Business Mathematics

Business Mathematics

Ameet Kumar Banerjee
Faculty
Institute of Development Studies,
Lucknow University and IIPM (Lucknow)
Lucknow

BSP BS Publications
A unit of **BSP Books Pvt., Ltd.**
4-4-309, Giriraj Lane, Sultan Bazar,
Hyderabad - 500 095 - A.P.
Phone : 040 - 23445605, 23445688

Published by :

BSP BS Publications
A unit of **BSP Books Pvt., Ltd.**
4-4-309, Giriraj Lane, Sultan Bazar,
Hyderabad - 500 095 - A.P.
Phone : 040 - 23445605, 23445688
e-mail : info@bspbooks.net
www.bspublications.net

Printed at

Aditya Offset Process (I) Pvt., Ltd.
Hyderabad.

Price: Rs. 525.00

ISBN: 978-93-81075-94-4

Preface

This book is dedicated to my late father Sri. Ajoy Kumar Banerjee who has been a constant source of strength, guidance and inspiration. His godly blessing has given worlds strength to start this work. Students also played supporting role in writing of the book, when I found that the students especially from non mathematical stream expressed great discomfort for the subject.

The objective of this book was primarily meant to demystify the subject of mathematics and reach students from various backgrounds to cherish the richness of the subject and its application in diverse stream of management and economics. The book is presented in simple language with elaborate examples in order represent the idea clearly to understand the concepts in 22 chapters. Each chapter as such is presented in manner to appeal one and all from BBA, B. Com (Regular/ Hons), BA (Eco), MA (Eco), MBA and other allied courses.

My best wishes to all my students and teachers fraternity who have directly or indirectly have participated in my work. A special mention is required to be mentioned for my students Animesh, Praveen and Devesh who have relentlessly typed the manuscripts of the book, Mr. Vasudev, Mr. Avadesh of sales department and Mr. Naresh of production department of BSP Publication, who needs to receive a note of appreciation for his excellent coordination during the execution of the project. Without mentioning two more important personalities it will be highly ungrateful in my part my wife, Sumita who has exercised great support and patience and my pet Ruff – the great, Labrador retriever always into pranks in order to divert my attention towards him.

I will gratefully acknowledge valuable suggestions from all quarters from students and teachers who are teaching courses to both undergraduate and post graduate level in several management schools and universities in pan India basis in order to improve the book further and incorporate the same in the subsequent editions.

Every possible effort has been made to bring out the book error free and without any discrepancies. If you need any clarification please do write to me which will be thankfully received and reverted.

At last I express my deep sense of gratitude to all core team members of BSP Publication by bring the book in such a short duration.

Ameet Kumar Banerjee

email: ameet.banerjee@rediffmail.com

Contents

CHAPTER 3

Limits and Continuity

CHAPTER 4

Derivatives and its Applications

CHAPTER 5

Maxima and Minima and its Application

CHAPTER 6

Partial Derivatives and its Application

CHAPTER 7

Integration and its Application

CHAPTER 8

Differential Equations and its Application

CHAPTER 9

Theory of Matrices-I

CHAPTER 14

Progression

CHAPTER 15

Permutation and Combination

CHAPTER 16

Theory of Probability-I

CHAPTER 17

Theory of Probability - II

CHAPTER 18

Linear Programming Formulation and Graphical Method

CHAPTER 19

The Simplex Method and Duality

CHAPTER 20

Decision Theory and Decision Tree

CHAPTER 21

Game Theory

CHAPTER 22

Mathematics of Finance

CHAPTER 1

SET THEORY

Introduction

In this chapter we will understand the various concepts in relation to set theory from the basic understanding, what is a set? In daily life we often used the word set in various situations, but in mathematical terminology it represents a group or a collection of books, toys or numbers etc., these objects which are the part of the set are called as elements of the set. Some of the key contributors are Georg Cantor (1845-1918) from whose work the modern set theory was largely originated and other prominent contributors are Ernst Zermelo, Abraham Fraenkel and John Von Neumann.

Definition of Set

A set is a well defined collection of objects and these objects are termed as the members or elements of the set. But we need to elaborate on the term "well-defined" it means that each element bears certain characteristics with which it can be identified under a particular head.

For example

(i) The set {a, e, i} belongs to family of vowels

(ii) The set {1, 2, 3, 4} belongs to natural numbers (N).

Set Notation

The two most common way of expressing a set are:

(i) *Roster, Tabular or Enumeration Form*

In this method all the elements are listed within braces { } or brackets [ ] or parentheses ( ) separated by commas.

For example

All natural numbers less than six can be written as {1, 2, 3, 4, 5}

(ii) *Set builder Form*

In this method all the elements are listed according to characteristics or properties.

For example

(i) The Set A = {a, e, i, o, u} can be written as

A = {x | x is vowel in English alphabet} or

A = {x : x is vowel in English alphabet}

(ii) The Set B = {1, 3, 5, 7..........} can be written as B = {x | x is a odd number integer}

Note: A Colon (:), a vertical line (|) or a semi colon (;) can be used after x and read as "such that"

Types of Set

(i) *Finite set*

A set in countable form is a finite set and it means each element can be counted physically.

For example

Set A = {1, 2, 3, 4, 5} is a finite set with five elements.

(ii) *Infinite set*

A set in uncountable form is an infinite set. The elements in the set cannot be counted.

For example

The set of natural numbers (N)

A = {1, 2, 3,... }

(iii) *Null, empty or void set*

A set which has no elements is called as null, empty or void set denoted by {Ø} and read as phi in Greek and Latin.

For example

Set A = {Ø} is a null set.

(iv) *Singleton set*

A set containing only one element is called as singleton set.

For example

Set A = {1}

(v) ***Equality of sets***

Two sets A and B are termed equal if only if every element of Set A is also an element of Set B and also every element of Set B is an element of Set A i.e. A = B if only if {$x \in A$ and $x \in B$}. Also called as Axiom of Extension or Axiom of Identity

For example

If Set A = {2, 3, 4} and Set B = {4, 3, 2}. Here Set A = Set B because both sets have common and equal numbers of elements.

(vi) ***Equivalent set***

If the elements of one set can be set to one-one correspondence with the elements of other set, then the sets are called equivalent set denoted by the symbol ~. Here we mean by one–one correspondence is that each element in Set A can be matched with one element in B and vice-versa.

For example

Set A = {a, b, c} and Set B = {x, y, z} , here set A is not equal to set B but the elements of Set A can be put into one–one correspondence with each elements of Set B then Set A ~ Set B.

(vii) ***Subset***

If A and B are two sets such that every element of Set A is also an element of Set B then Set A is said to be a subset of Set B or read as "Set A is contained in Set B" or "Set A is a subset of Set B"

We can write this relationship as $A \subseteq B$ or $B \supseteq A$, i.e. it means if $x \in A$ and $A \subseteq B$ then $x \in B$

For example

Set A = {1, 2, 3} and Set B = {1, 2, 3, 4, 5} then all elements of Set A are also elements of Set B which means $A \subseteq B$

Properties of Subsets

(i) If Set A is subset of Set B then Set B is called the super set of Set A.

(ii) If Set A ⊆ Set B and Set B ⊆ Set A then Set A = Set B.

(iii) If Set A ⊆ Set B, Set B ⊆ Set C then Set A ⊆ Set C

(viii) ***Proper Subset***

A Set A is called proper subset of a Set B, if each and every element of Set A are contained in Set B and there exists at least one element in Set B such that it is not an element of Set A. Symbolically denoted as $A \subset B$

For example

If Set A = {1, 2, 3, 4, 5} and Set B = {1, 2, 3, 4, 5, 6, 7} here Set A is a proper subset of Set B.

(ix) ***Comparable Sets***

Two set A and B are said to be comparable if one of them is the subset of the other i.e. $A \subseteq B$ or $B \subseteq A$.

(x) ***Non Comparable Sets***

Two set A and B are said to be non-comparable, if there exists and one element in Set A which is not in Set B and one element in Set B which is not in Set A .

Symbolically denoted by $A \not\subset B$ and $B \not\subset A$.

For example

Set A = {1, 2, 3} and Set B = {3, 4, 5} then both the sets are said to be non-comparable.

(xi) ***Disjoint sets***

Set A and Set B are said to be disjoint. If no element of set A is in B and no element of set B is in A.

For example

If Set A = {1, 2, 3} and Set B = {4, 5, 6}, then there is no common elements between Set A and Set B.

(xii) ***Set of sets or family of sets***

If the elements of a set are sets themselves then it is called set of sets or family of sets.

For example

If Set A = {1, 2} then,

The Set = {Ø, {1}, {2}, {1, 2}} is a family of sets whose elements are subset of A.

(xiii) ***Power set***

If for a given Set A, a set consisting of all the subsets of A is called the power of the set. The power set this is denoted by P(A).

For example

If Set A = {1, 2} then P (A) = {Ø}, {1}, {2}, {1, 2}.

***Note**: Suppose a set contains 'n' elements then 2^n subsets can be formed. The set consisting of these 2^n subsets is called power set.*

(xiv) *Universal Set*

If each set is a subset of some other set. A set which is superset of all the sets under consideration is called as Universal set and is denoted by U.

For example

If Set A = {1, 2, 3, 4} and set B = {5, 6, 7} then both sets A and B are subsets of the universal set of natural number.

Union of Set (Set Operation)

If A and B are two sets then the union of Set A and Set B is the set consisting of either all the elements of Set A or Set B or both and denoted by "A ∪ B" and read as A union B or A cup B

i.e. $A \cup B = \{x : x \in A \text{ or } x \in B \text{ or } x \in \text{ both A and B}\}$

For example

If Set A = {1, 2, 3, 4} and Set B = {4, 5, 6, 7} then

$A \cup B = \{1, 2, 3, 4, 5, 6, 7\}$

And hence the union of two Sets A and B is the logical sum of A and B where each element is written only once.

Properties of Union of Sets

(i) If Set A and Set B are two sets then A ∪ B is also a unique set

For example

If Set A = {1, 2, 3} and Set B = {3, 4, 5}

Therefore, A ∪ B = {1, 2, 3, 4, 5} is also a set.

(ii) *Commutative Property*

Union of set is commutative i.e. if Set A and Set B are two sets then A ∪ B = B ∪ A

For example

If Set A = {1, 2, 3} and Set B = {3, 4, 5}, then

$A \cup B = \{1, 2, 3, 4, 5\}$.

Similarly

$B \cup A = \{3, 4, 5, 1, 2\}$

$= \{1, 2, 3, 4, 5\}$

$= A \cup B$

Clearly $A \cup B = B \cup A$

Alternative Proof

Union of sets is Communicative i.e. if Set A and Set B are two sets then $A \cup B = B \cup A$.

In order to prove $A \cup B = B \cup A$,

We first prove that $A \cup B \subseteq B \cup A$ and then $B \cup A \subseteq A \cup B$, if both of them holds good then $A \cup B = B \cup A$.

Let x be any element belongs to $A \cup B$.

So $x \in A \cup B \Rightarrow x \in A$ or $x \in B$

$\Rightarrow x \in B$ or $x \in A$

$\Rightarrow x \in B \cup A$

So $A \cup B \subseteq B \cup A$ .... (1)

Again let's assume y be an element such that

$y \in B \cup A \Rightarrow y \in B$ or $y \in A$

$\Rightarrow y \in A$ or $y \in B$

$\Rightarrow y \in A \cup B$

So $B \cup A \subseteq A \cup B$ .... (2)

From (1) and (2) we get $A \cup B \subseteq B \cup A$ and $B \cup A \subseteq A \cup B$, therefore we get $A \cup B = B \cup A$.

(iii) Union of sets is associative i.e. If Set A, Set B and Set C are three sets then $A \cup (B \cup C) = (A \cup B) \cup C$

For example

If Set A = {1, 2, 3}, Set B = {3, 4, 5} and Set C = {5, 6}, then

$(A \cup B) = \{1, 2, 3, 4, 5\}$

Therefore

$(A \cup B) \cup C = \{1, 2, 3, 4, 5\} \cup \{5, 6\}$

$= \{1, 2, 3, 4, 5, 6\}$

Similarly,

$(B \cup C) = \{3, 4, 5, 6\}$

Therefore,

$A \cup (B \cup C) = \{1, 2, 3\} \cup \{3, 4, 5, 6\}$

$= \{1, 2, 3, 4, 5, 6\}$

and hence,

$A \cup (B \cup C) = (A \cup B) \cup C$

Alternate Proof

If Set A, Set B and Set C are three sets then

$A \cup (B \cup C) = (A \cup B) \cup C$

Let x be an element which belongs to $A \cup (B \cup C)$

Then $x \in A \cup (B \cup C)$,

$\Rightarrow x \in A$ or $x \in (B \cup C)$

$\Rightarrow x \in A$ or ($x \in B$ or $x \in C$)

$\Rightarrow$ ($x \in A$ or $x \in B$) or $x \in C$

$\Rightarrow x \in (A \cup B) \cup C$

Thus, $A \cup (B \cup C) = (A \cup B) \cup C$.

(iv) If Set A is a set, then $A \cup \emptyset = A$ where $\emptyset$ is a null set

For example

If Set A = {1, 2, 3}, then $A \cup \emptyset = \{1, 2, 3\} \cup \{\emptyset\}$

$= \{1, 2, 3\}$

$= A$

Alternate Proof

If Set A is any set then $A \cup \emptyset = A$

Let x be any element such that $x \in A \cup \emptyset$ then $x \in A$ or $x \in \emptyset$ but $\{\emptyset\}$ being null set therefore $x \notin \{\emptyset\}$, thus $x \in A$.

Hence $A \cup \emptyset = A$

(v) *Union of sets is idempotent*

If Set A is any set then $A \cup A = A$.

For example

If A = {1, 2, 3} then $A \cup A = \{1, 2, 3\} \cup \{1, 2, 3\}$

$= \{1, 2, 3\}$

$= A$

Alternate Proof

If Set A is any set then $A \cup A = A$

Let x be any element so that $x \in A \cup A$

$\Rightarrow x \in A$ or $x \in A$

$\Rightarrow x \in A$ thus $A \cup A = A$

(vi) If Set A is a subset of universal Set U then $A \cup U = U$

For example

If Set U = {1, 2, 3, 4, 5, 6, 7} and Set A = {1, 2, 3}

$A \cup U = \{1, 2, 3\} \cup \{1, 2, 3, 4, 5, 6, 7\}$

$= \{1, 2, 3, 4, 5, 6, 7\}$

$= U.$

(vii) If Set A and Set B are two sets such that $A \subseteq B$ then $A \cup B = B$ and if $B \subseteq A$ then $A \cup B = A$

For example

If Set A = {1, 2, 3} and Set B = {1, 2, 3, 4, 5}

Then $A \subseteq B$

$A \cup B = \{1, 2, 3\} \cup \{1, 2, 3, 4, 5\}$

$= \{1, 2, 3, 4, 5\} = B$

Similarly If A = {1, 2, 3, 4, 5}

B = {1, 2, 3}

In this case $B \subseteq A$

$A \cup B = \{1, 2, 3, 4, 5\} \cup \{1, 2, 3\}$

$A \cup B = \{1, 2, 3, 4, 5\} = A$

Alternate Proof

If A and B are two sets then $A \subseteq A \cup B$ and $B \subseteq A \cup B$

let x be any element of the Set A such that

$x \in A \Leftrightarrow x \in A$ or $x \in B$

$x \in (A \cup B) \Leftrightarrow A \subseteq A \cup B$

Similarly $B \subseteq A \cup B$ can be proved.

Intersection of Sets

Let A and B are two sets then intersection set of Set A and Set B is the set which consist of common elements which belongs to both A and B denoted by $A \cap B$ and read as " A Cap B " or "A intersection B". Symbolically represented as

$A \cap B = \{x : x \in A \text{ and } x \in B\}$

Otherwise if $x \in A \cap B \Rightarrow x \in A$ and $x \in B$

For example

If Set A = {1, 2, 3} and Set B = {3, 4, 5} then $A \cap B$ = {3}

The Venn diagram representation is shown below as:

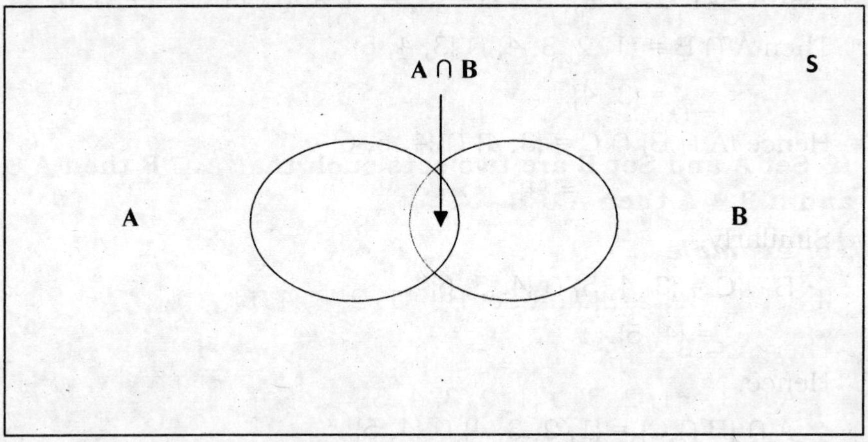

Properties of Intersection of Sets

The following are the properties which hold with respect to intersection of sets

(i) ***Communicative Property***

Intersection of sets is communicative i.e. if Set A and Set B are two sets then $A \cap B = B \cap A$

For example

If Set A = {1, 2, 3} and Set B = {3, 4, 5}

then $A \cap B$ = {1, 2, 3} ∩ {3, 4, 5}

= {3}.

Similarly,

$B \cap A$ = {3, 4, 5} ∩ {1, 2, 3}

= {3}.

Therefore $A \cap B = B \cap A$.

Alternative Proof

Let $x \in A \cap B$ then $x \in A$ and $x \in B \Leftrightarrow x \in B$ and $x \in A$ i.e. $x \in B \cap A$.

Hence $A \cap B = B \cap A$.

(ii) *Associative Property*

The intersection of sets are associative i.e. if Set A, Set B and Set C are three sets then $(A \cap B) \cap C = A \cap (B \cap C)$.

For example

If Set A = {1, 2, 3, 4}, Set B = {3, 4, 5} and Set C = {4, 5, 6}

Then $A \cap B = \{1, 2, 3, 4\} \cap \{3, 4, 5\}$

$= \{3, 4\}$.

Hence $(A \cap B) \cap C = \{3, 4\} \cap \{4, 5, 6\}$

$= \{4\}$.

Similarly,

$B \cap C = \{3, 4, 5\} \cap \{4, 5, 6\}$

$= \{4, 5\}$.

Hence,

$A \cap (B \cap C) = \{1, 2, 3, 4\} \cap \{4, 5\}$

$= \{4\}$

Therefore $(A \cap B) \cap C = A \cap (B \cap C)$

Alternative Proof:

Let $x \in (A \cap B) \cap C \Leftrightarrow$ then $x \in (A \cap B)$ and $x \in C$

$\Leftrightarrow (x \in A$ and $x \in B)$ and $x \in C$

$\Leftrightarrow x \in A$ and $(x \in B$ and $x \in C)$.

$\Leftrightarrow x \in A \cap (B \cap C)$.

Hence $(A \cap B) \cap C = A \cap (B \cap C)$.

(iii) *Idempotent Property*

The intersection of sets is idempotent i.e. if Set A is any set, then $A \cap A = A$

For example

If Set A = {1, 2, 3, 4} then

$A \cap A = \{1, 2, 3, 4\} \cap \{1, 2, 3, 4\}$

$= \{1, 2, 3, 4\}$

$= A$

Alternative Proof

Let x be any element such that $x \in A \cap A$, then $\Leftrightarrow x \in A$ and $x \in A$

$\Leftrightarrow x \in A$

Hence $A \cap A = A$.

(iv) If Set A is any set then $A \cap \emptyset = \emptyset$, $\emptyset$ is the null set

For example

If Set A = {1, 2, 3, 4} and $\emptyset = \{\ \}$ then,

$A \cap \emptyset = \{1, 2, 3, 4\} \cap \{\ \}$

$= \{\ \}$

$= \emptyset$.

(v) If Set A is any set subset of an Universal Set U then $A \cap U = A$

For example

If Set U = {1, 2, 3, 4, 5, 6, 7} and Set A = {1, 2, 3, 4}

then $A \cap U = \{1, 2, 3, 4, 5, 6, 7\} \cap \{1, 2, 3, 4\}$

$= \{1, 2, 3, 4\}$

$= A$

Alternative Proof

Let x be any element such that $x \in A \Rightarrow x \in A$ and $x \in U$ since $A \subseteq U$

Therefore $x \in A \cap U$,

And hence $A \subseteq A \cap U$ 1

But $A \cap U \subseteq A$ 2

From 1 and 2, we get,

$A = A \cap U$.

(vi) If A and B are disjoint sets then $A \cap B = \emptyset$

For example

if Set A = {1, 2, 3, 4} and Set B = {5, 6, 7}

then $A \cap B = \{1, 2, 3, 4\} \cap \{5, 6, 7\}$

$= \{\emptyset\}$

(vii) If Set A and Set B are two sets then $A \cap B \subseteq A$ and $A \cap B \subseteq B$

As $A \cap B$ contains only those elements which are in common A as well as in B. Therefore $A \cap B \subseteq A$ and $A \cap B \subseteq B$.

For example

If Set A = {1, 2, 3, 4} and Set B = {3, 4, 5, 6}

$A \cap B = \{1, 2, 3, 4\} \cap \{3, 4, 5, 6\}$

$= \{3, 4\}$ which is a subset of Set A and Set B

Alternative Proof

Let x be any element such that $x \in A \cap B$ then $x \in A$ and $x \in B$

$\Leftrightarrow x \in A$ as $A \cap B \subseteq A$. Similarly the other part $A \cap B \subseteq B$ can also be proved in similar manner.

Distributive Laws of Unions and Intersections

Result 1

If Set A, Set B and Set C are three sets, then

$A \cap (B \cup C) = (A \cap B) \cup (A \cap C)$

If Set A = {1, 2, 3, 4}, Set B = {3, 4, 5} and Set C = {4, 5, 6}

Then

$(B \cup C) = \{3, 4, 5\} \cup \{4, 5, 6\}$

$= \{3, 4, 5, 6\}.$

$A \cap (B \cup C) = \{1, 2, 3, 4\} \cap \{3, 4, 5, 6\}$

$= \{3, 4\}$

Similarly,

$(A \cap B) = \{1, 2, 3, 4\} \cap \{3, 4, 5\}$

$= \{3, 4\}$

and

$(A \cap C) = \{1, 2, 3, 4\} \cap \{4, 5, 6\}$

$= \{4\}$

Thus,

$(A \cap B) \cup (A \cap C) = \{3, 4\} \cup \{4\}$

$= \{3, 4\}$

Therefore $A \cap (B \cup C) = (A \cap B) \cup (A \cap C)$

Alternative Proof

Let x be any element belonging to $A \cap (B \cup C)$, then,

$x \in A \cap (B \cup C) \Leftrightarrow x \in A$ and $x \in (B \cup C)$

$\Leftrightarrow x \in A$ and ($x \in B$ or $x \in C$)

$\Leftrightarrow$ ($x \in A$ and $x \in B$) or ($x \in A$ and $x \in C$)

$\Leftrightarrow x \in (A \cap B)$ or $x \in (A \cap C)$

$\Leftrightarrow x \in (A \cap B) \cup (A \cap C)$

Hence $A \cap (B \cup C) = (A \cap B) \cup (A \cap C)$

Result 2

If Set A, Set B and Set C are three sets, then

$A \cup (B \cap C) = (A \cup B) \cap (A \cup C)$

If Set A = {1, 2, 3, 4}, Set B = {3, 4, 5} and Set C = {4, 5, 6}

Then

$(B \cap C) = \{3, 4, 5\} \cap \{4, 5, 6\}$

$= \{4, 5\}$

and

$A \cup (B \cap C) = \{1, 2, 3, 4\} \cup \{4, 5\}$

$= \{1, 2, 3, 4, 5\}$

Similarly

$(A \cup B) = \{1, 2, 3, 4\} \cup \{3, 4, 5\}$

$= \{1, 2, 3, 4, 5\}$

$(A \cup C) = \{1, 2, 3, 4\} \cup \{4, 5, 6\}$

$= \{1, 2, 3, 4, 5, 6\}$

Then $(A \cup B) \cap (A \cup C) = \{1, 2, 3, 4, 5\} \cap \{1, 2, 3, 4, 5, 6\}$

$= \{1, 2, 3, 4, 5\}$

Therefore $A \cup (B \cap C) = (A \cup B) \cap (A \cup C)$

Alternative Proof

Let x be any element belonging to $A \cup (B \cap C)$, then,

$x \in A \cup (B \cap C) \quad \Leftrightarrow x \in A$ or $x \in (B \cap C)$

$\Leftrightarrow x \in A$ or ($x \in B$ and $x \in C$)

$\Leftrightarrow$ ($x \in A$ or $x \in B$) and ($x \in A$ or $x \in C$)

$\Leftrightarrow x \in (A \cup B)$ and $x \in (A \cup C)$

$\Leftrightarrow x \in (A \cup B) \cap (A \cup C)$

Hence $A \cup (B \cap C) = (A \cup B) \cap (A \cup C)$.

Complement of a Set

The complement of a Set A is that set which contains all those elements of the universal set U which are not in A. The complement of set A is the set U – A and is denoted by A^c, A', $\bar{A}$ or ~ A. It can symbolically written as

$A' = U - A = \{x : x \in U \text{ and } x \notin A\}$

For example

If Set U = {1, 2, 3, 4, 5, 6, 7, 8} and the Set A = {1, 2, 3, 4, 5} then

A′ = U – A = {6, 7, 8}.

Properties of the Complement of Set

(i) The intersection of Set A and its complement A′ are disjoint sets i.e. $A \cap A'$ is a null set {Ø}

For example

If the Set U = {1, 2, 3, 4, 5, 6, 7, 8, 9} and Set A = {1, 2, 3, 4, 5} then

A′ = U – A = {6, 7, 8, 9}

Therefore $A \cap A'$ = {1, 2, 3, 4, 5} ∩ {6, 7, 8, 9}

= {Ø}

Alternative Proof

Let x be an element such that

$x \in A \cap A' \Rightarrow x \in A$ and $x \in A' \Rightarrow x \in \{Ø\}$

(ii) The union of Set A and its complement is the universal set i.e. $A \cup A' = U$, the universal set.

For example

If the Set U = {1, 2, 3, 4, 5, 6, 7, 8, 9} and Set A = {1, 2, 3, 4, 5} then

$A' = U - A = \{6, 7, 8, 9\}$

Therefore $A \cup A' = \{1, 2, 3, 4, 5\} \cup \{6, 7, 8, 9\}$

$= \{1, 2, 3, 4, 5, 6, 7, 8, 9\}$

= Set U.

Alternative Proof

As every set is the subset of the universal set, therefore $A \cup A' \subseteq U$. Let $x \in U$ that implies $x \in A$ or $x \in A' \Rightarrow x \in A \cup A'$. Therefore $U \subseteq A \cup A'$. Hence $A \cup A' = U$.

(iii) Complement of complement a Set A is the set itself i.e. $(A')' = A$

For example

If the Set U = {1, 2, 3, 4, 5, 6, 7, 8, 9} and Set A = {1, 2, 3, 4, 5} then

$A' = U - A = \{6, 7, 8, 9\}$

Therefore $(A')' = U - A' = \{1, 2, 3, 4, 5\}$ = Set A

Alternative Proof

Let $x \in (A')' \Leftrightarrow x \notin A' \Leftrightarrow x \in A$.

(iv) If the Set A is equal to the universal Set U then $A' = \{\emptyset\}$.

(v) If Set A and Set B are two sets then $A - B = A \cap B'$

For example

If the Set U = {1, 2, 3, 4, 5, 6, 7, 8, 9} , Set A = {1, 2, 3, 4, 5} and Set B = {4, 5, 6, 7} then

$A - B = \{1, 2, 3, 4, 5\} - \{4, 5, 6, 7\}$

$= \{1, 2, 3\}$.

But $B' = U - B = \{1, 2, 3, 4, 5, 6, 7, 8, 9\} - \{4, 5, 6, 7\}$

$= \{1, 2, 3, 8, 9\}$

Thus $A \cap B' = \{1, 2, 3, 4, 5\} \cap \{1, 2, 3, 8, 9\}$

$= \{1, 2, 3\}$

$= A - B$

Alternate Proof

Let $x \in A - B \Rightarrow x \in A$ and $x \notin B \Leftrightarrow x \in A$ and $x \in B' \Leftrightarrow x \in (A \cap B')$ and hence $A - B = A \cap B'$.

(vi) If $A \subseteq B$ then $A \cup (B - A) = B$

For example

If the Set B = {1, 2, 3, 4, 5, 6, 7, 8, 9} and Set A = {1, 2, 3, 4, 5}

B – A = {1, 2, 3, 4, 5, 6, 7, 8, 9} – {1, 2, 3, 4, 5}

= {6, 7, 8, 9}.

Therefore A ∪ (B – A) = {1, 2, 3, 4, 5} ∪ {6, 7, 8, 9}.

= {1, 2, 3, 4, 5, 6, 7, 8, 9}

= B

Alternate Proof

As given $A \subseteq B$ let $x \in A \cup (B - A) \Leftrightarrow x \in A$ or $x \in (B - A)$

$\Leftrightarrow (x \in A$ or $x \in B)$ and $(x \in A$ or $x \in A')$

$\Leftrightarrow x \in (A \cup B)$ and $x \in B$ as $(A \cup A' = B$ as $A \subseteq B)$

$\Leftrightarrow x \in (A \cup B)$

$\Leftrightarrow x \in B$, hence $A \cup (B - A) = B$

Difference of Sets

Let Set A and Set B are two sets then the difference of Set A and B is the set which consist of those elements which belongs to A but does not belong to B denoted by A – B and read as "A difference B "or A minus B" and also denoted by A ~ B. The symbolical representation is

$A - B = (x : x \in A$ and $x \notin B)$ similarly

$B - A = (x : x \in B$ and $x \notin A)$.

For example

If Set A = {1, 2, 3, 4}, Set B = {3, 4, 5} then

A – B = {1, 2, 3, 4} – {3, 4, 5}

= {1, 2}.

Similarly

B – A = {3, 4, 5} – {1, 2, 3, 4}

= {5}

Properties of Difference of Sets

(i) $A - A = \emptyset$.

For example

If Set A = {1, 2, 3, 4, 5) then

$A - A = \{1, 2, 3, 4, 5\} - \{1, 2, 3, 4, 5\}$

$= \emptyset$

Alternate Proof

Let $x \in A - A$ then $x \in A$ and $x \notin A$ but there is no such element satisfying both the conditions and hence, there is no element belonging to $A - A$ i.e $A - A = \emptyset$.

(ii) $A - \emptyset = A$.

For example

If Set A = {1, 2, 3, 4, 5} and Ø be a null set then

$A - \emptyset = \{1, 2, 3, 4, 5\} - \{\emptyset\}$.

$= \{1, 2, 3, 4, 5\} = A$, hence $A - \emptyset = A$.

Alternate Proof

Let $x \in A - \emptyset$ then $x \in A$ and $x \notin \emptyset$ which mean x belongs to A, since there is no element belonging to Ø.

Conversely, $x \in A$ then $x \in A - \emptyset$ i.e. $A - \emptyset = A$.

(iii) $A - B$, $A \cap B$ and $B - A$ are mutually disjoint.

For example

If Set A = {1, 2, 3, 4, 5} and Set B = {4, 5, 6, 7} then

$A - B = \{1, 2, 3, 4, 5\} - \{4, 5, 6, 7\}$

$= \{1, 2, 3\}$

$A \cap B = \{1, 2, 3, 4, 5\} \cap \{4, 5, 6, 7\}$

$= \{4, 5\}$

and

$B - A = \{4, 5, 6, 7\} - \{1, 2, 3, 4, 5\}$

$= \{6, 7\}$

Alternate Proof

We need to prove that $(A - B) \cap (A \cap B) = \emptyset$

Let $x \in (A - B) \cap (A \cap B)$ then $x \in (A - B)$ and $x \in (A \cap B)$

$\Leftrightarrow$ ($x \in A$ and $x \notin B$) and ($x \in A$ and $x \in B$)

$\Leftrightarrow$ $x \in A$ and $x \in \emptyset$ as there cannot be an element satisfying both the condition i.e. $x \in B$ and $x \notin B$)

$\Leftrightarrow$ $x \in \emptyset$.

Similarly other results can be equally proved in the same lines.

(iv) $(A - B) \cup A = A$

For example

If Set A = {1, 2, 3, 4, 5} and Set B = {4, 5, 6, 7} then

$A - B = \{1, 2, 3, 4, 5\} - \{4, 5, 6, 7\}$

$= \{1, 2, 3\}$.

Therefore,

$(A - B) \cup A = \{1, 2, 3\} \cup \{1, 2, 3, 4, 5\}$

$= \{1, 2, 3, 4, 5\}$.

Alternate Proof

Let x belongs to $(A - B) \cup A$ then $x \in (A - B)$ or $x \in A$

$\Leftrightarrow$ ($x \in A$ and $x \notin B$) or $x \in A$

$\Leftrightarrow$ $x \in A$ and $x \notin B$

$\Leftrightarrow$ $x \in A$

Hence, $(A - B) \cup A = A$

(v) $(A - B) \cap B = \emptyset$

For example

If Set A = {1, 2, 3, 4, 5} and Set B = {4, 5, 6, 7} then

$A - B = \{1, 2, 3\} = \{1, 2, 3, 4, 5\} - \{4, 5, 6, 7\}$

$= \{1, 2, 3\}$.

Therefore,

$(A - B) \cap B = \{1, 2, 3\} \cap \{4, 5, 6, 7\}$.

$= \emptyset$

Alternate Proof

Let x belongs to $(A - B) \cap B$ then $x \in (A - B)$ and $x \in B$

$$\Leftrightarrow (x \in A \text{ and } x \notin B) \text{ and } x \in B$$

Note there cannot be element satisfying both the condition i.e. $x \in B$ and $x \notin B)$

$$\Leftrightarrow x \in A \text{ and } x \in \emptyset$$

$$\Leftrightarrow x \in \emptyset, \text{ hence } (A - B) \cap B = \emptyset$$

De Morgan's Law

1st Law

Let Set A and Set B are two sets then $(A \cup B)' = A' \cap B'$

For example

Set U = {1, 2, 3, 4, 5, 6, 7, 8, 9},

Set A = {1, 2, 3, 4} and

Set B = {6, 7, 8, 9}

Then

$(A \cup B) = \{1, 2, 3, 4\} \cup \{6, 7, 8, 9\}$

$= \{1, 2, 3, 4, 6, 7, 8, 9\}$

Now,

$(A \cup B)' = U - (A \cup B)$

$= \{1, 2, 3, 4, 5, 6, 7, 8, 9\} - \{1, 2, 3, 4, 6, 7, 8, 9\}$

$= \{5\}$.

$A' = U - A = \{1, 2, 3, 4, 5, 6, 7, 8, 9\} - \{1, 2, 3, 4\}$

$= \{5, 6, 7, 8, 9\}$

$B' = U - B = \{1, 2, 3, 4, 5, 6, 7, 8, 9\} - \{6, 7, 8, 9\}$.

$= \{1, 2, 3, 4, 5\}$

Now,

$A' \cap B' = \{5, 6, 7, 8, 9\} \cap \{1, 2, 3, 4, 5\}$

$= \{5\}$

Therefore $(A \cup B)' = A' \cap B'$

Alternate Proof

$(A \cup B)' = A' \cap B'$

Let x be any element of $(A \cup B)'$,

Then $x \in (A \cup B)' \Leftrightarrow x \in U$ and $x \notin (A \cup B)$

$\Leftrightarrow x \in U$ and $(x \notin A$ or $x \notin B)$

$\Leftrightarrow (x \in U$ but $x \notin A)$ and $(x \in U$ but $x \notin B)$

$\Leftrightarrow x \in A'$ and $x \in B'$

$\Leftrightarrow x \in A' \cap B'$

Hence $(A \cup B)' = A' \cap B'$

2nd Law

Let Set A and Set B are two sets then $(A \cap B)' = A' \cup B'$

For example

Set U = {1, 2, 3, 4, 5, 6, 7, 8, 9},

Set A = {1, 2, 3, 4} and

Set B = {6, 7, 8, 9}

Then

$(A \cap B)$ = {1, 2, 3, 4} ∩ {6, 7, 8, 9}

= {Ø}

Now,

$(A \cap B)' = U - (A \cap B)$ = {1, 2, 3, 4, 5, 6, 7, 8, 9} – {Ø}

= {1, 2, 3, 4, 5, 6, 7, 8, 9}.

A' = U – A = {1, 2, 3, 4, 5, 6, 7, 8, 9} - {1, 2, 3, 4}

= {5, 6, 7, 8, 9}

B'= U – B = {1, 2, 3, 4, 5, 6, 7, 8, 9} - {6, 7, 8, 9}.

= {1, 2, 3, 4, 5}

Now,

$A' \cup B'$ = {5, 6, 7, 8, 9} ∪ {1, 2, 3, 4, 5}

= {1, 2, 3, 4, 5, 6, 7, 8, 9}

Therefore $(A \cap B)' = A' \cup B'$

Alternate Proof

$(A \cap B)' = A' \cup B'$

To prove the above result, let x be any element belonging to $(A \cap B)'$ then

$$x \in (A \cap B)' \Leftrightarrow x \notin (A \cap B)$$
$$\Leftrightarrow x \notin A \text{ or } x \notin B$$
$$\Leftrightarrow x \in A' \text{ or } x \in B'$$
$$\Leftrightarrow x \in A' \cup B'.$$

Hence, $(A \cap B)' = A' \cup B'$

De Morgan's Law on Difference of Sets

$A - (B \cup C) = (A - B) \cap (A - C)$

For example

Set A = {1, 2, 3, 4, 5, 6, 7, 8, 9},

Set B = {1, 2, 3, 4} and

Set C = {6, 7, 8, 9}

Then

$(B \cup C)$ = {1, 2, 3, 4, 6, 7, 8, 9}

And $A - (B \cup C)$ = {5}.

But $(A - B)$ = {5, 6, 7, 8, 9} and $(A - C)$ = {1, 2, 3, 4, 5}

Therefore $(A - B) \cap (A - C)$ = {5}

Hence $A - (B \cup C) = (A - B) \cap (A - C)$

Alternate Proof

Let x be any element such that

$$x \in A - (B \cup C) \Leftrightarrow x \in A \text{ and } x \notin (B \cup C)$$
$$\Leftrightarrow x \in A \text{ and } (x \notin B \text{ or } x \notin C)$$
$$\Leftrightarrow (x \in A \text{ but } x \notin B) \text{ and } (x \in A \text{ but } x \notin C)$$
$$\Leftrightarrow x \in (A - B) \text{ and } x \in (A - C)$$
$$\Leftrightarrow x \in (A - B) \cap (A - C)$$

Hence $A - (B \cup C) = (A - B) \cap (A - C)$

(ii) $A - (B \cap C) = (A - B) \cup (A - C)$

For Example

Set A = {1, 2, 3, 4, 5, 6, 7, 8, 9},

Set B = {1, 2, 3, 4} and

Set C = {6, 7, 8, 9}

Then

$(B \cap C) = \{\emptyset\}$

And $A - (B \cap C) = \{1, 2, 3, 4, 5, 6, 7, 8, 9\}$.

But $(A - B) = \{5, 6, 7, 8, 9\}$ and $(A - C) = \{1, 2, 3, 4, 5\}$

Therefore $(A - B) \cup (A - C) = \{1, 2, 3, 4, 5, 6, 7, 8, 9\}$

Hence $A - (B \cap C) = (A - B) \cup (A - C)$

Alternate Proof

Let x be any element such that

$$x \in A - (B \cap C) \Leftrightarrow x \in A \text{ and } x \notin (B \cap C)$$

$$\Leftrightarrow x \in A \text{ and } (x \notin B \text{ and } x \notin C)$$

$$\Leftrightarrow (x \in A \text{ and } x \notin B) \text{ or } (x \in A \text{ and } x \notin C)$$

$$\Leftrightarrow x \in (A - B) \text{ or } x \in (A - C)$$

$$\Leftrightarrow x \in (A - B) \cup (A - C)$$

Hence $A - (B \cap C) = (A - B) \cup (A - C)$.

Some Important results on Difference, Union and Intersection

(i) If Set A and Set B are two sets then $A \cup B = (A - B) \cup B$

For example

Set A = {1, 2, 3, 4, 5, 6, 7, 8, 9} and Set B = {1, 2, 3, 4} then

$A \cup B = \{1, 2, 3, 4, 5, 6, 7, 8, 9\}$

And

$(A - B) = \{5, 6, 7, 8, 9\}$, then,

$(A - B) \cup B = \{1, 2, 3, 4, 5, 6, 7, 8, 9\}$

Hence

$A \cup B = (A - B) \cup B$

(ii) If Set A and Set B are two sets then $A \cap (B - A) = \{\emptyset\}$

For example

Set $A = \{6, 7, 8, 9\}$ and Set $B = \{1, 2, 3, 4, 5\}$ then

$(B - A) = \{5\}$,

Then,

$$A \cap (B - A) = \{6, 7, 8, 9\} \cap \{5\}$$
$$= \{\emptyset\}$$

Hence proved.

(iii) If Set A and Set B are two sets then $(A - B) \cap B = \{\emptyset\}$

For example

Set $A = \{1, 2, 3, 6, 7, 8, 9\}$ and Set $B = \{1, 2, 3, 4\}$ then

$(A - B) = \{6, 7, 8, 9\}$

Therefore

$$(A - B) \cap B = \{6, 7, 8, 9\} \cap \{1, 2, 3, 4\}$$
$$= \{\emptyset\}$$

Hence proved

(iv) If Set A and Set B are two sets then $A - (A - B) = A \cap B$

For example

Set $A = \{1, 2, 3, 6, 7, 8, 9\}$ and Set $B = \{1, 2, 3, 4\}$ then

$(A - B) = \{6, 7, 8, 9\}$

Therefore

$$A - (A - B) = \{1, 2, 3, 6, 7, 8, 9\} - \{6, 7, 8, 9\}$$
$$= \{1, 2, 3\}$$

But

$$A \cap B = \{1, 2, 3, 6, 7, 8, 9\} \cap \{1, 2, 3, 4\}$$
$$= \{1, 2, 3\}.$$

Hence, $A - (A - B) = A \cap B$

(v) If Set A and Set B are two sets then

$(A - B) \cup (B - A) = (A \cup B) - (A \cap B)$.

For example

Set A = {1, 2, 3, 6, 7, 8, 9} and Set B = {1, 2, 3, 4} then

$(A - B) = \{6, 7, 8, 9\}$ and $(B - A) = \{4\}$

Therefore,

$(A - B) \cup (B - A) = \{4, 6, 7, 8, 9\}$.

Similarly,

$(A \cup B) = \{1, 2, 3, 4, 6, 7, 8, 9\}$ and $(A \cap B) = \{1, 2, 3\}$

Therefore

$(A \cup B) - (A \cap B) = \{1, 2, 3, 4, 6, 7, 8, 9\} - \{1, 2, 3\}$.

$= \{4, 6, 7, 8, 9\}$, hence proved.

(vi) If Set A, Set B and Set C are three sets then

$(A \cap B) - C = (A - C) \cap (B - C)$.

For example

Set A = {1, 2, 3, 4, 5, 6, 7, 8, 9},

Set B = {1, 2, 3, 4, 5, 6} and

Set C = {3, 5, 6, 7, 9}, then,

$(A \cap B) = \{1, 2, 3, 4, 5, 6\}$.

Therefore

$(A \cap B) - C = \{1, 2, 3, 4, 5, 6\} - \{3, 5, 6, 7, 9\}$

$= \{1, 2, 4\}$.

Similarly

$(A - C) = \{1, 2, 4, 8\}$ and $(B - C) = \{1, 2, 4\}$.

Therefore

$(A - C) \cap (B - C) = \{1, 2, 4, 8\} \cap \{1, 2, 4\}$

$= \{1, 2, 4\}$.

And hence $(A \cap B) - C = (A - C) \cap (B - C)$

(vii) If Set A, Set B and Set C are three sets then

$A \cap (B - C) = (A \cap B) - (A \cap C)$.

For example

Set A = {1, 2, 3, 4, 5, 6, 7, 8, 9},

Set B = {1, 2, 3, 4, 5, 6} and

Set C = {3, 5, 6, 7, 9}, then,

(B – C) = {1, 2, 4}.

Therefore

A ∩ (B – C) = {1, 2, 3, 4, 5, 6, 7, 8, 9} ∩ {1, 2, 4},

= {1, 2, 4}.

Similarly

(A ∩ B) = {1, 2, 3, 4, 5, 6} and (A ∩ C) = {3, 5, 6, 7, 9}.

Therefore (A ∩ B) – (A ∩ C) = {1, 2, 3, 4, 5, 6} – {3, 5, 6, 7, 9}

= {1, 2, 4}.

And hence A ∩ (B – C) = (A ∩ B) – (A ∩ C)

(viii) If Set A and Set B are two sets then (A – B) = A ∩ B′

For example

If Set U = {1, 2, 3, 4, 5, 6, 7, 8, 9}, Set A = {1, 2, 3, 6, 7, 8, 9} and Set B = { 6, 7, 8, 9} then

(A – B) = {1, 2, 3}

And B′ = U – B = {1, 2, 3, 4, 5, 6, 7, 8, 9} – {6, 7, 8, 9}.

= {1, 2, 3, 4, 5}.

A ∩ B′ = {1, 2, 3, 6, 7, 8, 9} ∩ {1, 2, 3, 4, 5}

= {1, 2, 3}.

Hence (A – B) = A ∩ B′

(ix) If Set A and Set B are two sets then (A – B) = B′ – A′

If Set U = {1, 2, 3, 4, 5, 6, 7, 8, 9}, Set A = {1, 2, 3, 6, 7, 8, 9} and Set B = {6, 7, 8, 9} then

(A – B) ={1, 2, 3, 6, 7, 8, 9} – {6, 7, 8, 9} = {1, 2, 3}

And A′ = U – A = {1, 2, 3, 4, 5, 6, 7, 8, 9} – {1, 2, 3, 6, 7, 8, 9}

= {4, 5}.

B′ = U – B = {1, 2, 3, 4, 5, 6, 7, 8, 9} – {6, 7, 8, 9}.

= {1, 2, 3, 4, 5}.

Therefore B′ – A′ = {1, 2, 3, 4, 5} – {4, 5}.

= {1, 2, 3}.

Hence (A – B) = B′ – A′.

(x) If Set A, Set B and Set C are three sets then A ∪ (B – C) ≠ (A ∪ B) – (A ∪ C)

For example

If Set A = {1, 2, 3, 4, 5, 6, 7, 8, 9}, Set B = {1, 2, 3, 4, 5 ,6} and Set C ={3, 5, 6, 7, 9} , then,

(B – C) = {1, 2, 3, 4, 5, 6 } - {3, 5, 6, 7, 9} = {1, 2, 4}

Therefore,

A ∪ (B – C) = {1, 2, 3, 4, 5, 6, 7, 8, 9} ∪ {1, 2, 4}

= {1, 2, 3, 4, 5, 6, 7, 8, 9}.

Similarly

(A ∪ B) = {1, 2, 3, 4, 5, 6, 7, 8, 9} and

(A ∪ C) = {1, 2, 3, 4, 5, 6, 7, 8, 9}

Therefore

(A ∪ B) – (A ∪ C) = {Ø}

Hence A ∪ (B – C) ≠ (A ∪ B) – (A ∪ C)

Illustration No. 1

If Set A = {1, 2, 3, 4, 5, 6}, Set B = {4, 5, 6, 7}

and Set C = { 3, 5, 6, 7, 9}, then, find

(i) A – (B ∩ C) = (A – B) ∪ (A – C)

(ii) A ∪ (B ∩ C) = (A ∪ B) ∩ (A ∪ C)

(iii) A ∩ (B – C) = (A ∩ B) – (A ∩ C)

Solution:

(B ∩ C) = {4, 5, 6, 7} ∩ {3, 5, 6, 7, 9}

= {5, 6, 7}

Therefore

A – (B ∩ C) = {1, 2, 3, 4, 5, 6} – {5, 6, 7}

= {1, 2, 3, 4}

Similarly

$(A - B) = \{1, 2, 3\}$

and

$(A - C) = \{1, 2, 4\}$

Therefore $(A - B) \cup (A - C) = \{1, 2, 3, 4\}$

Hence, $A - (B \cap C) = (A - B) \cup (A - C)$

(ii) $(B \cap C) = \{4, 5, 6, 7\} \cap \{3, 5, 6, 7, 9\}$

$= \{5, 6, 7\}$

Therefore

$A \cup (B \cap C) = \{1, 2, 3, 4, 5, 6\} \cup \{5, 6, 7\}$

$= \{1, 2, 3, 4, 5, 6, 7\}$

Similarly,

$(A \cup B) = \{1, 2, 3, 4, 5, 6\} \cup \{4, 5, 6, 7\}$

$= \{1, 2, 3, 4, 5, 6, 7\}$

And

$(A \cup C) = \{1, 2, 3, 4, 5, 6\} \cup \{3, 5, 6, 7, 9\}$,

$= \{1, 2, 3, 4, 5, 6, 7, 9\}$

Therefore,

$(A \cup B) \cap (A \cup C) = \{1, 2, 3, 4, 5, 6, 7\} \cap \{1, 2, 3, 4, 5, 6, 7, 9\}$,

$= \{1, 2, 3, 4, 5, 6, 7\}$.

Hence $A \cup (B \cap C) = (A \cup B) \cap (A \cup C)$.

(iii) $(B - C) = \{4\}$

Therefore

$A \cap (B - C) = \{4\}$

But $(A \cap B) = \{4, 5, 6\}$

And

$(A \cap C) = \{3, 5, 6\}$

Therefore,

$(A \cap B) - (A \cap C) = \{4\}$

Hence $A \cap (B - C) = (A \cap B) - (A \cap C)$.

Illustration No. 2

If Set U = {1, 2, 3, 4, 5, 6, 7, 8, 9}, Set A = {1, 2, 3, 4, 5}, Set B = {4, 5, 6, 7} and Set C = {5, 6, 7, 8} then find

(i) $(A \cup B) \cap (A \cup C)$

(ii) $(A \cap B) \cup (A \cap C)$

(iii) $(A \cup B \cup C)'$

(iv) $(A \cup B') \cap (A' \cup B)$

Solution:

(i) $(A \cup B) = \{1, 2, 3, 4, 5\} \cup \{4, 5, 6, 7\}$

$= \{1, 2, 3, 4, 5, 6, 7\}.$

And

$(A \cup C) = \{1, 2, 3, 4, 5\} \cup \{5, 6, 7, 8\}$

$= \{1, 2, 3, 4, 5, 6, 7, 8\}.$

Therefore,

$(A \cup B) \cap (A \cup C) = \{1, 2, 3, 4, 5, 6, 7\}$

(ii) $(A \cap B) = \{1, 2, 3, 4, 5\} \cap \{4, 5, 6, 7\}.$

$= \{4, 5\}$

And

$(A \cap C) = \{1, 2, 3, 4, 5\} \cap \{5, 6, 7, 8\}$

$= \{5\}$

Therefore,

$(A \cap B) \cup (A \cap C) = \{4, 5\} \cup \{5\}.$

$= \{4, 5\}.$

(iii) $(A \cup B \cup C)' = U - (A \cup B \cup C)$

But,

$(A \cup B \cup C) = \{1, 2, 3, 4, 5, 6, 7, 8\}$

Therefore,

$(A \cup B \cup C)' = \{1, 2, 3, 4, 5, 6, 7, 8, 9\} - \{1, 2, 3, 4, 5, 6, 7, 8\}$

$= \{9\}.$

(iv) $(A \cup B') = \{1, 2, 3, 4, 5\} \cup \{1, 2, 3, 8, 9\}$,

$= \{1, 2, 3, 4, 5, 8, 9\}$.

And

$(A' \cup B) = \{6, 7, 8, 9\} \cup \{4, 5, 6, 7, 8\}$,

$= \{4, 5, 6, 7, 8, 9\}$.

Therefore,

$(A \cup B') \cap (A' \cup B) = \{1, 2, 3, 4, 5, 8, 9\} \cap \{4, 5, 6, 7, 8, 9\}$

$= \{4, 5, 8, 9\}$.

Illustration No. 3

If Set $U = \{1, 2, 3, 4, 5, 6, 7, 8\}$, Set $A = \{2, 3, 4\}$, Set $B = \{1, 4, 5, 6\}$ and Set $C = \{1, 2, 4, 5, 7\}$ then find

(i) $(A \cap B)' = A' \cup B'$.

(ii) $A \cup (B \cap C) = (A \cup B) \cap (A \cup C)$.

Solution:

(i) $(A \cap B) = \{2, 3, 4\} \cap \{1, 4, 5, 6\}$.

$= \{4\}$

Therefore $(A \cap B)' = U - (A \cap B)$

$= \{1, 2, 3, 4, 5, 6, 7, 8\} - \{4\}$

$= \{1, 2, 3, 5, 6, 7, 8\}$.

Similarly

$A' = U - A = \{1, 2, 3, 4, 5, 6, 7, 8\} - \{2, 3, 4\}$,

$= \{1, 5, 6, 7, 8\}$.

And

$B' = U - B = \{1, 2, 3, 4, 5, 6, 7, 8\} - \{1, 4, 5, 6\}$,

$= \{2, 3, 7, 8\}$

Therefore,

$A' \cup B' = \{1, 5, 6, 7, 8\} \cup \{2, 3, 7, 8\}$,

$= \{1, 2, 3, 5, 6, 7, 8\}$.

Hence $(A \cap B)' = A' \cup B'$.

(ii) $(B \cap C) = \{1, 4, 5, 6\} \cap \{1, 2, 4, 5, 7\}$
$= \{1, 4, 5\}$.

Therefore,

$A \cup (B \cap C) = \{2, 3, 4\} \cup \{1, 4, 5\}$.
$= \{1, 2, 3, 4, 5\}$.

Similarly,

$A \cup B) = \{2, 3, 4\} \cup \{1, 4, 5, 6\}$
$= \{1, 2, 3, 4, 5, 6\}$

And

$(A \cup C) = \{2, 3, 4\} \cup \{1, 2, 4, 5, 7\}$
$= \{1, 2, 3, 4, 5, 7\}$.

Therefore,

$(A \cup B) \cap (A \cup C) = \{1, 2, 3, 4, 5, 6\} \cap \{1, 2, 3, 4, 5, 7\}$.
$= \{1, 2, 3, 4, 5\}$

Hence, $A \cup (B \cap C) = (A \cup B) \cap (A \cup C)$

Symmetric Difference of Two Sets

If Set A and Set B are two sets then the difference set is called symmetric difference of two sets if it contains all those elements which belongs to A or to B. Symmetric difference of Set A and Set B is symbolically represented as

$A \Delta B = (A - B) \cup (B - A)$

For example

Set A = {1, 2, 3, 5, 6, 7,} and Set B = {1, 3, 4, 8, 9} then

$A \Delta B = \{2, 5, 6, 7\} \cup \{4, 8, 9\}$
$= \{2, 4, 5, 6, 7, 8, 9\}$

Some Properties of Symmetric Difference

(i) Commutative Property $A \Delta B = B \Delta A$.

(ii) Associative Property $(A \Delta B) \Delta C = A \Delta (B \Delta C)$.

(iii) $A \Delta A = \{\varnothing\}$.

(iv) $A \Delta \varnothing = A$

(v) $A \Delta (A \cap B) = A - B$.

(vi) $(A \Delta B) \cup (A \cap B) = A \cup B$.

(vii) $A \Delta B = (A - B) \cup (B - A)$

(viii) $A \Delta B = (A \cup B) - (A \cap B)$.

(ix) $A \Delta B = \{\emptyset\} \Leftrightarrow A = B$.

(x) $(A \Delta B) \cup (A \cap B) = A \cup B$.

Illustration No. 1

If Set A = {1, 2, 3, 5, 6} and Set B = {1, 3, 4, 8, 9} then verify that $A \Delta B = B \Delta A$ and also prove that $A \Delta B = (A \cup B) - (A \cap B)$?

Solution:

$A \Delta B = (A - B) \cup (B - A)$

$= \{2, 5, 6\} \cup \{4, 8, 9\}$

$= \{2, 4, 5, 6, 8, 9\}$

And

$B \Delta A = (B - A) \cup (A - B)$

$= \{4, 8, 9\} \cup \{2, 5, 6\}$.

$= \{2, 4, 5, 6, 8, 9\}$

Similarly, in order to prove that $A \Delta B = (A \cup B) - (A \cap B)$

As $(A \cup B) = \{1, 2, 3, 5, 6\} \cup \{1, 3, 4, 8, 9\}$.

$= \{1, 2, 3, 4, 5, 6, 8, 9\}$.

And

$(A \cap B) = \{1, 3\}$.

$(A \cup B) - (A \cap B) = \{1, 2, 3, 4, 5, 6, 8, 9\} - \{1, 3\}$.

$= \{2, 4, 5, 6, 8, 9\}$.

Hence

$A \Delta B = (A \cup B) - (A \cap B)$

Illustration No. 2

If Set A = {1, 3, 6, 9}, Set B = {3, 4, 7, 9} and Set C = {2, 4, 5, 7} then find

(i) $(A \Delta B) \Delta C = A \Delta (B \Delta C)$

(ii) $A \Delta (A \cap B) = A - B$

(iii) $(A \Delta B) \cup (A \cap B) = A \cup B$.

Solution:

$(A \Delta B) = (A - B) \cup (B - A)$

$= \{1, 6\} \cup \{4, 7\}$

$= \{1, 4, 6, 7\}$

Therefore,

$(A \Delta B) \Delta C = \{1, 6\} \cup \{2, 5\}$

$= \{1, 2, 5, 6\}$.

Similarly,

$(B \Delta C) = \{3, 9\} \cup \{2, 5\}$

$= \{2, 3, 5, 9\}$.

Therefore,

$A \Delta (B \Delta C) = \{1, 6\} \cup \{2, 5\}$

$= \{1, 2, 5, 6\}$

Hence,

$(A \Delta B) \Delta C = A \Delta (B \Delta C)$

(ii) $A \Delta (A \cap B) = A - B$

$(A \cap B) = \{3, 9\}$

And

$A \Delta (A \cap B) = \{1, 6\} \cup \{\emptyset\}$

$= \{1, 6\}$.

And

$A - B = \{1, 3, 6, 9\} - \{3, 4, 7, 9\}$.

$= \{1, 6\}$.

And hence,

$A \Delta (A \cap B) = A - B$.

(iii) $(A \Delta B) = (A - B) \cup (B - A)$

$= \{1, 6\} \cup \{4, 7\}$.

$= \{1, 4, 6, 7\}$.

And $(A \cap B) = \{3, 9\}$

Therefore,

$(A \Delta B) \cup (A \cap B) = \{1, 4, 6, 7\} \cup \{3, 9\}$
$= \{1, 3, 4, 6, 7, 9\}$.

And

$A \cup B = \{1, 3, 4, 6, 7, 9\}$.

And hence $(A \Delta B) \cup (A \cap B) = A \cup B$.

Cardinal Number of a Set

The numbers of elements in a finite set say Set A is called as cardinal number of A and symbolically represented as n (A).

For example

If Set A = {1, 2, 3, 4, 5, 6, 7} then n(A) = 7 as A contains only seven elements.

Some Important Results

(i) $n(A \cup B) = n(A) + n(B) - n(A \cap B)$.

Note: If Set A and Set B are disjoint sets then $(A \cap B) = \{\emptyset\}$.

Therefore $n(A \cup B) = n(A) + n(B)$

(ii) $n(A \cup B \cup C) = n(A) + n(B) + n(C) - n(A \cap B) - n(B \cap C)$
$- n(A \cap C) + n(A \cap B \cap C)$.

(iii) $n(A') = n(U) - n(A)$.

(iv) $n(A) = n[(A \cap B') \cup (A \cap B)]$
$= n[(A - B) \cup (A \cap B)] = n(A - B) + n(A \cap B)$.

(v) $n(A \cup B) = n(A - B) + n(A \cap B) + n(B - A)$.

Illustration No. 1

In a class of 30 students, 14 has taken mathematics, 10 has taken mathematics but not economics. Find the number of students who had taken mathematics and economics also find the number of students who had taken economics but not mathematics?

Solution:

Let

A = Set of students who have taken mathematics as subject.

B = Set of students who have taken economics as subject.

Then given are $n(A \cup B) = 30$, $n(A) = 14$ and $n(A \cap B') = 10$.

Now we need to find the students who have taken both the subjects i.e. $n(A \cap B)$ and the number of students who have taken economics as subject but not mathematics i.e. $n(B \cap A')$

But

$n(A) = n(A \cap B') + n(A \cap B)$

$14 = 10 + n(A \cap B)$.

$n(A \cap B) = 14 - 10$.

$= 4$

But

$n(A \cup B) = n(A) + n(B) - n(A \cap B)$.

$30 = 14 + n(B) - 4$.

$n(B) = 30 - 10 = 20$.

Therefore,

$n(B) = n(B \cap A') + n(A \cap B)$

$20 = n(B \cap A') + 4$.

$n(B \cap A') = 20 - 4 = 16$.

Illustration No. 2

Out of 200 students in a management school 120 students read Indian Economic Review and 100 read Harvard Business Review if the number of students who read neither of the journals is 40 then find the number of students who read both them?

Solution:

Let

A = Set of students who read Indian Economic Review.

B = Set of students who read Harvard Business Review.

Then given are $n(U) = 200$, $n(A) = 120$, $n(B) = 100$ and $n(A' \cap B') = 40$.

We need to find the number of students who read both the journals i.e. Indian Economic Review and Harvard Business Review i.e. $n(A \cap B)$

But,

$A' \cap B' = (A \cup B)'$ therefore $n(A \cup B)' = 40$.

Since $n(A \cup B)' = n(U) - n(A \cup B)$ or $n(A \cup B) = 200 - 40 = 160$

But as we know,

$$n(A \cup B) = n(A) + n(B) - n(A \cap B).$$

$$160 = 120 + 100 - n(A \cap B).$$

$$n(A \cap B) = 220 - 160 = 60.$$

Thus there are 60 students who read both the journals.

Illustration No. 3

In an examination 100 students secured 80% and more marks in Economics or Accounts. Out of these 70 obtained 80% and more marks in Economics and 20 in both Economics and Accounts. Then find how many of them have secured 80% and more marks in Accounts only?

Solution:

Let

A = Set of students who scored over 80% in Economics.

B = Set of students who scored over 80% in Accounts.

Then given are $n(A \cup B) = 100$, $n(A) = 70$, and $n(A \cap B) = 20$.

We need to find the number of students who scored over 80% in Accounts.

But we know,

$$n(A \cup B) = n(A) + n(B) - n(A \cap B)$$

$$100 = 70 + n(B) - 20.$$

$$n(B) = 100 - 50 = 50.$$

As we need to find the students who scored over 80% in Accounts are

$$N(A' \cap B) = n(B) - n(A \cap B)$$

$$= 50 - 20 = 30.$$

Illustration No. 4

In a pollution study of 200 cities has revealed the following report, 70 were polluted by Sulphur compounds, 60 were polluted by Lead and 80 were polluted by Fly Ash, 40 were polluted by Sulphur and Fly Ash, 30 were polluted by Sulphur and Lead , 35 were polluted by Lead and Fly ash and 10 cities were polluted by all three compounds. Then find

(i) How many were polluted by one of three impurities?

(ii) How many cities are only polluted by Sulphur, Lead and Fly Ash?

(iii) How many of the cities are not polluted at all?

Solution:

Let

S = City polluted by Sulphur compounds.

L = City polluted by Lead.

F = City polluted by Fly Ash.

(i) Since we need to find the number of cities polluted by at least one of the pollutants is given by

$$n(S \cup L \cup F) = n(S) + n(L) + n(F) - n(S \cap L) - n(L \cap F) - n(S \cap F) + n(S \cap L \cap F)$$

$$= 70 + 60 + 80 - 40 - 30 - 35 + 10$$

$$= 115.$$

Thus 115 cities are polluted by atleast one of the pollutants.

(ii) Cities polluted by only one pollutant are

Case I: By Sulphur Compounds,

$$n(S \cap L' \cap F') = n[S \cap (L \cup F)']$$

$$= n(S) - n(S \cap L) - n(S \cap F) + n(S \cap L \cap F)$$

$$= 70 - 30 - 40 + 10.$$

$$= 10$$

Case II By Lead,

$$n(L \cap S' \cap F') = n[L \cap (S \cup F)']$$

$$= n(L) - n(L \cap S) - n(L \cap F) + n(S \cap L \cap F)$$

$$= 60 - 30 - 35 + 10.$$

$$= 5$$

Case III : By Fly Ash,

$$n(F \cap S' \cap L') = n[F \cap (S \cup L)']$$

$$= n(F) - n(F \cap S) - n(F \cap L) + n(S \cap L \cap F)$$

$$= 80 - 35 - 40 + 10$$

$$= 15$$

(iii) To find the number of cities which are not polluted were

$$= n(U) - n((S \cup L \cup F)$$

$$= 200 - 115 = 85$$

Ordered Pairs

An ordered pair consists of two elements, say a and b represented within parenthesis as (a, b) where a is the first member and b the second member.

For example: the odd numbers and their squares can be represented in the form of ordered pairs as

(1, 1); (3, 9); (5, 25); (7, 49)............................

Note

Two ordered pair (a, b) and (c, d) are said to be equal if and only if a = c and b = d. In other words (a, b) = (c, d) $\Leftrightarrow$ a = c and b = d.

Cartesian Products

If Set A and Set B are two sets then the set of all ordered pairs whose first member belongs to Set A and the second member belongs to Set B is the Cartesian product of A and B in that order and read as A cross B and symbolically represented as

$A \times B = \{ (x, y) ; x \in A \text{ and } y \in B\}$

Note

(i) If $A \times B$ and $B \times A$ have same number of elements then $A \times B \neq B \times A$ unless and until A = B . Therefore the Cartesian product of two sets is commutative if the two sets are equal.

(ii) If Set A has m elements and Set B has n elements then $A \times B$ has mn elements.

(iii) If Set A and Set B are disjoint then the Cartesian product $A \times B$ is also disjoint.

(iv) If either of Set A or Set B is a null set then the set $A \times B$ is also a null set.

(v) If either of Set A or Set B is a infinite set and the other is a non empty set then the set $A \times B$ is also an infinite set.

(vi) If Set A or Set B are finite sets then $n(A \times B) = n(A) \times n(B)$.

Some Important Properties

(i) If Set A, Set B and Set C are three sets and $A \subseteq B$, then $(A \times C) \subseteq (B \times C)$

For example

Let $\{1, 2\} \in A \times C$ then $1 \in A$ and $2 \in C$

$\Leftrightarrow$ 1 $\in$ B and 2 $\in$ C (Since A $\subseteq$ B)

$\Leftrightarrow$ {1, 2} $\in$ (B × C)

Hence (A × C) $\subseteq$ (B × C)

Alternate Proof

Let { x, y } $\in$ A × C then x $\in$ A and y $\in$ C

$\Leftrightarrow$ x $\in$ B and y $\in$ C (Since A $\subseteq$ B)

$\Leftrightarrow$ { x, y} $\in$ (B × C)

Hence (A × C) $\subseteq$ (B × C)

(ii) If Set A, Set B and Set C are three sets and A $\subseteq$ B , and C $\subseteq$ D then (A × C) $\subseteq$ (B × D)

For example

let {1, 2} $\in$ A × C then 1 $\in$ A and 2 $\in$ C

$\Leftrightarrow$ 1 $\in$ B and 2 $\in$ D (Since A $\subseteq$ B and C $\subseteq$ D)

$\Leftrightarrow$ {1, 2} $\in$ (B × D)

Hence (A × C) $\subseteq$ (B × D)

Alternate Proof

let {x, y} $\in$ A × C then x $\in$ A and y $\in$ C

$\Leftrightarrow$ x $\in$ B and y $\in$ D (Since A $\subseteq$ B and C $\subseteq$ D)

$\Leftrightarrow$ {x, y} $\in$ (B × D)

Hence (A × C) $\subseteq$ (B × D)

(iii) If Set A and Set B are two sets and A $\subseteq$ B , then (A × A) = (A × B) $\cap$ (B × A).

For example

Let {1, 2} $\in$ A × A then 1 $\in$ A and 2 $\in$ A

$\Leftrightarrow$ (1 $\in$ A and 2 $\in$ A) and (1 $\in$ A and 2 $\in$ A)

$\Leftrightarrow$ (1 $\in$ A and 2 $\in$ B) and (1 $\in$ B and 2 $\in$ A) (Since A $\subseteq$ B)

$\Leftrightarrow$ {1, 2} $\in$ (A × B) and {1, 2} $\in$ (B × A)

$\Leftrightarrow$ {1, 2} $\in$ (A × B) $\cap$ (B × A)

Hence (A × A) = (A × B) $\cap$ (B × A)

Alternate Proof:

Let $\{x, y\} \in A \times A$ then $x \in A$ and $y \in A$

$\Leftrightarrow$ ($x \in A$ and $y \in A$) and ($x \in A$ and $y \in A$)

$\Leftrightarrow$ ($x \in A$ and $y \in B$) and ($x \in B$ and $y \in A$) { Since $A \subseteq B$ }

$\Leftrightarrow$ $\{x, y\} \in (A \times B)$ and $\{x, y\} \in (B \times A)$

$\Leftrightarrow$ $\{x, y\} \in (A \times B) \cap (B \times A)$

Hence $(A \times A) = (A \times B) \cap (B \times A)$.

(iv) If Set A, Set B and Set C are three sets then

$A \times (B - C) = (A \times B) - (A \times C)$

For example

Let $(1, 2) \in A \times (B - C)$ then $1 \in A$ and $2 \in (B - C)$

$\Leftrightarrow$ $1 \in A$ and $2 \in B$ and $2 \notin C$

$\Leftrightarrow$ ($1 \in A$ and $2 \in B$) and ($1 \in A$ and $2 \notin C$)

$\Leftrightarrow$ $\{1, 2\} \in (A \times B)$ and $\{1, 2\} \notin (A \times C)$

$\Leftrightarrow$ $\{1, 2\} \in [(A \times B) - (A \times C)]$

Hence $A \times (B - C) = (A \times B) - (A \times C)$.

Alternate Proof

Let $(x, y) \in A \times (B - C)$ then $x \in A$ and $y \in (B - C)$

$\Leftrightarrow$ $x \in A$ and $y \in B$ and $y \notin C$

$\Leftrightarrow$ ($x \in A$ and $y \in B$) and ($x \in A$ and $y \notin C$)

$\Leftrightarrow$ $\{x, y\} \in (A \times B)$ and $\{x, y\} \notin (A \times C)$

$\Leftrightarrow$ $\{x, y\} \in [(A \times B) - (A \times C)]$

Hence $A \times (B - C) = (A \times B) - (A \times C)$.

Illustration No. 1

If Set $A = \{1, 2, 3\}$ and Set $B = \{1, 2\}$ then prove that $A \times B \neq B \times A$

Solution:

The Cartesian product $A \times B = \{(1, 1), (1, 2), (2, 1), (2, 2), (3, 1) (3, 2)\}$

Similarly $B \times A = \{(1, 1), (1, 2), (1, 3), (2, 1), (2, 2), (2, 3)\}$

Thus (3, 1) and (3, 2) are elements of A × B are not elements B × A. Similarly (2,3) is a element in B × A but not A × B, therefore A ×B ≠ B × A

Illustration No. 2

If Set A = {1, 2} , Set B = {2, 3} and Set C {3, 4} then find

(i) A × (B ∪ C)

(ii) (A × B) ∩ (A × C).

(iii) A × (B ∩ C).

Solution:

(i) (B ∪ C) = {2, 3} ∪ {3, 4}

= {2, 3, 4}.

Therefore,

A × (B ∪ C) = {1, 2} × {2, 3, 4}.

= { (1, 2), (1, 3), (1, 4), (2, 2), (2, 3) ,(2, 4) }.

(ii) (A × B) = { 1, 2} × { 2, 3}

= { (1,2) , (1,3), (2,2), (2,3) }.

And (A × C) = { 1, 2} × { 3, 4}

= { (1, 3) , (1, 4), (2, 3), (2, 4) }

Therefore

(A × B) ∩ (A × C) = { (1, 3), (2, 3) }.

(iii) (B ∩ C) = { 2, 3 } ∩ {3, 4}

= { 3 }.

Therefore A × (B ∩ C) = { 1, 2} × { 3 }.

= { (1, 3) , (2, 3) }.

Illustration No. 3

If Set A = {1, 2} , Set B = {1, 2, 3} and Set C {1, 2, 3, 4} then verify whether A × (B ∩ C) = (A × B) ∩ (A × C) ?

Solution:

(B ∩ C) = {1, 2, 3} ∩ {1, 2, 3, 4}

= {1, 2, 3}

Therefore

$A \times (B \cap C) = \{1, 2\} \times \{1, 2, 3\}$.

$= \{(1, 1), (1, 2), (1, 3), (2, 1), (2, 2), (2, 3)\}$

But,

$(A \times B) = \{1, 2\} \times \{1, 2, 3\}$

$= \{(1, 1), (1, 2), (1, 3), (2, 1), (2, 2), (2, 3)\}$

And

$(A \times C) = \{(1, 1), (1, 2), (1, 3), (1, 4), (2, 1), (2, 2), (2, 3), (2, 4)\}$.

Therefore,

$(A \times B) \cap (A \times C) = \{(1, 1), (1, 2), (1, 3), (2, 1), (2, 2), (2, 3)\}$.

Hence $A \times (B \cap C) = (A \times B) \cap (A \times C)$

Summary

A set is a well defined collection of objects and these objects are tended as the members or elements of the set.

A set which has no elements is called as null, empty or void set denoted by {Ø} and read as phi in Greek and Latin.

If for a given set A, a set consisting of all the subsets of A is called the power of the set. The power set this is denoted by P (A).

If A and B are two sets then the union of Set A and Set B is the set consisting of either all the elements of Set A or Set B or both and denoted by "A ∪ B" and read as A union B or A cup B

i. e. A ∪ B = {x : x €A or x € B or x € both A and B}

The complement of a set A is that set which contains all those elements of the universal set U which are not in A. The complement of set A is the set U – A and is denoted by A^c, A′, A or ~ A. It can symbolically written as

$A' = U - A = \{x : x \in U \text{ and } x \notin A\}$

The number of elements in a finite set say Set A is called as cardinal number of A and symbolically represented as n (A).

If Set A and Set B are two sets then the set of all ordered pairs whose first member belongs to Set A and the second member belongs to Set B is the Cartesian product of A and B in that order and read as A cross B and symbolically represented as

$A \times B = \{(x, y) ; x \in A \text{ and } y \in B\}$

Self Practice Exercises

1. If Set A = {1, 2, 3, 4} and Set B = {3, 4, 5, 6} then find
   (i) $A \cup B$ (ii) $A \cap B$
2. If Set A = {1, 2, 3, 4, 5}, Set B = {4, 5, 6, 7} and
   Set C = {2, 3, 5, 6, 7} then find
   (i) $A \cap (B \cup C)$. (ii) $A \cup (B \cap C)$.
3. If Set A ={ 1, 2, 3 }, Set B = { 2, 3, 4, 5} and Set C = {3, 5, 6} then find
   (i) $(A \cup B) \cap (B \cup C)$ (ii) $(A \cap B) \cup (B \cap C)$
4. If Set A ={1, 2, 3, 4}, Set B = {2, 3, 4, 5} and Set C = {3, 4, 5, 6} then find
   (i) $(A - B) \cup C$ (ii) $A - (B \cap C)$
   (iii) $(A - B) \cup (B - C)$ (iv) $(A \cap B) - (B \cap C)$
5. If Set A ={ 1, 2, 3, 4, 5 } and Set B = { 3, 4, 5, 6, 7, 8 } then find
   (i) $A - (A - B)$ (ii) $(A - B) \cap (B - A)$ (iii) $(A - B) \cup (B - A)$
6. If Set U = {0, 1, 2, 3, 4, 5}, Set A = {0, 1, 2, 3} and
   Set B = {2, 3, 4} then find,
   (i) $(A' \cup B) \cap (B' \cup A)$ (ii) $(A' \cap B) \cup (B' \cap A)$.
7. If Set A = {1, 2, 3, 4, 5}, Set B = {3, 4, 5, 6, 7} and Set C = {5, 6, 7, 8} then find
   (i) $A \cup (B \cap C)$ (ii) $A \cap (B \cup C)$ (iii) $A \cap (B \cap C)$
8. If Set A = {1, 2, 3}, Set B = {2, 3, 4, 5} and Set C ={2, 4, 5, 6, 8} then prove that
   (i) $A \cap (B - C) = (A \cap B) - (A \cap C)$
   (ii) $(A \cap B) - C = (A - C) \cap (B - C)$
9. If Set U = {1, 2, 3, 4, 5, 6, 7, 8, 9}, Set A = {4, 5, 7}, Set B = {1, 2, 3, 4} and Set C = {3, 4, 5, 7} then find
   (i) $A' \cup B'$ (ii) $A' \cap B'$. (iii) $(A \cup B \cup C)'$.
10. if Set U = {1, 2, 3, 4, 5, 6, 7}, Set A = {1, 2, 3}, Set B = {3, 4, 5} and Set C = {4, 5, 7} then find
    (i) $A' \cap (B \cup C)$ (ii) $(A' \cap B) \cup (A \cap B')$
11. If Set A = {1, 2, 3, 4, 5} and Set B = {3, 4, 5, 6, 7} then find
    (i) $(A - B) \cup (B - A)$ (ii) $(A - B) \cap (B - A)$

12. If Set A = {1, 2, 3, 4, 5}, Set B = {3, 4, 5, 6} and Set C = {3, 5, 7, 8} then find
    (i) (A ∩ B) ∪ (B – C) (ii) (A U B) ∩ (B – C)

13. In a survey conducted in a town of 2000 population to see the readership of newspaper, it has revealed that 1000 read Hindustan Times, 800 read Times of India and 700 read Indian Express. If there was 400 who read both Hindustan Times and Times of India, 300 who read both Times of India and Indian Express, 500 read only Hindustan times and 100 read all the three papers. Then find,
    (i) How many read both Hindustan Times and Indian Express?
    (ii) How of them read only one paper?
    (iii) How many of them read none of the newspaper?

14. In a class of management course of 120 students appeared for their tri semester examinations of Business Mathematics, Business Statistics and Business Economics. After the results were declared it was found that 50 failed in Business Mathematics, 48 failed in Business Statistics and 64 failed in Business Economics. Moreover 18 failed in Business Mathematics alone, 12 failed Business Statistics alone and 10 in Business Statistics and Business Economics then find
    (i) How many failed in all the subjects?
    (ii) How many passed in all the subjects?
    (iii) How many passed in Business Economics alone?

15. In an examination 86 passed in marketing paper, 96 passed in managerial economics paper and 104 has passed in quantitative techniques paper. Only 16 students could manage to pass all the three, 28 passed in marketing and managerial economics paper and 42 passed in marketing and quantitative techniques paper and 40 passed in managerial economics paper and quantitative techniques paper then find,
    (i) How many students passed in only marketing paper?
    (ii) Find the ratio of students passing in managerial economics and quantitative techniques paper?
    (iii) If a student is declared pass in the examination provided he or she clears at least two subjects. Find how many were declared pass?

16. In a picnic party it was found that 20 people like all three beverages tea, coffee and cold drink. It was found that 40 people like coffee and tea, 50 like coffee and cold drink and 50 like tea and cold drink. It was noticed that 110 liked coffee, 100 tea and 100 cold drink. From the above mentioned information find
    (i) How many picnic goers preferred only coffee, only tea and only cold drink?
    (ii) How many of the picnic goers liked both tea and cold drink but not coffee?
    (iii) How many of the picnic goers liked at least two of the beverages?
17. The combined membership of Accounts Society and Mathematical society is 100. Find the number of members in Mathematical society knowing that 60 members are the members of accounts society and 25 are the members of both the institutions?
18. If Set A = { 1, 3 } , Set B = { 2, 4 } and Set C = { 3, 4} then find
    (i) $A \times (B \cup C)$ (ii) $A \times (B \cap C)$ (iii) $(A \times B) \cap (A \times C)$
19. If Set A ={ a, b} , Set B = { c, d } and Set C = { e, f } show that $A \times B \neq B \times A$ and also find $(A \times B) \cup (A \times C)$?
20. If Set A = { 1, 2, 3 } , Set B = { 1, 2 } and Set C = { 2, 4} then find
    (i) $A \times (B \cap C)$ (ii) $(A \times B) \cap (A \times C)$

Answers

1. (i) {1, 2, 3, 4, 5, 6}
   (ii) {3, 4}
2. (i) {2, 3, 4, 5}
   (ii) {1, 2, 3, 4, 5, 6, 7}
3. (i) {2, 3, 4, 5}
   (ii) {2, 3, 5}
4. (i) {1, 3, 4, 5, 6}
   (ii) {1, 2}
   (iii) {1, 2}
   (iv) { 2 }
5. (i) {3, 4, 5} , (ii) { Ø }, (iii) {1, 2, 6, 7, 8}

6. (i) {2, 3, 5} (ii) {0, 1, 4}
7. (i) {1, 2, 3, 4, 5, 6, 7}, (ii) {3, 4, 5} and (iii) { 5 }
9. (i) {1, 2, 3, 5, 6, 7, 8, 9}, (ii) {6, 8, 9} and (iii) {6, 8, 9}
10. (i) {4, 5, 7} and (ii) {1, 2, 4, 5}
11. (i) {1, 2, 6, 7} and (ii) { Ø }
12. (i) {3, 4, 5, 6} and (ii) {4, 6}
13. (i) 100, (ii) 1000 and (iii) 300
14 (i) 20, (ii) 20 and (iii) 28.
15. (i) 32, (ii) 22: 19 and (iii) 78
16. (i) 90, (ii) 30 and (iii) 100
17. 65.
18. (i) { (1, 2), (1, 3), (1, 4), (3, 2), (3, 2), (3, 4) }
(ii) { (1, 4), (3, 4)}
(iii) { (3, 4) }.
19. { (a, c), (a, d), (a, e), (a, f), (b, c), (b, d), (b, e), (b, f)}.
20. (i) { (1, 2),(2, 2), (3, 2) }
(ii) { (1, 2),(2, 2), (3, 2) }

CHAPTER 2

Functions and its Applications

Brief History and Introduction

The word function first appeared in Latin Manuscript "Methodus Tangentium Inversa, Seu De Functionibus "by Leibnitz in the year 1673. Later on Bernoulli was the first to assign a specialized use of the term function in an analytical sense. Today we use function in general sense to understand the interrelationship among variables which are often expressed in the form of equations or inequalities showing that how a function is affected by the movement of the variables by which it is associated. For example demand is a function of price, income, taste and preferences etc. So the demand function can be written as a function of

$$D = f(P, I, T \ldots)$$

If any change takes place in any of these variables is likely to affect demand. In this chapter we will highlight the fundamental concepts of functions and its usage in business and economics.

Function

Let x and y are real valued variables such that for every value of x in a given set X, we get a definite value of y then y is termed as a function of x which can be written as $y = f(x)$, where x is the independent variable and y is the dependent variable.

For example,

(i) $y = 5x + 4$ where $2 \leq x \leq 4$,

(ii) $y = 3x^2 + 4x + 5$ where $1 \leq x \leq 3$.

Definition of a Function or Mapping

Let X and Y be two non-empty sets with 'x' being a variable whose domain is set X. If there exists a rule say f which associates each element x of X with a unique element y in Y. Then f is termed as a function of x over the set X or f maps X to Y. Here X is called as the domain of the function f and Y is the co-domain.

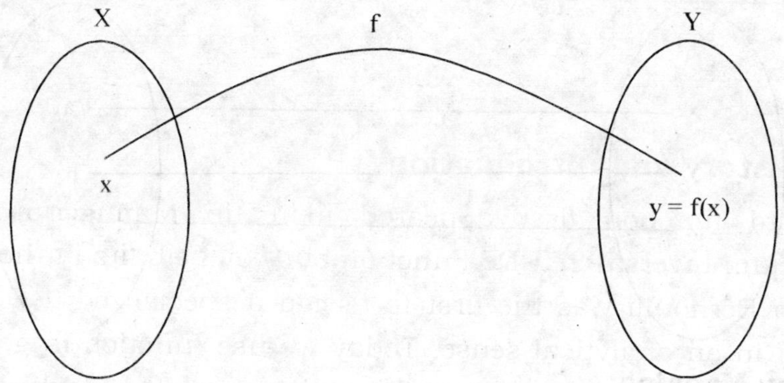

The element y which corresponds to a given x written in the form of f(x) is called the image of x under f while x is called the pre-image of f(x).

Note:

1. If an element x of the set X having one and only one image in Y under f then the function is a single valued function of x. But for every x in X there exists two or more images in Y then the function is a multi valued function of x

For example,

(i) $y = 5x + 4$ where $2 \leq x \leq 4$, is a single valued functions of x.

(ii) $y^2 = 9x$ where $1 \leq x \leq 9$ is a multi valued function.

(iii) Each element of X must have a unique image in Y but all elements of Y may not appear as images under f.

(iv) Range of f is defined to consist those elements in Y which appear as the image of at least one element in X. Thus, there can be more than one element in set X having the same image in Y and written as

$$f(X) = \{f(x) : x \in X\} = \{y : y \in Y, y = f(x) \text{ for some } x \in X\}$$

then the image set f(X) is called the range of f.

(v) The rule f which associates an element in X with a unique element of Y may be given by a relation or by a set of order pairs.

Types of Mapping

One–One Mapping (or Injection)

If each element in X have different f-image in Y then the mapping is one–one mapping i.e f(x) = f(a) then x = a.

One–Many Mapping

If each element in X has more than one f–image in Y then the mapping is called as one to many mapping i.e f(x) = f(a) = f(b) even if a ≠ b for all a, b ∈ Y

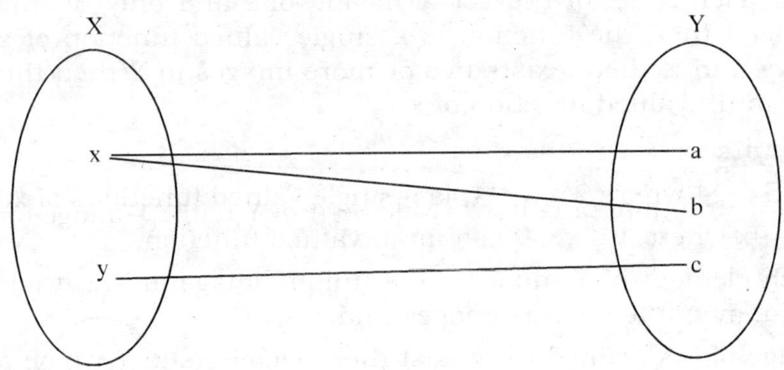

Many–One Mapping

If two or more different elements in X have same f-image in Y then the mapping is called many to one mapping i.e f(x) = f(y) = f(a) even if x ≠ y for all x, y ∈X

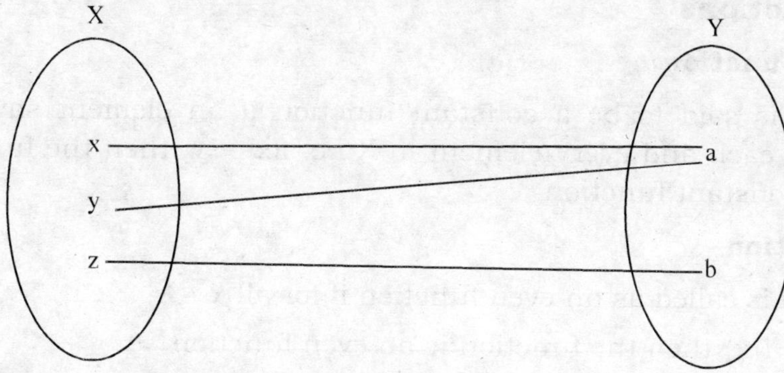

Into–Mapping

A mapping is into mapping if there exists at least one element in Y which is not the f–image of any element in X, but all the elements in X has f–image in Y.

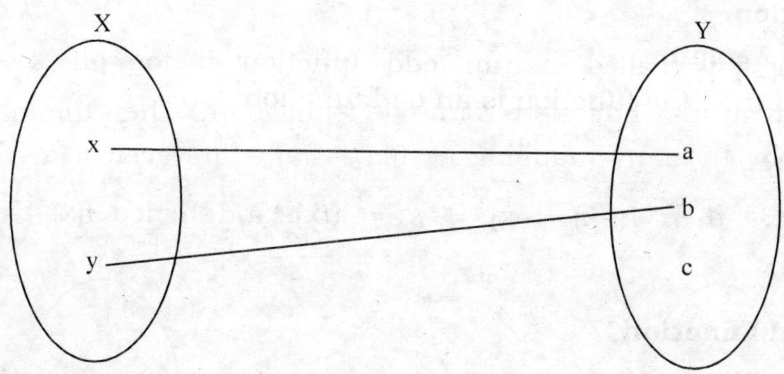

Onto–Mapping

A mapping is onto mapping if every element of Y is the f–image of at least one element in X.

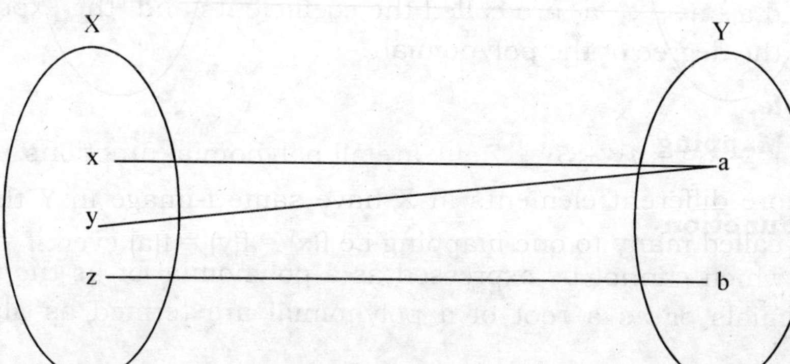

Some Functions

Constant Function

A function is said to be a constant function if an element say $y \in Y$ is assigned to each and every element in X i.e. f(x) = y, then the function is said to be constant function.

Even Function

A function f is called as an even function if for all $x \in X$,

if $f(-x) = f(x)$ then the function is an even function.

For example,

If $f(x) = x^2$ then $f(-x) = (-x)^2 = x^2 = f(x)$ and hence f(x) is an even function of x.

Odd Function

A function f is called as an odd function if for all $x \in X$, if $f(-x) = -f(x)$ then the function is an odd function.

For example,

If $f(x) = x^3$ then $f(-x) = (-x)^3 = -x^3 = -f(x)$ and hence f(x) is an odd function of x.

Polynomial Function

A function which can be written in the general form of

$f(x) = a_n x^n + a_{n-1} x^{n-1} + a_{n-2} x^{n-2} + \dots a_0$ where a_is (i = 0, 1, 2, ... n) are real constant, $a_n \neq 0$ and n is a positive integer called the polynomial in x of degree n and $a_n, a_{n-1}, \dots a_0$ are called the coefficients and the exponent n is called as the degree of the polynomial.

For example,

$2x^2 + 4x + 5$, $3x^3 + 4x^2 + 5x + 7$ etc. are all polynomial functions.

Algebraic Function

A function which cannot be expressed as a polynomial or as the ratio of two polynomials or as a root of a polynomial are termed as algebraic functions.

For example,

$\sqrt{x+5}$ *and* $\dfrac{2x+3}{\sqrt{3x+5}}$ are algebraic functions

Rational Function

The ratio of two polynomial functions say P(x) and Q(x) of a variable x such that $R(x) = \dfrac{P(x)}{Q(x)}$ where $Q(x) \neq 0$ for any value of x is called as a rational function of x.

For example,

$\dfrac{3x+5}{2x-6}$ for $x \neq 3$ is a rational function

Irrational Function

The algebraic functions which are not rational functions are irrational functions.

For example,

$\sqrt{2x+3}, \sqrt{x^2+3x+4}$ and $\dfrac{x}{\sqrt{3x+7}}$ are all irrational functions

Explicit Function

A function which can expressed in terms of an independent variable say x then it is called as explicit function

For example,

$y = \dfrac{4(x-3)}{5}$ is an explicit function.

Implicit Function

A function which is not an explicit function is an implicit function.

For example,

$3x + 4y + 2xy = 0$ is an implicit function

Bounded Function

If f(x) is a function defined in the domain of X and two finite numbers a and b can be found such that $a \leq f(x) \leq b$ for all x belongs to X, then f(x) is said to be bounded both above and below in the domain of X and we say that f(x) is bounded in X.

For example,

If $f(x) = x^2 + 3x + 2$ defined in the domain of $1 \le x \le 3$

Inverse Function

If a function $y = f(x)$ is defined in an interval such that we can express x as a function of y say $x = g(x)$ then $g(x)$ is called the inverse of $f(x)$.

For example,

If $y = \frac{ax+b}{cx+d}$ then $x = \frac{b-dy}{cy-a}$ is the inverse of the function y.

Exponential Function

A function having a constant base and a variable exponent is termed as an exponential function

For example,

$y = 5^x$, $y = 3^{x+2}$ are some exponential functions.

Logarithmic Function

A function which can be written in the form of $y = \log_c x$ where $c > 0$ and $c \neq 1$ and read as y is the log to the base c of x is called a logarithmic function.

The above function $y = \log_c x$ can also be rewritten as $x = c^y$ which is an exponential function. Thus logarithmic and exponential functions are inverse functions.

For example,

$y = \log_2 x$ and $y = \log_3 x^2$ are some logarithmic functions.

Incommensurable Power Function

A function which has a variable base and a constant exponent are termed as incommensurable power function.

For example,

$y = x^{\frac{4}{3}}$ and $y = x^{\sqrt{2}}$ are some incommensurable power functions.

Trigonometric Function

A trigonometric function is generally written in the form $y = \sin\theta$ or any other trigonometric function.

Note: The exponential, logarithmic, incommensurable power function and trigonometric function are examples of transcendental functions which are not algebraic functions.

Periodic Function

A function f is called as periodic function if $f(x + \alpha) = f(x)$ where α is a real number then f(x) is called as periodic function.

For example,

$\sin(2\pi + x) = \sin x$ for all x is periodic function with period as 2π radians.

Composite Function

If $y = f(x)$ and $x = g(z)$ then $y = f\{g(z)\}$ is called as function of function or a composite function.

Absolute Value Function

The functional relationship of the given form $y = f(x) = |x|$ is called as absolute value function where $|x|$ represents the absolute value of x and is written as

$$|x| = \begin{cases} x & \text{when } x \geq 0 \\ 0 & \text{when } x = 0 \\ -x & \text{when } x \leq 0 \end{cases}$$

Graphically presented as

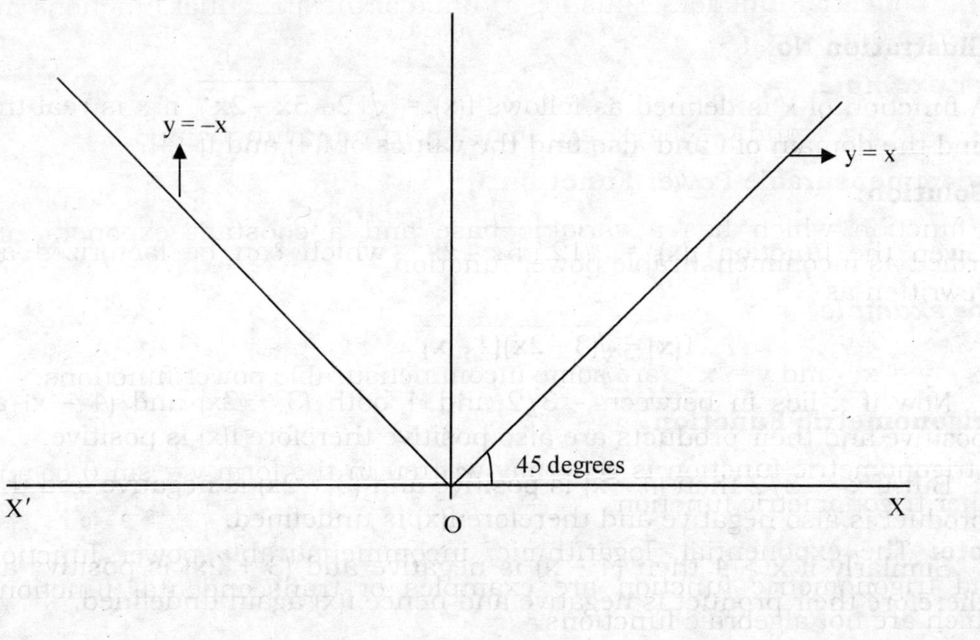

Step Function

If a function is defined under a closed interval say [a, b] and takes a constant value within the interior of each sub interval $[a, x_1]$, $[x_1, x_2]$ ... $[x_n, b]$ of [a, b] where $a < x_1 < x_2 < \ldots x_n < b$, then such a function is called as step function. Mathematically can be written as $f(x) = k_i$ for all values of x in the ith sub interval.

Graphically presented as shown below

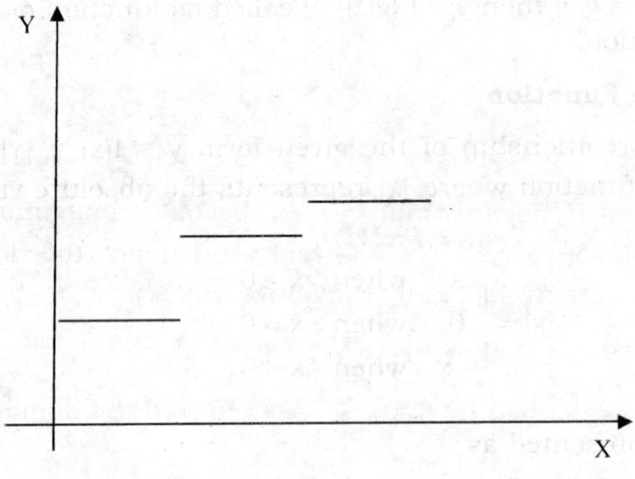

Illustration No. 1

A function of x is defined as follows $f(x) = \sqrt{12+5x-2x^2}$ if x is real then find the domain of f and also find the values of f(4) and f(– 4)?

Solution:

Given the function $f(x) = \sqrt{12+5x-2x^2}$ which can be factorized and rewritten as

$$f(x) = \sqrt{(3+2x)(4-x)}$$

Now if x lies in between – 3/2 and 4 both (3 + 2x) and (4 – x) are positive and their products are also positive therefore f(x) is positive.

But if x < –3/2 then (4 – x) is positive and (3 + 2x) is negative and their product is also negative and therefore f(x) is undefined.

Similarly if x > 4 then (4 – x) is negative and (3 + 2x) is positive and therefore their product is negative and hence f(x) again undefined.

Therefore the domain of the function f(x) is $-3/2 \le x \le 4$.

Now we need to find the value of f(4) and f(– 4)

Hence,

(i) $f(4) = \sqrt{(3+2x)(4-x)} = \sqrt{(3+2.4)(4-4)} = 0$ and

(ii) $f(-4) = \sqrt{(3+2x)(4-x)} = \sqrt{(3+2(-4))(4-(-4))} = \sqrt{-40}$ which is not real.

Illustration No. 2

Find the domain for the following functions $f(x) = \sqrt{7-x}$

Solution:

Since we have,

$$f(x) = \sqrt{7-x}$$

when $x > 7$, $7 - x$ is negative and $\sqrt{7-x}$ becomes imaginary and when $x \le 7$, $7 - x$ is positive so $\sqrt{7-x}$ is real. And hence the domain of the function of the function f(x) is all real values of $x \le 7$.

Illustration No. 3

If a function f(x) is defined by $f(x) = \frac{x}{x+6}$ then find the domain and range of the function?

Solution:

We have $$f(x) = \frac{x}{x+6}$$

clearly f(x) is defined only when $x + 6 \neq 0$ i.e. $x \neq -6$,

Therefore the domain of the function $f(x) = R - \{-6\}$

Let $$y = f(x) = \frac{x}{x+6}$$

Therefore $$y = \frac{x}{x+6}$$

$$y(x + 6) = x$$

$$yx + 6y = x$$

$$yx - x = -6y$$

$$x(y - 1) = -6y$$

$$x = \frac{6y}{1-y}$$

From above its clear that x is not defined for y = 1

Therefore the range of f(x) = R – {1}.

Illustration No. 4

If a function f(x) is defined by $f(x) = \sqrt{4-x^2}$ then find the domain and range of the function?

Solution

We have

$$f(x) = \sqrt{4-x^2}$$

Clearly f(x) is defined when $4 - x^2 \geq 0$ or $x^2 - 4 \leq 0$,

$$(x - 2)(x + 2) \leq 0,$$

$-2 \leq f(x) \leq 2$ within this range f(x) is non-negative, therefore $f(x) \in [0, 2]$

Hence the range of f(x) = [0, 2]

Illustration No. 5

Find the domain of the function defined by $f(x) = |x|$?

Solution

As we know that

$$f(x)\begin{cases} = x & \text{if } x > 0 \\ = 0 & \text{if } x = 0 \\ = -x & \text{if } x < 0 \end{cases}$$

From above we see that for every value of x in R, we have a unique value for f(x) and hence the domain of x is the set of all real numbers R.

Illustration No.6

If y = f(x) is defined as $y = f(x) = \frac{x+2}{x+3}$ then find f(1) and f(–1)? Then also find the value of f(y)?

Solution:

Given

$$y = f(x) = \frac{x+2}{x+3}$$

then $$f(1) = \frac{1+2}{1+3} = \frac{3}{4}$$

and $$f(-1) = \frac{-1+2}{-1+3} = \frac{1}{2}$$

But $$f(y) = \frac{y+2}{y+3} = \frac{\left(\frac{x+2}{x+3}\right)+2}{\left(\frac{x+2}{x+3}\right)+3} \quad \left\{\text{Since } y = f(x) = \frac{x+2}{x+3}\right\}$$

$$= \frac{\left\{\frac{(x+2)+2(x+3)}{(x+3)}\right\}}{\left\{\frac{(x+2)+3(x+3)}{(x+3)}\right\}}$$

$$= \frac{3x+8}{4x+11}$$

Illustration No. 7

If $y = f(x) = \dfrac{px+q}{qx-p}$ then show that $f(y) = x$?

Solution:

Given

$$y = f(x) = \frac{px+q}{qx-p}$$

then

$$f(y) = \frac{py+q}{qy-p} = \frac{p\left(\frac{px+q}{(qx-p)}\right)+q}{q\left(\frac{px+q}{(qx-p)}\right)-p}$$

$$= \frac{\dfrac{p(px+q)+q(qx-p)}{(qx-p)}}{\dfrac{q(px+q)-p(qx-p)}{(qx-p)}}$$

$$= \frac{\dfrac{(p^2x+pq)+(q^2x-pq)}{(qx-p)}}{\dfrac{(qpx+q^2)-(pqx-p^2)}{(qx-p)}}$$

$$= \frac{p^2x+q^2x}{p^2+q^2}$$

$$= \frac{(p^2+q^2)x}{(p^2+q^2)} = x$$

Illustration No.8

If $f(x + 2) = x^2 + 2x + 1$ then find $f(x - 1)$?

Solution:

Let $f(x) = ax^2 + bx + c$

Therefore

$$f(x + 2) = a(x + 2)^2 + b(x + 2) + c$$
$$= a(x^2 + 4x + 4) + b(x + 2) + c$$
$$= ax^2 + 4ax + 4a + bx + 2b + c$$
$$= ax^2 + x(4a + b) + 4a + 2b + c \quad .....(1)$$

Since,

$$f(x + 2) = x^2 + 2x + 1 \quad .....(2)$$

Comparing both the equation 1 and 2

$$a = 1,\ 4a + b = 2 \text{ and } 4a + 2b + c = 1$$

therefore substituting the value of a, in the equation $(4a + b = 2)$, we get,

$$4 + b = 2 \text{ or } b = -2$$

Similarly substituting the value of a and b in $4a + 2b + c = 1$

We get,

$$4 - 4 + c = 1 \text{ or } c = 1$$

Therefore

$$f(x) = x^2 - 2x + 1$$

and hence

$$f(x - 1) = (x - 1)^2 - 2\,(x - 1) + 1$$
$$= x^2 - 2x + 1 - 2x + 2 + 1$$
$$= x^2 - 4x + 4$$

Illustration No.9

If two functions f(x) and g(x) is defined as $f(x) = \frac{2}{2-x}$ and $g(x) = \frac{x-2}{x}$ then find whether f {g(x)} is equal to g {f(x)}?

Solution:

Given $\qquad f(x) = \frac{2}{2-x}$ and $g(x) = \frac{x-2}{x}$

$$f\{g(x)\} = f\left(\frac{x-2}{x}\right) \qquad \left\{\text{since } g(x) = \frac{x-2}{x}\right\}$$

therefore

$$= \left\{\frac{2}{2-\left(\frac{x-2}{x}\right)}\right\}$$

$$= \left\{\frac{2x}{2x-(x-2)}\right\}$$

$$= \left\{\frac{2x}{x+2}\right\}$$

Similarly

$$g\{f(x)\} = g\left(\frac{2}{2-x}\right) \qquad \left\{\text{since } f(x) = \frac{2}{2-x}\right\}$$

therefore

$$= \left\{\frac{\left(\frac{2}{2-x}\right)-2}{\left(\frac{2}{2-x}\right)}\right\} \qquad \left\{\text{since } g(x) = \frac{x-2}{x}\right\}$$

$$= \left\{ \frac{\dfrac{2-2(2-x)}{2-x}}{\left(\dfrac{2}{2-x}\right)} \right\}$$

$$= \left\{ \frac{2-4+2x}{2} \right\}$$

$$= \left\{ \frac{2x-2}{2} \right\} = x-1 \text{, hence } f\{g(x)\} \neq g\{f(x)\}.$$

Illustration No. 10

If two functions f(x) and g(x) is defined as $f(x) = x^2 + 1$ and $g(x) = 2x + 3$ then find f{g(x)} and g{f(x)} ?

Solution:

Given that

$$f(x) = x^2 + 1 \text{ and } g(x) = 2x + 3$$

Then

$$f\{g(x)\} = f(2x + 3) \qquad \{\text{since } g(x) = 2x + 3\}$$

$$= (2x + 3)^2 + 1$$

$$= 4x^2 + 12x + 9 + 1$$

$$= 4x^2 + 12x + 10$$

Similarly

$$g\{f(x)\} = g(x^2 + 1) \qquad \{\text{since } f(x) = x^2 + 1\}$$

$$= 2(x^2 + 1) + 3$$

$$= 2x^2 + 2 + 3$$

$$= 2x^2 + 5$$

Illustration No. 11

If the function f(x) is defined as $f(x) = \dfrac{3x+4}{2x-5}$ then find the inverse of the function?

Solution:

Since $f(x) = \frac{3x+4}{2x-5}$

Let $f(x) = y$ then $y = f(x) = \frac{3x+4}{2x-5}$

or $(2x - 5)\, y = (3x + 4)$

or $2xy - 5y = 3x + 4$

or $x(2y - 3) = 4 + 5y$

therefore $x = \frac{5y+4}{2y-3}$

And hence inverse of the function $f(x) = \frac{5y+4}{2y-3} = f(y)$

Illustration No. 12

If a function is defined over R such that $f(x) = 2x^2 + 3x - 5$ then find $\frac{f(2+h)-f(2)}{h}$?

Solution:

We have $f(x) = 2x^2 + 3x - 5$

Then

$$\begin{aligned} f(2 + h) &= 2(2 + h)^2 + 3(2 + h) - 5 \\ &= 2(4 + 4h + h^2) + (6 + 3h) - 5 \\ &= 8 + 8h + 2h^2 + 6 + 3h - 5 \\ &= 9 + 11h + 2h^2. \end{aligned}$$

And

$$\begin{aligned} f(2) &= 2(2)^2 + 3(2) - 5 \\ &= 8 + 6 - 5 = 9. \end{aligned}$$

Therefore,

$$\frac{f(2+h)-f(2)}{h} = \frac{9+11h+2h^2-9}{h}$$

$$= \frac{11h + 2h^2}{h}$$

$$= \frac{h(11 + 2h)}{h}$$

$$= 11 + 2h.$$

Illustration No. 13

If a function is defined as $f(x) = x^2 - x$ then prove that $f(1 + h) = f(-h)$?

Solution:

Given $f(x) = x^2 - x$

Therefore,

$$f(1+h) = (1 + h)^2 - (1 + h)$$
$$= 1 + 2h + h^2 - 1 - h$$
$$= h + h^2$$

Similarly

$$f(-h) = (-h)^2 - (-h) = h + h^2$$

and hence

$$f(1 + h) = f(-h)$$

Illustration No. 14

If two functions f(x) and g(x) are defined as $f(x) = 6x^6 + 3x^4 + 2x^2$ and $g(x) = 3x^3 + 4x$ then prove that f(x) is an even function and that g(x) is an odd function?

Solution:

We have $f(x) = 6x^6 + 3x^4 + 2x^2$ and $g(x) = 3x^3 + 4x$

Therefore

$$f(-x) = 6(-x)^6 + 3(-x)^4 + 2(-x)^2$$
$$= 6x^6 + 3x^4 + 2x^2 = f(x)$$

Hence f(x) is an even function of x

Similarly

$$g(x) = 3x^3 + 4x$$

therefore

$$g(-x) = 3(-x)^3 + 4(-x)$$
$$= -3x^3 - 4x$$
$$= -(3x^3 + 4x)$$
$$= -f(x)$$

Hence g(x) is an odd function of x

Illustration No. 15

If $f(x) = x\left(\frac{x-2}{h-2}\right) + x\left(\frac{x-h}{2-h}\right)$ prove that f(2) + f(h) = f(2 + h)?

Solution:

Given the function f(x) is defined as

$$f(x) = x\left(\frac{x-2}{h-2}\right) + x\left(\frac{x-h}{2-h}\right)$$

Then we have,

$$f(2) = 2\left(\frac{2-2}{h-2}\right) + 2\left(\frac{2-h}{2-h}\right) = 2(0) + 2(1) = 2$$

Similarly,

$$f(h) = h\left(\frac{h-2}{h-2}\right) + h\left(\frac{h-h}{2-h}\right) = h(1) + h(0) = h$$

And,

$$f(2+h) = (2+h)\left(\frac{2+h-2}{h-2}\right) + (2+h)\left(\frac{2+h-h}{2-h}\right)$$
$$= (2+h)\left(\frac{h}{h-2}\right) + (2+h)\left(\frac{2}{2-h}\right)$$
$$= (2+h)\left(\frac{h}{h-2}\right) - (2+h)\left(\frac{2}{h-2}\right)$$
$$= (2+h)\left(\frac{h-2}{h-2}\right)$$
$$= (2+h)$$

But $\quad L.H.S = f(2) + f(h) = 2 + h$

And $\quad R.H.S = f(2 + h) = 2 + h$

Hence $\quad L.H.S = R.H.S$

Illustration No. 16

If a function is defined as $f(x) = \frac{2x+3}{3x+2}$ then prove that $f(x).\ f\left(\frac{1}{x}\right) = 1$?

Solution:

We have

$$f(x) = \frac{2x+3}{3x+2}$$

then

$$f\left(\frac{1}{x}\right) = \frac{2\left(\frac{1}{x}\right)+3}{3\left(\frac{1}{x}\right)+2}$$

$$= \frac{\frac{2+3x}{x}}{\frac{3+2x}{x}}$$

$$= \frac{2+3x}{3+2x}$$

Therefore,

$$f(x).f\left(\frac{1}{x}\right) = \left(\frac{2x+3}{3x+2}\right)\left(\frac{2+3x}{3+2x}\right) = 1$$

Illustration No. 17

If a function is defined as $f(x) = x^2 - \frac{1}{x^2}$ then find the value of $f(x) + f\left(\frac{1}{x}\right)$?

Solution:

Given $\quad f(x) = x^2 - \frac{1}{x^2}$

And hence

$$f\left(\frac{1}{x}\right) = \frac{1}{x^2} - x^2$$

therefore

$$f(x) + f\left(\frac{1}{x}\right) = x^2 - \frac{1}{x^2} + \frac{1}{x^2} - x^2 = 0$$

Illustration No. 18

If a function is defined as $f(x) = \log\left(\frac{1-x}{1+x}\right)$ then show that $f\left(\frac{2x}{1+x^2}\right) = 2f(x)$?

Solution:

Given the function $f(x) = \log\left(\frac{1-x}{1+x}\right)$

Therefore,

$$f\left(\frac{2x}{1+x^2}\right) = \log\left(\frac{1-\frac{2x}{1+x^2}}{1+\frac{2x}{1+x^2}}\right)$$

$$= \log\left(\frac{\frac{1+x^2-2x}{1+x^2}}{\frac{1+x^2+2x}{1+x^2}}\right)$$

$$= \log\left\{\frac{(1-x)^2}{(1+x)^2}\right\}$$

$$= \log\left(\frac{1-x}{1+x}\right)^2$$

$$= 2\ \log\left(\frac{1-x}{1+x}\right)$$

$$= 2\ f(x)$$

Illustration No. 19

Given f(x) = |x + 2| and g(x) = 3x – 5 then find

(i) f[g(3)]

(ii) f[g(–1)]

(iii) f[g(2)]

Solution:

Given $f(x) = |x + 2|$ and $g(x) = 3x - 5$

Therefore,

$$f[g(x)] = f(3x - 5) = |3x - 5 + 2| = |3x - 3|$$

and hence

(i) $f[g(3)] = |3(3) - 3| = |6| = 6$

(ii) $f[g(-1)] = |3(-1) - 3| = |-6| = 6$

(iii) $f[g(2)] = |3(2) - 3| = |3| = 3$

Illustration No. 20

If two functions are defined as

$$f(x) = 4x + 3, \quad x \in R,$$

$$g(x) = -x, \quad x \in R,$$

then find f[g(x)] and g[f(x)]

Solution:

Since

$$f(x) = 4x + 3, \quad x \in R,$$

$$g(x) = -x, \quad x \in R,$$

therefore

$$f[g(x)] = f(-x)$$

$$= 4(-x) + 3 \qquad \text{(Since } f(x) = 4x + 3\text{)}$$

$$= 3 - 4x.$$

and

$$g[f(x)] = g(4x + 3) \qquad \text{(Since } f(x) = 4x + 3\text{)}$$

$$= 4(-x) + 3 \qquad \text{(Since } g(x) = -x\text{)}$$

$$= -4x - 3.$$

Graph of Functions

Illustration No. 1

Draw the graphs of the following functions with definition of domain being the set of all real numbers

(i) $x = 2$

(ii) $y = 4$

(iii) $\frac{x}{4} + \frac{y}{9} = 1$

Solution:

(i) Given x = 2, then y can assume any value between $-\infty < y < \infty$, so we have,

X	2	2	2	2	2	2	2
Y	-3	-2	-1	0	1	2	3

When the corresponding values of x and y are plotted against each in the XY plane, we get a straight line parallel to Y–axis at a distance of 2 units from it. Therefore the required graph is

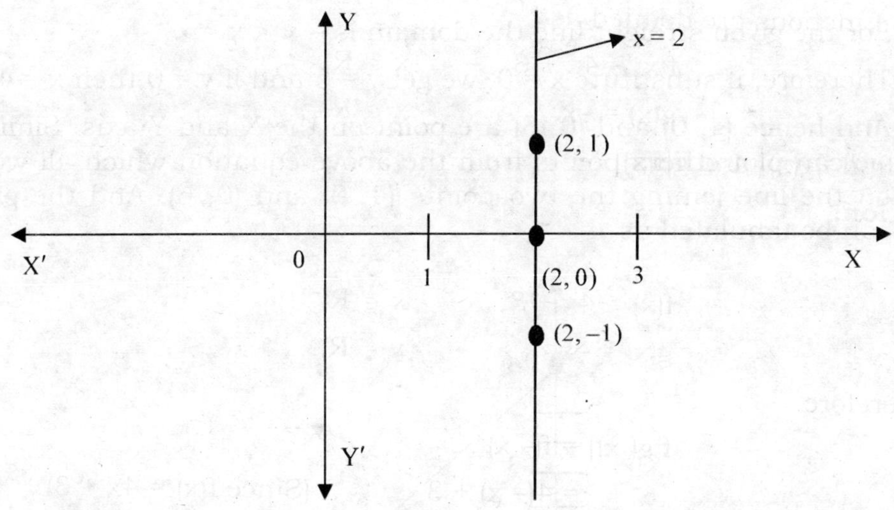

(ii) Given y = 4, then x can assume any value between $-\infty < x < \infty$, so we have,

X	–3	–2	–1	0	1	2	3
Y	4	4	4	4	4	4	4

When the corresponding values of x and y are plotted against each in the XY plane, we get a straight line parallel to x–axis at a distance of 4 units from it. Therefore the required graph is

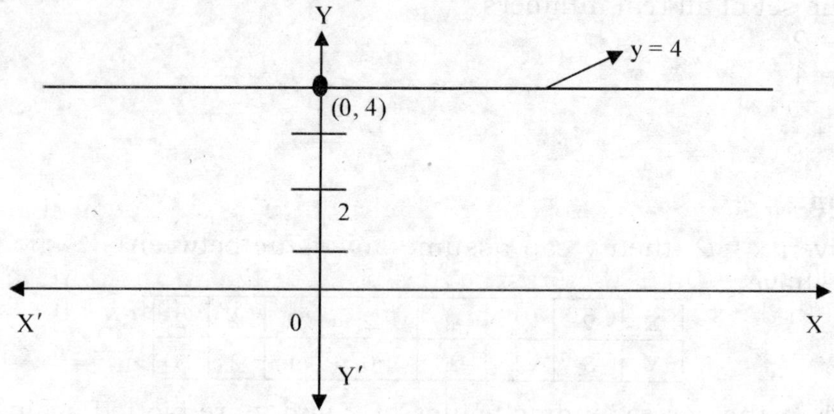

(iii) Given $\frac{x}{4}+\frac{y}{6}=1$

For the given straight line the domain is $-\infty < x < \infty$

Therefore, if substitute x = 0, we get y = 6 and if y = 0 then x = 4.

And hence (4, 0) and (0, 6) are point on the X and Y axis. Similarly we can plot others points from the above equation which all will lie on the line joining the two points (4, 0) and (0, 6). And the graph can be tabulated as

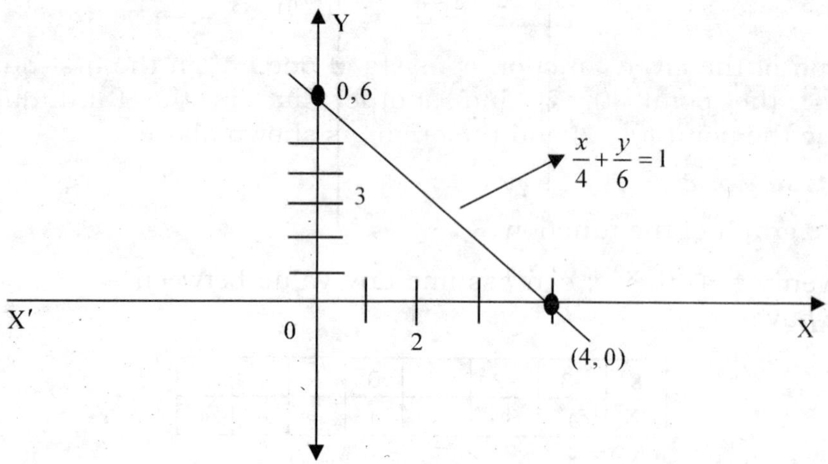

Illustration No. 2

Draw the graph of the following function

$$y \begin{cases} = 2, & \text{when } x > 0 \\ = 0, & \text{when } x = 0, \\ = -2, & \text{when } x < 0 \end{cases}$$

Solution:

The given function y is defined as such that, it is a straight line in first quadrant parallel to X–axis at a distance of 2 units above it when $x > 0$, $y = 0$ when $x = 0$ i.e. the origin and a straight line in the third quadrant parallel to the x axis at distance of 2 units below it, when $x < 0$.

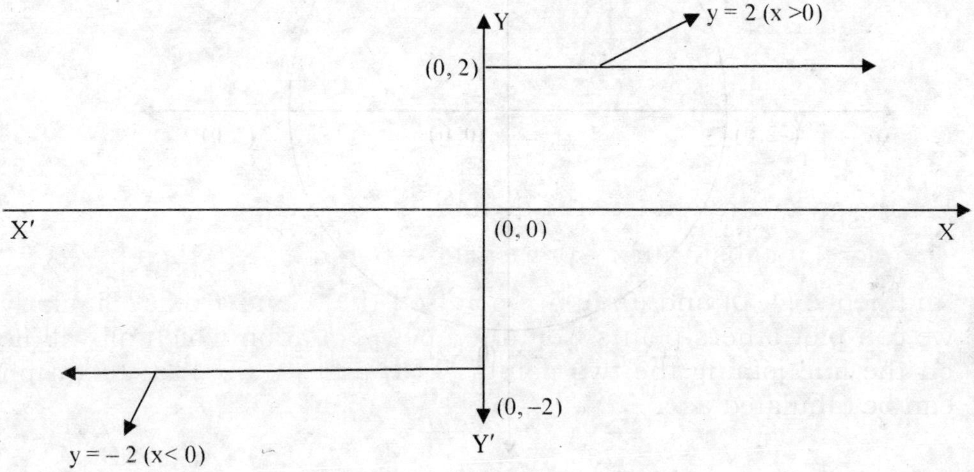

The graph of the given function consists of one line in the first quadrant excluding the point (0, 2) and another line in the third quadrant excluding the point (0, –2) and the origin as shown above.

Illustration No: 3

Draw the graph of the function

$$x^2 + y^2 = 4$$

Solution:

Given

$$x^2 + y^2 = 4$$

or $$y^2 = 4 - x^2$$

$$y = \pm\sqrt{4 - x^2}$$

For different values of x the corresponding values of y can be tabulated as

X	2	1	0	1	2
Y	0	±1.732	2	±1.732	0

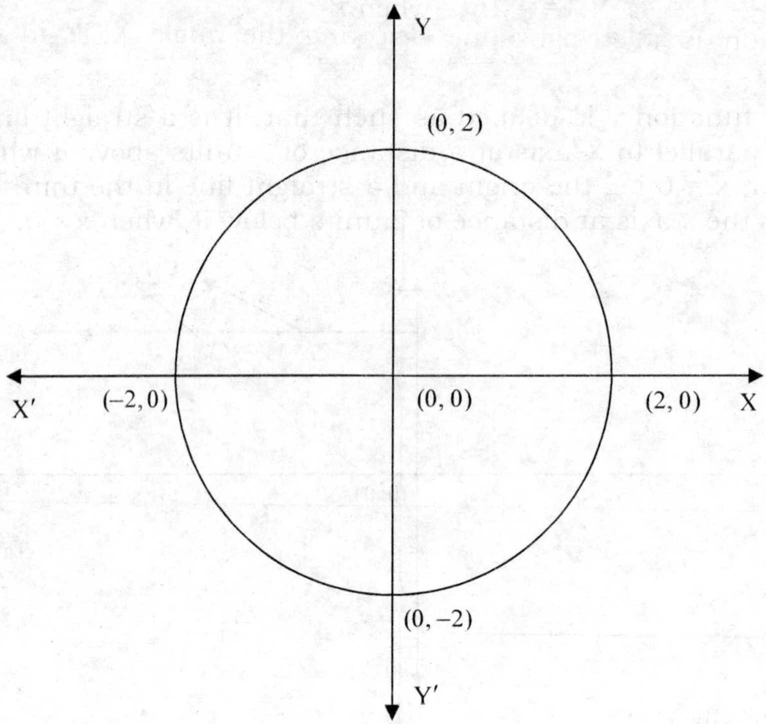

Illustration No. 4

Draw the graph of the function

$$f(x) \begin{cases} = x, & \text{when } x > 0 \\ = 0, & \text{when } x = 0, \\ = -x, & \text{when } x < 0 \end{cases}$$

Solution:

Let y = f(x) = x when x > 0

x	0.5	1	1.5	2	2.5	3
y	0.5	1	1.5	2	2.5	3

The graph is a straight line bisecting the angle XOY in the first quadrant

Whereas, y = f(x) = 0 when x = 0 and y = f(x) = –x when x < 0.

x	– 0.5	–1	–1.5	– 2	– 2.5	–3
y	0.5	1	1.5	2	2.5	3

The graph is a straight line bisecting the angle X'OY in the first quadrant

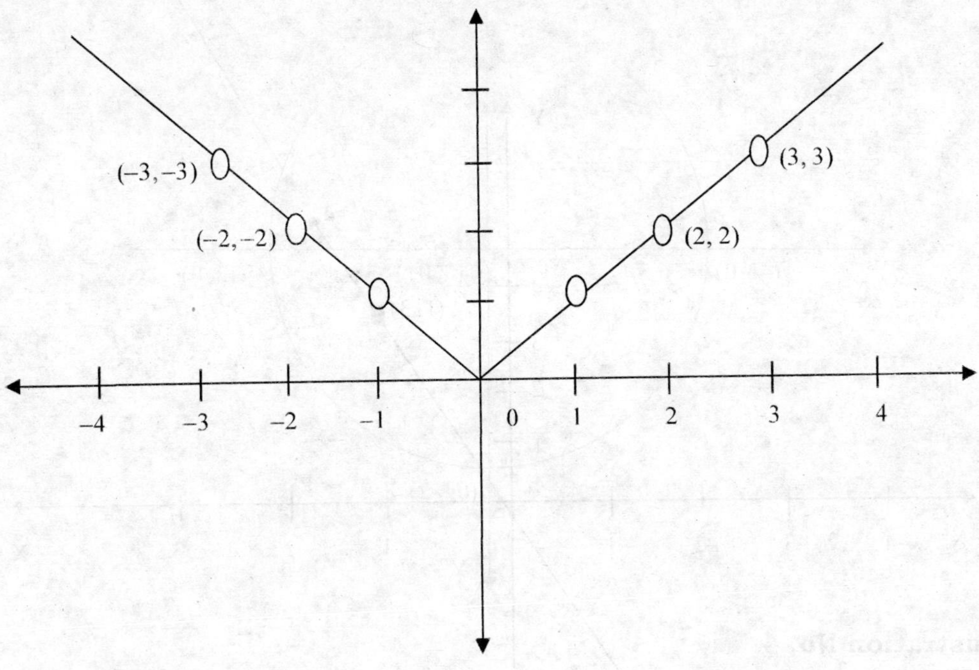

Illustration No. 5

Draw the graph of the function

$$y = f(x) = \frac{x^2 - 4}{x - 2}$$

Solution:

Given $y = f(x) = \dfrac{x^2 - 4}{x - 2}$

The given function is not defined at the point x = 2

The corresponding value of the function y is obtained by substituting various values of x

x	– 3	– 2	– 1	0	1	2	3
y	– 1	0	1	2	3	–	5

By plotting various points of x and corresponding values of y, we get the following graph which given as

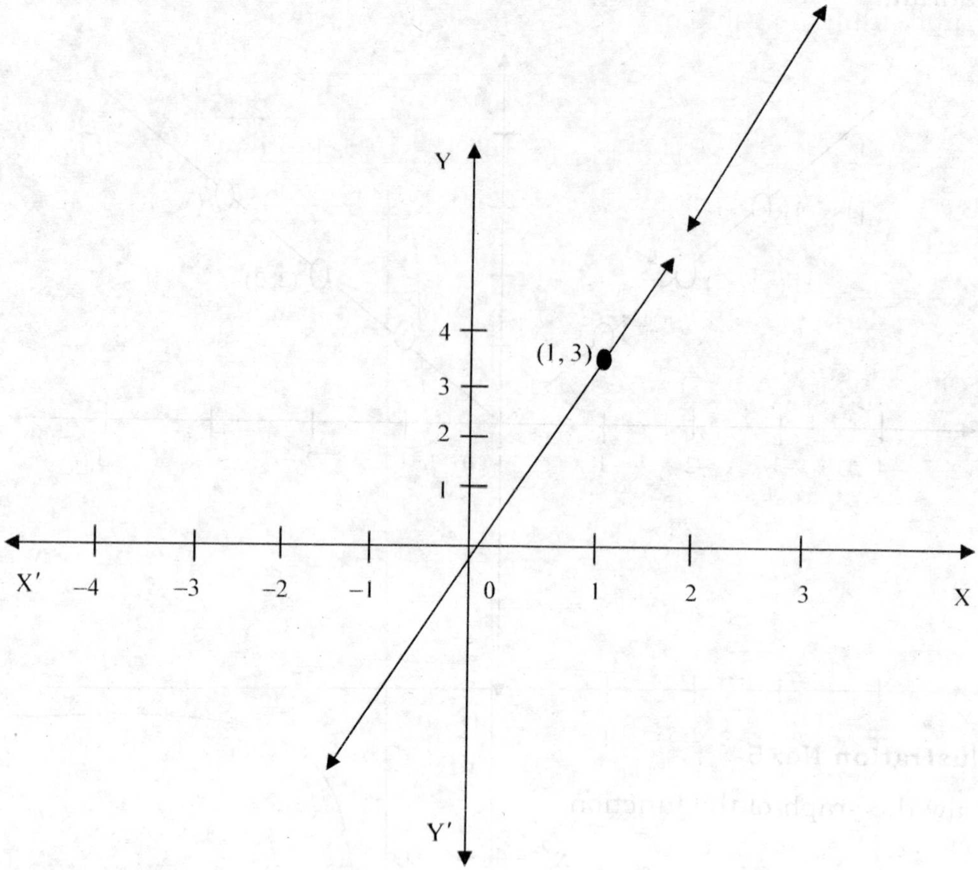

Illustration No. 6

Draw the graph of the function

$$f(x) = \frac{1}{1-x}$$

Solution:

The given function is $f(x) = \frac{1}{1-x}$

Let us assume $y = f(x) = \frac{1}{1-x}$, the function as visible is not defined for the value of x = 1 i.e at x = 1, y approaches infinite value i.e $y = \infty$. And when $x = -\infty$ then y = 0 also when $x = \infty$ then also y = 0.

Tabulating the value of y against the value of x, we get,

x	$-\infty$	-2	-1	0	1	2	3	∞
y	0	1/3	1/2	1	∞	-1	-2	0

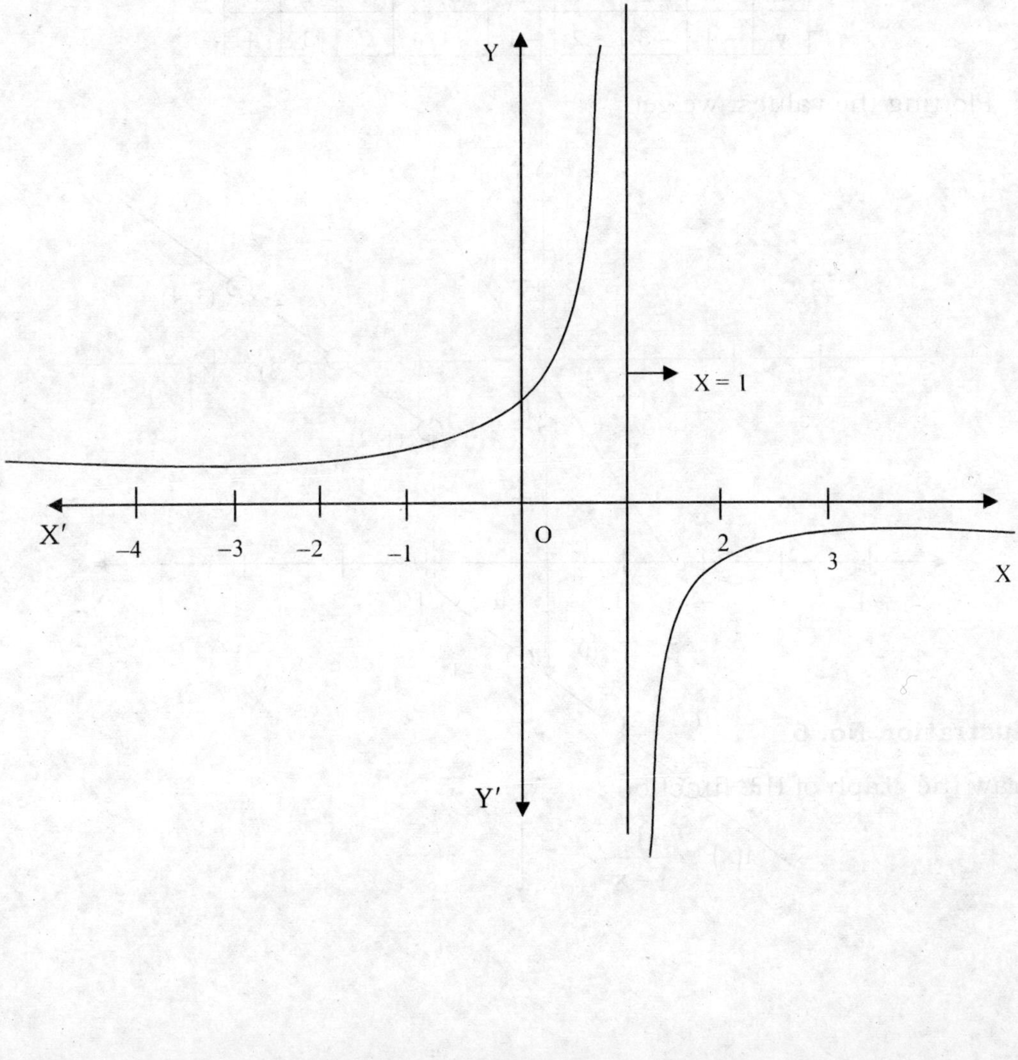

Illustration No. 7

Draw the graph of the function

$$f(x) = x + 2 \qquad \text{for } x \geq 1$$
$$= x - 2 \qquad \text{for } x < 1$$

Solution:

Given the function is defined as

$$f(x) = x + 1 \qquad \text{for } x \geq 1$$
$$= x - 1 \qquad \text{for } x < 1$$

Let us assume y = f(x) then tabulating some of the values of x and y, we get,

x	− 3	− 2	− 1	0	3/4	1	2	3
y	− 4	− 3	− 2	− 1	− 1/4	2	3	4

Plotting the values, we get

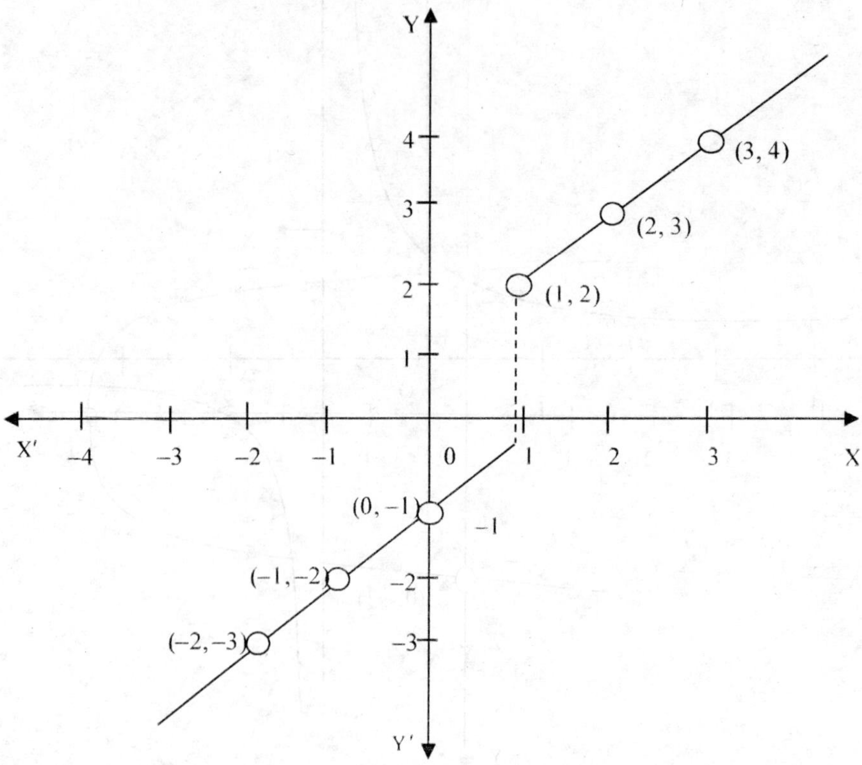

Illustration No. 8

Draw the graph of the function $y^2 = 4 - x$

Solution:

Since the function is defined as $y^2 = 4 - x$

or $$y = \pm\sqrt{4 - x}$$

For different values of x, we find the corresponding values of y as given in the following table.

x	−3	−2	−1	0	1	2	3
y	$\pm\sqrt{7}$	$\pm\sqrt{6}$	$\pm\sqrt{5}$	±2	$\pm\sqrt{3}$	$\pm\sqrt{2}$	±1

Plotting the values in the XY–Plane, we get,

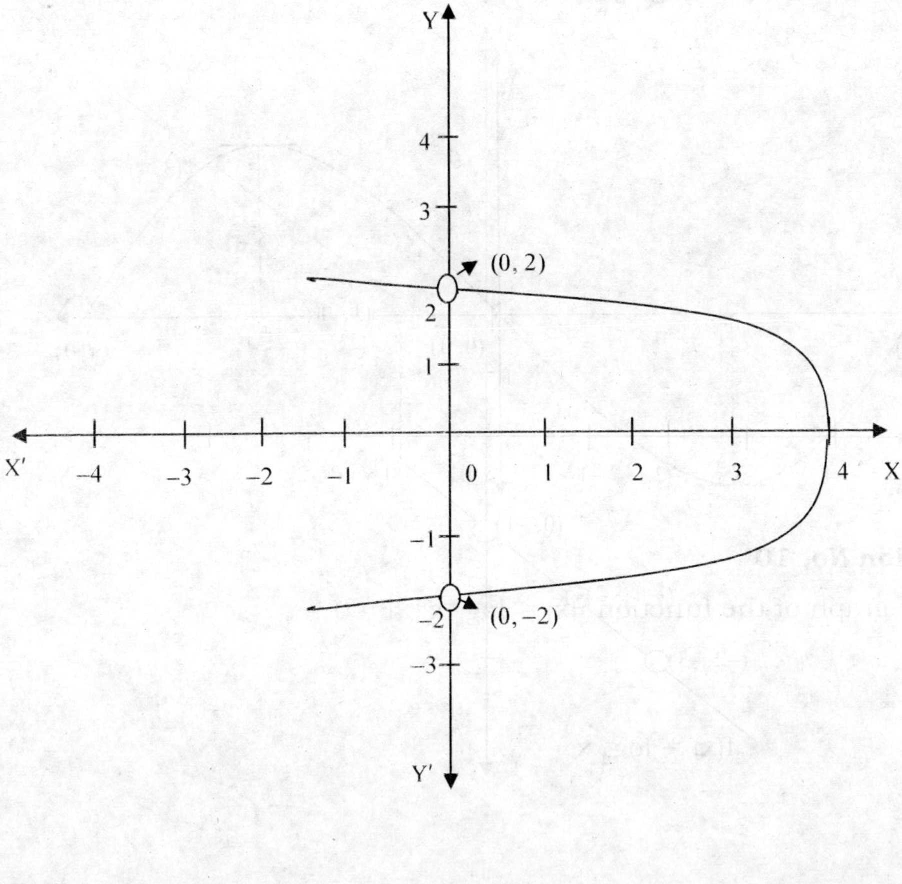

Illustration No. 9

Draw the graph of the function f(x) = sinx, where x ∈ (−π,π)

Solution:

We have f(x) = sin x, x defined in the interval (−π,π)

By substituting π different values of x between the range of the interval (−π,π)we get various values of sin x which are shown below in the table.

X	$-\pi$	$-\frac{\pi}{6}$	$-\frac{\pi}{3}$	$-\frac{\pi}{2}$	$-\frac{2\pi}{3}$	$-\frac{5\pi}{6}$	0	$\frac{\pi}{6}$	$\frac{\pi}{3}$	$\frac{\pi}{2}$	$\frac{2\pi}{3}$	$\frac{5\pi}{6}$	π
f(x)	0	$-\frac{1}{2}$	$-\frac{\sqrt{3}}{2}$	-1	$-\frac{\sqrt{3}}{2}$	$-\frac{1}{2}$	0	$\frac{1}{2}$	$\frac{\sqrt{3}}{2}$	1	$\frac{\sqrt{3}}{2}$	$\frac{1}{2}$	0

Therefore the required graph is shown as

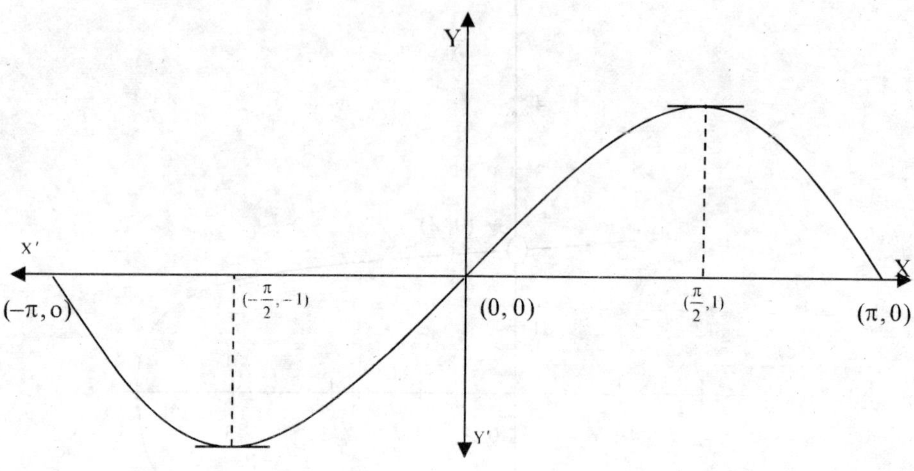

Illustration No. 10

Draw the graph of the function f(x) = $\log_a$ x , a > 0 ?

Solution:

We have

$$f(x) = \log_a x$$

the domain of the function is $0 < x < \infty$ and as x increase from $0 < x < \infty$ the value of f(x) also increases from $-\infty$ to $+\infty$. Therefore the graph is shown below for all values a > 0 which passes through the point (1, 0)

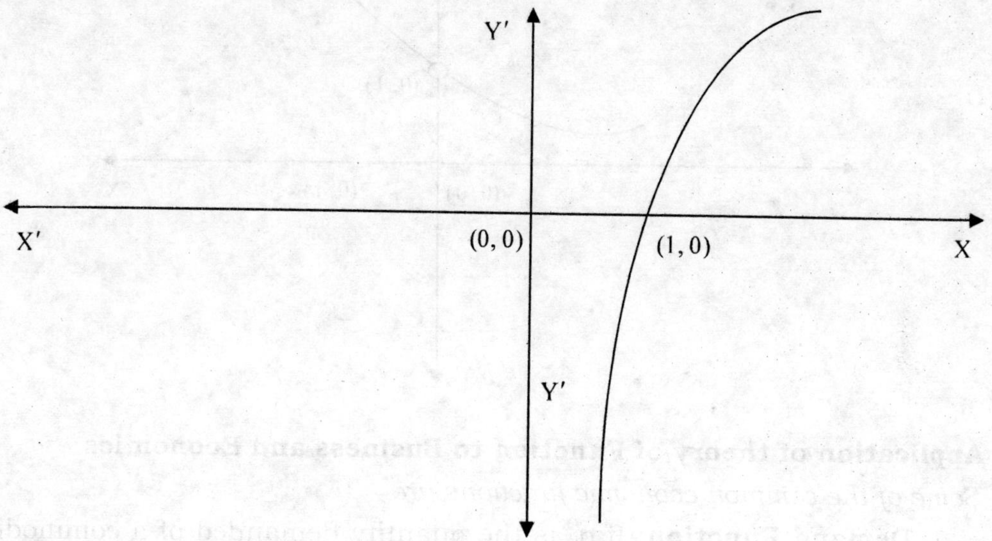

Illustration No. 11

Draw the graph of the function $f(x) = e^x$?

Solution:

Since $f(x) = e^x$

Let us assume that $y = f(x) = e^x$.

But we know that $e^x = 1 + x + \frac{x^2}{2!} + \ldots$

Now when x = 0 then y = 1 and as x increase tends towards $+\infty$ then even y tends towards infinity. And for negative values of x, y decreases from 1 to 0 as x decreases from 0 to infinity.

Then plotting the graph for varied values of x, we get,

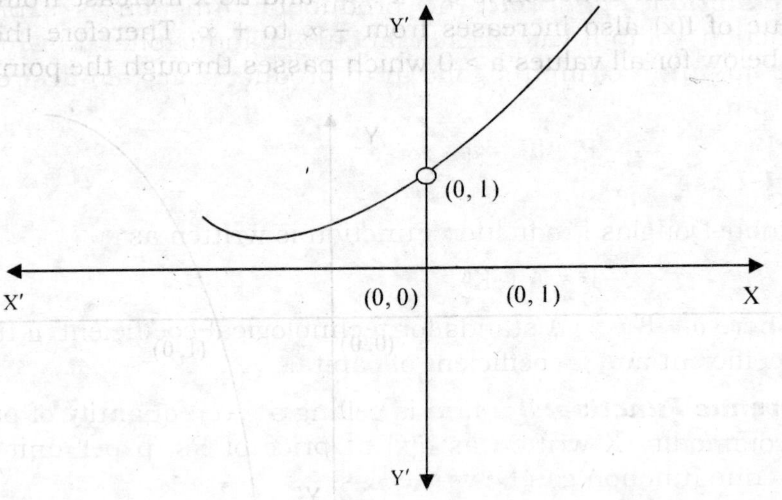

Application of theory of Function to Business and Economics

Some of the common economic functions are

1. **Demand Function**: If q_d is the quantity demanded of a commodity X and p is the price at which the commodity is purchased then the demand function can be expressed as

   $q_d = a - b\,p$ where a and b are constants.

2. ***Supply Function:*** If q_s is the quantity supplied of a commodity X and p is the price at which the commodity is offered for sale then the supply function can be expressed as

   $q_s = c + d\,p$ where c and d are constants.

3. ***Consumption Function:*** The consumption function can be written as $C = a + bY$ where a stands for autonomous consumption (consumption not dependent on income), b stands for marginal propensity to consume (i.e. the change in consumption brought about by change in income and Y is the national income.

4. ***Cost Function***: A cost function of a commodity X can be expressed as $C = C(x)$ where x is the quantity produced and C is the total cost incurred.

   Total Cost can be further expressed in terms of fixed cost F (independent of the number of units produced) and variable cost V(x) which varies with output thus total cost can be written as

   $$C(x) = F + V(x).$$

5. ***Production Function:*** The production function is expressed as function of input variables of labour (L) and capital (K) which shows the quantum of inputs gone in to produce a certain desired level of output.

$$P = f(L, K)$$

Spl. Case.

Cobb–Douglas Production Function is written as

$$P = A L^{\alpha} K^{\beta}$$

where $\alpha + \beta = 1$, A stands for technological coefficient, α the labour coefficient and β coefficient of capital.

6. ***Revenue Function:*** If a firm is selling a given quantity of produce of a commodity X written as q(x) at price of Rs. p per unit then the revenue function can be written as

$$R(x) = p \times q(x)$$

7. ***Investment Function:*** A simple investment function can be written as

$$I = a + br$$

where a and b are constants such that $a > 0$ and $b < 0$ and r is the rate of interest.

Illustration No. 1

A market supply function of a firm is given by $q = a + bp$ where p is the market price per unit of the item produced. If it was found that the number of items supplied at a pricing of Rs. 10 is 2000 units and 2400 at a price of Rs. 9 then find the values of the constants a and b and interpret your result?

Solution:

We have the supply function of the firm as

$$q = a + bp$$

But we are provided with the information that the quantity supplied to the market under two different pricing levels. Therefore substituting the values in the supply function we get,

$$2000 = a + 10b \quad .....(1)$$

Similarly, we have,

$$2400 = a + 9\,b \quad .....(2)$$

Solving both equations 1 and 2, we get,

$$a = 6000 \text{ and } b = -400$$

And hence the supply function can be generally written as

$$q = 6000 - 400p$$

So, we can conclude that from the negative sign in the above equation that increase of pricing by the firm will lead to fall in sales.

Illustration No. 2

Timkin bearings a manufacturer of bearings have a fixed cost of manufacturing as Rs 2000 and a variable cost of production as Rs. 5 per unit. If the supply function of the firm is given by $q = 500 + 50p$ where p is the market price of the bearings and q is the quantity supplied, then find the cost function in terms of price p. Also calculate what should be the price charged in order to earn a profit of Rs. 48,000?

Solution:

Given the demand function as $q = 500 + 50p$

Therefore

The total cost function C(x) = Fixed Cost (F) + Variable Cost {V(q)}

$$= 2000 + 5.q$$

$$= 2000 + 5(500 + 50p)$$

{Since $q = 500 + 50p$}

$$= 2000 + 2500 + 250p.$$

$$= 4500 + 250p.$$

But Total sales revenue (TR) = $p.\, q = p\,(500 + 50p)$

$$= 500p + 50\,p^2.$$

And hence,

Total Profit = Total Revenue (TR) – Total Cost (C(x))

$$= (500p + 50p^2) - (4500 + 250p).$$

$$= 50p^2 + 250p - 4500.$$

But the profit earned is Rs. 48,000 therefore, we have,

$$48{,}000 = 50p^2 + 250p - 4500 \text{ or } 50p^2 + 250p - 52500 = 0$$

or $\quad p^2 + 5p - 1050 = 0$

or $\quad p^2 - 30p + 35p - 1050 = 0$

or $\quad p(p - 30) + 35\,(p - 30) = 0$

or $(p + 35)(p - 30) = 0$

or $p = 30$ or -35

Therefore the firm must charge a price of Rs. 30 in order to earn a profit of Rs. 48,000

Illustration No. 3

A fruit vendor sells fresh strawberries which he purchases for Rs. p_1 and sells them at a price of Rs. p_2 per piece where $p_2 > p_1$; at the end of the day the unsold strawberries are sold at lower pricing p_3 such that $p_3 < p_1$ in order dispose the stock. Find the profit function of the fruit vendor per day?

Solution:

The profit of the fruit vendor depends on the number of strawberries he sells in relation to the amount of purchase he makes at the beginning of each day.

Let D be the demand for the strawberries and Q be the stocking,

Case I

When $D < Q$

Then the Profit = p_2 (Number of Strawberries sold) + p_3 (Number of Strawberries sold at lower pricing) – p_1 (quantity purchased)

$= p_2 D + p_3(Q - D) - p_1 Q$

$= (p_2 - p_3) D + (p_3 - p_1) Q$

Case II

When $D \geq Q$

Then the profit $= p_2 Q - p_1 Q$.

$= (p_2 - p_1)Q$.

General Condition for Market Equilibrium in an Economic System

A market is said to be in equilibrium when demand in the market is matched by the equivalent supply. Otherwise a market is in equilibrium when quantity demanded is equal quantity supplied

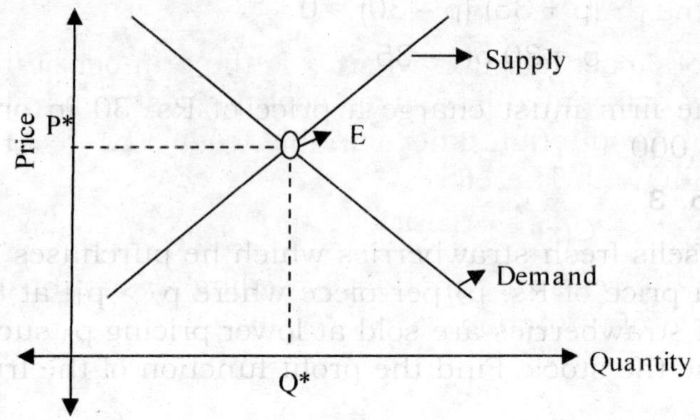

Market Equilibrium

Quantity Demanded = Quantity Supplied i.e. $Q_d = Q_s$.

Here E is the point of equilibrium, P^* and Q^* are the equilibrium price and quantity.

Illustration No. 4

For a given firm the demand and supply function is given by q = 25 – 3p and q = 5 + 2p where q is the quantity demanded in units and p is the price in rupees. Then find the equilibrium quantity and price?

Solution:

The general condition for market equilibrium

Quantity Demanded = Quantity Supplied

i.e.

$$25 - 3p = 5 + 2p$$

therefore,

$$20 = 5p \text{ or } p = 4$$

Substituting the value of p in any one of the equation we get,

$$q = 25 - 12 = 13.$$

Hence the equilibrium quantity and price are 13 units and Rs. 4

Break Even Analysis

Break Even Analysis commonly known form of Cost–Volume–Profit analysis is a specific technique in order to study the inter relationship between costs, volume and profit. It provides information regarding the level of sales at which costs and revenue are in equilibrium. The equilibrium point is called as break even point a point of sales volume at which total revenue is equal to total costs.

Let us assume that,

C(x) = The total cost function where x is the number of units produced. The total cost is bifurcated into, fixed cost (F) (cost remaining constant at any volume of production and variable cost V(x) (cost varying with volume of production). Therefore,

$$C(x) = F + V(x)$$

And R(x) be the revenue function by selling x amount of commodities and P(x) be the profit function,.

As we know that,

$$\text{Profit} = \text{Revenue} - \text{Cost}$$

$$P(x) = R(x) - C(x)$$

$$= R(x) - \{F + V(x)\}$$

But from above, we know that at break event point profit is zero i.e. P(x) = 0 or R(x) = F + V(x)

In order to determine the break-even point in terms of units, we assume that p is the selling price per unit and v be the variable cost per unit and x be the number of units sold.

Thus,

$$\text{Revenue } R(x) = \text{Price} \times \text{Quantity Sold.}$$

$$= p \times x.$$

$$\text{Cost } C(x) = F + V(x)$$

$$= F + V \times x$$

But we know,

Revenue = Cost at BEP point, therefore,

$$px = F + vx$$

$$px - vx = F \text{ or } x = \frac{F}{p - v}$$

$$\text{BEP (units)} = \frac{\text{Fixed Cost}}{\text{Selling Price} - \text{Variable cost per unit}}$$

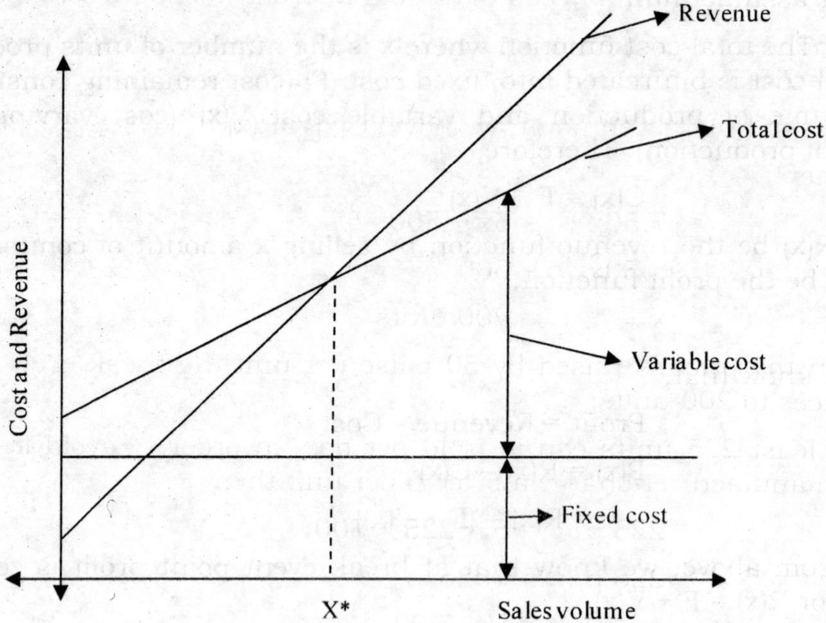

Illustration No. 5

A electrical bulb manufacturer has a daily cost of production of x units given by C(x) = 5x + 500. If the bulbs are sold at Rs. 7 per unit then,

(i) find what should be the minimum number of bulbs be sold daily to ensure that the manufacturer does not incurs any loss?

(ii) If the selling price is raised to Rs. 7.50 per unit, then find the break-even level of sales in units?

(iii) If it is found that at least 225 units can be sold daily then find what should be the price charged per unit in order to avoid loss?

Solution:

Let us assume that the manufacturer sells x units per day then the revenue generated by selling at Rs 7 per unit is

$$R(x) = \text{No. of units Sold} \times \text{Price per unit}$$

$$= 7x$$

At break – even point, we have, C(x) = R(x)

$$5x + 500 = 7x$$

or $$2x = 500$$

or $$x = 250.$$

Thus at least 250 units needs to be sold in order to ensure zero loss.

(ii) As the price is raised from Rs. 7 to Rs. 7.50 the revenue raised is R(x) = 7.50x

But at break-even point, we have, C(x) = R(x)

Therefore,

$$7.50x = 5x + 500$$

or $$2.50x = 500$$

or $$x = 200 \text{ units}$$

When the price is raised by 50 paise per unit the break even sales reduces to 200 units.

(iii) If at least 225 units can be sold per day, in order to avoid loss let the manufacturer charge a price p per unit then,

$$225\,p = 5 \times 225 + 500$$

or $$225\,p = 1625$$

or $$p = \text{Rs. } 7.22$$

Illustration No. 6

A manufacturer has installed a new manufacturing plant to produce bearings. The total cost incurred for initial set up (fixed cost) is Rs.10 lakhs and additional cost for producing each bearing is Rs. 500. If the bearings are sold at Rs. 1000 and during the first month operation 1500 units can be manufactured and sold then find

(i) The cost function C(x) for producing x units of bearings?

(ii) The revenue function from the sale of x units?

(iii) Profit function from the sale of x units?

(iv) Also find the break even sales?

(v) And what will be the profit or loss the manufacturer will incur during the first month when all that is produced are sold?

Solution:

Total Cost Function C(x) = Fixed Cost + Variable cost for producing x units

= 10,00,000 + 500x.

Revenue Function R(x) = Revenue from the sales of x units

= No. of units sold × Price

$$= 1000x$$

$$\text{Profit Function } P(x) = \text{Revenue Function} - \text{Cost Function}$$

$$= R(x) - C(x).$$

$$= 1000x - (10,00,000 + 500x).$$

$$= 500x - 10,00,000.$$

At Break Even Sales Profit = zero

Therefore,

$$500x - 10,00,000 = 0$$

or $$x = 2000.$$

And hence at 1500 level of sales the loss incurred by the firm is

$$= 500 \times 1500 - 10,00,000 = -2,50,000.$$

Summary

Definition of a Function or Mapping

Let X and Y be two non-empty sets with 'x' being a variable whose domain is set X. If there exists a rule say f which associates each element x of X with a unique element y in Y. Then f is termed as a function of x over the set X or f maps X to Y. Here X is called as the domain of the function f and Y is the co-domain

Into-Mapping

A mapping is into mapping if there exists at least one element in Y which is not the f-image of any element in X, but all the elements in X has f-image in Y.

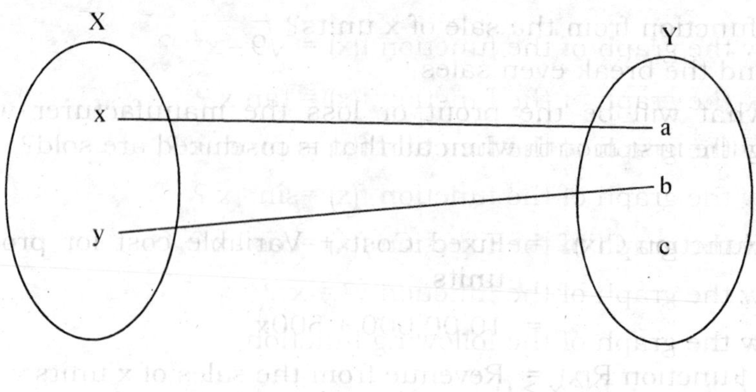

Algebraic Function

A function which cannot be expressed as a polynomial or as the ratio of two polynomials or as a root of a polynomial are termed as algebraic functions

For example,

$$\sqrt{x+5} \text{ and } \frac{2x+3}{\sqrt{3x+5}} \text{ are algebraic functions}$$

Rational Function

The ratio of two polynomial functions say P(x) and Q(x) of a variable x such that $R(x) = \frac{P(x)}{Q(x)}$ where $Q(x) \neq 0$ for any value of x is called as a rational function of x

A market is said to be in equilibrium when demand in the market is matched by the equivalent supply. Otherwise a market is in equilibrium when quantity demanded is equal quantity supplied

Break Even Analysis commonly known form of Cost – Volume – Profit analysis is a specific technique in order to study the inter relationship between costs, volume and profit.

Self Practice Exercise

1. Find the domain and range of the function $f(x) = \sqrt{9-x^2}$?
2. Find the domain and range of the function $f(x) = \frac{x}{x+7}$?
3. Find the domain and range of the function $f(x) = \frac{1}{\sqrt{x+2}}$?
4. Draw the graph of the function $f(x) = \sqrt{9-x^2}$?
5. Draw the graph of the function f(x) = tan x ?
6. Draw the graph of the function f(x) = cosec x ?
7. Draw the graph of the function $f(x) = \sin^{-1} x$?
8. Draw the graph of the function $f(x) = \cos^{-1} x$?
9. Draw the graph of the function $y^2 = x^3$?
10. Draw the graph of the following function

$$y = \begin{cases} 2 & \text{for } x \geq 0, \\ -2 & \text{for } x < -2 \end{cases}$$

11. Draw the graph of the following function

$$y = \begin{cases} 2x + 1 & \text{for } x \geq 1, \\ 2x - 1 & \text{for } x < 1 \end{cases}$$

12. Draw the graph of the function $f(x) = 2 - 3x^2$?

13. If $f(x) = 3x^4 + 2x^2$ and $g(x) = 3x + 5x^3$, then prove that f(x) is an even function and g(x) is an odd function?

14. If $f(x) = 2x^2 - 4x + 3$ then find the value of $\frac{f(3+h) - f(3)}{h}$?

15. If $f(x) = \frac{x-1}{x+1}$ then prove that $\frac{f(x) - f(y)}{1 + f(x)f(y)} = \frac{x - y}{xy + 1}$?

16. If $f(x) = ax^2 + bx + c$ and $f(1) = 4$, $f(2) = 10$ and $f(3) = 20$ then find a, b and c ?

17. If $f(x) = \frac{2x+3}{2x-3}$ then find $\frac{f(x)+2}{f(x)-2}$?

18. If $f(x) = x^2 + x$ then find $f(2 + h) - f(2 - h)$?

19. If $f(x) = x^2 - 2x + 3$ then find f(0), f(1), f(2) and f(– 2) ?

20. Find the nature of the function $f(x) = x\left(\frac{5^x - 1}{5^x + 1}\right)$?

21. Find the domain and range of the function $f(x) = \frac{x+2}{x^2 - 8x - 4}$?

22. If $f(x) = e^{px + q}$ then prove that f(a). f(b). f(c) = f(a + b + c) . e^{2q} ?

23. If $y = f(x) = \frac{3x-5}{4x-3}$ then prove that f(y) = x ?

24. A Electrical manufacturer has installed a new plant to produce electrical heaters. The total cost incurred for initial set up (fixed cost) is Rs. 5 lakhs and additional cost for producing each heater is Rs. 750. If the heater are sold at Rs. 1250 then find
    (i) The cost function C(x) for producing x units of heaters?
    (ii) The revenue function from the sale of x units?
    (iii) Profit function from the sale of x units?
    (iv) Also find the break even sales?

(v) And what will be the profit the manufacturer will make during the first month of operation if 1200 units are manufactured and sold?

25. A toy manufacturer has a daily cost of production of x units given by C(x) = 30x + 5000. If the toys are sold at Rs. 40 per unit then,

(i) Find what should be the minimum number of toys needs to be sold daily to ensure that the manufacturer does not incurs any loss?

(ii) If the selling price is raised to Rs. 50 per unit, then find the break - even level of sales in units?

(iii) If it is found that at least 400 units can be sold daily then find what should be the price charged per unit in order to avoid loss?

Answers

1. Domain: $(-3 \le x \le 3)$ and Range: $[0, 3]$
2. Domain: R – [– 7] and Range: R – {1}
3. Domain: $(-2, \infty)$ and Range: $(0, \infty)$
14. 8 + 2h
16. a = 1, b = 4 and c = –1
17. $\dfrac{6x-3}{-2x+9}$
18. 8h + 4
20. Even function.
21. Domain: $[R - \{4 \pm 2\sqrt{5}\}]$ and Range = $(-\infty, -1/4] \cup [-1/20, \infty]$
24. (i) 750 x,
    (ii) 1250x.
    (iii) 500x – 5,00,000
    (iv) 1,000 units
    (v) Rs. 1,00,000
25.
    (i) 500 units.
    (ii) 250 units.
    (iii) Rs 42.50

CHAPTER 3

Limits and Continuity

Brief Overview and Introduction

The credit of developing the subject of calculus goes to Isaac Newton and Leibnitz who are said to have worked on the subject independently during 17th century. The other pioneers are Cauchy, Lagrange and Weierstrass. Today calculus finds wide application in Mathematics, Physics, Chemistry, Business and Economics.

Limit of a Function

Let us assume that f be a real valued function defined for all real values of x in the neighborhood of a fixed number say 'a' such that f(x) approaches a fixed number 'l' as x approaches 'a' from either side of a, then l is the limit of the function f as x tends to a or written as $x \to a$, mathematically written as,

$$\lim_{x \to a} f(x) = l$$

where l is called as the limiting value of f(x) as $x \to a$

Note: If $f(x) \to l$ as $x \to a$ then l is the limit of f as $x \to a$ and if no such value l exists then $\lim_{x \to a} f(x)$ does not exist or the limit at a does not exists.

Meaning of x tends to a

Let x be a real variable and a being a fixed number, then x tends to a written as ($x \to a$) we mean that x assumes successive values either greater than a or lesser than a, whose numerical differences form a i.e. $|x - a|$ become smaller and smaller such that it can taken to be any small positive quantity say δ. In this case we can say that $x \neq a$ but approaches a.

Otherwise by $x \to a$, we mean given a positive number δ however small, the successive values of x satisfying the relation

$$0 < |x - a| < \delta \text{ or } a - \delta < x < a + \delta \ (x \neq a).$$

If x approaches a from values greater than 'a' such that the difference x – a becomes smaller and smaller then we say x tends to a from the right hand side and written as $x \to a^+$, here x is very close to a but greater than a.

If x approaches a from values lesser than 'a' such that the difference x – a becomes smaller and smaller then we say x tends to a from the left hand side and written as $x \to a^-$, here x is very close to a but lesser than a.

For Example

When we say x a real variable assuming successive value 1.1, 1.01, 1.001, 1.0001............. such that the difference of x – 1 becomes smaller and smaller such that x comes closer and closer to 1 then we can say that $x \to 1$ from the right hand side and written as $x \to 1^+$ or $x \to 1+$

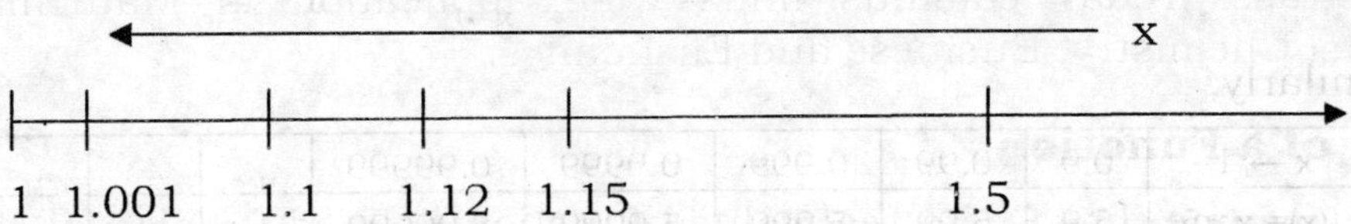

Similarly when x a real variable assuming successive value 0.9, 0.99, 0.999............. such that the difference of x – 1 becomes smaller and smaller such that x comes closer and closer to 1 then we can say that $x \to 1$ from the left hand side and written as $x \to 1^-$ or $x \to 1-$

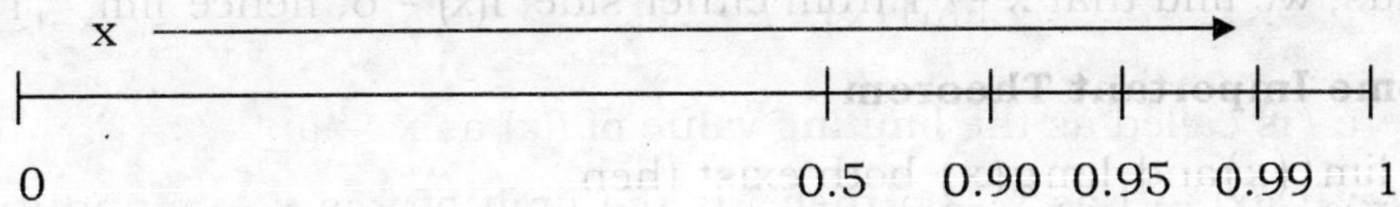

Some Other Types of Limits

(i) What do you understand by a function tending towards infinity i.e. $f(x) \to \infty$ as $x \to a$?

When $f(x) \to \infty$ as $x \to a$, we mean that when x approaches a from either side then f(x) becomes larger and larger finally exceeding any positive number known. In this case we can write

$$\lim_{x \to a} f(x) = \infty$$

(ii) By $f(x) \to l$ as $x \to \infty$, we mean that as x becomes larger and larger f(x) becomes closer and closer to 'l'.

(iii) As $x \to \infty$ then $\frac{1}{x} \to 0$.

(iv) $\lim_{x \to \infty} f(x) = \infty$ as $x \to \infty$, we mean that as x becomes larger and larger f(x) also becomes larger and larger.

Illustration No1

Evaluate $\lim\limits_{x \to 1} f(x)$ where f(x) = x + 5

Solution:

When we need to calculate the $\lim_{x \to 1} f(x)$ where f(x) = x + 5

$x \to 1^+$	1.1	1.01	1.001	1.0001	1.00001	....	....	....
f(x) = x +5	6.1	6.01	6.001	6.0001	6.00001	....	....	....

From the above table, we see that as $x \to 1^+$, $f(x) \to 6$. Here 6 is called the right hand limit of f(x) as $x \to 1^+$, written as $\lim\limits_{x \to 1^+} f(x) = 6$.

Similarly,

$x \to 1^-$	0.9	0.99	0.999	0.9999	0.99999	....	....	....
f(x)= x +5	5.9	5.99	5.999	5.9999	5.99999	....	....	....

From the above table, we see that as $x \to 1^-$; $f(x) \to 6$. Here 6 is called the left hand limit of f(x) as $x \to 1^-$ written as $\lim\limits_{x \to 1^-} f(x) = 6$

Thus, we find that $x \to 1$ from either side, f(x) = 6, hence $\lim_{x \to 1} f(x) = 6$.

Some Important Theorem

If $\lim\limits_{x \to a} f(x)$ and $\lim\limits_{x \to a} g(x)$ both exist then

(i) $\lim\limits_{x \to a} \{k f(x)\} = k \lim\limits_{x \to a} \{f(x)\}$

(ii) $\lim\limits_{x \to a} \{f(x) \pm g(x)\} = \lim\limits_{x \to a} \{f(x)\} \pm \lim\limits_{x \to a} \{g(x)\}$

(iii) $\lim\limits_{x \to a} \{f(x) \times g(x)\} = \lim\limits_{x \to a} \{f(x)\} \times \lim\limits_{x \to a} \{g(x)\}$

(iv) $\lim\limits_{x \to a} \dfrac{f(x)}{g(x)} = \dfrac{\lim\limits_{x \to a} f(x)}{\lim\limits_{x \to a} g(x)}$

(v) $\lim\limits_{x \to a} f\{g(x)\} = f\{\lim\limits_{x \to a} g(x)\}$

(vi) $\lim_{x\to a}\left(\frac{k}{f(x)}\right) = k\lim_{x\to a}\left(\frac{1}{f(x)}\right)$

(vii) $\lim_{x\to a} x\sqrt{f(x)} = x\sqrt{\lim_{x\to a} f(x)}$

for any positive integer.

Some Important Results

1. $\lim_{x\to a}\frac{x^n - a^n}{x-a} = na^{n-1}$ for all a > 0 and n is positive integer
2. $\lim_{x\to 0}\frac{a^x - 1}{x} = \log a (a > 0)$
3. $\lim_{x\to 0}\left[\frac{(1+x)^n - 1}{x}\right] = n$
4. $\lim_{x\to 0}\frac{e^x - 1}{x} = 1$
5. $\lim_{x\to 0}\frac{1}{x}[\log(1+x)] = 1$
6. $\lim_{x\to 0}(1+x)^{\frac{1}{x}} = e$
7. $\lim_{x\to 0} x^n = 0 (-1 < x < 1)$
8. $\lim_{x\to 0}\frac{x^n}{n!} = 0$
9. $\lim_{x\to 0} e^x = 1$
10. $\lim_{x\to\infty}\left(x + \frac{1}{x}\right)^x = e$

Methods of Evaluation of Limits

Case I: When the limits of the function f(x) as $x \to a$, is a finite quantity.

Factorization Method

If the function is expressed in the form of $\frac{f(x)}{g(x)}$ where $g(x) \neq 0$ and by substituting the value of x = a the value of the function is not indeterminate, then it is the limit value at x = a. Otherwise spilt the function f(x) and g(x) into their factors and cancel out the common factor and substitute the value of x.

Rationalization Method

If the function is expressed as an irrational fraction, then the limit value of the function is obtained by first rationalizing either the numerator or denominator as per the form of the function and then proceed in the manner discussed earlier.

Substitution Method

The one sided limits both left hand or right hand limits are evaluated by substituting $x = a \pm h$ where h is small and $h \neq 0$, so that when $x \to a$, $h \to 0$.

Case II

When the limits of the function f(x) as $x \to \infty$, an infinite quantity

For evaluation $\lim_{x\to\infty}\frac{f(x)}{g(x)}$ divide the terms of the ratio $\frac{f(x)}{g(x)}$ by the highest degree of the polynomial and then substitute the value of x.

Illustration No. 1

Find $\lim_{x\to 0}\left(5x^2-3x+2\right)=?$

Solution:

Given $\lim_{x\to 0}\left(5x^2-3x+2\right)$

$$= \lim_{x\to 0}\left(5x^2\right)-\lim_{x\to 0}3x+2 \to \text{Note} \lim_{x\to 0}\{k\,f(x)\} = \lim_{x\to 0}\{k\,f(x)\}$$

$$= 5\lim_{x\to 0}x^2-3\lim_{x\to 0}x+2 = 5.0^2-3.0+2=2$$

Illustration No. 2

Find $\lim_{x\to 0}\frac{\left(2x^3+6x+4\right)}{\left(x^2-2x+2\right)}$

Solution:

Since $\lim_{x\to 0}\left\{\frac{f(x)}{g(x)}\right\}$

$$= \frac{\lim_{x\to 0}\{f(x)\}}{\lim_{x\to 0}\{g(x)\}}$$

Applying the above Rule:

$$= \frac{\lim_{x\to 0}\left(2x^3+6x+4\right)}{\lim_{x\to 0}\left(x^2-2x+2\right)}$$

$$= \frac{\left\{\lim_{x\to 0} 2x^3\right\}+\left\{\lim_{x\to 0} 6x\right\}+4}{\left\{\lim_{x\to 0} x^2\right\}-\left\{\lim_{x\to 0} 2x\right\}+2}$$

$$= \frac{2\left\{\lim_{x\to 0} x^3\right\}+6\left\{\lim_{x\to 0} x\right\}+4}{\left\{\lim_{x\to 0} x^2\right\}-2\left\{\lim_{x\to 0} x\right\}+2}$$

$$= \frac{2\times 0+6\times 0+4}{0-2\times 0+2} = \frac{4}{2} = 2$$

Illustration No. 3

Find $\lim_{x\to 0}\frac{x^2-16}{x-4}$

Solution:

Given $\lim_{x\to 0}\frac{x^2-16}{x-4}$

$$= \frac{\lim_{x\to 0}\left(x^2-16\right)}{\lim_{x\to 0}\left(x-4\right)}$$

$$= \frac{\lim_{x\to 0}(x-4)(x+4)}{\lim_{x\to 0}(x-4)} = \lim_{x\to 0}(x+4) = 0+4 = 4$$

Note: $x^2-a^2=(x-a)(x+a)$

Illustration No.4

Find $\lim_{x\to 2}\frac{x^3-2^3}{x-2}$

Solution:

Given $\lim\limits_{x\to 2}\dfrac{x^3-2^3}{x-2}$

Therefore,

$$= \lim_{x\to 2}\frac{(x-2)\left(x^2+2x+4\right)}{x-2}$$

Since $x^3-a^3=(x-a)(x^2+ax+a^2)$

$$= \left\{\lim_{x\to 2}x^2\right\}+2\left\{\lim_{x\to 2}x\right\}+4$$

$$= 2^2 + 2 \times 2 + 4$$

$$= 4 + 4 + 4 = 12.$$

Illustration No. 5

Find $\lim\limits_{x\to 1}\dfrac{\left(x^2-5x+4\right)}{\left(x^2-3x+2\right)}$

Solution:

Given $\lim\limits_{x\to 1}\dfrac{\left(x^2-5x+4\right)}{\left(x^2-3x+2\right)}$

$$= \frac{\lim\limits_{x\to 1}\left(x^2-5x+4\right)}{\lim\limits_{x\to 1}\left(x^2-3x+2\right)}$$

As $(x^2 - 5x + 4)$ and $(x^2 - 3x + 2)$ can be split into factors i.e. $x^2 - 5x + 4 = (x - 1)(x - 4)$

Similarly

$x^2 - 3x + 2 = (x - 1)(x - 2)$

$$= \lim_{x\to 1}\frac{(x-1)(x-4)}{(x-1)(x-2)}$$

$$= \lim_{x\to 1}\frac{(x-4)}{(x-2)}$$

$$= \frac{\lim_{x\to 1}(x-4)}{\lim_{x\to 1}(x-2)}$$

$$= \frac{1-4}{1-2} = \frac{-3}{-1} = 3$$

Illustration No. 6

Find $\lim_{x\to 0}\frac{2^x-1}{3^x-1}$

Solution:

Given $\lim_{x\to 0}\frac{2^x-1}{3^x-1}$

Multiply and divide numerator and denominator by x, so we get,

$$= \frac{\lim_{x\to 0}\frac{2^x-1}{x}}{\lim_{x\to 0}\frac{3^x-1}{x}}$$

Note: $\lim_{x\to 0}\frac{a^x-1}{x} = \log a$

$$= \frac{\log 2}{\log 3}$$

Illustration No. 7

Find $\lim_{x\to 3}\left(\frac{6}{(x^2-9)}+\frac{1}{3-x}\right)$

Solution:

We have,

$$\lim_{x\to 3}\left(\frac{6}{(x^2-9)}+\frac{1}{3-x}\right)$$

Splitting $x^2-9 = x^2-3^2 = (x-3)(x+3)$

We get,

$$= \lim_{x\to 3}\left(\frac{6}{(x-3)(x+3)}+\frac{1}{3-x}\right)$$

$$= \lim_{x\to 3}\frac{1}{x-3}\left(\frac{6}{x+3}-1\right)$$

$$= \lim_{x\to 3}\left(\frac{1}{x-3}\right)\left(\frac{6-x-3}{x+3}\right)$$

$$= \lim_{x\to 3}\frac{-(x-3)}{(x-3)(x+3)}$$

$$= \lim_{x\to 3}\frac{-1}{x+3}$$

$$= -\frac{1}{3+3} = -\frac{1}{6}$$

Illustration No. 8

Find $\lim\limits_{x\to 0}\dfrac{e^{5x}-e^{3x}}{x}$

Solution:

$$\lim_{x\to 0}\frac{e^{5x}-e^{3x}}{x}$$

Adding and subtracting 1 in the numerator, we get

$$= \lim_{x\to 0}\frac{e^{5x}-e^{3x}+1-1}{x}$$

$$= \lim_{x\to 0}\left\{\frac{\left(e^{5x}-1\right)-\left(e^{3x}-1\right)}{x}\right\}$$

$$= \lim_{x\to 0}\left\{\frac{e^{5x}-1}{x}\right\}-\lim_{x\to 0}\left\{\frac{e^{3x}-1}{x}\right\}$$

Multiply and Divide 5 in first part and 3 in the second part

We get,

$$= \lim_{x\to 0} 5\left\{\frac{e^{5x}-1}{5x}\right\} - \lim_{x\to 0} 3\left\{\frac{e^{3x}-1}{3x}\right\}$$

$$= 5\lim_{x\to 0}\left\{\frac{e^{5x}-1}{5x}\right\} - 3\lim_{x\to 0}\left\{\frac{e^{3x}-1}{3x}\right\}$$

$$= 5 \times 1 - 3 \times 1 \qquad \text{since } \lim_{x\to 0}\frac{e^x-1}{x} = 1$$

$$= 2$$

Illustration No. 9

Find $\lim_{x\to 0}\frac{\sqrt{2+x}-\sqrt{2-x}}{x}$

Solution:

Given $\lim_{x\to 0}\frac{\sqrt{2+x}-\sqrt{2-x}}{x}$

Rationalize the Numerator with rationalizing factor,

$$= \lim_{x\to 0}\frac{\sqrt{2+x}-\sqrt{2-x}}{x} \times \frac{\sqrt{2+x}+\sqrt{2-x}}{\sqrt{2+x}+\sqrt{2-x}}$$

$$= \lim_{x\to 0}\frac{2x}{x\sqrt{2+x}+\sqrt{2-x}}$$

$$= \lim_{x\to 0}\frac{2}{\sqrt{2+x}+\sqrt{2-x}}$$

$$= \frac{2}{\sqrt{2}+\sqrt{2}}$$

$$= \frac{2}{2\sqrt{2}} = \frac{1}{\sqrt{2}}$$

Illustration No. 10

Find $\lim_{x\to 0}\frac{e^x - e^{-x}}{x}$

Solution:

We have
$$\lim_{x\to 0}\frac{e^x - e^{-x}}{x}$$

$$= \lim_{x\to 0}\frac{\left(1+x+\frac{x^2}{2!}+......\right)-\left(1-x+\frac{x^2}{2!}-\frac{x^3}{3!}+......\right)}{x}$$

$$= \lim_{x\to 0}\frac{2\left(x+\frac{x^3}{3!}+......\right)}{x}$$

$$= \lim_{x\to 0} 2\left(1+\frac{x^2}{3!}+......\right)$$

$$= 2 \times 1$$

$$= 2$$

Illustration No. 11

Find $\lim_{x\to 0}\frac{x}{\sqrt{5+x}-\sqrt{5-x}}$

Solution:

We have,
$$\lim_{x\to 0}\frac{x}{\sqrt{5+x}-\sqrt{5-x}}$$

$$= \lim_{x\to 0}\frac{x}{\sqrt{5+x}-\sqrt{5-x}} \times \frac{\sqrt{5+x}+\sqrt{5-x}}{\sqrt{5+x}+\sqrt{5-x}}$$

$$= \lim_{x\to 0}\frac{x\sqrt{5+x}+\sqrt{5-x}}{(5+x)-(5-x)}$$

$$= \lim_{x\to 0}\frac{x\sqrt{5+x}+\sqrt{5-x}}{2x}$$

$$= \lim_{x\to 0}\frac{\sqrt{5+x}+\sqrt{5-x}}{2}$$

$$= \frac{\sqrt{5}+\sqrt{5}}{2} = \sqrt{5}$$

Illustration No. 12

Find $\lim\limits_{x\to 0}\dfrac{e^x-1-x}{x}$

Solution:

$$\text{Given}\quad \lim_{x\to 0}\frac{e^x-1-x}{x}$$

$$=\lim_{x\to 0}\frac{\left(1+x+\dfrac{x^2}{2!}+\dfrac{x^3}{3!}+....\right)-1-x}{x}$$

$$=\lim_{x\to 0}\frac{\left(\dfrac{x^2}{2!}+\dfrac{x^3}{3!}+....\right)}{x}$$

$$=\lim_{x\to 0}\left(\frac{x}{2!}+\frac{x^2}{3!}....\right)$$

$$=0$$

Illustration No. 13

Find $\lim\limits_{x\to 0}\dfrac{x}{\sqrt{1+ax}-\sqrt{1-ax}}$

Solution:

$$\text{We have,}\quad \lim_{x\to 0}\frac{x}{\sqrt{1+ax}-\sqrt{1-ax}}$$

$$=\lim_{x\to 0}\frac{x}{\sqrt{1+ax}-\sqrt{1-ax}}\times\frac{\sqrt{1+ax}+\sqrt{1-ax}}{\sqrt{1+ax}+\sqrt{1-ax}}$$

$$=\lim_{x\to 0}\frac{x\sqrt{1+ax}+\sqrt{1-ax}}{(1+ax)-(1-ax)}$$

$$=\lim_{x\to 0}\frac{x\sqrt{1+ax}+\sqrt{1-ax}}{2ax}$$

$$= \lim_{x \to 0} \frac{\sqrt{1+ax} + \sqrt{1-ax}}{2a}$$

$$= \frac{\sqrt{1} + \sqrt{1}}{2a} = \frac{2}{2a} = \frac{1}{a}$$

Illustration No. 14

Find $\lim_{x \to 0} \frac{x^p - a^p}{x^q - a^q}$

Solution:

Multiply and divide by (x – a)

$$= \frac{\lim_{x \to a} \frac{x^p - a^p}{x - a}}{\lim_{x \to a} \frac{x^q - a^q}{x - a}} = \frac{p \times a^{p-1}}{q \times a^{q-1}} = \frac{pa^{p-q}}{q}$$

Illustration No. 15

Find $\lim_{x \to 0} \frac{3^x - 1}{\sqrt{1+x-1}}$

Solution:

$$\lim_{x \to 0} \frac{3^x - 1}{\sqrt{1+x} - 1}$$

Multiply and Divide by 'x', we get,

$$\lim_{x \to 0} \left(\frac{\frac{3^x - 1}{x}}{\frac{\sqrt{1+x-1}}{x}} \right)$$

$$= \frac{\lim_{x \to 0} \frac{3^x - 1}{x}}{\lim_{x \to 0} \frac{\sqrt{1+x} - 1}{x}}$$

$$= \frac{\log 3}{\lim_{x \to 0} \frac{\sqrt{1+x-1}}{x}}$$

Rationalizing the Denominator

$$= \frac{\log 3}{\lim\limits_{x\to 0}\frac{\left(\sqrt{1+x}-1\right)\left(\sqrt{1+x}+1\right)}{x\left(\sqrt{1+x}+1\right)}}$$

$$= \frac{\log 3}{\lim\limits_{x\to 0}\frac{\{(1+x)-1\}}{x\left(\sqrt{1+x}+1\right)}}$$

$$= \frac{\log 3}{\lim\limits_{x\to 0}\frac{1+x-1}{x\left(\sqrt{1+x}+1\right)}}$$

$$= \frac{\log 3}{\lim\limits_{x\to 0}\left\{\frac{x}{x\left(\sqrt{1+x}+1\right)}\right\}} = \frac{\log 3}{\frac{1}{2}}$$

$$= 2\log 3$$

$$= \log 3^2 = \log 9$$

Illustration No. 16

Find $\lim\limits_{x\to 0}\frac{x}{(1+x)^{1/3}-1}$

Solution:

Given $\lim\limits_{x\to 0}\frac{x}{(1+x)^{1/3}-1}$

$$= \frac{x}{\lim\limits_{x\to 0}\left\{1+\frac{x}{3}+\frac{\frac{1}{3}}{\left(\frac{1}{2}-2\right)2!}x^2+.....\right\}-1}$$

$$= \frac{x}{\lim_{x\to 0}\left\{\frac{x}{3} + \frac{\frac{1}{3}}{\left(\frac{1}{3}-2\right)2!}x^2 + \ldots..\right\}}$$

$$= \frac{x}{x\lim_{x\to 0}\left\{\frac{1}{3} + \frac{\frac{1}{3}}{\left(\frac{1}{3}-2\right)2!}x + \ldots..\right\}}$$

$$= \frac{1}{\frac{1}{3}} = 3$$

Illustration No. 17

Find $\lim_{x\to 0}\frac{\log\left(1+\frac{x}{3}\right)}{x}$

Solution: $\lim_{x\to 0}\frac{\log\left(1+\frac{x}{3}\right)}{x}$

Expanding by log $\left(1+\frac{x}{3}\right)$ we get,

$$= \lim_{x\to 0}\frac{\left(\frac{x}{3} - \frac{1}{2}\left(\frac{x}{3}\right)^2 + \frac{1}{3}\left(\frac{x}{3}\right)^3 + \ldots..\right)}{x}$$

$$= \lim_{x\to 0}\frac{x\left\{\frac{1}{3} - \frac{1}{2}\left(\frac{x}{9}\right) + \ldots..\right\}}{x}$$

$$= \frac{1}{3} - 0\ldots..$$

$$= \frac{1}{3}$$

Illustration No.18

Find $\lim_{x\to 0}\frac{4e^x-4}{x}$

Solution:

$$\lim_{x\to 0}\frac{4e^x-4}{x}$$

Expanding e^x, we get

$$=\lim_{x\to 0}\frac{4\left(1+x+\frac{x^2}{2!}+.....-1\right)}{x}$$

$$=\lim_{x\to 0}4\left(\frac{x+\frac{x^2}{2!}+.....}{x}\right) \quad \text{(Taking x common from N}^r\text{)}$$

$$=\lim_{x\to 0}4x\left(\frac{1+\frac{x}{2!}+........}{x}\right)$$

$$=4\lim_{x\to 0}\left(1+\frac{x}{2!}+........\right)$$

$$=4$$

Illustration No. 19

Find $\lim_{x\to\infty}\frac{1^2+2^2+3^2+.....x^2}{x^3}$

Solution:

$$\lim_{x\to\infty}\frac{1^2+2^2+3^2+......x^2}{x^3}$$

$$=\lim_{x\to\infty}\frac{x(x+1)(2x+1)}{6x^3}$$

$$= \lim_{x\to\infty} \frac{2x^3\left(1+\frac{1}{2x}\right)\left(1+\frac{1}{x}\right)}{6x^3}$$

As $x \to \infty$, $1/x \to 0$

$$= \lim_{\frac{1}{x}\to 0} \frac{\left(1+\frac{1}{x}\right)\left(1+\frac{1}{2x}\right)}{3} = \frac{1}{3}$$

Illustration No. 20

Find $\lim_{x\to 0} \frac{\sqrt{4+x}-2}{x}$

Solution:

$$\lim_{x\to 0} \frac{\sqrt{4+x}-2}{x}$$

Rationalizing the Numerator using rationalizing factor, we get,

$$= \lim_{x\to 0} \frac{\sqrt{4+x}-2}{x} \times \frac{\sqrt{4+x}+2}{\sqrt{4+x}+2}$$

$$= \lim_{x\to 0} \frac{4+x-4}{x\left(\sqrt{4+x}+2\right)}$$

$$= \lim_{x\to 0} \frac{x}{x\left(\sqrt{4+x}+2\right)}$$

$$= \frac{1}{2+2} = \frac{1}{4}$$

The Concept of Right and Left hand Limit

Right hand Limit

If f(x) is a function such that f(x) tends to value l_1 when x tends to a $(x \to a)$ from the right i.e. the values greater than a, then l_1 is called as the limiting value and the right hand limit of f(x). Symbolically written as R.H.L $= \lim_{x\to a^+} f(x) = f(a+0)$

Similarly

Left hand Limit

If f(x) is a function such that f(x) tends to value l_2 when x tends to a ($x \to a$) from the left i.e. the values lesser than a, then l_2 is called as the limiting value and the left hand limit of f(x). Symbolically written as L.H.L $= \lim_{x \to a^-} f(x) = f(a-0)$

Existence of $\lim_{x \to a} f(x)$

If the right hand limit i.e. $\lim_{x \to a^+} f(x)$ and the left hand limit i.e. $\lim_{x \to a^-} f(x)$ if both exists and $\lim_{x \to a^+} f(x) = \lim_{x \to a^-} f(x)$ then the $\lim_{x \to a} f(x)$ exists and is equal to the common value of the above two limits i.e.

$$\lim_{x \to a} f(x) = \lim_{x \to a^+} f(x) = \lim_{x \to a^-} f(x)$$

Difference between Limit and Value of a Function

When we directly substitute the value of x say (= a) in the given function f(x), then we get the functional value f(a), whereas $\lim_{x \to a} f(x)$ is the limiting value of f(x) when x approaches a from either side of a i.e. when $|x-a|$ can be made smaller by assigning a positive number say δ, however small but $x \neq a$. Therefore in general $\lim_{x \to a} f(x) \neq f(a)$.

Note: f(a) may be undefined but the limiting value of f(x) when $x \to a$ may be a definite number and vice–versa.

Illustration No. 1

Find the existence of $\lim_{x \to 0} \frac{|x|}{x}$

Solution:

Since $|x| = x$ for all $x \geq 0$

$= -x$ for all $x < 0$

Therefore, R.H.L $= \lim_{x \to 0^+} \frac{|x|}{x} = \lim_{x \to 0^+} \frac{x}{x} = \lim_{h \to 0} (1) = 1$

Similarly, L.H.L $= \lim_{x\to 0^-} \frac{|x|}{x} = \lim_{x\to 0^-} = \frac{-x}{x} = \lim_{h\to 0}(-1) = -1$

Therefore $\lim_{x\to 0^+} \frac{|x|}{x} \neq \lim_{x\to 0} \frac{|x|}{x}$

Hence $= \lim_{x\to 0} \frac{|x|}{x}$ does not exist

Illustration No. 2

If $f(x) = \begin{cases} x & x \le 3 \\ 6 - x & x > 3 \end{cases}$

Find out whether $\lim_{x\to 3} f(x)$ exist, then find its value

Solution:

We have

R.H.L $= \lim_{x\to 3^+} f(x)$

$= \lim_{h\to 0} f(3+h)$ (substituting x = 3 + h as x → 3, h → 0)

$= \lim_{h\to 0} \{6 - (3 + h)\}$

$= \lim_{h\to 0} \{(3 - h)\}$

= 3

L.H.L $= \lim_{x\to 3^-} f(x)$

$= \lim_{h\to 0} f(3-h)$ (similarly x = 3 - h as x → 3, h → 0)

= 3

Hence the R.H.L = L.H.L Therefore, the limit exits and the value of the function is 3.

Illustration No. 3

If f(x) is defined by

$$f(x)=\begin{cases} x & \text{when } 0 \le x < \frac{1}{2} \\ 0 & \text{when } x = \frac{1}{2} \\ 1-x & \text{when } \frac{1}{2} < x \le 1 \end{cases}$$

Find whether $\lim_{x\to 1/2} f(x)$ exists then what is its value?

Solution:

$$\text{R.H.L} = \lim_{x\to \frac{1}{2}^+} f(x)$$

$$= \lim_{h\to 0} f(1/2 + h)$$

$$= \lim_{h\to 0} [1 - (1/2 + h)]$$

$$= \lim_{h\to 0} [1/2 - h]$$

$$= 1/2$$

$$\text{L.H.L} = \lim_{x\to \frac{1}{2}^-} f(x)$$

$$= \lim_{h\to 0} f(1/2 - h)$$

$$= \lim_{h\to 0} (1/2 - h)$$

$$= \lim_{h\to 0} (1/2 - h) = 1/2$$

Thus L.H.L = R.H.L = $1/2$

Hence $\lim_{x\to 1/2} f(x)$ exists and its value is equal to $1/2$

Illustration No. 4

If $f(x) = \dfrac{|x-1|}{x-1}$, find whether $\lim_{x\to 0} f(x)$ exits or not?

Solution:

$$\text{R.H.L} = \lim_{x \to 0^+} f(x)$$

$$= \lim_{h \to 0} \frac{|1+h-1|}{1+h-1}$$

$$= \lim_{h \to 0} \frac{|h|}{h}$$

$$= \lim_{h \to 0} \frac{h}{h} = 1$$

$$\text{L.H.L} = \lim_{x \to 1^-} f(x)$$

$$= \lim_{h \to 0} \frac{|1-h-1|}{1-h-1}$$

$$= \lim_{h \to 0} \frac{|-h|}{-h} = -1$$

Since R.H.L $\neq$ L.H.L

Therefore $\lim f(x)$ does not exist.

Illustration No.5

If $f(x) = \begin{cases} 4 + x^2 & 0 \leq x < 2 \\ 4 - x & \text{where } x > 2 \end{cases}$

Prove that f(x) doesn't exist?

Solution:

$$\text{R.H.L} = \lim_{x \to 2} f(x)$$

$$= \lim_{h \to 0} f(2+h)$$

$$= \lim_{h \to 0} [4 - (2+h)]$$

$$= \lim_{h \to 0} [2 - h]$$

$$= 2$$

Also,

$$\text{L.H.L} = \lim_{x \to 2^-} f(x)$$

$$= \lim_{h \to 0} f(2-h)$$

$$= \lim_{h \to 0} \left[4 + (2 - h)^2\right]$$

$$= \lim_{h \to 0} \left[4 + 4 - 4h + h^2\right]$$

$$= \lim_{h \to 0} \left[8 - 4h + h^2\right] = 8$$

Since R.H.L ≠ L.H.L

Therefore $\lim_{x \to 2} f(x)$ doesn't exit

Continuity

A function f(x) is said to be continuous at a point x = a if and only if

(i) f(a) is defined and

(ii) $\lim_{x \to a^+} f(x) = \lim_{x \to a^-} f(x) = f(a)$

Thus f(x) is continuous at x = a if $\lim_{x \to a} f(x) = f(a)$ provided f(a) is a finite and the limit exists.

Cauchy's Definition of Continuity

A function is said to be continuous at x = a if f(a) exists corresponding to an arbitrary number ϵ, however small there exists a number δ such that

$$|f(x) - f(a)| < \epsilon, \text{ whenever } |x - a| < \delta$$

Continuity in an interval

A function is said to be continuous in the interval a < x < b, if f(x) is continuous at every point in the interval. And further $\lim_{x \to a^+} f(x) = f(a)$ and $\lim_{x \to b^-} f(x) = f(b)$ then f(x) is said to be continuous in the closed interval $a \le x \le b$.

Illustration No.1

If f (x) is defined as

$$f(x) = 4x + 2 \quad \text{when } x \geq 2$$
$$= 3x^2 - 2 \quad \text{when } x < 2$$

Show that whether function is continuous at x = 2?

Solution:

We calculate the R.H.L

$$\text{R.H.L} = \lim_{x\to 2^+} f(x) = \lim_{h\to 0} f(2+h)$$

$$= \lim_{h\to 0} \{4(2+h)+2\}$$

$$= \lim_{h\to 0} \{10+h\}$$

$$= 10$$

$$\text{L.H.L} = \lim_{x\to 2^-} f(x) = \lim_{h\to 0} f(2-h)$$

$$= \lim_{h\to 0} \{3(2-h)^2 - 2\}$$

$$= \lim_{h\to 0} \{3(4-4h+h^2) - 2\}$$

$$= \lim_{h\to 0} \{10 - 12h + 3h^2\}$$

$$= 10$$

Also f (2) = 4(2) + 2 = 10

R.H.L = L.H.L = f(2)

Hence the function is continuous at x = 2

Illustration No. 2

A function is defined as

$$f(x) = \begin{cases} 3x+3 & \text{for } x \leq 0 \\ 3 & \text{for } x = 0 \\ x^2+3 & \text{for } x > 0 \end{cases}$$

Examine the continuity at x = 0

Solution:

We calculate R.H.L

$$\lim_{x\to0^+} f(x) = \lim_{h\to0} f(0+h)$$

$$= \lim_{h\to0} (0+h)^2 + 3$$

$$= \lim_{h\to0} \left(h^2 + 3\right)$$

$$= 0^2 + 3$$

$$= 3$$

$$\text{L.H.L} = \lim_{x\to0^-} f(x) = \lim_{h\to0} f(0-h)$$

$$= \lim_{h\to0} \{3\ (0-h) + 3\}$$

$$= \lim_{h\to0} (-\ 3h + 3)$$

$$= -3\ (0) + 3$$

Also f(x) = 3 at x = 0

Hence R.H.L = L.H.L = f(0)

Therefore the function is continuous at x = 0

Illustration No. 3

The function f(x) is define as

$$\begin{aligned} f(x) &= x + 1 && \text{when } x < 1 \\ &= x^2 + 1 && \text{when } 1 \le x < 2 \\ &= x - 3 && \text{when } x \ge 2 \end{aligned}$$

Examine the continuity of the function at x = 1 and x = 2

Solution:

We calculate **R.H.L** at x = 1

$$\lim_{x\to1^+} f(x) = \lim_{h\to0} f(1+h)$$

$$= \lim_{h\to0} \left\{(1+h)^2 + 1\right\}$$

$$= \lim_{h\to0} \{h^2 + 2h + 1 + 1\}$$

$$= \lim_{h \to 0} \{h^2 + 2h + 2\}$$

$$= 0 + 0 + 2$$

L.H.L

$$\lim_{x \to 1^-} f(x) = \lim_{h \to 0} f(1 - h)$$

$$= \lim_{h \to 0} \{(1 - h) + 1\}$$

$$= \lim_{h \to 0} \{2 - h\}$$

$$= 2$$

And $f(1) = 1^2 + 1 = 2$

Since R.H.L = L.H.L = f(1)

Therefore the function f(x) is continuous at x = 1

And At x = 2

$$\text{R.H.L} = \lim_{x \to 2^+} f(x) = \lim_{h \to 0} f(2 + h)$$

$$= \lim_{h \to 0} [(2 + h) - 3]$$

$$= \lim_{h \to 0} [2 + h - 3]$$

$$= \lim_{h \to 0} [h - 1]$$

$$= -1$$

$$\text{L.H.L} = \lim_{x \to 2^-} f(x) = \lim_{h \to 0} f(2 - h)$$

$$= \lim_{h \to 0} \{(2 - h)^2 + 1\}$$

$$= \lim_{h \to 0} \{4 - 4h + h^2 + 1\}$$

$$= \lim_{h \to 0} \left[5 - 4h + h^2\right]$$

$$= 5$$

f(2) = 2 – 3 = –1

Since R.H.L ≠ L.H.L and hence f(x) discontinuous at x = 2

Illustration No. 4

If f(x) is defined as follows

$$f(x) = x - 1 \text{ for } x < 3$$
$$= 3x - 7 \text{ for } x \geq 3$$

show that f(x) is continues everywhere

Solution:

We find the R.H.L and L.H.L

$$\text{R.H.L} = \lim_{x \to 3^+} f(x)$$

$$= \lim_{h \to 0} \{3(3 + h) - 7\}$$

$$= \lim_{h \to 0} \{9 + 3h - 7\}$$

$$= \lim_{h \to 0} \{2 + 3h\}$$

$$= 2 + 0$$

$$= 2$$

$$\text{L.H.L} = \lim_{x \to 3^-} f(x)$$

$$= \lim_{h \to 0} f(3 - h)$$

$$= \lim_{h \to 0} \{(3 - h) - 1\}$$

$$= \lim_{h \to 0} \{2 - h\}$$

$$= 2$$

We observe that R.H.L = L.H.L

Also f(3) = 3x – 7

$$= 3 \times 3 - 7$$

$$= 9 - 7$$

$$= 2$$

And Hence

R.H.L = L.H.L = f(3)

There the given function f(x) is continuous at x = 3

Illustration No. 5

Show that the function $f(x) = x^2 - 4x + 3$ is continuous at the point $x = 1$

Solution:

At $x = 1$, we have

$$\text{R.H.L} = \lim_{x \to 1^+} f(x)$$

$$= \lim_{h \to 0} (1 + h)$$

$$= \lim_{h \to 0} \{(1 + h)^2 - 4(1 + h) + 3\}$$

$$= 1 + 2h + h^2 - 4 - 4h + 3$$

$$= h^2 - 2h$$

$$= 0$$

$$\text{L.H.L} = \lim_{x \to 1^-} f(x)$$

$$= \lim_{h \to 0} f(1 - h)$$

$$= \lim_{h \to 0} \{(1 - h)^2 - 4(1 - h) + 3\}$$

$$= \lim_{h \to 0} \{h^2 + 2h\}$$

$$= 0$$

And the value of f(x) at $x = 1$

i.e. $f(0) = 1^2 - 4(0) + 3$

$= 1 - 4 + 3 = 0$

Hence the R.H.L = L.H.L = f(1) hence the given function is continuous at $x = 1$

Illustration No. 6

Find the values of a and b. So that function f(x) is defined as

$$f(x) = 2 \quad \text{for } x \le 2$$

$$= ax + b \text{ for } 2 < x < 4$$

$$= 8 \quad \text{for } x \le 4$$

is continuous?

Solution:

As the given function is constant for all values x < 2 and x > 4. This shows that f(x) is continuous for all $x \in (2, 4)$

Therefore at, x = 2, we have f(2) = 2

$$\text{R.H.L} = \lim_{x\to 2^+} f(x) = \lim_{h\to 0} \{f(2+h)\}$$

$$= \lim_{h\to 0} \{a(2+h)+b\}$$

$$= 2a + b$$

$$\text{L.H.L} = \lim_{x\to 2^-} f(x)$$

$$= \lim_{h\to 0} f(2-h)$$

$$= \lim_{h\to 0} \{2\} = 2$$

Since f(x) is continuous at x = 2

Therefore we have, 2a + b = 2 ..... (3.1)

Similarly at x = 4

We have f(4) = 8 and

$$\text{R.H.L} = \lim_{x\to 4^+} f(x)$$

$$= \lim_{h\to 0} f(4+h)$$

$$= \lim_{h\to 0} (8)$$

$$= 8$$

$$\text{L.H.L} = \lim_{x\to 4^-} f(x)$$

$$= \lim_{h\to 0} f(4-h)$$

$$= \lim_{h\to 0} \{a(4-h)+b\}$$

$$= 4a + b$$

Since the function is continuous at x = 4

Since 4a +b = 8 ..... (3.2)

Solving the (1) and (2)

$$\left.\begin{aligned} 2a + b &= 2 \\ 4a + b &= 8 \end{aligned}\right\}$$

We get a = 3 and b = – 4

Illustration No. 7

Examine the continuity of the function f(x) at x = 1 and x = 2 where

$$f(x) = 4 - 2x \quad \text{if} \quad x \le 1$$
$$= 2 \quad \text{if} \quad 1 \le x \le 2$$
$$= 4x - 6 \quad \text{if} \quad x \ge 2$$

Solution:

Let us check the continuity of the function at x = 1

$$\text{R.H.L} = \lim_{x\to 1^+} f(x) = \lim_{h\to 0} f(1+h)$$
$$= \lim_{h\to 0} (2) = 2$$
$$\text{L.H.L} = \lim_{x\to 1^-} f(x) = \lim_{h\to 0} f(1-h)$$
$$= \lim_{h\to 0} \{4 - 2(1-h)\}$$
$$= \lim_{h\to 0} [4 - 2 + 2h] = 2$$

Since x = 1, R.H.L = L.H.L = f(1) = 2 hence f(x) is continuous at x = 1

(ii) Let us check the continuity at x = 2

$$\text{R.H.L} = \lim_{x\to 2^+} f(x)$$
$$= \lim_{h\to 0} f(2+h)$$
$$= \lim_{h\to 0} \{(4(2+h) - 6\}$$
$$= \lim_{h\to 0} [8 + 4h - 6]$$
$$= 2$$

$$\text{L.H.L} = \lim_{x \to 2^-} f(x)$$

$$= \lim_{h \to 0} f(2-h)$$

$$= \lim_{h \to 0} [2] = 2$$

The value of the function f(x) at x = 2 is

f(2) = 4(2) – 6

= 8 – 6

= 2

Since the R.H.L = L.H.L = f(2)

And hence the function is continuous at x = 2

Illustration No. 8

Examine the continuity of f(x) at x = 1 for a function defined as

$f(x) = |x - 1|$

Solution:

We have

$$\text{R.H.L} = \lim_{x \to 1^+} f(x)$$

$$= \lim_{h \to 0} f(1+h)$$

$$= \lim_{h \to 0} |(1+h) - 1|$$

$$= \lim_{h \to 0} |1 + h - 1|$$

$$= 0$$

$$\text{L.H.L} = \lim_{x \to 1^-} f(x)$$

$$= \lim_{h \to 0} f(1-h)$$

$$= \lim_{h \to 0} |(1-h) - 1|$$

$$= \lim_{h \to 0} |-h|$$

$$= 0$$

Also, $f(1) = |1 - 1| = |0| = 0$

Therefore R.H.L = L.H.L = f(1)

Hence the function is continuous at x = 1

Application of Limits and Continuity in Business and Economics

Illustration No. 1

A toy manufacturer charges at the rate of Rs 50 per unit for purchase of 50 or less in numbers. He charges at discounted price of Rs 45 if the number of toys purchased varies between above 50 and less than or equal 100 and the charges are Rs. 40 for any purchase above 100 units. From the cost function, find at what points the cost function is not continuous?

Solution:

Let 'x' be the quantity purchased and f (x) be the cost function.

Therefore

$$f(x) = \begin{cases} 50x & \text{for } x \le 50 \\ 45x & \text{for } 50 < x \le 100 \\ 40x & \text{for } x > 100 \end{cases}$$

At x = 50

$$\text{R.H.L} = \lim_{x \to 50^+} f(x)$$

$$= \lim_{h \to 0} f(50 + h)$$

$$= \lim_{h \to 0} \{45(50 + h)\}$$

$$= \lim_{h \to 0} \{2250 + 45h\}$$

$$= 2250$$

$$\text{L.H.L} = \lim_{x \to 50^-} f(x)$$

$$= \lim_{h \to 0} f(50 - h)$$

$$= \lim_{h\to 0} 50\,(50 - h)$$

$$= \lim_{h\to 0} (2500 - 50\,h)$$

$$= 2500$$

Since R.H.L ≠ L.H.L therefore the function is not continuous at x = 50

Similarly at x = 100

$$\text{R.H.L} = \lim_{x\to 100^+} f(x)$$

$$= \lim_{h\to 0} f(100 + h)$$

$$= \lim_{h\to 0} 40(100 + h)$$

$$= \lim_{h\to 0} (4000 + 40h)$$

$$= 4000$$

$$\text{L.H.L} = \lim_{h\to 100^-} f(x)$$

$$= \lim_{h\to 0} f(100 + h)$$

$$= \lim_{h\to 0} \{45\,(100 - h)\}$$

$$= \lim_{h\to 0} (4500 + 45h)$$

$$= 4500$$

Therefore R.H.L ≠ L.H.L and hence the function is discontinuous at x = 100

Illustration No.2

A TV manufacturer pay his employee at the rate Rs 100 per hour for at least 8 hrs of working in a day and anything above 8 hours but less than 10 hrs as paid at additional Rs 20 over above normal rate. And hrs is paid at the rate of Rs 40 over normal rate for working hours from 10 hours and above. Find the wage function and explain the parts of discontinuity?

Solution:

Let f (x) be the cost function & 'x' be the amount of services given

Therefore

$$f(x) = \begin{cases} 100x & \text{for } x \le 8 \\ 120x & \text{for } 8 < x < 10 \\ 140x & \text{for } x \ge 10 \end{cases}$$

At x = 8

$$\text{R.H.L} = \lim_{x \to 8^-} f(x)$$

$$= \lim_{h \to 0} \{100(8 - h)\}$$

$$= 800$$

$$\text{L.H.L} = \lim_{x \to 8^+} f(x)$$

$$= \lim_{h \to 0} \{120(8 + h)\}$$

$$= 960$$

Hence f(x) discontinous at x = 8 as R.H.L ≠ L.H.L

And at x = 10

$$\text{L.H.L} = \lim_{x \to 10^-} f(x)$$

$$= \lim_{x \to 0} 120(10 - h)$$

$$= 1200$$

$$\text{R.H.L} = \lim_{x \to 10^+} f(x)$$

$$= \lim_{h \to 0} 140(10 + h)$$

$$= 1400$$

f(x) is also discontinuous at x = 10 as R.H.L ≠ L.H.L

Brief Summary

(i) If be a real valued function defined for all values of x in the neighborhood of a fixed number say 'a' such that f(x) approaches a fixed number 'l' as x approaches 'a' from either side of a, then l is the limit of the function f as x tends to a or written as $x \to a$, mathematically written as,

$\lim_{x\to a} f(x) = l$ where l is called as the limiting value of $f(x)$ as $x \to a$

(ii) $\lim_{x\to a} \{f(x) \times g(x)\} = \lim_{x\to a} \{f(x)\} \times \lim_{x\to a} \{g(x)\}$

(iii) $\lim_{x\to a} \frac{f(x)}{g(x)} = \frac{\lim_{x\to a} f(x)}{\lim_{x\to a} g(x)})$

(iv) $\lim_{x\to 0} (1+x)^{\frac{1}{x}} = \in$

(v) $\lim_{x\to 0} x^n = 0$ for $(-1 < x < 1)$

(vi) **Existence of** $\lim_{x\to a} f(x)$

If the right hand limit i.e. $\lim_{x\to a^+} f(x)$ and the left hand limit i.e. $\lim_{x\to a^-} f(x)$ if both exists and $\lim_{x\to a^+} f(x) = \lim_{x\to a^-} f(x)$ then the $\lim_{x\to a} f(x)$ exists and is equal to the common value of the above two limits i.e.

$$\lim_{x\to a} f(x) = \lim_{x\to a^+} f(x) = \lim_{x\to a^-} f(x)$$

(vii) **Cauchy's Definition of Continuity**

A function is said to be continuous at x = a if f(a) exists corresponding to an arbitrary number ∈, however small there exists a number δ such that

$$|f(x) - f(a)| < \in, \text{ whenever } |x - a| < \delta$$

Self Practice Exercises

1. $\lim_{x\to 3} \sqrt{x+3-x}$

2. $\lim_{x\to \infty} \dfrac{4x^2+7x-3}{7x^2-6x+4}$

3. $\lim_{x\to 0} \dfrac{x-|x|}{2x-|x|}$

Hint $|x| = \begin{Bmatrix} x & x>0 \\ -x & x<0 \end{Bmatrix}$

4. $\lim_{x\to 0} \dfrac{5-\sqrt{5^2-x^2}}{x^2}$

5. $\lim_{x\to 0} \dfrac{\sqrt{3+5x}-\sqrt{3-7x}}{4x}$

6. $\lim_{x\to 0} \dfrac{2x}{\sqrt{7+x}-\sqrt{7-x}}$

7. $\lim_{x\to \infty} \dfrac{1^2+2^2+3^3+ \text{-------} x^2}{x^3}$

8. $\lim_{x\to 0} \dfrac{11^x-13^x}{x}$

9. $\lim_{x\to 5} \dfrac{x^5-5^5}{x^4-5^4}$

10. $\lim_{x\to 0} \dfrac{\sqrt{1+4^x}-\sqrt{1-5^x}}{x}$

11. $\lim_{x\to 0} \dfrac{\sqrt{x-2}+\sqrt{x}-\sqrt{2}}{\sqrt{x^2-2^2}}$

12. $\lim_{x\to 0} \dfrac{\sqrt{1+3x}-\sqrt{1-3x}}{x}$

13. $\lim_{x\to 1} \dfrac{x^2-1}{\sqrt{5x-1}-\sqrt{3x+1}}$

14. $\lim_{x\to 5} \dfrac{\sqrt{x-3}-\sqrt{2}}{x^2-5^2}$

15. $\lim_{x \to \sqrt{3}} \frac{x^2 - 3}{x - \sqrt{3}}$

16. If f(x) is defined as

$$f(x) = 2x - 1 \text{ if } x < 1$$
$$= 3x - 2 \text{ if } x \geq 2$$

find whether the function f(x) is continuous at x = 1

17. Show that the function $f(x) = x^2 - 2x + 3$ is continuous at the point x = 0

18. Find the value of p and q so that the function f(x) defined as 5 for $x \leq 3$

$$f(x) = px + q \text{ for } 3 < x < 5$$
$$= 11 \text{ for } 5 \leq x$$

19. Examine the continuity of f(x) at x = 3 for a function defined as $f(x) = |x - 3|$

20. Find the value of A & B so that function (x) where f(x) is defined as

$$f(x) \begin{cases} 3 & \text{if } x \leq 5 \\ ax + b & \text{if } 5 < x\ 7 \\ 13 & \text{if } x > 7 \end{cases}$$

21. If f(x) is defined as $f(x) = |x - 5|$ find the whether function is continuous at x = 5?

22. If $f(x) = \begin{cases} x^2 + a^2 & \text{when } x \geq 0 \\ x^2 - a^2 & \text{when } x < 0 \end{cases}$

If f(x) is continuous at x = 0

find the value of 'a'?

23. If f(x) is defined by

$$f(x) = \begin{cases} 2 + x & x \leq 2 \\ 6 - x & \text{when } x > 2 \end{cases}$$

Find:

(i) Whether f(x) is discontinuous at x = 2

(ii) Is f(x) continuous at x = 2

(iii) None of these.

24. Show that the function f(x) is continuous at x = 1 if $f(x) = x^2 + 8x + 7$

25. Prove that if

$$f(x) = \frac{x^2 - x - 4}{x - 4} \quad \text{When } x \neq 6$$

$$= 2 \quad \text{When } x = 6$$

Calculate the continuity at x = 6?

Answers

1. $\frac{\sqrt{3}}{\sqrt{2} + \sqrt{3}}$

2. $\frac{4}{7}$

4. $\frac{1}{10}$

5. $\frac{1}{\sqrt{3}}$

6. $2\sqrt{7}$

7. $\frac{1}{3}$

8. $-\log_{13} 11$

9. $\frac{25}{4}$

10. $\frac{9}{2}$

11. $\frac{1}{2}$

12. 3

13. -8

14. $\frac{1}{20\sqrt{2}}$

15. $2\sqrt{3}$.

18. p = 3 and q = – 4

19. Continuous at x = 3

20. a = 5 and b = 22

21. Continuous at x = 5

22. a = 0

23. Continuous at x = 2

25. Continuous at x = 6

Appendix

Some Important Proofs

1. $\lim_{x \to 0} \frac{x^n - a^n}{x - \alpha} = na^{n-1}$

Proof:

$$\lim_{x\to 0} \frac{x^n - a^n}{x - a}$$

Substituting x = a + h, therefore as x → a, then h → 0

And hence

$$\lim_{x\to 0} \frac{x^n - a^n}{x - a} = \lim_{h\to 0} \frac{(a+h)^n - a^n}{a+h-a}$$

$$= \lim_{x\to 0} \frac{a^n\left[\left\{1+\frac{h}{a}\right\}^n - 1\right]}{h} \quad \text{(taking } a^n \text{ common from the numerator)}$$

$$= \lim_{x\to 0} \frac{a^n\left[1 + \left(\frac{n\,h}{a}\right) + \left(\frac{n\,(n-1)}{2!}\right)\left(\frac{h}{a}\right)^2 + \ldots\ldots -1\right]}{h}$$

$$= \lim_{x\to 0} \frac{a^n\left[n\left\{\frac{h}{a}\right\} + \left(\frac{n\,(n-1)}{2!}\right)\left\{\frac{h}{a}\right\}^2 + \ldots\ldots\right]}{h}$$

Taking h as common factor from the numerator, we get

$$= \lim_{x\to 0} \frac{a^n h\left[n\left\{\frac{1}{a}\right\} + \frac{n\,(n-1)}{2!}\left\{\frac{h}{a^2}\right\} + \text{------}\right]}{h}$$

$$= \lim_{x\to 0} a^n\left[n\left\{\frac{1}{a}\right\} + \frac{n\,(n-1)}{2!}\left\{\frac{h}{a^2}\right\} + \text{------}\right]$$

$$= a^n\left[n\left(\frac{1}{a}\right)\right]$$

$$= n\,a^{n-1}$$

2. $\lim_{x\to 0} \frac{a^x - b^x}{x} = \log\left(\frac{a}{b}\right)$

Proof:

$$\lim_{x\to 0} \frac{a^x - b^x}{x}$$

$$= \lim_{x\to 0} \frac{1}{x}\left[\left\{1 + x \log a + \frac{x^2 (\log a)^2}{2!} + \ldots\right\} - \left\{1 + x \log b + \frac{x^2 (\log b)^2}{2!} + \ldots\right\}\right]$$

$$= \lim_{x\to 0} \frac{1}{x}\left\{x (\log a - \log b) + \frac{x^2}{2!}\{(\log a)^2 - (\log b)^2\} + \text{-----------}\right\}$$

Taking x common and applying limits we get all other parts zero,

Therefore, we get

$$= \log a - \log b = \log\left(\frac{a}{b}\right)$$

3. $\lim_{x\to\infty} x\left(a^{\frac{1}{x}} - 1\right) = \log a$

Proof:

$$\lim_{x\to\infty} x\left(a^{\frac{1}{x}} - 1\right)$$

as $x \to \infty, \frac{1}{x} \to 0$

Substitute, $\frac{1}{x} = y$

Therefore,

$$\lim_{y\to 0} \frac{1}{y}\left(a^y - 1\right)$$

$$= \lim_{y\to 0} \frac{1}{y}\left[1 + y (\log a) + \frac{y^2}{2!} (\log a)^2 + \ldots\ldots -1\right]$$

$$= \lim_{y\to 0} \frac{1}{y}\left[(y\log a) + \frac{y^2}{2!}(\log a)^2 + \ldots\ldots\right]$$

Taking y as a common factor we get,

$$= \lim_{y\to 0} \frac{1}{y} \times y\left[(\log a) + \frac{y}{2!}(\log a)^2 + \ldots\ldots\ldots\right]$$

Applying the limit, we get

$$= \log a$$

4. $\lim_{x\to 0} \frac{(1+x)^n - 1}{x} = n$

Proof:

$$\lim_{x\to 0} \frac{(1+x)^n - 1}{x}$$

$$= \lim_{x\to 0}\left[\frac{\left\{\left(1 + nx + \frac{n(n-1)x^2}{2!} + \ldots\ldots\right)\right\} - 1}{x}\right]$$

$$= \lim_{x\to 0}\left(\frac{\{nx + \frac{n(n-1)x^2}{2!} + \ldots\ldots\ldots\}}{x}\right)$$

$$= \lim_{x\to 0} \frac{x}{x}\left[n + \frac{n(n-1)}{2!}x + \ldots\ldots\ldots\right]$$

taking x common from N^r

$$= \lim_{x\to 0}\left[n + \frac{n(n-1)}{2!}x + \ldots\ldots\ldots\right]$$

$$= n + 0$$

$$= n$$

5. $\lim_{x\to\infty}\left[1 + \frac{1}{x}\right]^x = e$

Proof:

$$\lim_{x\to\infty}\left[1+\frac{1}{x}\right]^x$$

Using Binomial Expansion

$$(1+x)^n = {}^nC_0(1)^n + {}^nC_1\ (1)^{n-1}\,x + {}^nC_2\ (1)^{n-2}\,x^2 + \ldots\ldots\ldots$$

Therefore,

$$\left[1+\frac{1}{x}\right]^x = {}^xC_0\ (1)^n + {}^xC_1\ (1)^{x-1}\left[\frac{1}{x}\right] + \ldots\ldots\ldots\ldots$$

And hence,

$$\lim_{x\to\infty}\left[1+\frac{1}{x}\right]^x = \lim_{x\to\infty}\left[{}^xC_0\,(1)^n + {}^xC_1\,(1)^{x-1}\left\{\frac{1}{x}\right\} + \ldots\ldots\ldots\right]$$

$$= \left[\lim_{x\to\infty}\left\{1 + x\,\frac{1}{x} + \frac{x(x-1)}{2!}\,\frac{1}{x^2} + \text{-----}\right\}\right]$$

Note: ${}^xC_o = 1$, ${}^xC_1 = x$ and ${}^xC_2 = \dfrac{x\ (x-1)}{2!}$

$$= \left[\lim_{x\to\infty}\left\{1 + 1 + \frac{\left(1-\frac{1}{x}\right)}{2!} + \text{-----}\right\}\right]$$

As $\quad x \to \infty,\ \dfrac{1}{x} \to 0$

$$1+1+\frac{1}{2!}+\text{----}$$

$$= e^1 = e$$

Note: $e^x = 1 + x + \dfrac{x^2}{2!} + \ldots\ldots\ldots$

CHAPTER 4

Derivatives and its Applications

Concept and Definition

Derivative is a measure of rate of change of given function with respect to its independent variable. Let us understand the concept

Let us assume y = f(x) be a function defined under a given interval and Δx (or h) be the increment of x such that $x + \Delta x$ (or x + h) lies within that interval and the corresponding increment of y be Δy or $f(x + h) - f(x)$ then the incremental ratio is given by

$$\frac{\Delta y}{\Delta x} = \frac{f(x+\Delta x)-f(x)}{\Delta x} \quad \text{or} \quad \frac{f(x+h)-f(x)}{h}$$

And if the ratio $\frac{\Delta y}{\Delta x}$ tends to definite quantity as Δx (or h) $\rightarrow 0$ from either end then the definite quantity is termed as derivative of the function with respect to x and is denoted by $f'(x)$ or $\frac{d}{dx}f(x)$ or $\frac{dy}{dx}$

Thus

$$f'(x) \text{ or } \frac{dy}{dx} = \lim_{\Delta x \to 0} \frac{\Delta y}{\Delta x} = \lim_{h \to 0} \frac{f(x+h)-f(x)}{h}$$

provided the limit exists and for the derivative to exist both the Right hand derivative and the left hand derivative should exist and be equal.

i.e

$$Rf'(x) = \lim_{\Delta h \to 0^+} \frac{f(x+h)-f(x)}{h} \text{ the right hand derivative}$$

and the

$$Lf'(x) = \lim_{h \to 0^-} \frac{f(x+h) - f(x)}{h}$$ the left hand derivative

such that $Rf'(x) = Lf'(x)$

Note

Meaning of Increment

If x is a variable function which changes from one value say x_1 to another x_2 then the difference $x_2 - x_1$ i.e final value minus the initial value is called the increment of x and is denoted by Δx and read as delta x or h. And hence $\Delta x = x_2 - x_1$ or $x_2 = \Delta x + x_1$ and if x increases from x_1 to x_2 where $x_2 > x_1$ is called positive increment otherwise negative increment.

And if y a function is defined in terms of x i.e. $y = f(x)$ were x is independent in nature then a given increment in x from a given value $x = x_1$ then the corresponding increment of y will be denoted by Δy or $\Delta f(x)$ and we have

$$y + \Delta y = f(x_1 + \Delta x) \text{ where } y = f(x_1)$$

Therefore

$$\Delta y = f(x_1 + \Delta x) - f(x_1)$$

And the ratio $\frac{\Delta y}{\Delta x}$ is called the incremental ratio.

Remarks

The derivative of f(x) at any fixed value a is given by

$$f'(a) = \lim_{h \to 0} \frac{f(a+h) - f(a)}{h}$$

provided that the limit exists.

Some Important Rules

Rule 1

If u(x) is a function such that defined as $u(x) = c\,f(x)$, where f(x) is a derivative function and c is a constant then,

$$\frac{d}{dx}u(x) = \frac{d}{dx}\{cf(x)\} = c\frac{d}{dx}\{f(x)\}$$

Proof

Since $u(x) = c\, f(x)$, then

$u(x + h) = c\, f(x + h)$

hence

$$\frac{d}{dx}\{u(x)\} = \lim_{h\to 0}\frac{u(x+h)-u(x)}{h}$$

$$= \lim_{h\to 0}\frac{cf(x+h)-cf(x)}{h}$$

$$= c\lim_{h\to 0}\left\{\frac{f(x+h)-f(x)}{h}\right\}$$

$$= c\left\{\frac{d}{dx}\{f(x)\}\right\}$$

Hence Proved

Rule 2

If $f(x)$ is a function defined as $f(x) = u(x) \pm v(x)$

Then $$\frac{d}{dx}\{f(x)\} = \frac{d}{dx}\{u(x)\} \pm \frac{d}{dx}\{v(x)\}$$

Proof

Since $f(x) = u(x) \pm v(x)$

$f(x) = u(x + h) \pm v(x + h)$

Since $$\frac{d}{dx}f(x) = \lim_{h\to 0}\frac{\{u(x+h)\pm v(x+h)\}-\{u(x)\pm v(x)\}}{h}$$

$$= \lim_{h\to 0}\frac{\{u(x+h)-u(x)\pm\{v(x+h)-v(x)\}}{h}$$

$$= \lim_{h\to 0}\frac{\{u(x+h)-u(x)\}}{h} \pm \lim_{h\to 0}\frac{\{v(x+h)-v(x)\}}{h}$$

$$= \frac{d}{dx}\{u(x)\} \pm \frac{d}{dx}\{v(x)\}$$

Hence proved

Rule 3

If f(x) is a function defined as the product of two differentiable functions u(x) and v(x). Then the derivative of f(x) denoted by $\frac{d}{dx}\{f(x)\}$ can be expressed as

$$\frac{d}{dx}\{f(x)\} = v(x)u'(x) + u(x)v'(x)$$

where $\frac{d}{dx}u(x) = u'(x)$, $\frac{d}{dx}v(x) = v'(x)$ are the respective first derivative of u(x) and v(x)

Proof:

Since $f(x) = u(x) \times v(x)$

Then $f(x + h) = u(x + h) \times v(x + h)$

Therefore $f(x + h) - f(x) = u(x + h)\, v(x + h) - u(x)\, v(x)$

Applying the Rule of 1st principle we get:

$$\frac{d}{dx}f(x) = \lim_{h \to 0}\frac{f(x+h) - f(x)}{h} = \lim_{h \to 0}\frac{u(x+h)v(x+h) - u(x)v(x)}{h}$$

Adding and Subtracting u(x+h) v(x) in the numerator we get

$$= \lim_{h \to 0}\frac{u(x+h)v(x+h) - u(x+h)v(x) + u(x+h)v(x) - u(x)v(x)}{h}$$

$$= \lim_{h \to 0}\frac{u(x+h)\{v(x+h) - v(x)\} + v(x)\{u(x+h) - u(x)\}}{h}$$

$$= \lim_{h \to 0}\frac{u(x+h)\{v(x+h) - v(x)\}}{h} + \lim_{h \to 0}\frac{v(x)\{u(x+h) - u(x)\}}{h}$$

Applying limits as $h \to 0$

We get, $u(x + h) \to u(x)$

Therefore,

$$\lim_{h \to 0}\frac{u(x)\{v(x+h) - v(x)\}}{h} + \lim_{h \to 0}\frac{v(x)\{u(x+h) - u(x)\}}{h}$$

$$= v(x)\, u'(x) + u(x)\, v'(x)$$

where

$$u'(x) = \lim_{h\to 0}\frac{\{u(x+h)-u(x)\}}{h} \text{ and } v'(x) = \lim_{h\to 0}\frac{\{v(x+h)-v(x)\}}{h}$$

and hence,

Differential Coefficient of the product of two functions = Second function × Derivative of the first function + first function × Derivative of the second function

Rule 4

If f(x) is a function defined as a function expressed in the following form

$$f(x) = \frac{u(x)}{v(x)}$$

where u(x) and v(x) are function of x then $\frac{d}{dx}\{f(x)\}$ the derivative of f(x) can be expressed as

$$\frac{d}{dx}\{f(x)\} = \frac{v(x)\frac{d}{dx}\{u(x)\} - u(x)\frac{d}{dx}\{v(x)\}}{v(x)^2}$$

Proof:

Since

$$f(x) = \frac{u(x)}{v(x)}$$

Therefore (x + h) can be written as

$$f(x+h) = \frac{u(x+h)}{v(x+h)}$$

$$f(x+h) - f(x) = \frac{u(x+h)}{v(x+h)} - \frac{u(x)}{v(x)}$$

$$= \frac{u(x+h)v(x) - v(x+h)u(x)}{v(x+h)v(x)}$$

Apply 1st Principle, we get,

$$\frac{d}{dx}\{f(x)\} = \lim_{h\to 0}\frac{u(x+h)v(x) - v(x+h)u(x)}{h\,v(x+h)v(x)}$$

By adding and subtracting u(x) v(x) in the numerator ,we get

$$=\lim_{h\to 0}\frac{u(x+h)v(x)+u(x)v(x)-u(x)v(x)-v(x+h)u(x)}{hv(x+h)v(x)}$$

$$=\lim_{h\to 0}\frac{v(x)\{u(x+h)-u(x)\}-u(x)\{v(x+h)-v(x)\}}{hv(x+h)v(x)}$$

$$=\frac{v(x)\lim\limits_{h\to 0}\dfrac{\{u(x+h)-u(x)\}}{h}-u(x)\lim\limits_{h\to 0}\dfrac{\{v(x+h)-v(x)\}}{h}}{\lim\limits_{h\to 0}v(x+h)v(x)}$$

As $h\to 0$, $v(x + h) = v(x)$

$$=\frac{v(x)\dfrac{d}{dx}\{u(x)\}-u(x)\dfrac{d}{dx}\{v(x)\}}{\{v(x)\}^2}$$

where

$$u'(x)=\lim_{h\to 0}\frac{\{u(x+h)-u(x)\}}{h} \text{ and } v'(x)=\lim_{h\to 0}\frac{\{v(x+h)-v(x)\}}{h}$$

and hence proved.

Some Important Standard Derivative

Standard Derivative 1

The derivative of any Constant 'c' is zero i.e $\frac{d}{dx}\{c\}=0$

Proof

Let us define f(x) = c; then f(x + h) = c

$$\therefore \quad \frac{d}{dx}\{f(x)\}=\lim_{h\to 0}\frac{f(x+h)-f(x)}{h}$$

$$=\lim_{h\to 0}\frac{c-c}{h}=0$$

Hence the derivative of a constant is zero

Standard Derivative 2

If $f(x) = x^n$ then $\frac{d}{dx}\{f(x)\}=nx^{n-1}$

Proof

By first principle: we know,

$$\frac{d}{dx}\{f(x)\} = \lim_{h\to 0}\frac{f(x+h)-f(x)}{h}$$

But $f(x) = x^n$

Since $f(x + h) = (x + h)^n$

Substituting the values of f(x) and f(x + h) we get,

$$\frac{d}{dx}\{f(x)\} = \lim_{h\to 0}\frac{f(x+h)^n - x^n}{h}$$

$$= \lim_{h\to 0}\frac{\{{}^nc_0\, x^n + {}^nc_1\, x^{n-1} h^2 + \ldots\ldots + {}^n c_n x^0 h^n\} - x^n}{h}$$

Note: By Binomial Theorem

$$(x + h)^n = {}^nc_0\, x^n + {}^nc_1 \times x^{n-1} h + \ldots.. + {}^nc_n\, x^0 h^n$$

Therefore

$$= \lim_{h\to 0}\frac{\{x^n + n\times x^{n-1}\times h + \ldots.. + h^n\} - x^n}{h}$$

$$= \lim_{h\to 0}\frac{h\{nx^{n-1} + \ldots. + h^{n-1}\}}{h}$$

$$= \lim_{h\to 0}\{nx^{n-1} + \ldots\ldots + h^{n-1}\}$$

$= nx^{n-1} + 0,$ (as all other terms are zero)

$= nx^{n-1}$

Hence Proved

Standard Derivative 3

If $f(x) = e^x$ then $\frac{d}{dx}\{f(x)\} = e^x$

Proof

Since $f(x) = e^x$

Therefore, $f(x + h) = e^{x+h}$

Using 1st principle, we get

$$\frac{d}{dx}\{f(x)\} = \lim_{h\to0}\frac{e^{x+h}-e^{x}}{h}$$

$$= \lim_{h\to0}\frac{e^{x}\times e^{h}-e^{x}}{h}$$

$$= \lim_{h\to0}\frac{e^{x}\left[e^{h}-1\right]}{h}$$

Since $e^h = 1 + 1 + \frac{h}{1!} + \frac{h^2}{2!} + \frac{h^3}{3!} + ...$

Substituting the value of e^h we get,

$$= \lim_{h\to0}\frac{e^{x}\left[1+h+\frac{h^2}{2!}+\frac{h^3}{3!}+....-1\right]}{h}$$

$$= \lim_{h\to0}\frac{e^{x}\left[h+\frac{h^2}{2!}+\frac{h^3}{3!}+....\right]}{h}$$

Taking h as common factor from numerator we get,

$$= \lim_{h\to0}\frac{e^{x}\,h\left[1+\frac{h}{2!}+\frac{h^2}{3!}+....\right]}{h}$$

Applying $h\to0$

$$= e^x \lim_{h\to0}\left[1+\frac{h^2}{2!}+\frac{h^2}{3!}+....\right]$$

$$= e^x\left[1+\left\{\lim_{h\to0}\frac{h}{2!}+\frac{h^2}{3!}+....\right\}\right]$$

$$= e^x$$

Hence Proved

Standard Derivative 4

If $f(x) = a^x$ then $\frac{d}{dx}\{a^x\} = a^x \log a$

Proof

Since $f(x) = a^x$ then $f(x + h) = a^{x+h}$

Applying 1st principle

$$\frac{d}{dx}\{f(x)\} = \lim_{h\to 0}\frac{f(x+h)-f(x)}{h} = \lim_{h\to 0}\frac{a^{x+h}-a^x}{h}$$

$$= \lim_{h\to 0}\frac{a^x a^h - a^x}{h}$$

$$= \lim_{h\to 0}\frac{a^x(a^h-1)}{h}$$

Since $a^n = 1 + h \log a + \frac{(h\log a)^2}{2!} + \ldots\ldots$

Therefore substituting the value a^n, we get

$$= \lim_{h\to 0} a^x \times \left[\frac{(1+h\log a+h^2\frac{(\log a)^2}{2!}+\ldots\ldots)-1}{h}\right]$$

taking 'h' as common factor in the numerator, we get

$$= \lim_{h\to 0} a^x \times h \times \frac{\left(\log a + \frac{h(\log a)^2+\ldots\ldots}{2!}\right)}{h}$$

$$= \lim_{h\to 0} a^x\left(\log a + h\frac{(\log a)^2}{2!}+\ldots\ldots\right)$$

$$= a^x \times \log a + 0$$

$$= a^x \log a$$

Hence proved

Standard Derivative 5

If $f(x) = \log x$ then $\frac{d}{dx}\{f(x)\} = \frac{1}{x}$

Proof

Since f(x) = log (x)

then f(x + h) = log (x + h)

$$\frac{d}{dx}f(x) = \lim_{h\to 0}\frac{\log(x+h)-\log x}{h}$$

$$= \lim_{h\to 0}\frac{\log\left(\frac{x+h}{x}\right)}{h}$$

$$= \lim_{h\to 0}\frac{1}{h}\log\left(1+\frac{h}{x}\right)$$

$$= \lim_{h\to 0}\frac{1}{h}\left\{\left(\frac{h}{x}\right)-\frac{1}{2}\left(\frac{h}{x}\right)^2+\frac{1}{3}\left(\frac{h}{x}\right)^3+\ldots\ldots\right\}$$

Taking 'h' as a common factor in the N^r

$$= \frac{1}{h}\lim_{h\to 0}h\left\{\frac{1}{x}-\frac{1}{2}\left(\frac{h}{x^2}\right)+\frac{1}{2}\left(\frac{h^2}{x^3}\right)+\ldots\ldots\right\}$$

Applying $h\to 0$

$$= \frac{1}{x}-0$$

$$= \frac{1}{x}$$

Hence Proved

Some Solved Illustrations using 1st principle

Illustration No. 1

If $f(x) = x^4$ find $\frac{d}{dx}\{f(x)\}$ using 1st principle?

Solution:

Since $f(x) = x^4$

Therefore $f(x + h) = (x + h)^4$

Since $\frac{d}{dx}f(x) = \lim_{h\to 0}\frac{f(x+h)-f(x)}{h}$

$$= \lim_{h\to 0}\frac{(x+h)^4 - x^4}{h}$$

$$= \lim_{h\to 0}\frac{\{x^4 + 4x^3h + 6x^2h^2 + 4xh^3 + h^4\} - x^4}{h}$$

$$= \lim_{h\to 0}\frac{\{x^4 + 4x^3h + 6x^2h^2 + 4xh^3 + h^4 - x^4\}}{h}$$

$$= 4x^3 + 6x^2 \times 0 + 4x \times 0 + (0)$$

$$= 4x^3$$

Illustration No. 2

If f(x) = $\sqrt{x}$, find $\frac{d}{dx}\{f(x)\} = ?$

Solution:

Since f (x) = $\sqrt{x}$

Therefore f(x + h) = $\sqrt{x+h}$

Since $\frac{d}{dx}\{f(x)\} = \lim_{h\to 0}\frac{f(x+h)-f(x)}{h}$

$$= \lim_{h\to 0}\frac{\sqrt{x+h}-\sqrt{x}}{h}$$

Rationalize the numerator with Rationalizing factor

$$= \lim_{h\to 0}\frac{\sqrt{x+h}-\sqrt{x}}{h} \times \frac{\sqrt{x+h}+\sqrt{x}}{\sqrt{x+h}+\sqrt{x}}$$

$$= \lim_{h\to 0}\frac{(x+h)-x}{h\sqrt{x+h}+\sqrt{x}}$$

$$= \lim_{h\to 0}\frac{x+h-x}{h\sqrt{x+h}+\sqrt{x}}$$

$$= \lim_{h\to 0}\frac{h}{h\sqrt{x+h}+\sqrt{x}}$$

$$= \frac{1}{\sqrt{x}+\sqrt{x}} = \frac{1}{2\sqrt{x}}$$

Illustration No 3.

If f(x) = $\sqrt{x+2}$ then $\frac{d}{dx}f(x) = ?$

Solution:

Since f(x) = $\sqrt{x+2}$

Since f(x + h) = $\sqrt{x+h+2}$

Since $\frac{d}{dx}\{f(x)\} = \lim_{h\to0}\frac{f(x+h)-f(x)}{h}$

$$= \lim_{h\to0}\frac{\sqrt{x+h+2}-\sqrt{x+2}}{h}$$

$$= \lim_{h\to0}\frac{\sqrt{x+h+2}-\sqrt{x+2}}{h}\times\frac{\sqrt{x+h+2}+\sqrt{x+2}}{\sqrt{x+h+2}+\sqrt{x+2}}$$

$$= \lim_{h\to0}\frac{(x+h+2)-(x+2)}{\sqrt{x+h+2}+\sqrt{x+2}}$$

$$= \lim_{h\to0}\frac{x+h+2-x-2}{h\sqrt{x+h+2}+\sqrt{x+2}}$$

$$= \lim_{h\to0}\frac{h}{h\sqrt{x+h+2}+\sqrt{x+2}}$$

$$= \frac{1}{\sqrt{x+2}+\sqrt{x+2}} = \frac{1}{2\sqrt{x+2}}$$

Illustration No. 4

If f(x) = e^{3x} then find $\frac{d}{dx}f(x) = ?$

Solution:

Since f(x) = e^{3x}

Then f(x + h) = $e^{3(x+h)}$

Since $\frac{d}{dx}f(x) = \lim_{h\to0}\frac{f(x+h)-f(x)}{h}$

$$= \lim_{h\to0}\frac{e^{3x+3h}-e^{3x}}{h}$$

$$=\lim_{h\to 0}\frac{e^{3x}\left(e^{3h}-1\right)}{h}$$

$$=\lim_{h\to 0}\frac{e^{3x}\left(1+3h+\frac{(3h)^2}{2!}+\frac{(3h)^3}{3!}+....-1\right)}{h}$$

$$=\lim_{h\to 0}\frac{3h\times e^{3x}\left(\frac{(3h)}{2!}+\frac{(3h)^2}{3!}+....\right)}{h}$$

$$=\lim_{h\to 0}\frac{3\times e^{3x}\left(\frac{3h}{2!}+\frac{(3h)^2}{3!}+....\right)}{h}$$

Applying limits we get,

$$= 3 \times e^{3x} = 3e^{3x}$$

Illustration No. 5

If $f(x) = \frac{1}{\sqrt{x+1}}$, find $\frac{d}{dx}f(x) = ?$

Solution:

As $\quad f(x) = \frac{1}{\sqrt{x+1}}$

$$f(x + h) = \frac{1}{\sqrt{x+h+1}}$$

Since $\quad \frac{df(x)}{dx} = \lim_{h\to 0}\frac{f(x+h)-f(x)}{h}$

$$= \lim_{h\to 0}\frac{\frac{1}{\sqrt{x+h+1}}-\frac{1}{\sqrt{x+1}}}{h}$$

$$= \lim_{h\to 0}\frac{\sqrt{x+1}-\sqrt{x+h+1}}{h\times\sqrt{x+h+1}\sqrt{x+1}}$$

Using rationalizing factor

$$= \lim_{h\to 0}\frac{\sqrt{x+1}-\sqrt{x+h+1}}{h\sqrt{x+1}\sqrt{x+h+1}}\times\frac{\sqrt{x+1}+\sqrt{x+h+1}}{\sqrt{x+1}+\sqrt{x+h+1}}$$

$$= \lim_{h\to 0}\frac{(x+1)-(x+h+1)}{h\sqrt{x+1}\sqrt{x+h+1}\left(\sqrt{x+1}+\sqrt{x+h+1}\right)}$$

$$= \lim_{h\to 0}\frac{x+1-x-h-1}{h\times\sqrt{x+1}\sqrt{x+h+1}\left(\sqrt{x+1}+\sqrt{x+h+1}\right)}$$

$$= \lim_{h\to 0}\frac{-h}{h\sqrt{x+1}\sqrt{x+h+1}\left(\sqrt{x+1}+\sqrt{x+h+1}\right)}$$

$$= \frac{-1}{\sqrt{x+1}\sqrt{x+1}\left(\sqrt{x+1}+\sqrt{x+1}\right)}$$

$$= \frac{-1}{2(x+1)\left(\sqrt{x+1}\right)}$$

Illustration No. 6

If $f(x) = \cos x^2$, find $\frac{d}{dx}\{f(x)\} = ?$

Solution:

As $f(x) = \cos x^2$

therefore $\quad f(x + h) = \cos(x + h)^2$

$$\frac{df(x)}{dx} = \lim_{h\to 0}\frac{f(x+h)-f(x)}{h}$$

$$= \lim_{h\to 0}\frac{\cos(x+h)^2-\cos x^2}{h}$$

Since $\cos A - \cos B = 2\sin\left(\frac{A+B}{2}\right)\sin\left(\frac{A-B}{2}\right)$

$$= \lim_{h\to 0}\frac{2\sin\left\{\frac{(x+h)^2+x^2}{2}\right\}\sin\left\{\frac{(x+h)^2-x^2}{2}\right\}}{h}$$

$$= \lim_{h\to 0}\frac{2\sin\left\{\frac{x^2+2hx+h^2+x^2}{2}\right\}\sin\left\{\left(\frac{x^2+2hx+h^2-x^2}{2}\right)\right\}}{h}$$

$$= \lim_{h\to 0}\frac{2\sin\left\{\frac{2x^2+2hx+h^2}{2}\right\}\sin\left\{\left(\frac{2hx+h^2}{2}\right)\right\}}{h}$$

$$= 2\lim_{h\to 0}\sin\left\{\frac{2x^2+2hx+h^2}{2}\right\}\lim_{h\to 0}\sin\frac{\left\{\left(\frac{h(2x+h)}{2}\right)\right\}}{h}$$

$$= \lim_{h\to 0}\sin\left\{\frac{2x^2+2hx+h^2}{2}\right\}\lim_{h\to 0}\sin\left\{\frac{\frac{h(2x+h)}{2}}{\frac{h(2x+h)}{2}}\right\}(2x+h)$$

$$= \lim_{h\to 0}\sin\left\{x^2+hx+\frac{h^2}{2}\right\}\lim_{h\to 0}\sin\left\{\frac{\frac{h(2x+h)}{2}}{\frac{h(2x+h)}{2}}\right\}\lim_{h\to 0}(2x+h)$$

$= \sin x^2 \times 2x$

$= 2x \sin x^2$

Note: $\lim_{h\to 0}\frac{\sin x}{x} = 1$

Illustration No. 7

If $f(x) = \sin 5x$

$$\frac{df(x)}{dx} = ?$$

Solution

Since $f(x) = \sin 5x$

Then $f(x + h) = \sin 5(x + h)$

Since $$\frac{df(x)}{dx} = \lim_{h\to 0}\frac{f(x+h)-f(x)}{h}$$

$$= \lim_{h\to 0}\frac{\sin 5(x+h)-\sin 5x}{h}$$

$$= \lim_{h\to 0}\frac{2\cos\left\{\left(\frac{5x+5h+5x}{2}\right)\right\}\sin\left\{\frac{(5x+5h-5x)}{2}\right\}}{h}$$

$$= \lim_{h\to 0}\frac{2\cos\left(\frac{10x+5h}{2}\right)\sin\left\{\frac{5h}{2}\right\}}{h}$$

$$= \lim_{h\to 0}\frac{\cos\left(5x+\frac{5h}{2}\right)\sin\left\{\frac{5h}{2}\right\}}{\left(h/2\right)}$$

Multiply and divide by 5 we get,

$$= \lim_{h\to 0}\frac{\cos\left(5x+\frac{5h}{2}\right)\sin\left(\frac{5h}{2}\right)5}{\left(5h/2\right)}$$

$$= 5\lim_{h\to 0}\cos\left(5x+\frac{5h}{2}\right)\lim_{\frac{5h}{2}\to 0}\frac{\sin\left(\frac{5h}{2}\right)}{\left(\frac{5h}{2}\right)}$$

$$= 5 \times \cos 5x \times 1$$

$$= 5 \cos 5x$$

Illustration No. 8

If f(x) = x sin x, then $\frac{df(x)}{dx} = ?$

Solution:

As f(x) = x sin x

f(x + h) = (x + h) sin(x + h)

Since $$\frac{d}{dx}\{f(x)\} = \lim_{h\to 0}\frac{(x+h)\sin(x+h) - x\sin x}{h}$$

$$= \lim_{h\to 0}\frac{x\sin(x+h) + h\sin(x+h) - x\sin x}{h}$$

$$= \lim_{h\to 0}\frac{x\{\sin(x+h) - \sin x\} + h\sin(x+h)}{h}$$

$$= \lim_{h\to 0}\frac{x\left[2\cos\left(x+\frac{h}{2}\right)\sin\frac{h}{2}\right] + h\sin(x+h)}{h}$$

$$= \lim_{h\to 0}\frac{2x\left[\cos\left(x+\frac{h}{2}\right)\sin\frac{h}{2}\right]}{h} + \lim_{h\to 0}\frac{h\sin(x+h)}{h}$$

$$= \lim_{h\to 0}\frac{x\left[\cos\left(x+\frac{h}{2}\right)\sin\frac{h}{2}\right]}{h} + \lim_{h\to 0}\sin(x+h)$$

$$= x\cos x \lim_{\frac{h}{2}\to 0}\frac{\sin\frac{h}{2}}{\frac{h}{2}} + \sin x$$

$$= x\cos x \times 1 + \sin x$$

$$= x\cos x + \sin x$$

Hence Proved

Illustration No 9

If f(x) = log sin x, $\frac{dy}{dx} = ?$

Solution:

Let y = sin x and y + k = sin (x + h) or k = sin (x + h) – sin x

Since h $\to$ 0, then k $\to$ 0

But $$\frac{df(x)}{dx} = \lim_{h\to 0}\frac{f(x+h) - f(x)}{h}$$

$$= \lim_{h\to 0} \frac{\log \sin(x+h) - \log \sin x}{h}$$

$$= \lim_{h\to 0} \frac{\log(y+k) - \log y}{h}$$

$$= \lim_{h\to 0} \frac{\log \frac{y+k}{y}}{h}$$

$$= \lim_{h\to 0} \frac{\log\left(1+\frac{k}{y}\right) \times \frac{k}{h} \times \frac{1}{y}}{\frac{k}{y}}$$

$$= \frac{1}{y} \times \lim_{h\to 0} \frac{\log\left(1+\frac{k}{y}\right) \times \frac{\sin(x+h) - \sin x}{h}}{\frac{k}{y}}$$

$$= \frac{1}{\sin x} \lim_{k\to 0} \frac{\log\left(1+\frac{k}{y}\right)}{\frac{k}{y}} \times \lim_{h\to 0} \frac{\sin(x+h) - \sin x}{h}$$

$$= \frac{1}{\sin x} \times 1 \times \lim_{h\to 0} \frac{\sin(x+h) - \sin x}{h}$$

$$= \frac{1}{\sin x} \times \lim_{h\to 0} \frac{2\cos\left(x+\frac{h}{2}\right)\sin\left(\frac{h}{2}\right)}{h}$$

$$= \frac{1}{\sin x} \times \cos \times \lim_{\frac{h}{2}\to 0} \frac{\sin\left(\frac{h}{2}\right)}{\frac{h}{2}}$$

$$= \frac{\cos x}{\sin x} = \cot x$$

Illustration No. 10

If $y = \sin^{-1}\left(\frac{x}{a}\right)$ then find $\frac{dy}{dx}$

Solution:

Let $y = \sin^{-1}\left(\frac{x}{a}\right)$ and $y + k = \sin^{-1}\left(\frac{x+h}{a}\right)$

where $k \to 0$ or $h \to 0$

Since $\frac{x}{a} = \sin y$ and $\frac{x+h}{a} = \sin(y + k)$

Therefore,

$$\frac{dy}{dx} = \lim_{h\to 0}\frac{\sin^{-1}\left(\frac{x+h}{a}\right) - \sin^{-1}\left(\frac{x}{a}\right)}{h}$$

$$= \lim_{h\to 0}\frac{(y+k)-y}{h}$$

$$= \lim_{h\to 0}\frac{k}{h}$$

$$= \lim_{h\to 0}\frac{k}{\frac{[(x+h)-x]}{a}}$$

$$= \frac{1}{a}\lim_{k\to 0}\frac{k}{\sin(y+k)-\sin y}$$

$$= \frac{1}{a}\lim_{k\to 0}\frac{k}{2\cos\left(y+\frac{k}{2}\right)\sin\left(\frac{k}{2}\right)}$$

$$= \frac{1}{a}\frac{1}{\cos(y)\lim_{k\to 0}\left\{\frac{\sin\left(\frac{k}{2}\right)}{\left(\frac{k}{2}\right)}\right\}}$$

$$= \frac{1}{a \cos y}$$

$$= \frac{1}{a\sqrt{1-\sin^2 y}}$$

$$= \frac{1}{a\sqrt{1-\left(\frac{x}{a}\right)^2}} = \frac{1}{\sqrt{a^2-x^2}}$$

Illustration No. 11

If $y = e^{\sin x}$ then $\frac{d}{dx}\left\{e^{\sin x}\right\} = ?$

Solution:

Let sin x = y then sin(x + h) = y + k

Then k = sin(x + h) – sin x

where h $\rightarrow$ 0, k $\rightarrow$ 0

Therefore, we have

$$\frac{d}{dx}\left(e^{\sin x}\right) = \lim_{h\to 0} \frac{e \sin(x+h) - e \sin x}{h}$$

$$= \lim_{h\to 0} \frac{e^{y+k} - e^y}{k} \frac{k}{h}$$

$$= \lim_{h\to 0} \frac{e^y (e^k - 1)}{k} \frac{k}{h}$$

$$= \lim_{k\to 0} \frac{e^y\left(1+\frac{k}{1}+\frac{k^2}{2!}+ \ldots\ldots -1\right)}{k} \frac{k}{h}$$

$$= e^y \lim_{k\to 0} \frac{\left(k + \frac{k^2}{2!} + ...\right)}{k}$$

$$= e^y \lim_{k\to 0} \frac{k\left(1 + \frac{k}{2!} + ...\right)}{k} \lim_{h\to 0} \frac{k}{h}$$

$$= e^{\sin x} \lim_{h\to 0} \frac{\sin(x+h) - \sin x}{h}$$

$$= e^{\sin x} \times \lim_{h\to 0} \frac{2\cos\left(x + \frac{h}{2}\right)\sin\left(\frac{h}{2}\right)}{h}$$

$$= e^{\sin x} \times \cos x \times \lim_{\frac{h}{2}\to 0} \frac{\sin\left(\frac{h}{2}\right)}{\frac{h}{2}}$$

$$= e^{\text{Sinx}} \cos x$$

Some Important Standard Derivatives of Trigonometric Functions

Standard Derivative 1

$$\frac{d}{dx}\{\sin x\} = \cos x$$

Proof:

Let us define

$$f(x) = \sin x,$$

Then $f(x + h) = \sin(x + h)$

Therefore,

Note: $\sin A - \sin B = 2\cos\left(\frac{A+B}{2}\right) \times \sin\left(\frac{A-B}{2}\right)$

$$= \lim_{h\to 0} \frac{2\times\cos\left(\frac{x+h+x}{2}\right)\sin\left(\frac{x+h-x}{2}\right)}{h}$$

$$= 2\lim_{h\to 0} \frac{\cos\left(\frac{2x+h}{2}\right)\sin\left(\frac{h}{2}\right)}{h}$$

$$= \lim_{h\to 0}\cos\left(x+\frac{h}{2}\right) \lim_{\frac{h}{2}\to 0} \frac{\sin\left(\frac{h}{2}\right)}{\left(\frac{h}{2}\right)}$$

$$= \cos x \times 1$$

$$= \cos x$$

Standard Derivative 2

$$\frac{d}{dx}\{\cos x\} = -\sin x$$

Proof:

Let f(x) = cos x

then f(x + h) = cos(x + h)

Therefore,

$$\frac{d}{dx}\{f(x)\} = \lim_{h\to 0}\frac{f(x+h)-f(x)}{h}$$

$$= \lim_{h\to 0}\frac{\cos(x+h)-\cos x}{h}$$

$$= \lim_{h\to 0}(-)\frac{2\times\sin\left(\frac{2x+h}{2}\right)\times\sin\left(\frac{h}{2}\right)}{h}$$

$$=(-)\lim_{h\to 0}\sin\left(x+\frac{h}{2}\right)\lim_{\frac{h}{2}\to 0}\sin\left(\frac{\frac{h}{2}}{\frac{h}{2}}\right)$$

As $h \to 0 \to h/2 \to 0$

$$= -\sin x \times 1$$

$$= -\sin x$$

Hence Proved

Standard Derivative 3

$$\frac{d}{dx}(\tan x) = \sec^2 x$$

Proof:

Let us define f(x) = tan(x)

Since tan x = sin x/cos x

Therefore f(x) = sin x/cos x

$$\frac{d}{dx}f(x) = \frac{d}{dx}\left\{\frac{\sin x}{\cos x}\right\}$$

{Using u/v Rule}

$$= \frac{\cos x \times \left\{\frac{d}{dx}(\sin x)\right\} - \sin x \times \left\{\frac{d}{dx}(\cos x)\right\}}{\cos^2 x}$$

$$= \frac{\cos x \times \cos x - \sin x \times (-\sin x)}{\cos^2 x}$$

$$= \frac{\cos^2 x + \sin^2 x}{\cos^2 x}$$

$$= \frac{1}{\cos^2 x} \qquad \{\text{since } \cos^2 + \sin^2 x = 1\}$$

$$= \sec^2 x$$

Note

Similarly the derivative of cot x i.e. $\frac{d}{dx}\{\cot x\} = -\operatorname{cosec}^2 x$.

(Left to students to solve. Hint apply 1st principle)

Standard Derivative 4

$\frac{d}{dx}\{\sec x\} = \sec x \tan x$

Proof:

Let f(x) = sec x or f(x) can be written as f(x) = 1/cos x

$$= \frac{d}{dx}\{1/\cos x\}$$

$$= \frac{d\,f(x)}{dx} = \frac{d}{dx}\left\{\frac{1}{\cos x}\right\}$$

$$= \frac{\cos x \frac{d}{dx}(1) - 1 \frac{d}{dx}(\cos x)}{\cos^2 x}$$

$$= \frac{0 - 1(-\sin x)}{\cos^2 x}$$

$$= \frac{\sin x}{\cos^2 x} = \frac{\sin x}{\cos x} \frac{1}{\cos x}$$

$$= \tan x \sec x$$

Hence Proved

Note: Similarly $\frac{d}{dx}(\text{cosec } x) = -\text{cosec } x \cot x$

(Left to students to solve. *Hint*: apply 1st principle)

Illustration No. 12

Find the derivative

(i) $x^5 + 3e^x \log x + 6$

(ii) $x^{3/4} + 4a^x + 3$

Solution:

(i) Since $y = x^5 + 3e^x \log x + 6$

$$\frac{dy}{dx} = \frac{d}{dx}(x^5 + 3\,e^x \log x + 6)$$

$$= \frac{d}{dx}(x^5) + 3 \times \frac{d}{dx}\{e^x \log x\} + \frac{d}{dx}(6)$$

Note: $\frac{d}{dx}(u \times v) = v \times \frac{d}{dx}(u) - u \times \frac{d}{dx}(v)$

$$= 5x^4 + 3 \times \{\frac{d}{dx}(e^x) \log x + \frac{d}{dx}(\log x)(e^x)\} + 0$$

$$= 5x^4 + 3\{e^x \log x + e^x/x\}$$

$$= 5x^4 + 3e^x \log x + 3\{e^x/x\}$$

(ii) $y = x^3/4 + 4a^x + 3$

$$\frac{dy}{dx} = \frac{d}{dx}(x^3/4 + 4a^x + 3)$$

$$= \frac{d}{dx}(x^3/4) + \frac{d}{dx}(4a^x) + \frac{d}{dx}(3)$$

$$= \frac{3}{4}x^2 + 4\{\frac{d}{dx}(a^x)\} + 0$$

$$= \frac{3}{4}x^2 + 4\{a^x \log a\}$$

$$= \frac{3}{4}x^2 + 4a^x \log a.$$

Illustration No. 13

Find the derivative of the following function:

(a) $4^x \log x + 5x^3$

(b) $\frac{4x^2 + 9}{\sqrt{x}}$

(c) $\frac{8}{1 + 6x}$

Solution:

(a) Let $y = 4^x \log + 5x^3$

$$\frac{dy}{dx} = \frac{d}{dx}(4^x \log x + 5x^3)$$

$$= \frac{d}{dx}(4^x \log x) + \frac{d}{dx}(5x^3)$$

$$= \frac{d}{dx}(4^x) \times \log x + \frac{d}{dx}(\log x) \times (4^x) + 5 \times 3 \times x^2$$

$$= (4^x \log 4) \log x + \frac{4^x}{x} + 15\, x^2$$

(b) Since $y = \dfrac{4x^2 + 9}{\sqrt{x}}$

$$\frac{dy}{dx} = \frac{d}{dx} \times \left\{\frac{4x^2 + 9}{\sqrt{x}}\right\}$$

$$= \frac{\frac{d}{dx}(4x^2 + 9) \times \sqrt{x} - \frac{d}{dx}\left(\sqrt{x}\right)\left(4x^2 + 9\right)}{\left(\sqrt{x}\right)^2}$$

$$= \frac{8x \times \sqrt{x} - \frac{1}{2\sqrt{x}}\left(4x^2 + 9\right)}{x}$$

$$= \frac{16x^2 - \left(4x^2 + 9\right)}{2 \times x \times \sqrt{x}}$$

$$= \frac{12x^2 - 9}{2x\sqrt{x}}$$

(c) $y = \dfrac{8}{1 + 6x}$

$$\frac{dy}{dx} = \frac{d}{dx}\left\{\frac{8}{1 + 6x}\right\}$$

$$= \frac{(1 + 6x)\frac{d}{dx}(8) - 8\frac{d}{dx}(1 + 6x)}{(1 + 6x)^2}$$

$$= \frac{(1 + 6x) \times (0) - 8 \times (0 + 6)}{(1 + 6x)^2}$$

$$= \frac{0 - 48}{(1 + 6x)^2} = \frac{-48}{(1 + 6x)^2}$$

Illustration No.14

If $y = \frac{\sqrt{1-x}}{\sqrt{1+x}}$ find $\frac{dy}{dx} = ?$

Solution:

Since $y = \frac{\sqrt{1-x}}{\sqrt{1+x}}$

$$\frac{dy}{dx} = \frac{\left\{\frac{d}{dx}\left(\sqrt{1-x}\right)\right\} \times \left(\sqrt{1+x}\right) - \left\{\frac{d}{dx}\left(\sqrt{1+x}\right)\right\} \times \left(\sqrt{1-x}\right)}{\left(\sqrt{1+x}\right)^2}$$

$$= \frac{\frac{1}{2\sqrt{1-x}}\sqrt{1+x} - \frac{1}{2\sqrt{1+x}}\sqrt{1-x}}{\left(\sqrt{1+x}\right)^2}$$

$$= \frac{(1+x) - (1-x)}{2\sqrt{1-x^2}\ (1+x)}$$

$$= \frac{2x}{2\sqrt{1-x^2}\ (1+x)}$$

$$= \frac{x}{\sqrt{1-x^2}\ (1+x)}$$

Illustration No.15

If $y = \log\,[1 + \sqrt{x^2 + a^2}]$ find $\frac{dy}{dx} = ?$

Solution

$y = \log\,[1 + \sqrt{x^2 + a^2}]$

Let us assume that $1 + \sqrt{x^2 + a^2} = t$

We can write $y = \log t$

$$\frac{dy}{dx} = \frac{1}{t}\frac{dt}{dx}$$

But $t = 1 + \sqrt{x^2 + a^2}$

$$\frac{dt}{dx} = 0 + \frac{1 \times 2x}{2\sqrt{x^2 + a^2}}$$

$$= \frac{x}{\sqrt{x^2 + a^2}}$$

Substituting the value of t and $\frac{dt}{dx}$ in $\frac{dy}{dx}$ therefore,

we get

$$\frac{dy}{dx} = \frac{1}{1 + \sqrt{x^2 + a^2}} \times \frac{x}{\sqrt{x^2 + a^2}}$$

$$= \frac{x}{\left(x^2 + a^2\right) + \sqrt{x^2 + a^2}}$$

Illustration No.16

If $y = \frac{4x^2 + 3x - 2}{3x^2 + 4}$ then find $\frac{dy}{dx}$

Solution:

Since $y = \frac{4x^2 + 3x - 2}{3x^2 + 4}$

$$\frac{dy}{dx} = \frac{\left(3x^2 + 4\right)\frac{d}{dx}\left(4x^2 + 3x - 2\right) - \left(4x^2 + 3x - 2\right)\frac{d}{dx}\left(3x^2 + 4\right)}{\left(3x^2 + 4\right)^2}$$

$$= \frac{\left(3x^2 + 4\right) \times (8x + 3) - \left(4x^2 + 3x - 2\right) \times (6x)}{\left(3x^2 + 4\right)^2}$$

$$= \frac{24x^3 + 9x^2 + 32x + 12 - 24x^3 - 18x^2 + 12x}{\left(3x^2 + 4\right)^2}$$

$$= \frac{-9x^2 + 44x + 12}{\left(3x^2 + 4\right)^2}$$

Illustration No.17

If $y = a^{2x} + \sqrt{x-5}$ then find $\frac{dy}{dx} = ?$

Solution:

Since

$$y = a^{2x} + \sqrt{x-5}$$

Differentiating w.r.t 'x'

$$\frac{dy}{dx} = \frac{d}{dx}\left(a^{2x} + \sqrt{x-5}\right)$$

$$= \frac{d}{dx}\left(a^{2x}\right) + \frac{d}{dx}\left(\sqrt{x-5}\right)$$

$$= a^{2x} \times 2 \times \log a + \frac{1}{2\sqrt{x-5}}$$

$$= 2a^{2x}\log a + \frac{1}{2\sqrt{x-5}}$$

Illustration No.18

If $y = \frac{1}{\sqrt{x+2} + \sqrt{x-2}}$ then find $\frac{dy}{dx} = ?$

Solution:

Since

$$y = \frac{1}{\sqrt{x+2} + \sqrt{x-2}}$$

Rationalizing the denominator with the rationalizing factor we get

$$= \frac{1}{\sqrt{x+2} + \sqrt{x-2}} \times \frac{\sqrt{x+2} - \sqrt{x-2}}{\sqrt{x+2} - \sqrt{x-2}}$$

$$= \frac{\sqrt{x+2} - \sqrt{x-2}}{\left(\sqrt{x+2}\right)^2 - \left(\sqrt{x-2}\right)^2}$$

$$= \frac{\sqrt{x+2} - \sqrt{x-2}}{(x+2) - (x-2)}$$

$$= \frac{\sqrt{x+2} - \sqrt{x-2}}{4}$$

$$= \frac{\sqrt{(x+2)}}{4} - \frac{\sqrt{(x-2)}}{4}$$

$$= \frac{1}{4}\left(\sqrt{x+2} - \sqrt{x-2}\right)$$

$$= \frac{dy}{dx} = \frac{1}{4}\frac{d}{dx}\left(\sqrt{x+2} - \sqrt{x-2}\right)$$

$$= \frac{1}{4}\frac{d}{dx}\left(\sqrt{x+2}\right) - \frac{d}{dx}\left(\sqrt{x-2}\right)$$

$$= \frac{1}{4}\left(\frac{1}{2\sqrt{x+2}} - \frac{1}{2\sqrt{x-2}}\right)$$

$$= \frac{1}{8}\left(\frac{1}{\sqrt{x+2}} - \frac{1}{\sqrt{x-2}}\right)$$

Illustration No.19

If $y = \dfrac{a^2 + x^2}{a^2 - x^2}$ then find $\dfrac{dy}{dx} = ?$

Solution:

$$y = \frac{a^2 + x^2}{a^2 - x^2}$$

therefore

$$\frac{dy}{dx} = \frac{d}{dx}\left(\frac{a^2 + x^2}{a^2 - x^2}\right)$$

$$= \frac{\left(a^2 - x^2\right) \times \frac{d}{dx}\left(a^2 + x^2\right) - \left(a^2 + x^2\right) \times \frac{d}{dx}\left(a^2 - x^2\right)}{\left(a^2 - x^2\right)^2}$$

$$= \frac{\left(a^2 - x^2\right)(0 + 2x) - \left(a^2 + x^2\right)(0 - 2x)}{\left(a^2 - x^2\right)^2}$$

$$= \frac{2a^2x - 2x^3 + 2a^2x + 2x^3}{\left(a^2 - x^2\right)^2}$$

$$= \frac{4a^2x}{\left(a^2 - x^2\right)^2}$$

Illustration No. 20

If $y = x^x$ then find $\frac{dy}{dx}$

Solution:

$$y = x^x$$

Taking log both side

$\log y = x \log x$ {Since $\log x^n = n \log x$}

Now, differentiate with respect to x,

We get $\frac{1}{y}\frac{dy}{dx} = x \times \frac{d}{dx}(\log x) + \log x \times \frac{d}{dx}(x)$

$$= x \times \frac{1}{x} + \log x \times 1$$

$$= (1 + \log x)$$

Since $\frac{dy}{dx} = y\left(1 + \log x\right)$

$= x^x (1 + \log x)$ [Since $y = x^x$]

Illustration No. 21

If $y = e^{x^2+2x}$ then find $\frac{dy}{dx} = ?$

Solution:

Let $x^2 + 2x = t$

$$y = e^t$$

Differentiating w.r.t 'x'

Therefore $\frac{dy}{dx} = e^t \frac{dt}{dx}$

But $x^2 + 2x = t$

Now differentiate w. r. t. 'x',

we get, $$\frac{dt}{dx} = 2x + 2$$

(or)

$$\frac{dt}{dx} = 2(x+1)$$

Replace t, $\frac{dt}{dx}$

We get, $$\frac{dy}{dx} = e^{x^2+2x} \times 2(x+1)$$

$$= 2(x+1)\, e^{x^2+2x}$$

Illustration No. 22

If $y = 5\sqrt{x} + 7^{3x}$ find $\frac{dy}{dx} = ?$

Solution :

Since

$$y = 5\sqrt{x} + 7^{3x}$$

$$\frac{dy}{dx} = 5 \times \frac{d}{dx}\left(\sqrt{x}\right) + \frac{d}{dx}\left(7^{3x}\right) \quad (1)$$

Let, $z = 7^{3x}$ and $3x = t$

So, we can write

$$z = 7^t$$

Differentiating both side w.r.t x

We get,

$$\frac{dz}{dx} = 7^t \log 7 \frac{dt}{dx}$$

Since, $t = 3x$

$$\frac{dt}{dx} = 3$$

therefore $\frac{dz}{dx} = 3 \times 7^{3x} \log 7$

(Or)

$$\frac{dz}{dx} = 3(7^{3x} \log 7)$$

Replacing the value of the derivative of 7^{3x} in (1)

We get, $\frac{dy}{dx} = 5 \times \frac{1}{2} \times x^{-1/2} + 3 \,.\, 7^{3x} \log 7$

$$= \frac{5}{2\sqrt{x}} + 3(\,7^{3x} \log 7)$$

Some Standard Derivative of Inverse Trigonometric Functions

1. Derivative of $\sin^{-1}(x)$

i.e $\frac{d}{dx}\{\sin^{-1} x\} = \frac{1}{\sqrt{1-x^2}}$

Proof:

Let

$y = \sin^{-1}x$ then $\sin y = x$

Therefore differentiating w.r.t 'x'

We get, $\cos y \times \frac{dy}{dx} = 1$

$$\frac{dy}{dx} = \frac{1}{\cos y} = \frac{1}{\sqrt{1-\sin^2 y}} \quad \{\text{Since } \cos^2 x + \sin^2 x = 1\}$$

$$= \frac{1}{\sqrt{1-x^2}}$$

$$\frac{d}{dx}\{\sin^{-1} x\} = \frac{1}{\sqrt{1-x^2}}$$

2. $\frac{d}{dx}\{\cos^{-1} x\} = \frac{-1}{\sqrt{1-x^2}}$

Proof:

Let us assume $\cos^{-1}x = y$

Therefore $\cos y = x$

Differentiating w. r. t x, we get $(-) \sin y \dfrac{dy}{dx} = 1$

(Or) $$\frac{dy}{dx} = \frac{-1}{\sin y}$$

$$= \frac{-1}{\sqrt{1-\cos^2 y}}$$

$$= \frac{-1}{\sqrt{1-x^2}} \qquad \{\text{since } \cos y = x\}$$

$$\frac{d}{dx}\{\cos^{-1}x\} = \frac{-1}{\sqrt{1-x^2}}$$

3. $\dfrac{d}{dx}\{\tan^{-1}x\} = \dfrac{1}{1+x^2}$

Proof

Let $y = \tan^{-1}x$

Therefore $\tan y = x$

Differentiating w. r. t 'x'

$$\sec^2 y \frac{dy}{dx} = 1$$

$$\frac{dy}{dx} = \frac{1}{\sec^2 x}$$

$$= \frac{1}{1+\tan^2 y} \qquad \text{Since } \{\sec^2 x - \tan^2 y = 1\}$$

or $\{\sec^2 y = 1 + \tan^2 y\}$

$$= \frac{1}{1+x^2}$$

$$\frac{d}{dx}\{\tan^{-1}x\} = \frac{1}{1+x^2}$$

4. $\dfrac{d}{dx}\{\cot^{-1}x\} = \dfrac{-1}{1+x^2}$

Proof:

Let $\cot^{-1}x = y$

Or $\cot y = x$

Differentiating w. r. t 'x'

We get $(-)\ \text{cosec}^2 y \dfrac{dy}{dx} = 1$

$$\frac{dy}{dx} = \frac{-1}{\text{cosec}^2 y} \qquad \{\text{Since } \text{cosec}^2 y - \cot^2 y = 1\}$$

$$= \frac{-1}{1+\cot^2 y}$$

$$= \frac{-1}{1+x^2}$$

$$\frac{d}{dx}\{\cot^{-1}x\} = \frac{-1}{1+x^2}$$

5. $\dfrac{d}{dx}\{\sec^{-1}x\} = \dfrac{1}{x\sqrt{x^2-1}}$

Proof:

Let $\sec^{-1} x = y$

OR $\sec y = x$

Differentiating w.r.t 'x'

$\sec y \times \tan y \dfrac{dy}{dx} = 1$

$$\frac{dy}{dx} = \frac{1}{\sec y \times \tan y} \qquad \{\text{Since } \tan^2 y = \text{Sec}^2 y - 1\}$$

$$= \frac{1}{\sec y\sqrt{\sec^2 y - 1}}$$

$$= \frac{1}{x\sqrt{x^2-1}}$$

therefore $$\frac{d}{dx}\{\sec^{-1} x\} = \frac{1}{x\sqrt{x^2-1}}$$

6. $$\frac{d}{dx}\{\operatorname{cosec}^{-1} x\} = \frac{-1}{x\sqrt{x^2-1}}$$

Proof:

Let $\operatorname{cosec}^{-1} x = y$

OR $\operatorname{cosec} y = x$

Differentiating w.r.t x we get, $(-)\operatorname{cosec} y \cot y \frac{dy}{dx} = 1$

$$\frac{dy}{dx} = \frac{-1}{\operatorname{cosec} y \times \cot y} \qquad \{\text{Since } \cot^2 y = \operatorname{cosec}^2 y - 1\}$$

$$= \frac{-1}{\operatorname{cosec} y \sqrt{\operatorname{cosec}^2 y^{-1}}}$$

$$= \frac{-1}{x\sqrt{x^2-1}}$$

Therefore $$\frac{d}{dx}\{\operatorname{cosec}^{-1} x\} = \frac{-1}{x\sqrt{x^2-1}}$$

Illustration No.23

If $y = \log(x \sin x)$, find $\frac{dy}{dx} = ?$

Solution:

Since $y = \log(x \sin x)$

Let $x \sin x = t$

Differentiating w. r. t 'x',

We get $$\frac{dt}{dx} = x\frac{d}{dx}\{\sin x\} + \sin x \frac{d}{dx}\{x\}$$

$$= x \cos x + \sin x$$

But y = log t, now differentiating with respect to 'x' on both side we get

$$\frac{dy}{dx} = \frac{1}{t}\frac{dt}{dx} \quad .....(1)$$

$$\{t = x \sin x\} \text{ and } \frac{dt}{dx} \{x \cos x + \sin x\}$$

Substituting the value of 't' and $\frac{dt}{dx}$ in (1) we get

$$\frac{dy}{dx} = \frac{1\{x \cos x + \sin x\}}{x \sin x}$$

$$= \frac{x \cos x + \sin x}{x \sin x}$$

$$= \frac{x \cos x}{x \sin x} + \frac{\sin x}{x \sin x}$$

$$= \cot x + \frac{1}{x}$$

Illustration No. 24

If $y = a^x \cos x + \log(\sin^{-1} x)$,then find $\frac{dy}{dx} = ?$

Solution:

Since $y = a^x \cos x + \log(\sin^{-1} x)$

Differentiating w.r.t x

$$\frac{dy}{dx} = \frac{d}{dx}\{a^x \cos x\} + \frac{d}{dx}\{\log(\sin^{-1} x)\}$$

$$= a^x \times \frac{d}{dx}\{\cos x\} + \cos x \times \frac{d}{dx}\{a^x\} + \frac{d}{dx}\{\log(\sin^{-1} x)\}$$

$$= a^x \times (-\sin x) + \cos x \times \{a^x \log a\} + \frac{1}{\sin^{-1} x} \times \frac{1}{\sqrt{1 - x^2}}$$

$$= -a^x \times \sin x + a^x \log a \cos x + \frac{1}{\sqrt{1 - x^2}\ \sin^{-1} x}$$

Illustration No. 25

If $y = \log (xe^x) + \tan \frac{x}{x^2}$, find $\frac{dy}{dx}$

Solution:

Since $y = \log (xe^x) + \tan \frac{x}{x^2}$

$$\frac{dy}{dx} = \frac{d}{dx} \{\log \left(xe^x\right)\} + \frac{d}{dx} \frac{\tan x}{x^2}$$

Let us bifurcate the above into 2 parts; first let's find the derivative of $\log x\, e^x$ then $\tan \frac{x}{x^2}$

Part 1 $\frac{d}{dx} \{\log x e^x\}$

$$= e^x \{\log x + 1/x\}$$

Part 2 $$\frac{d}{dx} \frac{\tan x}{x^2} = \frac{x^2 \times \frac{d}{dx} (\tan x) - \tan x \frac{d}{dx} \{x^2\}}{\left(x^2\right)^2}$$

$$= \frac{x^2 \sec^2 x - 2x \tan x}{x^4}$$

$$= \frac{x \sec^2 x - 2 \tan x}{x^3}$$

Hence $$\frac{dy}{dx} = e^x \left(\log x + \frac{1}{x}\right) + \frac{x \sec^2 x - 2 \tan x}{x^3}$$

Illustration No. 26

If $y = \cos^{-1} (x^2 + 4x + 3)$, find $\frac{dy}{dx} = ?$

Solution:

Let $x^2 + 4x + 3 = t$

Differentiating w.r.t 'x'

$$\frac{dt}{dx} = 2x + 4$$

But $y = \cos^{-1}t$

Since differentiating w.r.t 't', both side we get

$$\frac{dy}{dx} = \frac{-1}{\sqrt{1-t^2}} = \frac{dt}{dx} \quad .....(1)$$

Substituting the value of t and dt in (1) we get,

$$= \frac{-1}{\sqrt{1-(x^2+4x+3)}} \times (2x+4)$$

$$= \frac{-(2x+4)}{\sqrt{1-(x^2+4x+3)}}$$

Application of Concepts of Differentiation to Business and Economics

Concept of Elasticity

Elasticity measures the degree of sensitivity or responsiveness to change in quantity demanded or sold of a particular commodity to a change brought about in a particular determining factor may be price.

Alternatively, elasticity measures the percentage change in one variable (x) in relation to percentage change in another variable (y) expressed as

For example

$$e = \frac{\text{Percentage Change in x}}{\text{Percentage Change in y}} = \frac{\left(\frac{dx}{x}\right) \times 100}{\left(\frac{dy}{y}\right) \times 100}$$

Elasticity of Demand

Ceteris paribus (i.e. keeping other thing constant) the elasticity of demand can be defined as the measure of degree of sensitivity or responsiveness of a consumer to quantity of commodity demanded to change in price. Or it refers to percentage change in quantity demanded of a commodity because of change in the price of the commodity.

Mathematically expressed as

$$e_d = \frac{\text{Percentage Change in Qty. demanded}}{\text{Percentage Change in price}}$$

$$= \frac{\frac{\text{Change in Qty.}}{\text{Initial Qty.}} \times 100}{\frac{\text{Change in Price.}}{\text{Initial Price.}} \times 100} = (-)\frac{\frac{dq}{q} \times 100}{\frac{dp}{p} \times 100}$$

$$= (-)\ \frac{dq}{dp} \times \frac{p}{q}$$

Note: Elasticity is expressed in only absolute terms as its mere number but the negative sign is affixed as the price and quantity are inversely related (law of demand). With the introduction of (–) sign we get a positive value of e_d.

Some Important cases of e_d

(i) If the value of $e_d = \infty$, then demand is said to be perfectly elastic i.e., change in price brings about infinite change in quantity demanded

(ii) If the value of $e_d > 1$, then demand is said to be elastic i.e. change in price brings about more than proportionate change in quantity demanded.

(iii) If the value of $e_d = 1$, then the demand is said to be unitary elastic i.e change in price brings about proportionate change in quantity demanded.

(iv) If the value lies between $0 < e_d < 1$, then demand is said to be in-elastic i.e change in price brings about less than proportionate change in quantity demanded.

(v) If the value of $e_d = 0$, then the demand is perfectly inelastic i.e change in price no change in quantity demanded.

Note:

The above formula is applicable to small change in price but if the price change is appreciable then the elasticity is measured using arc elasticity of demand which is given by.

$$e_d = \frac{\Delta q \times (p_1 + p_2)}{\Delta P \times (q_1 + q_2)}$$

Cross Elasticity of Demand refers to percentage change in quantity demanded of a commodity 'X' because of change in the price of the commodity 'Y'.

Mathematically expressed as

$$e_c = \frac{\text{Percentage Change in Qty. demanded of Commodity X}}{\text{Percentage Change in price of Commodity Y}}$$

$$= \frac{\dfrac{\text{Change in Qty }(q_x)}{\text{Initial Qty }(q_x)} \times 100}{\dfrac{\text{Change in Price }(p_y)}{\text{Initial Price }(P_y)} \times 100} = \frac{\dfrac{dq_x}{q_x} \times 100}{\dfrac{dp_y}{p_y} \times 100}$$

$$= \frac{dq_x}{dp_y} \times \frac{p_y}{q_x}$$

Elasticity of Supply

Ceteris paribus (i.e. keeping other thing constant) the elasticity of supply can be defined as the measure of degree of sensitivity or responsiveness of a supplier to quantity of commodity supplied to change in price. Or it refers to percentage change in quantity supplied of a commodity because of change in the price of the commodity.

Mathematically expressed as

$$e_s = \frac{\text{Percentage Change in Qty. Supplied}}{\text{Percentage Change in price}}$$

$$= \frac{\dfrac{\text{Change in Qty Supplied.}}{\text{Initial Qty.}}}{\dfrac{\text{Change in Price.}}{\text{Initial Price.}} \times 100} = \frac{\dfrac{dq_s}{q_s} \times 100}{\dfrac{dp}{p} \times 100}$$

$$= \frac{dq_s}{dp} \times \frac{p}{q_s}$$

Some Important cases of e_s

(i) If the value of $e_s = \infty$, then supply is said to be perfectly elastic i.e change in price brings about infinite change in quantity supplied.

(ii) If the value of $e_s > 1$, then demand is said to be elastic i.e. change in price brings about more than proportionate change in quantity supplied.

(iii) If the value of $e_s = 1$, then the supply is said to be unitary elastic i.e change in price brings about proportionate change in quantity supplied.

(iv) If the value lies between $0 < e_s < 1$, then supply is said to be in-elastic i.e., change in price brings about less than proportionate change in quantity supplied.

(v) If the value of $e_s = 0$, then the demand is perfectly inelastic i.e change in price no change in supplied.

Illustration No.27

Find the elasticity of demand for the following function $p = x^3 e^{-2x}$ where p = price per unit output and x is the output produced. Also calculate the elasticity at output level of 10 units?

Solution:

Since $p = x^3 e^{-2x}$

$$\text{Elasticity of demand } (e_d) = (-)\frac{dx}{dp} \times \frac{p}{x} \qquad \text{.....(i)}$$

therefore,

$$\frac{dp}{dx} = \frac{d}{dx}\left(x^3 e^{-2x}\right)$$

$$= 3x^2 e^{-2x} - 2x^3 e^{-2x}$$

$$= x^2 e^{-2x} (3 - 2x)$$

Substituting the value in (i) we get,

$$e_d = \frac{(-)1}{x^2 e^{-2x} (3 - 2x)} \times \frac{x^3 e^{-2x}}{x}$$

$$= \frac{-1}{(3 - 2x)}$$

$$= \frac{1}{(2x - 3)}$$

And hence the value of elasticity of demand (e_d) at output level of x = 10 units is

$$e_d = \frac{1}{(2x - 3)} = \frac{1}{17}$$

Illustration No.28

Vimal Confectioners a biscuit manufacturer knows that the quantity demanded and supplied at a price of Rs. 10 is 30 units. And the demand falls to 20 units if the price is raised to Rs. 15 per unit. But the manufacturer is comfortable in supplying 50 units at a price – level of Rs. 20 per unit. Find the price elasticity of demand and supply at equilibrium pricing assuming linear relationship between price and quantity?

Solution:

Let the demand function is written as

$$p = a + bx$$

Demand Condition

Substituting the value of price and corresponding quantity demanded in demand function we get,

$$10 = a + 30\,b \quad \text{.....(i)}$$

$$15 = a + 20\,b \quad \text{.....(ii)}$$

Therefore solving i and ii we get

$$a = 25 \text{ and } b = -\frac{1}{2}$$

Therefore $$p = 25 - \frac{x}{2}$$

Supply Condition

Similarly the supply function is given by

$$10 = a + 30\,b \quad \text{......(iii)}$$

$$20 = a + 50\,b \quad \text{.....(iv)}$$

Solving (iii) and (iv) we get

$$a = -5 \text{ and } b = \frac{1}{2}$$

Therefore $$p = -5 + \frac{x}{2}$$

But at equilibrium condition Demand = Supply

And hence equating the demand and supply conditions,

We get,

$$25 - \frac{x}{2} = -5 + \frac{x}{2}$$

$$x = 30 \text{ units.}$$

Elasticity of Demand

$$\text{Elasticity of demand } (e_d) = (-)\frac{dx}{dp} \times \frac{p}{x}$$

$$= (-)\frac{\left(25 - \frac{30}{2}\right)}{30} \times (-2)$$

$$= \frac{2(25 - 15)}{30} = \frac{2}{3}$$

Elasticity of Supply

$$\text{Elasticity of supply } (e_s) = \frac{dx}{dp} \times \frac{p}{x}$$

$$= \frac{\left(-5 + \frac{x}{2}\right)}{x} \times (2)$$

$$= \frac{(-5 + 15)}{30} \times (2) = \frac{2}{3}$$

Illustration No. 29

If the demand function is expressed as $p = x^a e^{bx+c}$ then show that the elasticity of demand is given by $e_d = -\{1/ a + bx\}$?

Solution:

Since the demand function is expressed as

$$p = x^a e^{bx+c}$$

therefore,

$$\frac{dp}{dx} = \frac{d}{dx}(x^a)\, e^{bx+c} + x^a \frac{d}{dx}(e^{bx+c})$$

$$= a\, x^{a-1} e^{bx+c} + x^a\, e^{bx+c}\, b$$

$$= x^{a-1} e^{bx+c} (a + bx)$$

But elasticity of demand $(e_d) = (-)\frac{dx}{dp} \times \frac{p}{x}$

$$= (-)\frac{1}{x^{a-1}\, e^{bx+c}\, (a + bx)} \times \frac{x^a e^{bx+c}}{x}$$

$$= (-)\frac{1}{(a + bx)}$$

Illustration No. 30

If the supply function of a firm is given as $p = \sqrt{b + (x/a)}$, then calculate the elasticity of supply of the firm at x = 2a?

Solution:

As we know,

Elasticity of supply $(e_s) = \frac{dx}{dp} \times \frac{p}{x}$

therefore $\frac{dp}{dx} = \frac{1}{2\sqrt{b + \left(\frac{x}{a}\right)}} \times a$

Substituting the value of the derivative we get,

$$= \frac{2 \times \sqrt{b + \left(\frac{x}{a}\right)} \times a \times \sqrt{b + \left(\frac{x}{a}\right)}}{x}$$

$$= \frac{2a\,(b + x/a)}{x}$$

$$= \frac{2ab}{x} + 2$$

In order calculate the elasticity of supply at $x = 2a$ we substitute the value of x in the above equation then we get,

$$= \frac{2ab}{2a} + 2$$

$$= b + 2$$

Illustration No. 31

If e_1 and e_2 be the price elasticities of the demand curves $p = e^{ax}$ and $p = p = \frac{e^{ax}}{x}$, then show that $e_1 e_2 = e_1 - e_2$?

Solution:

$$p = e^{ax} \text{ and } p = p = \frac{e^{ax}}{x}$$

Finding the elasticity in both the context,

But elasticity $(e) = (-)\frac{dx}{dp} \times \frac{p}{x}$

Therefore, $(e_1) = (-)\frac{dx}{dp} \times \frac{p}{x}$

$$= (-)\frac{e^{ax}}{ae^{ax} \times x}$$

$$= \frac{-1}{ax} \qquad \left\{\text{Since } \frac{dp}{dx} = ae^{ax}\right\}$$

Similarly $(e_2) = (-)\frac{dx}{dp} \times \frac{p}{x}$

$$= (-)\frac{x^2 \times e^{ax}}{(ax-1) \times x \times x}$$

$$= \frac{-1}{ax-1} \qquad \left\{\text{Since } \frac{dp}{dx} = (-)\frac{(ax-1)e^{ax}}{x^2}\right\}$$

Now $e_1 e_2 = \frac{-1}{ax} \times \frac{-1}{ax-1}$

$$= \frac{1}{ax\,(ax-1)}$$

Similarly, $$e_1 - e_2 = \frac{-1}{ax} + \frac{1}{ax-1}$$

$$= \frac{ax - ax + 1}{ax\ (ax-1)}$$

$$= \frac{1}{ax\ (ax-1)}$$

And hence $e_1 e_2 = e_1 - e_2$

Concept of Total Cost, Average Cost and Marginal Cost

Let C(x) be the cost function also called as total cost function representing the total cost required in producing 'x' units of goods or service. As per traditional theory total cost is split into two groups: the Fixed Cost and the Variable Cost

i.e TC {C (x)} = FC + TVC.

Fixed Cost (FC): It is a cost which remains constant irrespective to the level of production e.g. Depreciation of plants and machineries, Salaries of Administrative Staff etc.

Variable Cost (VC): It is cost which depends on the level of production as production increases so does variable cost.

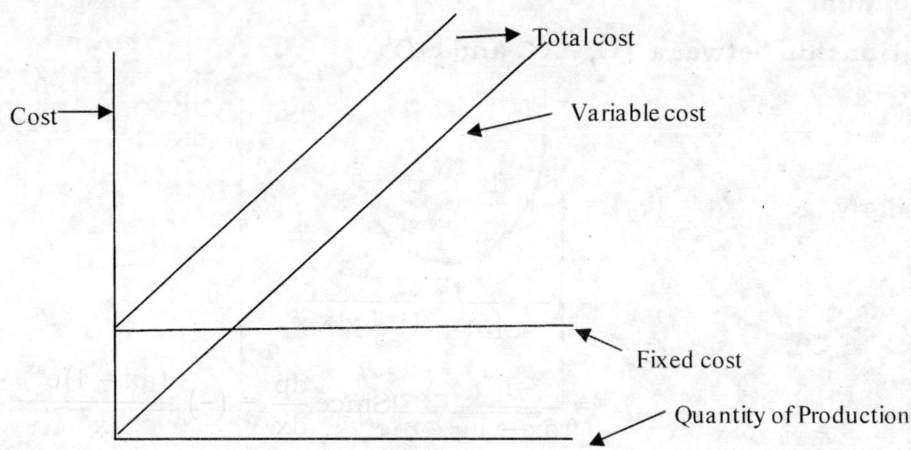

Average Total Cost (ATC) or Average Cost (AC) = It is the ratio of total cost to number of units produced i.e. it represents cost per unit of production.

$$\textbf{ATC} \text{ or } \textbf{AC} = \frac{\text{Total Cost}}{\text{No of Units produced}} = \frac{C(x)}{x}$$

Note: for convenience we will write ATC as AC in all future references.

Average Fixed Cost (AFC): It is the ratio of fixed cost to number of units produced i.e. it represents fixed cost per unit of production.

$$\textbf{AFC} = \frac{\text{Fixed Cost}}{\text{No. of Units produced}} = \frac{FC}{x}$$

Average Variable Cost (AVC): It is the ratio of variable cost to number of units produced i.e. it represents variable cost per unit of production.

$$\textbf{AVC} = \frac{\text{Variable Cost}}{\text{No. of Units produced}} = \frac{VC}{x}$$

Marginal Cost: Refers to change in total cost brought about by a change in production. If dC represent change in production cost and dx representing change in production then mathematically Marginal Cost can written as

$$MC = \frac{\text{Change in Total Cost}}{\text{Change in Production}} = \frac{dc}{dx}$$

Addendum

Relationship between TC, ATC and MC

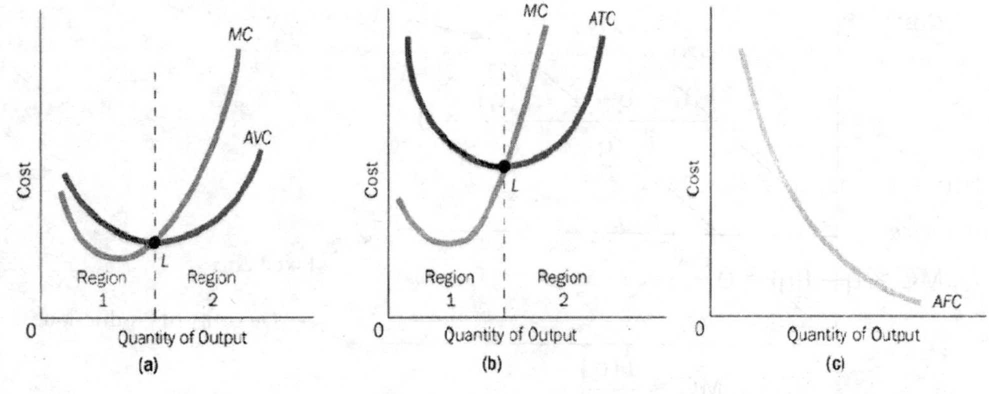

MC curve cuts both AVC and ATC curves at their respective low points

(i) Why MC cuts ATC at minimum point?

Let us assume

C = f(q)

Divide both side by 'q'

We get,

$$\frac{C}{q} = \frac{f(q)}{q}$$

i.e. $$AC = \frac{f(q)}{q}$$

Finding the 1st order condition,

$$\frac{dAC}{dq} = \frac{\frac{d}{dq} f(q) \times q - \frac{d}{dq}(q) \times f(q)}{q^2}$$

$$= \frac{MC \times q - 1 \times f(q)}{q^2}$$

Since $$MC = \frac{dc}{dq} = \frac{\text{Change in Cost}}{\text{Change in Output}}$$

Applying maximizing and minimizing conditions

We get,

$$\frac{dAC}{dq} = 0$$

$$\Rightarrow \frac{MC \times q - 1 \times f(q)}{q^2} = 0$$

Since $q \neq 0$, $q^2 \neq 0$

Therefore

$MC \times q - f(q) = 0$

Or

$$MC = \frac{f(q)}{q}$$

$$= AC \quad \{\text{Since } AC = \frac{f(q)}{q}\}$$

Hence proved

(ii) Why MC is below AC when AC falls

since,

$$AC = \frac{f(q)}{q}$$

As AC is decreasing then

$$\frac{dAC}{dq} < 0$$

$$\frac{dAC}{dq} = \frac{\frac{d}{dq} f(q) \times q - \frac{d}{dq}(q) \times f(q)}{q^2} < 0$$

$$= \frac{\frac{d}{dq} f(q) \times q - \{f(q)\}}{q^2} < 0 \quad (\text{since } dq/dq = 1)$$

$$\Rightarrow \quad \frac{MC \times q - 1 \times f(q)}{q^2} < 0$$

Since $q > 0$, $q^2 > 0$

Therefore

$$MC \times q - f(q) < 0$$

Or

$$MC < \frac{f(q)}{q}$$

$$< AC \qquad \{\text{Since } AC = \frac{f(q)}{q}\}$$

Hence proved.

(iii) Why MC is above AC when AC rising

since,

$$AC = \frac{f(q)}{q}$$

As AC is increasing then $\frac{dAC}{dq} > 0$

$$\frac{dAC}{dq} = \frac{\frac{d}{dq} f(q) \times q - \frac{d}{dq}(q) \times f(q)}{q^2} > 0$$

$$\frac{\frac{d}{dq} f(q) \times q - \{f(q)\}}{q^2} > 0 \qquad \text{(since } dq/dq = 1)$$

$$\Rightarrow \qquad \frac{MC \times q - 1 \times f(q)}{q2} > 0$$

Since $q > 0$, $q^2 > 0$

Therefore

$$MC \times q - f(q) > 0$$

Or

$$MC > \frac{f(q)}{q}$$

$$> AC \qquad \{\text{Since } AC = \frac{f(q)}{q}\}$$

Hence proved

Illustration No. 32

If the total cost function of a firm is expressed as $C(x) = 2500 + 3x^2 + 5x^3$

Find

(i) Fixed Cost (FC)

(ii) Variable Cost (VC)

(iii) Slope of MC Curve

(iv) Average Variable Cost (AVC)

Solution:

Since

Total Cost (TC) = Fixed Cost (FC) + Variable Cost (VC)

As we have the cost function $C(x) = 2500 + 3x^2 + 5x^3$

Therefore the Fixed Cost component (FC) = 2500

And the Variable Cost Component (VC) = $3x^2 + 5x^3$

But Marginal Cost (MC) = $\frac{dc}{dx} = \frac{d}{dx}\{2500 + 3x^2 + 5x^3\}$

$$= 6x + 15x^2$$

Therefore slope of Marginal Cost Curve

$$= \frac{d}{dx}(MC) = \frac{d}{dx}(6x + 15x^2).$$

$$= 6 + 30x.$$

$$\text{Average Variable Cost (AVC)} = \frac{\text{Variable Cost (VC)}}{\text{No. units Produced (x)}}$$

$$= \frac{3x^2 + 5x^3}{x}$$

$$= 3x + 5x^2$$

Illustration No.33

If the cost function is given by $C(x) = 5x^3 + 4x^2 + 3x + 7$ then find Average cost (AC), Marginal Cost (MC), Slope of AC and slope of MC?

Solution:

Since the total function $C(x) = 5x^3 + 4x^2 + 3x + 7$

Therefore the Average Cost (AC) = $C(x)/x$

$$AC = \frac{5x^3 + 4x^2 + 3x + 7}{x}$$

$$= 5x^2 + 4x + 3 + \frac{7}{x}.$$

But Marginal Cost (MC) = $\frac{d}{dx}\{C(x)\}$

$$= 15x^2 + 8x + 3.$$

And Slope of the Average Cost = $\frac{d}{dx}\{AC\} = \frac{d}{dx}[5x^2 + 4x + 3 + \frac{7}{x}]$

$$= 10x + 4 - \frac{7}{x^2}$$

Similarly slope of the Average Cost = $\frac{d}{dx}$ {MC} = $\frac{d}{dx}$ $[15x^2 + 8x + 3]$

$$= 30x + 8.$$

Illustration No. 34

The average cost function of a toy manufacturer is given by AC (x) is given by AC (x) = 5x + 7. + 125/x where 'x' represents the output produced. Find for what value of x for which Average Cost (AC) is increasing and for which the Average Cost (AC) is decreasing?

Solution:

Since

$$AC(x) = 5x + 7 + 125/x.$$

For Average Cost (AC) to be increasing, we need to have the slope of AC(x) i.e.

$$\frac{d}{dx}\{AC(x)\} > 0.$$

Therefore

$$\frac{d}{dx} = \frac{d}{dx}\left\{5x + 7 + 125/x\right\}$$

$$= 5 - \frac{125}{x^2}$$

But

$$5 - \frac{125}{x^2} > 0$$

Therefore $x > 5$

Similarly for Average Cost (AC) to be decreasing, we need to have d/dx {AC (x)} < 0

And hence

$$5 - \frac{125}{x^2} < 0 \text{ i.e. } x < 5.$$

But as $x \neq 0$ therefore $0 < x < 5$

Illustration No.35

If the cost function of a firm $C(x) = ax^2 + bx + c$ were x stands for the level of output then show that d/dx {AC (x)} = $\frac{1}{x} \times$ {MC(x) – AC (x)} where MC stands for Marginal Cost and AC stands for Average Cost?

Solution:

Since the cost function of the firm is given by

$$C(x) = ax^2 + bx + c \qquad \{\text{where } x = \text{output produced}\}$$

$$\text{But Average Cost } \{AC(x)\} = \frac{\text{Total Cost}\{C(x)\}}{\text{No. of units produced}}$$

$$= \frac{ax^2 + bx + c}{x}$$

$$= ax + b + \frac{c}{x}$$

$$\text{and} \qquad \frac{d}{dx}\{AC(x)\} = \frac{d}{dx}\left\{ax + b + \frac{c}{x}\right\}$$

$$= a - \frac{c}{x^2}$$

$$\text{And Marginal Cost } \{MC(x)\} = \frac{\text{Change in Total Cost }\{C(x)\}}{\text{Change in Output }(x)}$$

$$= \frac{d\{C(x)\}}{dx}$$

$$= \frac{d}{dx}\{ax^2 + bx + c\}$$

$$= 2ax + b.$$

But

$$\frac{1}{x} \times \{MC(x) - AC(x)\} = \{2ax + b - ax - b - \frac{c}{x}\}$$

$$= \frac{1}{x}\left\{ax - \frac{c}{x}\right\}$$

$$= \left(a - \frac{c}{x^2}\right)$$

$$= \frac{d}{dx}\{AC(x)\}$$

Hence proved.

Illustration No.36

The cost function of a firm producing electrical switches is given by

$$C(x) = \left(\frac{x^2+4x}{x+3}\right)+20$$

show that with increase in output level results in fall of Marginal Cost?

Solution:

Since

$$C(x) = \left(\frac{x^2+4x}{x+3}\right)+20$$

Marginal cost is the rate of change of cost resulting from change in output level

i.e $$MC(x) = \frac{d}{dx}\{C(x)\}$$

$$= \frac{d}{dx}\left\{\left(\frac{x^2+4x}{x+3}\right)+20\right\}$$

$$= \frac{(x+3)\frac{d}{dx}(x^2+4x)-(x^2+4x)\frac{d}{dx}(x+3)}{(x+3)^2}$$

$$= \frac{(x+3)(2x+4)-(x^2+4x)(1)}{(x+3)^2}$$

$$= \frac{2x^2+6x+4x+12-x^2-4x}{(x+3)^2}$$

$$= \frac{x^2+6x+12}{(x+3)^2}$$

$$= \frac{(x+3)^2+3}{(x+3)^2}$$

$$= 1+ + \frac{3}{(x+3)^2}$$

Finding the differentiation of MC (x)

We get,

$$\frac{d}{dx}\{MC(x)\} = (-)\frac{3(x+3)}{(x+3)^4}$$

$$= (-)\frac{3}{(x+3)^3}$$

Which is (–)ve for $x \geq 0$,

Hence as MC falls as the output level increases

Concept of Total Revenue, Average Revenue and Marginal Revenue

Total Revenue (TR) is the amount generated by selling a given level of output (q) at a given rate or price say p.

Therefore Total Revenue (TR) = Quantity Sold (q) × Price (p).

$$TR = q \times p.$$

$$\text{Average Revenue} \quad (AR) = \frac{\text{Total Revenue}}{\text{No. of units sold}} = \frac{q \times p}{q} = p$$

$$\text{Marginal Revenue} \quad (MR) = \frac{\text{Change in Total Revenue}}{\text{Change in No. of units sold}} = \frac{dTR}{dq}$$

Addendum

Relationship between TR, AR and MR Curves

Since,

$$TR = q \times p.$$

And

$$MR = \frac{dTR}{dq} = \frac{d}{dq} = (q \times p)$$

$$= p \times \left(\frac{d}{dq}(q)\right) + q \times \left(\frac{d}{dp}(p)\right)$$

$$= p + q \times \frac{dp}{dq}$$

Assuming at any given level of output q, MR lies below AR

Therefore it implies $p + q \times \frac{dp}{dq} < AR$

or $$p + q \times \frac{dp}{dq} < p \qquad \{\text{since } AR = p\}$$

Since q > 0, we get $\frac{dp}{dq} < 0$ or $\frac{d(AR)}{dq} < 0$

Hence in order MR to lie below AR the necessary condition is

$\frac{d}{dq}(AR)$ should be less than zero

Or AR curve should be declining which gives us the general relationship between marginal- average revenue relationship that MR lies below when AR is downward sloping.

The relationship between Marginal Revenue and Price Elasticity

$$MR = p\left(1 - \frac{1}{e}\right)$$

Proof

Let us assume that demand function as

P = f(q)

Therefore the Total Revenue (TR) = Price (p) × Quantity Sold (q)

$$= p \times q$$

$$= f(q) \times q$$

But Marginal Revenue (MR) = $\frac{d}{dq}(p \times q)$

$$= p\left(\frac{dq}{dq}\right) + q\left(\frac{dp}{dq}\right)$$

$$= p + q\left(\frac{dq}{dq}\right)$$

But the price elasticity of demand

$$e = (-)\frac{dq}{dp} \times \frac{p}{q}$$

$$-e(\frac{q}{p}) = (\frac{dq}{dp})$$

or $$-(\frac{p}{eq}) = \frac{dp}{dq}$$

And hence substituting the value of (dp/dq) in the above equation we have

$$MR = p + q\left(\frac{dp}{dq}\right)$$

$$= p - q\left(\frac{p}{eq}\right)$$

$$= p - \left(\frac{p}{e}\right)$$

$$MR = P\left(1 - \frac{1}{e}\right)$$

The Relationship between TR, MR and Price Elasticity

As TR reaches maximum level at e = 1 as at this point its slope is zero or MR = 0 and hence at e = 1

$$MR = p\left(1 - \frac{1}{1}\right) = 0$$

And if e <1 then TR curve will have negative slope i.e. given p > 0 and (1 – 1/e) < 0 and hence MR < 0.

Similarly if e >1 then TR curve will have positive slope i.e. given p > 0 and (1 – 1/e) > 0 and hence MR > 0.

And hence we conclude that

(i) If demand is inelastic i.e. (e < 1) price rise will lead to rise in TR and fall in price will resulting in fall in TR.

(ii) If demand is elastic i.e. (e > 1) price rise will lead to fall in TR and fall in price will resulting in rise in TR.

(iii) If demand is unitary elastic then TR remains unaffected by change in price as e = 1 resulting in MR = 0.

Relation between AR, MR and Price Elasticity of Demand

Let us find out the mathematical relationship between AR, MR and e_d.

Since we know,

$$\text{Total Revenue (TR)} = \text{Price (p)} \times \text{Qty (q)}$$

$$TR = p \times q$$

As $$\text{Average Revenue} = \frac{\text{Total Revenue (TR)}}{\text{No. of Quantity sold (q)}} = \frac{TR}{q}$$

$$= \frac{p \times q}{q} = p$$

But $$MR = p\left(1 - \frac{1}{e}\right) \quad \text{Since } AR = p$$

Therefore,

$$MR = AR\left(1 - \frac{1}{e}\right)$$

$$= \frac{e \times AR - AR}{e}$$

$$e\,MR = e \times AR - AR$$

$$e(MR - AR) = -AR$$

$$e\,(AR - MR) = AR$$

$$e = \frac{AR}{AR - MR}$$

Illustrations No.37

Animesh Electrical Limited manufacturer of Cool Breeze brand of electrical fans has invested an amount of Rs. 2 lakhs to set up a production unit. A variable cost of Rs. 150 was also estimated per unit of production whereas each unit of Cool Breeze fans selling price was marked Rs. 175. Write down the cost function C(x), revenue function and the profit function for 'x' units of production. Also find at what level of production the company reaches the breakeven point (BEP)?

Solution:

Given the fixed cost is Rs. 200000

Variable Cost per unit = 150

And Selling Price = 175

Therefore the Total Cost = Fixed Cost + Variable Cost

$$C(x) = 200000 + 150\,x$$

But Revenu $= \text{Price} \times \text{Qty. sold}$

$$= 175 \times x = 175x$$

As the Profit (π) = Revenue – Total Cost

$$= 175\,x - (200000 + 150\,x)$$

$$= 25\,x - 200000$$

But as we know Breakeven Point (BEP) where the company makes neither profit or loss. In other words at BEP profit (π) = 0

Therefore we have,

$$25\,x - 200000 = 0$$

Or $x = 8000$

Illustration No. 38

If the demand function is given by $p = a - bx$ then show that $e = \dfrac{AR}{AR - MR}$ at $p = k$?

Solution:

Since

$$p = a - bx$$

Therefore substituting $p = k$ we have,

$$k = a - bx$$

rearranging the equation we get,

$$x = \frac{a - k}{b}$$

But Total Revenue = Price × Qty

$$= (a - bx) \times x \qquad \{\text{since } p = a - bx\}$$

$$= ax - bx^2$$

As,

$$MR = \frac{dR}{dx} = \frac{d\left(ax - bx^2\right)}{dx} = a - 2bx \qquad .....(i)$$

Substituting the value of 'x' in the above equation, we get

$$= a - 2b \times \frac{(a-k)}{b}$$

$$= a - 2a + 2k.$$

$$= 2k - a.$$

$$AR = \frac{TR}{x}$$

$$= \frac{ax - bx^2}{x} = a - bx$$

Substituting the value of x in eq. (i), we get

$$= a - b \times \frac{(a-k)}{b}$$

$$= a - a + k$$

$$= k$$

But elasticity $\qquad (e) = (-)\,\frac{dx}{dp} \times \frac{p}{x}$

As $\qquad \frac{dp}{dx} = -b \qquad$ {since p = a – bx}

$$e = \frac{(a-bx)}{bx} = \frac{1}{b}\left(\frac{a}{x} - b\right) \qquad ......(ii)$$

substituting the value of x in the equation (ii) we get,

$$= \frac{1}{b}\left(\frac{a}{x} - b\right)$$

$$= \frac{1}{b}\left(\frac{ab}{(a-k)} - b\right)$$

$$= \left(\frac{a}{(a-k)} - 1\right)$$

$$= \frac{k}{a-k}$$

Similarly finding the value of $\frac{AR}{AR - MR}$

$$= \frac{k}{k-(2k-a)}$$

$$= \frac{k}{k-2k+a}$$

$$= \frac{k}{k-a}$$

Hence proved.

Summary

- Derivative is a measure of rate of change of given function with respect to its independent variable

$$\frac{\Delta y}{\Delta x} = \frac{f(x+\Delta x) - f(x)}{\Delta x} \quad \text{or} \quad \frac{f(x+h) - f(x)}{h}$$

- If u(x) is a function such that defined as u(x) = c f(x) , where f(x) is a derivative function and c is a constant then ,

$$\frac{d}{dx}u(x) = \frac{d}{dx}\left\{(c f(x) = c\frac{d}{dx} f(x)\right\}$$

- If f(x) is a function defined as f(x) = u(x) ± v(x)

Then $$\frac{d}{dx}\{f(x)\} = \frac{d}{dx}\{u(x)\} \pm \frac{d}{dx}\{v(x)\}$$

If f(x) is a function defined as the product of two differentiable f(x) u(x) and v(x). Then the derivative of f(x) denoted by $\frac{d}{dx}$\{ f(x)\} can be expressed as

$$\frac{d}{dx}\{f(x)\} = v(x)\, u'(x) + u(x)\, v'(x)$$

where $\frac{d}{dx}$ u(x) = $u'(x)$, $\frac{d}{dx}v(x)$= v'(x) are the respective first derivative of u(x) and v(x)

Differential Coefficient of the product of two functions = Second function × Derivative of the first function + first function × Derivative of the second function

- If f(x) is a function defined as a function expressed in the following form

$$f(x)=\frac{u(x)}{v(x)}$$

where u(x) and v(x) are function of x then $\frac{d}{dx}\{f(x)\}$ the derivative of f(x) can be expressed as

$$\frac{d}{dx}\{f(x)\}=\frac{v(x)\frac{d}{dx}\{u(x)\}-u(x)\frac{d}{dx}\{v(x)\}}{v(x)^2}$$

- $\frac{d}{dx}\{x^n\}=nx^{n-1}$
- $\frac{d}{dx}\{\tan x\}=\sec^2 x$
- $\frac{d}{dx}\{\operatorname{cosec} x\}=-\operatorname{cosec} x\cot x$
- Ceteris paribus (i.e. keeping other thing constant) the elasticity of demand can be defined as the measure of degree of sensitivity or responsiveness of a consumer to quantity of commodity demanded to change in price. Or it refers to percentage change in quantity demanded of a commodity because of change in the price of the commodity.

  Mathematically expressed as

$$e_d=\frac{\text{Percentage Change in Qty. demanded}}{\text{Percentage Change in price}}$$

Self Practice Exercises

Section No 1

Find the limit of the function using 1st principle

1. $y = x^2 + 5x + 4.$
2. $y = \sqrt{3x}$
3. $y = 2x^3 + 4x + 7$
4. $y = \frac{1}{\sqrt{x}}$
5. $y = e^{\sqrt{x}}$
6. $y = \log \cos x$
7. $y = x + \frac{1}{x} + 2$
8. $y = a^{5x}$
9. $y = \log \sqrt{x}$
10. $y = \sin 4x$
11. $y = \tan \sqrt{x}$
12. $y = x^2 \sin x$
13. $y = e^{\sec x}$
14. $y = \sqrt{\tan x}$
15. $y = \tan^{-1}\left(\frac{x}{5}\right)$

Section 2

Find the derivative of the following function

(a) $y = x^4 + \frac{7}{x} + e^x$

(b) $y = a^x + e^x + x^a$

(c) $y = e^x \sin x$

(d) $y = x \log x.$

(e) $y = \frac{x+1}{x^2 - 4x + 3}$

Section 3

(a) $y = 2x^{5/2} + 3 \log x + 5$

(b) $y = \frac{2^2 - x^2}{2^2 + x^2}$

(c) $y = \frac{3x^2 + 5}{\sqrt{x}}$

(d) $y = \frac{5 + 3x}{5 - 3x}$

(e) $y = e^{4x^2} + Bx + 6$

(f) $y = \sqrt{x} + 5^{5x}$

(g) $y = \frac{1 - x^2}{1 + x^2}$

Section 4

(a) $y = \sqrt{\dfrac{3-x}{3+x}}$ (b) $y = \dfrac{\sqrt{7}+\sqrt{x}}{\sqrt{7}-\sqrt{x}}$

(c) $y = x\log\left(\dfrac{x-5}{x+5}\right)$ (d) $y = x^{x^{x}}$

(e) $y = x^{a^{x}}$

Section 5

(a) $y = \log\sqrt{x+4}+\sqrt{x+3}$ (b) $y = \dfrac{(x-6)(x-4)}{\sqrt{x-5}}$

(c) $y = \sin^{-1}\left(\dfrac{5+4\cos x}{4+5\cos x}\right)$, *Hint*: substitute $t = \left(\dfrac{5+4\cos x}{4+5\cos x}\right)$

(d) $y = (x)^{\log x} + (\log x)^{x}$ (e) $x^{y} = e^{(x-y)}$

(f) $\sqrt{1-x^2}+\sqrt{1-y^2} = 9(x-y)$ (g) $\cos y = x\cos(3+y)$

Section No 6

(a) $y = \tan^{-1}\dfrac{\sqrt{1+5x^2x^2}-1}{5x}$ *Hint*: Substitute $5x = t$.

(b) $y = \log\left(x+\sqrt{x^2+7^2}\right)$ *Hint*: Substitute $=\left(x+\sqrt{x^2+7^2}\right)$

(c) $e^{y} = \left(\dfrac{5+4\tan x}{5-4\tan x}\right)$

Hint: convert $\tan x = \dfrac{\sin x}{\cos x}$ and then apply log both sides

(d) $y = \dfrac{1}{\sqrt{x+7}-\sqrt{x+5}}$

(e) $y = \sin^{-1}\left(\dfrac{\sqrt{x+1}-\sqrt{1-x}}{2}\right)$ *Hint*: substitute $x = \cos x$

(f) $y = (\log x)^{\text{Sinx}}$

Section No 7

1. If the total cost function of a firm is defined as $C(q) = 25{,}000 + 7q + 5q^2 + 3q^4$ then find
   - (a) Fixed Cost (FC)
   - (b) Variable Cost (VC)
   - (c) Average Cost (AC)
   - (d) Marginal Cost (MC)
   - (e) Slope of Marginal Cost
   - (f) Average Variable Cost (AVC)
2. If the demand function of the monopolist is given by $x = 50 - 10p$ then find the marginal revenue function of the monopolist?
3. The cost function of a manufacturer is given by

   $C = 3x^3 - 18x^2 + 100x$ then find
   - (a) Average Cost (AC).
   - (b) Marginal Cost (MC).
   - (c) Find at what value of x is the slope of the Average Cost is zero?
4. A toy manufacturer can sell 500 units of toys at Rs. 8 per unit and if the price is reduced by half the number units sold just doubles. Assuming linear demand function, find the revenue and demand function of the toy manufacturer?
5. If the total cost function of a firm is given by $TC = 1000\ e^{3x}$, then find the elasticity of the average cost?
6. A widget manufacturer knows that he can sells 50 units at the rate of Rs.10 and the demand falls to 25 units if the price is raised to Rs.15. But the manufacturer is happy to supply total of 70 units if the price is raised to Rs. 20. Find the manufacturers price elasticity of demand and supply at equilibrium· price assuming linear relationship between price and quantity in each case?
7. The demand function of a product is given by $p = 150 - 3x$ where p is the price unit and x is the quantity demand. Find
   - (a) Marginal Revenue.
   - (b) Relationship between Average and Marginal Revenue.
   - (c) Also the price at which Marginal Revenue is zero.

Answers

Section No 1

1. $2x + 5$
2. $\frac{\sqrt{3x}}{2}$
3. $6x^2 + 4$
4. $\frac{-1}{2x\sqrt{x}}$
5. $\frac{e^{\sqrt{x}}}{2\sqrt{x}}$
6. $-\tan x.$
7. $1-\frac{1}{x^2}$
8. $5a^{5x}\log a$
9. $\frac{1}{2x}$
10. $4\cos 4x.$
11. $\frac{\sec^2\sqrt{x}}{2\sqrt{x}}$
12. $-x^2\sin x + 2x\cos x$
13. $e^{\sec x}\sec x\tan x$
14. $\frac{\sec^2 x}{2\sqrt{\tan x}}$
15. $\frac{5}{x^2+5^2}$

Section No. 2

(a) $4x^3 - 7/x^2 + e^x$

(b) $a^x\log a + e^x + ax^{a-1}$

(c) $e^x(\sin x + \cos x)$

(d) $1 + \log x$

(e) $-(x^2 + 2x - 7)/(x^2 - 4x + 3)^2$

Section 3

(a) $5x^{3/2} + \frac{3}{x}$

(b) $\frac{-16x}{(4+x^2)^2}$

(c) $\frac{9x^2-5}{2x\sqrt{x}}$

(d) $\frac{30}{(5-3x)^2}$

(e) $8(x+1)\,e^{4x^2+8x+1}$

(f) $\frac{1}{2\sqrt{x}} + 5\,5^{5x}\log 5$

(g) $\dfrac{-4x}{\left(1+x^2\right)^2}$

Section 4

(a) $\dfrac{-3}{(3+x)\sqrt{1-x^2}}$

(b) $\sqrt{\dfrac{7}{x}}\dfrac{1}{\left(\sqrt{7}-\sqrt{x}\right)^2}$

(c) $\dfrac{10x}{x^2-5^2}+\log\left(\dfrac{x-5}{x+5}\right)$

(d) $x^{x^x}\left[x^x+x^x(1+\log x)\log x\right]$

(e) $x^{a^x}\left[\dfrac{1}{x}+\log x\log a\right]$

Section 5

(a) $\dfrac{1}{2}\left(\dfrac{1}{\sqrt{x+4}+\sqrt{x+3}}\right)\left(\dfrac{1}{\sqrt{x+4}}+\dfrac{1}{\sqrt{x+3}}\right)$

(b) $\dfrac{(x-6)(x-4)}{\sqrt{x-5}}\left\{\dfrac{1}{x-6}+\dfrac{1}{x-4}-\dfrac{1}{2(x-5)}\right\}$

(c) $-\dfrac{\sqrt{4^2-5^2}}{4+5\cos x}$

(d) $x^{\log x}\,\dfrac{2\log x}{x}+(\log x)^x$

(e) $\dfrac{\log x}{(1+\log x)^2}$

(f) $\sqrt{\dfrac{1-y^2}{1-x^2}}$

(g) $\dfrac{\cos^2(3x+y)}{\sin 3}$

Section 6

(a) $\dfrac{5}{2\left(1+5^2x^2\right)}$

(b) $\dfrac{1}{\sqrt{x^2+7^2}}$

(c) $\dfrac{40}{25\cos^2 x-16\sin^2 x}$

(d) $\dfrac{1}{4}\left\{\dfrac{1}{\sqrt{x+7}}+\dfrac{1}{\sqrt{x+5}}\right\}$

(e) $-\dfrac{1}{2\sqrt{1-x^2}}$

(f) $(\log x)^{\sin x}\left(\frac{\sin x}{x(\log x)} + \cos x \log\log x\right)$

Section No 7

1.

(a) FC = 25,000.

(b) $VC = 7q + 5q^2 + 3q^4$

(c) $AC = \frac{25000}{q} + 7 + 5q + 3q^3$

(d) $MC = 7 + 10q + 12q^3$

(e) $AVC = 7 + 5q + 3q^3$

2. $MR = 5 - \frac{x}{5}$

3.

(a) $AC = 3x^2 - 18x + 100$

(b) $MC = 9x^2 - 36x + 100$

(c) Slope of AC = 0 at x = 2

4. $12 - \frac{1}{125}x,\ 12x - \frac{1}{125}x^2$

5. $3x - 1$

6. 1.5, 0.4

7.

(a) MR = 150 – 6x

(b) Slope of MR = 2 Slope of AR.

(c) Rs.100.

Appendix

The Concept of Elasticity of Cost

If output produced be 'q' and the total cost function be TC such that TC = f(q),

Then the Elasticity of Cost is given by the ratio of proportional change in total cost to proportional change in the output.

Therefore,

$$\text{Elasticity of Cost } e_k = \frac{\text{Proportionate change in Total Cost}}{\text{Proportionate Change in output}}$$

$$= \frac{\frac{dTC}{TC}}{\frac{dq}{q}} = \frac{dTC}{TC} \times \frac{q}{dq}$$

$$= \frac{dTC}{dq} \times \frac{q}{TC}$$

$$= \frac{MC}{AC} \text{ since } \frac{dTC}{dq} = MC \text{ and } \frac{TC}{q} = AC$$

Thus elasticity of cost is the ratio of marginal cost to average cost.

Note:

(i) If MC > AC then $e_k > 1$

(ii) If MC < AC then $e_k < 1$ and

(iii) If MC = AC then $e_k = 1$

Elasticity of Average Cost

Since Average Cost (AC) $= \frac{TC}{q}$

Therefore

Elasticity of Average Cost $\left(\frac{TC}{q}\right) = \frac{d\left(\frac{TC}{q}\right)}{dq} \times \frac{q}{\frac{TC}{q}}$

$$= \frac{d\left(\frac{TC}{q}\right)}{dq} \times \frac{q^2}{TC}$$

$$= \frac{q^2}{TC} \times \left(\frac{q\frac{dTC}{dq} - TC}{q^2} \right)$$

$$= \frac{q^2}{TC} \times \frac{1}{q^2} \times \left(q\frac{dTC}{dq} - TC \right) = \frac{q}{TC}\frac{dTC}{dq} - 1$$

$$= e_k - 1$$

The following important results can be drawn

(i) If e_k the elasticity of cost is greater than or equal to or less than unity then the elasticity of average cost is greater than or equal to or less than zero.

(ii) The elasticity of total cost exceeds the elasticity of average cost by unity.

Elasticity of Marginal Cost

$$\text{Marginal Cost} = \frac{\text{Change in Total cost}}{\text{Change in output}}$$

$$MC = \frac{dTC}{dq}$$

Therefore the elasticity of marginal cost (MC) = $\dfrac{d\dfrac{dTC}{dq}}{dq} \times \dfrac{q}{\dfrac{dTC}{dq}}$

But $\quad e_k = \dfrac{dTC}{dq} \times \dfrac{q}{TC}$ or $\dfrac{dTC}{dq} = e_k \times \dfrac{TC}{q}$

And hence we get,

The elasticity of Marginal Cost (MC) = $\dfrac{d}{dq}\left(\dfrac{TC\,e_k}{q}\right)\dfrac{q^2}{TC\,e_k}$

CHAPTER 5

Maxima and Minima and its Application

Introduction

In previous all chapters we were discussing on various concepts of limits and differentiation in understanding the relationship between the functions. But in this chapter we will not only determine relationship but we will go one step ahead to understand the behavior pattern of the function against the other i.e. what will be the impact on a function defined in terms of another function at a particular point whether the function is increasing or decreasing or determine the point of inflexion all these concepts will be dealt in this chapter. The concept of maxima and minima finds wide application in business and economics to understand vital parameters like sales on revenue, cost on production and effects of taxes and subsidy in business profitability etc.

Increasing Function

Let $y = f(x)$ is a function of x in a given interval say $a \leq x \leq b$ if x increases resulting in increase in y then the function is said to be a increasing function i.e. if $a \leq x_1 < x_2 \leq b$ then $f(x_2) > f(x_1)$.

Geometrically it can be interpreted as we move along the curve from left to right in the interval $a \leq x \leq b$ the curve is raising. And in this case the increment in *x* i.e. Δx and the corresponding increment in y i.e. Δy have the same sign i.e both are either positive or negative such that the ratio $\frac{\Delta y}{\Delta x}$ is always positive

Therefore when the limit $\Delta x \to 0$, $\frac{\Delta y}{\Delta x} \to \frac{dy}{dx}$ so that the derivative at any point say P of the rising curve is positive.

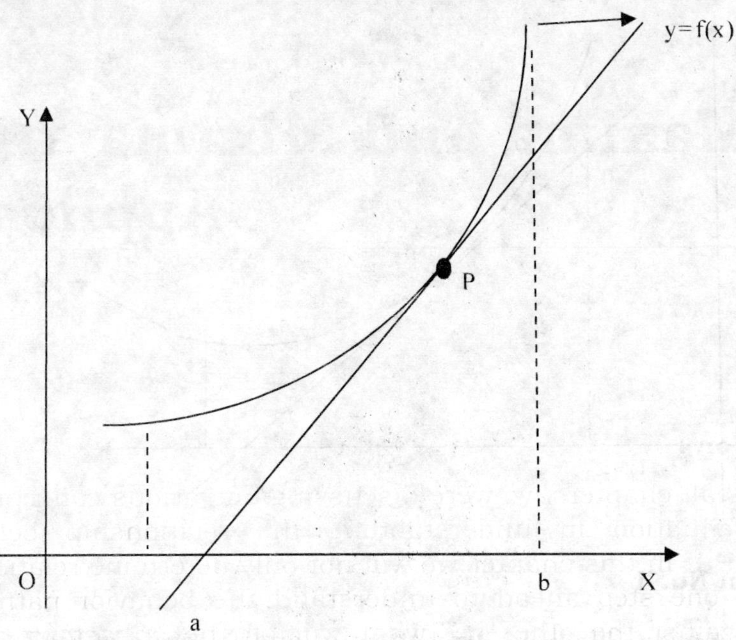

Decreasing Function

Let y = f(x) is a function of x in a given interval say $a \leq x \leq b$ if x increases resulting in decrease in y then the function is said to be a decreasing function i.e. if $a \leq x_1 < x_2 \leq b$ then $f(x_2) < f(x_1)$.

Geometrically it can be interpreted as we move along the curve from left to right in the interval $a \leq x \leq b$ the curve is falling. And in this case the increment in x i.e. Δx and the corresponding increment in y i.e. Δy have opposite sign such that the ratio $\frac{dy}{dx}$ is always negative.

Therefore when the limit $\Delta x \to 0$, $\frac{\Delta y}{\Delta x} \to \frac{dy}{dx}$ so that the derivative at any point say P of the falling curve is negative.

Note:

A function y = f(x) may neither be an increasing nor a decreasing function x at a certain interval [a, b]. In this, case, y = f(x) is stationary at that point and at that point the derivative $f'(x) = 0$.

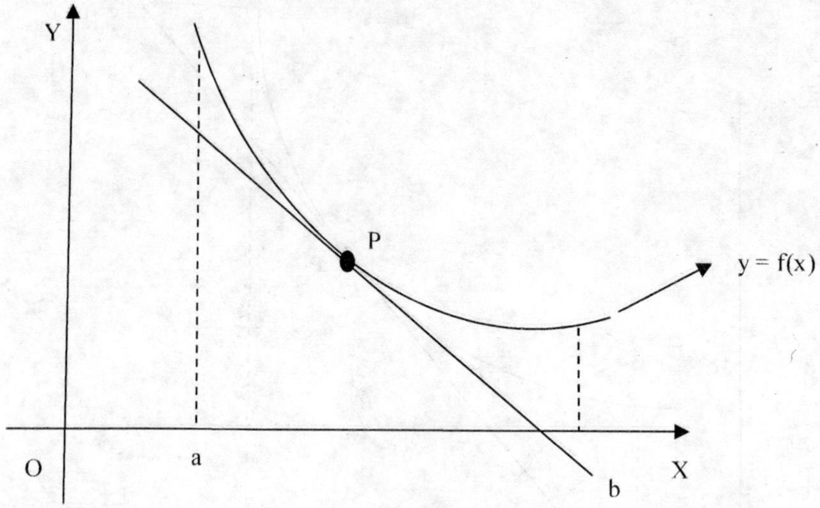

Illustration No. 1

If $f(x) = 4x^3 + 6x^2 - 24x + 14$, find the stationary points of f(x)?

Solution:

We have $f(x) = 4x^3 + 6x^2 - 24x + 14$

Therefore,

$$f'(x) = 12x^2 + 12x - 24.$$

But at stationary point of f(x), we have $f'(x) = 0$

And hence,

$$12x^2 + 12x - 24 = 0$$

i.e. $12(x^2 + x - 2) = 0$ or $x^2 + x - 2 = 0$

Or $x^2 + 2x - x - 2 = 0$

Or $x(x + 2) - 1(x + 2) = 0$

Or $(x - 1)(x + 2) = 0$

Therefore,

$$x = -2, 1.$$

Illustration No. 2

Show that the function $y = a + \frac{a}{x}$ is always a decreasing function for all real of values of x and a > 0?

Solution:

We have $$y = a + \frac{a}{x}$$

Therefore,

$$\frac{dy}{dx} = 0 - \frac{a}{x^2} = -\frac{a}{x^2} < 0$$ for all values of x, except at x = 0 provided (a > 0).

Hence y is always a decreasing function of x for all values of x, except at x = 0 and a > 0.

Maxima and Minima of a Function

A function y = f(x) is said to be relative maximum or simply maximum at x = a if f(a) is the greatest of all the values of f(x) in the vicinity of a i.e. we can find a positive number δ such that

$$f(a) > f(x) \text{ for all x in } a - \delta < x < a + \delta \ (x \neq a)$$

A function y = f(x) is said to be relative minimum or simply minimum at x = a′ if f(a′) is the least of all the values of f(x) in the vicinity of a′ i.e. we can find a positive number δ such that

$$f(a') < f(x) \text{ for all x in } a' - \delta < x < a' + \delta \ (x \neq a')$$

Extreme or Turning Values

By extreme or turning value of a function we mean the maximum and the minimum value of a function. The point at which the function has a maximum or minimum value is called as extremum or turning points.

Necessary Condition for Maxima and Minima

If f(x) has an maximum value at x = a and that f ′(a) exists then f ′(a) = 0 .

Proof:

Let us assume that f(x) is maximum at x = a and that f ′(a) exists

Now, if $f'(a) \neq 0$ then either $f'(a) > 0$ or $f'(a) < 0$

If $f'(a) > 0$ then f(x) is increasing at a and hence the values of f(x) > f(a) to the right of a which contradiction to the assumption.

Similarly,

If $f'(a) < 0$ then f(x) is decreasing at a and hence the values of f(x) < f(a) to the left of a which is also a contradiction to the assumption.

And hence $f'(a)$ can only be zero i.e. $f'(a) = 0$.

Note:

We can similarly proof that If f(x) has an minimum value at x = a and that $f'(a)$ exists then $f'(a) = 0$.

First Order Derivative test for Maxima and Minima

If f(x) is a function such that $f'(a) = 0$, then f(a) will be the maximum value of f(x) if its first derivative $\frac{dy}{dx}$ changes sign from positive to the left of a to negative to the right of a.

Similarly, f(a) will be the minimum value of f(x) if its first derivative $\frac{dy}{dx}$ changes sign from negative to the left of a to positive to the right of a.

And, has neither a maximum nor minimum value if its first derivative $\frac{dy}{dx}$ has the same sign on both the side of a. Such points are called as point of inflexion.

Second Order Derivative test for Maxima and Minima

A function f(x) has a maximum value at x = a if $f'(a) = 0$ and $f''(a) < 0$ if and only if the second derivative $\{f''(a)\}$ exists.

Similarly,

A function f(x) has a minimum value at x = a if $f'(a) = 0$ and $f''(a) > 0$ if and only if the second derivative $\{f''(a)\}$ exists.

Illustration No. 3

For what values of x, the following function has maximum or minimum value: $f(x) = 2x^3 - 9x^2 + 12x + 9$?

Solution:

Given $f(x) = 2x^3 - 9x^2 + 12x + 9$

Then finding the 1st derivative we get,

$$f'(x) = 6x^2 - 18x^2 + 12$$

Using 1st order condition for maxima and minima, we have,

$$f'(x) = 0 \text{ or } 6x^2 - 18x^2 + 12 = 0$$

Or $$6(x^2 - 3x + 2) = 0$$

Or $$(x^2 - 3x + 2) = 0$$

$$x^2 - 2x - x + 2 = 0$$

$$(x - 1)(x - 2) = 0$$

$$x = 1, 2.$$

Now, finding the 2nd order derivative i.e.

$$f''(x) = 12x - 18$$

At x = 1 we have,

$$f''(x)_{x=1} = 12(1) - 18 = -6 < 0$$

And hence f(x) is maximum at x = 1 and the maximum value is

$$f(1) = 2(1)^3 - 9(1)^2 + 12(1) + 9$$

$$= 2 - 9 + 12 + 9.$$

$$= 14.$$

At x = 2 we have,

$$f''(x)_{x=2} = 12(2) - 18 = 6 > 0$$

And hence f(x) is minimum at x = 2 and the minimum value is

$$f(2) = 2(2)^3 - 9(2)^2 + 12(2) + 9$$

$$= 16 - 36 + 24 + 9.$$

$$= 13.$$

Illustration No. 4

For what values of x, the following function has maximum or minimum value $y = \frac{2x}{x^2 + 4}$?

Solution:

Given, $$y = \frac{2x}{x^2 + 4}$$

Therefore,

$$\frac{dy}{dx} = \frac{2(4+x^2)-(2x)(2x)}{(4+x^2)^2}$$

$$= \frac{[(8+2x^2)-(4x^2)]}{(4+x^2)^2}$$

$$= \frac{(8-2x^2)}{(4+x^2)^2}$$

Using 1st order condition for maxima and minima, we have,

$$\frac{dy}{dx} = 0, \quad \frac{(8-2x^2)}{(4+x^2)^2} = 0$$

$(8 - 2x^2) = 0$ or $4 - x^2 = 0$

$x = \pm 2.$

Now, finding the 2nd order derivative i.e.

$$\frac{d^2x}{dy^2} = \left(\frac{(-4x)\ (4+x^2)^2 - (8-2x^2).2.(4+x^2).2x}{(4+x^2)^4}\right)$$

$$= (4+x^2)\left(\frac{(-4x)\ (4+x^2) - 4x\ (8-2x^2)}{(4+x^2)^4}\right)$$

$$= (-4x)\left(\frac{(4+x^2) + (8-2x^2)}{(4+x^2)^3}\right)$$

$$= (-4x)\left(\frac{12-x^2}{(4+x^2)^3}\right)$$

At $x = 2$, we have,

$$\frac{d^2x}{dy^2} = -(4 \times 2)\left(\frac{12-2^2}{(4+2^2)^3}\right)$$

$$= -\frac{1}{8} < 0$$

Therefore the given function is maximum at $x = 2$

Similarly,

At x = – 2, we have,

$$\frac{d^2x}{dy^2} = -[4 \times (-2)]\left(\frac{12-(-2)^2}{[4+(-2)^2]^3}\right)$$

$$= \frac{1}{8} > 0$$

Therefore the 2nd derivative is positive. And hence the function is minimum at x = – 2.

The maximum value of the function at x = 2 is $\frac{1}{2}$ and the minimum value of the function at x = – 2 is $-\frac{1}{2}$.

Illustration No. 5

Find the maximum and minimum value of the function

$$y = 2x + \frac{3}{x}$$

Solution:

Given

$$y = 2x + \frac{3}{x}$$

Finding the 1st derivative of the function, we get,

$$\frac{dy}{dx} = 2 - \frac{3}{x^2}$$

Now using the 1st order condition for maxima and minima, we have,

$$\frac{dy}{dx} = 0 \text{ or } 2 - \frac{3}{x^2} = 0$$

or $$x^2 = \frac{3}{2} \rightarrow x = \pm\sqrt{\frac{3}{2}}$$

Now, finding the 2nd derivative, we get,

$$\frac{d^2x}{dy^2} = 0 - \frac{3\times(-2)}{x^3} = \frac{6}{x^3}$$

Now at $x = \sqrt{\frac{3}{2}}$, we have,

$\frac{d^2x}{dy^2} = \frac{6}{\left(\sqrt{\frac{3}{2}}\right)^3}$, which is a positive value, therefore the function is minimum at $x = \sqrt{\frac{3}{2}}$ and the minimum value is given by,

$$y = 2\left(\sqrt{\frac{3}{2}}\right) + \frac{3}{\left(\sqrt{\frac{3}{2}}\right)}$$

$$= \sqrt{6} + \sqrt{6} = 2\sqrt{6}$$

Similarly,

Now at $x = -\sqrt{\frac{3}{2}}$, we have, $\frac{d^2x}{dy^2} = \frac{6}{\left(-\sqrt{\frac{3}{2}}\right)^3}$, which is a negative value, therefore the function is maximum at $x = -\sqrt{\frac{3}{2}}$ and the maximum value is given by,

$$y = 2\left(-\sqrt{\frac{3}{2}}\right) + \frac{3}{\left(-\sqrt{\frac{3}{2}}\right)}$$

$$= -\sqrt{6} - \sqrt{6} = -2\sqrt{6}$$

Illustration No.6

If $y = 4x^3 - 18x^2 - 48x + 24$ find for what value is the function is maximum or minimum?

Solution:

$$y = 4x^3 - 18x^2 - 48x + 24$$

diff w. r. t 'x'

$$dy/dx = 12x^2 - 36x - 48 + 0$$

$$= 12x^2 - 36x - 48$$

.....(5.1)

In order to find the maximum or minimum equate $\frac{dy}{dx} = 0$

We get, $12(x^2 - 3x - 4) = 0.$

(or) $x^2 - 3x - 4 = 0.$

Solving the quadric equation we get,

$$x^2 - 4x + x - 4 = 0$$

$$x(x-4) + 1(x-4) = 0$$

$$(x+1)(x-4) = 0$$

$$x = -1, 4$$

Now find the 2nd derivative (1), we get

$$\frac{d^2y}{dx^2} = 12(2x - 3)$$

Substituting the values of $x = -1, 4$

When $x = -1$

$$\left\{\frac{d^2y}{dx^2}\right\}_{x=-1} = 12\{2(-1) - 3\} = 12(-5) = -60$$

Which is negative value and hence the function is maximum at $x = -1$

And when $x = 4$

$$\left\{\frac{dy}{dx}\right\}_{x=4} = 12\{2(4) - 3\} = 12(5) = 60$$

Which is a positive value Therefore the function is minimum at $x = 4$

Illustration No.7

Show that the function $2x^3 - 6x^2 + 6x + 14$ has neither maximum nor minimum value?

Solution:

Let us assume that $f(x) = 2x^3 - 6x^2 + 6x + 14$, then, finding the 1st derivative we have

$$f'(x) = 6x^2 - 12x + 6$$

Using 1st order condition for maxima and minima, we have,

$$f'(x) = 0 \text{ or } 6x^2 - 12x + 6 = 0$$

Or $\qquad 6(x^2 - 2x + 1) = 0$

Or $\qquad (x^2 - 2x + 1)$

$$(x-1)^2 = 0$$

$$x = 1.$$

Now, finding the 2nd order derivative i.e.

$$f''(x) = 12x - 12$$

At x =1 we have

$$f''(x) = 12(1) - 12 = 0$$

$$= 0.$$

Thus nothing can be said about maximum and minimum at x =1. And hence the given function has neither maximum nor minimum at x =1.

Illustration No. 8

Find the maximum or the minimum value of the function $f(x) = x^{1/x}$?

Solution:

Given $f(x) = x^{1/x}$

Let us assume $y = f(x) = x^{1/x}$

Taking log on both side, we get,

$$\log y = \frac{1}{x}\log x$$

Finding the 1st derivative, we have

$$\frac{1}{y}\frac{dy}{dx}=\frac{1}{x}.\frac{1}{x}-\frac{1}{x^2}\log x$$

$$=\frac{1}{x^2}-\frac{1}{x^2}\log x$$

$$=\frac{1}{x^2}\ (1-\log x).$$

For maximum and minimum value of y, we equate $\frac{dy}{dx}=0$

Or $\frac{1}{x^2}\ (1-\log x)=0$

Or $\log x=1$

Or $x=e.$

Now, finding the 2nd derivative, we get,

$$\frac{1}{y}\ \frac{d^2x}{dy^2}-\frac{1}{y^2}\ \frac{dy}{dx}=-\left(\frac{2}{x^3}\right)\ (1-\log x)+\left(\frac{1}{x^2}\right)\left(-\frac{1}{x}\right)$$

$$\frac{1}{y}\ \frac{d^2x}{dy^2}-\frac{1}{y^2}\ \frac{dy}{dx}=-\left(\frac{2}{x^3}\right)\ (1-\log x)-\left(\frac{1}{x^3}\right)$$

Now at x = e, we have,

$$\frac{1}{e}\ \frac{d^2x}{dy^2}-\frac{1}{e^2}\ \frac{dy}{dx}=-\left(\frac{2}{e^3}\right)\ (1-\log e)-\left(\frac{1}{e^3}\right)$$

$$\frac{1}{e}\ \frac{d^2x}{dy^2}-\frac{1}{e^2}.0=-\left(\frac{2}{e^3}\right)\ (1-\log e)-\left(\frac{1}{e^3}\right)$$

(Since $\frac{dy}{dx}=0$ at $x=e$ and $\log e=1$)

Therefore,

$$\frac{1}{e}\ \frac{d^2x}{dy^2}=-\left(\frac{1}{e^3}\right)$$

$$\frac{d^2x}{dy^2}=-\left(\frac{1}{e^2}\right)<0$$

And hence the function is maximum at the point x = e and the maximum value is $e^{\frac{1}{e}}$.

Illustration No. 9

Find whether the function is maximum or minimum if the function is defined by $4e^{2x} + 9e^{-2x}$?

Solution:

Let $y = 4e^{2x} + 9e^{-2x}$

Therefore, finding the 1st derivative of the function, we get,

$$\frac{dy}{dx} = 2.4.e^{2x} - 2.9.e^{-2x}$$

$$\frac{dy}{dx} = 8e^{2x} - 18e^{-2x}$$

Using 1st order condition for maxima and minima, we have, $\frac{dy}{dx} = 0$

Or $$8e^{2x} - 18e^{-2x} = 0$$

Or $$(e^{2x})2 = \frac{18}{8} = \frac{9}{4} \text{ or } e^{2x} = \frac{3}{2}$$

Now finding the 2nd order derivative, we get

$$\frac{d^2x}{dy^2} = 16e^{2x} + 36e^{-2x}$$

Now at $$e^{2x} = \frac{3}{2},$$

$$\frac{d^2x}{dy^2} = 16\left(\frac{3}{2}\right) + 36\left(\frac{2}{3}\right) = 48 > 0$$

And hence the function is minimum and the minimum value is

$$4\left(\frac{3}{2}\right) + 9\left(\frac{2}{3}\right) = 12$$

Illustration No. 10

If the function is defined as $x+\frac{1}{x}$ then find the maximum and minimum value of the function?

Solution:

Given the function is of the form $x+\frac{1}{x}$

Let us assume $y = x+\frac{1}{x}$

Therefore the 1st derivative

$$\frac{dy}{dx} = 1 - \frac{1}{x^2}$$

For maximum and minimum value we equate, $\frac{dy}{dx} = 0$

Or $\quad 1 - \frac{1}{x^2} = 0$ or $x^2 = 1$ or $x = \pm 1$

Now, finding the 2nd order derivative, we get,

$$\frac{d^2x}{dy^2} = \frac{2}{x^3}$$

At $x = 1$, $\frac{d^2x}{dy^2} = 2 > 0$ and so y is minimum at $x = 1$ and the minimum value is $1+\frac{1}{1} = 1+1= 2$.

At $x = -1$, $\frac{d^2x}{dy^2} = -2 < 0$ and so y is maximum at $x = -1$ and the maximum value is $-1 - \frac{1}{1} = -2$.

Illustration No. 11

If $\frac{x}{3} + \frac{y}{5} = 1$ then find whether the function defined as $x^2 + y^2$ is minimum function or maximum function?

Solution:

Since $\frac{x}{3} + \frac{y}{5} = 1$ rearranging the equation, we have,

$$\frac{y}{5} = 1 - \frac{x}{3} = \left(\frac{3-x}{3}\right) \text{ or } y = \frac{5}{3}(3-x)$$

Let us assume that $z = x^2 + y^2$

By substituting the value of y from the above equation, we get,

$$z = x^2 + \left[\frac{5}{3}(3-x)\right]^2$$

Now finding the 1st order derivative, we get,

$$\frac{dy}{dx} = 2x + \frac{25}{9}.2.(3-x)(-1)$$

Using 1st order condition for maxima and minima, we have, $\frac{dy}{dx} = 0$

Therefore,

$$2x = \frac{50}{9}.(3-x) \text{ or } x = \frac{66}{25}$$

Now finding the 2nd order derivative, we get,

$$\frac{d^2x}{dy^2} = 2 + \frac{50}{9} > 0 \text{ for all value of x}$$

So, the function z is minimum and the minimum value at $x = \frac{66}{25}$ is

Illustration No. 12

If the function is defined as $\left(\frac{1}{4} - x\right)^2 + 2x^3$ prove that the function has maximum value at $x = -\frac{1}{2}$ and minimum value at $x = \frac{1}{6}$?

Solution:

Let us assume that $y = \left(\frac{1}{4} - x\right)^2 + 2x^3$ then finding the derivative, we get,

$$\frac{dy}{dx} = 2\left(\frac{1}{4} - x\right)(-1) + 6x^2$$

Using 1st order condition for maxima and minima, we have, $\frac{dy}{dx} = 0$

$$2\left(\frac{1}{4} - x\right)(-1) + 6x^2 = 0$$

or $$\left(2x - \frac{1}{2}\right) + 6x^2 = 0$$

or $$(4x - 1) + 12x^2 = 0$$

Solving the quadratic, we get,

$$(2x + 1)(6x - 1) = 0 \text{ or } x = -\frac{1}{2} \text{ or } \frac{1}{6}$$

Now finding the 2nd order derivative,

$$\frac{d^2x}{dy^2} = 2 + 12x$$

Therefore $$\frac{d^2x}{dy^2} \text{ at } x = -\frac{1}{2},\ 2 + 12\left(-\frac{1}{2}\right) = 2 - 6 = -4 < 0$$

And hence the function is maximum at $= -\frac{1}{2}$.

Similarly,

$\frac{d^2x}{dy^2}$ at $x = \frac{1}{6}$, $2 + 12\left(\frac{1}{6}\right) = 2 + 2 = 4 > 0$ and hence the function has minimum value at $x = \frac{1}{6}$.

Convexity and Concavity

Let P(x, y) be any point on the curve y = f(x). Then the slope of the tangent at P varies with P (x, y). And if the tangent line is below the curve then the curve is concave upward or convex downward. Similarly when the tangent is above the curve then the curve is concave downward or convex upward. See the graphical representation below

Case 1: Concave Downward or Convex Upward

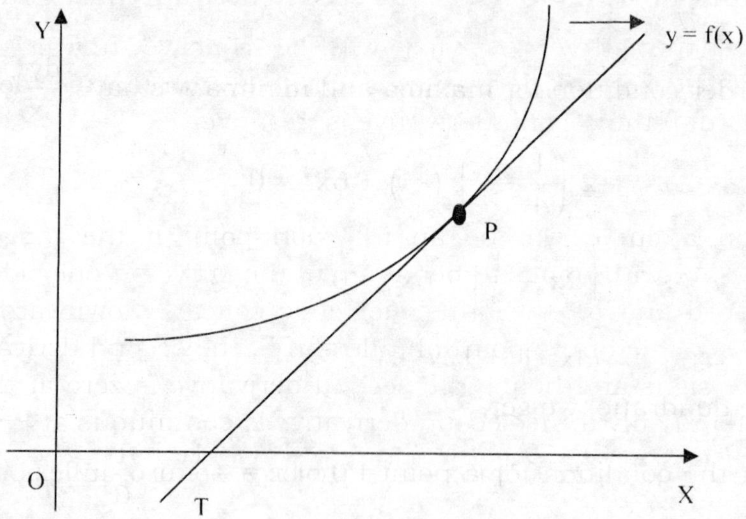

Case 2 :Concave upward or Convex downward

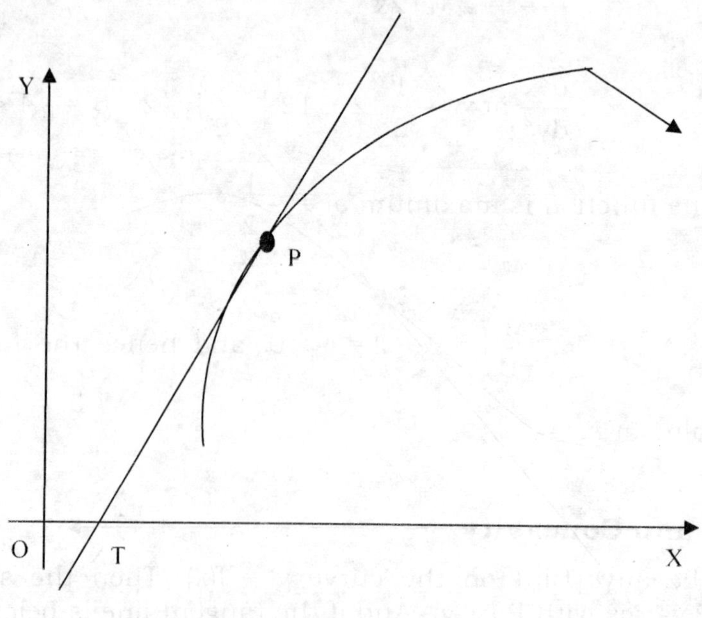

Note:

In case 1 as x increases the slope also increases i.e. the first derivative is an increasing function of x and hence the second derivative is positive. In Case 2 as x increases the slope decreases i.e. the first derivative is a decreasing function of x and hence the second derivative is negative.

And hence the curve y = f(x) will be concave upward (convex downward) if the second derivative is positive and concave downward (convex upward) if the second derivative is negative.

Point of Inflexion

A point P on a curve will be an inflexion point if the tangent line separates the arc with opposite bending i.e. the curve on one side of P is concave upward and on the other side it is concave downward or vice versa. On either side of the point of inflexion P, the second derivative will have opposite signs and hence the second derivative is zero at the point of inflexion if and only if the second derivative is continuous at P.

Therefore, the condition for a point P to be a point of inflexion on the curve y = f(x) is that, at the point P the second derivative i.e. $\frac{d^2y}{dx^2} = 0$ and $\frac{d^3y}{dx^3} \neq 0$.

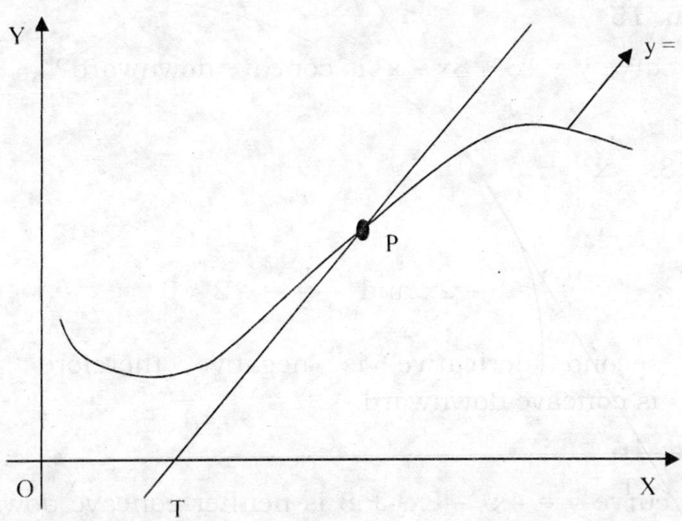

Illustration No. 13

Show that the curve $y = 2x^2 + 4$ is concave upward?

Solution:

Since $y = 2x^2 + 4$,

Therefore,

$$\frac{dy}{dx} = 4x \text{ and } \frac{d^2y}{dx^2} = 4 > 0$$

Since the second derivative is positive therefore the curve $y = 2x^2 + 4$ is concave upward.

Illustration No.14

Show that the curve $y = 8 - 4x - 3x^2$ is concave downward?

Solution:

Since $y = 8 - 4x - 3x^2$,

Therefore,

$$\frac{dy}{dx} = -4 - 6x \text{ and } \frac{d^2y}{dx^2} = -6 < 0$$

Since the second derivative is negative therefore the curve $y = 8 - 4x - 3x^2$ is concave downward.

Illustration No. 15

Show that the curve $y = 15 + 8x - x^2$ is concave downward?

Solution:

Since $y = 15 + 8x - x^2$,

Therefore,

$$\frac{dy}{dx} = 8 - 2x \text{ and } \frac{d^2y}{dx^2} = -2 < 0$$

Since the second derivative is negative therefore the curve $y = 15 + 8x - x^2$ is concave downward.

Illustration No. 16

Show that the curve $y = 4x^4 - 8x^3 + 8$ is neither concave downward nor upward?

Solution:

Since $y = 4x^4 - 8x^3 + 8$

Therefore, we have,

$$\frac{dy}{dx} = 16x^3 - 24x^2$$

Then finding the second order derivative we get, $\frac{d^2y}{dx^2} = 48x^2 - 48x$

But for the point of inflexion we have, $\frac{d^2y}{dx^2} = 0$ i.e. $48x^2 - 48x = 0$ or $48x(x - 1) = 0$

Therefore $x = 0, 1$.

Explanation

When $x > 0$ but close to 0, then $\frac{d^2y}{dx^2}$ is negative i.e. the curve is concave downward.

And when $x < 0$ but close to 0, then $\frac{d^2y}{dx^2}$ is positive i.e. the curve is concave upward.

Hence $x = 0$ is a point of inflexion.

Similarly

When $x > 1$ but close to 1, then $\frac{d^2y}{dx^2} = 48x(x - 1)$ is positive i.e. the curve is concave upward.

And When $x < 1$ but close to 1, then $\frac{d^2y}{dx^2} = 48x(x - 1)$ is negative i.e. the curve is concave downward.

Hence at $x = 1$ is the point of inflexion of the curve.

Illustration No. 17

Find the two numbers such that the sum of the two number is 8 and their product is maximum?

Solution:

Let one of the number be x, then the other number is (8 – x). Assuming that the product of the numbers be Z then, we have

$$Z = x(8 - x) \text{ or } Z = 8x - x^2.$$

Therefore,

$$\frac{dz}{dx} = 8 - 2x$$

Using 1st order condition for maxima and minima, we get,

$$\frac{dz}{dx} = 0 \text{ or } 8 - 2x = 0$$

Or $x = 4.$

Now finding the 2nd order derivative we get,

$\frac{d^2z}{dx^2} = -2$ which is negative. And hence the product of the number is maximum at x = 4 and the maximum value is 16.

Illustration No. 18

Divide a number 6 into two parts such that the product of the square of the first part and the fourth power of the second part is maximum at 2 and 4?

Solution:

Let us assume that one of the numbers be x then the second part is (6 – x),

Therefore,

The square of the first part = x^2.

And the fourth power of the second part = $(6 - x)^4$.

Let the product of the numbers be equal $Z = x^2(6 - x)^4$

And hence,

$$\frac{dz}{dx} = x^2 . 4(6 - x)^3 (-1) + 2x(6 - x)^4$$

$$= 2x\,(6-x)^3\,\{2x\,(-1) + (6-x)\}$$

$$= 2x\,(6-x)^3\,[\{-2x + 6 - x\}]$$

$$= 2x\,(6-x)^3\,[6-3x]$$

Using 1st order condition for maxima and minima, we get,

$$\frac{dz}{dx} = 0$$

Or $\quad 2x\,(6-x)^3\,[6-3x] = 0,$

Therefore x = 0, 2 and 4.

Now finding the 2nd order derivative we have,

$$\frac{d^2z}{dx^2} = 2[\frac{dz}{dx}\,(6-x)^3\,(6x-3x^2)]$$

$$= 2\,[3\,(6-x)^2\,(6x-3x^2) + (6-x)^3\,(6-6x)]$$

$$= 2\,(6-x)^2\,[\,(18x-9x^2) + (36+6x^2-42x)]$$

$$= 2\,(6-x)^2\,[\,36-60x-6x^2]$$

Therefore at x = 0, we get,

$\frac{d^2z}{dx^2} = 2592 > 0$, which is positive hence the product is minimum at x = 0,

Similarly, at x = 2, we get

$\frac{d^2z}{dx^2} = -1184 < 0$ which is negative hence the product is maximum at x = 2.

And at x= 4, we have,

$\frac{d^2z}{dx^2} = -2400 < 0$ which is negative hence the product is maximum at x = 4.

Illustration No. 19

Show that the height of the cylinder of maximum volume that can be curved out of a given sphere of radius r is $\frac{2r}{\sqrt{3}}$?

Solution:

Let us assume that the radius of the base of the cylinder be 'a' and height be x and AC the diameter of the sphere.

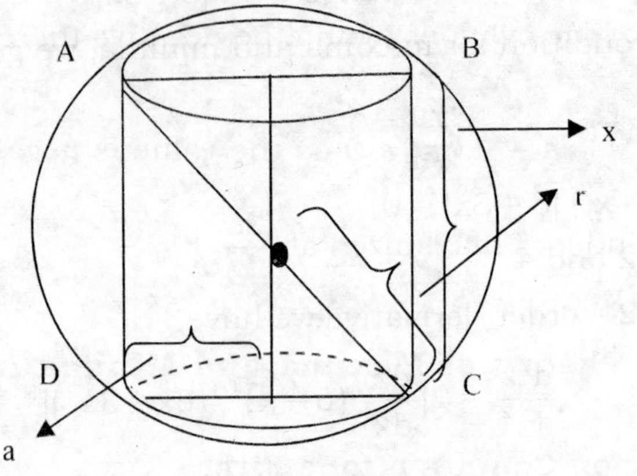

Since

$AB^2 + BC^2 = AC^2$

$(2a)^2 + x^2 = (2r)^2$

$4a^2 + x^2 = 4r^2$

If V is the volume of the cylinder, then

$V = \pi a^2 x.$

$= \pi x \{\frac{1}{4}\ 4r^2 - x^2.)\}$

$= \frac{\pi}{4}\left(4r^2 x - x^3\right)$

Therefore,

$$\frac{dv}{dx} = \frac{\pi}{4}\left(4r^2 - 3x^2\right)$$

For maxima and minima we equate $\frac{dv}{dx} = 0,$

Therefore,

$$\frac{\pi}{4}\left(4r^2 - 3x^2\right) = 0 \text{ or } x = \pm\frac{2r}{\sqrt{3}}$$

Now, finding

$$\frac{d^2V}{dx^2} = \frac{\pi}{4}(-6x)$$

Since the height of the cylinder cannot be negative therefore $x = \frac{2r}{\sqrt{3}}$

Now $\frac{d^2V}{dx^2}$ at $x = \frac{2r}{\sqrt{3}}$ is $-\sqrt{3}\,\pi r < 0$ as the value is negative, therefore the volume of the cylinder is maximized at $\frac{2r}{\sqrt{3}}$.

Application of Theory of Maxima and Minima in Business and Economics

Profit Maximization Condition for a Firm

As we know,

Profit (π) = Total Revenue (TR) – Total Cost (TC).

Now for profit maximization condition, Equate 1st Order derivative to zero.

Therefore,

$$\frac{d(\pi)}{dq} = \frac{dTR}{dq} - \frac{dTC}{dq} = 0$$

Or $\quad \frac{dTR}{dq} - \frac{dTC}{dq} \Rightarrow MR = MC \quad \{\text{since } MR = \frac{dTR}{dq},\ MC = \frac{dTC}{dq}\}$

Hence the above states the 1st order profit maximizing output level is where Marginal Revenue equals Marginal Cost i.e. MR = MC.

To check the viability of the above condition, find the 2nd order derivative

We have,

$$\frac{d^2(\pi)}{d^2q} = \frac{dMR}{dq} - \frac{dMC}{dq} < 0$$

i.e $\quad \frac{dMR}{dq} < \frac{dMC}{dq}$

i.e, Slope of MR curve < Slope of MC curve at profit maximizing output level.

Note: *Rate of change in Revenue is less than the rate of change in cost.*

Illustration No. 1

A manufacturer produces x units per day at a total cost of Rs. $(x^3 - 3x^2 + 15x + 9)$. Find at what level the average variable cost and the marginal cost is at their respective minimum?

Solution:

Let us assume that C(x) be the total cost of producing x unit per day in rupees, then we have,

$$C(x) = x^3 - 3x^2 + 15x + 9$$

And V(x) the average variable cost (AVC) in rupees, then

$$AVC(x) = \frac{\text{variable Cost}}{x} = \frac{x^3 - 3x^2 + 15x}{x} = x^2 - 3x + 15$$

Finding the 1st derivative i.e.

$$AVC'(x) = 2x - 3 + 0$$

Using the 1st order condition for maxima and minima, we get

$$AVC'(x) = 0 \text{ or } 2x - 3 = 0 \text{ or } x = \frac{3}{2}$$

Now finding the 2nd derivative, we get,

$AVC''(x) = 2 > 0$, hence the average variable function is minimum at output level $x = \frac{3}{2}$

Since Marginal cost M(x) is the 1st derivative of the total cost function C (x), hence

$$M(x) = \frac{dc(x)}{dx} = \frac{d}{dx}(x^3 - 3x^2 + 15x + 9)$$

$$= 3x^2 - 6x + 15$$

Now finding the 1st order derivative i.e.

$$M'(x) = 6x - 6$$

Using 1st order condition for maxima and minima, we get,

$$M'(x) = 0 \text{ or } 6x - 6 = 0 \text{ or } x = 1.$$

Now finding the 2nd order derivative

$M''(x) = 6 > 0$ a positive value thus the marginal cost attains its minimum value at $x = 1$.

Illustration No. 2

A manufacturer of ball bearing with an average cost of production $AC = 7 + \frac{x}{14}$ and a demand function $x = 52 - p$ where p the price per unit and x is the number of units produced. Find the most profitable output and the maximum profit?

Solution:

The average cost of production (AC) $= 7 + \frac{x}{14}$

And since $$AC = \frac{\text{Total Cost (TC)}}{\text{No. of units produced}}$$

Therefore, $$TC = x \times AC.$$

$$= x \times \left(7 + \frac{x}{14}\right)$$

$$= 7x + \frac{x^2}{14}$$

But

Profit (π) = Total Revenue (TR) – Total Cost (TC)

Since Total Revenue (TR) = Price × Quantity Sold.

$= p \times x$ {since $x = 52 - p$ or $p = 52 - x$}

$= (52 - x)\,x = 52x - x^2$.

Therefore

$$\text{Profit } (\pi) = [52x - x^2] - \left[7x + \frac{x^2}{14}\right]$$

$$= 45x - \frac{15x^2}{14}$$

For maxima and minima condition find $\frac{d\pi}{dx} = 0$

$$\frac{d\pi}{dx} = 45 - \{30x / 14\} = 0 \text{ or } x = 21 \text{ units.}$$

To check maximizing condition

i.e. $\frac{d^2\pi}{dx^2} < 0,$

But

$$\frac{d^2\pi}{dx^2} = -\frac{15}{7} < 0 \text{, a negative value.}$$

Therefore the profit (π) is maximized at x = 21 units and the maximum profit is

$$\text{profit } (\pi) = 45 \times 21 - \{15 \times \frac{21 \times 21}{14}\}.$$

$$= \text{Rs.}472.50.$$

Illustration No. 3

A TV manufacturer with a demand function as p = 114 – 3x and total cost function as TC (x) = 210 + 30x + x^2? Find the output that he should produce in order to maximize his profit?

Solution:

Since the total cost function

TC(x) = 210 + 30x + x^2

and the demand function p = 114 – 3x

Since Profit (π) = Total Rev – Total Cost = TR – TC

But Total Revenue (TR) = Price (p) × Qty. (x)

$$= (114 - 3x) \times x$$

$$= 114x - 3x^2$$

Therefore,

$$\text{Profit } (\pi) = (114x - 3x^2) - (210 + 30x + x^2)$$

$$= 114\,x - 3x^2 - 210 - 30x - x^2.$$

$$= 96x - 4x^2 - 210.$$

Differentiate the profit function with respect to 'x', we get

$$\frac{d\pi}{dx} = 96 - 8x$$

Now for maxima or minima condition, we equate $\frac{d\pi}{dx} = 0$.

We get,

$$96 - 8x = 0 \text{ or } x = 12$$

Now find the 2nd derivative of the profit function,

We get,

$$\left(\frac{d^2\pi}{dx^2}\right) = -8 < 0$$

Hence the profit is maximized at output level of 12 units.

Illustration No. 4

A travel and tour operator charge Rs 50 as ticket charge per passenger and gives Rs 2 as discount for passenger over and above 50. Find the number of passenger which will maximize his revenue?

Solution:

Let 'x' be the number of passengers who avail the operator services

Since the operator will give a discount of Rs. 2 for passengers over and above 50

Therefore the ticket charged by the operator per passenger for x > 50 is

$$y = 50 - \frac{2}{5}(x - 50)$$

Explanation: A passenger received Rs. $\frac{2}{5}$ of the discount amount per ticket if the number exceeds 50 (x > 50). If the number of tickets sold is 'x' then the total discount received is 2/5 (x – 50). And hence the price charged per ticket after discount 50 – 2/5 (x – 50)

Therefore the revenue generated by ticket sales is

Revenue = price × No. of tickets Sold.

Revenue (R) = p × x.

$$= [50 - \frac{2}{5}(x - 50)] \times x.$$

$$= 50x - \frac{2}{5}x^2 + 50x$$

$$= 100x - \frac{2}{5}x^2$$

For Maximum Revenue – find the 1st derivative

$$\frac{dR}{dx} = 100 - \frac{4}{5}x$$

Equating $\frac{dR}{dx} = 0$,

We get, $100 - \frac{4}{5}x = 0$ or $x = 125$.

Now finding the 2nd derivative, we have

$$\left(\frac{d^2\pi}{dx^2}\right) = -\frac{4}{5} < 0$$

And hence the revenue is maximized at x = 125 passenger and the maximum revenue is Rs. 2500.

Illustration No. 5

A manufacturer produces ball bearings at the rate of x units per week at a total cost of Rs = $\frac{x^3}{3} - 4x^2 + 7x + 9$. Find the output level at which the marginal cost is minimized and also find at what level the AVC is minimized?

Solution:

Since, we know

$$T(x) = \frac{x^3}{3} - 4x^2 + 7x + 9$$

As Marginal cost

$$\{MC(x)\} = \frac{d}{dx}\{TC\ (x)\}$$

$$= \frac{d}{dx}\{\frac{x^3}{3} - 4x^2 + 7x + 9\}$$

$$= x^2 - 8x + 7$$

For Maxima or Minima Condition, we equate MC(x) = 0

$$x^2 - 8x + 7 = 0 \text{ or } x = 1, 7$$

Now, finding the 2nd derivative, we get,

$$\frac{d}{dx}\{MC(x)\} = 2x - 8$$

Now at x = 1, we have

$$\frac{d}{dx}\{MC(x)\} = 2.1 - 8 = -6 < 0$$

Therefore at x = 1 the marginal cost is maximized

Now let us find $\frac{d}{dx}\{MC(x)\}$ at x = 7

$$\frac{d}{dx}\{MC(x)\} = 2.7 - 8 = 6 > 0$$

Therefore at x = 7 the marginal cost is minimized

For the 2nd part as the variable cost component in the total cost function is

Variable Cost $VC(x) = \frac{x^3}{3} - 4x^2 + 7x$

And Average Variable Cost

$$= \frac{\text{Variable cost}}{\text{No. of Units Produced}} = \frac{VC(x)}{x}$$

$$AVC(x) = \frac{\frac{x^3}{3} - 4x^2 + 7x}{x} = \frac{x^2}{3} - 4x + 7$$

Now finding the 1st derivative of AVC(x), we get,

$$\frac{d}{dx}\{AVC(x)\} = \frac{2x}{3} - 4$$

For Maxima or Minima Condition equating $\frac{d}{dx}\{AVC(x)\} = 0$

$$\frac{2x}{3} - 4 = 0 \text{ or } x = 6$$

Now, finding the 2nd derivative and substituting x = 6, we get,

$$\left(\frac{d^2\{AVC(x)\}}{dx^2}\right) = \frac{2}{3} > 0 \text{ a positive number}$$

And hence, the AVC is minimized at an output level of 6 units.

Illustration No. 6

If the total revenue and total cost function of a firm are TC = $35x - x^2$ and TC = 7 + 3x, were x is the output produced. Then find the equilibrium output level of the firm?

Solution:

Since Total Rev = TR (x) = 35 x – x^2 and Total Cost = TC (x) = 7 + 3x

Then Marginal Revenue MR = $\frac{d}{dx}\{TR(x)\} = 35 - 2x.$

and MC = $\frac{d}{dx}\{TC(x)\} = 3$

But the Equilibrium output level is achieved at a point where MR = MC

And hence,

$$35 - 2x = 3$$

$$32 = 2x \text{ or } x = 16$$

Therefore at 16 units equilibrium is reached.

Illustration No. 7

A Firm producing widgets has following cost of manufacture of Rs 900 as overhead cost, raw material cost is Rs. 5 per widgets and labor cost is $x^2/9$. Find at what level is the average cost is minimum?

Solution:

Total Cost function can be written as,

$TC(x) = 900 + 5x + \frac{x^2}{9}$ where x is the number of units produced.

Since $AC(x) = \frac{TC(x)}{x}$ as Average Cost $= \frac{\text{Total Cost}}{\text{No. of units Produced}}$

$$AC(x) = \frac{900 + 5x + \frac{x^2}{9}}{x}$$

$$= \frac{900}{x} + 5 + \frac{x}{9}$$

Finding the 1st order derivative, we get,

$$\frac{d}{dx}\{AC(x)\} = -\frac{900}{x^2} + \frac{1}{9}$$

For Maxima or Minima condition, we equate $\frac{d}{dx}\{AC(x)\} = 0$

$$-\frac{900}{x^2} + \frac{1}{9} = 0 \text{ or } x = 90$$

Now finding the 2nd derivative and substituting the value of x = 90, we get

$$\left(\frac{d^2\{AC(x)\}}{dx^2}\right) = \frac{1800}{x^3} = \frac{1800}{(90)^3} > 0$$

a positive number

Hence the AC is minimized at level of output of 90 units.

Illustration No. 8

A demand function of a firm is p = 225 – 3x and the cost function is c = 15x + 4000 (where p is the price and c the cost). Find the profit maximizing level of output and also find the price charged?

Solution:

Since the profit maximizing output condition is MC = MR

$$MC(x) = \frac{d}{dx}\{TC(x)\} = \frac{d}{dx}\{15x + 4000\} = 15 \qquad \ldots..1$$

But Total Revenue = (Price Charged × No. units produced)

$TR(x) = p \times x = x \times (225 - 3x).$

$= 225x - 3x^2.$

But Marginal Revenue (MR) = $\frac{d}{dx}\{TR(x)\}$

Therefore differentiating w.r.t 'x', we get

$$MR(x) = \frac{d}{dx}\{225x - 3x^2\} = 225 - 6x \quad .....2)$$

Equating (1) & (2) for profit maximizing condition, i.e. MC = MR

$$225 - 6x = 15$$

$$225 - 15 = 6x \text{ or } x = 35$$

Since the profit maximized at output of x = 35 and then the price charged at x = 35 is

$$P = 225 - 3 \times 35$$

$$= 225 - 105 = 120.$$

Thus at a price of Rs.120 the profit is maximized.

Illustration No. 9

If the production function of a automobile manufacturer is $P(k) = 30k + \frac{k^2}{2} - \frac{k^3}{3}$ where P stands for total output generated by 'k' units of capital utilized then

1. Find the number of units of capital required to maximize output?
2. Also find the maximum value of Marginal productivity of capital?

Solution:

Given the production function as

$$P(k) = 30k + \frac{k^2}{2} - \frac{k^3}{3}$$

Differentiating the production function w.r.t 'k', we get,

$$\frac{d}{dx}\{P(k)\} = 30 + k - k^2$$

In order to check the maxima and minima condition, we equate $\frac{d}{dx}\{P(k)\} = 0$

Therefore,

$$30 + k - k^2 = 0$$

$$30\,k + 6k - 5k - k^2 = 0$$

$$6(5 + k) - k\,(5 + k) = 0$$

$$(6 - k)\ (5 + k) = 0$$

or $$k = 6 \text{ or } - 5$$

Since there cannot be negative units of capital, therefore k = 6

Now finding the 2nd order derivative of the production function, we have,

$$\frac{d^2}{dx^2}\{P(k)\} = 1 - 2k$$

Substituting the value of k, we get

$$\frac{d^2}{dx^2}\{P(k)\} = 1 - 2 \times 6 = -11 < 0$$

Hence, the output is maximized at 6 units of capital.

Since $$MP_k = \frac{d}{dx}\{P(k)\} = 30 + k - k^2$$

Differentiating w.r.t 'k' to obtain the 1st order derivative, we get,

$$\frac{d}{dx}\{MP_k\} = 1 - 2k$$

For maxima and minima condition equating $\frac{d}{dx}\{MP_k\} = 0$, we get,

$$1 - 2k = 0 \text{ or } k = \frac{1}{2} \text{ units.}$$

Now finding the 2nd derivative, we get,

$$\frac{d^2}{dx^2}\{MP_k\} = -2 < 0 \text{ which is negative number}$$

And hence MP_k is maximum at $k = \frac{1}{2}$ units.

Effect of Taxation and Subsidy on Profitability

Let us understand the cumulative effect of taxes and subsidies received on the profitability of the firm

Assume that C(x) be the cost of production, were 'x' denotes the no. of units produced.

And p is the price charged per unit and price is determined by a demand function. In other words price is a function of demand i.e.

$$p = f(x).$$

Since Revenue (R) = Price charged (p) × No. of Units Sold (x).

$$R = p \times x.$$

Therefore, Profit (π) = Revenue (R) – Cost (C)

$$= (p \times x) - \{C(x)\}$$

Effect of Taxation

Let us assume tax at the rate of 't' is imposed per unit of output. Therefore the cost is increased by a factor of t × x.

Now the Total cost after tax imposition $C_t(x) = C(x) + t \times x$

But the profit maximization conditions is MC = MR

In order to determine Profit with taxation

$$= \text{Revenue (R)} - \text{Tax Adjusted Cost } \{C_t(x)\}$$

$$\text{Profit } (\pi) = R(x) - \{C_t(x)\}$$

$$= R(x) - \{C(x) + t \times x\}$$

For Maxima and Minima condition finding $\frac{d}{dx}\{\pi\}$

$$= \frac{d}{dx}\{R(x)\} - \frac{d}{dx}\{C_t(x)\}$$

$$= \frac{d}{dx}\{R(x)\} - \frac{d}{dx}\{C(x) + t\,x\}$$

$$= \frac{d}{dx}\{R(x)\} - \frac{d}{dx}\{C(x)\} - t$$

Equating $\frac{d}{dx}\{\pi\} = 0$

we get,

$$MR - MC - t = 0 \qquad \{\text{Since } MR = \frac{d}{dx}\{R(x)\},\ MC = \frac{d}{dx}\{C(x)\}$$

or $MR = MC + t$.

and necessary condition for maximization is

$$\frac{d^2}{dx^2}\{\pi\} < 0,\ \frac{d}{dx}\{MR\} - \frac{d}{dx}\{MC\} < 0$$

Hence the tax collected T = tx. Similarly the maximum value can be calculated by finding the 1st and 2nd derivative of 'T'.

it observed that MR reduces by an amount of 't'

Effect of Subsidy

In order to understand the effect of subsidy, consider subsidy as a negative tax thus the total cost can be written as

$$TC = \{C_t(x)\} = \{C(x) - t\,x\}$$

But

$$\text{Profit } (\pi) = R(x) - \{C_t(x)\}$$
$$= R(x) - \{C(x) - t\,x\}$$

For Maxima and Minima condition finding $\frac{d}{dx}\{\pi\}$

$$= \frac{d}{dx}\{R(x)\} - \frac{d}{dx}\{C_t(x)\}$$

$$= \frac{d}{dx}\{R(x)\} - \frac{d}{dx}\{C(x) - t\,x\}$$

$$= \frac{d}{dx}\{R(x)\} - \frac{d}{dx}\{C(x) + t$$

Equating $\frac{d}{dx}\{\pi\} = 0$

we get,

$MR - MC + t = 0$ {Since MR = d/ dx {R(x)}, MC = d/dx {C(x)}

or $MR = MC - t$, therefore MR requires by an amount of t

Illustration No. 10

A monopolist with cost function as C (x) = ax + b and demand function of the form p = c – dx where a, b, c and d are all positive numbers. If the government has imposed a tax at the rate of Rs. t per unit of output, find out the maximum tax paid by the monopolist also the output at which profit is maximized?

Solution:

Given $C(x) = ax + b$

But a tax is imposed on the output per unit at the rate of Rs. t.

Therefore the total cost $C(x) = ax + b + tx$

i.e. $C(x) = (a + t)x + b$

$$\text{But Revenue} = \text{Price} \times \text{Quantity Sold} = p \times x$$

$$= x \times (c - dx)$$

$$= cx - dx^2.$$

As profit (π) is defined as difference of revenue and cost.

$$\text{Profit } (\pi) = \text{Total Rev} - \text{Total Cost}$$

$$= \{cx - dx^2\} - \{(a + t)x + b\}$$

$$= \{(c - a - t)x - dx^2 - b\}$$

Using Maxima and Minima Rule

We have

$$\frac{d^2}{dx^2}(\pi) = (c - a - t) - 2dx \qquad \ldots\ldots \text{(i)}$$

Equating $\frac{d}{dx}(\pi) = 0$

We get $(c - a - t)x - 2dx = 0$

so, $x = \frac{(c - a - t)}{2d}$

By calculating the 2nd derivative of equation (i) we have

$$\frac{d^2}{dx^2}(\pi) = -2d < 0 \text{ as d is positive number}$$

So the profit is maximized at an output level of

$$x = \frac{(c-a-t)}{2d}$$

Now in order to determine the Tax on the monopolist which is depended on the output level therefore?

Tax paid (T) = t x

$$= \frac{t\,(c-a-t)}{2d}$$

$$= \frac{tc - ta - t^2}{2d}$$

Diff w.r.t 't', we get,

$$\frac{dT}{dt} = \frac{c-a-2t}{2d}$$

Equating $\frac{dT}{dt} = 0$ for maximum and minimum condition, we have,

$$\frac{c-a-2t}{2d} = 0 \text{ or } t = \frac{c-a}{2}$$

Now finding the 2nd order derivative, we get,

$$\frac{d^2T}{dt^2} = -\frac{2}{2d} = \frac{1}{d} < 0$$

Hence the tax paid is maximized at $t = \frac{c-a}{2}$ and the maximum tax paid is $T = t\,x = \frac{c-a}{2} \times \frac{c-a}{4d}$

$\{x = \frac{c-a-t}{2d}$ but as $t = \frac{c-a}{2}$ substituting t in x we get $x = \frac{c-a}{4d}$

$$T = tx.$$

$$= \left(\frac{c-a}{2}\right)\left(\frac{c-a}{4d}\right)$$

$$= \frac{(c-a)^2}{8d}$$

Therefore the tax revenue is maximized at an output level of $x = \frac{c-a}{4d}$.

Illustration No: 11

Given the demand function and cost function of a suitcase manufacturer are p = 30 – 5x and c = 10x. then

(i) Find the optimizing quantity and price?

(ii) Calculate the profit

(iii) Also find the rate of tax which will maximize tax revenues?

Solution:

(i) Since the demand function is p = 30 – 5x and cost function is C(x) = 10x

But total revenue TR(x) = px

$= (30 - 5x)x$

$= 30x - 5x^2$

Now, diff. with respect to x, we get,

$$MR = \frac{d}{dx}\{TR(x)\} = 30 - 10x \qquad (i)$$

But Cost Function C(x) = 10x, therefore

$$MC(x) = \frac{d}{dx}\{C(x)\} = 10.$$

But we know that profit maximum is condition is in Marginal Revenue = Marginal Cost i.e. MC =MR.

Therefore

$10 = 30 - 10x$

$10x = 20$ or $x = 2$.

And hence the price charged is $p = 30 - 5x = 30 - 5\times 2 = 20$.

(ii) Let 't' be the tax imposed per unit of output

Now the new cost function $C_N(x) = 10x + tx$

$$MC(x) = \frac{d}{dx}\{C(x)\} = 10 + t$$

But we know, MR = MC (Profit to be maximized)

Therefore,

$$10 + t = 30 - 10x$$

Or $$x = \frac{20 - t}{10}$$

(iii) But Tax Revenue T (t) = tx

$$= t\,x = \frac{20 - t}{10}$$

$$= 2t - \frac{t^2}{10}$$

Differentiating w.r.t to find maximum or minimum condition

$$\frac{d}{dt}\{T(t)\} = 2 - \frac{2t}{10}$$

$$= 2 - \frac{t}{5}$$

Now equate $\frac{d}{dt}$ {T(t)} = 0 for 1st order condition.

Therefore $2 - \frac{t}{5} = 0$ or t = 10

Now finding the 2nd derivative we get,

$$\frac{d^2T}{dt^2} = -\frac{1}{5} < 0 \text{ a negative number}$$

And hence tax revenue is maximized at t = 10

And the tax revenue is = $T(x) = 10 \times \left(\frac{20 - 10}{10}\right) = 10$

Illustration No.12

If the cost function of a manufacturing firm is $C(x) = 600x - 20x^2 + 2x^3/3$ where C (x) stands for the total cost function and x the output. Find the output level at which the cost is minimized?

Solution:

Given

$$C(x) = 600\,x - 20\,x^2 + \frac{2x^3}{3}$$

But,

$$MC(x) = \frac{d}{dx}\{C(x)\} = 600 - 40x + 2x^2.$$

In order to determine the minimum marginal cost we differentiate the marginal cost function MC(x).

$$\frac{d}{dx}\{MC(x)\} = -40 + 4x.$$

Now equating $\frac{d}{dx}\{MC(x)\} = 0$, we get

$$4x - 40 = 0 \text{ or}$$

$$x = 10.$$

Finding 2nd order derivative of the Marginal cost function, we have

$$\frac{d^2}{dt^2}\{MC(x)\} = 4 > 0$$

Therefore the marginal cost is minimized at 10 units.

Summary

- If $y = f(x)$ is a function of x in a given interval say $a \le x \le b$ if x increases resulting in increase in y then the function is said to be a increasing function i.e. if $a \le x_1 < x_2 \le b$ then $f(x_2) > f(x_1)$.
- If $y = f(x)$ is a function of x in a given interval say $a \le x \le b$ if x increases resulting in decrease in y then the function is said to be a decreasing function i.e. if $a \le x_1 < x_2 \le b$ then $f(x_2) < f(x_1)$.
- A function $y = f(x)$ is said to be relative maximum or simply maximum at $x = a$ if f(a) is the greatest of all the values of f(x) in the vicinity of a i.e. we can find a positive number δ such that $f(a) > f(x)$ for all x in $a - \delta < x < a + \delta$ $(x \ne a)$
- A function $y = f(x)$ is said to be relative minimum or simply minimum at $x = a'$ if f(a') is the least of all the values of f(x) in the vicinity of a' i.e. we can find a positive number δ such that $f(a') < f(x)$ for all x in $a' - \delta < x < a' + \delta$ $(x \ne a')$
- If f(x) is a function such that $f'(a) = 0$, then f(a) will be the maximum value of f(x) if its first derivative $\frac{dy}{dx}$ changes sign from positive to the left of a to negative to the right of a.

- Similarly, f(a) will be the minimum value of f(x) if its first derivative $\frac{dy}{dx}$ changes sign from negative to the left of a to positive to the right of a.

Self Practice Exercises

1. Find the maximum and minimum value of the function given by $2x^3 - 36x^2 + 192x + 148$?
2. Find the maximum and minimum value of the function $(1 - x)^2 e^x$?
3. Show that the function $x e^x$ has a minimum value at $x = -1$?
4. Show that the function $3x^3 - 9x^2 + 9x + 21$ has neither maximum nor minimum value at $x = 1$?
5. Show that the function $xy(y - x) = 16$ is having a minimum value at $x = 4$?
6. Find the maximum or minimum value of the function $9x + \frac{4}{x}$?
7. If $f(x) = |x|$ and $c = 0$, show that f(x) assume a maximum or a minimum value if $f'(c)$ does not exist?
8. Find the minimum value of the function $5^2e^{2x} + 7^2e^{2x}$?
9. If the function is defined as $y = \frac{7x}{7^2 + x^2}$ then show that the function is maximum at $x = 7$ and minimum at $x = -7$?
10. Show that the function $2x + \frac{2}{x}$ has only one maximum value and one minimum value?
11. Show that the curve $y = x^2(6 - x)$ has a point of inflexion at $x = 2$?
12. Show that the curve $y = 3x^2$ is concave upward?
13. Show that the curve $y = 2x^3 - 12x^2 + 24x - 10$ has a point of inflexion at $x = 2$?
14. Show that the curve $y = 9 - 5x - 2x^2$ is concave downward?
15. A manufacturer produces x units per day incurring a cost of Rs.$(2x^3 - 12x^2 + 18x + 12)$. Find at what level the average variable cost and the marginal cost is at their respective minimum?
16. If the demand function of a firm is given as $p = \sqrt{27 - x}$, find at what output level at which total revenue is maximized and calculate the total revenue?

17. A demand function of a confectioner is $p = 75 - 5x$ and the cost function is $c = 25x + 500$ (where p is the price and c the cost). Find the profit maximizing level of output of the confectioner, the price charged and also the total revenue generated?

18. It was found that for a product of a manufacturer the relationship between tax applicable and the output level is given by $t = -3x + 18$. Find what level of output the tax revenue is maximized, the tax rate applicable and the maximum tax revenue generated?

19. If the quantity sold and the price of a firm are related as $q = 5e^{-3p}$ and the cost of production is $c = 2 + 4q$ then find the optimal price which maximizes the profit?

20. A copper wire manufacturer produces two types of copper wires Grade I of high quality and Grade II of low quality. If the manufacturer has have capacity to produce x tones of Grade I and y tones of Grade II copper wires where x and y are related as $y = \frac{80 - 10x}{20 - 2x}$ then show that in order to maximize revenue 5.5 tones of Grade I copper wire are produced where the market price of Grade I is half that of the Grade II copper wire?

21. A Firm producing chocolates has following cost of manufacture of Rs 1000 as overhead cost, raw material cost is Rs 4 per unit and labor cost is $x^2/10$. Find at what level is the AC is minimum?

22. A cycle manufacturer produces 10,000 cycles per year. The product is carried on in lots. If the cost incurred to make the production unit ready irrespective of the number of unit produced is Rs. 100 and the storage cost is Rs. ½ then find the economic order quantity of production? *(Hint Inventory Cost = Ordering Cost + Holding Cost)*

23. Hansung a LCD television manufacturer charges Rs.15,000 for a LCD television set on orders of 100 or less and the charge is reduced by Rs. 125 per set if the order size increases in excess of 100. Find the largest order size so that Hansung revenue is maximized?

24. Given the demand function & cost function of a manufacturer are $p = 30 - 3x$ and $c = 2x^2 + 10x$ then
    (i) Find the optimizing quantity and price that maximizes profit?
    (ii) What will happen to pricing if a tax of Rs. 2 per unit is imposed on the output?

Answers

1. 308, 276.
2. $\frac{4}{e}, 0$
6. – 27.
7. f(0) = 0 is the minimum value of f(x).
8. Minimum value = 70
10. – 4, 4
15. AVC is minimum at output level of 3 units and Marginal cost is minimum at output level of 2 units.
16. At output level of x = 18 units and Total Revenue (TR) = 54.
17. Price Charged = Rs. 25, Out Level = 10 units and Total Revenue = Rs. 250.
18. Out – level = 3 units, tax applicable = Rs 9 and Total Tax Revenue = Rs. 27.
19. Price = Rs. $\frac{13}{3}$
21. At output level of 100 units.
22. 2000 units.
23. 10.
24. (i) x = 2 and p= 24, (ii) x = 1.8 and p = 24.6.

CHAPTER 6

Partial Derivatives and its Application

Concept and Definition

Let f(x, y) be a function with two independent variables x and y then the ordinary derivative of f(x, y) with respect to x assuming y to be a constant, is called the partial derivative of f(x, y) with respect to x and is denoted by the symbol $\frac{\partial f(x,y)}{\partial x}$ or $f_x(x,y)$. In the same lines the partial derivative of f(x, y) can also be defined.

Thus we have,

$$\frac{\partial f(x,y)}{\partial x} = \lim_{h\to 0}\frac{f(x+h,y)-f(x,y)}{h}$$

provided the limit exists and h is the incremental value of x such that the new value of x is x + h and the corresponding value of f(x, y) = f(x + h, y) keeping y fixed.

Similarly

$$\frac{\partial f(x,y)}{\partial x} = \lim_{k\to 0}\frac{f(x,y+k)-f(x,y)}{k}$$

provided the limit exists and k is the incremental value of y such that the new value of y is y + k and the corresponding value of f(x, y) = f(x, y + k) keeping x fixed.

The Concept of Successive Partial Derivatives

If f(x, y) has partial derivatives with respect to two independent functions x and y denoted by $\frac{\partial f(x,y)}{\partial x}$ or $f_x(x,y)$ and $\frac{\partial f(x,y)}{\partial y}$ or $f_y(x,y)$ then each of these partial derivatives may have partial derivatives with respect to x and y called the second order partial derivatives of f(x, y) and are denoted by

$$\frac{\partial}{\partial x}\left(\frac{\partial f(x,y)}{\partial x}\right)=\frac{\partial^2 f(x,y)}{\partial x^2}=f_{xx}(x,y)$$

$$\frac{\partial}{\partial y}\left(\frac{\partial f(x,y)}{\partial y}\right)=\frac{\partial^2 f(x,y)}{\partial y^2}=f_{yy}(x,y)$$

The Concept of Cross Partial Derivatives

If f(x, y) has partial derivatives with respect to two independent functions x and y then the second order partial derivatives $f_{xy}(x,y)$ or $f_{yx}(x,y)$ are called as the cross partial derivatives and represented by $\frac{\partial^2 y(x,y)}{\partial x\partial y}=f_{xy}(x,y)$ or $\frac{\partial^2 y(x,y)}{\partial x\partial y}=f_{yx}(x,y)$ which means that they are only different in the order in which differentiated partially.

$\frac{\partial^2 f(x,y)}{\partial x\partial y}=f_{xy}(x,y)$ shows that the given function f(x, y) is first partially differentiated with respect to y and then with respect to x i.e. $\frac{\partial}{\partial x}\left(\frac{\partial f(x,y)}{\partial y}\right)$.

Similarly $\frac{\partial^2 f(x,y)}{\partial y\partial x}=f_{yx}(x,y)$ shows that the given function f(x,y) is first partially differentiated with respect to x and then with respect to y i.e. $\frac{\partial}{\partial y}\left(\frac{\partial f(x,y)}{\partial x}\right)$.

Under certain conditions the cross partial derivatives are identical in value i.e

$$\frac{\partial}{\partial y}\left(\frac{\partial f(x,y)}{\partial x}\right)=\frac{\partial}{\partial x}\left(\frac{\partial f(x,y)}{\partial y}\right) \text{ i.e. } f_{yx}(x,y)=f_{xy}(x,y)$$

Illustration No 1

If f(x, y) = $x^2 + y^2 + 2xy$ then find $\frac{\partial f(x,y)}{\partial x},\frac{\partial f(x,y)}{\partial y},\frac{\partial^2 f(x,y)}{\partial x^2},\frac{\partial^2 f(x,y)}{\partial y^2}$ and $\frac{\partial^2 f(x,y)}{\partial y\partial x}$.

Solution:

Since we have $f(x, y) = x^2 + y^2 + 2xy$

Therefore,

$$\frac{\partial f(x,y)}{\partial x} = 2x + 0 + 2y$$

$$= 2(x + y)$$

$$\frac{\partial^2 f(x,y)}{\partial x^2} = \frac{\partial}{\partial x}\left(\frac{\partial f(x,y)}{\partial x}\right) = \frac{\partial}{\partial x}(2x + 2y)$$

$$= 2$$

Similarly

$$\frac{\partial f(x,y)}{\partial y} = 0 + 2y + 2x$$

$$= 2(x + y)$$

$$\frac{\partial^2 f(x,y)}{\partial x^2}$$

$$\frac{\partial^2 f(x,y)}{\partial y^2} = \frac{\partial}{\partial y}\left(\frac{\partial f(x,y)}{\partial y}\right) = \frac{\partial}{\partial y}(2y + 2x)$$

$$= 2$$

Finally,

$$\frac{\partial^2 f(x,y)}{\partial y \partial x} = \frac{\partial}{\partial y}\left(\frac{\partial f(x,y)}{\partial x}\right) = \frac{\partial}{\partial y}(2x + 2y) = 2.0 + 2$$

$$= 2$$

Illustration No. 2

If $f(x,y) = 2x^3 + 4x^2y + y^3$ then show that $x\frac{\partial f(x,y)}{\partial x} + y\frac{\partial f(x,y)}{\partial y} = 3f(x,y)$

Solution:

Given

$$f(x,y) = 2x^3 + 4x^2y + y^3$$

therefore

$$\frac{\partial f(x,y)}{\partial x} = 6x^2 + 8xy$$

Similarly $$\frac{\partial f(x,y)}{\partial y} = 4x^2 + 3y^2$$

And hence $$x\frac{\partial f(x,y)}{\partial x} = 6x^3 + 8x^2y \text{ and } y\frac{\partial f(x,y)}{\partial y} = 4x^2y + 3y^3$$

Therefore $$x\frac{\partial f(x,y)}{\partial x} = y\frac{\partial f(x,y)}{\partial y} = 6x^3 + 8x^2y + 4x^2y + 3y^3$$

$$= 6x^3 + 12x^2y + 3y^3$$

$$= 3(2x^3 + 4x^2y + y^3)$$

$$= 3\,f(x, y)$$

Illustration No.3

If $z = x^3 + 3x^2y + y^3$ then show that $\frac{\partial^2 z}{\partial y \partial x} = \frac{\partial^2 z}{\partial x \partial y}$

Solution:

Since $$z = x^3 + 3x^2y + y^3$$

Therefore $$\frac{\partial z}{\partial x} = 3x^2 + 6xy \text{ and } \frac{\partial^2 z}{\partial y \partial x} = \frac{\partial}{\partial y}\left(\frac{\partial z}{\partial x}\right) = \frac{\partial}{\partial y}(3x^2 + 6xy)$$

$$= 0 + 6x = 6x.$$

Similarly

$$\frac{\partial z}{\partial y} = 3y^2 + 3x^2 \text{ and } \frac{\partial^2 z}{\partial x \partial y} = \frac{\partial}{\partial x}\left(\frac{\partial z}{\partial y}\right) = \frac{\partial}{\partial x}(3y^2 + 3x^2)$$

$$= 0 + 6x = 6x.$$

Hence $$\frac{\partial^2 z}{\partial y \partial x} = \frac{\partial^2 z}{\partial x \partial y} = 6x$$

Illustration No. 4

If $f(x, y) = \log(x^2 + y^2)$, then prove that $\frac{\partial^2 f(x,y)}{\partial x^2} + \frac{\partial^2 f(x,y)}{\partial y^2} = 0$

Solution:

Given $f(x, y) = \log(x^2 + y^2)$

Therefore

$$\frac{\partial f(x,y)}{\partial x} = \frac{1}{x^2+y^2} \times (2x+0) = \frac{2x}{x^2+y^2}$$

$$\therefore \quad \frac{\partial^2 f(x,y)}{\partial x^2} = 2\left[\frac{(x^2+y^2).1 - x.2x}{(x^2+y^2)^2}\right] = 2\left[\frac{-x^2+y^2}{(x^2+y^2)^2}\right]$$

Similarly $$\frac{\partial f(x,y)}{\partial y} = \frac{1}{x^2+y^2} \times (2y+0) = \frac{2y}{x^2+y^2}$$

$$\therefore \quad \frac{\partial^2 f(x,y)}{\partial y^2} = 2\left[\frac{(x^2+y^2).1 - y.2y}{(x^2+y^2)^2}\right] = 2\left[\frac{x^2-y^2}{(x^2+y^2)^2}\right]$$

And hence,

$$\frac{\partial^2 f(x,y)}{\partial x^2} + \frac{\partial^2 f(x,y)}{\partial y^2} = 2\left\{\frac{-x^2+y^2}{(x^2+y^2)^2} + \frac{x^2-y^2}{(x^2+y^2)^2}\right\}$$

$$= 2\left\{\frac{-x^2+y^2+x^2-y^2}{(x^2+y^2)^2}\right\} = 0$$

Illustration No. 5

If $f(x,y) = x^3 + y^3 + 2x^2y$ then prove that $x\frac{\partial f(x,y)}{\partial x} + y\frac{\partial f(x,y)}{\partial y} = 3f(x,y)$

Solution:

Given

$$f(x, y) = x^3 + y^3 + 2x^2y$$

therefore $$\frac{\partial f(x,y)}{\partial x} = 3x^2 + 4xy$$

Similarly $$\frac{\partial f(x,y)}{\partial y} = 3y^2 + 2x^2$$

And hence $$x\frac{\partial f(x,y)}{\partial x} = 3x^3 + 4x^2y \text{ and } y\frac{\partial f(x,y)}{\partial y} = 2x^2y + 3y^3$$

Therefore $$x\frac{\partial f(x,y)}{\partial x} + y\frac{\partial f(x,y)}{\partial y} = 3x^3 + 4x^2y + 2x^2y + 3y^3$$

$$= 3x^3 + 6x^2y + 3y^3$$

$$= 3(x^3 + 2x^2y + y^3)$$

$$= 3\ f(x, y)$$

Illustration No. 6

If $f(x,y) = \frac{3x + 4y}{5y}$ then prove that $x\frac{\partial f(x,y)}{\partial x} + y\frac{\partial f(x,y)}{\partial y} = 0$

Solution:

Given $\quad f(x, y) = \frac{3x + 4y}{5y}$

therefore $\quad \frac{\partial f(x,y)}{\partial x} = \frac{1}{5y} \times 3 = \frac{3}{5y}$

and hence $\quad x\frac{\partial f(x,y)}{\partial x} = \frac{3}{5y} \times x = \frac{3x}{5y}$

$$\frac{\partial f(x,y)}{\partial y} = \frac{4(5y) - 5(3x + 4y)}{(5y^2)} = -\frac{15x}{(5y)^2} = -\frac{3x}{5y^2}$$

$\therefore \quad y\frac{\partial f(x,y)}{\partial y} = -\frac{3x}{5y^2} \times y = -\frac{3x}{5y}$

$\therefore \quad x\frac{\partial f(x,y)}{\partial x} + y\frac{\partial f(x,y)}{\partial y} = \frac{3x}{5y} - \frac{3x}{5y} = 0$

Total Differentials

The partial derivative techniques has measured the rate of change in f(x, y) a function with two variable x and y, when small change was made in x or in y keeping either of y (or x) as constant. But in case where there is simultaneous change in x and y, the corresponding change in the given function f(x,) is given by finding the total differentials

As we know that $\frac{\partial f(x,y)}{\partial x}$ is the partial derivative of f(x, y) with respect to x, explaining the change in f(x, y) when there is a small change in x holding y constant.

Let us assume that x changes by an amount of dx, then the change in f(x, y) will be $\left(\frac{\partial f(x,y)}{\partial x}dx\right)$

Similarly, $\frac{\partial f(x,y)}{\partial y}$ is the partial derivative of f(x,y) with respect to y, explaining the change in f(x,y) when there is a small change in y holding x constant.

And hence $\left(\frac{\partial f(x,y)}{\partial y}dy\right)$ will be change in f(x,y) with change in y by amount of dy.

Therefore the total change in f(x,y) due to small change in x and y equal to dx and dy is

$$df(x,y) = \left(\frac{\partial f(x,y)}{\partial x}dx\right) + \left(\frac{\partial f(x,y)}{\partial y}dy\right)$$

Also written as

$$df(x,y) = f_x dx + f_y dy$$

Illustration No.7

If $f(x, y) = 2x^3 + xy - 3y^2$ then find the total differential of f(x,y)?

Solution:

Since,

$$f(x, y) = 2x^3 + xy - 3y^2$$

therefore $\frac{\partial f(x,y)}{\partial x} = 6x^2 + y$

and $\frac{\partial f(x,y)}{\partial y} = y - 6y$

therefore

$$df(x,y) = \left(\frac{\partial f(x,y)}{\partial x}dx\right) + \left(\frac{\partial f(x,y)}{\partial y}dy\right)$$

$$= \left(6x^2 + y\right)dx + (y - 6y)dy$$

Illustration No.8

If $f(x,y) = \frac{y}{x+y}$ then find the total differential of f(x, y)?

Solution:

Given $f(x, y) = \frac{y}{x+y}$

Therefore $$\frac{\partial f(x,y)}{\partial x} = \frac{[0.(x+y)-1.(y)]}{(x+y)^2} = -\frac{y}{(x+y)^2}$$

And $$\frac{\partial f(x,y)}{\partial y} = \frac{1.(x+y)-1.(y)}{(x+y)^2} = \frac{x}{(x+y)^2}$$

therefore, the total differentials

$$df(x,y) = \left(\frac{\partial f(x,y)}{\partial x}dx\right) + \left(\frac{\partial f(x,y)}{\partial y}dy\right)$$

$$= \left(-\frac{y}{(x+y)^2}\right)dx + \left(\frac{x}{(x+y)^2}\right)dy$$

Illustration No.9

If $f(x,y) = y^x + x^y$ then find the total differential of f(x,y)?

Solution:

Given $f(x, y) = y^x + x^y$

And hence $$\frac{\partial f(x,y)}{\partial x} = y\ x^{y-1} + y^x \log y$$

Similarly $$\frac{\partial f(x,y)}{\partial y} = x^y \log x + xy^{x-1}$$

Therefore

$$df(x, y) = \left(\frac{\partial f(x,y)}{\partial x}dx\right) + \left(\frac{\partial f(x,y)}{\partial y}dy\right)$$

$$= (yx^{y-1} + y^x \log y)dx + (x^y \log x + xy^{x-1})dy$$

Illustration No. 10

If f(x, y) = $a^x e^y$ then find the total differential of f(x, y)?

Solution:

Since

$$f(x, y) = a^x e^y$$

$$\frac{\partial f(x,y)}{\partial x} = (a^x \log a)e^y$$

And $$\frac{\partial f(x,y)}{\partial x} = a^x e^y$$

$$\text{Therefore, } df(x,y) = \left(\frac{\partial f(x,y)}{\partial x}dx\right) + \left(\frac{\partial f(x,y)}{\partial y}dy\right)$$

$$= \left\{\left(a^x \log a\right)e^y\right\}dx + \left(a^x e^y\right)dy$$

Second Order Total Differential

Since we have seen that for a given function f(x,y) the total differential can be written in the form

$$df(x,y) = \left(\frac{\partial f(x,y)}{\partial x}dx\right) + \left(\frac{\partial f(x,y)}{\partial y}dy\right)$$

the right hand side of the expression of total differential is 1st order partial differentials

Now let us find out the second order total differential in terms of second order partial derivatives as

$$d\{df(x,y)\} = d\left\{\left(\frac{\partial f(x,y)}{\partial x}dx\right)\right\} + d\left\{\left(\frac{\partial f(x,y)}{\partial y}dy\right)\right\}$$

since $$d\left\{\left(\frac{\partial f(x,y)}{\partial y}dy\right)\right\} = d\left(\frac{\partial f(x,y)}{\partial y}dx\right) = d(f_x)dx = \left(\frac{\partial}{\partial x}f_x dx + \frac{\partial}{\partial y}f_x dy\right)dx$$

$$= (f_{xx}dx + f_{xy}dy)dx$$

Similarly

$$d\left\{\left(\frac{\partial f(x,y)}{\partial y}dy\right)\right\} = d\left(\frac{\partial f(x,y)}{\partial y}\right)dy = d(f_y)dy = \left(\frac{\partial}{\partial x}f_y dx + \frac{\partial}{\partial y}f_y dy\right)dy$$

$$= (f_{xy}dx + f_{yy}dy)dy$$

Thus

$$d\{d(f(x, y)\} = d^2f(x, y) = (f_{xx}dx + f_{xy}dy)\,dx + (f_{xy}dx + f_{yy}dy)dy$$

$$= f_{xx}(dx)^2 + f_{xy}dxdy + f_{xy}dxdy + f_{yy}(dy)^2$$

$$= f_{xx}(dx)^2 + 2f_{xy}dxdy + f_{yy}(dy)^2$$

$$= (f_x dx + f_y dy)^2$$

Thus an nth order total differential can be written in the same lines as

$$d^nf(x,y) = (f_x dx + f_y dy)^n$$

In case if the function has more than two variables the same rule will be applied i.e.

$$d^2f(x, y, z) = (f_x dx + f_y dy + f_z dz)^2$$

and the nth order total differential can be written as

$$d^nf(x, y, z) = (f_x dx + f_y dy + f_z dz)^n$$

Illustration No.11

If $f(x, y) = 2x^2 + 4xy + y^2$ then find the first and second order total differentials?

Solution:

Given $f(x, y) = 2x^2 + 4xy + y^2$

therefore

$$f_x = 4x + 4y$$

$$f_{xx} = 4$$

Similarly

$$f_y = 4x + 2y$$

$$f_{yy} = 2$$

We have,

$$f_{xy} = f_{yx} = 4$$

And hence the first order total differential

$$dz = f_x dx + f_y dy$$

$$= (4x + 4y)\ dx + (4x + 2y)\ dy$$

We know second order total differential

$$d^2f(x,y) = \left(f_x dx + f_y dy\right)^2$$

$$= f_{xx}(dx)^2 + 2f_{xy} dx\ dy + f_{yy}(dy)^2$$

$$= 4(dx)^2 + 2 \times 4(dx\ dy) + (dy)^2$$

$$= 4(dx)^2 + 8\,dx\ dy + 2(dy)^2$$

Illustration No.12

If f(x, y) = x {log (2 + y)} then find the first and second order total differentials?

Solution:

Given

$$f(x, y) = x\ \{\ \log\ (\ 2 + y\ )\}$$

and hence

$$f_x = \log(2 + y)$$

$$f_{xx} = 0$$

Similarly $$f_y = \frac{x}{2 + y}$$

$$f_{yy} = -\frac{x}{(2 + y)^2}$$

We have, $f_{xy} = f_{yx} = \frac{1}{2+y}$

And hence the first order total differential

$$dz = f_x dx + f_y dy$$

$$= \log(2+y)dx + \left(\frac{x}{2+y}\right)dy$$

We know second order total differential

$$d^2f(x,y) = (f_x dx + f_y dy)^2$$

$$= f_{xx}(dx)^2 + 2f_{xy}dxdy + f_{yy}(dy)^2$$

$$= 0.(dx)^2 + 2\left\{\frac{1}{2+y}\right\}dxdy + \left\{-\frac{x}{(2+y)^2}\right\}(dy)^2$$

$$= \left\{\frac{2}{2+y}\right\}dx\ dy + \left\{-\frac{x}{(2+y)^2}\right\}(dy)^2$$

Illustration No.13

If $f(x, y) = x\,e^y$ then find the second order total differentials?

Solution:

Given $f(x, y) = x\,e^y$

and hence $f_x = e^y$

$f_{xx} = 0$

Similarly $f_y = x\,e^y$

$f_{yy} = x\,e^y$

We have, $f_{xy} = f_{yx} = e^y$

Then the second order total differentials

$$d^2f(x,y) = (f_x dx + f_y dy)^2$$

$$= f_{xx}(dx)^2 + 2f_{xy}dx\ dy + f_{yy}(dy)^2$$

$$= 0.(dx)^2 + 2e^y(dx\ dy) + xe^y(dy)^2$$

$$= 2e^y(dx\ dy) + xe^y(dy)^2$$

Euler's Theorem

A function f (x, y) of two independent variables of x and y is said to be homogenous in x and y of degree n if it can be represented in the following form $x^n\phi\left(\frac{y}{x}\right)$ or $y^n\ \phi\left(\frac{x}{y}\right)$.

Alternatively

A function f(x, y) of two independent variable x any y is said to be homogenous in x and y of degree n if and only if

$$f(zx, zy) = z^n f(x, y)$$

Illustration No. 14

If $f(x,y) = \dfrac{x^2 + 2xy + y^2}{x + y}$ then express f(x, y) as homogenous function in x and y of degree one?

Solution:

Given

$$f(x,y) = \frac{x^2 + 2xy + y^2}{x + y} = \frac{x^2\left\{1 + 2\left(\frac{y}{x}\right) + \left(\frac{y}{x}\right)^2\right\}}{x\left(\frac{y}{x}\right)}$$

$$= \frac{x\left\{1 + 2\left(\frac{y}{x}\right) + \left(\frac{y}{x}\right)^2\right\}}{\left(\frac{y}{x}\right)} = x \times \phi\left(\frac{y}{x}\right) = x\phi\left(\frac{y}{x}\right)$$

Therefore f(x, y) is a homogenous function in x and y of degree one.

Illustration No.15

If f(x, y) $= \dfrac{x^{3/4} + y^{5/4}}{x^{1/2} + y^{1/2}}$ then prove that f(x, y) as homogenous function in x and y of degree 3/2?

Solution:

Since

$$f(x,y)=\frac{x^{5/4}+y^{5/4}}{x^{1/2}+y^{1/2}}=\frac{x^{\frac{5}{4}}}{x^{\frac{1}{2}}}\left\{\frac{1+\left(\frac{y}{x}\right)^{5/4}}{1+\left(\frac{y}{x}\right)^{1/2}}\right\}$$

$$=x^{3/2}\left\{\frac{1+\left(\frac{y}{x}\right)^{5/4}}{1+\left(\frac{y}{x}\right)^{1/2}}\right\}$$

$$=x^{3/2}\times\text{a function of}\left\{\frac{y}{x}\right\}=x^{3/2}\phi\left(\frac{y}{x}\right)$$

Hence f(x, y) as homogenous function in x and y of degree 3/2

Euler's Theorem on Homogenous Functions of Two Variables

If f(x, y) be a homogenous function of x and y of degree n having partial derivatives $\frac{\partial f}{\partial x}$ and $\frac{\partial f}{\partial y}$, then

$$x\frac{\partial f}{\partial y}+y\frac{\partial f}{\partial y}=nf(x,y).$$

Proof:

Since f(x, y) is a homogenous function of degree n, we can write

$$f(x,y)=x^n z\left(\frac{y}{x}\right)$$

therefore

$$\frac{\partial f}{\partial x}=nx^{n-1}z\left(\frac{y}{x}\right)+x^n z'\left(\frac{y}{x}\right)\frac{\partial}{\partial x}\left(\frac{y}{x}\right)$$

where $z'\left(\frac{y}{x}\right)$ is the derivative of $z\left(\frac{y}{x}\right)$

$$\frac{\partial f}{\partial x}=nx^{n-1}z\left(\frac{y}{x}\right)+x^n z'\left(\frac{y}{x}\right)\times\left(-\frac{y}{x^2}\right)$$

$$x\frac{\partial f}{\partial x} = nx^n z\left(\frac{y}{x}\right) + x^n z'\left(\frac{y}{x}\right) \times \left(-\frac{y}{x}\right)$$

i.e.
$$x\frac{\partial f}{\partial x} = nx^n z\left(\frac{y}{x}\right) - x^n z'\left(\frac{y}{x}\right) \times \left(\frac{y}{x}\right) \qquad .....(1)$$

Similarly

$$\frac{\partial f}{\partial y} = x^n z'\left(\frac{y}{x}\right)\frac{\partial}{\partial y}\left(\frac{y}{x}\right)$$

$$\frac{\partial f}{\partial y} = x^n z'\left(\frac{y}{x}\right)\left(\frac{1}{x}\right)$$

$$y\frac{\partial f}{\partial y} = x^n z'\left(\frac{y}{x}\right)\left(\frac{y}{x}\right) \qquad .....(2)$$

Adding eq.(1) and (2)

We get,

$$x\frac{\partial f}{\partial x} + y\frac{\partial f}{\partial y} = n\ f(x,y)$$

Hence proved

Note: *The theorem can be extended to a homogenous function of three or more variables. For e.g. if f(x, y, z) be a homogenous function of degree n, then*

$$x\frac{\partial f}{\partial x} + y\frac{\partial f}{\partial y} + z\frac{\partial f}{\partial x} = n\ f(x,y)$$

Illustration No. 16

Verify Euler's theorem for the function $f(x , y) = x^4 + 2x^3y + y^4$

Solution:

Since

$$f(x, y) = x^4 + 2x^3y + y^4 = x^4 \left\{1 + 2\left(\frac{y}{x}\right) + \left(\frac{y}{x}\right)^4\right\} = x^4 z\left(\frac{y}{x}\right)$$

therefore f(x, y) is a homogenous function of degree 4

In order to verify

$$f(x, y) = x^4 + 2x^3y + y^4$$

therefore $$\frac{\partial f(x,y)}{\partial x} = 4x^3 + 6x^2y$$

Similarly $$\frac{\partial f(x,y)}{\partial y} = 4y^3 + 2x^3$$

And hence $$x\frac{\partial f(x,y)}{\partial x} = 4x^4 + 6x^3y \text{ and } y\frac{\partial f(x,y)}{\partial y} = 2x^3y + 4y^4$$

Therefore $$x\frac{\partial f(x,y)}{\partial x} + y\frac{\partial f(x,y)}{\partial y} = 4x^4 + 6x^3y + 2x^3y + 4y^4$$

$$= 4x^4 + 8x^3y + 4y^4$$

$$= 4(x^4 + 2x^3y + y^4)$$

$$= 4f(x,y)$$

Hence Euler's theorem is verified

Illustration No.17

If $f(x, y, z) = x^3 + y^3 + z^3 + 4xyz$, then prove that $x\frac{\partial f}{\partial x} + y\frac{\partial f}{\partial y} + z\frac{\partial f}{\partial z} = 3f(x,y)$

Solution:

We have

$$f(x, y, z) = x^3 + y^3 + z^3 + 4xyz$$

therefore $$\frac{\partial f(x,y,z)}{\partial x} = 3x^2 + 4yz$$

$$\frac{\partial f(x,y,z)}{\partial y} = 3y^2 + 4xz$$

And $$\frac{\partial f(x,y,z)}{\partial z} = 3z^2 + 4xy$$

And hence

$$x\frac{\partial f(x,y)}{\partial x} = 3x^3 + 4xyz, y\frac{\partial f(x,y)}{\partial y} = 3y^3 = 4xyz \text{ and } z\frac{\partial f(x,y,z)}{\partial z} + 3z^3 + 4xyz$$

Therefore

$$x\frac{\partial f(x,y)}{\partial x}+y\frac{\partial f(x,y)}{\partial y}+z\frac{\partial f(x,y,z)}{\partial z}=3x^3+4xyz+3y^3+4xyz+3z^3+4xyz$$

$$=3x^3+3y^3+3z^3+12xyz$$

$$=3\{x^3+y^3+z^3+4xyz\}$$

$$=3\ f(x,\ y).$$

Hence Euler's theorem is verified

Illustration No.18

If $f(x,y)=\log\frac{x^3-y^3}{x^2-y^2}$ prove by Euler's theorem that $x\frac{\partial f(x,y)}{\partial x}+y\frac{\partial f(x,y)}{\partial y}=1$

Solution:

We have,

$$f(x,y)=\log\frac{x^3-y^3}{x^2-y^2};\ \text{therefore } e^{f(x,y)}=\frac{x^3-y^3}{x^2-y^2}$$

$$\text{Let } z=e^{f(x,y)}\rightarrow z=\frac{x^3-y^3}{x^2-y^2}=\frac{x^3}{x^2}\times\frac{\left(1-\frac{y^3}{x^3}\right)}{\left(1-\frac{y^2}{x^2}\right)}=x\left\{\frac{\left(1-\frac{y^3}{x^3}\right)}{\left(1-\frac{y^2}{x^2}\right)}\right\}$$

$\therefore$ z is a homogenous function in x, y of degree 1

Hence by Euler's Theorem

$$x\frac{\partial z}{\partial x}+y\frac{\partial z}{\partial y}=1.z=z$$

As $z=e^{f(x,y)}$

$$\frac{\partial z}{\partial x}=e^{f(x,y)}\frac{\partial f(x,y)}{\partial x}\ \text{and}\ \frac{\partial z}{\partial y}=e^{f(x,y)}\frac{\partial f(x,y)}{\partial y}$$

Therefore substituting the value of $\frac{\partial z}{\partial x}$ and $\frac{\partial z}{\partial y}$ in the above equation

we get

$$xe^{f(x,y)}\frac{\partial f(x,y)}{\partial x}+ye^{f(x,y)}\frac{\partial f(x,y)}{\partial y}=e^{f(x,y)}$$

$$e^{f(x,y)}\left\{x\frac{\partial f(x,y)}{\partial x}+y\frac{\partial f(x,y)}{\partial y}\right\}=e^{f(x,y)}$$

$$=x\frac{\partial f(x,y)}{\partial x}+y\frac{\partial f(x,y)}{\partial y}=1$$

Hence proved

Illustration No.19

If f(x, y)= g(x – ay) + h (x + ay) where g and h are any two functions and a is a constant then show that

$$a^2\frac{\partial^2 f(x,y)}{\partial x^2}=\frac{\partial^2 f(x,y)}{\partial y^2}$$

Solution:

Since we have

$$f(x, y)= g(x - ay) + h (x + ay)$$

therefore

$$\frac{\partial f(x,y)}{\partial x}=g'(x-ay)\frac{\partial}{\partial x}(x-ay)+h'(x+ay)\frac{\partial}{\partial x}(x+ay)$$

$$=g'(x-ay)\times(1-0)+h'(x+ay)\times(1+0)$$

$$=g'(x-ay)+h'(x+ay)$$

Finding the second order partial derivatives w.r.t x, we get

$$\frac{\partial^2 f(x,y)}{\partial x^2}=\frac{\partial}{\partial x}\left(\frac{\partial f(x,y)}{\partial x}\right)=g''(x-ay)+h''(x+ay)$$

Similarly we find the 1st and 2nd order derivative with respect to y, we get

$$\frac{\partial f(x,y)}{\partial y}=g'(x-ay)\frac{\partial}{\partial y}(x-ay)+h'(x+ay)\frac{\partial}{\partial y}(x+ay)$$

$$=g'(x-ay)\times(0-1.a)+h'(x+ay)\times(0.1.a)$$

$$=-ag'(x-ay)+ah'(x+ay)$$

$$\frac{\partial^2 f(x,y)}{\partial y^2} = \frac{\partial}{\partial y}\left(\frac{\partial f(x,y)}{\partial y}\right) = -a \times -ag''^{(x-ay)} + a \times ah''(x+ay)$$

$$= a^2 g''(x-ay) + a^2 h''(x+ay)$$

$$= a^2\left[g''(x-ay) + h''(x+ay)\right]$$

$$= a^2 \frac{\partial^2 f(x,y)}{\partial x^2}$$

Hence proved

Derivative of Implicit Functions

Let f(x,y) is a function then it's a implicit function if there is a change in x should be accompanied by change in y such that the f(x, y)= 0

For example if 4y – 2xy + 3x = 0 is an implicit function then if x changes should be accompanied by change in y in order to make the expression equal to zero.

General Rule for finding the first order derivative of Implicit Function

The general rule for finding the derivative $\left(\frac{dy}{dx}\right)$ for an implicit function of a single variable is given by f (x, y)= 0

First find the total differentials of f(x, y)= 0

i.e. $f_x\, dx + f_y\, dy = d(0) = 0$

which is equal to

$$f_x\, dx = -\, f_y\, dy$$

So that we get

$$\left(\frac{dy}{dx}\right) = -\frac{f_x}{f_y}$$

Thus we see that the first order derivative $\left(\frac{dy}{dx}\right)$ of the implicit function f(x, y) = 0 is the negative ratio of two partial derivatives f_x and f_y.

Illustration No.20

If f (x, y) = 4y – 2xy + 3x = 0 then find the derivative of the implicit function?

Solution:

Given

$$f(x, y) = 4y - 2xy + 3x = 0$$

Therefore $f_x = 3 - 2y$ and $f_y = 4 - 2x$

And hence $\left(\frac{dy}{dx}\right) = -\frac{f_x}{f_y} = -\frac{(3-2y)}{(4-2x)} = \frac{2y-3}{4-2x}$

Illustration No.21

Find the $\left(\frac{dy}{dx}\right)$ of the function $f(x, y) = 2x^3 + 8xy + y^3 = 0$?

Solution:

Given

$$f(x, y) = 2x^3 + 8xy + y^3$$

Therefore

$$f_x = 6x^2 + 8y$$

And $f_y = 3y^2 + 8x$

Since $\left(\frac{dy}{dx}\right) = -\frac{f_x}{f_y}$

Hence $\left(\frac{dy}{dx}\right) = -\frac{(6x^2 + 8y)}{(3y^2 + 8x)}$

Illustration No. 22

Find the $\left(\frac{dy}{dx}\right)$ of the function $f(x,y) = x^y + y^x = 0$

Solution:

Given

$f(x, y) = x^y + y^x$ finding the partial derivative we get,

$$f_x = yx^{y-1} \text{ and}$$

$$f_y = xy^{x-1}$$

$$\left(\frac{dy}{dx}\right) = -\frac{f_x}{f_y}$$

Hence

$$\left(\frac{dy}{dx}\right) = \frac{yx^{y-1}}{xy^{x-1}} = \frac{x^{y-2}}{y^{x-2}}$$

Application of the Concepts of Partial Derivatives to Business and Economics

Let f(x, y, z) be a function of three independent variable x, y, and z then the partial derivative $\frac{\partial f}{\partial x}$ measures the change in the function f when there is marginal change in x keeping y and z constant. Similarly $\frac{\partial f}{\partial y}$ and $\frac{\partial f}{\partial z}$ measures the change if f with marginal change in y and z keeping other two variables constant.

We will utilize the above concept in understanding

(i) Demand Analysis.
(ii) Utility Analysis.
(iii) Production Analysis.

Demand Analysis

Let us assume that X_1, X_2, X_3 ... X_n be n commodities available in the market at the price of p_1, p_2, p_3 ... p_n. Then the demand for good X_i will depend on the pricing of the good sold in the market keeping all other factors constant (*law of demand*). Hence the quantity demanded say x_i can be written as function of pricing i.e.

$$x_i = f(p_i)$$

then the partial derivative $\frac{\partial x_i}{\partial x}$ indicates the variation in demand with change in pricing assuming other factors remaining constant.

Now we restrict the market with two goods X_1 and X_2 with pricing p_1 and p_2.

Then the quantity demand for both goods X_1 and X_2 say x_1 and x_2 can be expressed as a function of pricing

$$x_1 = f(p_1, p_2) \text{ and } x_2 = f(p_1, p_2)$$

Then the partial derivative of $\frac{\partial x_1}{\partial p_1}$ shows the marginal change in the quantity demanded of x_1 with change in pricing in p_1. Similarly $\frac{\partial x_2}{\partial p_2}$ shows the marginal change in the quantity demanded of x_2 with change in pricing in p_2.

Nature of Commodities whether substitutes (competitive) or Complementary

The commodities are said to be substitutes if the increase in demand for one commodity corresponds to the decrease in the demand of the other. i.e. if the demand for one commodity rises then the demand for the other will fall. And the commodities are said to be complementary if the increase in demand of one commodity will also result in the increase of demand for the other.

Substitute and Partial Differentiation

If two commodities X and Y with price p_1 and p_2 and the quantity demanded be x_1 and x_2 then if the increasing in pricing of p_1 will result in fall in demand for X and increase in demand of Y then the two commodities are competitive or substitutes of each other.

i.e. $\frac{\partial x_2}{\partial p_1} > 0$ as x_2 and p_1 change in the same direction and $\frac{\partial x}{\partial p} < 0$.

Similarly if the increasing in pricing of p_2 will result in fall in demand for Y and increase in demand of X then the two commodities are competitive or substitutes of each other.

i.e. $\frac{\partial x_1}{\partial p_2} > 0$ as x_1 and p_2 change in the same direction and $\frac{\partial x_2}{\partial p_2} < 0$.

Complementary and Partial Differentiation

If the increasing in pricing of p_1 will not only result in fall in demand for X and but also demand for Y then the two commodities are complementary to each other.

i.e. $\frac{\partial x_1}{\partial p_1} < 0$ as well as $\frac{\partial x_2}{\partial p_1} < 0$ i.e. the movement of x_2 and p_1 in opposite direction

Similarly

If the increasing in pricing of p_2 will not only result in fall in demand for Y and but also demand for X then the two commodities are complementary to each other.

i.e. $\frac{\partial x_2}{\partial p_2} < 0$ as well as $\frac{\partial x_1}{\partial p_1} < 0$

Partial Elasticity of Demand

Elasticity measures the rate of change in demand because rate change in pricing

Say two commodities X and Y with pricing p_1 and p_2 and the quantity demanded x_1 and x_2 then the elasticity of demands measured as

The partial elasticity of demand for X with respect to its own pricing p_1

$$= \frac{\partial x_1}{\partial p_1} \times \frac{p_1}{x_1}$$

Similarly

The partial elasticity of demand for Y with respect to its own pricing p_2

$$= \frac{\partial x_2}{\partial p_2} \times \frac{p_2}{x_2}$$

Again,

The cross elasticity of demand for X with respect to pricing p_2 of Y is given by

$$= \frac{\partial x_1}{\partial p_2} \times \frac{p_2}{x_1}$$

Similarly

The cross elasticity of demand for Y with respect to pricing p_1 of X is given by

$$= \frac{\partial x_2}{\partial p_1} \times \frac{p_1}{x_2}$$

Illustration No. 23

The demand function for a commodity x is given by

$$D(x) = 60 - 3p_x + 2p_y$$

Where D(x) is the quantity demanded p_x is the price of x, and p_y is the price of the other commodity y, then

(i) Calculate the price elasticity of demand of x?

(ii) Find the price elasticity at $p_x = 5$ and $p_y = 2$?

(iii) And also calculate the cross elasticity of demand for x with respect to price p_y?

Solution:

Given that

$$D(x) = 60 - 3p_x + 2p_y, \text{ therefore,}$$

$$\frac{\partial D(x)}{\partial x} = -3 \text{ and } \frac{\partial D(x)}{\partial y} = 2$$

(i) the price elasticity of demand for x is given by

$$e_{x_p} = \frac{p_x}{D(x)} \times \frac{\partial D(x)}{\partial x} = \frac{p_x}{60 - 3p_x + 2p_y} \times -3 = -\frac{3p_x}{60 - 3p_x + 2p_y}$$

(ii) The price elasticity of demand for x at $p_x = 5$ and $p_y = 2$

$$e_{x_p} = -\frac{3 \times 5}{60 - 3 \times +5 - 2 \times 2} = -\frac{15}{49}$$

(iii) the cross elasticity of demand for x with respect to price p_y is given by

$$= \frac{p_y}{D(x)} \times \frac{\partial D(x)}{\partial y} = \frac{p_y}{60 - 3p_x + 2p_y} \times 2 = \frac{2p_y}{60 - 3p_x + 2p_y}$$

Illustration No.24

If the demand function for two commodities x and y are given as $x_1 = 4p_1^{0.5}\ p_2^{0.7}$ and $x_2 = 3p_1^{0.7}\ p_2^{-0.4}$ then determine the four partial elasticities of demand with respect to price where x_1 and x_2 are the quantity demanded of x and y? Also find whether the commodities are substitutes or compliments?

Solution:

Given are the demand function

$$x_1 = 4p_1^{0.5}\ p_2^{0.7} \text{ and } x_2 = 3p_1^{0.7}\ p_2^{-0.4}$$

Let eij where (i = 1, 2 and j =1, 2) be the elasticity of demand of xi with respect to the price p_j. Now the four partial elasticity of demand is given by

$$e_{11} = \frac{p_1}{x_1} \times \frac{\partial x_1}{\partial p_1} = \frac{p_1}{4p_1^{-0.5} p_2^{0.7}} \times 4 \times (-0.5) \times p_1^{-1.5} p_2^{0.7} = -0.50$$

$$e_{12} = \frac{p_2}{x_1} \times \frac{\partial x_1}{\partial p_2} = \frac{p_1}{4p_1^{-0.5} p_2^{0.7}} \times 4 \times (0.7) \times p_1^{-0.5} p_2^{-0.3} = 0.70$$

$$e_{21} = \frac{p_1}{x_2} \times \frac{\partial x_2}{\partial p_1} = \frac{p_1}{3p_1^{0.7} p_2^{-0.4}} \times 3 \times (0.7) \times p_1^{-0.3} p_2^{-0.4} = 0.70$$

$$e_{22} = \frac{p_2}{x_2} \times \frac{\partial x_2}{\partial p_2} = \frac{p_2}{3p_1^{0.7} p_2^{-0.4}} \times 3 \times (-0.4) \times p_1^{0.7} p_2^{-1.4} = -0.40$$

Since the values of e_{12} and e_{21} are positive and greater than zero therefore the commodities are substitutes.

Illustration No. 25

If the demand function for two commodities x and y are given as $x_1 = 5p_1^{-1.2} p_2^{0.9}$ and $x_2 = p_1^{0.6} p_2^{-0.8}$ then determine whether the commodities are substitutes or complements where x_1 and x_2 are the quantity demanded of x and y?

Solution:

Given the demand functions are $x_1 = 5p_1^{-1.2} p_2^{0.9}$ and $x_2 = p_1^{0.6} p_2^{-0.8}$

Therefore in order to determine whether the commodities are compliments or substitutes we find the partial derivatives $\frac{\partial x_1}{\partial p_2}$ and $\frac{\partial x_2}{\partial p_1}$.

And hence

$$\frac{\partial x_1}{\partial p_2} = 5 \times 0.9 \times p_1^{-1.2} p_2^{-0.1} = 4.5 p_1^{-1.2} p_2^{-0.1} > 0,$$

which is also a positive quantity

Similarly

$$\frac{\partial x_1}{\partial p_2} = 0.6 \times p_1^{-0.4} p_2^{-0.8} > 0$$

which is also positive quantity

Therefore the commodities x and y are competitive or substitutes.

Illustration No.26

A monopolist has different demand functions for two different markets given by

$$q_x^d = a - b\, p_x \text{ and } q_y^d = c - d\, p_y$$

and his cost function is given by c(q). Find under what conditions is the price of the first market is higher than that of the second?

Solution:

Given that

$$q_x^d = a - b\, p_x \text{ and } q_y^d = c - d\, p_y$$

Therefore

$$q_x^d = a - b\, p_x \Rightarrow p_x = \frac{1}{b}(a - q_x^d) \text{ and}$$

$$q_y^d = c - d\, p_y \Rightarrow p_y = \frac{1}{d}(c - q_y^d)$$

Since Revenue = Price × Quantity Demanded

Therefore $R_x = q_x^d \times \frac{1}{b}\left(a - q_x^d\right) = \frac{1}{b}\left\{a q_x^d - \left(q_x^d\right)^2\right\}$

And $R_y = q_y^d \times \frac{1}{d}\left(c - q_y^d\right) = \frac{1}{d}\left\{c q_y^d - \left(q_y^d\right)^2\right\}$

The necessary condition for profit maximization is:

$$MR_1 = MR_2 = MC$$

Thus

$$MR_1 = \frac{\partial R_1}{\partial x} = \frac{1}{b}\left(a - 2q_x^d\right) = MC \text{ and } MR_2 = \frac{\partial R_1}{\partial x} = \frac{1}{d}\left(c - 2q_y^d\right) = MC$$

Therefore

$$\frac{1}{b}\left(a-2q_x^d\right)=\frac{1}{d}\left(c-2q_y^d\right)$$

$$\frac{1}{b}\{a-2\,(a-bp_x)\}=\frac{1}{d}\{c-2(c-dp_y)\}$$

$$\frac{1}{b}\left\{\frac{a}{b}-\frac{2a}{b}+2p_x\right\}=\frac{1}{d}\left\{\frac{c}{d}-\frac{2c}{d}+2p_y\right\}$$

$$-\frac{a}{b}+\frac{c}{d}=2(p_y-p_x)$$

$$(p_y-p_x)=\frac{ad-bc}{2bd}$$

Thus if $(p_y-p_x)>0$ or $p_y > p_x$ if $ad-bc>0$. Hence the price in the first market to be higher than the second market the necessary condition is ad > bc.

Utility Analysis

If we assume that U is the utility derived by the consumption of n goods $X_1, X_2, X_3 \ldots X_n$ then the total utility derived can be expressed as

$U = f\,(x_1, x_2, x_3 \ldots x_n)$ where $x_1, x_2, x_3 \ldots x_n$ are the amount of goods demanded of $X_1, X_2, X_3 \ldots X_n$ at any time period.

Thus the partial derivative $\frac{\partial U}{\partial x_i}$ will reflect the rate of change in total utility U with respect to change in the amount consumption of good X_i keeping the other consumption unchanged. And hence $\frac{\partial U}{\partial x_i}$ where (i = 1, 2, 3 ... n) will give the marginal utilities of various goods of an individual for a given set of purchases.

Marginal Rate of Substitution

Let us assume that a consumer is consuming two goods X and Y and the amount of goods consumed be x and y. And if the utility function is continuous then the consumer can derive a given level of satisfaction or utility by various combinations of quantities consumed of X and Y. However all such combination gives the indifference curve which is the combination of bundles of two commodities giving equal level of satisfaction?

Some of the properties of Indifference Curve are

1. Indifference curve are downward sloping to the right.
2. Indifference Curve are convex to origin.
3. Two different indifference curves do not intersect.

But the total differential of the utility function U = f(X, Y) is

$$dU = \frac{\partial f}{\partial x}dx + \frac{\partial f}{\partial y}dy = f_x dx + f_y dy$$

Since the change of U i.e. dU along the indifference curve is zero therefore dU = 0 that implies

$$f_x dx + f_y dy = 0$$

And hence

$$f_x dx = -f_y dy$$

$$\frac{dy}{dx} = -\frac{f_x}{f_y}$$

The negative value of $\frac{dy}{dx}$ is the marginal rate of substitution of commodity X for Y or vice versa. Otherwise stated that it is the rate at which a consumer substitutes X for Y or vice versa.

Illustration No. 27

Suppose the utility function of a consumer is given by U = ax + by + cxy where X and Y are the two commodities consumed by a consumer in the process deriving satisfaction level of U then find the ratio of their marginal utilities?

Solution:

Given U = ax + by + cxy

Therefore

$$f_x = \frac{\partial U}{\partial x} = a + cy \text{ and } f_y = \frac{\partial U}{\partial y} = b + cx$$

And hence the ratio of their marginal utilities is given by

$$\frac{f_x}{f_y} = \frac{\frac{\partial U}{\partial x}}{\frac{\partial U}{\partial y}} = \frac{a + cy}{b + cx}$$

Illustration No.28

If the utility function of a consumer is of the form given by U = log (x + y + xy) where X and Y are the two commodities consumed by the consumer and U the utility derived then find the marginal rate of substitution X for Y ?

Solution:

Given

$$U = \log (x + y + xy)$$

$$f_x = \frac{\partial U}{\partial x} = \frac{1}{x+y+xy}(1+y) \text{ and } f_y = \frac{\partial U}{\partial y} = \frac{1}{x+y+xy}(x+1)$$

Therefore the marginal rate of substitution of commodity X for Y is given by

$$\frac{dy}{dx} = -\frac{f_x}{f_y} = -\left\{\frac{\frac{1}{x+y+xy}(1+y)}{\frac{1}{x+y+xy}(x+1)}\right\} = -\left\{\frac{1+y}{1+x}\right\}$$

Some Additional Concepts

Constrained Utility Maximization

Let us assume U = f(x, y) be the utility function and B = $p_x x + p_y y$ be the budget constraint where p_x and p_y are the prices of two commodities X and Y respectively and $p_x x$ and $p_y y$ are the apportionment of total income B into commodities X and Y.

The budget constraint can be written as $y = \dfrac{B - p_x x}{p_y}$ 1

Thus the utility function can be expressed as

$$U = f\left(x, \frac{B - p_x x}{p_y}\right) \quad .....2$$

Differentiating 1 and 2 with respect to x, we get,

$$\frac{dy}{dx} = -\frac{p_x}{p_y}$$

And

$$\frac{\partial U}{\partial x} = f_x + f_y\left(-\frac{p_x}{p_y}\right)$$

For maximization condition $\frac{\partial U}{\partial x} = 0$ or $f_x + f_y\left(-\frac{p_x}{p_y}\right) = 0$ or $\frac{f_x}{f_y} = \frac{p_x}{p_y}$

Since $f_x = \frac{\partial U}{\partial x} = MU_x$ and $f_y = \frac{\partial U}{\partial y} = MU_y$

And hence

$$\frac{MU_x}{MU_y} = \frac{p_x}{p_y} \text{ or } \frac{MU_x}{p_x} = \frac{MU_y}{p_y}$$

Thus the maximization condition is the ratio of marginal utilities must be equal to the ratio of the prices.

But for the second order condition we have,

$$\frac{d^2U}{dx^2} = \frac{d}{dx}\left(\frac{dU}{dx}\right) = \frac{d}{dx}\left\{f_x + f_y\left(-\frac{p_x}{p_y}\right)\right\} = \left\{\frac{\partial}{\partial x}\left(\frac{dU}{dx}\right)\left(\frac{\partial x}{\partial x}\right)\right\} + \left\{\frac{\partial}{\partial y}\left(\frac{dU}{dx}\right)\left(\frac{\partial y}{\partial x}\right)\right\}$$

$$= \left\{\frac{\partial}{\partial x}\left(f_x + f_y\left(-\frac{p_x}{p_y}\right)\right)\right\} + \left\{\frac{\partial}{\partial y}\left((f_x + f_y)\left(-\frac{p_x}{p_y}\right)\right)\left(-\frac{p_x}{p_y}\right)\right\}$$

$$= \left(f_{xx} + f_{xy}\left(-\frac{p_x}{p_y}\right)\right) + \left(f_{yx} + f_{yy}\left(-\frac{p_x}{p_y}\right)\right)\left(-\frac{p_x}{p_y}\right)$$

$$= \left[f_{xx} + 2f_{xy}\left(-\frac{p_x}{p_y}\right) + f_{yy}\left(-\frac{p_x}{p_y}\right)^2\right]$$

Thus U will be maximum only when $\frac{d^2U}{dx^2} < 0$ i.e. the sign of the expression inside the bracket must be negative.

Production Analysis

Production of a product or commodity requires factor inputs like land, labour, raw materials and capital etc. which goes into the system in different proportions to determine a given level of output. Alternatively production is the transformation of factor inputs into outputs in a given period of time. Thus a function which connects various factor inputs is a production function.

Let us assume that X and Y are two factors gone into the production of a certain commodity and the amount produced be 'q' using x amount of X and y amount of Y then 'q' can be expressed mathematically as

$$q = f(x, y)$$

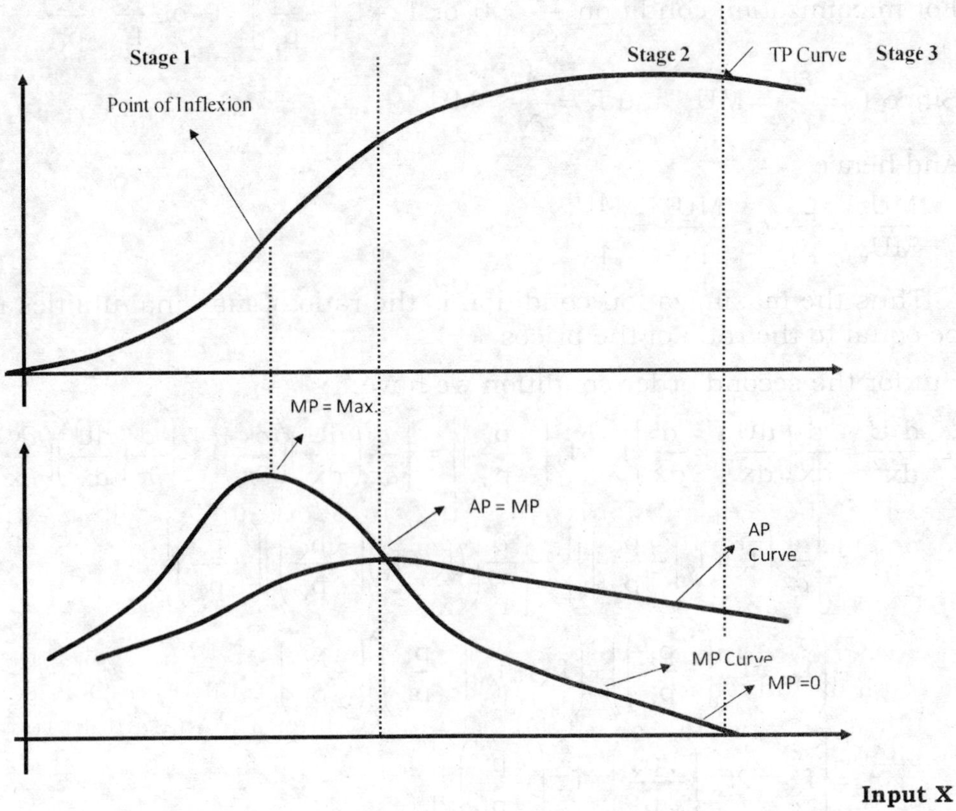

Now if we assume that out of the both the factors X and Y, let X be the variable factor and Y be the constant factor then the ratio $\left(\frac{q}{x}\right)$ is the average product of the factor X and the partial derivative $\frac{\partial q}{\partial x}$ will show the rate of change in the amount of production brought about by marginal change in x. Otherwise $\frac{\partial q}{\partial x}$ is the marginal product of the factor X.

The above diagrammatic presentation shows the total product curve (TP), the average product curve (AP) and the marginal product curve (MP) with a single factor input X.

The production process goes through 3 stages which can be summarized as

Stage 1

Marginal product curve starts falling and average product curve rises and reaches its maximum. At maximum point marginal product cuts average product curve i.e. MP = AP.

Stage 2

Both Marginal and average product are falling and marginal product turns zero i.e. MP = 0 at this point the total product is maximum.

Stage 3

Marginal product turns negative, and both average product and total product are both falling.

Note: The above discussion is restricted to Short Run condition where only one factor input is considered as variable factor.

Illustration No. 29

If the production function of a firm is given by $Q = AL^aK^b$ (L : Labour and K = Capital) and where a + b = 1. Then show that total product is equal to L times marginal product of labour plus K times the marginal product of capital?

Solution:

Given the production of the firm as $Q = AL^aK^b$ where a + b = 1

Therefore the marginal product of labour is given by

$$MP_L = \frac{\partial Q}{\partial L} = A\frac{\partial}{\partial x}\left(L^aK^b\right)$$

$$= Aa\left(L^{a-1}K^b\right)$$

And $$LMP_L = LAa\left(L^{a-1}K^b\right) = Aa\left(L^aK^b\right)$$

Similarly

$$MP_k = \frac{\partial Q}{\partial K} = A\frac{\partial}{\partial x}\left(L^a K^b\right)$$

$$= Ab\left(L^a K^{b-1}\right)$$

And $$KMP_K = KAb\left(L_a K^{b-1}\right) = Ab\left(L^a K^b\right)$$

Therefore $$LMP_L + KMP_K = Aa(L^a K^b) + Ab(L^a K^b) = (a+b)A\ (L^a K^b)$$

{Since (a + b) = 1}

$$= A(L^a K^b) = Q$$

Illustration No. 30

For the production function of a manufacturer is given by q (l, k) = $\frac{1}{l+k}(2hlk - al^2 - bk^2)$. Show that the marginal product of the both factor inputs is depends on the ratio of the factors. Also prove the degree of homogeneity and Euler's theorem?

Solution:

Given the product function

$$q\ (l,\ k) = \frac{1}{l+k}(2hlk - al^2 - bk^2)$$

Thus the marginal product of the factor l is MP_l is given by

$$MP_l = \frac{\partial q}{\partial l} = \frac{(l+k)(2hk - 2al) - 1.(2hlk - al^2 - bk^2)}{(l+k)^2}$$

$$= \frac{2hk - 2al}{(l+k)} - \frac{(2hlk - al^2 - bk^2)}{(l+k)^2}$$

$$= \frac{2h\left(\frac{k}{l}\right) - 2a}{1+\left(\frac{k}{l}\right)} - \frac{\left\{2h\left(\frac{k}{l}\right) - a - b\left(\frac{k}{l}\right)^2\right\}}{\left(1+\frac{k}{l}\right)^2}$$

This shows that marginal product of labour MP_l depends only on the ratio of $\left(\frac{k}{l}\right)$ of factors.

Similarly, the marginal product of the factor k MP_k is given by

$$MP_k = \frac{\partial q}{\partial k} = \frac{(l+k)(2hk-2al)-1.(2hlk-al^2-bk^2)}{(l+k)^2}$$

$$= \frac{2hl-2bk}{(l+k)} - \frac{(2hlk-al^2-bk^2)}{(l+k)^2}$$

$$= \frac{2h\left(\frac{l}{k}\right)-2b}{\left(\frac{l}{k}+1\right)} - \frac{\left\{2h\left(\frac{l}{k}\right)-a\left(\frac{l}{k}\right)^2-b\right\}}{\left(\frac{l}{k}+1\right)}$$

This shows that marginal product of capital MP_k depends only on the ratio of $\left(\frac{l}{k}\right)$ of factors.

In order to establish the degree of homogeneity and verifying Eulers theorem multiply MP_l by l and MP_k by k and then adding them, we have,

$$lMP_l + kMP_k = l\left\{\frac{(l+k)(2hk-2al)-(2hlk-al^2-bk^2)}{(l+k)^2}\right\}$$

$$+k\left\{\frac{(l+k)(2hk-2al)-1.(2hlk-al^2-bk^2)}{(l+k)^2}\right\}$$

$$= l\left\{\frac{2hk-2al}{(l+k)} - \frac{(2hlk+-al^2-bk^2)}{(l+k)^2}\right\} + k\left\{\frac{2hl-2bk}{(l+k)} - \frac{(2hlk-al^2-bk^2)}{(l+k)^2}\right\}$$

Solving the above we get,

$$= \frac{2\{2hlk-al^2-bk^2\}}{(l+k)} - \frac{(l+k)\{2hlk-al^2-bk^2\}}{(l+k)^2}$$

$$= \frac{2(l+k)\{2hlk-al^2-bk^2\}-(l+k)\{2hlk-al^2-bk^2\}}{(l+k)^2}$$

$$= \frac{(l+k)\{2hlk - al^2 - bk^2\}}{(l+k)^2}$$

$$= \frac{\{2hlk - al^2 - bk^2\}}{(l+k)} = q(l,k)$$

Hence q(l, k) is a homogenous function of degree 1

Illustration No.31

For a given firm if the production function is given by $q = 5L^3K^2 - 3L^2K^3$ where L and K are the labor and capital inputs. Find the marginal product of labor and the marginal product of capital. If we assume that capital k to be constant then find for what value of L for which AP is maximum?

Solution:

Given the production function as $q = 5L^3K^2 - 3L^2K^3$

Therefore

$$MP_L = \frac{\partial q}{\partial L} = 15L^2K^2 - 6LK^3$$

Similarly

$$MP_K = \frac{\partial q}{\partial K} = 10L^3K - 9L^2K^2$$

Since we have assumed capital K to be constant, therefore the Average Product of Labor is given by

$$AP_L = \frac{q}{L} = 5L^2K^2 - 3LK^3$$

But we know for the maximum value of AP_L the condition is that $AP_L = MP_L$.

And hence,

$$15L^2K^2 - 6LK^3 = 5L^2K^2 - 3LK^3$$

$$10L^2K^2 - 3LK^3 = 0 \text{ or } L = \frac{3K}{10}$$

And hence for a value of $L = \frac{3K}{10}$, AP_L is maximum.

Illustration No.32

Following is the linear homogenous production function $q = \sqrt{al^2 - 2hlk + bk^2}$ where q,l and k represent the output , labour and capital respectively. Then show that 1 times the marginal product of labour and k times the marginal product of capital equals total product?

Solution:

Given the production function as $q = \sqrt{al^2 - 2hlk + bk^2}$

Therefore

$$MP_l = \frac{\partial q}{\partial l} = \frac{1}{2\sqrt{al^2 - 2hlk + bk^2}}(2al - 2hk) = \frac{al - hk}{\sqrt{al^2 - 2hlk + bk^2}}$$

Similarly

$$MP_k = \frac{\partial q}{\partial k} = \frac{1}{2\sqrt{al^2 - 2hlk + bk^2}}(2bk - 2hl) = \frac{bk - hl}{\sqrt{al^2 - 2hlk + bk^2}}$$

Therefore

$$lMP_l = l\left(\frac{al - hk}{\sqrt{al^2 - 2hlk + bk^2}}\right) = \frac{al^2 - hlk}{\sqrt{al^2 - 2hlk + bk^2}} \quad ......(1)$$

And

$$kMP_k + k = \left(\frac{bk - hl}{\sqrt{al^2 - 2hlk + bk^2}}\right) = \frac{bk^2 - hlk}{\sqrt{al^2 - 2hlk + bk^2}} \quad ......(2)$$

Adding eq, 1 and 2, we get,

$$lMP_l + kMP_k = \frac{al^2 - hlk}{\sqrt{al^2 - 2hlk + bk^2}} + \frac{bk^2 - hlk}{\sqrt{al^2 - 2hlk + bk^2}}$$

$$= \frac{al^2 - 2hlk + bk^2}{\sqrt{al^2 - 2hlk + bk^2}}$$

$$= \sqrt{al^2 - 2hlk + bk^2} = q$$

the total product.

The Concept of Maxima and Minima of a function of two variables

Let f(x, y) be a function of two variables x and y then in order to derive the condition for deciding the maximum or minimum value of f(x, y) the following rules are used

(i) Find the first order partial derivative of function f with respect to x and y and equate to zero i.e.

$$\frac{\partial f}{\partial x} = f_x = 0 \text{ and } \frac{\partial f}{\partial y} = f_y = 0$$

Then

(ii) Then find the second order partial derivative of the function f

i.e $$f_{xx} = \frac{\partial^2 f}{\partial x^2} \text{ and } f_{yy} = \frac{\partial^2 f}{\partial y^2}$$

if f_{xx}, $f_{yy} < 0$ then the function is maximum

Otherwise if

f_{xx}, $f_{yy} > 0$ then the function is minimum

(iii) f_{xx}, $f_{yy} > (f_{xy})^2$

Illustration No.33

Find the maximum and minimum value of a function given by $f(x,y) = \frac{1}{3}x^3 + y^2 - 9x + 8y$??

Solution:

Given

$$f(x,y) = \frac{1}{3}x^3 + y^2 - 9x + 8y$$

Therefore

$$\frac{\partial f}{\partial x} = f_x = x^2 - 9 = 0 \qquad \text{1st order condition}$$

And hence, we get $3x^2 = 9$ or $x = \pm 3$

Similarly,

$$\frac{\partial f}{\partial y} = f_y = 2y + 8 = 0 \text{ or } y = -4$$

Thus we have obtained two points (3, –4) and (–3, –4) where the function take maximum or minimum value.

Now finding the 2nd order condition, we have

$$f_{xx} = \frac{\partial^2 f}{\partial x^2} = 2x \text{ and } f_{yy} = \frac{\partial^2 f}{\partial y^2} = 2 \text{ and } f_{xy} = 0$$

At the point (3, –4), we have,

$$f_{xx} = 6 \text{ and } f_{yy} = 2 \text{ and } f_{xy} = 0$$

Therefore $f_{xx} > 0$, $f_{yy} > 0$ and $f_{xx}\, f_{yy} = 12 > 0$

Thus satisfying the minimum value condition at the point (3, –4) and the minimum value is

$$f(x, y) = \frac{1}{3} \times 3^3 + (-4)^2 - 9 \times 3 + 8 \times (-4)$$

$$= -34$$

At the point (– 3, – 4), we have,

$$f_{xx} = 6 \text{ and } f_{yy} = 2 \text{ and } f_{xy} = 0$$

Therefore $f_{xx} < 0$, $f_{yy} > 0$ therefore there exist a point of inflexion at the point (– 3, – 4).

Illustration No. 34

Find the maximum and minimum value of a function given by $f(x, y) = x^3 + x^2 - xy + y^2 + 6$.

Solution:

Given

$$f(x, y) = x^3 + x^2 - xy + y^2 + 6$$

Therefore

$$\frac{\partial f}{\partial x} = f_x = 3x^2 + 2x - y = 0 \qquad \text{1st order condition}$$

Similarly,

$$\frac{\partial f}{\partial x} = f_y = 2y - x = 0$$

Solving the above two equations we get, $x = \frac{1}{2}$ and $y = \frac{1}{4}$

Now finding the 2nd order condition, we have

$$f_{xx} = \frac{\partial^2 f}{\partial x^2} = 6x + 2 \text{ and } f_{yy} = \frac{\partial^2 f}{\partial y^2} = 2 \text{ and } f_{xy} = 0$$

At the point $\left(\frac{1}{2}, \frac{1}{4}\right)$, we have,

$$f_{xx} = 6\left(\frac{1}{2}\right) + 2 = 5 \text{ and } f_{yy} = 2$$

Therefore $\quad f_{xx} > 0,\ f_{yy} > 0$

Thus satisfying the minimum value condition at the point $\left(\frac{1}{2}, \frac{1}{4}\right)$ and the minimum value is

$$f(x, y) = \left(\frac{1}{2}\right)^3 + \left(\frac{1}{2}\right)^2 - \frac{1}{2} \times \frac{1}{4} + 6 = \frac{25}{4}$$

Summary

- If f(x, y) be a function two independent variables x and y then the ordinary derivative of f(x, y) with respect to x assuming y to be a constant, is called the partial derivative of f(x, y) with respect to x and is denoted by the symbol $\frac{\partial f(x, y)}{\partial x}$ or $f_x(x, y)$.
- **Euler's Theorem**

  A function f (x, y) of two independent variables of x and y is said to be homogenous in x and y of degree n if it can represented in the following form

  $$x^n \phi\left(\frac{y}{x}\right) \text{ or } y^n \phi\left(\frac{x}{y}\right).$$

- Elasticity measures the rate of change in demand because rate change in pricing

Self Review Questions

1. If $f(x,y) = x^4 + y^4 + 3x^2y + 3xy^2$ then find $\frac{\partial f(x, y)}{\partial x}$, $\frac{\partial f(x, y)}{\partial y}$, $\frac{\partial^2 f(x, y)}{\partial x^2}$, and $\frac{\partial^2 f(x, y)}{\partial y^2}$?

2. If $f(x, y) = 3x^2 - 5y^2 + 4xy$ then find $\frac{\partial f(x,y)}{\partial x}$, $\frac{\partial f(x,y)}{\partial y}$ and $\frac{\partial^2 f(x,y)}{\partial y \partial x}$.
3. If $f(x,y) = 9x^3 + 6y^3 - 15x^2y$
   then show that $x\frac{\partial f(x,y)}{\partial x} + y\frac{\partial f(x,y)}{\partial y} = 3f(x,y)$.
4. Verify the Euler theorem for the function $f(x, y) = 3x^2 + 3y^2 + 5xy$?
5. Verify the Euler theorem for the function $f(x, y) = x^5 + 4x^3y^2 + y^5$?
6. Find the first order total differentials of $f(x, y) = \sqrt{2x^2 + y^2}$
7. Find the first order total differentials of $f(x, y) = \log(x - 2y^2)$
8. Find the first order total differentials of $f(x, y) = \frac{x^2 + y^2}{x + y}$
9. Find the first order total differentials of $f(x, y) = \sqrt{x^2 + 3y}$
10. Find the first order total differentials of $f(x, y) = \sqrt{x - y}$
11. Find the second order total differentials of $f(x, y) = x^2 + 2xy$.
12. Find the second order total differentials of $f(x, y) = 2xy$.
13. Find the second order total differentials of $f(x, y) = y\log(1 + x)$.
14. Find the second order total differentials of $f(x, y) = xy^2 + x^2y$.
15. Find the derivatives of the implicit function $f(x, y) = x^y y^x$
16. Find the derivatives of the implicit function $f(x, y) = xy - x$
17. Find the derivatives of the implicit function $f(x,y) = a^x e^y$
18. Find the derivatives of the implicit function $f(x,y) = a^x y + x^2y + y^3$.
19. Find whether the function is having maximum or minimum value if $f(x, y) = 4/3\, x^3 + y^2 - 4x + 8y$.
20. The demand function for a commodity x is given by

$$D(x) = 30 - 4p_x + 3p_y$$

where D(x) is the quantity demanded, p_x is the price of x, and p_y is the price of the other commodity y, then

(i) Calculate the price elasticity of demand of x?

(ii) And also calculate the cross elasticity of demand for x with respect to price p_y?

Answers

1. (i) $4x^3 + 6xy + 3y^2$, (ii) $4y^3 + 3x^2 + 6xy$, (iii) $12x^2 + 6y$, (iv) $12y^2 + 6x$
2. (i) $6x + 4y$, (ii) $-10y + 4x$, (iii) 4
4. $2f(x, y)$
5. $5f(x, y)$
6. $d\,f(x, y) = \frac{1}{2\sqrt{2x^2 + y^2}}\{4xfx + 2ydy\}$
7. $d\,f(x, y) = (1/x - 2y^2)\{dx - 4ydy\}$
8. $d\,f(x, y) = \{(x^2 + 2xy - y^2)/(x + y)^2\}dx + \{(y^2 + 2xy - x^2)/(x + y)^2\}dy$
9. $d\,f(x, y) = 1/(x^2 + 3y)^{1/2}\{6xdx + 3dy\}$
10. $d\,f(x, y) = \frac{1}{2\sqrt{x - y}}(dx - dy)$
11. $d^2f(x, y) = 2(dx)^2 + 2(dx\,dy)$
12. $d^2f(x, y) = 2(dx\,dy)$
13. $d^2f(x, y) = -\frac{y}{(1 + x)^2}(dx)^2 + \left\{\frac{1}{1 + x}\right\}(dx\,dy)$
14. $d^2f(x, y) = (2y)(dx)^2 + (2x)(dxdy) + (2x)(dy)^2$
15. $\frac{dy}{dx} = -\left\{\frac{y^2(y + x^2 \log y)}{x^2(x + y^2 \log x)}\right\}$
16. $\frac{dy}{dx} = \frac{1 - y}{x}$
17. $\frac{dy}{dx} = -(\log a)$
18. $\frac{dy}{dx} = -\left\{\frac{a^x y \log a + 2xy}{a^x + x^2 + 3y^2}\right\}$
19. Minimum value at (–1, –4)
20. (i) $-\frac{4p_x}{60 - 3p_x + 2p_y}$, (ii) $-\frac{3p_y}{60 - 3p_x + 2p_y}$

Appendix

i) Maximization of Output subject to Cost Constraint (Budget Constraint)

Maximize

$$X = f(L, K) \qquad .....(1)$$

Subject to

$$C = wL + r K \qquad ..... 2. \text{(Constraint Equation)}$$

Since this is a case of constrained maximization and the condition for the equilibrium of the firm may be obtained by using Lagrangian Multipliers.

Rewriting the constraint equation we get,

$$C - wL - r K = 0$$

We will multiply the constraint by a constant λ the Lagrangian multiplier

$$\lambda (C - wL - r K) = 0$$

Lagrangian multiplier are undefined constant used for solving maxima and minima constrains.

Framing the composite function, we get,

$$\Phi = X + \lambda (C - wL - r K)$$

Thus it can be shown that the maximization of the function Φ implies maximization of the output.

Applying the concept of Maxima and minima for which we obtain the 1st order derivative of the above composite function with respect to L, K and λ.

Thus, we get

$$\frac{\partial}{\partial L} = \frac{\partial X}{\partial L} + \lambda(-w) = 0 \qquad .....(3)$$

$$\frac{\partial \phi}{\partial K} = \frac{\partial X}{\partial K} + \lambda(-r) = 0 \qquad .....(4)$$

$$\frac{\partial \phi}{\partial x} = C - wL - rK = 0 \qquad .....(5)$$

Solving the equations 3 and 4 for λ we get,

$$\frac{\partial X}{\partial L} = \lambda \text{ w or } \lambda = \frac{\partial X / \partial L}{w} = \frac{MP_L}{w}$$

$$\frac{\partial X}{\partial K} = \lambda \text{ r or } \lambda = \frac{\partial X / \partial K}{r} = \frac{MP_K}{r}$$

Equating both equations in terms of λ hence we get,

$$\frac{MP_L}{w} = \frac{MP_K}{r} \text{ or } \frac{MP_L}{MP_K} = \frac{w}{r}$$

The firm is in equilibrium when the ratio of marginal productivities of factors are equal to the ratio of their prices.

The second order condition for the firm to be in equilibrium requires that the slope of marginal product of both the factors have a negative slope.

Therefore the second order derivative of marginal product of labour i.e.

$$\text{Slope of } MP_L = \frac{\partial^2 x}{\partial L^2}$$

Similarly

$$\text{Slope of } MP_K = \frac{\partial^2 x}{\partial K^2}$$

The second order conditions are

$$\frac{\partial^2 x}{\partial L^2} < 0 \text{ and } \frac{\partial^2 x}{\partial K^2} < 0$$

And

$$\left(\frac{\partial^2 X}{\partial L^2}\right)\left(\frac{\partial^2 X}{\partial K^2}\right) > \left(\frac{\partial^2 X}{\partial L \partial K}\right)^2$$

ii) Minimization of Cost for a given level of Output

Minimize

$$C = wL + r\,K$$

Subject to

$$X = f(L,K) \quad \text{(Constraint Equation)}$$

Rewriting the constraint equation, we get,

$$X - f(L,K) = 0$$

Forming the composite function using Lagrangian Multiplier λ

$$\Phi = C - \lambda\,[X - f(L,K)]$$

Or

$$\Phi = (wL + r\,K) - \lambda\,[X - f(L,K)]$$

Finding the partial derivatives of Φ with respect to L, K and λ and equating to zero

$$\frac{\partial \phi}{\partial L} = w - \lambda \frac{\partial f(L.K)}{\partial L} = 0 = w - \lambda \frac{\partial X}{\partial L} \quad .....(6)$$

$$\frac{\partial \phi}{\partial K} = r - \lambda \frac{\partial f(L.K)}{\partial K} = 0 = r - \lambda \frac{\partial X}{\partial K} \quad .....(7)$$

$$\frac{\partial \phi}{\partial \lambda} = -[X - f(L,K)] = 0 \quad .....(8)$$

From equation 6 and 7 we get,

$$\lambda\left(\frac{\partial X}{\partial L}\right) = w \text{ and } \lambda\left(\frac{\partial X}{\partial K}\right) = r$$

Dividing these equations we get,

$$\frac{w}{r} = \frac{\left(\frac{\partial X}{\partial L}\right)}{\left(\frac{\partial X}{\partial K}\right)} = MRS_{L,K}$$

The second order condition for the firm to be in equilibrium requires that the slope of marginal product of both the factors have a negative slope.

Therefore the second order derivative of marginal product of labour i.e.

$$\text{Slope of } MP_L = \frac{\partial^2 X}{\partial L^2}$$

Similarly

$$\text{Slope of } MP_K = \frac{\partial^2 X}{\partial K^2}$$

The second order conditions are

$$\frac{\partial^2 X}{\partial L^2} < 0 \text{ and } \frac{\partial^2 X}{\partial K^2} < 0$$

And

$$\left(\frac{\partial^2 X}{\partial L^2}\right)\left(\frac{\partial^2 X}{\partial K^2}\right) > \left(\frac{\partial^2 X}{\partial L \partial K}\right)^2$$

Chapter 7

Integration and its Application

Definition

Integration is reverse or inverse of Differentiation.

If f(x) is the derivative of a function F(x) for all real x in a given interval i.e. $\frac{d}{dx}\{F(x)\} = f(x)$ or $d\{F(x)\} = f(x)\,dx$, then F(x) is called the integral or to be more specific the indefinite integral of f(x) with respect to x and mathematically written as

$$\int f(x)dx = F(x)$$

Here the symbol $\int$ stands for integration, dx after f(x) indicates the integration to be performed with respect to variable x or termed as variable of integration, f(x) is the integrand and F(x) the anti derivative of f(x).

However if we write $\frac{d}{dx}\{F(x)+c\} = f(x)$ where c is an arbitrary constant such that $\frac{d}{dx}(c) = 0$ then we have

$$\frac{d}{dx}\{F(x)+c\} = f(x)$$

Or $$\int f(x)dx = F(x) + c$$

where c is the constant of integration.

Some Standard Fundamental Integrals

1. $\int x^n \, dx = \frac{x^{n+1}}{n+1} + c$

Proof

By differentiating $\frac{x^{n+1}}{n+1} + c$ we have,

$$\frac{d}{dx}\left\{\left(\frac{x^{n+1}}{n+1}\right) + c\right\} = \frac{1}{n+1}\frac{d}{dx}\left(x^{n+1}\right) + \frac{d}{dx}(c)$$

$$\left\{\text{since } \frac{d}{dx}(c) = 0\right\}$$

$$= \frac{1}{n+1}(n+1)\, x^{n+1-1}$$

$$= x^n$$

Therefore by the definition of integral, we have

$$\int x^n \, dx = \frac{x^{n+1}}{n+1} + c$$

2. $\int \frac{1}{x} \, dx = \log x + c$

Proof:

By differentiating log x + c we have

$$\frac{d}{dx}(\log x) = \frac{1}{x} \qquad \left\{\text{since } \frac{d}{dx}(c) = 0\right\}$$

Therefore by the definition of integration, we have

$$\int \frac{1}{x} \, dx = \log x + c$$

Similarly

$$\int \frac{1}{x \pm a} \, dx = \log(x \pm a) + c$$

3. $\int e^x dx = e^x + c$

Differentiating $e^x + c$

i.e. $\frac{d}{dx}\left(e^x + c\right) = e^x$ $\left\{\text{since } \frac{d}{dx}(c) = 0\right\}$

Therefore by the definition of integration, we have

$$\int e^x dx = e^x + c$$

Similarly

$$\int e^{mx} dx = \frac{e^{mx}}{m} + c$$

$$\left\{\text{since } \frac{d}{dx}\left(\frac{e^{mx}}{m} + c\right) = \frac{1}{m} \times m \times e^{mx} = e^{mx}\right\}$$

4. $\int a^x\, dx = \frac{a^x}{\log a} + c \ (a > 0)$

$$\left[\left\{\text{since } \frac{d}{dx}\left(\frac{a^x}{\log a}\right) + c\right\} = \frac{1}{\log a} a^x \log a = a^x\right]$$

Similarly

$$\int a^{mx}\, dx = \frac{a^{mx}}{m \log a} + c$$

$$\left[\left\{\text{since } \frac{d}{dx}\left(\frac{a^{mx}}{m \log a}\right) + c\right\} = \frac{1}{m \log a} a^{mx}\, m \log a = a^{mx}\right]$$

5. $\int \sin x\, dx = -\cos x + c$

$$\left[\left\{\text{since } \frac{d}{dx}(-\cos x) + c\right\} = -(-\sin x) = \sin x\right]$$

Similarly

$$\int \cos x\, dx = \sin x + c$$

6. $\int \sec^2 x dx = \tan x + c$

$$\left[\left\{\text{since } \frac{d}{dx} \tan x + c\right\} = \sec^2 x\right]$$

Similarly

$$\left[\int \sec x \tan x \, dx = \sec x + c\right.$$

7. $\int \operatorname{cosec}^2 x dx = -\cot x + c$

$$\left[\left\{\text{Since } \frac{d}{dx} (-\cot x + c)\right\} = \operatorname{cosec}^2 x\right]$$

Similarly

$$\left[\int \operatorname{cosec} x \cot x \, dx = -\operatorname{cosec} x + c\right.$$

8. $\int \frac{1}{\sqrt{1-x^2}} dx = \sin^{-1} x + c$

$$\left[\left\{\text{Since } \frac{d}{dx} \left(\sin^{-1} x + c\right)\right\} = \frac{1}{\sqrt{1-x^2}}\right]$$

Similarly

$$\left[\int \frac{-1}{\sqrt{1-x^2}} dx = \cos^{-1} x + c\right.$$

9. $\int \frac{1}{1+x^2} dx = \tan^{-1} x + c$

$$\left[\left\{\text{Since } \frac{d}{dx} \left(\tan^{-1} x + c\right)\right\} = \frac{1}{1+x^2}\right]$$

Similarly

$$\left[\int \frac{-1}{1+x^2} dx = \cot^{-1} x + c\right.$$

10. $\int \frac{1}{x\sqrt{x^2-1}}\,dx = \sec^{-1}x + c$

$$\left[\left\{\text{Since } \frac{d}{dx}\left(\sec^{-1}x + c\right)\right\} = \frac{1}{x\sqrt{x^2-1}}\right]$$

Similarly

$$\left[\int \frac{-1}{x\sqrt{x^2-1}}\,dx = \text{cosec}^{-1}x + c\right.$$

Rules of Integration

1. $\int c\, f(x)dx = c \int f(x)dx$

Proof:

Let $\left[\int f(x)dx = F(x) \text{ then } \frac{d}{dx}F(x) = f(x)\right.$

By differentiating c f(x)

i.e $\frac{d}{x}\{cf(x)\} = c\frac{d}{dx}F(x)$

$= c\, f(x)$

Therefore by the definition of integral, we have

$$\left[\begin{array}{l}\int c\, f(x)dx = x\,F(x)\\ \int c\, f(x)dx = c\int f(x)dx\end{array}\right.$$

2. $\int\{f(x) \pm g(x)dx = \int f(x)dx \pm \int g(x)dx$

Proof:

Let $\int f(x)dx = F(x)$ then $\frac{d}{dx}F(x) = f(x)$

And $\int g(x)dx = G(x)$ then $\frac{d}{dx}G(x) = g(x)$

Then

$$\left[\frac{d}{dx}\{f(x) + G(x)\} = \frac{d}{dx}\{F(x)\} \pm \frac{d}{dx}\{G(x)\}\right.$$

$$= f(x) \pm g(x)$$

And hence by the definition of integration, we have

$$\left[\begin{array}{l}\int\{f(x) \pm g(x)\}\,dx = F(x) \pm G(x)\\ \int\{f(x) \pm g(x)\}\,dx = \int f(x)dx \pm \int g(x)dx\end{array}\right.$$

Illustration No.1

Find $\int x^6\,dx$

Solution:

$$\left[\begin{array}{l}\int x^6\,dx\\ = \dfrac{x^{6+1}}{6+1} + c\\ = \dfrac{x^7}{7} + c\end{array}\right.$$

Illustration No. 2

Find $\int 3 \log x\,dx$

Solution:

$$\left[\begin{array}{l}\int 3 \log x\,dx = 3 \times \int \log x\,dx\\ = 3 \times \dfrac{1}{x} + c\\ = \dfrac{3}{x} + c\end{array}\right.$$

Illustration No. 3

Find the $\int 4a^x dx$

Solution:

$$\left[\int 4a^x dx = 4 \int a^x\,dx\right.$$

$$= 4 \times \frac{a^x}{\log a} + c$$

$$= \frac{4a^x}{\log a} + c$$

Illustration No. 4

Find the $\int (x^3 + \frac{2}{x} + e^x + 5)dx$

Solution:

$$\int (x^3 + \frac{2}{x} + e^x + 5)\,dx = \int x^3\,dx + \int \frac{2}{x}\,dx + \int e^x\,dx + \int 5dx$$

$$= \frac{x^{3+1}}{3+1} + 2 \times \log x + e^x + 5 \times x + c$$

$$= \frac{x^4}{4} + 2\log x + e^x + 5x + c$$

Illustration No. 5

Find the $\int (1 - x)\, x^{3/2}\, dx$

Solution:

$$\int (1 - x)\, x^{3/2}\, dx = \int (x^{3/2} - x^{5/2})\, dx$$

$$= \int x^{3/2}\, dx - \int x^{5/2}\, dx$$

$$= \frac{x^{\frac{3}{2}+1}}{\frac{3}{2}+1} - \frac{x^{\frac{5}{2}+1}}{\frac{5}{2}+1} + c$$

$$= \frac{2}{5}x^{5/2} - \frac{2}{7}x^{7/2} + c$$

$$= [2\{\frac{1}{5}x^{\frac{5}{2}} - \frac{1}{7}x^{7/2}\}] + c$$

Illustration No. 6

Find the $\int \left(x + \frac{1}{x}\right)^2 dx$

Solution:

$$\int\left(x + \frac{1}{x}\right)^2 dx = \int(x^2 + \frac{1}{x^2} + 2)dx$$

$$= \int x^2\, dx + \int\frac{1}{x^2}\, dx + \int 2\, dx$$

$$= \frac{x^3}{3} + \frac{x^{-1}}{-1} + 2x + c$$

$$= \frac{x^3}{3} + \frac{-1}{x^{-1}} + 2x + c$$

Illustration No. 7

Find $\int\frac{(x-1)^2}{\sqrt{x}}dx$

Solution:

$$\int\frac{(x-1)^2}{\sqrt{x}}dx = \int\frac{(x^2 - 2x + 1)}{\sqrt{x}}dx$$

$$= \int\{x^{3/2} - 2x^{1/2} + x^{-1/2}\}\, dx$$

$$= \int x^{-1/2}\, dx + \int x^{3/2}\, dx - 2\int x^{1/2}\, dx$$

$$= \frac{x^{1/2}}{1/2} + \frac{x^{5/2}}{5/2} - \frac{2\,x^{3/2}}{3/2} + c$$

$$= 2x^{1/2} + \frac{2}{5}x^{5/2} - \frac{4}{3}x^{3/2} + c$$

Illustration No. 8

Find $\int\left(x^a + a^x + a^a\right)dx$

Solution:

$$\left[\int\left(x^a + a^x + a^a\right)dx = \int x^a dx + \int a^x dx + \int a^a dx\right.$$

$$= \frac{x^{a+1}}{a+1} + \frac{a^x}{\log a} + a^a x + c$$

Illustration No. 10

$$\int\left(e^{5x}+\frac{3}{x}\right)dx$$

Solution:

$$\int\left(e^{5x}+\frac{3}{x}\right)dx = \int e^{5x}dx + \int\frac{3}{x}dx$$

$$= \frac{e^{5x}}{5} + 3\log x + c$$

Illustration No. 11

Find $\int(e^x - \frac{1}{e^x})^2 dx$

Solution:

$$\int\left(e^x-\frac{1}{e^x}\right)^2 dx = \int\{e^{2x} - 2 + \frac{1}{e^{2x}}\}dx$$

$$= \int e^{2x}dx - \int 2dx + \int\frac{1}{e^{2x}}\, dx$$

$$= \int e^{2x}\, dx - 2\int dx + \int e^{-2x}\, dx$$

$$= \frac{e^{2x}}{2} - 2x - \frac{e^{-2x}}{2} + c$$

Illustration No. 12

Find $\int\frac{x^2}{(x+3)}dx$

Solution:

$$\int\frac{x^2}{(x+3)}dx = \int\frac{x^2+9-9}{(x+3)}dx$$

$$= \int\left\{(x-3)+\frac{9}{(x+3)}\right\} dx$$

$$= \int x\,dx - \int 3dx + \int \frac{9}{(x+3)}\,dx$$

$$= \frac{x^2}{2} - 3x + 9\log(x+3) + c$$

Illustration No. 13

Find $\int \left(x + \frac{1}{x}\right)^3 dx$

Solution:

$$\int \left(x + \frac{1}{x}\right)^3 dx = \int \left(x^3 + 3x^2\frac{1}{x} + 3x\frac{1}{x^2} + \frac{1}{x^3}\right) dx$$

$$= \int x^3\,dx + \int 3x\,dx + \int 3\frac{1}{x}\,dx + \int \frac{1}{x^3}\,dx$$

$$= \frac{x^4}{4} + 3\frac{x^2}{2} + 3\log x - \frac{1}{2x^2} + c$$

Methods of Integration

Some of the common useful techniques of integration are

1. Integration by Substitution
2. Integration by Parts
3. Integration by Partial Fraction

Integration by Substitution

This method is utilized when the basic rule of integration as discussed in the previous section are not directly applicable. In such cases the given integral is converted into known standard form by substitution or changing the independent variable in terms of some other variable in order to set the integral into a basic form so that rules of integration are made applicable. This method of changing the variable is called as method of substitution. Lot many illustration are discussed below in this section for the convenience of the students.

Some Standard Integration

(i) $\int \frac{dx}{a^2 + x^2} = \frac{1}{a} \tan^{-1} \frac{x}{a}$

Proof:

$$\int \frac{dx}{a^2 + x^2}$$

Let x = at

Therefore we have,

dx = a dt, Replacing the value of x and dx

$$= \int \frac{a\,dt}{a^2 + (at)^2}$$

$$= \frac{a}{a^2} \int \frac{dt}{1+t^2}$$

$$= \frac{1}{a} \int \frac{dt}{1+t^2}$$

$$= \frac{1}{a} \tan^{-1}\left(\frac{x}{a}\right) + c$$

Alternatively

$$\int \frac{dx}{a^2 + x^2}$$

Put x = a tan t

Therefore dx = a $\sec^2$ t dt

And $a^2 + x^2 = a^2 + a^2 \tan^2 t = a^2(1 + \tan^2 t) = a^2 \sec^2 t$

And hence

$$\int \frac{dx}{a^2 + x^2} = \int \frac{a \sec^2 t dt}{a^2 + a^2 \tan^2 t}$$

$$= \int \frac{a \sec^2 t dt}{a^2 \sec^2 t}$$

$$= \frac{1}{a} \int dt$$

$$= \frac{1}{a} t + c$$

{Since x = a tan t replacing t in terms of x}

We get,

$$= \frac{1}{a} \tan^{-1}\left(\frac{x}{a}\right) + c$$

(ii) $\int \frac{dx}{\sqrt{a^2 - x^2}} = \sin^{-1}\left(\frac{x}{a}\right) + c$

Proof:

$$\int \frac{dx}{\sqrt{a^2 - x^2}}$$

Let $x = at$; $dx = a\,dt$

Replacing 'x' and dx values, we get,

$$\int \frac{dx}{\sqrt{a^2 - x^2}} = \int \frac{a\,dt}{\sqrt{a^2 - (at)^2}}$$

$$= \frac{a}{a} \int \frac{dt}{\sqrt{1 - t^2}}$$

$$= \sin^{-1} t \qquad \text{\{Since x = at\}}$$

$$= \sin^{-1} (x/a)$$

Alternatively

$$\int \frac{dx}{\sqrt{a^2 - x^2}}$$

Let $x = a \sin t$; $dx = a \cos t\, dt$

Replacing 'x' and dx values, we get,

$$\int \frac{dx}{\sqrt{a^2 - x^2}} = \int \frac{a \cos t\, dt}{\sqrt{a^2 - a^2 \sin^2 t}}$$

$$= \frac{a}{a} \int \frac{\cos\, dt}{\sqrt{1 - \sin^2 t}}$$

$$= \int \frac{\cos t \, dt}{\cos t}$$

$$= \int dt$$

$$= t + c \qquad \{\text{Since } x = a \sin t \rightarrow t = \sin^{-1}(x/a)\}$$

Therefore

$$\int \frac{dx}{\sqrt{a^2 - x^2}} = \sin^{-1}(x/a) + c$$

(iii) $\int \frac{dx}{x\sqrt{x^2 - a^2}} = \frac{1}{a} \sec^{-1}\left(\frac{x}{a}\right)$

Proof:

$$\int \frac{dx}{x\sqrt{x^2 - a^2}}$$

Substitute x = at; dx = a dt

$$\int \frac{dx}{x\sqrt{x^2 - a^2}} = \int \frac{a \, dt}{a \, t \sqrt{(at)^2 - a^2}}$$

$$= \int \frac{dt}{t\sqrt{(at)^2 - a^2}}$$

$$= \frac{1}{a} \int \frac{dt}{t\sqrt{t^2 - 1}}$$

$$= \frac{1}{a} \sec^{-1} t \qquad \left\{\text{As } t = \left(\frac{x}{a}\right)\right\}$$

Therefore

$$\int \frac{dx}{x\sqrt{x^2 - a^2}} = \frac{1}{a} \sec^{-1}\left(\frac{x}{a}\right)$$

Alternative Proof

$$\int \frac{dx}{x\sqrt{x^2 - a^2}} = \frac{1}{a} \sec^{-1}\left(\frac{x}{a}\right)$$

$$\int \frac{dx}{x\sqrt{x^2 - a^2}}$$

Put x = a sec t → dx = a sect tan t dt

$$\int \frac{dx}{x\sqrt{x^2 - a^2}} = \int \frac{a \sec \tan t\, dt}{a \sec t \sqrt{a^2 \sec^2 t - a^2}}$$

$$= \frac{1}{a}\int \frac{\tan t\, dt}{\tan t}$$

$$= \frac{1}{a}\int dt$$

$$= \frac{1}{a} t + c$$

$$= \frac{1}{a} \sec^{-1}\left(\frac{x}{a}\right) + c$$

{Since x = a sect → t = $\sec^{-1}\left(\frac{x}{a}\right)$}

(iv) $\int \frac{dx}{x^2 - a^2}\ (x > a) = \frac{1}{2a} \log\left[\frac{x-a}{x+a}\right] + c$

Proof

$$\int \frac{dx}{x^2 - a^2} = \int \left\{\frac{1}{2a}\left(\frac{1}{x-a} - \frac{1}{x+a}\right)\right\} dx$$

(By splitting using Partial Fraction)

$$= \frac{1}{2a}\left\{\int \frac{dx}{x-a} - \int \frac{dx}{x+a}\right\}$$

$$= \frac{1}{2a} \{\log(x-a) - \log(x+a) + c$$

$$= \frac{1}{2a} \log\left[\frac{x-a}{x+a}\right] + c$$

Alternative Proof

$$\left[\int \frac{dx}{x^2 - a^2}\right.$$

Put x = a sect → dx = a sec t tan t dt

$$\left[\int \frac{dx}{x^2 - a^2} = \int \frac{a \text{ sect tant dt}}{(a^2 \text{ sec}^2 t - a^2)}\right.$$

$$= \int \frac{a \text{ sect tant dt}}{(a^2 \tan^2 t)}$$

$$= \frac{1}{a}\int \frac{\text{sect dt}}{\tan t}$$

$$= \frac{1}{a}\int \frac{1 \times \cos t \text{ dt}}{\cos t \times \sin t}$$

$$= \frac{1}{a}\int \frac{1}{\sin t} \text{ dt}$$

$$= \frac{1}{a}\int \text{cosec t dt}$$

$$= \frac{1}{2a} \log \{\tan^2\left(\frac{t}{2}\right)\} + c$$

$$= \frac{1}{2a} \log \left[\frac{x-a}{x+a}\right] + c$$

{Since $\cos t = \left(\dfrac{1-\tan^2 \frac{t}{2}}{1+\tan^2 \frac{t}{2}}\right)$, $\dfrac{x}{a} = \text{sect} = \dfrac{1+\tan^2 \frac{t}{2}}{1-\tan^2 \frac{t}{2}} = \dfrac{x-a}{x+a} = \tan^2\left(\dfrac{t}{2}\right)$}

Similarly

$$\int \frac{dx}{a^2 - x^2} \ (a > x) = \frac{1}{2a} \log \left[\frac{a+x}{a-x}\right] + c$$

(Proof is left for students to practice)

Hint: $\dfrac{1}{a^2 - x^2} = \dfrac{1}{2a}\left(\dfrac{1}{x+a} + \dfrac{1}{a-x}\right)$ or put x = a sint $\rightarrow$ dx = a cost dt

(v) $\displaystyle\int \frac{1}{\sqrt{x^2 + a^2}} dx = \log\left(x + \sqrt{x^2 + a^2}\right) + c$

Proof:

$$\int \frac{1}{\sqrt{x^2+a^2}}\,dx$$

Put x = a tan t → dx = a sec^2 t dt

$$\int \frac{1}{\sqrt{x^2+a^2}}\,dx = \int \frac{a\sec^2 t}{\sqrt{a^2\ \tan^2 t + a^2}}\,dt$$

$$= \int \frac{\sec^2 t}{\sec t}\,dt$$

$$= \int \sec t\ dt$$

$$= \log\ (\sec t + \tan t) + c$$

{Since x = a tan t → tan t = $\frac{x}{a}$ and sec t = $\sqrt{\tan^2 t + 1} = \frac{1}{a}\sqrt{x^2+a^2}$}

Therefore the values we get,

$$= \log\left(x + \sqrt{x^2+a^2}\right) + c$$

Similarly

$$\int \frac{1}{\sqrt{x^2-a^2}}\,dx = \log\left(x + \sqrt{x^2-a^2}\right) + c$$

proof of the above is left to students to practice

(*Hint* = Put x = a sec t ↔ dx = a sect tantdt)

Illustration No. 1

$$\int \frac{e^{2x}+e^{-2x}}{e^{2x}-e^{-2x}}\,dx$$

Solution:

$$\int \frac{e^{2x}+e^{-2x}}{e^{2x}-e^{-2x}}\,dx$$

Let

$$e^{2x} - e^{-2x} = t$$

Since $2(e^{2x} + e^{-2x})\ dx = dt$

and hence substituting the values we get,

$$= \int \frac{dt}{2t}$$

$$= \frac{1}{2} \log t + c \qquad \{\text{Since } t = e^{x} - e^{-x}\}$$

Therefore replacing t, we get

$$= \frac{1}{2} \log (e^{2x} - e^{-2x}) + c$$

Illustration No. 2

$$\int \frac{\sqrt{5 + \log x}}{x} dx$$

Solution:

$$\int \frac{\sqrt{5 + \log x}}{x} dx$$

Put 5 + log x = t, differentiating with respect to x we get $\frac{1}{x} dx = dt$

$$= \int \sqrt{t}\, dt$$

$$= \int t^{1/2}\, dt$$

$$= \frac{2}{3} t^{3/2} + c \qquad \{\text{Since } t = 5 + \log x\}$$

$$= \frac{2}{3} (5 + \log x)^{3/2} + c$$

Illustration No. 3

$$\int \frac{1 + \cos x}{\sqrt{x + \sin x}} dx$$

Solution:

$$\int \frac{1 + \cos x}{\sqrt{x + \sin x}} dx$$

Let x + sin x = t differentiating with respect to we get,

$$(1 + \cos x)\, dx = dt$$

Replacing the values we get,

$$= \int \frac{dt}{\sqrt{t}}$$

$$= \int t^{-1/2}\, dt$$

$$= \frac{t^{\frac{1}{2}+1}}{\frac{1}{2}+1} + c$$

$$= 2t^{1/2} + c \qquad \text{[Since } t = (x + \sin x)]$$

$$= 2(x + \sin x)^{1/2} + c$$

$$= 2\sqrt{x + \sin x} + c$$

Illustration No. 4

$$\int \frac{\log \sqrt{x+5}}{(x+5)}\, dx$$

Solution:

$$\int \frac{\log \sqrt{x+5}}{(x+5)}\, dx$$

$$= \int \frac{\log (x+5)^{1/2}}{(x+5)}\, dx$$

$$= \frac{1}{2}\int \frac{\log (x+5)}{(x+5)}\, dx$$

Let $\log (x + 5) = t$; $\frac{dx}{(x+5)} = dt$

Substituting the values x and dx we get

$$\frac{1}{2}\int t\, dt = \frac{1}{2} \times \frac{t^2}{2} + c$$

$$= \frac{t^2}{4} + c \qquad \{\text{since } t = \log (x + 5)\}$$

$$= \frac{[\log(x+5)]^2}{4} + c$$

Illustration No. 5

$$\int \frac{\cos ecx}{\log(\cos ecx + \cot x)} dx$$

Solution:

$$\int \frac{\cos ecx}{\log(\cos ecx + \cot x)} dx$$

Let log [cosec x + cot x] = t

$$dt = \frac{1}{(\cos ecx + \cot x)}\left(- \text{cosecx} \cot x - \text{cosec}^2 x\right) dx$$

$$dt = - \text{cosec } x\, dx$$

Replacing the values of by t and dt we get

$$= \int \frac{-1}{t} dt = - \log t + c$$

$$= - \log [\cot x + \text{cosec } x] + c$$

(since log (cosec x + cot x) = t)

Illustration No. 6

$$\int \frac{1}{1+9x^2} dx$$

Solution:

$$\int \frac{1}{1+9x^2} dx$$

Rewriting the above integrals, we get

$$= \frac{1}{9}\int \frac{1}{\left(\frac{1}{9}\right) + x^2} dx$$

$$= \frac{1}{9}\int \frac{1}{\left(\frac{1}{3}\right)^2 + x^2} dx$$

$$= \frac{1}{9} \times \frac{1}{1/3} \times \tan^{-1}\frac{x}{1/3} + c$$

$$= \frac{1}{3} \tan^{-1}(3x) + c$$

Illustration No. 7

$$\int \frac{1}{(x-2)\left(\sqrt{x^2-2x+3}\right)} dx$$

Solution:

$$\int \frac{1}{(x-2)\left(\sqrt{x^2-2x+3}\right)} dx$$

$$= \int \frac{1}{(x-2)\ (\sqrt{x^2-2x+4-1}} dx$$

$$= \int \frac{1}{(x-2)\ (\sqrt{(x-2)^2-1}} dx$$

Let (x – 2) = t; dx = dt

$$= \int \frac{1}{t\sqrt{(t)^2-1}} dx$$

$$= \sec^{-1} t + c$$

Replacing t by (x – 2) we get

$$= \sec^{-1} (x-2) + c$$

Illustration No. 8

$$\int \frac{\cos\ (\cot^{-1}x)}{1+x^2} dx$$

Solution:

$$\int \frac{\cos\left(\cot^{-1} x\right)}{1+x^2} dx$$

Let $\cot^{-1} x = t;\ \frac{-1}{1+x^2} dx = dt$

$$= \int \cos\ t \times (-dt)$$

$$= -\int \cos\ t\ dt$$

$$= -[\sin t] + c$$

Replacing the value of t by $\cot^{-1}$ x we get,

$$= -\sin[\cot^{-1} x] + c$$

Illustration No. 9

$$\int\sqrt{\frac{1+\sin 2x}{1-\sin 2x}}$$

Solution:

$$\int\sqrt{\frac{1+\sin 2x}{1-\sin 2x}}$$

$$= \int\sqrt{\frac{1+\sin 2x}{1-\sin 2x}}$$

$$= \int\sqrt{\frac{\sin^2 x+\cos^2 x+\sin 2x}{\sin^2 x+\cos^2 x-\sin 2x}}\,dx$$

$$= \int\sqrt{\frac{\sin^2 x+\cos^2 x+2\sin x\cos x}{\sin^2 x+\cos^2 x-2\sin x\cos x}}\,dx$$

$$= \int\sqrt{\frac{(\sin x+\cos x)^2}{(\sin x-\cos x)^2}}\,dx$$

$$= \int\frac{(\sin x+\cos x)}{(\sin x-\cos x)}\,dx$$

Let sin x – cos x = t; (cos x + sin x) dx = dt

$$= \int\frac{1}{t}\,dt$$

$$= \log t + c \qquad \text{(Since sin x – cos x = t)}$$

$$= \log[\cos x - \sin x] + c$$

Illustration No.10

$$\int\frac{e^{\sin^{-1} x}}{\sqrt{1-x^2}}\,dx$$

Solution:

$$\int \frac{e^{\sin^{-1} x}}{\sqrt{1-x^2}}\,dx$$

Let $\sin^{-1} x = t$; $\dfrac{1}{\sqrt{1-x^2}} = dt$

Replacing t and dt in the original integral we get

$$\int e^t\,dt$$

$= e^t + c$ (Since $t = \sin^{-1} x$)

$= e^{\sin^{-1} x} + c$

Illustration No. 11

$$\int \frac{1}{\sin^{-1} x\sqrt{1-x^2}}\,dx$$

Solution:

Let $\sin^{-1} x = t$; $\dfrac{1}{\sqrt{1-x^2}} = dt$

Replacing the values of x and dx we get,

$$= \int \frac{1}{t}\,dt$$

$= \log t + c$ (Since $t = \sin^{-1} x$)

$= \log \sin^{-1} x + c$

Illustration No. 12

$$\int \left(1+\frac{1}{x^2}\right) e^{x-\frac{1}{x}} dx$$

Solution:

$$\int \left(1+\frac{1}{x^2}\right) e^{x-\frac{1}{x}} dx$$

Put $x - \dfrac{1}{x} = t$; $\left(1+\dfrac{1}{x^2}\right) dx = dt$

By replacing the values of x and dx we get,

$$\int\left(1+\frac{1}{x^2}\right)e^{x-\frac{1}{x}}\,dx = \int e^t dt$$

$$= e^t + c$$

$$= e^{x-\frac{1}{x}} + c$$

Illustration No. 13

$$\int\frac{1}{\sqrt{9x^2+16}}\,dx$$

Solution:

$$\int\frac{1}{\sqrt{9x^2+16}}\,dx = \int\frac{1}{3\sqrt{x^2+16/9}}\,dx$$

$$= \frac{1}{3}\int\frac{1}{\sqrt{x^2+\left(\frac{4}{3}\right)^2}}$$

$$= \frac{1}{3}\log\left[x+\sqrt{x^2+\left(\frac{4}{3}\right)^2}\right] + c$$

Illustration No. 14

$$\int\sqrt{\frac{3-x}{3+x}}\,dx$$

Solution:

$$\int\sqrt{\frac{3-x}{3+x}}\,dx$$

$$= \int\sqrt{\frac{3-x}{3+x}} \times \sqrt{\frac{3+x}{3+x}}\,dx$$

$$= \int\frac{3-x}{\sqrt{9-x^2}}\,dx$$

$$= \int\frac{3}{\sqrt{3^2-x^2}}\,dx - \int\frac{x}{\sqrt{9-x^2}}\,dx$$

$$= 3\sin^{-1}\left(\frac{x}{3}\right) - \int\frac{x}{\sqrt{9-x^2}}\,dx$$

For the second part of the integral, we use substitution method

Let $9 - x^2 = t^2$

$- 2x\,dx = 2t\,dt$ Or $x\,dx = - t\,dt$

$$\int \frac{x}{\sqrt{9-x^2}}\,dx = \int \frac{-t}{t}\,dt$$

$$= - \int dt$$

$$= - t + c$$

{Replacing t in terms of x as $t = \sqrt{9-x^2}$ }

$$= -\left(\sqrt{9-x^2}\right) + c$$

Therefore

$$\int \sqrt{\frac{3-x}{3+x}}\,dx = 3\sin^{-1}\left(\frac{x}{3}\right) - \left(\sqrt{9-x^2}\right) + c$$

Illustration No. 15

$$\int \frac{1}{\sqrt{(5-x)^2 + 2^2}}\,dx$$

Solution:

$$\int \frac{1}{\sqrt{(5-x)^2 + 2^2}}\,dx$$

Let $5 - x = t$

$-dx = dt$ or $dx = -dt$

$$= \int \frac{1}{\sqrt{t^2 + 2^2}}\,dx$$

$$= \log\left[t + \sqrt{t^2 + 2^2}\right] + c \quad \text{\{Replacing } t = 5 - x\text{\}}$$

$$= \log\left[(5-x) + \sqrt{(5-x)^2 + 4}\right] + c$$

Integration by Parts

This method is based on the product rule of differentiation and is utilized when the integrand is product of two functions.

According to this method, if u and v be two functions of x such that u is differentiable and v is integrable, then

$$\int (u \cdot v)dx = u \int v\, dx - \int \left\{ \frac{du}{dx} \int v\, dx \right\} dx$$

i.e the integral of the product of two functions = 1st Function × the Integral of the 2nd – Integral of {the derivative of the 1st × Integral of the Second}.

Proof:

Since we have $\frac{d}{dx}(u \cdot z) = \frac{du}{dz} z + u \frac{dz}{dx}$, where u, z are two differentiable function of x.

Integrating both sides w.r.t x, we get

$$uz = \int \left(\frac{du}{dx} z \right) dx + \int \left(u . \frac{dz}{dx} \right) dx$$

or

$$\int \left(u . \frac{dz}{dx} \right) dx = uz - \int \left(\frac{du}{dx} z \right) dx \qquad .....(1)$$

Let $\frac{dz}{dx} = v$, then $dz = vdx$

Therefore $z = \int v\, dx$

Hence substituting the value in (1)

$$\int (u \cdot v)dx = u \int v\, dx - \int \left\{ \frac{du}{dx} \int v\, dx \right\} dx$$

Illustration No. 1

$\int x^2 e^x dx$

Solution:

$$\int x^2 e^x\,dx = x^2 \int e^x\,dx - \int\left\{\frac{d}{dx}x^2 \int e^x\,dx\right\}dx$$

$$= x^2 e^x - \int 2x\,e^x\,dx$$

$$= x^2 e^x - 2\int x\,e^x\,dx$$

$$= x^2 e^x - 2\,\{xe^x - \int 1.\,e^x dx\}$$

$$= x^2 e^x - 2\,\{xe^x - e^x\} + c$$

$$= e^x\,(x^2 - 2x + 1) + c$$

Illustration No. 2

$\int x^n \log x\,dx$

Solution:

$$\int x^n \log x\,dx = \log x \int x^n\,dx - \int\left\{\left(\frac{d}{dx}\log x\right)\int x^n\,dx\right\}dx$$

$$= \log x\,\frac{x^{n+1}}{n+1} - \int\frac{1}{x}\times\frac{x^{n+1}}{n+1}dx$$

$$= \frac{x^{n+1}}{n+1}\log x - \frac{1}{n+1}\int x^n\,dx$$

$$= \frac{x^{n+1}}{n+1}\log x - \frac{1}{n+1}\times\frac{x^{n+1}}{n+1} + c$$

$$= \frac{x^{n+1}}{n+1}\log x - \frac{x^{n+1}}{(n+1)^2} + c$$

$$= \frac{x^{n+1}}{n+1}\left(\log x - \frac{1}{n+1}\right) + c$$

Illustration No. 3

$\int\frac{\log x}{x}\,dx$

Solution:

$$I = \int \frac{\log x}{x} dx = \int \log x \times \frac{1}{x} dx$$

$$= \log x \int \frac{1}{x} dx - \int \left\{ \frac{d}{dx} \log x \int \frac{1}{x} dx \right\} dx$$

$$= \log x \times \log x - \int \frac{1}{x} \times \log x \, dx$$

$$I = (\log x)^2 - I + c$$

$$2I = (\log x)^2 + c$$

$$I = \frac{(\log x)^2}{2} + c$$

Illustration No. 4

$\int x \cos^2 x dx$

Solution:

$$\int x \cos^2 x dx = \int x \left\{ \frac{1 + \cos 2x}{2} \right\} dx$$

$$= \frac{1}{2}\left[\int x \, (1 + \cos 2x) dx \right]$$

$$= \frac{1}{2}\left[\int (x + x\cos 2x) dx \right]$$

$$= \frac{1}{2}\left[\int x \, dx + \int x \cos 2x \, dx \right]$$

$$= \frac{1}{2}\left[\frac{x^2}{2} + \left\{ \frac{x \sin 2x}{2} - \int 1 \times \frac{\sin 2x}{2} dx \right\} \right]$$

$$= \frac{1}{2}\left[\frac{x^2}{2} + \left\{ \frac{x \sin 2x}{2} - \frac{1}{2} \int \sin 2x dx \right\} \right]$$

$$= \frac{x^2}{4} + \frac{x \sin 2x}{4} + \frac{\cos 2x}{8} + c$$

Illustration No. 5

$$\int \frac{\log \log e^x}{e^x}\, dx$$

Solution:

$$\int \frac{\log \log e^x}{e^x}\, dx$$

Now let $\log e^x = t \rightarrow \frac{1}{e^x} dx = dt$

Therefore

$$\int \frac{\log \ \log e^x}{e^x}\, dx = \int \log t \ dt$$

$$= \int 1 \times \log t \, dt$$

$$= \log \int 1 \ dt - \int \left\{ \frac{d}{dt} \log t \int 1 \ dt \right\} dt$$

$$= t \ \log t - \int \frac{1}{t} t \ dt$$

$$= t \log t - t + c$$

{Replacing t by log e^x) we get

$$= \log e^x \log \log e^x - \log e^x + c$$

$$= \log e^x (\log \log e^x - 1) + c$$

Illustration No. 6

$$\int \tan^{-1} \frac{x}{\sqrt{1-x^2}}\, dx$$

Solution:

$$\int \tan^{-1} \frac{x}{\sqrt{1-x^2}}\, dx$$

Put $x = \sin t \rightarrow dx = \cos t \ dt$

Replacing the value of x, we get,

$$\int \tan^{-1} \frac{x}{\sqrt{1-x^2}}\, dx = \int \tan^{-1} \frac{\sin t}{\sqrt{1-\sin^2 t}}\, dt$$

$$= \int \tan^{-1} (\tan t) \cos t \, dt$$

$$= \int t \cos t \, dt$$

$$= t \sin t - \int 1 \times \sin t \, dt$$

$$= t \sin t + \cos t + c$$

{Replacing t by $\sin^{-1} x$, we get}

$$= x \sin^{-1} x + \sqrt{1 - x^2} + c$$

{since $\cos t = \sqrt{1 - \sin^2 t}$ }

Illustration No. 7

$$\int \frac{\sin \sqrt{x}}{\sqrt{x}} dx$$

Solution:

$$\int \frac{\sin \sqrt{x}}{\sqrt{x}} dx$$

Put $\sqrt{x} = t \rightarrow \frac{1}{2\sqrt{x}} dx = dt$

Therefore

$$\int \frac{\sin \sqrt{x}}{\sqrt{x}} dx = \int \sin t \times 2 \, dt$$

$$= 2 \int \sin t \, dt$$

$$= -2 \cos t + c$$

Illustration No. 8

$$\int \frac{1}{x^3} e^{\frac{1}{x}} dx$$

Solution:

$$\int \frac{1}{x^3} e^{\frac{1}{x}} dx = \int \frac{1}{x^2} \times \frac{1}{x} \times e^{\frac{1}{x}} dx$$

Put $$\frac{1}{x} = t \rightarrow \frac{-1}{x^2}dx = dt$$

Replacing with values to t terms, we get,

$$\int \frac{1}{x^3} e^{\frac{1}{x}} dx = \int \frac{1}{x^2} \times \frac{1}{x} \times e^{\frac{1}{x}} dx = \int t\, e^t \left(-\frac{dt}{2}\right)$$

$$= -\int t\, e^t\, dt$$

$$= -\left[t\, e^t - \int 1 \times e^t dt\right]$$

$$= -\left[t\, e^t - e^t\right] + c$$

$$= e^t\, [1 - t] + c$$

Replacing t in terms of x we get,

$$= e^{1/x}\,[1 - 1/x] + c$$

Some Important Rule

Rule 1:

$$\int e^x \{f(x) + f'(x)\} dx = e^x\, f(x) + c$$

Proof:

$$\int e^x \{f(x) + f'(x)\} dx = \int e^x\, f(x)\, dx + \int e^x\, f'(x) dx$$

$$= f(x) \int e^x\, dx + f'(x) \int e^x\, dx$$

$$= f(x) \int e^x\, dx - \int\left\{\frac{d}{dx}\, f(x) \int e^x\, dx\right\} dx + \int e^x\, f'(x) dx$$

By integrating the 1st component by parts

$$= f(x)\, e^x - \int e^x\, f'(x) dx + \int e^x\, f'(x) dx + c$$

$$= e^x f(x) + c$$

Rule 2

$$\int x\, \{f(x) + f'(x)\} dx = x f(x) + c$$

Proof:

$$\int x \{f(x) + f'(x)\}dx = \int x\, f(x)dx + \int x\, f'(x)dx$$

$$= x\int f(x)dx + x \int f'(x)\, dx$$

$$= x\int f'(x)\, dx - \int\left\{\frac{d}{dx} x \int f'(x)\, dx\right\}dx + \int x\, f(x)dx$$

By integrating the 1st component by parts

$$= f(x)\, x - \int x f(x)dx + \int x\, f(x)dx + c$$

$$= x\, f(x) + x$$

Illustration No. 9

$$\int\frac{e^x}{x}\ (x \log x + 1)\, dx$$

Solution:

$$\int\frac{e^x}{x}\ (x \log x + 1)\, dx = \int e^x \log x\, dx + \int\frac{e^x}{x}\, dx$$

$$= \log \times \int e^x\, dx - \int\{\frac{d}{dx}\log x \int e^x\, dx\}dx + \int\frac{e^x}{x}dx$$

$$= e^x \log x - \int\frac{e^x}{x}dx + \int\frac{e^x}{x}\, dx + c$$

$$= e^x \log x + c$$

Some Standard Integrals

$$\int e^{ax} \sin bx\, dx = \frac{e^{ax}}{a^2 + b^2}\ (a \sin bx - b \cos bx) + c$$

Proof:

Let $I = \int e^{ax} \sin bx\, dx$

$$= \int e^{ax} \sin bx\, dx - \int\left\{\frac{d}{dx} e^{ax} \int \sin\ bx\, dx\right\}dx$$

$$= e^{ax}\ \frac{-(\cos bx)}{b} - \int\left\{a\, e^{ax}\ \frac{-(\cos bx)}{b}\right\}dx$$

$$= -\ \frac{e^{ax}(\cos bx)}{b} + \frac{a}{b}\int e^{ax}\cos bx\, dx$$

$$= -\frac{e^{ax}(\cos bx)}{b} + \frac{a}{b}\ \{e^{ax}\int \cos bx\, dx - \int \frac{d}{dx}\, e^{ax}\int \cos bx\}\, dx$$

$$= -\frac{e^{ax}(\cos bx)}{b} + \frac{a}{b}\ \{e^{ax}\ \frac{(\sin bx)}{b} - \frac{a}{b}\int e^{ax}\sin bx\, dx\}$$

$$= \frac{e^{ax}(\cos bx)}{b} + \frac{a}{b^2}\ e^{ax}\sin bx - \frac{a^2}{b^2}\ I$$

$$I\left(1 + \frac{a^2}{b^2}\right) = \frac{e^{ax}(\cos bx)}{b} + \frac{a}{b^2}\ e^{ax}\sin bx$$

$$I\left(a^2 + b^2\right) = e^{ax}\ \left(-\, b\cos bx + a\sin bx\right)$$

$$I = \frac{e^{ax}}{\left(a^2 + b^2\right)}\ [a\sin bx + b\cos bx]$$

Similarly

$$\int e^{ax}\ \cos bx\ dx = \frac{e^{ax}}{a^2 + b^2}\ (a\cos bx + b\sin bx) + c$$

Proof:

Let $I = \int e^{ax}\ \cos bx\ bx\ dx$

$$= \int e^{ax}\cos bx\ dx - \int\left\{\frac{d}{dx}\, e^{ax}\int \cos bx\ dx\right\}dx$$

$$= e^{ax}\ \frac{(\sin bx)}{b} - \int\left\{a\, e^{ax}\,\frac{(\sin bx)}{b}\right\}dx$$

$$= e^{ax}\ \frac{(\sin bx)}{b} - \frac{a}{b}\int e^{ax}\sin bx\, dx$$

$$= \frac{e^{ax}(\sin bx)}{b} - \frac{a}{b}\{e^{ax}\int \sin bx\,dx - \int \frac{d}{dx}e^{ax}\int \sin bx\,dx$$

$$= \frac{e^{ax}(\sin bx)}{b} - \frac{a}{b}\{e^{ax}\frac{-(\cos bx)}{b} + \frac{a}{b}\int e^{ax}\frac{-(\cos bx)}{b}dx\}$$

$$= \frac{e^{ax}(\sin bx)}{b} - \frac{a}{b^2}e^{ax}\cos bx - \frac{a^2}{b^2}I$$

$$I\left(1+\frac{a^2}{b^2}\right) = \frac{e^{ax}(\sin bx)}{b} + \frac{a}{b^2}e^{ax}\cos bx$$

$$I\left(a^2+b^2\right) = e^{ax}\ (a\cos bx + b\sin bx)$$

$$I = \frac{e^{ax}}{\left(a^2+b^2\right)}[a\cos bx + b\sin bx]$$

Illustration No. 10

$\int e^{\cos x}\sin 2x\ dx$

Solution:

$\int e^{\cos x}\sin 2x\ dx$

$$= \int e^{\cos x}2\sin x\cos x\ dx$$

[Since $\sin 2x = 2\sin x\cos x$]

$$= 2\int e^{\cos x}\ x\sin x\cos x\ dx$$

Let $\cos x = t$ $\rightarrow$ $-\sin x\ dx = dt$

Therefore

$$= 2\int e^t \times t\ (-dt)$$

$$= -2\int e^t \times t\ dt$$

$$= -2[t\int e^t dt - \int\{\frac{d}{dt}t\int e^t dt\}\ dt$$

$$= -2[te^t - \int 1\times e^t\ dt]$$

$$= -2[te^t - \int e^t\,dt]$$

$$= -2[te^t - e^t] + c$$

Replacing t = cos x

$$= -2\,[\cos x\, e^{\cos x} - e^{\cos x}] + c$$

$$= -2\, e^{\cos x}\,[\cos x - 1] + c$$

$$= 2\, e^{\cos x}\,[1 - \cos x] + c$$

Illustration No. 11

$\int x^2 (\log x)^2 dx$

Solution:

$\int x^2 (\log x)^2 dx$

$$= (\log x)^2 \int x^2\,dx - \int \{\frac{d}{dx}(\log x)^2 \int x^2\,dx\}\,dx$$

$$= \frac{x^3}{3}(\log x)^2 - \int 2\log x \times \frac{1}{x} \times \frac{x^3}{3}\,dx$$

$$= \frac{x^3}{3}(\log x)^2 - \frac{2}{3}\int x^2 \;\log x\;dx$$

$$= \frac{x^3}{3}(\log x)^2 - \frac{2}{3}\left(\log x \int x^2\,dx - \int\left\{\frac{d}{dx}\log x \int x^2\,dx\right\}dx\right)$$

$$= \frac{x^3}{3}(\log x)^2 - \frac{2}{3}\left(\frac{x^3}{3}\log x - \int \frac{1}{x} \times \frac{x^3}{3}\,dx\right)$$

$$= \frac{x^3}{3}(\log x)^2 - \frac{2x^3}{9}\log x - \int \frac{x^2}{9}\,dx$$

$$= \frac{x^3}{3}(\log x)^2 - \frac{2x^3}{9}\log x - \frac{1}{9}\int x^2\,dx$$

$$= \frac{x^3}{3}(\log x)^2 - \frac{2x^3}{9}\log x - \frac{1}{9}\frac{x^3}{3}$$

$$= \frac{x^3}{3}(\log x)^2 - \frac{2x^3}{9}\log x - \frac{x^3}{27} + c$$

Illustration No. 12

$\int x^2 \cos x\, dx$

Solution:

$\int x^2 \cos x\, dx$

$$= \int x^2 \cos x\, dx - \int \left\{ \frac{d}{dx} x^2 \int \cos x\, dx \right\} dx$$

$$= x^2 \sin x - \int 2x \sin x\, dx$$

$$= x^2 \sin x - 2 \int x \sin x\, dx$$

$$= x^2 \sin x - 2 [x \int \sin x\, dx - \int \{\frac{d}{dx} x \int \sin x\, dx\} dx$$

$$= x^2 \sin x - 2 [x(-\cos x) - \int 1(-\cos x)\, dx$$

$$= x^2 \sin x - 2 [-x \cos x + \int \cos x\, dx]$$

$$= x^2 \sin x + 2x \cos x - 2 \sin x + c$$

Some Standard Integrals

(i) $\int \sqrt{x^2 + a^2}\, dx$

Proof:

Let $I = \int 1 \times \sqrt{x^2 + a^2}\, dx$

$$= \sqrt{x^2 + a^2} \int 1\, dx - \int \left\{ \frac{d}{dx} \sqrt{x^2 + a^2} \int 1\, dx \right\} dx$$

$$= x\sqrt{x^2 + a^2} - \int \frac{2x \times x\, dx}{2\sqrt{x^2 + a^2}}$$

$$= x\sqrt{x^2 + a^2} - \int \frac{x^2\, dx}{\sqrt{x^2 + a^2}}$$

Adding and subtracting a^2 in the Second Part

$$= x\sqrt{x^2 + a^2} - \int \frac{(x^2 + a^2 - a^2)\, dx}{\sqrt{x^2 + a^2}}$$

$$= x\sqrt{x^2+a^2} - \int\sqrt{x^2+a^2} + \int\frac{a^2\,dx}{\sqrt{x^2+a^2}}$$

$$= x\sqrt{x^2+a^2} + \int\frac{a^2\,dx}{\sqrt{x^2+a^2}} - I$$

$$2I = x\sqrt{x^2+a^2} + a^2\int\frac{dx}{\sqrt{x^2+a^2}}$$

$$2I = x\sqrt{x^2+a^2} + a^2\log[x+\sqrt{x^2+a^2}] + c$$

$$\int\sqrt{x^2+a^2}\,dx = \frac{1}{2}[x\sqrt{x^2+a^2} + a^2\log[x+\sqrt{x^2+a^2}] + c$$

(ii) $\int\sqrt{x^2-a^2}\,dx$

Proof:

Let $I = \int 1\times\sqrt{x^2-a^2}\,dx$

$$= \sqrt{x^2-a^2}\int 1\,dx - \int\left\{\frac{d}{dx}\sqrt{x^2-a^2}\int 1\,dx\right\}dx$$

$$= x\sqrt{x^2-a^2} - \int\frac{2x\times x\,dx}{2\sqrt{x^2-a^2}}$$

$$= x\sqrt{x^2-a^2} - \int\frac{x^2\,dx}{\sqrt{x^2-a^2}}$$

Adding and subtracting a^2 in the Second Part

$$= x\sqrt{x^2-a^2} - \int\frac{(x^2-a^2+a^2)\,dx}{\sqrt{x^2-a^2}}$$

$$= x\sqrt{x^2-a^2} - \int\sqrt{x^2-a^2} - \int\frac{a^2\,dx}{\sqrt{x^2-a^2}}$$

$$= x\sqrt{x^2-a^2} - \int\frac{a^2\,dx}{\sqrt{x^2-a^2}} - I$$

$$2I = x\sqrt{x^2-a^2} - a^2\int\frac{dx}{\sqrt{x^2-a^2}}$$

$$2I = x\sqrt{x^2 - a^2} - a^2 \log[x + \sqrt{x^2 - a^2}] + c$$

$$I = \frac{1}{2}[x\sqrt{x^2 - a^2} - a^2 \log[x + \sqrt{x^2 - a^2}] + c$$

(iii) $\int \sqrt{a^2 - x^2}\, dx$

Proof:

Let $I = \int 1 \times \sqrt{a^2 - x^2}\, dx$

$$= \sqrt{a^2 - x^2} \int 1 dx - \int \left\{ \frac{d}{dx} \sqrt{a^2 - x^2} \int 1\, dx \right\} dx$$

$$= x\sqrt{a^2 - x^2} - \int \frac{-2x \times x\, dx}{2\sqrt{a^2 - x^2}}$$

$$= x\sqrt{a^2 - x^2} + \int \frac{x^2\, dx}{\sqrt{a^2 - x^2}}$$

Adding and subtracting a^2 in the Second Part

$$= x\sqrt{a^2 - x^2} - \int \frac{-(a^2 - x^2) + a^2 dx}{\sqrt{a^2 - x^2}}$$

$$= x\sqrt{a^2 - x^2} - \int \sqrt{a^2 - x^2} + \int \frac{a^2\, dx}{\sqrt{a^2 - x^2}}$$

$$= x\sqrt{a^2 - x^2} + \int \frac{a^2 dx}{\sqrt{a^2 - x^2}} - I$$

$$2I = x\sqrt{a^2 - x^2} + a^2 \int \frac{dx}{\sqrt{a^2 - x^2}}$$

$$2I = x\sqrt{a^2 - x^2} + a^2 \sin^{-1}\left(\frac{x}{a}\right) + c$$

$$I = \frac{1}{2}\left[x\sqrt{x^2 - a^2} + a^2 \sin^{-1}\left(\frac{x}{a}\right)\right] + c$$

Illustration No. 13

$\int \sqrt{25 - x^2}\, dx$

Solution:

Let $I = \int\sqrt{25 - x^2}\ dx$

$$= \int\sqrt{5^2 - x^2}\ dx$$

$$= \frac{1}{2}\left[x\sqrt{5^2 - x^2}\right] + 5^2 \sin^{-1}\left\{\frac{x}{5}\right\}\} + c$$

$$= \frac{1}{2}\left[x\sqrt{25 - x^2}\right] + 25 \sin^{-1}\left\{\frac{x}{5}\right\}\} + c$$

Illustration No: 14

$\int\sqrt{9x^2 - 16}\ dx$

Solution:

Let $I = \int\sqrt{9x^2 - 16}\ dx$

Take 9 as common factor within the roots, then we have

$$= 3\int\sqrt{x^2 - \left(\frac{4}{3}\right)^2}\ dx$$

$$= 3\left[\frac{1}{2}\left\{x\sqrt{x^2 - \left(\frac{4}{3}\right)^2} - \left(\frac{4}{3}\right)^2\left(\log x + \sqrt{x^2 - \left(\frac{4}{3}\right)^2}\right)\right\}\right] + c$$

$$= \left[\frac{3}{2}\left\{x\sqrt{x^2 - \left(\frac{4}{3}\right)^2} - \frac{16}{9}\left(\log x + \sqrt{x^2 - \left(\frac{4}{3}\right)^2}\right)\right\}\right] + c$$

Illustration No.15

$\int x\ \sqrt{x^2 - 4}\ dx$

Solution:

Let $I = \int x\sqrt{x^4 - 4}\ dx$

Assume $x^2 = t$

Then $2x\ dx = dt \rightarrow x\ dx = \frac{1}{2}\ dt$

Therefore $\int x\sqrt{x^4-4}\,dx = \int\sqrt{t^2-2^2}\times\frac{1}{2}dt$

$$= \frac{1}{2}\int\sqrt{t^2-2^2}\times dt$$

$$= \frac{1}{2}\left[\frac{1}{2}\left\{t\sqrt{t^2-(2)^2}-(2)^2\left(\log t+\sqrt{t^2-(2)^2}\right)\right\}\right]+c$$

$$= \left[\frac{1}{4}\left\{t\sqrt{t^2-4}-4\left(\log t+\sqrt{t^2-4}\right)\right\}\right]+c$$

$$= \left[\frac{1}{4}\left\{x^2\sqrt{x^4-4}-4\left(\log x^2+\sqrt{x^4-4}\right)\right\}\right]+c$$

Illustration No. 16

$\int e^{3x}\left(e^{6x}+9\right)^{\frac{1}{2}}dx$

Solution:

$I = \int e^{3x}\left(e^{6x}+9\right)^{\frac{1}{2}}dx$

Put $e^{3x} = t \rightarrow 3e^{3x} = dt$

Therefore

$$\int e^{3x}\left(e^{6x}+9\right)^{\frac{1}{2}}dx = \int\sqrt{t^2+3^2}\times\frac{1}{3}dt$$

$$= \frac{1}{3}\int\sqrt{t^2+3^2}\,dt$$

$$= \frac{1}{3}\left[\frac{1}{2}\left\{t\sqrt{t^2+(3)^2}+(3)^2\left(\log t+\sqrt{t^2+(3)^2}\right)\right\}\right]+c$$

$$= \left[\frac{1}{6}\left\{t\sqrt{t^2+9}+9\left(\log t+\sqrt{t^2+9}\right)\right\}\right]+c$$

$$= \left[\frac{1}{6}\left\{e^{3x}\sqrt{e^{6x}+9}+9\left(\log e^{3x}+\sqrt{e^{6x}+9}\right)\right\}\right]+c$$

Illustration No. 17

$\int \sec x\tan x\sqrt{\sec^2 x+3}\,dx$

Solution:

$$I = \int \sec x \tan x \sqrt{\sec^2 x + 3}\, dx$$

Let $\sec^2 x = t$, $\sec x \tan x\, dx = dt$

$$= \int \sqrt{t^2 + 3}\, dt$$

$$= \int \sqrt{t^2 + \left(\sqrt{3}\right)^2}\, dt$$

$$= \left[\frac{1}{2}\left\{t\sqrt{t^2 + \left(\sqrt{3}\right)^2} + \left(\sqrt{3}\right)^2 \left(\log t + \sqrt{t^2 + \left(\sqrt{3}\right)^2}\right)\right\}\right] + c$$

Replacing t with sec x we get

$$= \left[\frac{1}{2}\left\{\sec x \sqrt{\sec^2 x + \left(\sqrt{3}\right)^2} + \left(\sqrt{3}\right)^2 \left(\log \sec x + \sqrt{\sec^2 x + \left(\sqrt{3}\right)^2}\right)\right\}\right] + c$$

Illustration No. 18

$$\int \frac{x \sin^{-1} x}{\sqrt{1 - x^2}}\, dx$$

Solution:

$$\int \frac{x \sin^{-1} x}{\sqrt{1 - x^2}}\, dx$$

Let $\sin^{-1} x = t$; $\frac{1}{\sqrt{1 - x^2}}\, dx = dt$

Since $\sin^{-1} x = t$ or $x = \sin t$

Therefore we get

$$\int \frac{x \sin^{-1} x}{\sqrt{1 - x^2}}\, dx = \int t \sin t\, dt$$

$$= -t \cos t - \int \left\{\frac{d}{dt} t \int \sin t\, dt\right\} dt$$

$$= -t \cos t - \int 1(-\cos t)\, dt$$

$$= -t \cos t + \int \cos t \, dt$$

$$= -t \cos t + \sin t + c$$

Replacing $t = \sin^{-1} x$

$$= -\sqrt{1-x^2} \sin^{-1} x + x + c$$

$\{$since $\sin \sin^{-1} t = x$ and $\cos t = \sqrt{1-\sin^2 t}$

$$= \{\sqrt{1-x^2}\}$$

Illustration No. 19

$\int \log (1+x) \, dx$

Solution:

$$\int \log (1+x) \, dx = \int 1 \times \log (1+x) \, dx$$

$$= \log (1+x) \int 1 \, dx - \int \left\{ \frac{d}{dx} \log (1+x) \int 1 \, dx \right\} dx$$

$$= x \log (1+x) - \int \frac{1}{(1+x)} \times x \, dx$$

Adding subtracting 1 in the numerator we get,

$$= x \log (1+x) - \int \frac{(x+1)-1}{(x+1)} \, dx$$

$$= x \log (1+x) - \int 1 \, dx + \int \frac{1}{(1+x)} \, dx$$

$$= x \log (1+x) - x + \log (1+x) + c$$

$$= (1+x) \log (1+x) - x + c$$

Integration by Partial Fraction

If u(x) and v(x) two functions which can be expressed in rational form i.e. $\frac{u(x)}{v(x)}$ where both u(x) and v(x) are polynomial functions. As all the polynomials with real coefficient can be split into product of real factors such that the rational algebraic function can be resolved into partial fractions. Once these partial fractions are obtained then each term can be integrated one by one. There exists three such cases while resolving a rational function into partial fraction.

1. The degree of the numerator is greater than the denominator i.e. the degree of u(x) is greater than the degree of v(x).
2. The degree of the numerator is equal to the denominator i.e. the degree of u(x) is equal to the degree of v(x).
3. The degree of the numerator is lesser than the denominator i.e. the degree of u(x) is less than the degree of v(x).

In the first two cases the fraction is an improper fraction and hence we divide the numerator until the fraction is converted into a proper form. And the partial fractions are written according to the following forms

Case I

When the denominator v(x) is expressed as the product of non repeating linear factors

Let us assume $u(x) = (x-a)(x-b)(x-c)\ ...$

Then

$$\frac{u(x)}{v(x)} = \frac{A}{(x-a)} + \frac{B}{(x-c)} + \frac{c}{(x-c)} + \ldots$$

Case II

When the denominator v(x) is of the form of $(ax + b)^n$

$$\frac{u(x)}{v(x)} = \frac{A}{(ax+b)} + \frac{B}{(ax+b)^2} + \frac{c}{(ax+b)^3} + \ldots$$

Case III

When the denominator v(x) is in quadratic form like $ax^2 + bx + c$ then the partial fraction can be expressed as

$$\frac{u(x)}{v(x)} = \frac{Ax+B}{ax^2+bx+c}$$

Case IV

When the denominator v(x) is of the form like $(ax^2 + bx + c)^n$ then the partial fraction can be expressed as

$$\frac{u(x)}{v(x)} = \frac{Ax + B}{ax^2 + bx + c} + \frac{Cx + D}{\left(ax^2 + bx + c\right)^2} + \ldots$$

In order to find the constant terms, we compare the coefficients of like powers of x on both sides to get a set of simultaneous equations which on solving we get the values of constant terms and hence the partial fractions are evaluated. The following solved illustrations will elaborate on the concepts.

Illustration No. 1

$$\int \frac{1}{x^2 - 3x - 4}\, dx$$

Solution:

$$\int \frac{1}{x^2 - 3x - 4}\, dx$$

$$\frac{1}{x^2 - 3x - 4} = \frac{1}{(x-4)(x+1)} = \frac{A}{(x+1)} + \frac{B}{x-4}$$

$$= \frac{A(x-4) + B(x+1)}{(x+1)\,(x-4)}$$

Equating coefficient of x & constant terms both side, we get,

$A + B = 0, - 4A + B = 1$

Solving both equations in A & B we get

$$A = \frac{-1}{5},\ B = \frac{-1}{5}$$

$$\frac{1}{x^2 - 3x - 4} = \frac{-1}{5(x+1)} + \frac{1}{5(x-4)}$$

$$\int \frac{1}{x^2 - 3x - 4}\, dx = \int \frac{-1}{5(x+1)}\, dx + \int \frac{1}{5(x-4)}\, dx$$

$$= \frac{-1}{5}\log(x+1) + \frac{1}{5}\log(x-4) + c$$

$$= \frac{1}{5}\log\frac{x-4}{x+1}$$

Illustration No. 2

$$\int \frac{1}{x^2-7}\,dx$$

Solution:

$$\int \frac{1}{x^2-7}\,dx = \int \frac{1}{x^2-\left(\sqrt{7}\right)^2}\,dx$$

$$= \frac{1}{2\sqrt{7}}\log\frac{x-\sqrt{7}}{x+\sqrt{7}} + c$$

Illustration No. 3

$$\int \frac{x-1}{x^2-5x+6}\,dx$$

Solution:

$$\int \frac{x-1}{x^2-5x+6}\,dx$$

As $\int \frac{x-1}{x^2-5x+6} = \frac{(x-1)}{(x-2)(x-3)} = \frac{A}{(x-2)} + \frac{B}{(x-3)}$

Since $(x-1) = A(x-3) + B(x-2)$

$$= x(A+B) - (3A+2B)$$

Equating the coefficient of x & constant term both sides, we get

$$A + B = 1,\ -(3A + 2B) = -1$$

Solving both equations in A & B

We get $A = -1$, $B = 2$

Therefore

$$\frac{x-1}{x^2-5x+6} = \frac{-1}{(x-2)} + \frac{2}{x-3}$$

$$\int \frac{x-1}{x^2-5x+6}\,dx = \int \frac{-1}{(x-2)}\,dx + \int \frac{2}{(x-3)}\,dx$$

$$= -\log(x-2) + 2\log(x-3)$$

$$= \log(x-3)^2 - \log(x-2) + c$$

$$= \log \frac{(x-3)^2}{(x-2)} + c$$

Illustration No. 4

$$\int \frac{1}{(x-1)(x^2-9)}\,dx$$

Solution:

$$\int \frac{1}{(x-1)(x^2-9)}\,dx$$

As $(x-1)(x^2-9) = (x-1)(x-3)(x+3)$

Therefore

$$\frac{1}{(x-1)\ (x-3)\ (x+3)} = \frac{A}{x-1} + \frac{B}{(x-3)} + \frac{C}{x+3}$$

$$= \frac{A(x-3)(x+3) + B(x-1)(x+3) + C(x-1)(x-3)}{(x-1)\ (x-3)(x+3)}$$

$$= \frac{Ax^2 - 9A + Bx^2 + 2Bx + 3B + Cx^2 - 4Cx + 3C}{(x-1)(x-3)(x+3)}$$

$$= \frac{(A+B+C)x^2 + (2B-4C)x + (-9A+3B+3C)}{(x-1)(x-3)(x+3)}$$

Equating the coefficient of x^2, x and Constant term where

$(A + B + C) = 0,$

$2B - 4C = 0,\ -9A - 3B + 3C = 1$

Solving for A, B, C we get $A = -1/8$, $B = 1/12$ and $C = 1/24$

$$= \int \frac{1}{(x-1)(x^{2-}9)} dx = \int \frac{-1}{8(x-1)} dx + \int \frac{1}{12(x-2)} dx + \int \frac{1}{24(x+3)} dx$$

$$= \frac{-1}{8} \log(x-1) + \frac{1}{12} \log(x-2) + \frac{1}{24} \log(x+3) + c$$

Illustration No. 5

$$\int \frac{1}{x^3 - 1} dx$$

Solution:

$$\int \frac{1}{x^3 - 1} dx$$

Since we know that $x^3 - 1 = (x - 1)(x^2 + x + 1)$

Let us assume that

$$\frac{1}{(x-1)(x^2+x+1)} = \frac{A}{x-1} + \frac{Bx+c}{x^2+x+1}$$

$\rightarrow 1 = A(x^2 + x + 1) + (Bx + C)(x - 1)$

$1 = Ax^2 + Ax + A + Bx^2 - Bx + Cx - C$

$1 = (A + B)x^2 + x(A - B + C) + A - C$

Equating the coefficient of x^2, x and Constant term we get,

$A + B = 0$

$A - B + C = 0$

$A - C = 1$

Solving the above three equations in A, B and C, we get

$A = 1/3$, $B = -1/3$ and $C = -2/3$

$$\int \frac{1}{x^3 - 1} dx = \int \frac{1}{3(x-1)} dx + \int \frac{(-x-2)}{3(x^2+x+1)} dx$$

$$= \int \frac{1}{3(x-1)} dx - \int \frac{(x+2)}{3(x^2+x+1)} dx$$

$$= \frac{1}{3}\int\frac{1}{(x-1)}\,dx - \frac{1}{6}\int\frac{(2x+4)}{(x^2+x+1)}\,dx$$

$$= \frac{1}{3}\log(x-1) - \frac{1}{6}\int\frac{(2x+1)+3}{(x^2+x+1)}\,dx$$

$$= \frac{1}{3}\log(x-1) - \frac{1}{6}\left\{\int\frac{(2x+1)}{(x^2+x+1)}\,dx + \int\frac{3}{(x^2+x+1)}\,dx\right\}$$

$$= \frac{1}{3}\log(x-1) - \frac{1}{6}\left\{\int\frac{(2x+1)}{(x^2+x+1)}\,dx + 3\int\frac{1}{(x^2+x+1)}\,dx\right\}$$

$$= \frac{1}{3}\log(x-1) - \frac{1}{6}\log\left(x^2+x+1\right) + \frac{1}{2}\int\frac{1}{\left(x+\frac{1}{2}\right)^2+\left(\frac{\sqrt{3}}{2}\right)^2}\,dx$$

$$= \frac{1}{3}\log(x-1) - \frac{1}{6}\log\left(x^2+x+1\right) + \frac{1}{2}\times\frac{2}{\sqrt{3}}\tan^{-1}\frac{\left(x+\frac{1}{2}\right)}{\frac{\sqrt{3}}{2}}$$

$$= \frac{1}{3}\log(x-1) - \frac{1}{6}\log\left(x^2+x+1\right) + \frac{1}{\sqrt{3}}\tan^{-1}\frac{(2x+1)}{\sqrt{3}} + c$$

Illustration No. 6

$$\int\frac{x^2}{(x-1)(x-2)(x-3)}\,dx$$

Solution:

$$\int\frac{x^2}{(x-1)(x-2)(x-3)}\,dx$$

Let

$$\frac{x^2}{(x-1)(x-2)(x-3)} = \frac{A}{(x-1)} + \frac{B}{(x-2)} + \frac{C}{(x-3)}$$

Therefore we have,

$$x^2 = A(x-2)(x-3) + B(x-1)(x-3) + C(x-1)(x-2)$$

Equating the x^2, x and constant terms both side we have,

$$A + B + C = 1$$

$$-5A - 4B - 3C = 0$$

$$6A + 3B + 2C = 0$$

Solving the above three equation, we get

$$A = \frac{1}{2},\ B = -\frac{4}{1} \text{ and } C = \frac{9}{2}$$

And hence

$$\int \frac{x^2}{(x-1)(x-2)(x-3)} dx = \int \frac{A}{(x-1)} dx + \int \frac{B}{(x-2)} dx + \int \frac{C}{(x-3)} dx$$

$$= A \int \frac{1}{(x-1)} dx + B \int \frac{1}{(x-2)} dx + C \int \frac{1}{(x-3)} dx$$

$$= \frac{1}{2} \int \frac{1}{(x-1)} dx - 4 \int \frac{1}{(x-2)} dx + \frac{9}{2} \int \frac{1}{(x-3)} dx$$

$$= \frac{1}{2} \log(x-1) - 4\log(x-2) + \frac{9}{2}\log(x-3) + c$$

Illustration No. 7

$$\int \frac{\sin x}{(1+\cos x)(2+\cos x)} dx$$

Solution:

$$\int \frac{\sin x}{(1+\cos x)(2+\cos x)} dx$$

Let cos x = t, –sin x dx = dt, therefore, we have,

$$\int \frac{\sin x}{(1+\cos x)(2+\cos x)} dx = \int \frac{-1}{(1+t)(2+t)} dt$$

$$= \frac{-1}{(1+t)(2+t)} = \frac{A}{(1+t)} + \frac{B}{(2+t)}$$

Equating the x and constant terms both side we have,

$$A + B = 0$$

And

$$2A + B = -1$$

And solving both the equations, we get, A = –1 and B = 1

$$\frac{-1}{(1+t)(2+t)} = \frac{-1}{(1+t)} + \frac{1}{(2+t)}$$

$$\int \frac{-1}{(1+t)(2+t)} dt = \int \frac{-1}{(1+t)} dt + \int \frac{1}{(2+t)} dt$$

$$= \log(2 + t) - \log(1 + t) + c$$

$$= \log\left(\frac{2+t}{1+t}\right) + c$$

$$= \log\left(\frac{2+\cos x}{1+\cos x}\right) + c$$

Illustration No. 8

$$\int \frac{1}{x(x^a+1)} dx$$

Solution:

$$\int \frac{1}{x(x^a+1)} dx$$

Let $x^a = t$ then, we have $a\, x^{a-1} = dt$ or $x^{a-1} = \frac{1}{a} dt$

Therefore,

$$\int \frac{1}{x(x^a+1)} dx$$

$$= \int \frac{x^{a-1}}{x^{a-1}\, x(x^a+1)} dx$$

$$= \int \frac{x^{a-1}}{x^a\left(x^a+1\right)}\,dx$$

$$= \int \frac{1}{t(t+1)} \times \frac{1}{n} \times dt$$

$$= \frac{1}{n}\int\left(\frac{1}{t} - \frac{1}{t+1}\right)dt$$

$$= \frac{1}{n}\left[\log t - \log(t+1)\right] + c$$

$$= \frac{1}{n}\log\left[\frac{t}{(t+1)}\right] + c$$

Replacing $t = x^a$,

We get,

$$= \frac{1}{n}\log\left[\frac{x^a}{\left(x^a+1\right)}\right] + c$$

Some Standard Integral Substitutions Forms

(i) $\int \frac{1}{ax^2+bx+c}\,dx$

Proof:

$$ax^2 + bx + c = a[x^2 + \frac{b}{a}x + \frac{c}{a}]$$

$$= a\ [x + \frac{b}{2a})^2 - \frac{b^2}{4a^2} - \frac{c}{a}]$$

$$= a\left[\left(x + \frac{b}{2a}\right)^2 - \frac{b^2-4ac}{4a^2}\right]$$

There exists two case

Case 1

When $b^2 - 4ac$ is positive then,

$$\int \frac{1}{ax^2+bx+c}\,dx = \frac{1}{a}\int \frac{1}{\left(x+\frac{b}{2a}\right)^2-\left(\frac{\sqrt{b^2-4ac}}{2a}\right)^2}\,dx \quad \left(\text{form } \int \frac{1}{x^2-a^2}\,dx\right)$$

$$= \frac{1}{a} \times \frac{1}{2\frac{\sqrt{b^2-4ac}}{2a}} \log \frac{\left(x+\frac{b}{2a}\right)-\frac{\sqrt{b^2-4ac}}{2a}}{\left(x+\frac{b}{2a}\right)+\frac{\sqrt{b^2-4ac}}{2a}} + c$$

$$= \frac{1}{\sqrt{b^2-4ac}} \log \frac{(2ax+b)-\sqrt{b^2-4ac}}{(2ax+b)+\sqrt{b^2-4ac}} + c$$

Case 2

When $b^2 - 4ac$ is negative then,

$$\int \frac{1}{ax^2+bx+c}\,dx = \frac{1}{a}\int \frac{1}{\left(x+\frac{b}{2a}\right)^2+\left(\frac{\sqrt{4ac-b^2}}{2a}\right)^2}\,dx \quad \left(\text{form } \int \frac{1}{x^2+a^2}\,dx\right)$$

$$= \frac{1}{a} \times \frac{1}{\frac{\sqrt{4ac-b^2}}{2a}} \tan^{-1} \frac{\left(x+\frac{b}{2a}\right)}{\left(\frac{\sqrt{4ac-b^2}}{2a}\right)} + c$$

$$= \frac{2}{\sqrt{4ac-b^2}} \tan^{-1} \frac{2ax+b}{\sqrt{4ac-b^2}} + c$$

(ii) $\int \frac{px+q}{ax^2+bx+c}\,dx$

Proof:

Let us write

$$px + q = M\frac{d}{dx}\left(ax^2+bx+c\right) + N$$

$$= M\,(2ax + b) + N$$

$$= 2Max + Mb + N$$

Comparing the Coefficient of x and constant term we have

$$M = \frac{p}{2a} \text{ and } N = q - \frac{p}{2a} b$$

Therefore,

$$\frac{px + q}{ax^2 + bx + c} = \frac{M\frac{d}{dx}\left(ax^2 + bx + c\right) + N}{ax^2 + bx + c}$$

$$= \log\left(ax^2 + bx + c\right) + N \int \frac{1}{ax^2 + bx + c} dx$$

The second component is the form of $\int \frac{1}{x^2 \pm a^2} dx$

Illustration No. 9

$$\int \frac{1}{x^2 + 6x + 13} dx$$

Solution:

$$= \int \frac{1}{x^2 + 6x + 13} dx$$

$$= \int \frac{1}{x^2 + 6x + 9 + 4} dx$$

$$= \int \frac{1}{(x+3)^2 + 4} dx$$

$$= \int \frac{1}{(x+3)^2 + 2^2} dx$$

$$= \frac{1}{2} \tan^{-1}\left(\frac{x+3}{2}\right) + c$$

Definite Integrals

Definite Integration as the Limit of a Sum

Let f(x) be single–valued continuous function defined in a closed interval $a \le x \le b$ where a and b are both finite i.e. f(x) is bounded within [a, b].

Let us divide the interval [a, b] into n equal sub-parts each of length h, such that

$a < a + h < a + 2h <\ldots < a +(r - 1)h < a + rh <\ldots a + (n - 1)h < a + nh = b$

Then the sum

$hf(a) + h f(a + h) + h f(a + 2h) + \ldots h f(a + rh) +\ldots h f\{ a + (n-1) h\}$

$= h [f(a) + f(a + h) + f(a + 2h) + \ldots f(a + rh) +\ldots f\{ a + (n - 1) h\}]$

$= h \sum_{r=0}^{n-1} f(a + rh)$ where a + nh = b or nh = b – a

Then the limit $\lim_{h\to 0} h\sum_{r=0}^{n-1} f(a+rh)$ if exists then called as definite integral of f(x) with respect to x in between the limits a and b and denoted by $\int_a^b f(x)dx$.

Therefore

$$\int_a^b f(x)dx = \lim_{h\to 0} h\sum_{r=0}^{n-1} f(a+rh) \qquad \text{where } nh = b - a$$

Remarks

Instead of the above form we may state the sum as

$hf(a) + h f(a + h) + h f(a + 2h) + \ldots h f(a + rh) +\ldots h f(a + n h)$

$$= h \sum_{r=0}^{n} f(a+rh) \qquad \text{where } a + nh = b \text{ or } nh = b - a$$

In this case

$$\int_a^b f(x)dx = \lim_{h\to 0} h\sum_{r=0}^{n} f(a+rh) \qquad \text{where } nh = b - a$$

Illustration No. 1

Calculate the integral as the limit of the sum

$$\int_1^2 2x dx$$

Solution:

Here f(x) = 2x, a = 1 and b = 2

Therefore f(a + rh) = 2(1 + rh) and nh = b – a = 2 – 1 = 1

By the definition definite integral

$$\int_1^2 5x\,dx = \lim_{h\to 0} h\sum_{r=1}^{n} f(a+rh) = \lim_{h\to 0} h\sum_{r=1}^{n} 2(1+rh)$$

$$\lim_{h\to 0}\ 2h\sum_{r=1}^{n} 1 + h\sum_{r=1}^{n} r = 2\lim_{h\to 0} h\left[n + h\,\frac{n(n+1)}{2}\right]$$

Since $\sum_{r=1}^{n} r = 1 + 2 + 3 + \dots + n = \frac{n(n+1)}{2}$

$$= 2\lim_{h\to 0}\left[nh + \frac{nh(nh+h)}{2}\right], \text{ where } nh = 1,$$

$$2\lim_{h\to 0}\left[1 + \frac{1(1+h)}{2}\right] = 2\left[1 + \frac{1(1+0)}{2}\right] = 2 \times \frac{3}{2} = 3$$

Illustration No. 2

Calculate the integral as the limit of the sum

$$\int_0^1 6x^2\,dx$$

Solution:

Here $f(x) = 6x^2$, $a = 0$ and $b = 1$

Therefore $f(a + rh) = 6(rh)^2$ and $nh = b - a = 1 - 0 = 1$

By the definition definite integral

$$\int_0^1 6x^2\,dx = \lim_{h\to 0} h\sum_{r=1}^{n} f(rh) = \lim_{h\to 0} h\sum_{r=1}^{n} 6(rh)^2$$

$$= 6\lim_{h\to 0} h\left[1^2h^2 + 2^2h^2 + \dots + n^2h^2\right]$$

$$= 6\lim_{h\to 0} h\,h^2\left[1^2 + 2^2 + \dots + n^2\right]$$

$$= 6\lim_{h\to 0} h^3\left[\frac{n(n+1)(2n+1)}{6}\right]$$

since $\left[1^2 + 2^2 + \ldots + n^2\right] = \dfrac{n(n+1)(2n+1)}{6}$

$$= 6 \lim_{h \to 0} \frac{nh\ (nh+h)(2nh+h)}{6}$$

$$= \lim_{h \to 0} 1(1 + h)(2 + h)$$

$$= 2$$

Fundamental Theorem of Integral Calculus – Valuation of Definite Integral

If $\int_a^b f(x)dx$ exists and there also exists a function F(x) such that $\dfrac{d}{dx} F(x) = f(x)$ for every x in the range $a \le x \le b$ then

$$\int_a^b f(x)dx = \left[F(x)\right]_a^b = F(b) - F(a)$$

Some Important Properties of Definite Integrals

(i) $\int_a^b f(x)dx = \int_a^b f(z)dz$

Proof

Let $\int_a^b f(x)dx = F(x)$ then $\int_a^b f(z)dz = F(z)$

As

$$\int_a^b f(x)dx = F(b) - F(a)$$

and $$\int_a^b f(z)dz = F(b) - F(a)$$

And hence

$$\int_a^b f(x)dx = \int_a^b f(z)\,dz$$

(ii) $\int_a^b f(x)dx = -\int_a^b f(x)\,dx$

Proof

Let $\int_a^b f(x)dx = F(x)$ then LHS $\int_a^b f(x)dx = F(b) - F(a)$

And RHS = $-[F(x)]_b^a = -[F(a) - F(b)] = F(b) - F(a)$

Hence LHS = RHS

(iii) $\int_a^b f(x)dx = \int_a^c f(x)\,dx + \int_c^b f(x)dx$

Proof

As $\int_a^b f(x)dx = F(x)$ then LHS $\int_a^b f(x)dx = F(b) - F(a)$

Therefore in RHS

$$[F(x)]_a^c = [F(c) - F(a)] \text{ similarly } [F(x)]_c^b = [F(b) - F(c)]$$

Adding the above two results, we have

$$[F(c) - F(a)] + [F(b) - F(c)] = F(b) - F(a)$$

(iv) $\int_0^a f(x)dx = \int_0^a f(a-x)dz$

Proof:

Let $a - x = z$ then $-dx = dz$

Now if $x = 0$ then $z = a$ and if $x = a$ then $z = 0$

The RHS

$$\int_0^a f(a-x)dz = \int_a^0 f(z)\,(-dz) = -\int_a^0 f(z)\,(dz) = \int_0^a f(z)\,(dz)$$

Therefore RHS = LHS

(v) $\int_{-a}^a f(x)dx = \int_{-a}^0 f(x)dx + \int_0^a f(x)dx$

Proof:

Put $x = -z$ then $dx = -dz$

Substitute when $x = -a$, $z = a$ and when $x = 0$ then $z = 0$

Therefore,

$$\int_{-a}^0 f(x)dx = \int_a^0 f(-z)(-dz)$$

$$= -\int_a^0 f(-z)(dz) = \int_0^a f(-z)(dz) = \int_0^a f(-x)(dx)$$

And hence

$$\int_{-a}^{0} f(x)dx = \int_{-a}^{a} f(-x)dx + \int_{0}^{a} f(x)dx = \int_{0}^{a} \{f(-x)dx + f(x)dx\}$$

(vi) $\int_{-a}^{a} f(x)dx = 2\int_{0}^{a} f(x)dx$ if $f(-x) = f(x)$

$= 0$ if $f(-x) = -f(x)$

Proof:

If $f(-x) = f(x)$ i,e $f(x)$ is an even function of them

$$\int_{-a}^{a} f(x)dx = \int_{0}^{a} \{f(-x)dx + f(x)dx\}$$

$$= \int_{0}^{a} \{f(x)dx + f(x)dx\} = 2\int_{0}^{a} f(x)\,dx$$

(vii) $\int_{0}^{2a} f(x)dx = \int_{0}^{a} f(x)\,dx + \int_{0}^{a} f(2a - x)\,dx$

Proof:

Let $2a - x = z$ then $dz = -dx$

Now substituting $x = 0$ we get $z = 2a$ and if $x = a$ then $z = a$

Thus

$$\int_{0}^{a} f(2a - x)dx$$

$$= \int_{2a}^{a} f(z)(-dz)$$

$$= -\int_{2a}^{a} f(z)(dz) = \int_{a}^{2a} f(z)(dz) = \int_{a}^{2a} f(x)(dx)$$

Therefore

$$\int_{0}^{a} f(x)dx + \int_{a}^{2a} f(x)(dx) = \int_{0}^{2a} f(x)(dx)$$

Illustration No. 1

Find $\int_{0}^{1} \frac{1}{1+x^2}\,dx$

Solution:

$$\int_{0}^{1} \frac{1}{1+x^2}\,dx = \left[\tan^{-1} x\right]_{0}^{1}$$

$$= \left[\tan^{-1} 1 - \tan^{-1} 0\right]$$

$$= \frac{\pi}{4} - 0$$

$$= \frac{\pi}{4}$$

Illustration No. 2

Find $\int_{-1}^{5} \frac{1}{x+3}\,dx$

Solution:

$$\int_{-1}^{5} \frac{1}{x+3}\,dx = \left[\log\ (x+3)\right]_{-1}^{5}$$

$$= \left[\log\ (5+3)\right] - \left[\log\ (-1+3)\right]$$

$$= \left[\log\ (8) - \log\ (2)\right]$$

$$= \log \left\{\frac{8}{2}\right\}$$

$$= \log 4$$

Illustration No. 3

Find $\int_{0}^{\infty} \frac{1}{3^2 + (2x)^2}\,dx$

Solution:

$$\int_{0}^{\infty} \frac{1}{3^2 + (2x)^2}\,dx = \frac{1}{2^2} \int_{0}^{\infty} \frac{1}{\left(\frac{3}{2}\right)^2 + (x)^2}\,dx$$

$$= \frac{1}{2^2} \times \frac{1}{\frac{3}{2}} \times \left[\tan^{-1} \frac{x}{\left(\frac{3}{2}\right)}\right]_{0}^{\infty}$$

$$= \frac{1}{6}\left[\tan^{-1}\frac{2x}{3}\right]_0^{\infty}$$

$$= \frac{1}{6}\left[\tan^{-1}\infty - \tan^{-1}0\right]$$

$$= \frac{1}{6} \times \left(\pi/2\right) = \pi/12$$

Illustration No. 4

Find $\int_2^5 (3x + 7)\, dx$

Solution:

$$\int_2^5 (3x+7)\, dx = \left[\frac{3}{2}x^2 + 7x\right]_2^5$$

$$= \left[\frac{3}{2}5^2 + 7 \times 5\right] - \left[\frac{3}{2}2^2 + 7 \times 2\right]$$

$$= \left[\frac{3}{2}(5^2 - 2^2) + 7 \times (5-2)\right]$$

$$= 63/2 + 21$$

$$= 105/2$$

Illustration No. 5

Find $\int_0^2 \frac{(3x+4)}{(x+1)}\, dx$

Solution:

$$\int_0^2 \frac{(3x+4)}{(x+1)}\, dx$$

$$= \int_0^2 \frac{(3x+3)+1}{(x+1)}\, dx$$

$$= \int_0^2 \frac{3(x+1)+1}{(x+1)}\, dx$$

$$= \int_0^2 3\,dx + \int_0^2 \frac{1}{(x+1)}\,dx$$

$$= 3\,[x]_0^2 - \left[\log\,(x+1)\right]_0^2$$

$$= 3[2-0] + [\log 3 - \log 1]$$

$$= 6 + \log 3$$

Illustration No. 6

$$\int_0^{\pi/4} \sqrt{(1+\sin 2x)}\,dx$$

Solution:

$$\int_0^{\pi/4} \sqrt{(1+\sin 2x)}\,dx$$

$$= \int_0^{\pi/4} \sqrt{(\sin^2 x + \cos^2 x + \sin 2x)}\,dx$$

$$= \int_0^{\pi/4} \sqrt{(\sin x + \cos x)^2}\,dx$$

$$= \int_0^{\pi/4} (\sin x + \cos x)\,dx$$

$$= \int_0^{\pi/4} \sin x\,dx + \int_0^{\pi/4} \cos x\,dx$$

$$= -[\cos x]_0^{\pi/4} + [\sin x]_0^{\pi/4}$$

$$= -[\cos(\pi/4) - \cos 0] + [\sin(\pi/4) - \sin 0]$$

$$= -\left[\frac{1}{\sqrt{2}} - 1\right] + \left[\frac{1}{\sqrt{2}} - 0\right]$$

$$= 1$$

Illustration No. 7

$$\int_0^3 \frac{dx}{(2+x)(3+x)}$$

Solution:

$$\int_0^3 \frac{dx}{(2+x)(3+x)}$$

Since

$$\frac{1}{(2+x)(3+x)} = \frac{A}{(2+x)} + \frac{B}{(3+x)}$$

$$= \frac{A(3+x)+B(2+x)}{(2+x)(3+x)}$$

$$= \frac{x(A+B)+(3A+2B)}{(2+x)(3+x)}$$

Comparing both side with x and Constant coefficient terms

A + B = 0 and 3A + 2B = 1

Solving both the equations, we have A =1, B = –1.

$$\left[\int_0^3 \frac{dx}{(2+x)(3+x)} = \int_0^3 \frac{1}{(2+x)}\,dx - \int_0^3 \frac{1}{(3+x)}\,dx\right.$$

$$= \left[\log(2+x)\right]_0^3 - \left[\log(3+x)\right]_0^3$$

$$= [\log 5 - \log 2] - [\log 6 - \log 3]$$

$$= [\log 5 + \log 3] - [\log 2 + \log 6]$$

$$= \log 15 - \log 12$$

$$= \log\frac{15}{12}$$

$$= \log\frac{5}{4}$$

Illustration No. 8

Find $\int_0^{\pi/4} \frac{\sin\left(\tan^{-1} x\right)}{1+x^2}\,dx$

Solution:

$$\int_0^{\pi/4} \frac{\sin\left(\tan^{-1} x\right)}{1+x^2}\,dx$$

Let $\tan^{-1}x = t$; $\frac{1}{1+x^2}dx = dt$

Therefore at $x = 0$, $\tan^{-1}(0) = t \leftrightarrow t = 0$

And at $x = \pi/4$, $t = 1$ {since, $\tan^{-1}(\pi/4) = 1$}

Hence the limit changes to

$$\int_0^1 \sin t\,dt = \left[-\cos t\right]_0^1$$

$$= -[\cos 1 - \cos 0]$$

$$= -[0 - \pi/2] = \pi/2$$

Illustration No. 9

$$\int_0^1 \left[xe^x + \cos\left(\frac{\pi}{4}\right)x\right]dx$$

Solution:

$$\int_0^1 \left[xe^x + \cos\left(\frac{\pi}{4}\right)x\right]dx$$

$$= \int_0^1 xe^x\,dx + \int_0^1 \cos\left(\frac{\pi}{4}\right)x\,dx$$

$$= \left[xe^x - e^x\right]_0^1 + \left[\sin\left(\frac{\pi x}{4}\right)\left(\frac{4}{\pi}\right)\right]_0^1$$

$$= \left[xe^x\right]_0^1 - \left[e^x\right]_0^1 + \frac{4}{\pi}\left[\sin\left(\frac{\pi}{4}\right)x\right]_0^1$$

$$= [1 \times e - 0] - [e - 0] + \frac{4}{\pi}\left[\sin\left(\frac{\pi}{4}\right) - \sin 0\right]$$

$$= +\frac{4}{\pi}\left[\frac{1}{\sqrt{2}}\right]$$

$$= \frac{2\sqrt{2}}{\pi}$$

Illustration No 10

$$\int_0^{\pi/4} \frac{\sin x}{1+\cos^2 x}\, dx$$

Solution:

$$\int_0^{\pi/4} \frac{\sin x}{1+\cos^2 x}\, dx$$

Let cos x = t

$\therefore$ sin x dx = dt

Substituting x = 0 and $\pi/4$, we get t = 1 and $\frac{1}{\sqrt{2}}$

$$= -\int_1^{\frac{1}{\sqrt{2}}} \frac{1}{1+t^2}\, dt$$

$$= -\left[\tan^{-1} x\right]_1^{\frac{1}{\sqrt{2}}}$$

$$= -\left[\tan^{-1}\frac{1}{\sqrt{2}} - \tan^{-1} 1\right]$$

$$= \left[\frac{\pi}{4} - \tan^{-1}\frac{1}{\sqrt{2}}\right]$$

Illustration No. 11

$$\int_0^{\infty} \frac{dx}{\left(x^2+5^2\right)\left(x^2+3^2\right)}$$

Solution:

$$\int_0^{\infty} \frac{dx}{\left(x^2+5^2\right)\left(x^2+3^2\right)}$$

Since

$$\frac{1}{\left(x^2+5^2\right)\left(x^2+3^2\right)} = \frac{1}{5^2-2^2}\left[\frac{\left(x^2+5^2\right)-\left(x^2+3^2\right)}{\left(x^2+5^2\right)\left(x^2+3^2\right)}\right]$$

$$= \frac{1}{5^2-3^2}\left[\frac{1}{x^2+5^2} - \frac{1}{x^2+3^2}\right]$$

Therefore

$$\int_0^\infty \frac{dx}{(x^2+5^2)(x^2+3^2)} = \frac{1}{5^2-3^2}\left[\frac{1}{x^2+5^2} - \frac{1}{x^2+3^2}\right]$$

$$= \frac{1}{5^2-3^2}\left[\left[\frac{1}{5}\tan^{-1}\frac{x}{5}\right]_0^\infty - \left[\frac{1}{3}\tan^{-1}\frac{x}{3}\right]_0^\infty\right]$$

$$= \frac{1}{5^2-3^2}\left[\frac{1}{5}\left(\frac{\pi}{2}-0\right)\right] - \frac{1}{3}\left[\left(\frac{\pi}{2}-0\right)\right]$$

$$= \frac{1}{5^2-3^2} \times \frac{\pi}{2} \times \left(\frac{1}{5}-\frac{1}{3}\right)$$

$$= \frac{\pi}{2} \times \left(\frac{3-5}{5\times 3}\right) \times \frac{1}{(5+3)(5-3)}$$

$$= -\frac{\pi}{2} \times \frac{1}{15\times 8}$$

$$= -\frac{\pi}{240}$$

Illustration No. 12

$$\int_0^{\frac{\pi}{2}} \frac{\sin^2 x}{\sin x + \cos x}\, dx$$

Solution:

Let $$I = \int_0^{\frac{\pi}{2}} \frac{\sin^2 x}{\sin x + \cos x}\, dx = \int_0^{\frac{\pi}{2}} \frac{\sin^2\left(\frac{\pi}{2}-x\right)}{\sin\left(\frac{\pi}{2}-x\right) + \cos\left(\frac{\pi}{2}-x\right)}\, dx$$

$$= \int_0^{\frac{\pi}{2}} \frac{\cos^2 x}{\sin x + \cos x}\, dx$$

Therefore

$$2I = I + I = \int_0^{\frac{\pi}{2}} \frac{\sin^2 x}{\sin x + \cos x}\, dx + \int_0^{\frac{\pi}{2}} \frac{\cos^2 x}{\sin x + \cos x}\, dx$$

$$= \int_0^{\frac{\pi}{2}} \frac{\sin^2 x + \cos^2 x}{\sin x + \cos x}\, dx$$

$$= \int_0^{\frac{\pi}{2}} \frac{1}{\sin x + \cos x} dx$$

$$= \int_0^{\frac{\pi}{2}} \frac{1}{\sqrt{2}\left(\frac{1}{\sqrt{2}}\sin x + \frac{1}{\sqrt{2}}\cos x\right)} dx$$

$$= 1/\sqrt{2} \int_0^{\frac{\pi}{2}} \frac{1}{\left(\cos\frac{\pi}{4}\sin x + \sin\frac{\pi}{4}\cos x\right)} dx$$

$$= 1/\sqrt{2} \int_0^{\frac{\pi}{2}} \frac{1}{\sin\left(\frac{\pi}{4} + x\right)} dx$$

$$= 1/\sqrt{2} \int_0^{\frac{\pi}{2}} \operatorname{cosec}\left(\frac{\pi}{4} + x\right) dx$$

$$= \frac{1}{\sqrt{2}}\left[\log\left\{\operatorname{cosec}\left(\frac{\pi}{4} + x\right) - \cot\left(\frac{\pi}{4} + x\right)\right\}\right]_0^{\frac{\pi}{2}}$$

$$= \frac{1}{\sqrt{2}}\left[\left(\operatorname{cosec}\frac{3\pi}{4} - \cot\frac{3\pi}{4}\right) - \left(\operatorname{cosec}\frac{\pi}{4} - \cot\frac{\pi}{4}\right)\right]$$

$$= \frac{1}{\sqrt{2}}\log\frac{\sqrt{2}+1}{\sqrt{2}-1}$$

Application of Integration in Business and Economics

Illustration No.1

The marginal Revenue function for an electric heater manufacturer is given by MR $= \frac{5}{(x-3)^2} + 3$. Find the total revenue function and demand function of the manufacturer?

Solution:

Since

$$MR = \frac{5}{(x-3)^2} + 3$$

Therefore Total Revenue (TR) = $\int MR\,dx$

$$= \int\left\{\frac{5}{(x-3)^2}+3\right\}dx$$

$$= \int\frac{5}{(x-3)^2}dx + \int 3dx$$

$$= 5\int\frac{1}{(x-3)^2}dx + 3\int dx$$

$$= 5\frac{(x-3)^{-2+1}}{-2+1} + 3x + c$$

$$= \frac{-5}{(x-3)} + 3x + c$$

But TR = 0 when x = 0,

$\therefore\ 0 = 3(0) - \dfrac{5}{0-3} + c \rightarrow c = -5/3$

Hence the Total Revenue function is given by

$$TR = 3x - \frac{5}{(x-3)} - \frac{5}{3}$$

But Total Revenue (TR) = Price (p) × Qty. (x)

$= p \times x$

Therefore $$p = \frac{3x - \dfrac{5}{(x-3)} - \dfrac{5}{3}}{x}$$

$$= 3 - \frac{5}{x(x-3)} - \frac{5}{3x}$$

$$= 3 - \frac{5[3-(x-3)]}{x(x-3)}$$

$$= 3 - \frac{5(6-x)}{x(x-3)}$$

Hence the demand function $p = 3 - \dfrac{5(6-x)}{x(x-3)}$

Illustration No. 2

Gaurav a toy manufacturer has a fixed cost of production as Rs.8000 and the marginal cost as a function of number of units produced as $MC(x) = 4000 + 60x + 3x^2$, then find the total cost function?

Solution :

Let TC(x) be the total cost function

Then $TC(x) = \int MC(x)\,dx$

$$= \int (4000 + 60x + 3x^2)\,dx$$

$$= \int 4000\,dx + \int 60x\,dx + \int 3x^2\,dx$$

$$= 4000x + 60\frac{x^2}{2} + 3\frac{x^3}{3} + c$$

where 'c' is the constant of integration

Therefore TC (x) $= 4000x + 60\frac{x^2}{2} + 3\frac{x^3}{3} + c$

But TC (x) at x = 0 is given by

TC (0) $= 4000(0) + 60\frac{0^2}{2} + 3\frac{0^3}{3} + c$

$= c$

But $[TC(x)]_{x=0}$ = 8000 { since Fixed Cost = 8000 }

And hence

$[TC(x)]_{x=0}$ $= c = 8000$

$\leftrightarrow$ $c = 8000$

Therefore

$$TC(x) = 8000 + 4000x + 60\frac{x^2}{2} + 3\frac{x^3}{3}$$

Illustration No. 3

Sanchit an electrical switch manufacturer, if it is known that his marginal revenue function is given by $MR(x) = 50 - 15x^2$. Then find his revenue and demand function?

Solution:

Let R(x) be the revenue function of Sanchit

Therefore

$$R(x) = \int MR(x)\, dx$$

$$= \int \left[50 - 15x^2\right] dx$$

$$= \int 50\, dx - \int \left(15x^2\right) dx$$

$$= 50x - 15\frac{x^3}{3} + c$$

$$= 50x - 5x^3 + c$$

Since revenue of Sanchit at zero level of production i.e. at $x = 0$ is $R(0) = 0$

Thus $[R(0)]_{x=0} = 50(0) - 50(0)^3 + c$

Therefore we get $c = 0$

Thus the required revenue function is given by

$$R(x) = 50\,x - 5x^3$$

But Revenue R(x) = Price (p) × Qty.(x) .

$$= p \times x$$

$$p = \frac{R(x)}{x}$$

$$\leftrightarrow p = \frac{50x - 5\,x^3}{x}$$

$$= 50 - 5x^2$$

Therefore the demand function is given by $p = 50 - 5x^2$

Illustration No. 4

Ramesh Automobiles the manufacturer of new breed electric scooters. If the additional cost per unit is given by $20 + 5x^4 + 2e^x$, where x is the number of scooters manufactured. Determine the total and the average cost of production if the there exist a fixed cost Rs. 30000?

Solution:

Let TC(x) be the total cost function of Ramesh Automobiles

Then $$TC(x) = \int MC(x)\,dx$$

$$= \int (20 + 5x^4 + 2e^x)\,dx$$

$$= \int 20\,dx + \int 5x^4\,dx + \int 2e^x\,dx$$

$$= 20x + 5\frac{x^5}{5} + 2e^x + c$$

$$= 20x + x^5 + 2e^x + c$$

where 'c' is the constant of integration

Therefore $TC(x) = 20x + x^5 + 2e^x + c$

But TC (x) at x = 0 is given by

$$TC(0) = 20(0) + (0)^5 + 2e^0 + c$$

$$= c + 2$$

But $[TC(x)]_{x=0} = 30000$ {since Fixed Cost = 30000}

And hence

$$[TC(x)]_{x=0} = c + 2 = 30000$$

$$\leftrightarrow c = 29998$$

Therefore

$$TC(x) = 20x + x^5 + 2e^x + 29998$$

But Average Cost $$AC(x) = \frac{\text{Total cost TC}}{\text{No. of Units Produced}}$$

$$= \frac{20x + x^5 + 2e^x + 29998}{x}$$

$$= 20 + x^4 + 2\frac{e^x}{x} + \frac{29998}{x}$$

Illustration No. 5

Harrier a leading air–conditioner manufacturer producer of *Kool Breeze* brand has an additional revenue per unit as 49 – 2x and additional cost per unit as 20 + 2x. The company has reputation of suffering loss of Rs. 50 when there is zero sales. With the available information calculate the total profit function and break-even points (BEP)?

Solution:

The company has additional revenue per unit i.e. Marginal Revenue (MR) = 49 – 2x and additional cost per unit Marginal Cost (MC) = 20 + 2x.

As we know that,

Profit (P) = Revenue – Cost

Therefore

$$\frac{dP}{dx} = \frac{dR}{dx} - \frac{dC}{dx} = MR - MC$$

$$\left(\text{Since } MR = \frac{dR}{dx},\ MC = \frac{dC}{dx}\right)$$

$$= (49 - 2x) - (20 + 2x)$$

$$= 29 - 4x$$

$$P = \int \frac{dP}{dx}\,dx = \int (29 - 4x)\,dx$$

$$= \int 29\,dx - 4\int x\,dx$$

$$= 29x - 4\frac{x^2}{2} + c$$

$$= 29x - 2x^2 + c$$

But additionally we knew that the firm suffers a loss Rs. 50 if there is no sales

Therefore at $x = 0, P = -50$

As $P = 29x - 2x^2 + c$

$$[P]_{x=0} = 29(0) - 2(0)^2 + c$$

$$-50 = 0 - 0 + c$$

$$\leftrightarrow c = -50$$

The profit function is given by $P = 29x - 2x^2 - 50$

As in break even sales the firm earns no profit or $P = 0$

Therefore $29x - 2x^2 - 50 = 0,$

Or $2x^2 - 29x + 50 = 0$

$$2x^2 - 4x - 25x + 50 = 0$$

Or $2x(x-2) - 25(x-2) = 0$

$(x-2)(2x-25) = 0$

Or $x = 2$ or $25/2$

Illustration No. 6

If the elasticity of demand for a hair–pin producer is $\frac{p}{(p-1)(p-2)}$ where p is the price then calculate the demand function knowing that the quantity demanded is 6 unit at a price p = 3.

Solution:

Given

$$e_d = \frac{p}{(p-1)(p-2)}$$

But elasticity of demand e_d is

$$e_d = \frac{-dx \times p}{x \times dp} = \frac{p}{(p-1)(p-2)}$$

$$\frac{-dx}{x} = \frac{-1}{(p-1)} + \frac{1}{(p-2)}$$

Intergrating both side, we have

$$-\log x = -\log(p-1) + \log(p-2) + \log c$$

substituting $x = 6$ and $p = 3$

$$-\log 6 = -\log 2 + \log 1 + \log c$$

$$\log c = \log 2 - \log 6$$
$$= \log(2/6)$$
$$= \log(1/3)$$

Therefore taking antilogs both side we have c = 1/3

And hence the demand function is given by

$$-\log x = -\log(p-1) + \log(p-2) + \log 1/3$$

$$\log x = \log(p-1) - \log(p-2) - \log 1/3$$

$$= \log \frac{(p-1)}{(p-2)/3}$$

$$= \log \frac{3(p-1)}{(p-2)}$$

Taking antilogs both side we have

$$x = \frac{3(p-1)}{(p-2)}$$

Illustration No. 7

Nitin Paints ltd., a paints manufacturing firm having the marginal revenue function as $MR = \frac{6}{(x+3)^2} + 8$ where x the output produced, find the firms total revenue function, demand function and also calculate the revenue generated at output level of 15 units.

Solution:

Since the firm's Marginal revenue function is given by

$$MR = \frac{6}{(x+3)^2} + 8$$

Therefore the total revenue is given by

$$R = \int MR\,dx = \int\left\{\frac{6}{(x+3)^2} + 8\right\} dx$$

$$= 6 \int (x+3)^{-2}\, dx + 8 \int dx$$

$$= \frac{6(x+3)^{-1}}{-1} + 8x + c$$

$$= \frac{-6}{(x+3)} + 8x + c$$

But at x = 0, Revenue (R) = 0 and substituting in the Revenue function we get,

$$0 = -2 + c \leftrightarrow c = 2$$

Therefore Revenue $(R) = \frac{-6}{(x+3)} + 8x + 2$

$$= \left(\frac{-6}{(x+3)} + 2\right) + 8x$$

$$= \frac{-6 + 2(x+3)}{(x+3)} + 8x$$

$$= \frac{2x}{(x+3)} + 8x$$

The demand function is given by as we know,

Revenue (R) = Price (p) × Qty.(x)

$$R = p \times x$$

$$p = \frac{R}{x}$$

$$= \frac{\frac{2x}{x+3} + 8x}{x}$$

$$= \frac{2}{(x+3)} + 8$$

And the Revenue of the firm at an output level of x = 15 units is

Since the Revenue $(R) = \frac{2x}{(x+3)} + 8x$

Therefore $[R]_{x=15} = \frac{2 \times 15}{(15+3)} + 8 \times 15$

$$= \text{Rs. } 121.66$$

Illustration No. 8

If the marginal cost of a firm MC = 60 $\sqrt{x}$ then calculate the total cost, average cost and the marginal cost of the firm at an output level of 9 units?

Solution:

As the Marginal Cost (MC) = $60\sqrt{x}$

$$TC = \int MC\, dx$$

$$= \int 60 \sqrt{x}\, dx$$

$$= 60 \int \sqrt{x}\, dx = 60 \frac{x^{\frac{3}{2}}}{\frac{3}{2}}$$

$$= 40x^{\frac{3}{2}}$$

Therefore $[TC]_{x=9}$ $= 40 \times 9^{\frac{3}{2}} = 1080$

And the Average Cost (AC) $= \frac{TC}{x}$

$$= 40x^{\frac{3}{2}} / x$$

$$= 40x^{\frac{1}{2}}$$

Therefore $[AC]_{x=9}$ $= 40 \times 9^{\frac{1}{2}} = 120.$

Similarly

Marginal Cost $[MC]_{x=9} = 60 \times 9^{\frac{1}{2}} = 180$

Illustration No. 9

Ganesh Hosieries ltd manufacturer of men's undergarments has the following marginal cost and marginal revenue given as Marginal Cost (MC) $= 15 + \frac{x}{15}$ and Marginal Revenue (MR) = 20. If the company has fixed cost component as Rs 100 then determine the profit maximizing level of output and the maximum profit?

Solution:

Let TC and TR be the Total Cost and Total Revenue of the firm

Then

$$TC = \int MC\, dx$$

$$= \int \left(15 + \frac{x}{15}\right) dx$$

$$= \int (15)\, dx + \int \left(\frac{x}{15}\right) dx$$

$$= 15x + \frac{x^2}{30} + c$$

The firm has fixed cost component even at zero-level of output

i.e. $[TC]_{x=0} = 100$

But $\quad [TC]_{x=0} = 15(0) + \frac{(0)^2}{30} + c$

Therefore, we have,

$$100 = c$$

And hence the Total Cost (TC) $= 15x + \frac{x^2}{30} + 100.$

Similarly

Total Revenue $\quad (TR) = \int MR\, dx$

$$= \int (20)\, dx = 20x + c'$$

As total revenue at zero sales x = 0 is TR = 0

And hence $[TR]_{x=0} = 20(0) + c'$

$$c' = 0$$

therefore (TR) = 20x

But as we know,

Profit (P) = Total Revenue (TR) – Total Cost (TC)

$$= 20x - \{15x + \frac{x^2}{30} + 100\}$$

$$= 5x - \frac{x^2}{30} - 100$$

In order to find the maximum profit and profit level, let us differentiate w.r.t 'x'

We get,

$$\frac{dp}{dx} = 5 - \frac{x}{15}$$

Now equating $\quad \frac{dp}{dx} = 0$, we get,

$$5 - \frac{x}{15} = 0, \ x = 75 \text{ units}$$

For maximizing condition we find the 2nd derivative we get

$$\frac{d^2P}{dx^2} = -\frac{1}{15} < 0$$

And hence the profit is maximum at x = 75 units of output level

The maximum profit is

$$\text{Profit } [P]_{x=75} = 5\,(75) - \frac{75^2}{30} - 100$$

$$= 87.5 \text{ (approx)}$$

Consumer and Producer Surplus

Let us assume p = f(x) as the demand function of a commodity where p the price that the consumer is prepared to pay for a given quantity x of the commodity. As per demand law, the given function is decreasing function as with increase in the price of x the demanded will decrease. Now p = g (x) be the supply function where p is the price that a producer is comfortable to make available quantity x of the commodity.

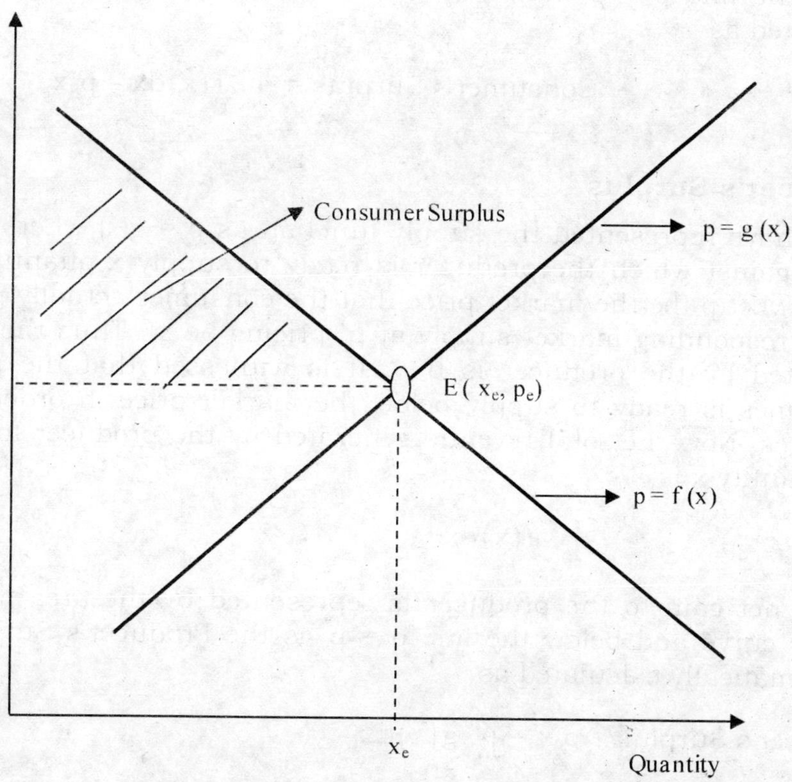

With the interaction of demand and supply curve gives the equilibrium point (E) the corresponding equilibrium pricing (p_e) and quantity (x_e). This point reflects both the consumers and producers acceptability in terms of pricing for benefits exchange. Sometimes the consumer are willing to pay more for the commodity than the prevailing market price that needs to be paid actually.

Consumer Surplus reflects the difference between what consumer is ready to pay and what actually pay in the absence of discrimination of price.

Let us formally derive Consumer Surplus mathematically,

Here consumer is more than willing to pay above the equilibrium pricing (p_e) thus the consumer expenditure to acquire x_e quantity is $\int_0^{x_e} f(x)dx$

But the consumer's actual expenditure will be $p_e x_e$. Thus the net gain to consumer is represented by the area under the demand curve and above the line $p = p_e$ and this area is the consumer surplus and is calculated as

$$\text{Consumer's Surplus} = \int_0^{x_e} f(x)dx - p_e x_e$$

Producer's Surplus

As we have represented the supply function as $p = g(x)$ where p is the price against which the producer is ready to supply x quantity to the market. Let p_e be the market price that the consumer actually pays and the corresponding market supply at p_e pricing be x_e. Thus the revenue generated by the producer is $p_e x_e$. It is witnessed that the producer sometimes is ready to supply below the market price in order to gain leverages. Now the total revenue generated by the producer for supply the quantity x is

$$\int_0^{x_e} g(x)dx$$

The net gain to the producer is represented by the area above the supply curve and below the line $p = p_e$ is the Producer's Surplus and mathematically calculated as

$$\text{Producer's Surplus} = p_e x_e - \int_0^{x_e} g(x)dx$$

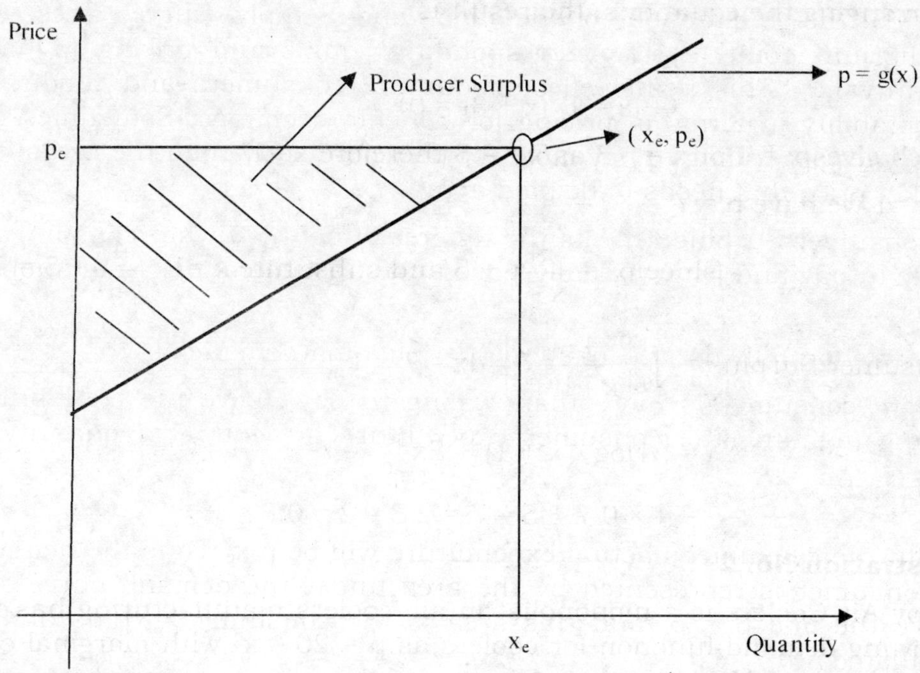

Illustration No.1

If a firm under perfect competition, the price of the quantity demanded of a product is determined by the demand and supply laws given by $p_d = \frac{4}{x+1} + 5$ and $p_s = 6 + x$ respectively then determine the consumer surplus associated?

Solution:

Since the equilibrium price and quantity is determined by the interaction of demand and supply where,

The demand function = $p_d = \frac{4}{x+1} + 5$

And the supply function = $p_s = 6 + x$

Equating Demand = Supply

We get,

$$\left(\frac{4}{x+1} + 5\right) = (6 + x)$$

Rearranging the equations, the result is

$$x^2 + 2x - 3 = 0$$

$$\leftrightarrow (x - 1)(x + 3) = 0$$

which gives x = 1 or x = – 3 as x ≠ – 3 therefore x = 1

At x = 1 we have p = 7

$$\left\{\text{since } p_d = \frac{4}{x+1} + 5 \text{ and substitute } x = 1, \frac{4}{1+1} + 5 = 7\right\}$$

$$\text{Consumer Surplus} = \int_0^1 \left\{\frac{4}{x+1} + 5\right\} dx - 7 \times 1$$

$$= \left[4\log\ (x + 1) + 5x\right]_0^1 - 7$$

$$= 4 \times 0.7 + 5 - 7 = 2.8 - 2 = 0.8$$

Illustration No. 2

Binoy Air Coolers is a monopolist in air–coolers manufacturing has the following demand function for coolers as $p = 20 - x^2$ with marginal cost of production (MC) = $x^2 + 4$. Determine the consumer surplus at the price Binoy Air Coolers will fix. And how will the scenario change if it is perfect competition?

Solution:

Given the demand function, $p = 20 - x^2$

But Revenue (R) = Price(p) × Qty (x)

$$= p \times x = (20 - x^2) \times x$$

$$= 20x - x^3$$

But in case of Monopoly the profit maximization condition is MR = MC

$$\text{As Marginal Revenue (MR)} = \frac{dR}{dx} = 20 - 3x^2$$

And hence $20 - 3x^2 = x^2 + 4$

$\rightarrow x = \pm 2$ as $x \neq -2$, therefore $x = 2$

At x = 2, we have $[p]_{x=2} = 20 - (2)^2$

$$= 20 - 4 = 16$$

$$\text{Calculating the consumer surplus (CS)} = \int_0^2 \left(20 - x^2\right) dx - 16 \times 2$$

$$= \left[20x - \frac{x^3}{3}\right]_0^2 - 32$$

$$= \left[20 \times 2 - \frac{2^3}{3}\right] - 32$$

$$= 16/3$$

Under perfectly competitive condition the supply function and the marginal cost function are equal and hence we have

$$20 - x^2 = x^2 + 4 \leftrightarrow x = 2\sqrt{2}$$

Therefore the Price $[p]_{x = 2\sqrt{2}} = 20 - 8 = 12$

The Consumer Surplus at Perfectly Competitive Condition

$$= \int_0^{2\sqrt{2}} (20 - x^2)\, dx - 32\sqrt{2}$$

$$= 8\sqrt{2}/3$$

Illustration No. 3

Abhishek Chemical works a monopolist in washing detergent cakes and has the following demand function $p = 15 - 3x^2$ where p is the price per unit and if the firm has a marginal cost (MC) = $9 + 3x$ then determine the consumer surplus?

Solution:

Let Revenue function of the firm be 'R'

Since Revenue (R) = Price (p) × Qty. (x)

$$= (15 - 3x^2) \times x = 15x - 3x^3$$

Now Marginal Revenue (MR) $= \frac{dR}{dx} = 15 - 9x^2$

But under monopoly

MC = MR

Therefore $15 - 9x^2 = 9 + 3x$

i.e. $9x^2 + 3x - 6 = 0$

$$3x^2 + x - 2 = 0$$

$$(3x - 2)(x + 1) = 0$$

$x = 2/3$ or -1 since $x \neq -1$

and hence $x = 2/3$, at $x = 2/3$ we have $p = 15 - 3\left(\frac{2}{3}\right)^2$

$$= 15 - 4/3 = 41/3$$

Therefore the Consumer Surplus = $\int_0^{2/3}(15 - 3x^2)\,dx - \frac{41}{3} \times 2/3$

$$= \left[15x - x^3\right]_0^{2/3} - \frac{82}{9}$$

$$= 15 \times \frac{2}{3} - \left(\frac{2}{3}\right)^3 - 82/9$$

$$= 16/27$$

Illustration No. 4

A firm operating in perfectly competitive market has the following demand and supply function $p = 16 - 3x - x^2$ and $p = 12 + x$ respectively. Find the consumer and producer surplus at equilibrium pricing?

Solution:

As the demand and supply function by $p = 16 - 3x - x^2$ and $p = 12 + x$

But under equilibrium condition Demand = Supply

And hence $16 - 3x - x^2 = 12 + x$

Therefore we get $(x - 2)^2 = 0 \rightarrow x = 2$

At $x = 2$ the pricing is given by $p = 16 - 6 - 4 = 6$

The Consumer Surplus (CS) = $\int_0^2 (16 - 3x - x^2)dx - 2 \times 6$

$$= \left[16x - 3\frac{x^2}{2} - \frac{x^3}{3}\right]_0^2 - 12$$

$$= \left[16 \times 2 - 3\frac{2^2}{2} - \frac{2^3}{3}\right] - 12$$

$$= 34/3$$

And now we find the producer surplus

At x = 2 the price $[p]_{x=2} = 12 + 2 = 14$.

Therefore the Producer Surplus (PS) $= 14 \times 2 - \int_0^2 (12 + x)dx$

$$= 28 - \left[12x + \frac{x^2}{2}\right]_0^2$$

$$= 28 - 24 - 4/2.$$

$$= 2$$

Learning Curve

In industrial production the line workers during the various working stage gains useful experience as a result the direct labour input per unit gradually reduces. The rate of reduction in direct labour requirement is represented by a curve known as learning curve. Basically learning curve is a graphical device used to show the rate of decline in direct labour requirement and mathematically represented as

$$f(x) = A\,x^{\alpha} \qquad \text{where } -1 \le \alpha \le 0,\ A > 0$$

where f(x) is the number of hours of direct labour required to produce x units of the product and graphically shown as

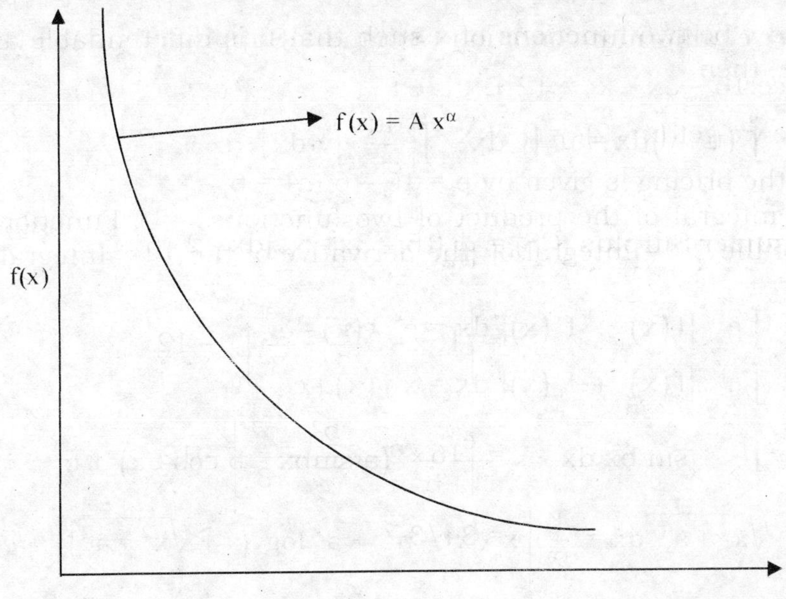

From the graph it can be deduced that as x increases results in f(x) to decrease. Thus the curve can be useful tool to predict the number of production hours required for future assignments.

Once the learning curve is determined then for a gross production process it is represented as

$$\int_a^b f(x)\,dx = \int_a^b Ax^{\alpha} dx$$

where a and b represents the number of unit produced.

Summary

Integration is reverse or inverse of Differentiation mathematically represented as.

$$\int f(x)dx = F(x)$$

$$\int c\,f(x)dx = c\int f(x)\,dx$$

$$\int \{f(x) \pm g(x)\}\,dx = \int f(x)\,dx \pm \int g(x)\,dx$$

$$\int \frac{dx}{a^2 + x^2} = \frac{1}{a}\tan^{-1}\frac{x}{a}$$

$$\int \frac{dx}{x\sqrt{x^2 - a^2}} = \frac{1}{a}\sec^{-1}\left(\frac{x}{a}\right)$$

If u and v be two functions of x such that u is differentiable and v is integrable, then

$$\int (u\,.\,v)dx = u\int v\,dx - \int\left\{\frac{du}{dx}\int v\,dx\right\}dx$$

i.e the integral of the product of two functions = 1st Function × the Integral of the 2nd – Integral of {the derivative of the 1st × Integral of the Second}.

$$\int e^x\,\{f(x) + f'(x)\}\,dx = e^x\,f(x) + c$$

$$\int e^x\,\{f(x) + f'(x)\}\,dx = x\,f(x) + c$$

$$\int e^{ax}\sin bx\,dx = \frac{e^{ax}}{a^2 + b^2}(a\sin bx - b\cos bx) + c$$

$$\sqrt{x^2 + a^2}\,dx = \frac{1}{2}\left[x\sqrt{x^2 + a^2} + a^2\log\,(x + \sqrt{x^2 + a^2})\right] + c$$

Fundamental Theorem of Integral Calculus – Valuation of Definite Integral

If $\int_a^b f(x)dx$ exists and there also exists a function F(x) such that

$\frac{d}{dx} F(x) = f(x)$ for every x in the range $a \le x \le b$ then

$$\int_0^{2a} f(x)\, dx = \int_0^a f(x)dx + \int_0^a f(2a - x)dx$$

Self Practice Exercises

Section 1

Evaluate the Integrals of the following

1. $\int (e^{2x} + 3^x)dx$
2. $\int \left(3x^2 + \frac{2}{x} - 3e^{3x} + 5x\right)dx$
3. $\int (x^3 + 5^x + 3^3)dx$
4. $\int \sqrt{x}\ (x + 5)\ dx$
5. $\int (x + 5)\ (3x + 2)dx$
6. $\int \frac{e^{3x} + e^{4x}}{e^x + e^{-x}}\ dx$
7. $\int \frac{2^x + 3^x}{2^x 3^x}\ dx$
8. $\int \left(\sqrt{\frac{5}{x}} + \sqrt{\frac{x}{5}}\right) dx$
9. $\int \frac{\sin x}{1 - \sin x} dx$
10. $\int \left(1 + x + \frac{x^2}{2!} + \frac{x^3}{3!} \ldots\right) dx$
11. $\int \frac{\cos x - \sin x}{\cos x + \sin x}\ (1 + \sin 2x)dx$
12. $\int \frac{a^x + b^x}{c^x}\ dx$
13. $\int \cos^{-1} x\ (\sin x)dx$
14. $\int \frac{e^{4\log x} - e^{3\log x}}{e^{3\log x} - e^{2\log x}} dx$

Section No. 2

1. $\int \frac{x}{(x^2 + 5^2)}\ dx$ {*Hint*: Substitute $x^2 + 5 = z$}

2. $\int \frac{(x+3)}{\sqrt{x-3}}\, dx$

3. $\int \frac{1}{\sqrt{3x+1}+\sqrt{3x-1}}\, dx$

4. $\int \frac{1}{(x-1)\sqrt{x^2-2x}}\, dx$

5. $\int \sqrt{\frac{7+x}{7-x}}\, dx$

6. $\int \frac{1}{\sqrt{25-9x^2}}\, dx$

7. $\int \frac{\left(3+4\sin^{-1}x\right)^3}{\sqrt{1-x^2}}\, dx$

8. $\int \frac{x}{\sqrt{9-x^4}}\, dx$

9. $\int \frac{\sin 2x}{4-\cos^2 x}\, dx$

10. $\int \frac{2x+3}{2x^2+6x+5}\, dx$

11. $\int \frac{1}{x(9+11\log x)}\, dx$

12. $\int \frac{3x^4}{x^5+7}\, dx$

13. $\int \frac{\cos x\ (1+\log x)}{x}\, dx$

Section No 3

1. $\int x^4 \log x\, dx$

2. $\int x^2 e^{3x}\, dx$

3. $\int x^2 \cos 2x\, dx$

4. $\int \frac{\log(\log x)}{x}\, dx$

5. $\int 6\sin^{-1} x\, dx$

6. $\int \tan^{-1} x\, dx$

7. $\int \cos^{-1}\left(\frac{1-x^2}{1+x^2}\right) dx$

8. $\int e^x \left(\frac{x^2-2x}{x^4}\right) dx$

9. $\int \left(\frac{\sin x\cos x+1}{\cos^2 x}\right) e^x dx$

10. $\int e^x \cos 2x\, dx$

11. $\int \frac{\log x}{(1+\log x)^2}\, dx$

12. $\int \sqrt{x^2-25}\, dx$

13. $\int \frac{1}{4x^2-9}\, dx$

14. $\int \sqrt{16-x^2}\, dx$

15. $\int \frac{1}{x\,[10+7\log x+(\log x)^2]}\, dx$

Section No. 4

1. $\int \frac{x^2}{(x^2+5^2)(x^2+3^2)} dx$ 2. $\int \frac{1}{(x-1)^2 (x+2)} dx$

3. $\int \frac{(x+1)}{(x-2)(x+3)} dx$ 4. $\int \frac{1}{x(x+2)^2} dx$

5. $\int \frac{x^3}{(x-2)(x-4)(x-6)} dx$ 6. $\int \frac{x}{(x+4)^2 (x+3)} dx$

7. $\int \frac{1}{(x^2+1)(x+1)} dx$ 8. $\int \frac{1}{\cos x\ (1+2\sin x)} dx$

9. $\int \frac{x}{x^4+3x^2+10} dx$ 10. $\int \frac{x}{2x^2+2x+3} dx$

Section No. 5

1. $\int_1^2 \frac{1}{(x+5)} dx$ 2. $\int_0^\infty \frac{1}{(5^2+7^2x^2)} dx$

3. $\int_0^2 \frac{x}{\sqrt{4-x^2}} dx$ 4. $\int_2^3 \frac{x^2}{\sqrt{x^2-1}} dx$

5. $\int_0^{\pi/2} \sqrt{1+\sin x}\ dx$ 6. $\int_0^1 \{xe^x + \sin(\pi x/3)\} dx$

7. $\int_0^\infty \frac{1}{(7^2+x^2)(3^2+x^2)} dx$ 8. $\int_0^\infty \frac{\cos(\tan^{-1} x)}{(1^2+x^2)} dx$

9. $\int_1^2 \frac{1}{(x+2)(x+3)} dx$ 10. $\int_1^{1/2} x \log(1-x)\ dx$

11. $\int_0^2 \sqrt{\frac{2+x}{2-x}} dx$ 12. $\int_0^{\pi/2} \frac{\sqrt{\sin x}}{\sin x + \sqrt{\cos x}} dx$

13. $\int_0^{\pi/2} \frac{1}{(2^2\cos^2 x + 3^2 \sin^2 x)^2} dx$

Section No. 6

1. Vineet Electricals producer of electric filaments has the following marginal revenue function (MR) $= \frac{xe^x}{(x+1)^2}$ where x is the output produced. Find the total revenue function (TR) of the firm?

2. If a firm has marginal cost function (MC) $= 10x + \frac{x}{10}$ and marginal revenue function as (MR) = 20 where x is output produced. Find the profit maximizing level of output and the profit generated if the firm has a fixed cost component (FC) = 100?

3. If a profit maximizing firm has the following marginal revenue function (MR) $= \frac{5}{(x+4)} + 7$, then find out the demand function?

4. Rylon tyres a leading manufacturer of motor cycle and scooter tyres has the following marginal cost function (MC) = 50 + 16x + $3x^2$ and fixed cost of production (FC) = 150. Calculate the total cost of production of the manufacturer?

5. If the elasticity of demand for a product $e_d = \frac{p}{(p-1)(p-2)}$ where p is the price per unit, find the demand function if the quantity demanded is 4 units at price p = 3.

6. A toy manufacturing firm's marginal cost is furnished as below MC = 3 + 6x + $5e^{-x}$ (lakhs of rupees) and where x is the number of units produced. Find the total cost of producing 10 units if the firm has a fixed cost of production as 5 lakhs?

7. The production manager of a widget producer has learned that after producing 10 units of widgets determines that the production facility has the following learning curve represented by f(x) = 100 – 70 e – x where f(x) is the rate of labour hours required to produce xth unit. Find how many labour hours will be required to produce additional 10 units?

Answers

Section No. 1

1. $\frac{e^{2x}}{2}+\frac{3^x}{\log 3}+c$

2. $x^2+2\log x-e^{3x}+5x+c$

3. $\frac{x^4}{4}+\frac{5^x}{\log 5}+27x+c$

4. $\frac{2}{5}x^{5/2}+\frac{10}{3}x^{3/2}+c$

5. $x^3+\frac{5}{2}x^2+2x+c$

6. $\frac{e^{4x}}{4}+c$

7. $\frac{(2/3)^x}{\log\left(\frac{2}{3}\right)}+\frac{(3/2)^x}{\log\left(\frac{3}{2}\right)}+c$

8. $2\sqrt{5x}+\frac{2}{3\sqrt{5}}x^{3/2}+c$

9. $(\tan x-\sec x)-x+c$

10. $x+\frac{x^2}{2!}+\frac{x^3}{3!}...+c$

11. $\frac{\sin 2x}{2}+c$

12. $\frac{(a/c)^x}{\log\left(\frac{a}{c}\right)}+\frac{(b/c)^x}{\log\left(\frac{b}{c}\right)}+c$

13. $\frac{\pi}{2}x-\frac{x^2}{2}+c$

14. $\frac{x^2}{2}+c$

Section No. 2

1. $-\frac{1}{2(x^2+5)}+c$

2. $\frac{2}{3}(x-3)^{3/2} + 12\sqrt{x-3} + c$

3. $\frac{1}{2}\left[\frac{(3x+1)^{\frac{3}{2}}}{3}-\frac{(3x-1)^{\frac{3}{2}}}{3}\right]+c$

4. $\sec^{-1}(x-1)+c$

5. $-14 \log(7-x) - x + c$

6. $\frac{1}{5}\sin^{-1}\frac{3x}{3}+c$

7. $\frac{1}{16}(3+4\sin^{-1}x)^4+c$

8. $\frac{1}{2}\sin^{-1}\frac{x^2}{3}+c$

9. $\log(4-\cos^2 x)+c$

10. $\frac{1}{2}\log(2x^2+6x+5)+c$

11. $\frac{1}{11}\log(9+11\log x)$

12. $\frac{3}{5}\log(x^5+7)+c$

13. $\sin(1+\log x)+c$

Section No. 3

1. $\frac{x^5}{5}\left(\log x-\frac{1}{5}\right)+c$

2. $\frac{1}{3}\left[e^{3x}\left(x^2-\frac{2}{3}x+\frac{2}{9}\right)\right]+c$

3. $\frac{x^2}{2}\sin 2x+\frac{x}{4}\cos 2x-\frac{1}{2}\sin 2x+c$

4. $\log x\,[\log(\log x)-1]+c$

5. $6x\sin^{-1}x+6\sqrt{1-x^2}+c$

6. $x\tan^{-1}x-\frac{1}{2}\log(1+x^2)+c$

7. $2\tan^{-1}x-2\log\sqrt{1+x^2}+c$

8. $\frac{e^x}{x^2}+c$

9. $e^x\tan x+c$

10. $\frac{2}{5}e^x\sin 2x+\frac{1}{5}e^x\cos 2x+c$

11. $\frac{x}{1+\log x}+c$

12. $\frac{1}{2}\left[\frac{x}{2}\sqrt{x^2-25}-25\log\left(x+\sqrt{x^2-25}\right)\right]+c$

13. $\frac{1}{12}\log\frac{2x-3}{2x+3}+c$

14. $\left[\left(\frac{x}{2}\right)\sqrt{4-x^2}+8\sin^{-1}\left(\frac{x}{4}\right)\right]+c$

15. $\frac{1}{3}\log\frac{\log(x+2)}{\log(x+5)}+c$

Section No. 4

1. $\frac{1}{16}[5\tan^{-1}\left(\frac{x}{5}\right)-3\tan^{-1}\left(\frac{x}{3}\right)+c$

2. $\frac{1}{x-1}+\log\left\{\frac{x-2}{x-1}\right\}+c$

3. $\log(x-2)+\frac{2}{5}\log(x+3)+c$

4. $\log\left(\frac{x}{x+2}\right)+\frac{1}{2(x+2)}+c$

5. $x+\log(x-2)-16\log(x-4)+27\log(x-6)+c$

6. $-\frac{4}{(x+4)}+3\log\left[\frac{x+4}{x+3}\right]+c$

7. $\frac{1}{2}\log(x+1)-\frac{1}{4}\log\left(x^2+1\right)+\frac{1}{2}\tan^{-1}x+c$

8. $\frac{1}{6}\log(1-\sin x)-\frac{1}{2}\log(1+\sin x)+\frac{4}{3}\log(1+2\sin x)+c$

9. $\frac{1}{2}\tan^{-1}\left(x^2+3\right)+c$

10. $\frac{1}{4}\log\left(2x^2+2x+3\right)-\frac{1}{2\sqrt{5}}\tan^{-1}\left(\frac{2x+1}{\sqrt{5}}\right)+c$

Section No. 5

1. $\log\left(\frac{7}{6}\right)$

2. $\frac{\pi}{70}$

3. 2

4. $1+\frac{1}{2}\log\left(\frac{3}{2}\right)$

5. $2-\sqrt{3}$

6. $\frac{3}{2\pi}+1$

7. $\frac{\pi}{420}$

8. 1

9. $\log\left(\frac{15}{16}\right)$

10. $\frac{3}{8}\log 2-\frac{5}{16}$

11. $\pi+2$

12. $\frac{\pi}{4}$

13. $\frac{13\pi}{864}$

Section No. 6

1. $-\frac{xe^x}{(x+1)}+e^x-1$

2. 100, 1600

3. $p=\frac{5}{x}\log\left(\frac{x+4}{4}\right)+7$

4. $TC = 50x + 8x^2 + x^3 + 150$

5. $y=8\left(\frac{x-2}{x-1}\right)$

6. $340-\frac{5}{e^{10}}$

7. $100 + 70\,(e^{-20} - e^{-10})$

CHAPTER 8

Differential Equations and its Application

Definitions

An equation having differentials or differential coefficients is termed as differential equation.

For Example

(i) $\frac{dy}{dx} = x \cos x$

(ii) $\frac{dy}{dx} = x + e^x$

(iii) $\frac{d^2y}{dx^2} = x\, e^x + 5$

Order of a Differential Equation

The order of the highest derivative or differential coefficient which appears in the equation is the order of the differential equation.

For Example

(i) $\frac{dy}{dx} = x \cos x$ is a differential equation of order 1, as the highest differential coefficient of y in the equation is $\frac{dy}{dx}$ which is the first derivative.

(ii) $\frac{d^2y}{dx^2} = x\, e^x + 5$ is a differential equation of order 2, as the highest differential coefficient of y in the equation is $\frac{d^2y}{dx^2}$ which is the second derivative.

Degree of the Differential Equation

The degree of a differential equation is the power of the highest derivative which appears in the differential equation, where the derivatives are rational and free from radicals and fractions.

For Example

(i) $\frac{dy}{dx} = x + \sin x$ is a differential equation of degree 1, as the power of the highest derivative $\frac{dy}{dx} = 1$

(ii) $\left(\frac{d^2y}{dx^2}\right)^2 = x \cos x + 7$ is a differential equation of degree 2, as the power of the highest derivative $\frac{d^2y}{dx^2} = 2$

Illustration No.1

Determine the order and degree of the following differential equations

(i) $\frac{dy}{dx} + \cos^2 x = 4$

(ii) $x\frac{d^2y}{dx^2} + 2\frac{dy}{dx} = xy$

Solution:

(i) The given differential equation is

$$\frac{dy}{dx} + \cos^2 x = 4$$

The highest order of the derivative in the equation is $\frac{dy}{dx}$ therefore the order of the equation is 1 and since the power of the derivative is also one hence the degree of the differential equation is 1.

(ii) The given differential equation is

$$x\frac{d^2y}{dx^2} + 2\frac{dy}{dx} = xy$$

The highest order of the derivative in the equation is $\frac{d^2y}{dx^2}$ therefore the order of the equation is 2 and since the power the derivative is one hence the degree is also 1.

Solution of a Differential Equation

A solution of a differential equation is any relation in between the variables involved which satisfies the differential equation. Such a solution is also called as the primitive as the differential equation can be regarded as a relation derived from it.

Illustration No. 2

Find the differential equation whose primitives is y = 4x + 5?

Solution:

Given $y = 4x + 5$ .....(1)

Differentiating eq.1 with respect to x, we get, $\frac{dy}{dx} = 4$ .....(2)

Now differentiating eq. 2 once again, we get, $\frac{d^2y}{dx^2} = 0$

Therefore, the required differential equation is $\frac{d^2y}{dx^2} = 0$

Illustration No. 3

Find the differential equation if the primitive is $xy = 2e^x + 3e^{-x}$?

Solution:

Given $xy = 2e^x + 3e^{-x}$ 1

Differentiating 1 with respect to x, we get,

$$x\frac{dy}{dx} + y = 2e^x - 3e^{-x} \quad .....2$$

Now differentiating 2 once again, we get,

$$x\frac{d^2y}{dx^2} + 1.\frac{dy}{dx} + \frac{dy}{dx} = 2e^x + 3e^{-x}$$

$$x\frac{d^2y}{dx^2} + 2\frac{dy}{dx} = xy$$

Illustration No. 4

Find the differential equation whose primitive is $4x^2 + 9y^2 = 36$?

Solution:

Given $4x^2 + 9y^2 = 36$ 1

Differentiating 1 with respect to x, we get,

$$8x + 18y\frac{dy}{dx} = 0 \quad .....2$$

Now differentiating 2 once again, we get,

$$8 + 18\left(\frac{dy}{dx}\right)\left(\frac{dy}{dx}\right) + 18y\frac{d^2y}{dx^2} = 0$$

$$18y\frac{d^2y}{dx^2} + 18\left(\frac{dy}{dx}\right)^2 + 8 = 0$$

Solution of First Order and First degree Differential Equation

Some of the methods of solving differential equation of first order belonging to some of the common standard forms are discussed below.

Solving Differential Equations of the form $\frac{dy}{dx} = f(x)$

Proof

The given differential equation is expressed as

$$\frac{dy}{dx} = f(x)$$

$$\Rightarrow \quad dy = f(x)dx$$

Now integrating both sides and adding an arbitrary constant of integration say 'c' to any one of the side, we get

$$\int dy = \int f(x)\, dx + c$$

Once we evaluate the value of $\int f(x)\, dx$, we get the required solution.

Illustration No. 5

Solve the differential equation $\frac{dy}{dx} = 5x$?

Solution:

Since $$\frac{dy}{dx} = 5x$$

$$\Rightarrow \quad dy = 5x\,dx$$

Now integrating both sides, we get

$$\int dy = \int 5x\,dx$$

$$y = \frac{5x^2}{2} + c$$

Illustration No. 6

Solve the differential equation $\frac{dy}{dx} = \sec^2 x$

Solution:

Since $$\frac{dy}{dx} = \sec^2 x$$

Therefore $dy = \sec^2 x\,dx$

Now integrating both sides, we get

$$\int dy = \int \sec^2 x\,dx$$

$$y = \tan x + c$$

Illustration No. 7

Solve the differential equation $\frac{dy}{dx} = x^3 + \sin 3x$

Solution:

The given equation is $\frac{dy}{dx} = x^3 + \sin 3x$

Therefore

$$dy = (x^3 + \sin 3x)\,dx$$

$$= x^3\,dx + \sin 3x\,dx$$

Now integrating both sides, we get

$$\int dy = \int x^3\ dx + \int \sin\ 3x\ dx$$

$$y = \frac{x^4}{4} - \frac{\cos\ 3x}{3} + c$$

Illustration No. 8

Solve the differential equation $e^x - e^{-x}\left(\frac{dy}{dx}\right) e^x + e^{-x}$?

Solution:

Since $$e^x - e^{-x}\left(\frac{dy}{dx}\right) = e^x + e^{-x}$$

Therefore $$\frac{dy}{dx} = \frac{e^x + e^{-x}}{e^x - e^{-x}}$$

Or $$dy = \left(\frac{e^x + e^{-x}}{e^x - e^{-x}}\right) dx$$

Now integrating both sides, we get

$$\int dy = \int\left(\frac{e^x + e^{-x}}{e^x - e^{-x}}\right) dx$$

Let $e^x - e^{-x} = t$

And hence, $$(e^x + e^{-x})\ dx = dt$$

Therefore, $$\int dy = \int \frac{1}{t} dt$$

Or $$y = \log t + c$$

But replacing t by $e^x - e^{-x}$, we get

$$y = \log\ (e^x - e^{-x}) + c$$

Illustration No. 9

Solve the differential equation $\frac{dy}{dx} = \frac{1+y}{y^2}$

Solution:

Since $\frac{dy}{dx} = \frac{1+y}{y^2}$.

Or rewriting the above equation as,

$$\frac{dx}{dy} = \frac{y^2}{1+y}$$

Or

$$dx = \left(\frac{y^2}{1+y}\right) dy$$

$$= \left(y - \frac{y}{1+y}\right) dy$$

Now integrating both sides, we get

$$\int dx = \int \left(y - \frac{y}{1+y}\right) dy$$

$$\int dx = \int y\ dy - \int \left(\frac{y}{1+y}\right) dy$$

$$\int dx = \int y\ dy - \int \left(\frac{1+y-1}{1+y}\right) dy$$

$$\int dx = \int y\ dy - \int dy + \int \frac{1}{1+y}\ dy$$

$$x = \frac{y^2}{2} - y + \log\ (1+y) + c$$

Equation with Variables are Separable

If an equation can be written as $f_1(x)\ dx = f_2(x)\ dy$ where $f_1(x)$ is a function of x and $f_2(x)$ is a function of y, then its solution is obtained by directly integrating as $\int f_1(x)\ dx = \int f_2(x)\ dy + c$ where c is arbitrary constant i.e., if an equation in x and y can be written down in such way that dx and all the terms containing x is on one side and that dy and the terms y on the other side then the equation can be termed as variables are separable and the solution of the equation obtained by direct integration.

Note:

Some of the different forms of differential equation that can be converted into separable form are

(i) $\frac{dy}{dx} = f(x)\, g(y)$ can be converted to $\frac{dy}{g(y)} = f(x)\, dx$

(ii) $f_1(x)\, g_1(y)\, dx + f_2(x)\, g_2(y)\, dy = 0$ can be converted to $\frac{f_1\, x}{f_2\, x} dx + \frac{g_2\, y}{g_1\, y}\, dy$

Illustration No. 10

Solve the differential equation $3\frac{dy}{dx} = \frac{y}{x}$?

Solution:

Since the equation is $3\frac{dy}{dx} = \frac{y}{x}$

Rearranging the equation, we get,

$$\frac{3dy}{y} = \frac{dx}{x}$$

Integrating both sides, we get,

$$3\int\frac{dy}{y} = \int\frac{dx}{x}$$

$$3 \log y = \log x + \log c$$

$$\log y^3 = \log cx$$

Therefore, $y^3 = cx$

Illustration No. 11

Solve $(5 + x)\frac{dy}{dx} = (y - 5y^2)$

Solution:

The given equation is

$$(5 + x)\frac{dy}{dx} = (y - 5y^2)$$

$$\frac{dy}{y(1-5y)} = \frac{dx}{(5+x)}$$

$$\left\{\frac{1}{y} + \frac{5}{(1-5y)}\right\} dy = \frac{dx}{(5+x)} \quad \text{Using partial fractions}$$

Integrating both sides, we get,

$$\int \left\{\frac{1}{y} + \frac{5}{(1-5y)}\right\} dy = \int \frac{dx}{(5+x)}$$

$$\int \frac{1}{y} dy + \int \frac{5}{(1-5y)} dy = \int \frac{dx}{(5+x)}$$

$$\log y - \log (1 - 5y) = \log (5 + x) + \log c$$

$$\log \frac{y}{(1-5y)} = \log\ c\ (5+x)$$

Taking antilogs, both sides, we get,

$$\frac{y}{(1-5y)} = c(5+x)$$

Or $$y = c\ (1 - 5\ y)\ (5+x)$$

Illustration No. 12

Solve the differential equation $(1 + x^2)\ dy - (1 + y^2)\ dx = 0$?

Solution:

The given equation is $(1 + x^2)\ dy - (1 + y^2)\ dx = 0$

Therefore,

$$(1 + x^2)\ dy = (1+ y^2)\ dx$$

$$\frac{dy}{1+y^2} = \frac{dx}{1+x^2}$$

Integrating both sides, we get,

$$\int \frac{dy}{1+y^2} = \int \frac{dx}{1+x^2}$$

Therefore,

$$\tan^{-1} y = \tan^{-1} x + \tan^{-1} c$$

$$\tan^{-1} y - \tan^{-1} x = \tan^{-1} c$$

$$\tan^{-1}\frac{y-x}{1+yx} = \tan^{-1} c$$

And hence,

$$\frac{y-x}{1+yx} = c$$

$$y - x = c\,(1 + yx)$$

$$y = x + c\,(1 + yx)$$

Illustration No.13

Solve the differential equation $\frac{dy}{dx} = \frac{1+\cos^2 y}{1-\cos^2 y}$?

Solution:

The given equation is

$$\frac{dy}{dx} = \frac{1+\cos^2 y}{1-\cos^2 y}$$

But,

$$\frac{1+\cos^2 y}{1-\cos^2 y} = \cot^2 y$$

since $1 - \cos^2 y = \sin^2 y$ and $1 + \cos^2 y = \cos^2 y$

Therefore,

$$\frac{dy}{dx} = \cot^2 y$$

Or, $$\frac{dy}{\cot^2 y} = dx$$

$$\tan^2 y\, dy = dx$$

Integrating both sides, we get,

$$\int \tan^2 y\, dy = \int dx$$

$$\int (\sec^2 y - 1)\, dy = \int dx$$

$$\int \sec^2 y\, dy - \int dy = \int dx$$

$$\tan y - y = x + c$$

Or

$$x - \tan y + y + c = 0$$

Illustration No.14

Solve the differential equation $(9 + y^2)x\,dx + (x^2 - 9)\,y\,dy = 0$?

Solution:

The given differential equation is $(9 + y^2)x\,dx + (x^2 - 9)\,y\,dy = 0$

Now dividing by $\frac{1}{2}(x^2 - 9)(y^2 + 9)$, we get,

$$\frac{2x\,dx}{x^2 - 9} + \frac{2y\,dy}{y^2 + 9} = 0$$

Now, integrating the above equation, we get,

$$\int \frac{2x\,dx}{x^2 - 9} + \int \frac{2y\,dy}{y^2 + 9} = 0$$

Therefore, we have,

$\log(x^2 - 9) + \log(y^2 + 9) = \log c$, where c is a constant

or $\log(x^2 - 9)(y^2 + 9) = \log c$

or $(x^2 - 9)(y^2 + 9) = c$

Homogenous Differential Equation

A function $f(x, y)$ is said to be homogenous function of degree n if $f(\alpha x, \alpha y) = \alpha^n f(x, y)$ and such that the function can be written as

$$f(x, y) = x^n f\left(\frac{y}{x}\right) \text{ or } y^n f\left(\frac{x}{y}\right)$$

For Example,

$f(x, y) = x^2 + y^2 + 3xy$ is a homogenous equation of degree 2

If a differential equation of 1st order and 1st degree can be expressed as

$$\frac{dy}{dx} = \frac{f(x,y)}{g(x,y)} = \frac{xf\left(\frac{y}{x}\right)}{xg\left(\frac{y}{x}\right)},$$

where f(x, y) and g(x, y) are homogenous equations. In order to solve such equations we substitute

$$y = vx \text{ where } \frac{dy}{dx} = v + x\frac{dv}{dx} \qquad .....1$$

Now, the given equation is

$$\frac{dy}{dx} = \frac{f(x,y)}{g(x,y)} = \frac{xf\left(\frac{y}{x}\right)}{xg\left(\frac{y}{x}\right)},$$

taking x common both numerator and denominator,

Or, $$\frac{dy}{dx} = \frac{f\left(\frac{y}{x}\right)}{g\left(\frac{y}{x}\right)}$$

Now substituting the values of y and $\frac{dy}{dx}$ from 1, we have,

$$f(v) = v + x\frac{dv}{dx}$$

Or $$x\frac{dv}{dx} = f(v) - v$$

Or $$\frac{dv}{f(v) - v} = \frac{dx}{x}$$

Now the variable can be separated and its solution is

$$\int\frac{dv}{f(v) - v} = \int\frac{dx}{x} = \log x + c \text{ where c is a constant.}$$

After integration v should be replaced by (y/x) to get the required solution.

Illustration No.15

Solve the differential equation $(x^2 - y^2)dx + 2xy\,dy = 0$

Solution:

The given equation is $(x^2 - y^2)\,dx + 2xy\,dy = 0$

The equation can be written as $\frac{dy}{dx} = \frac{y^2 - x^2}{2xy}$

Now the equation gets converted to a homogenous differential equation form as each term in the numerator and denominator on the R.H.S is of the second degree in x and y. Now substituting y = v x, we get,

$$v + x\frac{dv}{dx} = \frac{v^2x^2 - x^2}{2\,x\,vx}$$

$$x\,\frac{dv}{dx} = \frac{v^2 - 1}{2\,v} - v$$

$$x\,\frac{dv}{dx} = \frac{v^2 - 1 - 2v^2}{2\,v}$$

$$= \frac{-1 - v^2}{2\,v}$$

$$= \frac{2vdv}{-1 - v^2} = \frac{dx}{x}$$

Integrating both sides, we get,

$$\int\frac{2vdv}{-1 - v^2} = \int\frac{dx}{x}$$

Therefore,

$$-\log(1 + v^2) = \log x + \log c$$

$$\log x + \log(1 + v^2) = \log c$$

$$\log x(1 + v^2) = \log c$$

And hence,

$$x(1 + v^2) = c$$

Now substituting the value of v = y/x, we get,

$$[(y^2/x^2) + 1]\,x = c$$

$(x^2 + y^2) = cx$ is the required solution

Illustration No. 16

Solve the differential equation $(x^2 + y^2)dx + 3xy\,dy = 0$

Solution:

The given equation is $(x^2 + y^2)\,dx + 3xy\,dy = 0$

The equation can be written as $\frac{dy}{dx} = -\frac{y^2 + x^2}{3xy}$

Now the equation gets converted to a homogenous differential equation form as each term in the numerator and denominator on the R.H.S is of the second degree in x and y. Now substituting y = v x, we get,

$$v + x\frac{dv}{dx} = -\left(\frac{v^2x^2 + x^2}{3x\ vx}\right)$$

$$x\frac{dv}{dx} = -\left(\frac{v^2 + 1}{3v}\right) - v$$

$$x\frac{dv}{dx} = -\left(\frac{v^2 + 1 + 3v^2}{3v}\right)$$

$$= -\frac{\left(1 + 4v^2\right)}{3v}$$

$$\frac{3vdv}{1 + 4v^2} = -\frac{dx}{x}$$

Integrating both sides, we get,

$$\int\frac{3vdv}{1 + 4v^2} = -\int\frac{dx}{x}$$

Therefore,

$$(3/8)\log(1 + 4v^2) = -\log x + \log c$$

$$\log x + \log[(1 + 4v^2)]^{3/8} = \log c$$

$$\log x\,(1 + 4v^2)^{3/8} = \log c$$

And hence,

$$x(1 + 4v^2)^{3/8} = c$$

Now substituting the value of v = y/x, we get,

$$[4(y^2/x^2) + 1]^{3/8}\,x = c$$

$$[4\,(x^2 + y^2)^{3/8}]\,x^{1/4} = c \text{ is the required solution}$$

Illustration No. 17

Solve the differential equation $x\frac{dy}{dx} = y - x\sin^2\left(\frac{y}{x}\right)$

Solution:

The given equation is $x\frac{dy}{dx} = y - x\sin^2\left(\frac{y}{x}\right)$

Now the above equation can be rewritten as

$$\frac{dy}{dx} = \frac{y}{x} - \sin^2\left(\frac{y}{x}\right)$$

Now substituting y = vx, we get

$$v + x\frac{dv}{dx} = v - \sin^2 v$$

$$x\frac{dv}{dx} = -\sin^2 v$$

$$\frac{dv}{\sin^2 v} = -\frac{dx}{x}$$

$$\operatorname{cosec}^2 v \, dv = -\frac{dx}{x}$$

Integrating both sides, we get,

$$\int \operatorname{cosec}^2 v \, dv = -\int \frac{dx}{x}$$

$$-\cot v = -\log x + c$$

Or

$$-\cot v + \log x = c$$

Now replacing v by y/x, we get

$-\cot(y/x) + \log x = c$ is the required solution

An Equation Reducible to Homogenous Form

Consider the equation $\dfrac{dy}{dx} = \dfrac{a_1x + b_1y + c_1}{a_2x + b_2y + c_2}$ 1

where $\dfrac{a_1}{a_2} \neq \dfrac{b_1}{b_2}$

Now, if c_1 and c_2 are both zero then the equation reduces to homogenous form which can be solved by the method just discussed in the previous section but if c_1 and c_2 are both not equal to zero then, we change the variables so that the variables are no longer present by substituting x = X + h and y = Y + k where h and k are constant.

Now differentiating x = X + h and Y = y + k, we get dx = dX and dy = dY, so that $\frac{dy}{dx} = \frac{dY}{dX}$

So that eq. 1 can be rewritten as

$$\frac{dY}{dX} = \frac{a_1X + b_1Y + a_1h + b_1k + c_1}{a_2X + b_2Y + a_2h + b_2k + c_2}$$

h and k are chosen in a such way that,

$$a_1h + b_1k + c_1 = 0 \text{ and } a_2h + b_2k + c_2 = 0$$

thus the above equation reduces to homogenous form as

$$\frac{dY}{dX} = \frac{a_1X + b_1Y}{a_2X + b_2Y}$$

Which can be solved by substituting Y = vX and once the solution is obtained then X and Y are replaced by x – h and y – k to get the required solution.

Illustration No.18

Solve the differential equation (2x + y + 3) dy = (x + 2y + 3) dx

Solution:

The given equation is (2x + y + 3) dy = (x + 2y + 3) dx which can be rewritten as,

$$\frac{dy}{dx} = \frac{x + 2y + 3}{2x + y + 3} \qquad .....1$$

Now substituting x = X + h and y = Y + k

Hence the equation 1 gets converted into,

$$\frac{dY}{dX} = \frac{(X + h) + 2(Y + k) + 3}{2(X + h) + (Y + k) + 3}$$

$$\frac{dY}{dX} = \frac{X + 2Y + h + 2k + 3}{2X + Y + 2h + k + 3} \qquad .....2$$

Now h and k are chosen such way that $h + 2k + 3 = 0$ and 2h + k+ 3 = 0 solving both the equations, we get

$$h = k = -1$$

Therefore $\quad x = X - 1$ and $y = Y - 1$

Now eq. 2 gets reduced to,

$$\frac{dY}{dX} = \frac{X + 2Y}{2X + Y} \qquad .....3$$

Now, substituting Y = vX and then differentiating both side, we get, $\frac{dY}{dX} = v + X\frac{dv}{dx}$. Now replacing the value of $\frac{dY}{dX}$ in 3, we get,

$$v + X\frac{dv}{dX} = \frac{X + 2vX}{2X + vX}$$

$$X\frac{dv}{dX} = \frac{1 + 2v}{2 + v} - v$$

$$X\frac{dv}{dX} = \frac{1 + 2v - 2v - v^2}{2 + v} - \frac{1 - v^2}{2 + v}$$

$$\left(\frac{2 + v}{1 - v^2}\right) = \frac{dX}{X} \qquad ......4$$

Now resolving eq. 4 into partial fractions, we get,

$$\left[\frac{1}{2}\left(\frac{1}{v + 1}\right) + \frac{3}{2}\left(\frac{1}{1 - v}\right)\right] dv = \frac{dX}{X} \qquad .....5$$

Now integrating eq. 5, we get

$$\int\frac{1}{2}\left[\left(\frac{1}{v + 1}\right) + \frac{3}{2}\left(\frac{1}{1 - v}\right)\right] dv = \int\frac{dX}{X}$$

$$\frac{1}{2}\int\frac{1}{v + 1} dv + \frac{3}{2}\int\frac{1}{1 - v} dv = \int\frac{dX}{X}$$

$$\frac{1}{2}\log(v + 1) - \frac{3}{2}\log(1 - v) = \log X + \log C$$

where C is constant

$$\frac{1}{2}\log(v + 1) - \frac{1}{2}\log(1 - v)^3 = \log XC$$

$$\log\left\{\frac{v+1}{(1-v)^3}\right\} = 2\log XC$$

$$\log\left\{\frac{v+1}{(1-v)^3}\right\} = \log(XC)^2$$

$$\left\{\frac{v+1}{(1-v)^3}\right\} = (XC)^2$$

Now, substituting v = Y/X

$$\left\{\frac{\frac{Y}{X}+1}{\left(1-\frac{Y}{X}\right)^3}\right\} = (XC)^2$$

$$\left\{\frac{X+Y}{(X-Y)^3}\right\} = (C)^2$$

Now substituting X = x + 1 and Y = y + 1

We get,

$$\left\{\frac{x+1+y+1}{(x+1-y-1)^3}\right\} = (C)^2$$

$$\left\{\frac{x+y+2}{(x-y)^3}\right\} = (C)^2$$

$(x + y + 2) = (C)^2 (x - y)^3$ is the required solution.

Illustration No.19

Solve the differential equation (x + y – 1) dx = (x + y + 1) dy

Solution:

The given equation is (x + y – 1) dx = (x + y + 1) dy

Or the equation can be rewritten as

$$\frac{dy}{dx} = \frac{x+y-1}{x+y+1} \qquad .....1$$

Now, substituting x + y = v and differentiating, we get $1+\frac{dy}{dx}=\frac{dv}{dx}$

Or $$\frac{dv}{dx}-1=\frac{dy}{dx}$$

Substituting the changes in eq. 1, we get,

$$\frac{dv}{dx}-1=\frac{v-1}{v+1}$$

$$\frac{dv}{dx}=\frac{v-1}{v+1}+1$$

$$\frac{dv}{dx}=\frac{v-1+v+1}{v+1}$$

$$\frac{dv}{dx}=\frac{2v}{v+1}$$

Or $$\frac{(1+v)dv}{v}=2dx$$

$$\frac{1}{v}dv+dv=2dx$$

Now integrating, we get,

$$\log v+v=2x+c$$

Now substituting the value of v, we get

$$\log(x+y)+(x+y)=2x+c$$

Or $$x+y+\log(x+y)=c$$ is the required solution

Linear Differential Equation

A differential equation of the type $\frac{dy}{dx}+Py=Q$ where P and Q are constants or function of x alone is called as linear differential equation of the first order of y.

In order to solve linear differential equation of the above type multiply both sides of the equation by an integrating factor $e^{\int pdx}$,

Then we have,

$$\left(\frac{dy}{dx} + Py\right)e^{\int pdx} = Qe^{\int pdx} \qquad .....1$$

Now, we get,

$$\frac{dy}{dx}\left\{y\, e^{\int P\, dx}\right\} = y.\frac{d}{dx}\left[e^{\int P\, dx}\right] + e^{\int Pdx}.\frac{dy}{dx}$$

$$= y.\left[e^{\int P\, dx}\right].\frac{d}{dx}\left\{\int P\, dx\right\} + e^{\int P\, dx}.\frac{dy}{dx}$$

$$= y.\left[e^{\int P\, dx}\right].P + e^{\int P\, dx}.\frac{dy}{dx}$$

Or

$$\frac{d}{dx}\left[y\, e^{\int P\, dx}\right] = \left\{Py + \frac{dy}{dx}\right\}e^{\int P\, dx}$$

Substituting the value in eq.1, we get,

$$\frac{d}{dx}\left[y\, e^{\int P\, dx}\right] = Qe^{\int P\, dx}$$

Integrating both sides with respect to x , we get,

$$\left\{y\, e^{\int P\, dx}\right\} = c + \int Qe^{\int P\, dx}\, dx$$

where c is an arbitrary constant giving the complete solution of the differential equation.

Illustration No. 20

Solve the linear differential equation $\frac{dy}{dx} + y = e^x$?

Solution:

The given linear differential equation is $\frac{dy}{dx} + y = e^x$

Here P = 1 and Q = e^x

Therefore the integrating factor = $e^{\int pdx} = e^{\int 1dx} = e^x$

Now multiplying the integrating factor on both sides, we get,

$$e^x\frac{dy}{dx} + e^x y = e^x e^x$$

$$e^x \frac{dy}{dx} + e^x y = e^{2x}$$

$$\frac{d}{dx} + \left(e^x y\right) = e^{2x}$$

Integrating the above equation we get,

$$e^x y = \frac{e^{2x}}{2} + C$$

Or

$$2e^x y = e^{2x} + 2C$$

Illustration No. 21

Solve the linear differential equation $(1 + x^2)\frac{dy}{dx} + 2xy = 1$

Solution:

The given differential equation $(1 + x^2)\frac{dy}{dx} + 2xy = 1$

Now rearranging the equation, we have,

$$\frac{dy}{dx} + \frac{2xy}{1+x^2} = \frac{1}{1+x^2}$$

Here, $P = \frac{2x}{1+x^2}$ and $Q = \frac{1}{1+x^2}$

Now the integrating factor = $e^{\int P\,dx} = e^{\int \frac{2x}{1+x^2}dx} = e^{\log\left(1+x^2\right)} = 1 + x^2$

And hence multiplying the above equation by the integrating factor, we get,

$$\left(1+x^2\right)\frac{dy}{dx} + 2xy = 1$$

$$\frac{d}{dx}\{\left(1+x^2\right)y\} = 1$$

Now further integrating, we have,

$$\{(1 + x^2)y\} = x + c$$

Illustration No. 22

Solve the linear differential equation y dx – x dy + log x dx = 0?

Solution:

The given linear differential equation y dx – x dy + log x dx = 0 can be rewritten as

$$\frac{dy}{dx} - \frac{y}{x} = \frac{\log x}{x}$$

Here, $P = -\frac{1}{x}$ and $Q = \frac{\log x}{x}$

Therefore the integrating factor $= e^{\int P\,dx} = e^{\int -\frac{1}{x}dx} = e^{-\log x} = \frac{1}{x}$

And hence multiplying both sides by the integrating factor, we get,

$$\frac{1}{x}\left(\frac{dy}{dx}\right) - \frac{y}{x^2} = \frac{\log x}{x^2}$$

Now integrating the above equation, we get,

$$\left(\frac{1}{x}\right)y = C + \int \frac{\log x}{x^2}\,dx$$

Now putting $\log x = t \leftrightarrow \frac{1}{x} = e^{-t}$

Therefore $\frac{1}{x}dx = dt$

Hence,

$$\left(\frac{1}{x}\right)y = C + \int t\,e^{-t}\,dt$$

$$= C + \left(-te^{-t}\right) + \int e^{-t}\,dt \qquad \{\text{Using integration by parts}\}$$

$$= C + \left(-te^{-t}\right) - e^{-t}$$

$$= C - (1+t)e^{-t}$$

Replacing the values of t by x terms, we get the required solution i.e.

$$\left(\frac{1}{x}\right)y = C - (1 + \log x)\left(\frac{1}{x}\right) \text{ or } y = Cx - (1 + \log x)$$

Illustration No. 23

Solve the linear differential equation $(1 + x^2)\left(\frac{dy}{dx}\right) - y = \cot^{-1} x$?

Solution:

The given differential equation $(1 + x^2)\left(\frac{dy}{dx}\right) - y = \cot^{-1}x$ can be rewritten

$$\left(\frac{dy}{dx}\right) - \frac{y}{1+x^2} = \frac{\cot^{-1}x}{1+x^2}$$

Here, $P = -\frac{1}{1+x^2}$ and $Q = \frac{\cot^{-1}x}{1+x^2}$

Therefore the integrating factor = $e^{\int P\,dx} = e^{-\int \frac{1}{1+x^2}dx} = e^{\cot^{-1}x}$

Multiplying both sides by the integrating factor, we get,

$$e^{\cot^{-1}x}\left(\frac{dy}{dx}\right) - \frac{ye^{\cot^{-1}x}}{1+x^2} = \frac{\cot^{-1}x}{1+x^2}e^{\cot^{-1}x}$$

Now, integrating, we get,

$$\left(y\, e^{\cot^{-1}x}\right) = \int \frac{\cot^{-1}x}{1+x^2} e^{\cot^{-1}x} dx$$

Let $\cot^{-1} x = t$ then $-\frac{dx}{1+x^2} = dt$

Therefore, we have,

$$\left(y\, e^{\cot^{-1}x}\right) = -\int te^t\, dt$$

$$\left(y\, e^{\cot^{-1}x}\right) = -\left(te^t - \int e^t\, dt\right) + c$$

$$\left(y\, e^{\cot^{-1}x}\right) = -\left(te^t - e^t\right) + c$$

$$\left(y\, e^{\cot^{-1}x}\right) = -te^t + e^t + c$$

Now replacing t in x terms we get, the required solution,

$$\left(y\, e^{\cot^{-1}x}\right) = -\cot^{-1}x\, e^{\cot^{-1}x} + e^{\cot^{-1}x} + c$$

Equations Reducible to Linear Form (Bernoulli's Equation)

The equation of the form $\left(\frac{dy}{dx}\right) + Py = Qy^n$ where P and Q are constants or function of x and n is a constant other than zero and unity then the equation is the extended form of linear equation also called as Bernoulli's Equation. This type of equation can be reduced to linear form by dividing by y^n and substituting $1/y^{n-1}$ is equal to v

i.e. $$\left(\frac{dy}{dx}\right) + Py = Qy^n$$

$$\frac{1}{y^n}\left(\frac{dy}{dx}\right) + \frac{P}{y^{n-1}} = Q \qquad .....1$$

Now, substitute $\frac{1}{y^{n-1}} = v$

Differentiating both sides with respect to x, we get,

$$(-n+1)y^{-n}\left(\frac{dy}{dx}\right) = \left(\frac{dv}{dx}\right) \text{ or } \frac{1}{y^n}\left(\frac{dy}{dx}\right) = \frac{1}{1-n}\left(\frac{dv}{dx}\right)$$

Now replacing the value $\frac{1}{y^n}\left(\frac{dy}{dx}\right)$ in eq. 1, we have,

$$\frac{1}{1-n}\left(\frac{dv}{dx}\right) + Pv = Q$$

Or,

$$\left(\frac{dv}{dx}\right) + Pv(1-n) = Q(1-n)$$

which is a linear equation in v which can be solved as the procedure discussed in case of linear differential equation.

Illustration No. 24

Solve the differential equation $y^2\left(\frac{dy}{dx}\right) = x - y^3$?

Solution:

The given equation can be rewritten as

$$y^2\left(\frac{dy}{dx}\right) + y^3 = x \qquad .....1$$

putting $y^3 = v$ or $y^2 \frac{dy}{dx} = \frac{1}{3}\left(\frac{dv}{dx}\right)$

then the equation reduces to

$$\frac{1}{3}\left(\frac{dv}{dx}\right) + v = x$$

Or

$$\left(\frac{dv}{dx}\right) + 3v = 3x$$

which is of the standard form of linear equation in v

Therefore,

The integrating factor = $e^{\int 3\,dx} = e^{3x}$

And hence the solution is $ve^{3x} = \int 3x\, e^{3x}\, dx + c$

$$ve^{3x} = 3\left\{\frac{1}{3} x\, e^{3x} - \int 1.\ \frac{1}{3} e^{3x}\, dx\right\} + c \text{ \{By integration by Parts\}}$$

$$= x\, e^{3x} - \frac{1}{3} e^{3x} + c$$

$y^3e^{3x} = xe^{3x} - \frac{1}{3} e^{3x} + c$, is the required solution and c is an arbitrary constant.

Illustration No. 25

Solve the differential equation $x\,dy - y\,dx = y^2 \log x\, dx$?

Solution:

The given equation can be written as $x\left(\frac{dy}{dx}\right) - y = y^2 \log x$

Divide both side by y^2, we get,

$$\frac{1}{y^2}\left(\frac{dy}{dx}\right) - \frac{1}{x}, \frac{1}{y} = \frac{1}{x}\log x \qquad .....1$$

Now substituting $-\frac{1}{y} = v$ then $\frac{1}{y^2}\left(\frac{dy}{dx}\right) = \frac{dv}{dx}$ then replacing the value in eq.1, we get,

$$\frac{dv}{dx}+\frac{1}{x}v=\frac{1}{x}\log x$$

which is the standard form of the linear equation and hence the integrating factor is,

$$e^{\int \frac{1}{x}dx} = e^{\log x} = x$$

Therefore multiplying both sides by the integrating factor, we get,

$$x\frac{dv}{dx}+v=\log x$$

$$\frac{d}{dx}(vx) = \log x$$

Integrating both sides, we get,

$$vx = \int \log x \, dx + c$$

$$= x \log x - \int \frac{1}{x}.1dx+c$$

$$= x \log x - \log x + c$$

Now, replacing the value v by $-\frac{1}{y}$, we get,

$$-\frac{x}{y} = x \log x - \log x + c \text{ or}$$

$-x = xy \log x - y \log x + cy$ is the required solution and where c is a arbitrary constant.

Exact Differential Equation

A differential equation of the form M dx + N dy = 0 where M and N are functions of x and y or constants is said to be exact if the expression can be directly obtained by differentiating certain functions of x and y. If f(x, y) be a function such that

$$d\left[f(x,y)\right] = M\,dx + N\,dy$$

$$\text{or } \frac{\partial f}{\partial x}dx+\frac{\partial f}{\partial y}dy = Mdx+Ndy$$

Therefore the ***necessary condition*** that the equation M dx + N dy = 0 is exact if $M = \frac{\partial f}{\partial x}$ and $N = \frac{\partial f}{\partial y}$.

Also, differentiating M and N w. r. t y and x, we get $\frac{\partial M}{\partial y} = \frac{\partial^2 f}{\partial y \partial x}$ and $\frac{\partial N}{\partial x} = \frac{\partial^2 f}{\partial x \partial y}$.

As we know that $\frac{\partial^2 f}{\partial y \partial x} = \frac{\partial^2 f}{\partial x \partial y}$, therefore the above condition can be restated as $\frac{\partial M}{\partial y} = \frac{\partial N}{\partial x}$ in the assumption that $\frac{\partial^2 f}{\partial y \partial x} = \frac{\partial^2 f}{\partial x \partial y}$

Hence the necessary and sufficient condition that the differential equation to be exact is $\frac{\partial M}{\partial y} = \frac{\partial N}{\partial x}$

If we assume $P = \int M dx$ then $\frac{\partial P}{\partial x} = M$ so that $\frac{\partial^2 P}{\partial y \partial x} = \frac{\partial M}{\partial y}$

As, $\frac{\partial N}{\partial x} = \frac{\partial M}{\partial y} = \frac{\partial^2 P}{\partial y \partial x} = \frac{\partial^2 P}{\partial x \partial y}$ or $\frac{\partial N}{\partial x} = \frac{\partial}{\partial x} = \frac{\partial}{\partial x}\left(\frac{\partial P}{\partial y}\right)$

Now integrating both sides with respect to x, we get,

$$N = \frac{\partial P}{\partial y} + \varnothing(y), \text{ where } \varnothing(y) \text{ is a function of y alone}$$

Therefore,

$$\begin{aligned} M dx + N dy &= \frac{\partial P}{\partial x} dx + \left\{\frac{\partial P}{\partial y} + \varnothing(y)\right\} dy \\ &= \frac{\partial P}{\partial x} dx + \frac{\partial P}{\partial y} dy + \varnothing(y) dy \\ &= dP + dF(y) \text{ where } dF(y) = \varnothing(y) dy \\ &= d\,[P + F(y)] \end{aligned}$$

Showing that M dx + N dy is an exact differential

Working Rule:

(i) Write down the differential equation in Mdx + N dy form.

(ii) Find the partial derivatives of $\frac{\partial M}{\partial y}$ and $\frac{\partial N}{\partial x}$

(iii) If found that $\frac{\partial M}{\partial y} = \frac{\partial N}{\partial x}$ then the equation is exact.

(iv) Now integrate M with respect to x regarding y as constant.

(v) Then integrate N with respect to y only those terms which do not contain x.

(vi) Add the two expressions obtained in iv) and v) and equate the result to an arbitrary constant to give the solution.

Note:

Equations Reducible to Exact Differential Equation Form

A differential equation of the type Mdx + Ndy which is not exact can be made exact by multiplying the equation with certain function of x and y which are the integrating factors.

Some of the examples are given below used for finding the integrating factors in certain cases.

(i) $d\left(\frac{y}{x}\right) = \frac{xdy - ydx}{x^2}$

(ii) $d\left(\frac{x}{y}\right) = \frac{ydx - xdy}{y^2}$

(iii) $d(x, y) = xdy + ydx$

(iv) $d\left(\frac{x^2}{y}\right) = \frac{2xy\,dx - x^2\,dy}{y^2}$

(v) $d\left(\frac{y^2}{x}\right) = \frac{2xy\,dy - y^2\,dx}{x^2}$

(vi) $d\left(\frac{y^2}{x^2}\right) = \frac{2x^2y\,dy - 2xy^2\,dx}{x^4}$

(vii) $d\left[\tan^{-1}\left(\frac{x}{y}\right)\right] = \frac{y\,dx - x\,dy}{x^2 + y^2}$

(viii) $d\left[\tan^{-1}\left(\frac{y}{x}\right)\right] = \frac{x\,dy - y\,dx}{x^2 + y^2}$

(xi) $d\log\left(\frac{x}{y}\right) = \frac{x\,dy - y\,dx}{xy}$

(x) $d\log\left(\frac{y}{x}\right) = \frac{y\,dx - x\,dy}{xy}$

(xi) $d\left(-\frac{1}{xy}\right) = \frac{x\,dy + y\,dx}{x^2y^2}$

(xii) $d\left(\frac{e^x}{y}\right) = \frac{y\,e^x\,dx - e^x\,dy}{y^2}$

(xiii) $d[\frac{1}{2}\log\left(x^2 + y^2\right)] = \frac{x\,dx + y\,dy}{x^2 + y^2}$

Illustration No. 26

Solve $(x^2 - 3y)\,dx - (3x - y^2)dy = 0$?

Solution:

The given equation is $(x^2 - 3y)\,dx - (3x - y^2)\,dy = 0$

Here $M = x^2 - 3y$ and $N = -(3x - y^2)$

Therefore $\frac{\partial M}{\partial y} = -3$ and $\frac{\partial N}{\partial x} = -3$ and hence $\frac{\partial M}{\partial y} = \frac{\partial N}{\partial x}$ therefore the given equation is exact.

Now regarding y as constant integrate M w.r.t x

i.e $$\int M\,dx = \int\left(x^2 - 3y\right)dy = \frac{1}{3}x^3 - 3xy \quad .....1$$

And the only term we get by integrating N with respect to y

i.e. $$\int N\,dy = \int (y)^2 dy = \frac{1}{3}y^3 \quad .....2$$

Now from eq. 1 and eq. 2 we get,

$$\left(\frac{1}{3}x^3 - 3xy\right) + \frac{1}{3}y^3 = c$$ where c is arbitrary constant.

Illustration No. 27

Solve $(3x + 4y + 5)\,dx + (4x + 3y + 7)\,dy = 0$

Solution:

From the above equation we have,

$$M = 3x + 4y + 5 \text{ and } N = 4x + 3y + 7,$$

Therefore

$$\frac{\partial M}{\partial y} = 4 \text{ and } \frac{\partial N}{\partial x} = 4 \text{ and hence } \frac{\partial M}{\partial y} = \frac{\partial N}{\partial x}$$

therefore the given equation is exact

Now taking y as constant integrating M

i.e. $$\int M dx = \int (3x + 4y + 5)\,dx = \frac{3x^2}{2} + 4xy + 5x \quad \text{.....1}$$

Similarly integrating N w.r.t y by omitting x terms, we get,

i.e. $$\int N dy = \int (3y + 7)\,dx = \frac{3y^2}{2} + 7y \quad \text{.....2}$$

Adding the expression in eq. 1 and 2, we get,

$$\frac{3x^2}{2} + 4xy + 5x + \frac{3y^2}{2} + 7y = K \text{ where K is a arbitrary constant}$$

And hence, $$3x^2 + 8xy + 3y^2 + 10x + 14y = 2K$$

Or $$3x^2 + 8xy + 3y^2 + 10x + 14y - 2K = 0$$

Or $$3x^2 + 8xy + 3y^2 + 10x + 14y + C = 0 \text{ where } C = -2K$$

Illustration No.28

Solve $(x^2 - 2xy - y^2)dx - (x + y)^2 dy$

Solution:

From the given equation we have

$$M = x^2 - 2xy - y^2 \text{ and } N = -(x + y)^2 = -x^2 - 2xy - y^2$$

Hence $\frac{\partial M}{\partial y} = -2x - 2y$ and $\frac{\partial N}{\partial x} = -2x - 2y$ and hence $\frac{\partial M}{\partial y} = \frac{\partial N}{\partial x}$

therefore the given equation is exact

Integrating M w.r.t y, we get,

$$\int M\,dx = \int\left(x^2 - 2xy - y^2\right)dx = \frac{x^3}{3} - x^2y - y^2x \qquad .....1$$

Similarly integrating only term in N with respect to y,

$$\int N\,dx = \int -y^2 dy = -\frac{y^3}{3} \qquad .....2$$

Now adding eq. 1 and 2, we get,

$$\frac{x^3}{3} - x^2y - y^2x - \frac{y^3}{3} = C$$

$$x^3 - 3x^2y - 3y^2x - y^3 = 3C$$

Illustration No. 29

Solve $ydx - xdy = xy^2dx$?

Solution:

The given equation is not exact as it does not satisfies the necessary and sufficient condition i.e. $\frac{\partial M}{\partial y} \neq \frac{\partial N}{\partial x}$

Therefore dividing both sides of the equation by y^2, we get,

$$x\,dx = \frac{y\,dx - x\,dy}{y^2}$$

Or $$x\,dx - \frac{y\,dx - x\,dy}{y^2} = 0$$

But $$d\left(\frac{x}{y}\right) = \frac{y\,dx - x\,dy}{y^2}$$

Therefore $$x\,dx - d\left(\frac{x}{y}\right) = 0$$

Integrating we get, $\frac{x^2}{2} - \left(\frac{x}{y}\right) = C$ where C is an arbitrary constant

Or $$x^2y - 2x = 2Cy$$

Illustration No. 30

Solve $y(2xy + e^x)\,dx - e^x\,dy = 0$?

Solution:

The given equation is $y(2xy + e^x)\,dx - e^x\,dy = 0$ and since the condition of exactness is not satisfied, therefore let us rearrange the equation by dividing both the sides of the equation by y^2.

Or $$2xdx + \frac{e^x}{y}dx - \frac{e^x}{y^2}dy = 0 \text{ or } 2xdx + \left(\frac{y e^x\,dx - e^x\,dy}{y^2}\right) = 0$$

But $$d\left(\frac{e^x}{y}\right) = \frac{ye^x\,dx - e^x\,dy}{y^2}$$

Therefore $$2xdx + d\left(\frac{e^x}{y}\right) = 0$$

Integrating, we get $x^2 + \left(\frac{e^x}{y}\right) = C$ where C is an arbitrary constant

Or $$x^2y + e^x = Cy$$

Solving Second Order Differential Equations

The differential equation of the type $\frac{d^2y}{dx^2} = f(x)$ are solved by integrating twice with respect to x. The following examples will illustrate the procedure.

Illustration No. 31

Solve the differential equation $\frac{d^2y}{dx^2} = 2x$?

Solution:

Since the given differential equation is $\frac{d^2y}{dx^2} = 2x$

Therefore integrating with respect to x, we get,

$$\frac{dy}{dx} = 2\int x\,dx + c_1$$

$$= 2\,\frac{x^2}{2} + c_1$$

$$= x^2 + c_1$$

Integrating once again with respect to x, we get,

$$y = \int (x^2 + c_1)\, dx + c_2$$

$$= \frac{x^3}{3} + c_1 x + c_2$$

Hence the required solution is $y = \frac{x^3}{3} + c_1 x + c_2$

Illustration No. 32

Solve the differential equation $\frac{d^2y}{dx^2} = \sin x + e^x$?

Solution:

The given differential equation is $\frac{d^2y}{dx^2} = \sin x + e^x$

Integrating with respect to x,

$$\frac{dy}{dx} = \int (\sin x + e^x)\, dx + c_1$$

$$\frac{dy}{dx} = \int \sin x\, dx + \int e^x\, dx + c_1$$

$$\frac{dy}{dx} = -\cos x + e^x + c_1$$

Once again integrating with respect to x, we get,

$$y = \int (-\cos x + e^x + c_1)\, dx + c_2$$

$$= \int \left(-\cos x\, dx + \int e^x\, dx + \int c_1\right) dx + c_2$$

$$= -\sin x + e^x + c_1 x + c_2$$

Illustration No. 33

Solve the differential equation $\frac{d^2y}{dx^2} = x\, a^x$?

Solution:

The given differential equation is $\frac{d^2y}{dx^2} = x\, a^x$

Therefore,

Integrating with respect to x, we get,

$$\frac{dy}{dx} = \int (xa^x)\,dx + c_1$$

$$= \frac{xa^x}{\log a} - \int \left(\frac{a^x}{\log a}\right) dx + c_1$$

$$= \frac{xa^x}{\log a} - \frac{a^x}{(\log a)^2} + c_1$$

Once again, integrating with respect to x, we get,

$$y = \int \{\frac{xa^x}{\log a} - \frac{a^x}{(\log a)^2} + c_1\}\,dx + c_2$$

$$y = \int \left\{\frac{xa^x}{\log a}\right\} dx - \int \frac{a^x}{(\log a)^2} dx + \int c_1\,dx + c_2$$

$$= \frac{1}{\log a} \int xa^x\,dx - \frac{a^x}{(\log a)^3} + c_1 x + c_2$$

$$= \frac{1}{\log a}\left\{\frac{xa^x}{\log a} - \frac{a^x}{(\log a)^2}\right\} - \frac{a^x}{(\log a)^3} + c_1 x + c_2$$

$$= \frac{xa^x}{(\log a)^2} - \frac{a^x}{(\log a)^3} - \frac{a^x}{(\log a)^3} + c_1 x + c_2$$

$$= \frac{xa^x}{(\log a)^2} - \frac{2a^x}{(\log a)^3} + c_1 x + c_2,$$

is the required solution.

Illustration No. 34

Solve the differential equation $\frac{d^2y}{dx^2} = x\cos x + e^x$?

Solution:

The given differential equation is $\frac{d^2y}{dx^2} = x \cos x + e^x$

Therefore, integrating with respect to x, we get,

$$\frac{dy}{dx} = \int\left(x \cos x\, dx + e^x\right) dx + c_1$$

$$= \int x \cos x\, dx + \int e^x\, dx + c_1$$

$$= x \sin x - \int \sin x\, dx + e^x + c_1$$

$$= x \sin x + \cos x + e^x + c_1$$

Once again, integrating with respect to x, we get,

$$y = \int\left(x \sin x + \cos x + e^x + c_1\right) dx + c_2$$

$$y = \int x \sin x\, dx + \int \cos x\, dx + \int e^x\, dx + \int c_1\, dx + c_2$$

$$= -x \cos x + \int \cos x\, dx + \sin x + e^x + c_1 x + c_2$$

$$= -x \cos x + \sin x + \sin x + e^x + c_1 x + c_2$$

$$= -x \cos x + 2\sin x + e^x + c_1 x + c_2$$

Application of Differential Equation to Business and Economics

Illustration No. 35

Ganesh's individual preference for two goods x and y is defined by the marginal rate of substitution of y for x is equal to $\frac{x-3}{y-2}$ then show that $U = (y-2)^2 - (x-3)^2$ where U is the utility function?

Solution:

Since the marginal rate of substitution of y and x is given by

$$\frac{dy}{dx} = \frac{x-3}{y-2}$$

$$(y-2)dy = (x-3)dx$$

Now, integrating both sides, we get,

$$\int (y-2)dy = \int (x-3)dx$$

$$\frac{y^2}{2} - 2y = \frac{x^2}{2} - 3x + c$$

$$y^2 - 4y = x^2 - 6x + 2c$$

i.e.

$$(y-2)^2 = (x-3)^2 + 2c - 4 - 9,$$

Assuming $2c - 4 - 9 = U$

Hence we get, $U = (y-2)^2 - (x-3)^2$

Illustration No. 36

If the elasticity of demand of the demand curve of the form q = f(p) is 5 – 4p then find the demand curve?

Solution:

Given that the elasticity of demand = $\dfrac{\frac{dq}{q}}{\frac{dp}{p}} = \dfrac{dq}{dp}.\dfrac{p}{q}$

But,

$$\frac{dq}{dp}.\frac{p}{q} = 5 - 4p,$$

Therefore,

$$\frac{dq}{q} = \left(\frac{5-4p}{p}\right)dp$$

Now integrating both sides, we get,

$$\int\frac{dq}{q} = \int\left(\frac{5-4p}{p}\right)dp = \int\frac{5}{p}dp - \int 4\,dp$$

$$\log q = 5 \log p - 4p + c$$

$$q = e^{5 \log p - 4p + c}$$

$$= p^5.e^{-4p+c}$$

Hence the demand curve is given by $q = p^5.e^{-4p+c}$

Illustration No. 37

If the population of a city is growing by 10 % per annum then how long does it take for the population to double?

Solution:

Let the initial population be p_0 and the population after a time period be p_t.

But it is given that,

$$\frac{dp}{dt} = \frac{10p}{100} = \frac{p}{10}$$

Now rearranging the equation, we get,

$$\frac{dp}{p} = \frac{1}{10}dt$$

Now integrating both side we get,

$$\int\frac{dp}{p} = \frac{1}{10}\int dt$$

$$\log p = \frac{1}{10}t + c$$

But the initial population i.e at t= 0 is p

Therefore, C = $\log p_0$

And hence, $\log p = \frac{1}{10}t + \log\ p_o$

$$\frac{1}{10}t = \log\ \frac{p}{p_o}$$

$$t = 10\ \log\frac{p}{p_o}$$

When $p = 2p_0$ then,

$$t = 10\ \log\frac{2p_o}{p_o}$$

$$= 10\ \log\ 2$$

Hence the population will double in a time period of 10 log2 years.

Summary

- An equation having differentials or differential coefficients is termed as differential equation
- A function f(x, y) is said to be homogenous function of degree n if $f(\alpha x, \alpha y) = \alpha^n f(x, y)$ and such that the function can be written as $f(x,y) = x^n f\left(\frac{y}{x}\right)$ or $y^n f\left(\frac{x}{y}\right)$
- A differential equation of the type $\frac{dy}{dx} + Py = Q$ where P and Q are constants or function of x alone is called as linear differential equation of the first order of y.

Self Practice Questions

1. Determine the order and degree of the differential equation

$$\left(\frac{d^2y}{dx^2}\right)^3 + \left(\frac{d^3y}{dx^3}\right)^2 + xy = 0$$

2. Determine the order and degree of the differential equation

$$y = \frac{dy}{dx} + 4\sqrt{1 + \left(\frac{dy}{dx}\right)^2}$$

3. Find the differential equation whose primitive is given as $y = p\,e^x + q$?
4. Find the differential equation whose primitive is given as $y = a \sin(x + b)$
5. Solve the differential equation $\frac{dy}{dx} = x^4 + x^2 + 5e^x + 4$
6. Solve the differential equation $(x - y)^2 \frac{dy}{dx} = 9$
7. Solve the differential equation $(6x + x^2)\frac{dy}{dx} = 9 + 6x$
8. Solve the differential equation $\sin^2 x\, dy + y\, dx = \cot x\, dx$
9. Solve the differential equation $y\, dx - x dy = 6\,(y^2 dx + dy)$?
10. Solve the differential equation $-x\, dy + y\, dx = x^2 y^2 dx$

11. Solve the differential equation
$(x^3 + xy^2 + 16y)dx + (y^3 + yx^2 - 16x)\, dy = 0$

12. Solve the differential equation $x\,dy - ydx + 2x^3dx = 0$

13. Solve the differential equation $x\,dx + ydy + \frac{4(xdy - ydx)}{x^2 + y^2} = 0$

14. Solve the differential equation $x\,dx + y\,dy + (x^2 + y^2)\,dy = 0$

15. Solve the differential equation $x\,dy - y\,dx = 5\,(x^2 + y^2)\,dy$

16. Solve the differential equation $dy + \sec^2x\,dx = \tan x \sec^2x$

17. Solve the differential equation $(x + y)\,dy = 9\,dx$

18. Solve the differential equation $(x + ay + b)\,dy = dx$

19. Solve the differential equation $\frac{d^2y}{dx^2}$ $x \sin x$?

20. Solve the differential equation $\frac{d^2y}{dx^2} = \cos x - \sin x$

21. Solve the differential equation $\frac{d^2y}{dx^2}$ $3xe^x$?

22. Solve the differential equation $\frac{d^2y}{dx^2} = a^x + e^x$?

23. If the elasticity of demand of the demand curve of the form $q = f(p)$ is $2 - 3p$ then find the demand curve?

24. If the marginal cost of a suitcase manufacturer is $\frac{dc(x)}{dx} = 4 + 6x$ then find the total cost function $C(x)$, given that $C(0) = 50$?

25. If the population of a city is growing by 4% per annum then how long does it take for the population to double?

Answers

1. Order 3 and power 2
2. Order 1 and power 2

3. $y\left(\frac{d^2y}{dx^2}\right) = \left(\frac{dy}{dx}\right)^2$

4. $\left(\frac{d^2y}{dx^2}\right) + ay = 0$

5. $x^5/5 + x^3/3 + 5ex + 4x + c$

6. $2y + 2c = 3 \log \left\{\frac{x-y-3}{x-y+3}\right\}$

7. $3\log [x (x+ 6)^3] = 2y + 2c$

8. $y\, e^{-\cot x} = C + e^{-\cot x} (\cot x + 1)$

9. $x + 6 = y (C + 6x)$

10. $x = y (C + x^3/3)$

11. $x^2 + y^2 + 32\tan^{-1}(y/x) = 2C$

12. $y + x^3 = Cx$

13. $x^2 - 32 \tan^{-1}(y/x) + y^2 = C$

14. $x^2 + y^2 = e^{(C - 2y)}$

15. $\tan^{-1}(y/x) = 5y + C$

16. $ye^{\tan x} = c + e^{\tan x} (\tan x - 1)$

17. $x + y - 9 = c\, e^{(y/9)}$

18. $x + ay + b = ce^y$

19. $y = - x \sin x - 2\cos x + c_1x + c_2$

20. $y = - \cos x + \sin x + c_1x + c_2$

21. $y = 3xe^x - 6e^x + c_1x + c_2$

22. $y = \frac{a^x}{(\log a)^2} e^x + c_1x + c_2$

23. $q = p^2.e^{-3p+c}$

24. $C(x) = 4x + 3x^2 + 50$

25. 25 log 2 years

CHAPTER 9

Theory of Matrices-I

Brief History

The concept of matrix was conceived by Sir Arthur Cayley a British Mathematician in the year of 1858 is basically a rectangular array of rows and columns. The technique developed is a powerful tool to capture wide and disarray data into an organized form to facilitate easy handling for further treatments. Now-a-days matrices finds a wide application in Business, Social Sciences Sector, Quantum Theory, Computer Science and host of other subjects. In this chapter we will look into the various fundamental concepts and elaborate further in the following chapters.

Definition: Matrix is any rectangular arrangement of numbers both real or complex into rows and columns enclosed by a pair of brackets [ ] or a pair of parenthesis ( ) or a pair of double lines $\|\ \|$ and subject to certain rules of operation.

For example

(a)

$$\begin{array}{c} \\ \\ R_1 \rightarrow \\ R_2 \rightarrow \\ R_3 \rightarrow \end{array} \begin{array}{c} \begin{array}{ccc} C_1 & C_2 & C_3 \\ \downarrow & \downarrow & \downarrow \end{array} \\ \begin{pmatrix} 1 & 5 & 0 \\ 2 & 4 & 3 \\ 7 & 9 & 8 \end{pmatrix}_{3\times 3} \end{array}$$

the example (a) contains 3 rows (R_1, R_2 and R_3) and 3 – columns (C_1, C_2 and C_3).

(b)

$$\begin{array}{c} \\ \\ R_1 \rightarrow \\ R_2 \rightarrow \end{array} \begin{array}{c} \begin{array}{ccc} C_1 & C_2 & C_3 \\ \downarrow & \downarrow & \downarrow \end{array} \\ \begin{pmatrix} 5 & 1 & 0 \\ 4 & 3 & 2 \end{pmatrix} \end{array}$$

The Example (b) contains 2-rows (R_1 and R_2) and 3-columns (C_1, C_2 and C_3).

The matrix as in illustration (a) is a matrix of order 3 × 3 read as (3 by 3) which means that the matrix contains 3 rows and 3 columns. The illustration (b) is a matrix of order 2 × 3 which means that the matrix contains 2–rows and 3-columns. Therefore the order represents the number of rows and columns.

The same concept can be extended to m × n matrix case

For example

$$\text{Columns}$$
$$\begin{pmatrix} a_{11} & a_{12} & --- & a_{1j} & --- & a_{1n} \\ a_{21} & a_{22} & --- & a_{2j} & --- & a_{2n} \\ - & - & - & - & & - \\ a_{i1} & a_{i2} & & a_{ij} & & a_{in} \end{pmatrix}_{m \times n} \quad \text{Rows}$$

Each number a_{ij} where (i = 1, 2, 3, ... m) and (j =1, 2, 3, ...n) is called an element of a matrix which is distinctly identified with respect to a particular row or column in which it is situated. For instance the element a_{21} is located in 2nd row and 1st column.

In short form the matrix of the order m × n can be represented as

$$A = \left[a_{ij}\right]_{m \times n} \quad \text{or} \quad \left(a_{ij}\right)_{m \times n} \quad \text{where } i \le m \text{ and } j \le n$$

Forms of Matrix

1. ***Row Matrix***: A matrix having only one row is called as row matrix.

   For example

   $$\begin{pmatrix} 1 & -1 & 2 \end{pmatrix}_{1 \times 3}$$

   is 1 × 3 row matrix with a single row and 3 columns.

   $$\begin{pmatrix} a_{11} & a_{12} & a_{13} & --- & a_{1n} \end{pmatrix}_{1 \times n}$$

   is 1 × n row matrix with a single row and n –columns.

2. ***Column Matrix***: A matrix having a single column is called as a column matrix.

For example

$$\begin{pmatrix} 3 \\ 6 \\ 9 \end{pmatrix}_{3 \times 1}$$

is a 3 × 1 column matrix having 3 – rows and a single Column.

$$\begin{pmatrix} a_{11} \\ a_{12} \\ a_{13} \\ - \\ - \\ a_{1n} \end{pmatrix}_{n \times 1}$$

is a n × 1 column matrix having n – rows and a single column.

3. ***Null or Zero Matrix***: A matrix with all elements zero is called as Null or Zero matrix.

For example

$$\begin{pmatrix} 0 & 0 \\ 0 & 0 \end{pmatrix}_{2 \times 2} \text{ and } \begin{pmatrix} 0 & 0 & 0 \\ 0 & 0 & 0 \\ 0 & 0 & 0 \end{pmatrix}_{3 \times 3}$$

4. ***Square Matrix***: A matrix 'A' is said to be a square matrix if the number of rows is equal to number of columns i.e. if $A = [a_{ij}]_{m \times n}$ then A will be square matrix if and only if m = n.

For example

$$\begin{pmatrix} 1 & 3 & -1 \\ 6 & 8 & 5 \\ 0 & 2 & 4 \end{pmatrix}_{3 \times 3}$$

is a matrix with 3 rows and 3 columns. Therefore a square matrix is a matrix in which Number of Rows = Number of Columns.

5. ***Diagonal Matrix***: A matrix is called as diagonal matrix if its all non diagonal elements are zero.

For example $$\begin{pmatrix} 3 & 0 & 0 \\ 0 & 4 & 0 \\ 0 & 0 & 5 \end{pmatrix}_{3 \times 3}$$

Note: An element is a diagonal element if the subscript of the element has i = j.

$$\begin{pmatrix} a_{11} & a_{12} & a_{13} \\ a_{21} & a_{22} & a_{23} \\ a_{31} & a_{32} & a_{33} \end{pmatrix}$$

The elements a_{11}, a_{22} and a_{33} are diagonal as elements in each case where the value of i = j.

6. ***Scalar Matrix***: A diagonal matrix whose all diagonal elements are equal is called as a Scalar Matrix.

For example $\begin{pmatrix} 3 & 0 & 0 \\ 0 & 3 & 0 \\ 0 & 0 & 3 \end{pmatrix}$

7. ***Unit Matrix***: A scalar matrix with all elements as unity or one is called as unit matrix or an identity matrix.

For example

$$\begin{pmatrix} 1 & 0 & 0 \\ 0 & 1 & 0 \\ 0 & 0 & 1 \end{pmatrix}$$

the above example is a unit matrix of order 3×3.

8. ***Symmetric Matrix***: A matrix is said to be symmetric if and only if
   (i) It's a square matrix.
   (ii) $(a_{ij})^{th}$ element = $(a_{ji})^{th}$ element.

For example

$$\begin{pmatrix} 5 & 4 & 6 \\ 4 & 1 & 2 \\ 6 & 2 & 1 \end{pmatrix}_{3 \times 3} \Leftrightarrow \begin{pmatrix} a_{11} & a_{12} & a_{13} \\ a_{21} & a_{22} & a_{23} \\ a_{31} & a_{32} & a_{33} \end{pmatrix}_{3 \times 3}$$

Here $a_{12} = a_{21} = 4$, $a_{23} = a_{32} = 6$ and $a_{13} = a_{31} = 2$

9. ***Skew-symmetric Matrix***: A matrix is said to be skew-symmetric if and only if
   (i) It's a square matrix.

(ii) $(a_{ij})^{th}$ element = $-(a_{ji})^{th}$ element.

For example

$$\begin{pmatrix} 5 & 4 & 6 \\ 4 & 1 & 2 \\ 6 & 2 & 1 \end{pmatrix}_{3\times 3} \Leftrightarrow \begin{pmatrix} a_{11} & a_{12} & a_{13} \\ -a_{21} & a_{22} & a_{23} \\ -a_{31} & -a_{32} & a_{33} \end{pmatrix}_{3\times 3}$$

Here $a_{12} = -a_{21}$, $a_{23} = -a_{32}$ and $a_{13} = -a_{31}$

10. ***Upper Triangular Matrix***: A matrix is a upper triangular matrix if all the elements below the principal diagonal are zero.

For example

$$\begin{pmatrix} 4 & 4 & 3 \\ 0 & 1 & 5 \\ 0 & 0 & 1 \end{pmatrix}_{3\times 3}$$

11. ***Lower Triangular Matrix***: A matrix is a lower triangular matrix if all the elements above the principal diagonal are zero.

For example

$$\begin{pmatrix} 4 & 0 & 0 \\ 3 & 1 & 0 \\ 5 & 6 & 7 \end{pmatrix}_{3\times 3}$$

Illustration No.1

Find the value of 'a' and 'b' when

$$\begin{pmatrix} 5 & a-1 \\ b-2 & 7 \end{pmatrix} = \begin{pmatrix} a+1 & 3 \\ 2-3b & 8 \end{pmatrix}$$

Equating elements of both the matrices we have a+1 = 5 and a–1 = 3. Therefore from both the equations we get the value of a = 4.

Similarly, we have b – 2 = 2 – 3b → 4b = 4 or b =1. And hence a = 4 and b = 1

Rules of Matrix's Operation

Addition Rule

Two matrixes can only be added if there are of the same order i.e. each matrixes should have equal number of rows and columns.

Let $A = (a_{ij})_{m \times n}$ and $B = (b_{ij})_{m \times n}$ are two matrixes such that there sum is another matrix $C = (c_{ij})_{m \times n}$

i.e. $C = A + B = (a_{ij})_{m \times n} + (b_{ij})_{m \times n}$

$= (c_{ij})_{m \times n}$

where $c_{ij} = a_{ij} + b_{ij}$

For example

$$A = \begin{pmatrix} 2 & 4 & 6 \\ 5 & 3 & 2 \end{pmatrix} \text{ and } B = \begin{pmatrix} 5 & 3 & 1 \\ 1 & 6 & 2 \end{pmatrix}.$$

Then $$A + B = B = \begin{pmatrix} 2+5 & 4+3 & 6+1 \\ 5+1 & 3+6 & 2+2 \end{pmatrix}$$

$$= \begin{pmatrix} 7 & 7 & 7 \\ 6 & 9 & 4 \end{pmatrix}$$

Note: In matrix addition each elements of a matrix is added to the corresponding element in the other matrix

For example

$$A = \begin{pmatrix} a_{11} & a_{12} \\ a_{21} & a_{22} \end{pmatrix} \text{ and } B = \begin{pmatrix} b_{11} & b_{12} \\ b_{21} & b_{22} \end{pmatrix}$$

Then $$A + B = \begin{pmatrix} a_{11} + b_{11} & a_{12} + b_{12} \\ a_{21} + b_{21} & a_{22} + b_{22} \end{pmatrix}$$

Some of the Properties of Matrix Addition

(i) ***Matrix Addition is commutative:*** If A and B are two matrix of the same order. Then **A + B = B + A** (i.e addition of the matrix is commutative)

Proof

If $A = (a_{ij})_{m \times n}$ and $B = (b_{ij})_{m \times n}$

$A + B = (a_{ij})_{m \times n} + (b_{ij})_{m \times n}$

$= (b_{ij})_{m \times n} + (a_{ij})_{m \times n}$

$= B + A$

Hence matrix addition is commutative

For example

If $A = \begin{pmatrix} 2 & 4 \\ 5 & 6 \end{pmatrix}$ and $B = \begin{pmatrix} 7 & 3 \\ 2 & 2 \end{pmatrix}$

Then,

$$A + B = \begin{pmatrix} 2 & 4 \\ 5 & 6 \end{pmatrix} + \begin{pmatrix} 7 & 3 \\ 2 & 2 \end{pmatrix}$$

$$= \begin{pmatrix} 2+7 & 4+3 \\ 5+2 & 6+2 \end{pmatrix} = \begin{pmatrix} 9 & 7 \\ 7 & 8 \end{pmatrix}$$

Similarly,

$$B + A = \begin{pmatrix} 7 & 3 \\ 2 & 2 \end{pmatrix} + \begin{pmatrix} 2 & 4 \\ 5 & 6 \end{pmatrix}$$

$$= \begin{pmatrix} 7+2 & 3+4 \\ 2+5 & 2+6 \end{pmatrix} = \begin{pmatrix} 9 & 7 \\ 7 & 8 \end{pmatrix}$$

Hence, A + B = B + A.

(ii) *Matrix Addition is Associative*

If A, B and C are three matrices of the same order then **A + (B + C) = (A + B) + C** i.e. matrices addition is associative.

Proof:

If $A = (a_{ij})_{m \times n}$, $B = (b_{ij})_{m \times n}$ and $C = (c_{ij})_{m \times n}$

First of all we find $(B + C)_{m \times n} = (b_{ij} + c_{ij})_{m \times n}$

$$A + (B + C) = (a_{ij})_{m \times n} + (b_{ij} + c_{ij})_{m \times n}$$

$$= [(a_{ij})_{m \times n} + (b_{ij})_{m \times n}] + (c_{ij})_{m \times n}$$

$$= (A + B) + C$$

Hence Proved.

For example

If $A = \begin{pmatrix} 2 & 4 \\ 5 & 6 \end{pmatrix}$ $B = \begin{pmatrix} 7 & 3 \\ 2 & 2 \end{pmatrix}$ and $C = \begin{pmatrix} -1 & 3 \\ 2 & -2 \end{pmatrix}$

$$B + C = \begin{pmatrix} 7 & 3 \\ 2 & 2 \end{pmatrix} + \begin{pmatrix} -1 & 3 \\ 2 & -2 \end{pmatrix}$$

$$= \begin{pmatrix} 7-1 & 3+3 \\ 2+2 & 2-2 \end{pmatrix} = \begin{pmatrix} 6 & 6 \\ 4 & 0 \end{pmatrix}$$

Therefore

$$A + (B + C) = \begin{pmatrix} 2 & 4 \\ 5 & 6 \end{pmatrix} + \begin{pmatrix} 6 & 6 \\ 4 & 0 \end{pmatrix}$$

$$= \begin{pmatrix} 2+6 & 4+6 \\ 5+4 & 6-0 \end{pmatrix} = \begin{pmatrix} 8 & 10 \\ 9 & 6 \end{pmatrix}$$

Similarly,

$$A + B = \begin{pmatrix} 2 & 4 \\ 5 & 6 \end{pmatrix} + \begin{pmatrix} 7 & 3 \\ 2 & 2 \end{pmatrix}$$

$$= \begin{pmatrix} 2+7 & 4+3 \\ 5+2 & 6+2 \end{pmatrix} = \begin{pmatrix} 9 & 7 \\ 7 & 8 \end{pmatrix}$$

Therefore

$$(A + B) + C = \begin{pmatrix} 9 & 7 \\ 7 & 8 \end{pmatrix} + \begin{pmatrix} -1 & 3 \\ 2 & -2 \end{pmatrix}$$

$$= \begin{pmatrix} 9-1 & 7+3 \\ 7+2 & 8-2 \end{pmatrix} = \begin{pmatrix} 8 & 10 \\ 9 & 6 \end{pmatrix}$$

Hence A + (B + C) = (A + B) + C, therefore matrix addition is associative.

(iii) ***Existence of Additive Identity***

For a matrix A, there exists a matrix 'O' (a null matrix) of the same order such that, A + O = O + A = A, where 'O' is called as the additive identity matrix of A. Since $a_{ij} + 0 = 0 + a_{ij} = a_{ij}$.

For example

Let $A = \begin{pmatrix} 2 & 4 \\ 5 & 6 \end{pmatrix}$ and $O = \begin{pmatrix} 0 & 0 \\ 0 & 0 \end{pmatrix}$

then,

$$A + O = \begin{pmatrix} 2+0 & 4+0 \\ 5+0 & 6+0 \end{pmatrix}$$

$$A = \begin{pmatrix} 2 & 4 \\ 5 & 6 \end{pmatrix}$$

(v) ***Existence of Additive Inverse***

Any matrix A of order m × n is said to have additive inverse matrix i.e. (–A) if A + (–A) = (– A) + A = O where O is the additive identity.

For example

Let $A = \begin{pmatrix} 2 & 4 \\ 5 & 6 \end{pmatrix}$ and $(-A) = \begin{pmatrix} -2 & -4 \\ -5 & -6 \end{pmatrix}$

then,

$$A + O = \begin{pmatrix} 2-2 & 4-4 \\ 5-5 & 6-6 \end{pmatrix}$$

$$= \begin{pmatrix} 0 & 0 \\ 0 & 0 \end{pmatrix}$$

(vi) ***Cancellation Law:*** If A, B and C are three matrices of the same order then, A + C = B + C then A = B

Proof:

$$A + C = B + C$$

Adding both side (– C), we get,

$$A + C + (-C) = B + C + (-C).$$

$$A + (C - C) = B + (C - C)$$

$$A + O = B + O.$$

$$A = B.$$

Scalar Multiplication of Matrix

Let A be a matrix and k be a scalar (i.e. a real number), then the multiplication of k by A denoted by kA is a matrix obtained by multiplying each element of matrix A by k.

For example

Let $A = \begin{pmatrix} 1 & 3 & 4 \\ 5 & 2 & 6 \end{pmatrix}$ and k = 3, then,

$$kA = 3\begin{pmatrix} 1 & 3 & 4 \\ 5 & 2 & 6 \end{pmatrix}$$

$$= \begin{pmatrix} 3\times1 & 3\times3 & 3\times4 \\ 3\times5 & 3\times2 & 3\times6 \end{pmatrix}$$

$$= \begin{pmatrix} 3 & 9. & 12 \\ 15 & 6 & 18 \end{pmatrix}$$

Some of the Properties of Scalar Multiplication

If 'k' is any scalar A and B are any two matrix of order m × n. Then $k(A + B) = k(A) + k(B)$

Proof:

If $A = (a_{ij})_{m \times n}$ and $B = (b_{ij})_{m \times n}$

Since $kA = k(a_{ij})_{m \times n}$

Similarly $kB = k(b_{ij})_{m \times n}$

Therefore,

$$kA + kB = k(a_{ij})_{m \times n} + k(b_{ij})_{m \times n}$$

$$= k[(a_{ij})_{m \times n} + (b_{ij})_{m \times n}]$$

$$= k[A + B]$$

Hence Proved

For example

If $A = \begin{pmatrix} 1 & 2 \\ 2 & 1 \end{pmatrix}$ $B = \begin{pmatrix} 2 & 2 \\ 1 & 1 \end{pmatrix}$ and let k = 2

Then,

$$A + B = \begin{pmatrix} 1 & 2 \\ 2 & 1 \end{pmatrix} + \begin{pmatrix} 2 & 2 \\ 1 & 1 \end{pmatrix}$$

$$= \begin{pmatrix} 1+2 & 2+2 \\ 2+1 & 1+1 \end{pmatrix} = \begin{pmatrix} 3 & 4 \\ 3 & 2 \end{pmatrix}$$

Therefore

$$k\,(A + B) = 2\begin{pmatrix} 3 & 4 \\ 3 & 2 \end{pmatrix}$$

$$= \begin{pmatrix} 2\times 3 & 2\times 4 \\ 2\times 3 & 2\times 2 \end{pmatrix}$$

$$= \begin{pmatrix} 6 & 8 \\ 6 & 4 \end{pmatrix}$$

Similarly,

$$k\,A = 2\begin{pmatrix} 1 & 2 \\ 2 & 1 \end{pmatrix} = \begin{pmatrix} 2 & 4 \\ 4 & 2 \end{pmatrix}$$

and

$$k\,B = 2\begin{pmatrix} 2 & 2 \\ 1 & 1 \end{pmatrix} = \begin{pmatrix} 4 & 4 \\ 2 & 2 \end{pmatrix}$$

therefore,

$$= \begin{pmatrix} 2 & 4 \\ 4 & 2 \end{pmatrix} + \begin{pmatrix} 4 & 4 \\ 2 & 2 \end{pmatrix}$$

$$= \begin{pmatrix} 6 & 8 \\ 6 & 4 \end{pmatrix}$$

Hence $\qquad k\,(A + B) = k(A) + k(B)$

(ii) If k and l are two scalar and A is any matrix of order m × n, then $(k + l)\,A = kA + lA$

For example

Let $A = \begin{pmatrix} 1 & 2 \\ 2 & 1 \end{pmatrix}$ and k = 2, 1 = 3

Then $(k + l) = (2 + 3) = 5$

Hence

$$(k + l)\,A = 5\begin{pmatrix} 1 & 2 \\ 2 & 1 \end{pmatrix} = \begin{pmatrix} 5 & 10 \\ 10 & 5 \end{pmatrix}.$$

But

$$kA = 2\begin{pmatrix}1 & 2\\2 & 1\end{pmatrix} = \begin{pmatrix}2 & 4\\4 & 2\end{pmatrix}$$

Similarly,

$$lA = 3\begin{pmatrix}1 & 2\\2 & 1\end{pmatrix} = \begin{pmatrix}3 & 6\\6 & 3\end{pmatrix}$$

Thus

$$kA + lA = \begin{pmatrix}2 & 4\\4 & 2\end{pmatrix} + \begin{pmatrix}3 & 6\\6 & 3\end{pmatrix}$$

$$= \begin{pmatrix}2+3 & 4+6\\4+6 & 2+3\end{pmatrix} = \begin{pmatrix}5 & 10\\10 & 5\end{pmatrix}$$

Therefore,

$$(k + l)\, A = kA + lA$$

(iii) If k and l are two scalars and A is a matrix of order m × n, then $k \times (lA) = (k \times l)\, A$

For example

Let $A = \begin{pmatrix}1 & 2\\2 & 1\end{pmatrix}$ and k = 2, l = 3 Then

Then,

$$(lA) = 3\begin{pmatrix}1 & 2\\2 & 1\end{pmatrix} = \begin{pmatrix}3 & 6\\6 & 3\end{pmatrix}$$

therefore,

$$k\,(lA) = 2\begin{pmatrix}3 & 6\\6 & 3\end{pmatrix} = \begin{pmatrix}6 & 12\\12 & 6\end{pmatrix}$$

similarly

$$(k \times l)\, A = 6\begin{pmatrix}1 & 2\\2 & 1\end{pmatrix} = \begin{pmatrix}6 & 12\\12 & 6\end{pmatrix}$$

Hence proved that, $k \times (lA) = (k \times l)\, A$

(iv) If A is any Matrix of order m × n and k is a scalar, then $(-k)A = -(kA) = k(-A)$

For example

If $A = \begin{pmatrix} 1 & 1 \\ 2 & 1 \end{pmatrix}$ and k = 2

Then

$$(-k)A = (-2)\begin{pmatrix} 1 & 1 \\ 2 & 1 \end{pmatrix} = \begin{pmatrix} -2 & -2 \\ -4 & -2 \end{pmatrix}$$

Similarly

$$-(k \times A) = (-)\begin{pmatrix} 2 & 2 \\ 4 & 2 \end{pmatrix} = \begin{pmatrix} -2 & -2 \\ -4 & -2 \end{pmatrix}$$

And

$$k(-A) = 2\begin{pmatrix} -1 & -1 \\ -2 & -1 \end{pmatrix} = \begin{pmatrix} -2 & -2 \\ -4 & -2 \end{pmatrix}$$

Hence $(-k)A = -(kA) = k(-A)$

Multiplication Rule

If A and B are two matrix then product AB is defined only when the number of columns in A is equal to number of rows in B

Let $A = [a_{ij}]_{m \times n}$ and $B = [b_{ij}]_{n \times p}$

Then number of column's in A equal number of rows in B then there product AB is equal to, say C were $C = [c_{ij}]_{m \times p}$

Note

When the above condition is satisfied then only the matrix A and B are said to be conformable to multiplication.

For example

$$A = \begin{pmatrix} 4 & 5 \\ 6 & 2 \end{pmatrix}_{2 \times 2} \quad B = \begin{pmatrix} 1 \\ 5 \end{pmatrix}_{2 \times 1}$$

Here the number of columns in matrix A is '2' and number of rows in matrix B is 2 hence the matrix multiplication is conformable.

Therefore,

$$= \begin{pmatrix} 4\times1 + 5\times5 \\ 6\times1 + 2\times5 \end{pmatrix} = \begin{pmatrix} 4+25 \\ 6+10 \end{pmatrix} = \begin{pmatrix} 29 \\ 16 \end{pmatrix}_{2\times1}$$

And hence the product matrix is a of order (2 × 1)

Note:

Rules of Multiplication

If
$$A = \begin{matrix} & C_1 & C_2 \\ R_1 & a11 & a12 \\ R_2 & a21 & a22 \end{matrix}$$
and
$$B = \begin{matrix} & C_1 & C_2 \\ R_1 & b11 & b12 \\ R_2 & b21 & b22 \end{matrix}$$

Then AB is

$$\begin{pmatrix} R_1 C_1 & R_1 C_2 \\ R_2 C_1 & R_2 C_2 \end{pmatrix}$$

i.e.
$$\begin{pmatrix} a_{11} \times b_{11} + a_{12}\times b_{21} & a_{11} \times b_{12} + a_{12} \times b_{22} \\ a_{21} \times b_{11} + a_{22} \times b_{21} & a_{21} \times b_{12} + a_{22} \times b_{22} \end{pmatrix}$$

Note:

The same rule applies for a (3 × 3) matrix

Some of the Properties of Matrix Multiplication

(i) Matrix multiplication is not Commutative

If A and B are any two matrix, then AB ≠ BA i.e. matrix multiplication is not commutative.

For example

If $A = \begin{pmatrix} 1 & 2 \\ 2 & 1 \end{pmatrix}$ and $B = \begin{pmatrix} 1 & 2 \\ 1 & 1 \end{pmatrix}$

Then,

$$A \times B = \begin{pmatrix} 1 & 2 \\ 2 & 1 \end{pmatrix} \times \begin{pmatrix} 1 & 2 \\ 1 & 1 \end{pmatrix} = \begin{pmatrix} (1\times1) + (2\times1) & (1\times2) + (2\times1) \\ (2\times1) + (1\times1) & (2\times2) + (1\times1) \end{pmatrix}$$

$$= \begin{pmatrix} 3 & 4 \\ 3 & 5 \end{pmatrix}$$

(ii) Matrix Multiplication is Associative

If A, B and C are there matrix of the same order, then A × (B × C) = (A × B) × C i.e. matrix multiplication is associative.

For example

If $A = \begin{pmatrix} 1 & 2 \\ 2 & 1 \end{pmatrix}$, $B \begin{pmatrix} 1 & 2 \\ 1 & 1 \end{pmatrix}$ and $C = \begin{pmatrix} 1 & 1 \\ 1 & 0 \end{pmatrix}$

Then

$$B \times C = \begin{pmatrix} 1 & 2 \\ 1 & 1 \end{pmatrix} \times \begin{pmatrix} 1 & 1 \\ 1 & 0 \end{pmatrix} = \begin{pmatrix} (1\times1) + (2\times1) & (1\times1) + (2\times0) \\ (1\times1) + (1\times1) & (1\times1) + (1\times0) \end{pmatrix}$$

$$= \begin{pmatrix} 3 & 1 \\ 2 & 1 \end{pmatrix}$$

Therefore,

$$A \times (B \times C) = \begin{pmatrix} 1 & 2 \\ 2 & 1 \end{pmatrix} \times \begin{pmatrix} 3 & 1 \\ 2 & 1 \end{pmatrix}$$

$$= \begin{pmatrix} (1\times3) + (2\times2) & (1\times1) + (2\times1) \\ (2\times3) + (1\times2) & (2\times1) + (1\times1) \end{pmatrix}$$

$$= \begin{pmatrix} 7 & 3 \\ 8 & 3 \end{pmatrix}$$

Similarly

$$(A \times B) = \begin{pmatrix} 1 & 2 \\ 2 & 1 \end{pmatrix} \times \begin{pmatrix} 1 & 2 \\ 1 & 1 \end{pmatrix} = \begin{pmatrix} (1\times1) + (2\times1) & (1\times2) + (2\times1) \\ (2\times1) + (1\times1) & (2\times2) + (1\times1) \end{pmatrix}$$

$$= \begin{pmatrix} 3 & 4 \\ 3 & 5 \end{pmatrix}$$

Now,

$$(A \times B) \times C = \begin{pmatrix} 3 & 4 \\ 3 & 5 \end{pmatrix} \times \begin{pmatrix} 1 & 1 \\ 1 & 0 \end{pmatrix} = \begin{pmatrix} (3\times1) + (4\times1) & (3\times1) + (4\times0) \\ (3\times1) + (5\times1) & (3\times1) + (5\times0) \end{pmatrix}$$

$$= \begin{pmatrix} 7 & 3 \\ 8 & 3 \end{pmatrix}$$

Hence A × (B × C) = (A × B) × C

(iii) Distributive Law

If A, B and C are three matrix of the order m × n, n × p and p × q, then A × (B + C) = (A × B) + (A × C)

For example

If $A = \begin{pmatrix} 1 & 2 \\ 1 & 2 \end{pmatrix}$, $B = \begin{pmatrix} 1 & 1 \\ -1 & 0 \end{pmatrix}$ and $C = \begin{pmatrix} 1 & 0 \\ 1 & 1 \end{pmatrix}$

Then

$$(B + C) = \begin{pmatrix} 1 & 0 \\ -1 & 1 \end{pmatrix} + \begin{pmatrix} 1 & 0 \\ 1 & 1 \end{pmatrix} = \begin{pmatrix} 1+1 & 1+0 \\ -1+1 & 0+1 \end{pmatrix}$$

$$= \begin{pmatrix} 2 & 1 \\ 0 & 1 \end{pmatrix}$$

Therefore,

$$A \times (B + C) = \begin{pmatrix} 1 & 2 \\ 1 & 2 \end{pmatrix} \times \begin{pmatrix} 2 & 1 \\ 0 & 1 \end{pmatrix} = \begin{pmatrix} (1\times 2) + (2\times 0) & (1\times 1) + (2\times 1) \\ (1\times 2) + (2\times 0) & (1\times 1) + (2\times 1) \end{pmatrix}$$

$$= \begin{pmatrix} 2 & 3 \\ 2 & 3 \end{pmatrix}$$

$$A \times B = \begin{pmatrix} 1 & 2 \\ 1 & 2 \end{pmatrix} \times \begin{pmatrix} 1 & 1 \\ -1 & 0 \end{pmatrix} = \begin{pmatrix} (1\times 1) + (2\times -1) & (1\times 1) + (2\times 0) \\ (1\times 1) + (2\times -1) & (1\times 1) + (2\times 0) \end{pmatrix}$$

$$= \begin{pmatrix} -1 & 1 \\ -1 & 1 \end{pmatrix}$$

$$A \times C = \begin{pmatrix} 1 & 2 \\ 1 & 2 \end{pmatrix} \times \begin{pmatrix} 1 & 0 \\ 1 & 1 \end{pmatrix} = \begin{pmatrix} (1\times 1) + (2\times 1) & (1\times 0) + (2\times 1) \\ (1\times 1) + (2\times 1) & (1\times 0) + (2\times 1) \end{pmatrix}$$

$$= \begin{pmatrix} 3 & 2 \\ 3 & 2 \end{pmatrix}$$

Therefore,

$$AB + AC = \begin{pmatrix} -1 & 1 \\ -1 & 1 \end{pmatrix} + \begin{pmatrix} 3 & 2 \\ 3 & 2 \end{pmatrix}$$

$$= \begin{pmatrix} -1+3 & 1+2 \\ -1+3 & 1+2 \end{pmatrix}$$

$$= \begin{pmatrix} 2 & 3 \\ 2 & 3 \end{pmatrix}$$

Hence the result A (B + C) = AB + AC.

Illustration No.1

If $A = \begin{pmatrix} 1 & 4 \\ 2 & 1 \end{pmatrix}$ and $B = \begin{pmatrix} 1 & -1 \\ 2 & -1 \end{pmatrix}$

Then find

(i) A + B and B + A and check whether A + B = B + A

(ii) AB, BA and then find whether AB = BA

Solution:

Given

$$A = \begin{pmatrix} 1 & 4 \\ 2 & 1 \end{pmatrix} \text{ and } B = \begin{pmatrix} 1 & -1 \\ 2 & -1 \end{pmatrix}$$

Therefore,

$$A + B = \begin{pmatrix} 1 & 4 \\ 2 & 1 \end{pmatrix} + \begin{pmatrix} 1 & -1 \\ 2 & -1 \end{pmatrix}$$

$$= \begin{pmatrix} (1+1) & 4+(-1) \\ (2+2) & 1+(-1) \end{pmatrix}$$

$$= \begin{pmatrix} 2 & 3 \\ 4 & 0 \end{pmatrix}$$

Similarly,

$$B + A = \begin{pmatrix} 1 & -1 \\ 2 & -1 \end{pmatrix} + \begin{pmatrix} 1 & 4 \\ 2 & 1 \end{pmatrix}$$

$$= \begin{pmatrix} 1+1 & (-1)+4 \\ 2+2 & (-1)+1 \end{pmatrix}$$

$$= \begin{pmatrix} 2 & 3 \\ 4 & 0 \end{pmatrix}$$

Hence $\quad A + B = B + A$

(ii) Here

$$A = \begin{pmatrix} 1 & 4 \\ 2 & 1 \end{pmatrix} \text{ and } B = \begin{pmatrix} 1 & -1 \\ 2 & -1 \end{pmatrix}$$

$$AB = \begin{pmatrix} 1 & 4 \\ 2 & 1 \end{pmatrix} \times \begin{pmatrix} 1 & -1 \\ 2 & -1 \end{pmatrix}$$

$$= \begin{pmatrix} (1 \times 1) + (4 \times 2) & 1 \times (-1) + 4 \times (-1) \\ (2 \times 1) + (1 \times 2) & 2 \times (-1) + 1 \times (-1) \end{pmatrix}$$

$$= \begin{pmatrix} 1+8 & -1+(-4) \\ 2+2 & -2+(-1) \end{pmatrix}$$

$$= \begin{pmatrix} 9 & -5 \\ 4 & -3 \end{pmatrix}$$

Similarly,

$$BA = \begin{pmatrix} 1 & -1 \\ 2 & -1 \end{pmatrix} \times \begin{pmatrix} 1 & 4 \\ 2 & 1 \end{pmatrix}$$

$$= \begin{pmatrix} (1 \times 1) + (-1) \times 2 & (1 \times 4) + (-1) \times 1 \\ 2 \times 1 + (-1) \times 2 & 2 \times 4 + (-1) \times 1 \end{pmatrix}$$

$$= \begin{pmatrix} 1-2 & 4-1 \\ 2-2 & 8-1 \end{pmatrix} = \begin{pmatrix} -1 & 3 \\ 0 & 7 \end{pmatrix}$$

Hence $\quad AB \neq BA$

Illustration No. 2

What will be the product of two matrix A and B if

$$A = [2 \quad 1 \quad 4 \quad 3] \text{ and } B = \begin{pmatrix} 1 \\ 0 \\ 7 \\ 5 \end{pmatrix}$$

Solution:

Since A is matrix of order 1 × 4 and B is a matrix of order 4 × 1. Then the product of AB will be matrix of order 1 × 1, therefore

$$AB = [2 \quad 1 \quad 4 \quad 3]_{1 \times 4} \begin{pmatrix} 1 \\ 0 \\ 7 \\ 5 \end{pmatrix}_{4 \times 1}$$

$$= [(2 \times 1) + (1 \times 0) + (4 \times 7) + (3 \times 5)]$$

$$= [2 + 0 + 28 + 15]$$

$$= [45]_{1 \times 1}$$

Illustration No. 3

Matrix A and B is given by

$$A = \begin{pmatrix} 2 & 1 & 0 \\ 3 & 5 & 4 \\ 1 & 6 & 9 \end{pmatrix} \text{ and } B = \begin{pmatrix} 3 & 1 & 2 \\ 0 & 1 & 1 \\ 5 & 6 & 4 \end{pmatrix}$$

Then find

(i) 3A + 2B =?

(ii) If 4A – 2X = 5B. Then find the matrix 'X'?

Solution:

Here

$$A = \begin{pmatrix} 2 & 1 & 0 \\ 3 & 5 & 4 \\ -1 & 6 & 9 \end{pmatrix} \text{ and } B = \begin{pmatrix} 3 & 1 & 2 \\ 0 & 1 & 1 \\ 5 & 6 & 4 \end{pmatrix}$$

Then,

$$3A = 3 \begin{pmatrix} 2 & 1 & 0 \\ 3 & 5 & 4 \\ -1 & 6 & 9 \end{pmatrix}$$

$$= \begin{pmatrix} 6 & 3 & 0 \\ 9 & 15 & 12 \\ -3 & 18 & 27 \end{pmatrix}$$

And $$2B = 2\begin{pmatrix} 3 & 1 & 2 \\ 0 & 1 & 1 \\ 5 & 6 & 4 \end{pmatrix}$$

$$= \begin{pmatrix} 6 & 2 & 4 \\ 0 & 2 & 2 \\ 10 & 12 & 8 \end{pmatrix}$$

Thus,

$$3A + 2B = \begin{pmatrix} 6 & 3 & 0 \\ 9 & 15 & 12 \\ -3 & 18 & 27 \end{pmatrix} + \begin{pmatrix} 6 & 2 & 4 \\ 0 & 2 & 2 \\ 10 & 12 & 8 \end{pmatrix}$$

$$= \begin{pmatrix} 6+6 & 3+2 & 0+4 \\ 9+0 & 15+2 & 12+2 \\ -3+10 & 18+12 & 27+8 \end{pmatrix}$$

$$= \begin{pmatrix} 12 & 5 & 4 \\ 9 & 17 & 14 \\ 7 & 30 & 35 \end{pmatrix}$$

(ii) Since,

$$A = \begin{pmatrix} 2 & 1 & 0 \\ 3 & 5 & 4 \\ -1 & 6 & 9 \end{pmatrix} \text{ and } B = \begin{pmatrix} 3 & 1 & 2 \\ 0 & 1 & 1 \\ 5 & 6 & 4 \end{pmatrix}$$

Therefore,

$$A = 4\begin{pmatrix} 2 & 1 & 0 \\ 3 & 5 & 4 \\ -1 & 6 & 9 \end{pmatrix}$$

$$= \begin{pmatrix} 8 & 4 & 0 \\ 12 & 20 & 16 \\ -4 & 24 & 36 \end{pmatrix}$$

Similarly,

$$5B = 5\begin{pmatrix} 3 & 1 & 2 \\ 0 & 1 & 1 \\ 5 & 6 & 4 \end{pmatrix}$$

$$= \begin{pmatrix} 15 & 5 & 10 \\ 0 & 5 & 5 \\ 25 & 30 & 20 \end{pmatrix}$$

As, 4A – 2X = 5B

Therefore, rearranging the equation, we have,

$$2X = 4A - 5B \quad = \begin{pmatrix} 8 & 4 & 0 \\ 12 & 20 & 16 \\ -4 & 24 & 36 \end{pmatrix} - \begin{pmatrix} 15 & 5 & 10 \\ 0 & 5 & 5 \\ 25 & 30 & 20 \end{pmatrix}$$

$$= \begin{pmatrix} (8-15) & (4-5) & (0-10) \\ (12-0) & (20-5) & (16-5) \\ (-4-25) & (24-30) & (36-20) \end{pmatrix}$$

$$= \begin{pmatrix} -7 & -1 & -10 \\ 12 & 15 & 11 \\ -29 & -6 & 16 \end{pmatrix}$$

Therefore,

$$X = \frac{1}{2}\begin{pmatrix} -7 & -1 & -10 \\ 12 & 15 & 11 \\ -29 & -6 & 16 \end{pmatrix}$$

$$= \begin{pmatrix} -7/2 & -1/2 & 5 \\ 6 & 15/2 & 11/2 \\ -29/2 & -3 & 8 \end{pmatrix}$$

Illustration No. 4

Find x and y, if,

$$\begin{pmatrix} 3 & -2 \\ 2 & x \end{pmatrix}\begin{pmatrix} 6 & y \\ 3 & -1 \end{pmatrix} = \begin{pmatrix} 12 & 8 \\ 21 & 1 \end{pmatrix}$$

Solution:

Given

$$\begin{pmatrix} 3 & -2 \\ 2 & x \end{pmatrix} \begin{pmatrix} 6 & y \\ 3 & -1 \end{pmatrix} = \begin{pmatrix} 12 & 8 \\ 21 & 1 \end{pmatrix}$$

$$\begin{pmatrix} (3\times 6) + (-2\times 3) & (3\times y) + (-2)\times(-1) \\ (2\times 6) + 3\times x & (2\times y) + x\times(-1) \end{pmatrix} = \begin{pmatrix} 12 & 8 \\ 21 & 1 \end{pmatrix}$$

$$\begin{pmatrix} 18+(-6) & 3y+2 \\ 12+3x & 2y+x \end{pmatrix} = \begin{pmatrix} 12 & 8 \\ 21 & 1 \end{pmatrix}$$

Now, comparing both the matrices, we get,

$$3y + 2 = 8,\ 12 + 3x = 21 \text{ and } 2y - x = 1$$

Solving the above linear equations, we get

$$3y = 6 \text{ or } y = 2$$

And $3x = 9$ or $x = 3$

Illustration No.5

Find A and B if $A + B = \begin{pmatrix} 8 & 1 \\ 3 & 6 \end{pmatrix}$ and $A - B = \begin{pmatrix} 4 & 1 \\ 1 & 4 \end{pmatrix}$

Solution:

Since $A + B = \begin{pmatrix} 8 & 1 \\ 3 & 6 \end{pmatrix}$

And $A - B = \begin{pmatrix} 4 & 1 \\ 1 & 4 \end{pmatrix}$

Therefore,

$$(A + B) + (A - B) = \begin{pmatrix} 8 & 1 \\ 3 & 6 \end{pmatrix} + \begin{pmatrix} 4 & 1 \\ 1 & 4 \end{pmatrix}$$

$$= \begin{pmatrix} 8+4 & 1+1 \\ 3+1 & 6+4 \end{pmatrix}$$

$$= \begin{pmatrix} 12 & 2 \\ 4 & 10 \end{pmatrix}$$

$$2A = \begin{pmatrix} 12 & 2 \\ 4 & 10 \end{pmatrix}$$

or,

$$A = \begin{pmatrix} 6 & 1 \\ 2 & 5 \end{pmatrix}$$

Similarly,

$$(A + B) - (A - B) = \begin{pmatrix} 8 & 1 \\ 3 & 6 \end{pmatrix} - \begin{pmatrix} 4 & -1 \\ 1 & 4 \end{pmatrix}$$

$$= \begin{pmatrix} 8-4 & 1-1 \\ 3-1 & 6-4 \end{pmatrix}$$

$$= \begin{pmatrix} 4 & 0 \\ 2 & 2 \end{pmatrix}$$

Therefore

$$2B = \begin{pmatrix} 4 & 0 \\ 2 & 2 \end{pmatrix} \Rightarrow B = 1/2 \begin{pmatrix} 4 & 0 \\ 2 & 2 \end{pmatrix}$$

And Hence

$$B = \begin{pmatrix} 2 & 0 \\ 1 & 1 \end{pmatrix}$$

Illustration No. 6

If $A = \begin{pmatrix} 1 & 2 \\ 2 & 1 \end{pmatrix}$, $B = \begin{pmatrix} 1 & 2 \\ 3 & -1 \end{pmatrix}$ and $C = \begin{pmatrix} 1 & 1 \\ 1 & 0 \end{pmatrix}$

Then prove that A (B + C) = A B + A C

Solution:

Since

$$A = \begin{pmatrix} 1 & 2 \\ 2 & 1 \end{pmatrix}, B = \begin{pmatrix} 1 & 2 \\ 3 & -1 \end{pmatrix} \text{ and } C = \begin{pmatrix} 1 & 1 \\ 1 & 0 \end{pmatrix}$$

Therefore,

$$(B + C) = \begin{pmatrix} 1 & 2 \\ 3 & -1 \end{pmatrix} + \begin{pmatrix} 1 & 1 \\ 1 & 0 \end{pmatrix}$$

$$= \begin{pmatrix} 1+1 & 2+1 \\ 3+1 & -1+0 \end{pmatrix} = \begin{pmatrix} 2 & 3 \\ 4 & -1 \end{pmatrix}$$

Therefore,

$$A \times (B + C) = \begin{pmatrix} 1 & 2 \\ 2 & 1 \end{pmatrix} \times \begin{pmatrix} 2 & 3 \\ 4 & -1 \end{pmatrix}$$

$$= \begin{pmatrix} (1\times 2) + (2\times 4) & (1\times 3) + 2\times(-1) \\ (2\times 2) + (1\times 4) & (2\times 3) + (1\times -1) \end{pmatrix}$$

$$= \begin{pmatrix} 2+8 & 3-2 \\ 4+4 & 6-1 \end{pmatrix}$$

$$= \begin{pmatrix} 10 & 1 \\ 8 & 5 \end{pmatrix}$$

But,

$$AB = \begin{pmatrix} 1 & 2 \\ 2 & 1 \end{pmatrix} \times \begin{pmatrix} 1 & 2 \\ 3 & -1 \end{pmatrix}$$

$$= \begin{pmatrix} (1\times 1) + (2\times 3) & (1\times 2) + (2\times -1) \\ (2\times 1) + (1\times 3) & (2\times 2) + (1\times -1) \end{pmatrix}$$

$$= \begin{pmatrix} 1+6 & 2-2 \\ 2+3 & 4-1 \end{pmatrix} = \begin{pmatrix} 7 & 0 \\ 5 & 3 \end{pmatrix}$$

And

$$AC = \begin{pmatrix} 1 & 2 \\ 2 & 1 \end{pmatrix} \times \begin{pmatrix} 1 & 1 \\ 1 & 0 \end{pmatrix}$$

$$= \begin{pmatrix} 1\times 1 + 2\times 1 & 1\times 1 + 2\times 0 \\ 2\times 1 + 1\times 1 & 2\times 1 + 1\times 0 \end{pmatrix}$$

$$= \begin{pmatrix} 1+2 & 1+0 \\ 2+1 & 2+0 \end{pmatrix}$$

$$= \begin{pmatrix} 3 & 1 \\ 3 & 2 \end{pmatrix}$$

Therefore,

$$AB + AC = \begin{pmatrix} 7 & 0 \\ 5 & 3 \end{pmatrix} + \begin{pmatrix} 3 & 1 \\ 3 & 2 \end{pmatrix}$$

$$= \begin{pmatrix} 7+3 & 0+1 \\ 5+3 & 3+2 \end{pmatrix}$$

$$= \begin{pmatrix} 10 & 1 \\ 8 & 5 \end{pmatrix}$$

As the LHS and RHS are equal and hence, A (B + C) = AB + AC.

Illustration No. 7

Find a, b, c and d if

$$4\begin{pmatrix} a & b \\ c & d \end{pmatrix} + 3\begin{pmatrix} 2 & -2 \\ 0 & 4 \end{pmatrix} = 2\begin{pmatrix} 7 & 5 \\ 6 & 12 \end{pmatrix}$$

Solution:

Given

$$4\begin{pmatrix} a & b \\ c & d \end{pmatrix} + 3\begin{pmatrix} 2 & -2 \\ 0 & 4 \end{pmatrix} = 2\begin{pmatrix} 7 & 5 \\ 6 & 12 \end{pmatrix}$$

$$\begin{pmatrix} 4a & 4b \\ 4c & 4d \end{pmatrix} + \begin{pmatrix} 6 & -6 \\ 0 & 12 \end{pmatrix} = \begin{pmatrix} 14 & 10 \\ 12 & 24 \end{pmatrix}$$

$$\begin{pmatrix} 4a+6 & 4b-6 \\ 4c & 4d+12 \end{pmatrix} = \begin{pmatrix} 14 & 10 \\ 12 & 24 \end{pmatrix}$$

Now, comparing both sides, we get,

$4a + 6 = 14$, $4b - 6 = 10$, $4c = 12$ and $4d + 12 = 24$

Therefore solving the above set of equations we get $4a = 8$ or $a = 2$

$4b = 16$ or $b = 4$

$4c = 12$ or $c = 3$

$4d = 12$ or $d = 3$

Illustration No. 8

If $A = \begin{pmatrix} 3 & 2 \\ 4 & 6 \end{pmatrix}$, then show that $A^2 - 9A + 10I = 0$ where $I = \begin{pmatrix} 1 & 0 \\ 0 & 1 \end{pmatrix}$

Solution:

We have,

$$A = \begin{pmatrix} 3 & 2 \\ 4 & 6 \end{pmatrix}$$

Then

$$A^2 = A \times A = \begin{pmatrix} 3 & 2 \\ 4 & 6 \end{pmatrix} \times \begin{pmatrix} 3 & 2 \\ 4 & 6 \end{pmatrix}$$

$$= \begin{pmatrix} (3\times3) + (2\times4) & (3\times2) + (2\times6) \\ (4\times3) + (6\times4) & (4\times2) + (6\times6) \end{pmatrix}$$

$$= \begin{pmatrix} 9+8 & 6+12 \\ 12+24 & 8+36 \end{pmatrix}$$

$$= \begin{pmatrix} 17 & 18 \\ 36 & 44 \end{pmatrix}$$

And

$$9A = 9\begin{pmatrix} 3 & 2 \\ 4 & 6 \end{pmatrix} = \begin{pmatrix} 27 & 18 \\ 36 & 54 \end{pmatrix}$$

Similarly

$$10\,I = 10\begin{pmatrix} 1 & 0 \\ 0 & 1 \end{pmatrix} = \begin{pmatrix} 10 & 0 \\ 0 & 10 \end{pmatrix}$$

Then

$$A^2 - 9A + 10\,I = \begin{pmatrix} 17 & 18 \\ 36 & 44 \end{pmatrix} - \begin{pmatrix} 27 & 18 \\ 36 & 54 \end{pmatrix} + \begin{pmatrix} 10 & 0 \\ 0 & 10 \end{pmatrix}$$

$$= \begin{pmatrix} 17-27+10 & 18-18+0 \\ 36-36+0 & 44-54+10 \end{pmatrix}$$

$$= \begin{pmatrix} 0 & 0 \\ 0 & 0 \end{pmatrix}$$

Hence proved.

Transpose of a Matrix

If A is a matrix then the transpose of a matrix A denoted by A^T is obtained by interchanging rows into columns or columns into rows.

If $A = [a_{ij}]_{m \times n}$;

then $A^T = [a_{ji}]_{n \times m}$

For example

$$\text{If } A = \begin{pmatrix} 2 & 3 \\ 5 & 6 \end{pmatrix}$$

Then,

$$A^T = \begin{pmatrix} 2 & 5 \\ 3 & 6 \end{pmatrix}$$

Properties of Transpose of a Matrix

If A and B are any two matrix then

(i) The transpose of the transpose of a matrix is the original matrix i.e. $(A^T)^T = A$

(ii) The transpose of the sum or difference of two matrix is the sum or different of their transpose i.e. $(A + B)^T = A^T + B^T$ and $(A - B)^T = A^T - B^T$.

(iii) If A matrix of order $m \times n$ then $(k\ A)^T = k\ A^T$, where k is any number.

(iv) The transpose of the product of two matrices is the product of their transpose taken in the reserve order i.e. $(AB)^T = B^T A^T$.

Illustration No. 9

$$\text{If } A = \begin{pmatrix} 3 & -2 & 2 \\ 5 & 3 & 4 \end{pmatrix} \text{ and } B = \begin{pmatrix} 4 & -1 & 5 \\ 2 & 4 & -4 \end{pmatrix}$$

Then show that $(A + B)^T = A^T + B^T$

Solution:

$$\text{Given} \quad A = \begin{pmatrix} 3 & -2 & 2 \\ 5 & 3 & 4 \end{pmatrix} \text{ and } B = \begin{pmatrix} 4 & -1 & 5 \\ 2 & 4 & -4 \end{pmatrix}$$

Then,

$$A + B = \begin{pmatrix} 3 & -2 & 2 \\ 5 & 3 & 4 \end{pmatrix} + \begin{pmatrix} 4 & -1 & 5 \\ 2 & 4 & -4 \end{pmatrix}$$

$$= \begin{pmatrix} 3+4 & -2+(-1) & 2+5 \\ 5+2 & 3+4 & 4+(-4) \end{pmatrix}$$

$$=$$

Therefore,

$$(A + B)^T = \begin{pmatrix} 7 & 7 \\ -3 & 7 \\ 7 & 0 \end{pmatrix}$$

But,

$$(A)^T = \begin{pmatrix} 3 & 5 \\ -2 & 3 \\ 2 & 4 \end{pmatrix}$$

And,

$$(B)^T = \begin{pmatrix} 4 & 2 \\ -1 & 4 \\ 5 & -4 \end{pmatrix}$$

$$(A)^T + (B)^T = \begin{pmatrix} 3+4 & 5+2 \\ -2+(-1) & 3+4 \\ 2+5 & 4-4 \end{pmatrix}$$

$$= \begin{pmatrix} 7 & 7 \\ -3 & 7 \\ 7 & 0 \end{pmatrix}$$

Hence proved.

Illustration No. 10

If $A = \begin{pmatrix} 1 & 3 \\ 1 & 2 \end{pmatrix}$ and $B = \begin{pmatrix} 1 & 2 \\ 1 & 1 \end{pmatrix}$ then prove that $(AB)^T = B^T A^T$?

Solution:

$$AB = \begin{pmatrix} 1 & 3 \\ 1 & 2 \end{pmatrix} \times \begin{pmatrix} 1 & 2 \\ 1 & 1 \end{pmatrix}$$

$$= \begin{pmatrix} (1\times1) + (3\times1) & (1\times2) + (3\times1) \\ (1\times1) + (2\times1) & (1\times2) + (2\times1) \end{pmatrix}$$

$$= \begin{pmatrix} 1+3 & 2+3 \\ 1+2 & 2+2 \end{pmatrix}$$

$$= \begin{pmatrix} 4 & 5 \\ 3 & 4 \end{pmatrix}$$

$$(AB)^T = \begin{pmatrix} 4 & 3 \\ 5 & 4 \end{pmatrix}$$

But,

$$A^T = \begin{pmatrix} 1 & 1 \\ 3 & 2 \end{pmatrix}$$

$$B^T = \begin{pmatrix} 1 & 1 \\ 2 & 1 \end{pmatrix}$$

Therefore,

$$B^T A^T = \begin{pmatrix} 1 & 1 \\ 2 & 1 \end{pmatrix} \times \begin{pmatrix} 1 & 1 \\ 3 & 2 \end{pmatrix}$$

$$= \begin{pmatrix} (1\times1) + (1\times3) & (1\times1) + (1\times2) \\ (2\times1) + (1\times3) & (2\times1) + (1\times2) \end{pmatrix}$$

$$= \begin{pmatrix} 1+3 & 1+2 \\ 2+3 & 2+2 \end{pmatrix}$$

$$= \begin{pmatrix} 4 & 3 \\ 5 & 3 \end{pmatrix}$$

Hence $(AB)^T = B^T A^T$

Application of Theory of Matrix to Business and Economics

Illustration No. 11

Gaurav has Rs 20,000 to be invested in bank fixed deposits such that the annual interest earned at 10% on the first part and 8% on the second part amounts to Rs. 1840?

Solution:

Let the two parts are denoted by x and 20,000 –x respectively and I be the matrix representing the two parts as

$$I = [x \quad 20{,}000 - x]$$

Assuming that R represents the annual rate of interest on each part, therefore,

$$R = \begin{pmatrix} 0.10 \\ 0.08 \end{pmatrix}$$

And hence the interest earned on both the parts is given by

$$IR = [x \quad 20{,}000 - x] \begin{pmatrix} 0.10 \\ 0.08 \end{pmatrix} = [0.10x + 1600 - 0.08x]$$

$$= 0.02x + 1600.$$

But the annual interest earned is Rs. 1840

Therefore,

$$0.02x + 1600 = 1840$$

$$0.02x = 240,$$

$$x = 12{,}000.$$

Since the first part is Rs. 12000, the second part is Rs.8000.

Illustration No. 12

A company produces three types of bearings B_1, B_2 and B_3 and sells them in markets M_1, M_2 and M_3. The monthly sales in units are given by

Market	Product		
	B_1	**B_2**	**B_3**
M_1	100	40	20
M_2	140	60	40
M_3	120	100	80

If the selling price of B_1, B_2 and B_3 are Rs.100, Rs.80 and Rs.60 then find the profit earned from each market if it is known that each of B_1, B_2 and B_3 costs Rs.70, Rs.40 and Rs.20 to manufacture?

Solution:

Since Profit per unit (P)

= Selling Price per unit (S) – Cost price per unit (C)

But the selling price matrix (S) for the bearings is given by,

$$S = \begin{pmatrix} 100 \\ 80 \\ 60 \end{pmatrix}$$

Whereas the cost matrix (C) is given by

$$C = \begin{pmatrix} 70 \\ 40 \\ 20 \end{pmatrix}$$

Therefore

$$\text{Profit} \quad (P) = \begin{pmatrix} 100 \\ 80 \\ 60 \end{pmatrix} - \begin{pmatrix} 70 \\ 40 \\ 20 \end{pmatrix} = \begin{pmatrix} 30 \\ 40 \\ 40 \end{pmatrix}$$

Hence the total profit (T) earned by the firm is quantity sold of each bearing in three different markets i.e.

$$\begin{pmatrix} 100 & 40 & 20 \\ 140 & 60 & 40 \\ 120 & 100 & 80 \end{pmatrix} \begin{pmatrix} 30 \\ 40 \\ 40 \end{pmatrix} = \begin{pmatrix} 100 \times 30 + 40 \times 40 + 20 \times 40 \\ 140 \times 30 + 60 \times 40 + 40 \times 40 \\ 120 \times 30 + 100 \times 40 + 80 \times 40 \end{pmatrix}$$

$$\begin{pmatrix} 3000 + 1600 + 800 \\ 4200 + 2400 + 1600 \\ 3600 + 4000 + 3200 \end{pmatrix} = \begin{pmatrix} 5400 \\ 8200 \\ 10800 \end{pmatrix}$$

Illustration No. 13

If three individuals X, Y and Z purchased three types of investments A, B and C. The number of units purchased by X are 50 units of A, 40 units of B and 30 units of C. While Y has purchased 60 units of A, 40 units of B and 50 units of C whereas Z purchased 30 units of A,

40 units of B and 60 units C. If each of A, B and C per unit costs Rs. 50, Rs 70 and Rs. 80, then using matrix method find the total sum spend by each individual to purchase the share?

Solution:

Let S denotes the matrix representing the given information in which the 1st, 2nd and 3rd column represents the three different investments A, B and C whereas the row 1st, 2nd and 3rd represents the number of units purchased.

Therefore,

$$S = \begin{pmatrix} 50 & 40 & 30 \\ 60 & 40 & 50 \\ 30 & 40 & 60 \end{pmatrix}$$

Let P denote the individual prices of the investments A, B and C and hence,

$$P = \begin{pmatrix} 50 \\ 70 \\ 80 \end{pmatrix}$$

Therefore the product SP gives the total amount spend by the three individuals in acquiring A, B and C.

Hence,

$$SP = \begin{pmatrix} 50 & 40 & 30 \\ 60 & 40 & 50 \\ 30 & 40 & 60 \end{pmatrix} \begin{pmatrix} 50 \\ 70 \\ 80 \end{pmatrix} = \begin{pmatrix} 50 \times 50 + 40 \times 70 + 30 \times 80 \\ 60 \times 50 + 40 \times 70 + 50 \times 80 \\ 30 \times 50 + 40 \times 70 + 60 \times 80 \end{pmatrix}$$

$$= \begin{pmatrix} 2500 + 2800 + 2400 \\ 3000 + 2800 + 4000 \\ 1500 + 2800 + 4800 \end{pmatrix}$$

$$= \begin{pmatrix} 7700 \\ 9800 \\ 9100 \end{pmatrix}$$

Hence the amount spend by X = Rs. 7700, Y = 9800 and Z = 9100.

Summary

- Matrix is any rectangular arrangement of numbers both real or complex into rows and columns enclosed by a pair of brackets [ ] or a pair of parenthesis ( ) or a pair of double lines || || and subject to certain rules of operation.
- **Square Matrix**: A matrix 'A' is said to be a square matrix if the number of rows is equal to number of columns i.e. if $A = [a_{ij}]_{m \times n}$ then A will be square matrix if and only if m = n.
- **Symmetric Matrix** : A matrix is said to be symmetric if and only if

  (i) It's a square matrix.

  (ii) $(a_{ij})^{th}$ element = $(a_{ji})^{th}$ element.

  For example

$$\begin{pmatrix} 5 & 4 & 6 \\ 4 & 1 & 2 \\ 6 & 2 & 1 \end{pmatrix}_{3 \times 3} \Leftrightarrow \begin{pmatrix} a11 & a12 & a13 \\ a21 & a22 & a23 \\ a31 & a32 & a33 \end{pmatrix}_{3 \times 3}$$

  Here $a_{12} = a_{21} = 4$, $a_{23} = a_{32} = 6$ and $a_{13} = a_{31} = 2$.

- If A and B are two matrix then product AB is defined only when the number of columns in A is equal to number of rows in B. Let $A = [a_{ij}]_{m \times n}$ and $B = [b_{ij}]_{n \times p}$, Then number of column's in A equal number of rows in B then their product AB is equal to, say C were C $= [\, c_{ij} \,]_{m \times p}$
- If A is a matrix then the transpose of a matrix A denoted by A^T is obtained by interchanging the rows into columns or columns into rows. If $A = [a_{ij}]_{m \times n}$ then $A^T = [a_{ji}]_{n \times m}$

Self Practice Exercises

1. If $A = \begin{pmatrix} 1 & 2 \\ 4 & 3 \end{pmatrix}$ and $B = \begin{pmatrix} 1 & 1 \\ 5 & 6 \end{pmatrix}$ then find AB and BA? Also prove whether AB = BA?

2. If $A = \begin{pmatrix} 2 & 3 & 5 \\ 4 & 1 & 5 \end{pmatrix}$ and $B = \begin{pmatrix} 1 & 3 & 0 \\ 1 & 4 & 3 \end{pmatrix}$ then find 3A + 2B?

3. If $A = \begin{pmatrix} 2 & -1 \\ 1 & 3 \end{pmatrix}$, $B = \begin{pmatrix} 5 & 0 \\ -1 & 3 \end{pmatrix}$ and $C = \begin{pmatrix} 1 & -2 \\ 4 & 2 \end{pmatrix}$

   Find whether A (B + C) = AB + AC?

4. If $A = \begin{pmatrix} 1 & -1 \\ 2 & -1 \end{pmatrix}$ and $B = \begin{pmatrix} 2 & 1 \\ 3 & -1 \end{pmatrix}$ then find $A^2 + B^2 + 2(A + B)$?

5. If $A = \begin{pmatrix} 1 & 1 \\ 0 & 2 \end{pmatrix}$, $B = \begin{pmatrix} 1 & -1 \\ 1 & 1 \end{pmatrix}$ and $C = \begin{pmatrix} 0 & 1 \\ 1 & 1 \end{pmatrix}$ verify whether A(BC) = (AB)C?

6. If $A = \begin{pmatrix} 8 & 0 \\ -2 & 2 \end{pmatrix}$ and $B = \begin{pmatrix} 0 & 5 \\ 6 & -7 \end{pmatrix}$ find the matrix X such that 3A+ 4B + 2X = 0?

7. If $2A - B = \begin{pmatrix} 3 & -4 & 2 \\ 4 & 8 & 10 \\ 12 & 14 & 4 \end{pmatrix}$ and $A - 2B = \begin{pmatrix} 6 & 7 & -2 \\ -1 & -2 & -4 \\ 3 & -2 & -4 \end{pmatrix}$ find A and B?

8. If $A = \begin{pmatrix} 2 & 3 & -2 \\ 1 & 2 & 3 \\ 1 & 1 & 2 \end{pmatrix}$ and $B = \begin{pmatrix} 1 & -2 & 2 \\ 1 & 2 & -1 \\ 3 & 4 & 1 \end{pmatrix}$ find (A + B) and AB ?

9. If $A = \begin{pmatrix} 2 & 3 \\ 1 & -2 \end{pmatrix}$ and $B = \begin{pmatrix} 1 & 5 \\ 1 & -4 \end{pmatrix}$ then find A^TB^T and B^TA^T?

10. If $A = \begin{pmatrix} 2 & -2 & 0 \\ 4 & 6 & 8 \\ 0 & 2 & 4 \end{pmatrix}$ and $B = \begin{pmatrix} 4 & 4 & -8 \\ -8 & 4 & 8 \\ 4 & -2 & 10 \end{pmatrix}$ Then find A^T, B^T and $(A+B)^T$?

11. If $A = \begin{pmatrix} 1 & -1 \\ 2 & -1 \end{pmatrix}$ and $B = \begin{pmatrix} a & 1 \\ b & -1 \end{pmatrix}$ and if $(A + B)^2 = A^2 + B^2$, then find a and b?

12. If $A = \begin{pmatrix} 1 & 2 \\ 7 & 4 \end{pmatrix}$ and $B = \begin{pmatrix} 6 & 5 \\ 8 & 7 \end{pmatrix}$ then prove that $A^T + B^T = (A + B)^T$?

13. Prove that the matrix A given by $\begin{pmatrix} p & q \\ r & s \end{pmatrix}$ satisfying the following relation $A^2 - (p + s)A + (ps - rq)I = 0$ where I is the unit matrix of second order?

14. If $A = \begin{pmatrix} 3 & 4 \\ 5 & 6 \end{pmatrix}$ and $B = \begin{pmatrix} 1 & -2 \\ 4 & 2 \end{pmatrix}$ then find $A^T B$ and $B^T A$? Also check for whether $A^T B = (B^T A)^T$?

15. Find the value of x, y and z if the matrix A given by

$$A = \frac{1}{3}\begin{pmatrix} 2x & 4 & 4 \\ 4 & 2 & 2y \\ 4 & 2z & 2 \end{pmatrix}$$

Obeys the law that $A.\ A^T = 4I$ where I is a unit matrix of third order?

16. Ramesh buys 6 dozen of cups, 8 dozens of saucers and 4 dozens of glasses. If cups costs Rs 40 per dozen, saucers Rs. 24 per dozen and glasses Rs. 30 per dozen then represent the quantities by row matrix and the prices by column matrix and hence obtain the total cost incurred by Ramesh?

17. Two neighbors X and Y have the following of family members. There are 2 Men, 3 women and 2 child in X's family whereas 1 man, 2 women and 1 child in Y's family. If the recommended daily intake of food in calories for men is 3000 calories; women 2200 calories and children 1800 calories and proteins for Men is 150 grams, Women 120 grams and children 80 grams. Then represent the above information in matrix form? Also calculate the total requirement of calories and proteins of each family?

18. Dinesh has invested an amount of Rs. 60,000 in two different types of bonds A and B. The bond A earns 12% interest and whereas the bond B earns 9% interest per annum. Using matrix algebraic method determine the individual investments if the annual interest earned is Rs. 6,450?

19. An Electrical heater manufacturer produces two types of heater H_1 and H_2 for hyper markets M_1 and M_2. The monthly sales of these products are given as

Markets	**Products**	
	H_1	**H_2**
M_1	400	600
M_2	700	800

If the selling prices of the heaters are Rs. 550 and Rs. 400 then find the revenue generated per month by the firm?

20. A urea manufacturer uses three types of ingredients to manufacture two types of urea U_1 and U_2. The cost per kg of the ingredients is given by

$$C = [6 \quad 8 \quad 5]$$

In order to produce U_1 and U_2 the following are the requirements of the ingredients

$$\text{Requirement (R)} = \begin{pmatrix} 4 & 4 \\ 2 & 10 \\ 6 & 8 \end{pmatrix}$$

Find the cost of manufacturing of each unit?

Answers

1. $AB = \begin{pmatrix} 6 & 7 \\ 19 & 22 \end{pmatrix}$; $BA = \begin{pmatrix} 5 & 4 \\ 29 & 23 \end{pmatrix}$ and $AB \neq BA$

2. $3A + 2B = \begin{pmatrix} 8 & 15 & 15 \\ 14 & 11 & 21 \end{pmatrix}$

3. Yes, $A(B + C) = AB + AC$

4. $A^2 + B^2 + 2(A + B) = \begin{pmatrix} 12 & 1 \\ 15 & 1 \end{pmatrix}$

6. $X = \begin{pmatrix} -12 & 1 \\ -9 & 4 \end{pmatrix}$

7. $A = \begin{pmatrix} 4 & -5 & 6 \\ 3 & 6 & 8 \\ 7 & 10 & 4 \end{pmatrix}$ and $B = \begin{pmatrix} 5 & -6 & 4 \\ 2 & 4 & 6 \\ 2 & 6 & 4 \end{pmatrix}$

8. $A + B = \begin{pmatrix} 3 & 1 & 0 \\ 2 & 4 & 2 \\ 4 & 5 & 3 \end{pmatrix}$ and $AB = \begin{pmatrix} -1 & -6 & -1 \\ 12 & 14 & 3 \\ 8 & 8 & 3 \end{pmatrix}$

9. $A^TB^T = \begin{pmatrix} 7 & -2 \\ -7 & 11 \end{pmatrix}$ and $B^TA^T = \begin{pmatrix} 5 & -1 \\ -2 & 13 \end{pmatrix}$

10. $A^T = \begin{pmatrix} 2 & 4 & 0 \\ -2 & 6 & 2 \\ 0 & 8 & 4 \end{pmatrix}$, $B^T = \begin{pmatrix} 4 & -8 & 4 \\ 4 & 4 & -2 \\ -8 & 8 & 10 \end{pmatrix}$ and $(A + B)^T = \begin{pmatrix} 6 & -4 & 4 \\ 2 & 10 & 0 \\ -8 & 16 & 14 \end{pmatrix}$

11. a = 1 and b = 4.

14. $A^TB = \begin{pmatrix} 23 & 4 \\ 28 & 4 \end{pmatrix}$ and $B^TA = \begin{pmatrix} 23 & 28 \\ 4 & 4 \end{pmatrix}$. $A^TB = (B^TA)^T$

15. x = 1, y = –2 and z = –2.

16. $[6\ \ 8\ \ 4] \begin{pmatrix} 40 \\ 24 \\ 30 \end{pmatrix}$; Total Cost = Rs.552

17. $\begin{pmatrix} 2 & 3 & 2 \\ 1 & 2 & 1 \end{pmatrix} \begin{pmatrix} 3000 & 150 \\ 2200 & 120 \\ 1800 & 80 \end{pmatrix} = \begin{pmatrix} 16,200 & 820 \\ 9,200 & 470 \end{pmatrix}$

18. A = Rs.35,000 and B = Rs.25, 000.

19. M_1 = 4,60,000 and M_2 = 7,05,000.

20. U_1 = Rs.70 and U_2 = Rs.144.

CHAPTER 10

Determinants of Matrices

Definition

For every square matrix A of certain order, say n we can relate a number called the determinant of A denoted by det A or $|A|$ or Δ.

Definition of Determinant of Second Order

Let A be square matrix, such that $A = \begin{pmatrix} a_1 & b_1 \\ a_2 & b_2 \end{pmatrix}$

Then the symbol $|A| = \Delta = \begin{vmatrix} a_1 & b_1 \\ a_2 & b_2 \end{vmatrix}$

is the determinant of second order and its value is calculated by $a_1b_1 - a_2b_2$ otherwise the value of the determinant (product of elements of the principal diagonal) – (the product of the elements off the principle diagonal) where a_1, b_1, a_2 and b_2 are the elements of the matrix A , then

$$|A| = \begin{vmatrix} 2 & 3 \\ 1 & 4 \end{vmatrix}$$

$$= 2 \times 4 - 3 \times 1 = 8 - 3 = 5$$

Value of Determinant of Third Order

For a square matrix A the form $A = \begin{vmatrix} a_1 & b_1 & c_1 \\ a_2 & b_2 & c_2 \\ a_3 & b_3 & c_3 \end{vmatrix}$

Then the symbol $|A| = \Delta = \begin{vmatrix} a_1 & b_1 & c_1 \\ a_2 & b_2 & c_2 \\ a_3 & b_3 & c_3 \end{vmatrix}$

is called the determinant of third order and its value is calculated by

$$|A| = a_1 \begin{vmatrix} b_2 & c_2 \\ b_3 & c_3 \end{vmatrix} - b_1 \begin{vmatrix} a_2 & c_2 \\ a_3 & c_3 \end{vmatrix} + c_1 \begin{vmatrix} a_2 & b_2 \\ a_3 & b_3 \end{vmatrix}$$

For example

$$\begin{vmatrix} 1 & 3 & 2 \\ 4 & 2 & 1 \\ -2 & 1 & 3 \end{vmatrix} = 1 \begin{vmatrix} 2 & 1 \\ 1 & 3 \end{vmatrix} - 3 \begin{vmatrix} 4 & 1 \\ -2 & 3 \end{vmatrix} + 2 \begin{vmatrix} 4 & 2 \\ -2 & 1 \end{vmatrix}$$

$$= 1\,(6 - 1) - 3(12 + 2) + 2\,(4 + 4)$$

$$= 5 - 42 + 16 = -21$$

Minor and Cofactor of the Elements of a Determinant

Minors of the Elements

Let us consider a determinant of third order given by

$$\Delta = \begin{vmatrix} a_1 & b_1 & c_1 \\ a_2 & b_2 & c_2 \\ a_3 & b_3 & c_3 \end{vmatrix}$$

Then the minor of any element in Δ is a determinant of second order obtained by eliminating from the Δ the row and the column in which the element occurs and represented by M_{ij} where (ij) represents the ith row and j^{th} column.

For example

$$\Delta = \begin{vmatrix} a_1 & b_1 & c_1 \\ a_2 & b_2 & c_2 \\ a_3 & b_3 & c_3 \end{vmatrix}$$

The Minor of a_1 is obtain by deleting the row and the column in which a_1 is identified thus we obtain a determinant of second order denoted by M_{11}.

Therefore,

$$\Delta = \begin{vmatrix} a_1 & b_1 & c_1 \\ a_2 & b_2 & c_2 \\ a_3 & b_3 & c_3 \end{vmatrix}$$

i.e. $$M_{11} = \begin{vmatrix} b_2 & c_2 \\ b_3 & c_3 \end{vmatrix} = b_2\, c_3 - b_3\, c_2.$$

Similarly,

The minor of b_1 denoted by $$M_{12} = \begin{vmatrix} a_2 & c_2 \\ a_3 & c_3 \end{vmatrix} = a_2\, c_3 - c_2\, a_3.$$

Cofactors of the Elements

If any element a_{ij} occurs in the ith row and j^{th} column in Δ, then the cofactor of the element is obtained by multiplying the minor M_{ij} by $(-1)^{i+j}$ and denoted by A_{ij}. Otherwise the cofactor of any element is the minor of that element in Δ with appropriate sign depending on the row and the column in which the element is located.

Therefore, the cofactor of $a_{ij} = A_{ij} = (-1)^{i+j} M_{ij}$

For example

If $$|A| = \Delta = \begin{vmatrix} a_1 & b_1 & c_1 \\ a_2 & b_2 & c_2 \\ a_3 & b_3 & c_3 \end{vmatrix}$$

Therefore Cofactor of a_{11} is

$$A_{11} = (-1)^{1+1} \begin{vmatrix} b_2 & c_2 \\ b_3 & c_3 \end{vmatrix}$$

$$= (-1)^{1+1} M_{11} = M_{11}$$

Similarly, the cofactor of a_{12} *is*

$$A_{12} = (-1)^{1+2} \begin{vmatrix} a_2 & c_2 \\ a_3 & c_3 \end{vmatrix} = -M_{12}$$

Other cofactor elements can hence be calculated in same order.

Illustration No.1

Find the minor and the cofactor elements of the following determinant

$$\Delta = \begin{vmatrix} 2 & -1 & 1 \\ 3 & 4 & -2 \\ 0 & 5 & 6 \end{vmatrix}$$

Solution:

$$\Delta = \begin{vmatrix} 2 & -1 & 1 \\ 3 & 4 & -2 \\ 0 & 5 & 6 \end{vmatrix}$$

$$\text{Minor of } 2 = M_{11} = \begin{vmatrix} 4 & -2 \\ 5 & 6 \end{vmatrix} = 24 + 10 = 34$$

$$\text{Minor of } (-1) = M_{12} = \begin{vmatrix} 3 & -2 \\ 0 & 6 \end{vmatrix} = 18 - 0 = 18$$

$$\text{Minor of } 1 = M_{13} = \begin{vmatrix} 3 & 4 \\ 0 & 5 \end{vmatrix} = 15 - 0 = 15$$

$$\text{Minor of } 3 = M_{21} = \begin{vmatrix} -1 & 1 \\ 5 & 6 \end{vmatrix} = (-6) - 5 = -11$$

$$\text{Minor of } 4 = M_{22} = \begin{vmatrix} 2 & 1 \\ 0 & 6 \end{vmatrix} = 12 - 0 = 12.$$

$$\text{Minor of } (-2) = M_{23} = \begin{vmatrix} 2 & -1 \\ 0 & 5 \end{vmatrix} = 10 - 0 = 10$$

Minor of 0 = $M_{31} = \begin{vmatrix} -1 & 1 \\ 4 & -2 \end{vmatrix} = 2 - 4 = -2.$

Minor of 5 = $M_{32} = \begin{vmatrix} 2 & 1 \\ 3 & -2 \end{vmatrix} = -4 - 3 = -7.$

Minor of 6 = $M_{33} = \begin{vmatrix} 2 & -1 \\ 3 & 4 \end{vmatrix} = 8 + 3 = 11$

Now the cofactor of each element is given by:

Cofactor of $A_{11} = (-1)^{1+1} \times M_{11} = (1) \times 34 = 34$

Cofactor of $A_{12} = (-1)^{1+2} \times M_{12} = (-1) \times 18 = -18$

Cofactor of $A_{13} = (-1)^{1+3} \times M_{13} = (1) \times 15 = 15$

Cofactor of $A_{21} = (-1)^{2+1} \times M_{21} = (-1) \times (-11) = 11$

Cofactor of $A_{22} = (-1)^{2+2} \times M_{22} = (1) \times 12 = 12$

Cofactor of $A_{23} = (-1)^{2+3} \times M_{23} = (-1) \times 10 = -10$

Cofactor of $A_{31} = (-1)^{3+1} \times M_{31} = (1) \times (-2) = -2$

Cofactor of $A_{32} = (-1)^{3+2} \times M_{32} = (-1) \times (-7) = 7$

Cofactor of $A_{33} = (-1)^{3+3} \times M_{33} = (1) \times 11 = 11$

Some Important Properties of Determinants

The set of properties of determinants which holds good in any order form,

(i) If the row and the column of a determinant are interchanged then the value of the determinant remains unaltered, i.e.

$$\begin{vmatrix} a_1 & b_1 & c_1 \\ a_2 & b_2 & c_2 \\ a_3 & b_3 & c_3 \end{vmatrix} = \begin{vmatrix} a_1 & a_2 & a_3 \\ b_1 & b_2 & b_3 \\ c_1 & c_2 & c_3 \end{vmatrix}$$

Proof:

Determinant of the matrix = $\begin{vmatrix} a_1 & b_1 & c_1 \\ a_2 & b_2 & c_2 \\ a_3 & b_3 & c_3 \end{vmatrix}$

$$= a_1 \begin{vmatrix} b_2 & c_2 \\ b_3 & c_3 \end{vmatrix} - b_1 \begin{vmatrix} a_2 & c_2 \\ a_3 & c_3 \end{vmatrix} + c_1 \begin{vmatrix} a_2 & b_2 \\ a_3 & b_3 \end{vmatrix}$$

$= a_1 (b_2 c_3 - b_3 c_2) - b_1 (a_2 c_3 - a_3 c_2) + c_1 (a_2 b_3 - a_3 b_2).$

$= a_1 b_2 c_3 - a_1 b_3 c_2 - b_1 a_2 c_3 + b_1 a_3 c_2 + c_1 a_2 b_3 - c_1 a_3 b_2.$

$= a_1 (b_2 c_3 - b_3 c_2) - a_2 (b_1 c_3 - b_3 c_1) + a_3 (b_1 c_2 - b_2 c_1).$

$$= \begin{vmatrix} a_1 & a_2 & a_3 \\ b_1 & b_2 & b_3 \\ c_1 & c_2 & c_3 \end{vmatrix}$$

Hence Proved

(ii) If two adjacent rows or columns of a determinant are interchanged then the numerical value of the determinant doesn't change only sign of the determinant gets changed, i.e.

$$\begin{vmatrix} a_1 & b_1 & c_1 \\ a_2 & b_2 & c_2 \\ a_3 & b_3 & c_3 \end{vmatrix} = - \begin{vmatrix} a_1 & b_1 & c_1 \\ a_3 & b_3 & c_3 \\ a_2 & b_2 & c_2 \end{vmatrix}$$

Proof:

$$\begin{vmatrix} a_1 & b_1 & c_1 \\ a_2 & b_2 & c_2 \\ a_3 & b_3 & c_3 \end{vmatrix}$$

$$= a_1 \begin{vmatrix} b_2 & c_2 \\ b_3 & c_3 \end{vmatrix} - b_1 \begin{vmatrix} a_2 & c_2 \\ a_3 & c_3 \end{vmatrix} + c_1 \begin{vmatrix} a_2 & b_2 \\ a_3 & b_3 \end{vmatrix}$$

$= a_1 (b_2 c_3 - c_2 b_3) - b_1 (a_2 c_3 - c_2 a_3) + c_1 (a_2 b_3 - a_3 b_2)$

$= a_1 b_2 c_3 - a_1 c_2 b_3 - b_1 a_2 c_3 + b_1 c_2 a_3 + c_1 a_2 b_3 - c_1 a_3 b_2$

$= - [a_1 c_2 b_3 - a_1 c_3 b_2] + [b_1 c_2 a_3 - b_1 a_2 c_3] - [c_1 a_3 b_2 - c_1 a_2 b_3]$

$$= - a_1 \begin{vmatrix} b_3 & c_3 \\ b_2 & c_3 \end{vmatrix} + b_1 \begin{vmatrix} a_3 & c_3 \\ a_2 & c_2 \end{vmatrix} - c_1 \begin{vmatrix} a_3 & b_3 \\ a_2 & b_2 \end{vmatrix}$$

$$= -\left[a_1\begin{vmatrix} b_3 & c_3 \\ b_2 & c_2 \end{vmatrix} - b_1\begin{vmatrix} a_3 & c_3 \\ a_2 & c_2 \end{vmatrix} + c_1\begin{vmatrix} a_3 & b_3 \\ a_2 & b_2 \end{vmatrix}\right]$$

- Taking –ve sign as common factor

$$= -\begin{vmatrix} a_1 & b_1 & c_1 \\ a_3 & b_3 & c_3 \\ a_2 & b_2 & c_2 \end{vmatrix}$$

Hence proved.

(iii) If two rows or columns of a determinant are identical then the value of the determinants is zero i.e.

$$\begin{vmatrix} a_1 & b_1 & c_1 \\ a_1 & b_1 & c_1 \\ a_2 & b_2 & c_2 \end{vmatrix} = 0$$

Proof:

$$\begin{vmatrix} a_1 & b_1 & c_1 \\ a_1 & b_1 & c_1 \\ a_2 & b_2 & c_2 \end{vmatrix}$$

$$= a_1\begin{vmatrix} b_1 & c_1 \\ b_2 & c_2 \end{vmatrix} - b_1\begin{vmatrix} a_1 & c_1 \\ a_2 & c_2 \end{vmatrix} + c_1\begin{vmatrix} a_1 & b_1 \\ a_2 & b_2 \end{vmatrix}$$

$$= a_1 (b_1 c_2 - b_2 c_1) - b_1 (a_1 c_2 - a_2 c_1) + c_1 (a_1 b_2 - a_2 b_1)$$

$$= a_1 b_1 c_2 - a_1 b_2 c_1 - b_1 a_1 c_2 + b_1 a_2 c_1 + c_1 a_1 b_2 - c_1 a_2 b_1$$

$$= 0$$

Hence proved.

(iv) If all the elements of any row or column of a determinant is multiplied by the same constant say k then the value of the determinants gets multiplied by the same scalar, i.e.

$$\begin{vmatrix} ka_1 & kb_1 & kc_1 \\ a_2 & b_2 & c_2 \\ a_3 & b_3 & c_3 \end{vmatrix} = k\begin{vmatrix} a_1 & b_1 & c_1 \\ a_2 & b_2 & c_2 \\ a_3 & b_3 & c_3 \end{vmatrix}$$

Proof:

$$\begin{vmatrix} ka_1 & kb_1 & kc_1 \\ a_2 & b_2 & c_2 \\ a_3 & b_3 & c_3 \end{vmatrix} = k\,a_1 \begin{vmatrix} b_2 & c_2 \\ b_3 & c_3 \end{vmatrix} - k\,b_1 \begin{vmatrix} a_2 & c_2 \\ a_3 & c_3 \end{vmatrix} + k\,c_1 \begin{vmatrix} a_2 & b_2 \\ a_3 & b_3 \end{vmatrix}$$

$$= k \left[a_1 \begin{vmatrix} b_2 & c_2 \\ b_3 & c_3 \end{vmatrix} - b_1 \begin{vmatrix} a_2 & c_2 \\ a_3 & c_3 \end{vmatrix} + c_1 \begin{vmatrix} a_2 & b_2 \\ a_3 & b_3 \end{vmatrix} \right]$$

$$= k \begin{vmatrix} a_1 & b_1 & c_1 \\ a_2 & b_2 & c_2 \\ a_3 & b_3 & c_3 \end{vmatrix}$$

Hence proved.

(v) If the elements of any row or column of a determinant are multiplied in order by the cofactor elements of any other row or column then sum of the products thus obtained is zero i.e. $a_1 A_2 + b_1 B_2 + c_1 C_2 = 0$, $a_2 A_3 + b_2 B_3 + c_2 C_3 = 0$, $a_3 A_1 + b_3 B_1 + c_3 C_1 = 0$.

Proof:

$$\begin{vmatrix} a_1 & b_1 & c_1 \\ a_2 & b_2 & c_2 \\ a_3 & b_3 & c_3 \end{vmatrix}$$

$A_1\, B_1\, C_1$, $A_2\, B_2\, C_2$ and $A_3\, B_3\, C_3$ be the cofactor of (a_1, b_1, c_1), (a_2, b_2, c_2) and (a_3, b_3, c_3) respectively

Then,

$$A_2 = -\begin{vmatrix} b_1 & c_1 \\ b_3 & c_3 \end{vmatrix} = -(b_1 c_3 - b_3 c_1) = b_3 c_1 - b_1 c_3$$

$$B_2 = +\begin{vmatrix} a_1 & c_1 \\ a_3 & c_3 \end{vmatrix} = a_1 c_3 - a_3 c_1$$

$$C_2 = -\begin{vmatrix} a_1 & b_1 \\ a_3 & b_3 \end{vmatrix} = -(a_1 b_3 - a_3 b_1) = a_3 b_1 - a_1 b_3$$

Therefore,

$a_1 A_2 + b_1 B_2 + c_1 C_2$

$= a_1 (b_3 c_1 - b_1 c_3) + b_1 (a_1 c_3 - a_3 c_1) + c_1 (a_3 b_1 - a_1 b_3)$

$= a_1 b_3 c_1 - a_1 b_1 c_3 + b_1 a_1 c_3 - b_1 a_3 c_1 + c_1 a_3 b_1 - c_1 a_1 b_3$

$= 0.$

Hence proved.

(vi) If each element of any row or column is the sum of two numbers then the determinant can be written as the sum of the two determinants whose order is not changed i.e.

$$\begin{vmatrix} a_1+\alpha & b_1+\beta & c_1+\nu \\ a_2 & b_2 & c_2 \\ a_3 & b_3 & c_3 \end{vmatrix} = \begin{vmatrix} a_1 & b_1 & c_1 \\ a_2 & b_2 & c_2 \\ a_3 & b_3 & c_3 \end{vmatrix} + \begin{vmatrix} \alpha & \beta & \nu \\ a_2 & b_2 & c_2 \\ a_3 & b_3 & c_3 \end{vmatrix}$$

Proof:

If A_1, B_1, C_1 be the cofactor of the elements $a_1 + \alpha$, $b_1 + \beta$ and $c_1 + \nu$ of the 1st row of the determinant, then,

$$\begin{vmatrix} a_1+\alpha & b_1+\beta & c_1+\nu \\ a_2 & b_2 & c_2 \\ a_3 & b_3 & c_3 \end{vmatrix}$$

$= (a_1 + \alpha) A_1 + (b_1 + \beta) B_1 + (c_1 + \nu) C_1$

$= (a_1 A_1 + b_1 B_1 + c_1 C_1) + (\alpha A_1 + \beta B_1 + \nu C_1)$

$$= \begin{vmatrix} a_1 & b_1 & c_1 \\ a_2 & b_2 & c_2 \\ a_3 & b_3 & c_3 \end{vmatrix} + \begin{vmatrix} \alpha & \beta & \nu \\ a_2 & b_2 & c_2 \\ a_3 & b_3 & c_3 \end{vmatrix}$$

(vii) If all the elements of either row or column are added to the same multiples of the corresponding elements of any other row or column then the value of the determinant remains unaltered i.e.

$$\begin{vmatrix} a_1+\alpha a_2+\beta a_3 & b_1+\alpha b_2+\beta b_3 & c_1+\alpha c_2+\beta c_3 \\ a_2 & b_2 & c_2 \\ a_3 & b_3 & c_3 \end{vmatrix} = \begin{vmatrix} a_1 & b_1 & c_1 \\ a_2 & b_2 & c_2 \\ a_3 & b_3 & c_3 \end{vmatrix}$$

Proof:

$$= \begin{vmatrix} a_1 & b_1 & c_1 \\ a_2 & b_2 & c_2 \\ a_3 & b_3 & c_3 \end{vmatrix} + \begin{vmatrix} \alpha a_2 & \alpha b_2 & \alpha c_2 \\ a_2 & b_2 & c_2 \\ a_3 & b_3 & c_3 \end{vmatrix} + \begin{vmatrix} \beta a_3 & \beta b_3 & \beta c_3 \\ a_2 & b_2 & c_2 \\ a_3 & b_3 & c_3 \end{vmatrix}$$

$$= \begin{vmatrix} a_1 & b_1 & c_1 \\ a_2 & b_2 & c_2 \\ a_3 & b_3 & c_3 \end{vmatrix} + \alpha \begin{vmatrix} a_2 & b_2 & c_2 \\ a_2 & b_2 & c_2 \\ a_3 & b_3 & c_3 \end{vmatrix} + \beta \begin{vmatrix} a_3 & b_3 & c_3 \\ a_2 & b_2 & c_2 \\ a_3 & b_3 & c_3 \end{vmatrix}$$

$$= \begin{vmatrix} a_1 & b_1 & c_1 \\ a_2 & b_2 & c_2 \\ a_3 & b_3 & c_3 \end{vmatrix} + \alpha\ (0) + \beta\ (0)$$

$$= \begin{vmatrix} a_1 & b_1 & c_1 \\ a_2 & b_2 & c_2 \\ a_3 & b_3 & c_3 \end{vmatrix}$$

Hence proved.

Illustration No.2

Calculate the value of determinant without expansion

$$\begin{vmatrix} a-b & b-c & c-a \\ b-c & c-a & a-b \\ c-a & a-b & b-c \end{vmatrix} = 0$$

Solution:

$$\begin{vmatrix} a-b & b-c & c-a \\ b-c & c-a & a-b \\ c-a & a-b & b-c \end{vmatrix}$$

- By viewing the above matrix we use the following column operation i.e. adding column 2nd and 3rd column to column '1st ' i.e. $C_1 = C_1 + C_2 + C_3$

We get,

$$= \begin{vmatrix} a-b+b-c+c-a & b-c & c-a \\ b-c+c-a+a-b & c-a & a-b \\ c-a+a-b+b-c & a-b & b-c \end{vmatrix}$$

$$= \begin{vmatrix} 0 & b-c & c-a \\ 0 & c-a & a-b \\ 0 & a-b & b-c \end{vmatrix} = 0$$

Hence the value of the determinant is zero.

Illustration No.3

Calculate the determinant of

$$\begin{vmatrix} 1 & x & yz \\ 1 & y & zx \\ 1 & z & xy \end{vmatrix}$$

Solution:

$$\begin{vmatrix} 1 & x & yz \\ 1 & y & zx \\ 1 & z & xy \end{vmatrix}$$

$$= \begin{vmatrix} 1 & x & yz \\ 0 & y-x & zx-yz \\ 0 & z-x & xy-yz \end{vmatrix}$$

By using the following Row operation,

- $R_2 : R_2 - R_1$ and $R_3 : R_3 - R_1$.

$$= \begin{vmatrix} 1 & y & z \\ 0 & y-x & z(x-y) \\ 0 & z-x & y(x-z) \end{vmatrix}$$

$$= 1 \begin{vmatrix} (y-x) & z(x-y) \\ (z-y) & y(x-z) \end{vmatrix}$$

$$= [y\ (y-z)\ (x-y) - z\ (x-y)\ (z-x)]$$

$= (x - z)(x - y)(- y + z)$

$= (x - z)(x - y)(z - y)$

$= (x - y)(y - z)(z - x)$

[*Note*: By taking negative sign common from (z − y) (x − z)]

Illustration No. 4

Without direct expansion find the determinant

$$\begin{vmatrix} a-b-c & 2a & 2a \\ 2b & b-c-a & 2b \\ 2c & 2c & c-a-b \end{vmatrix}$$

Solution:

$$\begin{vmatrix} a-b-c & 2a & 2a \\ 2b & b-c-a & 2b \\ 2c & 2c & c-a-b \end{vmatrix}$$

$$= \begin{vmatrix} a+b+c & a+b+c & a+b+c \\ 2b & b-c-a & 2b \\ 2c & 2c & c-a-b \end{vmatrix}$$

Adding R_2 and R_3 to R_1 i.e. $R_1 = R_1 + R_2 + R_3$

$$= \begin{vmatrix} 1 & 1 & 1 \\ 2b & -b-c-a & 2b \\ 2c & 0 & -c-a-b \end{vmatrix} (a+b+c)$$

- By taking (a + b + c) as a common factor from R_1
- Subtracting 2nd Column from 1st Column i.e. C_2: $C_2 - C_1$.

$$= (a+b+c) \begin{vmatrix} 1 & 0 & 0 \\ 2b & -(b+c+a) & 0 \\ 2c & 0 & -(c+a+b) \end{vmatrix}$$

C_3: $C_3 - C_1$.

By expanding along the 1st Row, we get,

$= (a + b + c)\,[-(b + c + a) \times -(c + a + b)]$

$= (a + b + c)(a + b + c)^2$

$= (a + b + c)^3$

Illustration No.5

Prove that

$$\begin{vmatrix} 1 & a & a^2 \\ 1 & b & b^2 \\ 1 & c & c^2 \end{vmatrix} = (b-c)(c-a)(a-b)$$

Solution:

$$\begin{vmatrix} 1 & a & a^2 \\ 1 & b & b^2 \\ 1 & c & c^2 \end{vmatrix}$$

$$= \begin{vmatrix} 1 & a & a^2 \\ 0 & b-a & b^2-a^2 \\ 0 & c-a & c^2-a^2 \end{vmatrix}$$

- $R_2: R_2 - R_1$ and $R_3: R_3 - R_1$

Therefore by expanding along the 1st column, we get,

$$= 1 \begin{vmatrix} b-a & b^2-a^2 \\ c-a & c^2-a^2 \end{vmatrix}$$

$$= (b-c)(c^2-a^2) - (b^2-a^2)(c-a) \quad [\text{Since } a^2 - b^2 = (a+b)(a-b)]$$

The above expression can be written as,

$$= (b-a)(c+a)(c-a) - (b+a)(b-a)(c-b)$$

$$= (b-a)(c-a)[c+a-(b+a)]$$

$$= (b-a)(c-a)[c+a-b-a]$$

$$= (b-a)(c-a)(c-b)$$

$$= (a-b)(b-c)(c-a)$$

[By taking negative sign as common factor between $(c-b)$ and $(b-a)$]

Illustration No.6

Prove that

$$\begin{vmatrix} x+y+2z & x & y \\ z & y+z+2x & y \\ z & x & z+x+2y \end{vmatrix} = 2(x+y+z)^3$$

Solution:

$$\begin{vmatrix} x+y+2z & x & y \\ z & y+z+2x & y \\ z & x & z+x+2y \end{vmatrix}$$

$$\begin{vmatrix} 2(x+4+z) & x & y \\ 2(x+4+z) & y+z+2x & y \\ 2(x+4+z) & x & z+x+2y \end{vmatrix}$$

- C_1: $C_1 + C_2 + C_3$

$$= 2(x+y+z)\begin{vmatrix} 1 & x & y \\ 1 & y+z+2z & y \\ 1 & x & z+x+2y \end{vmatrix}$$

- By taking 2(x + y + z) as a common factor from Column 1

$$= 2(x+y+z)\begin{vmatrix} 1 & x & y \\ 0 & x+y+z & 0 \\ 0 & 0 & x+y+z \end{vmatrix}$$

R_2: $R_2 - R_1$ and R_3: $R_3 - R_1$

$$= 2(x+y+z)\,[1 \times (x+y+z) \times (x+y+z)]$$

$$= 2(x+y+z)(x+y+z)^2$$

$$= 2(x+y+z)^3$$

Hence Proved

Illustration No. 7

If $\begin{vmatrix} x+a & b & c \\ c & x+b & a \\ a & b & x+c \end{vmatrix} = 0$ and $x \neq 0$ then prove that $x = -(a+b+c)$?

Solution:

$$\begin{vmatrix} x+a & b & c \\ c & x+b & a \\ a & b & x+c \end{vmatrix} = 0$$

$$\begin{vmatrix} x+a+b+c & b & c \\ x+a+b+c & x+b & a \\ x+a+b+c & b & x+c \end{vmatrix} = 0$$

- $C_1 : C_1 + C_2 + C_3$

$$(x + a + b + c)\begin{vmatrix} 1 & b & c \\ 1 & x+b & a \\ 1 & b & x+c \end{vmatrix} = 0$$

- Taking (x + a + b + c) as a common factor from Column 1

$$(x + a + b + c)\begin{vmatrix} 1 & b & c \\ 0 & x & a-c \\ 0 & 0 & x \end{vmatrix} = 0$$

- R_2: $R_2 - R_1$ and R_3: $R_3 - R_1$

$$(x + a + b + c)\begin{vmatrix} x & a-c \\ 0 & x \end{vmatrix} = 0$$

$$(x + a + b + c)\,(x^2) = 0$$

i.e $x^2 = 0$ or $(x + a + b + c) = 0$ since $x \neq 0$

therefore,

$$x = -(a + b + c)$$

Illustration No. 8

Prove that $\begin{vmatrix} 1+x & 1 & 1 \\ 1 & 1+y & 1 \\ 1 & 1 & 1+z \end{vmatrix} = (x\,y\,z)\left(\frac{1}{x} + \frac{1}{y} + \frac{1}{z}\right)$

Solution:

$$\begin{vmatrix} 1+x & 1 & 1 \\ 1 & 1+y & 1 \\ 1 & 1 & 1+z \end{vmatrix}$$

$$= xyz \begin{vmatrix} \left(1+\frac{1}{x}\right) & \frac{1}{x} & \frac{1}{x} \\ \frac{1}{y} & \left(1+\frac{1}{y}\right) & \frac{1}{y} \\ \frac{1}{z} & \frac{1}{z} & \left(1+\frac{1}{z}\right) \end{vmatrix}$$

- By taking x, y, z as common factor from R_1, R_2 and R_3.

$$= x\,y\,z \begin{vmatrix} \left(1+\frac{1}{x}+\frac{1}{y}+\frac{1}{z}\right) & \left(1+\frac{1}{x}+\frac{1}{y}+\frac{1}{z}\right) & \left(1+\frac{1}{x}+\frac{1}{y}+\frac{1}{z}\right) \\ \frac{1}{y} & \left(1+\frac{1}{y}\right) & \frac{1}{y} \\ \frac{1}{z} & \frac{1}{z} & \left(1+\frac{1}{z}\right) \end{vmatrix}$$

By substituting $R_1 : R_1 + R_2 + R_3$

$$= x\,y\,z \left(1+\frac{1}{x}+\frac{1}{y}+\frac{1}{z}\right) \begin{vmatrix} 1 & 1 & 1 \\ \frac{1}{y} & \left(1+\frac{1}{y}\right) & \frac{1}{y} \\ \frac{1}{z} & \frac{1}{z} & \left(1+\frac{1}{z}\right) \end{vmatrix}$$

- By taking $\left(1+\frac{1}{x}+\frac{1}{y}+\frac{1}{z}\right)$ as a common factor from R_1.

$$= x\,y\,z \left(1+\frac{1}{x}+\frac{1}{y}+\frac{1}{z}\right) \begin{vmatrix} 1 & 0 & 0 \\ \frac{1}{y} & 1 & 0 \\ \frac{1}{z} & 0 & 1 \end{vmatrix}$$

By substituting C_2: $C_2 - C_1$ and C_3: $C_3 - C_1$

$$= x\,y\,z\left(1+\frac{1}{x}+\frac{1}{y}+\frac{1}{z}\right)\begin{vmatrix}1 & 0\\ 0 & 1\end{vmatrix}$$

$$= x\,y\,z\left(1+\frac{1}{x}+\frac{1}{y}+\frac{1}{z}\right), \text{ Hence proved.}$$

Illustration No. 9

If the determinant of the matrix

$$\begin{vmatrix}a & a^2 & 1+a^3\\ b & b^2 & 1+b^3\\ c & c^2 & 1+c^3\end{vmatrix} = 0, \text{ then, prove that } (1 + a\,b\,c) = 0$$

Solution:

$$\begin{vmatrix}a & a^2 & 1+a^3\\ b & b^2 & 1+b^3\\ c & c^2 & 1+c^3\end{vmatrix} = 0$$

$$\begin{vmatrix}a & a^2 & 1\\ b & b^2 & 1\\ c & c^2 & 1\end{vmatrix} + \begin{vmatrix}a & a^2 & a^3\\ b & b^2 & b^3\\ c & c^2 & c^3\end{vmatrix} = 0$$

- By using properties of determinants

$$\begin{vmatrix}a & a & 1\\ b & b^2 & 1\\ c & c^2 & 1\end{vmatrix} + abc\begin{vmatrix}1 & a & a^2\\ 1 & b & b^2\\ 1 & c & c^2\end{vmatrix} = 0$$

- By taking a, b and c as common factor from R_1, R_2 and R_3 from the 2nd component

$$(1 + abc)\;(1 + abc)\begin{vmatrix}1 & a & a^2\\ 1 & b & b^2\\ 1 & c & c^2\end{vmatrix} = 0$$

But we know that from solved illustration no.5 that

$$\begin{vmatrix} 1 & a & a^2 \\ 1 & b & b^2 \\ 1 & c & c^2 \end{vmatrix} = (a-b)\ (b-c)\ (c-a)$$

Hence,

$$(1 + a\,b\,c)\ [(a - b)\ (b - c)\ (c + a)] = 0$$

i.e.

Either $(1 + a\,b\,c) = 0$ or $[(a - b)\ (b - c)\ (c - a)] = 0$

Since the 2nd part $[(a - b)\ (b - c)\ (c - a)] \neq 0$, therefore, we have,

$(1 + a\,b\,c) = 0$, hence proved.

Illustration No.10

$$\begin{vmatrix} y+z & x & x \\ y & z+x & y \\ z & z & x+y \end{vmatrix} = 4xyz$$

Solution:

$$\begin{vmatrix} y+z & x & x \\ y & z+x & y \\ z & z & x+y \end{vmatrix}$$

$$= \begin{vmatrix} 0 & -2x & -2y \\ y & z+x & y \\ z & z & x+y \end{vmatrix}$$

- $R_1 = R_1 - (R_2 + R_3)$

$$= 2\begin{vmatrix} 0 & -x & -y \\ y & z+x & y \\ z & z & x+y \end{vmatrix}$$

- By taking 2 as a common factor from R_1 and substituting $R_2 = R_2 + R_1$

$$= 2\begin{vmatrix} 0 & -z & -y \\ y & z & 0 \\ z & z & x+y \end{vmatrix}$$

$$= 2 \begin{vmatrix} 0 & -z & -y \\ y & x & 0 \\ z & 0 & x \end{vmatrix}$$

- Substituting $R_3 = R_3 + R_1$

$$= 2\,[0 + z\,(xy - 0) - y\,(0 - xz)]$$

$$= 2\,[0 + xyz + xyz]$$

$$= 2\,(2\,x\,y\,z)$$

$$= 4xyz$$

Illustration No. 11

Prove that the determinant of

$$\begin{vmatrix} 0 & xy^2 & xz^2 \\ x^2y & 0 & yz^2 \\ x^2z & y^2z & 0 \end{vmatrix} = 2x^3y^3z^3$$

Solution:

$$\begin{vmatrix} 0 & xy^2 & xz^2 \\ x^2y & 0 & yz^2 \\ x^2z & y^2z & 0 \end{vmatrix}$$

- Taking x, y, z common from R_1, R_2 and R_3

we get,

$$= xyz \begin{vmatrix} 0 & y^2 & z^2 \\ x^2 & 0 & z^2 \\ x^2 & y^2 & 0 \end{vmatrix}$$

$$= x^3y^3z^3 \begin{vmatrix} 0 & 1 & 1 \\ 1 & 0 & 1 \\ 1 & 1 & 0 \end{vmatrix}$$

- By taking x^2, y^2 and z^2 common from C_1, C_2 and C_3

$$= x^3\,y^3\,z^3\,[0 - 1\,(0 - 1) + 1\,(1 - 0)]$$

$$= x^3\,y^3\,z^3\,(2)$$

$$= 2x^3y^3z^3,$$

hence Proved.

Illustration No. 12

Prove that

$$\begin{vmatrix} x-y & 1 & x \\ y-z & 1 & y \\ z-x & 1 & z \end{vmatrix} = \begin{vmatrix} x & 1 & y \\ y & 1 & z \\ z & 1 & x \end{vmatrix}$$

Solution:

$$\begin{vmatrix} x-y & 1 & x \\ y-z & 1 & y \\ z-x & 1 & z \end{vmatrix} = \begin{vmatrix} x & 1 & x \\ y & 1 & y \\ z & 1 & z \end{vmatrix} + \begin{vmatrix} -y & 1 & x \\ -z & 1 & y \\ -x & 1 & z \end{vmatrix}$$

- In the first part C_1 and C_3 are equal hence determinant is zero

$$= 0 + \begin{vmatrix} -y & 1 & x \\ -z & 1 & y \\ -x & 1 & z \end{vmatrix}$$

$$= -\begin{vmatrix} y & 1 & x \\ z & 1 & y \\ x & 1 & z \end{vmatrix}$$

- Taking (–)sign common from C_1

$$= \begin{vmatrix} x & 1 & y \\ y & 1 & z \\ z & 1 & x \end{vmatrix}$$

- By interchanging the position of C_1 by C_3 and C_3 by C_1

Illustration No. 13

$$\begin{vmatrix} \pi+x & \pi-x & \pi-x \\ \pi-x & \pi+x & \pi-x \\ \pi-x & \pi-x & \pi+x \end{vmatrix} = 0, \text{ if } x \neq 0, \text{ then prove that } x = 3\pi?$$

Solution:

$$\begin{vmatrix} \pi + x & \pi - x & \pi - x \\ \pi - x & \pi + x & \pi - x \\ \pi - x & \pi - x & \pi + x \end{vmatrix} = 0$$

$$\begin{vmatrix} 3\pi - x & \pi - x & \pi - x \\ 3\pi - x & \pi + x & \pi - x \\ 3\pi - x & \pi - x & \pi + x \end{vmatrix} = 0$$

- Applying $C_1 \rightarrow C_1 + C_2 + C_3$

$$(3\pi - x)\begin{vmatrix} 1 & \pi - x & \pi - x \\ 1 & \pi + x & \pi - x \\ 1 & \pi - x & \pi + x \end{vmatrix} = 0$$

- By taking $(3\pi - x)$ as common factor from C_1

$$(3\pi - x)\begin{vmatrix} 1 & \pi - x & \pi - x \\ 0 & 2x & 0 \\ 0 & 0 & 2x \end{vmatrix} = 0$$

- By applying $R_2 \rightarrow R_2 - R_1$ and $R_3 \rightarrow R_3 - R_1$

$$= (3\pi - x) \times 1 \times \begin{vmatrix} 2x & 0 \\ 0 & 2x \end{vmatrix} = 0$$

$$(3\pi - x) \times 1 \times 4x^2 = 0$$

Therefore, either $(3\pi - x) = 0$ or $4x^2 = 0$, since $x \neq 0$

Hence, $x = 0, 3\pi$.

Illustration No. 14

Prove that

$$\begin{vmatrix} \alpha & \beta & \alpha x + \beta y \\ \beta & \nu & \beta x + \nu y \\ \alpha x + \beta y & \beta x + \nu y & 0 \end{vmatrix} = (\beta^2 - \nu\alpha)(\alpha x^2 + 2\beta xy + \nu y^2)$$

Solution:

$$\begin{vmatrix} \alpha & \beta & \alpha x+\beta y \\ \beta & \nu & \beta x+\nu y \\ \alpha x+\beta y & \beta x+\nu y & 0 \end{vmatrix}$$

$$= \begin{vmatrix} \alpha & \beta & 0 \\ \beta & \nu & 0 \\ \alpha x+\beta y & \beta x+\nu y - x(\alpha x+\beta y) & -y(\beta x+\nu y) \end{vmatrix}$$

- By applying C_3: $C_3 - xC_1 - yC_2$

$$= -(\alpha x^2 + 2\beta xy + \nu y^2) \begin{vmatrix} \alpha & \beta \\ \beta & \nu \end{vmatrix}$$

$$= -(\alpha x^2 + 2\beta xy + \nu y^2)(\alpha\nu - \beta^2)$$

$$= (\beta^2 - \alpha\nu)(\alpha x^2 + 2\beta xy + \nu y^2)$$

Illustration No. 15

Prove that
$$\begin{vmatrix} 1+x^2-y^2 & 2xy & -2y \\ 2xy & 1+x^2-y^2 & 2x \\ 2y & -2x & 1+x^2-y^2 \end{vmatrix} = (1+x^2+y^2)^3$$

Solution:

$$\begin{vmatrix} 1+x^2-y & 2xy & -2y \\ 2xy & 1+x^2+y^2 & 2x \\ 2y & -2x & 1+x^2-y^2 \end{vmatrix}$$

$$\begin{vmatrix} 1+x^2+y^2 & 0 & -2y \\ 0 & 1+x^2+y^2 & 2x \\ y(1+x^2+y^2) & -x(1+x^2+y^2) & 1+x^2-y^2 \end{vmatrix}$$

- By applying C_1: $C_1 - yC_3$ and C_2: $C_2 + xC_3$

$$= (1+x^2+y^2)^2 \begin{vmatrix} 1 & 0 & -2y \\ 0 & 1 & 2x \\ y & -x & 1-x^2-y^2 \end{vmatrix}$$

By taking $(1 + x^2 + y^2)$ as common factor from C_1 and C_2

$$= (1 + x^2 + y^2)^2 \begin{vmatrix} 1 & 0 & -2y \\ 0 & 1 & 2x \\ 0 & 0 & 1+x^2+y^2 \end{vmatrix}$$

By applying $R_3 : R_3 - yR_1 + xR_2$

$$= (1 + x^2 + y^2)^2 \times 1 \times \begin{vmatrix} 1 & 2x \\ 0 & 1+x^2+y^2 \end{vmatrix}$$

$$= (1 + x^2 + y^2)^2 \times 1 \times (1 + x^2 + y^2)$$

$$= (1 + x^2 + y^2)^3$$

Illustration No. 16

Prove that $\begin{vmatrix} 1 & 1+x & 1+x+y \\ 2 & 3+2x & 1+3x+2y \\ 3 & 6+3x & 1+6x+3y \end{vmatrix} = 1$

Solution:

$$\Delta = \begin{vmatrix} 1 & 1+x & 1+x+y \\ 2 & 3+2x & 1+3x+2y \\ 3 & 6+3x & 1+6x+3y \end{vmatrix}$$

$$= \begin{vmatrix} 1 & 1 & 1+x \\ 2 & 3 & 1+3x \\ 3 & 6 & 1+6x \end{vmatrix}$$

By applying $C_2: C_2 - x\,C_1$ and $C_3: C_3 - y\,C_1$

$$= \begin{vmatrix} 1 & 1 & 1 \\ 2 & 3 & 1 \\ 3 & 6 & 1 \end{vmatrix} + x \begin{vmatrix} 1 & 1 & 1 \\ 2 & 3 & 3 \\ 3 & 6 & 6 \end{vmatrix}$$

$$= \begin{vmatrix} 1 & 1 & 1 \\ 2 & 3 & 1 \\ 3 & 6 & 1 \end{vmatrix} + x\,(0)$$

$$= \begin{vmatrix} 1 & 1 & 1 \\ 2 & 3 & 1 \\ 3 & 6 & 1 \end{vmatrix}$$

$$= \begin{vmatrix} 0 & 0 & 1 \\ 1 & 2 & 1 \\ 2 & 5 & 1 \end{vmatrix}$$

- By applying $C_1: C_1 - C_3$ and $C_2: C_2 - C_3$

$$= 0 \begin{vmatrix} 2 & 1 \\ 5 & 1 \end{vmatrix} - 0 \begin{vmatrix} 1 & 1 \\ 2 & 1 \end{vmatrix} + 1 \begin{vmatrix} 1 & 2 \\ 2 & 5 \end{vmatrix}$$

$$= 0 - 0 + (5 - 4) = 1$$

Illustration No. 17

Prove that

$$\begin{vmatrix} x & x+y & x+y+2 \\ 2x & 3x+2y & 4x+3y+2z \\ 3x & 6x+3y & 10x+6y+3z \end{vmatrix} = x^3$$

Solution:

$$\begin{vmatrix} x & x+y & x+y+2 \\ 2x & 3x+2y & 4x+3y+2z \\ 3x & 6x+3y & 10x+6y+3z \end{vmatrix}$$

- Since the elements of the second column is sum of two elements there we can write them into two separate determinants

$$= \begin{vmatrix} x & x & x+y+2 \\ 2x & 3x & 4x+3y+2z \\ 3x & 6x & 10x+6y+3z \end{vmatrix} + \begin{vmatrix} x & y & x+y+2 \\ 2x & 2y & 4x+3y+2z \\ 3x & 3y & 10x+6y+3z \end{vmatrix}$$

$$= \begin{vmatrix} x & x+y & x+y+2 \\ 2x & 3x+2y & 4x+3y+2z \\ 3x & 6x+3y & 10x+6y+3z \end{vmatrix} + xy \begin{vmatrix} 1 & 1 & x+y+z \\ 2 & 2 & 4x+3y+2z \\ 3 & 3 & 10x+6y+3z \end{vmatrix}$$

- Taking x and y common from 1st and 2nd Column of 2nd determinant

$$\begin{vmatrix} x & x+y & x+y+2 \\ 2x & 3x+2y & 4x+3y+2z \\ 3x & 6x+3y & 10x+6y+3z \end{vmatrix} + xy\ (0)$$

- Since C_1 and C_2 are identical in the 2nd determinant

$$= \begin{vmatrix} x & x & x \\ 2x & 3x & 4x \\ 3x & 6x & 10x \end{vmatrix} + \begin{vmatrix} x & x & y \\ 2x & 3x & 3y \\ 3x & 6x & 6y \end{vmatrix} + \begin{vmatrix} x & x & z \\ 2x & 3x & 2z \\ 3x & 6x & 3z \end{vmatrix}$$

- As 3rd column is a sum of three elements

$$= 0 + 0 + 0 = 0.$$

- Since C_1 and C_2 are identical in all the three determinant.

Illustration No. 18

Prove that $\begin{vmatrix} 2+x & y & z \\ x & 2+y & z \\ x & y & 2+z \end{vmatrix} = 4\ (2 + x + y + z)$

Solution:

$$\Delta = \begin{vmatrix} 2+x & y & z \\ x & 2+y & z \\ x & y & 2+z \end{vmatrix}$$

By Applying C_1: $C_1 + C_2 + C_3$

$$= (2 + x + y + z) \begin{vmatrix} 1 & y & z \\ 1 & 2+y & z \\ 1 & y & 2+z \end{vmatrix}$$

By taking (2 + x + y + z) as common from C_1

$$= (2 + x + y + z) \begin{vmatrix} 1 & y & z \\ 0 & 2 & 0 \\ 0 & 0 & 2 \end{vmatrix}$$

Applying $R_2:R_2 - R_1$ and $R_3:R_3 - R_1$

$$= (2 + x + y + z)\begin{vmatrix} 2 & 0 \\ 0 & 2 \end{vmatrix}$$

$$= 4\,(2 + x + y + z)$$

Hence proved

Illustration No. 19

Show that

$$\begin{vmatrix} b+c & a+c & a+b \\ q+r & r+p & p+q \\ y+z & z+x & x+y \end{vmatrix} = 2\begin{vmatrix} a & b & c \\ p & q & r \\ x & y & z \end{vmatrix}$$

Solution:

$$\Delta = \begin{vmatrix} b+c & a+c & a+b \\ q+r & r+p & p+q \\ y+z & z+x & x+y \end{vmatrix}$$

$$= \begin{vmatrix} 2(a+b+c) & a+c & a+b \\ 2(p+q+r) & r+p & p+q \\ 2(x+y+z) & z+x & x+y \end{vmatrix}$$

- By Applying $C_1: C_1 + C_2 + C_3$

$$= 2\begin{vmatrix} (a+b+c) & a+c & a+b \\ (p+q+r) & r+p & p+q \\ (x+y+z) & z+x & x+y \end{vmatrix}$$

By taking 2 common from C_1

$$\begin{vmatrix} a+b+c & -b & -c \\ p+q+r & -q & -r \\ x+y+z & -y & -z \end{vmatrix}$$

- By applying $C_2: C_2 - C_1$ and $C_3: C_3 - C_1$

$$= 2\begin{vmatrix} a & -b & -c \\ p & -q & -r \\ x & -y & -z \end{vmatrix}$$

By applying $C_1: C_1 + C_2 + C_3$

$$= 2\begin{vmatrix} a & b & c \\ p & q & r \\ x & y & z \end{vmatrix}$$

Illustration No. 20

Find 'π' if

$$\begin{vmatrix} \pi - 2 & 2\pi - 3 & 3\pi - 4 \\ \pi - 4 & 2\pi - 9 & 3\pi - 16 \\ \pi - 8 & 2\pi - 27 & 3\pi - 64 \end{vmatrix} = 0$$

Solution:

$$\Delta = \begin{vmatrix} \pi - 2 & 2\pi - 3 & 3\pi - 4 \\ \pi - 4 & 2\pi - 9 & 3\pi - 16 \\ \pi - 8 & 2\pi - 27 & 3\pi - 64 \end{vmatrix}$$

$$\begin{vmatrix} \pi - 2 & 1 & 2 \\ \pi - 4 & -1 & -4 \\ \pi - 8 & -11 & -40 \end{vmatrix}$$

By applying $C_2: C_2 - 2C_1$ and $C_3: C_3 - 3C_1$

$$\begin{vmatrix} -2 & 1 & 2 \\ -2 & -2 & -6 \\ -6 & -12 & -42 \end{vmatrix}$$

By applying $R_2: R_2 - R_1$ and $R_3: R_3 - R_1$

$$= (-2)(-6)\begin{vmatrix} \pi - 2 & 1 & 2 \\ 1 & 1 & 3 \\ 1 & 2 & 7 \end{vmatrix}$$

- By taking (–2) and (–6) Common from R_2 and R_3 respectively

$$= 12\begin{vmatrix} \pi - 2 & 1 & 2 \\ 1 & 1 & 3 \\ 1 & 2 & 7 \end{vmatrix}$$

$$= (12)\begin{vmatrix} \pi - 2 & 1 & 2 \\ 1 & 1 & 3 \\ 0 & 1 & 4 \end{vmatrix}$$

- By applying $R_3: R_3 - R_2$

$$= (12)\left[(\pi - 2)\begin{vmatrix}1 & 3\\1 & 4\end{vmatrix} - 1\begin{vmatrix}1 & 2\\1 & 4\end{vmatrix} + 0\begin{vmatrix}1 & 2\\1 & 3\end{vmatrix}\right]$$

$$= 12\,[(\pi - 2)(4 - 3) - (4 - 2)] = 0\ [\text{Since } \Delta = 0]$$

Therefore, $12(\pi - 4) = 0$ or $\pi = 4$

Illustration No. 21

Prove that

$$\begin{vmatrix}a & a^2 & 1+k\,a^3\\b & b^2 & 1+k\,b^3\\c & c^2 & 1+k\,c^3\end{vmatrix} = (1 + k\,abc)(a - b)(b - c)(c - a)$$

Solution:

$$\begin{vmatrix}a & a^2 & 1+k\,a^3\\b & b^2 & 1+k\,b^3\\c & c^2 & 1+k\,c^3\end{vmatrix}$$

$$\begin{vmatrix}a & a^2 & 1+k\,a^3\\b & b^2 & 1+k\,b^3\\c & c^2 & 1+k\,c^3\end{vmatrix} = \begin{vmatrix}a & a^2 & 1\\b & b^2 & 1\\c & c^2 & 1\end{vmatrix} + \begin{vmatrix}a & a^2 & ka^3\\b & b^2 & kb^3\\c & c^2 & kc^3\end{vmatrix}$$

- Using properties of determinants and splitting the determinant into two components.

$$\begin{vmatrix}a & a^2 & 1\\b & b^2 & 1\\c & c^2 & 1\end{vmatrix} + k\,abc\begin{vmatrix}1 & a & a^2\\1 & b & b^2\\1 & c & c^2\end{vmatrix} = (1 + k\,abc)\begin{vmatrix}1 & a & a^2\\1 & b & b^2\\1 & c & c^2\end{vmatrix}$$

- By taking k as common factor from third column in second component and a, b and c as common factor from R_1, R_2 and R_3 from 2nd component.

$$\begin{vmatrix}1 & a & a^2\\1 & b & b^2\\1 & c & c^2\end{vmatrix} = (a - b)(b - c)(a - c)$$

Therefore we have,

$$(1 + k\,abc)(a - b)(b - c)(c - a)$$

Illustration No. 22

Prove that

$$\begin{vmatrix} x & x+y & x+zy \\ x+zy & x & x+zy \\ x+y & x+zy & x \end{vmatrix} = 9y^2(x+y)^2$$

Solution:

$$\begin{vmatrix} x & x+y & x+zy \\ x+zy & x & x+zy \\ x+y & x+zy & x \end{vmatrix}$$

$$\begin{vmatrix} x & x+y & x+zy \\ x+zy & x & x+zy \\ x+y & x+zy & x \end{vmatrix} = \begin{vmatrix} 3(x+y) & x+y & x+zy \\ 3(x+y) & x & x+y \\ 3(x+y) & x+zy & x \end{vmatrix}$$

By applying $C_1 : C_1 + C_2 + C_3$

$$= 3\,(x+y) \begin{vmatrix} 1 & x+y & x+2y \\ 1 & x & x+y \\ 1 & x+2y & x \end{vmatrix}$$

- By taking 3(x + y) as common factor from 1st column.

$$= 3(x+y) \begin{vmatrix} 1 & x+y & x+2y \\ 0 & -y & -y \\ 0 & y & -2y \end{vmatrix}$$

By applying $R_2{:}R_2 - R_1$ and $R_3{:}R_3 - R_1$

$$= 3\,(x+y) \times 1 \times \begin{vmatrix} -y & -y \\ y & -2y \end{vmatrix}$$

$$= 3\,(x+y)\,(2y^2 + y^2)$$

$$= 3\,(x+y)\,(3y^2) = 9y^2\,(x+y)$$

Hence proved.

Illustration No. 23

Prove that

$$\begin{vmatrix} x & y & z \\ x-y & y-z & z-x \\ y+z & z+x & x+y \end{vmatrix} = x^3 + y^3 + z^3 - 3xyz$$

Solution:

$$\begin{vmatrix} x & y & z \\ x-y & y-z & z-x \\ y+z & z+x & x+y \end{vmatrix}$$

$$\begin{vmatrix} x & y & z \\ x-y & y-z & z-x \\ y+z & z+x & x+y \end{vmatrix} = \begin{vmatrix} x+y+z & y & z \\ 0 & y-z & z-x \\ 2(x+y+z) & (2+x) & x+y \end{vmatrix}$$

- By applying $C_1:C_1 + C_2 + C_3$

$$= (x+y+z)\begin{vmatrix} 1 & y & z \\ 0 & y-z & z-x \\ 2 & (z+x) & (x+y) \end{vmatrix}$$

By taking (x + y + z) as common factor from 1st column

$$= (x+y+z)\begin{vmatrix} 1 & y & z \\ 0 & y-z & z-x \\ 0 & (z+x)-2y & (x+y)-2z \end{vmatrix}$$

By applying R_3: $R_3 - 2R_1$

$= (x + y + z) [(y - z) \{(x + y) - 2z\} - (z - x) \{(z + x) - 2y\}]$

$= (x + y + z) [(2yx + y^2 - 2zy - 2x + 2z^2) - (z^2 + zx - 2zy - x^2 + 2yx)]$

$= (x + y + z) [yx + y^2 - 3zy - zx + 2z + 2z^2 - z^2 + 2zy + x^2 + xz - 2yx]$

$= (x + y + z) (x^2 + y^2 + z^2 - xy - yz - zx)$

$= (x^3 + y^3 + z^3 - xyz)$

Hence proved.

Illustration No. 24

Find the determinant without expansion

$$\begin{vmatrix} x & x+y & x+y+z \\ 2x & 3x+2y & 4x+3y+2z \\ 3x & 6x+3y & 10x+6y+3z \end{vmatrix}$$

Solution:

$$= \begin{vmatrix} x & x+y & x+y+z \\ 2x & 3x+2y & 4x+3y+2z \\ 3x & 6x+3y & 10x+6y+3z \end{vmatrix}$$

$$= \begin{vmatrix} x & x+y & x+y+z \\ 0 & x & 2x+y \\ 0 & 3x & 7x+3y \end{vmatrix}$$

- By Applying R_2: $R_2 - 2R_1$ and R_3: $R_3 - 3R_1$

$$= x \begin{vmatrix} x & 2x+y \\ 3x & 7x+3y \end{vmatrix}$$

$$= x\,(7x^2 + 3xy - 6x^2 - 3xy)$$

$$= x \times x^2$$

$$= x^3$$

Illustration No. 25

Prove that

$$\begin{vmatrix} p & q & r \\ q & r & p \\ r & p & q \end{vmatrix} = -(p^3 + q^3 + r^3 - pqr)$$

And hence prove

$$\begin{vmatrix} 2qr-p^2 & r^2 & q^2 \\ r^2 & 2qr-p^2 & p^2 \\ q^2 & p^2 & 2pq-r^2 \end{vmatrix} = (p^3 + q^3 + r^3 - 3pqr)^2$$

Solution:

We have

$$\begin{vmatrix} p & q & r \\ q & r & p \\ r & p & q \end{vmatrix} = \begin{vmatrix} p+q+r & q & r \\ p+q+r & q & r \\ p+q+r & q & r \end{vmatrix}$$

- By applying $C_1{:}C_1 + C_2 + C_3$

$$= (p + q + r) \begin{vmatrix} 1 & q & r \\ 1 & r & p \\ 1 & p & q \end{vmatrix}$$

- By taking $(p + q + r)$ as a common factor from C_1

$$= (p + q + r) \begin{vmatrix} 1 & q & r \\ 0 & r-q & p-r \\ 0 & p-q & q-r \end{vmatrix}$$

- Applying $R_2{:}\ R_2 - R_1$ and $R_3{:}\ R_3 - R_1$

$$= (p + q + r) \times 1 \times \begin{vmatrix} r-q & p-r \\ p-q & q-r \end{vmatrix}$$

$= (p + q + r) \times 1 \times \{(r - q)(q - r) - (p - r)(p - q)\}$

$= (p + q + r)(rq - r^2 - q^2 + rq - p^2 + pq + rp + rq)$

$= (p + q + r)(p^2 + q^2 + r^2 - pq - qr - rp)$

$= -(p^3 + q^3 + r^3 - 3pqr)$

As for the second part;

We know

$$(p^3 + q^3 + r^3 - 3pqr) = \begin{vmatrix} p & q & r \\ q & r & p \\ r & p & q \end{vmatrix}$$

Squaring the side, we get

$$(p^3 + q^3 + r^3 - 3pqr)^2 = \begin{vmatrix} p & q & r \\ q & r & p \\ r & p & q \end{vmatrix}^2$$

i.e

$$\begin{vmatrix} p & q & r \\ q & r & p \\ r & p & q \end{vmatrix}^2 = \begin{vmatrix} p & q & r \\ q & r & p \\ r & p & q \end{vmatrix}\begin{vmatrix} p & q & r \\ q & r & p \\ r & p & q \end{vmatrix}$$

$$\begin{vmatrix} p & q & r \\ q & r & p \\ r & p & q \end{vmatrix} \times \begin{vmatrix} -p & q & r \\ -q & r & p \\ -r & p & q \end{vmatrix}$$

- By interchanging 2nd and 3rd column

$$= \begin{vmatrix} -p^2 + qr + rq & -pq + pq + r^2 & -pr + q^2 + pr \\ -pq + r^2 + pq & -q^2 + pr + pr & -qr + qr + p^2 \\ -pr + pr + q^2 & -qr + p^2 + qr & -r^2 + pq + pq \end{vmatrix}$$

$$= \begin{vmatrix} 2qr - p^2 & r^2 & q^2 \\ r^2 & 2pr - q^2 & p^2 \\ q^2 & p^2 & 2pq - r^2 \end{vmatrix}$$

Hence proved.

Illustration No. 26

If $(p + q + r) = 0$ then prove that $\begin{vmatrix} p-r & s & q \\ s & q-x & p \\ q & p & r-x \end{vmatrix} = 0$

and then show that $x = 0$ or $x = \sqrt{\frac{3}{2}\left(p^2 + q^2 + r^2\right)}$

Solution:

Given

$$\Delta = \begin{vmatrix} p-x & s & q \\ s & q-x & p \\ q & p & r-x \end{vmatrix}$$

By applying $C_1: C_1 + C_2 + C_3$

$$\begin{vmatrix} p+q+r-x & r & q \\ p+q+r-x & q-x & p \\ p+q+r-x & p & r-x \end{vmatrix}$$

$$= (p+q+r-x)\begin{vmatrix} 1 & s & q \\ 0 & q-x-s & p-q \\ 0 & p-r & r-x-q \end{vmatrix}$$

- By taking (p + q + r − x) as a common factor from 1st column
- And then applying $R_2: R_2 - R_1$ and $R_3: R_3 - R_1$

$$= (p + q + r - x)\ [(q - x - s)\ (r - x - q) - (p - q)\ (p - r)]$$

$$= (p + q + r - x)\ [x^2 - (p^2 + q^2 + r^2 - pq - qr - pr)]$$

Since $\Delta = 0$,

now either $(p + q + r - x) = 0$

or

$[x^2 - (p^2 + q^2 + r^2 - pq - qr - pr)] = 0$

$\rightarrow x = p + q + r$ or $x = \pm\sqrt{p^2 + q^2 + r^2 - pq - qr - pr}$

But as $p + q + r = 0$

Therefore $x = 0$ or $x = \pm\sqrt{p^2 + q^2 + r^2 - pq - qr - pr}$

$$= \pm\sqrt{p^2 + q^2 + r^2 + \left(\tfrac{1}{2}\right)\left(p^2 + q^2 + r^2\right)}$$

$$= \pm\sqrt{\frac{3}{2}\left(p^2 + q^2 + r^2\right)}$$

Note

$$(p + q + r)^2 = p^2 + q^2 + r^2 + 2pq + 2qr + 2pr$$

Since $p + q + r = 0$

Therefore $p^2 + q^2 + r^2 = (2pq + 2qr + 2pr)$

$$(\tfrac{1}{2})\ (p^2 + q^2 + r^2) = -\ (pq + qr + pr)$$

Illustration No. 27

Prove that

$$\begin{vmatrix} 3+x & 3-x & 3-x \\ 3-x & 3+x & 3-x \\ 3-x & 3-x & 3+x \end{vmatrix} = 0$$

Solution:

Given

$$\Delta = \begin{vmatrix} 3+x & 3-x & 3-x \\ 3-x & 3+x & 3-x \\ 3-x & 3-x & 3+x \end{vmatrix} = 0$$

$$= \begin{vmatrix} 9-x & 3-x & 3-x \\ 9-x & 3+x & 3-x \\ 9-x & 3-x & 3+x \end{vmatrix}$$

Applying C_1: $C_1 + C_2 + C_3$

$$= (9-x) \begin{vmatrix} 1 & 3-x & 3-x \\ 1 & 3+x & 3-x \\ 1 & 3-x & 3+x \end{vmatrix}$$

By taking $(3 - x)$ common from C_1

$$= (9-x) \begin{vmatrix} 1 & 3-x & 3-x \\ 0 & 2x & 0 \\ 0 & 0 & 2x \end{vmatrix}$$

By applying R_2: $R_2 - R_1$ and R_3:$R_3 - R_1$

$$= (9-x) \times 1 \begin{vmatrix} 2x & 0 \\ 0 & 2x \end{vmatrix}$$

$$= (9-x)\,4x^2$$

Since $\Delta = 0$,

Hence

$$(9-x)\,4x^2 = 0$$

Therefore either $x = 0$ or 9

Summary

1. Consider a determinant of third order given by

$$\Delta = \begin{vmatrix} a_1 & b_1 & c_1 \\ a_2 & b_2 & c_2 \\ a_3 & b_3 & c_3 \end{vmatrix}$$

Then the minor of any element in Δ is a determinant of second order obtained by eliminating from the Δ the row and the column in which the element occurs and represented by M_{ij} where (ij) represents the ith row and j^{th} column.

2. If two adjacent rows or columns of a determinant are interchanged then the numerical value of the determinant doesn't change only sign of the determinant gets changed.

3. If all the elements of any row or column of a determinant is multiplied by the same constant say k then the value of the determinants gets multiplied by the same scalar.

4. If each element of any row or column is the sum of two numbers then the determinant can be written as the sum of the two determinants whose order is not changed.

Self Practice Exercise

1. Find the value of determinant

$$\begin{vmatrix} 3 & 4 & 7 \\ 6 & 3 & 9 \\ 2 & 5 & 8 \end{vmatrix}$$

2. Find the value of determinant

$$\begin{vmatrix} 1 & 0 & 7 \\ -3 & 5 & 4 \\ 0 & 8 & 1 \end{vmatrix}$$

3. Find the value of determinant

$$\begin{vmatrix} 7 & 1 & 6 \\ 11 & 2 & 4 \\ 15 & 3 & 2 \end{vmatrix}$$

4. Prove that without expansion

$$\begin{vmatrix} x+y+z & -z & -y \\ -z & x+y+z & -x \\ -y & -x & x+y+z \end{vmatrix} = 2\,(x+y)\,(y+z)\,(z+x)$$

5. Prove that

$$\begin{vmatrix} (x+y)^2 & x^2 & yz \\ (z+x)^2 & y^2 & zx \\ (z+y)^2 & z^2 & xy \end{vmatrix} = (x-y)(y-z)(z-x)(x+y+z)(x^2+y^2+z^2)$$

[Hint: Applying - C_1: $C_1 + C_2 + 2C_3$ and then take $(x^2 + y^2 + z^2)$ as common from C_1]

6. Prove that

$$\begin{vmatrix} 1 & a & a^2 \\ a^2 & 1 & a \\ a & a^2 & 1 \end{vmatrix} = (1 - a^3)^2$$

[*Hint*: Applying - C_1: $C_1 + C_2 + C_3$ and then take $(1 + a + a^2)$ common factor from C_1 and then apply: R_2: $R_2 - R_1$, R_3: $R_3 - R_1$]

7. Prove that

$$\begin{vmatrix} q+r & p & q \\ r+p & r & p \\ p+q & q & r \end{vmatrix} = (p+q+r)(p-r)^2$$

8. Show that

$$\begin{vmatrix} x+a & b & c \\ a & x+b & c \\ a & b & x+c \end{vmatrix} = 0$$

Hint: Apply $C_1 : C_2 + C_3$

9. Show that

$$\begin{vmatrix} 1 & 1 & 1 \\ 1 & 1+a & 1 \\ 1 & 1 & 1+b \end{vmatrix} = ab$$

10. Prove that

$$\begin{vmatrix} x^2+y^2 & xy & xz \\ yx & z^2+x^2 & yz \\ zx & zy & x^2+y^2 \end{vmatrix} = 4x^2 y^2 z^2$$

11. Prove that

$$\begin{vmatrix} q+r & p-q & p \\ r+p & q-r & q \\ p+q & r-p & r \end{vmatrix} = 3pqr - p^3 - q^3 - r^3$$

12. Show that

$$\begin{vmatrix} 0 & y^2x & z^2x \\ x^2y & 0 & z^2y \\ x^2z & y^2z & 0 \end{vmatrix} = 2x^3 y^3 z^3$$

13. Prove that

$$\begin{vmatrix} x^2 & x^2-(y-z)^2 & yz \\ y^2 & y^2-(z-x)^2 & zx \\ z^2 & z^2-(x-y)^2 & xy \end{vmatrix} = (x-y)(y-z)(z-x)(x+y+z)(x^2+y^2+z^2)$$

[*Hint*: $C_2 = C_2 - 2C_1 - 2C_3$ and then take $(x^2 + y^2 + z^2)$ as common factor]

14. Show that

$$\begin{vmatrix} -x(y^2+z^2-x^2) & 2y^3 & 2z^3 \\ 2x^3 & -y(z^2+x^2-y^2) & 2z^3 \\ 2x^3 & 2y^3 & -z(x^2+y^2-z^2) \end{vmatrix} = xyz(x^2+y^2+z^2)$$

[*Hint*: take x, y, z as common factor from C_1 C_2 and C_3. Then apply- R_1: $R_1 + R_3$, R_2: $R_2 + R_3$]

15. Show that

$$\begin{vmatrix} 1 & x & yz \\ 1 & y & zx \\ 1 & z & xy \end{vmatrix} = \begin{vmatrix} 1 & x & x^2 \\ 1 & y & y^2 \\ 1 & z & z^2 \end{vmatrix}$$

16. Prove that

$$\begin{vmatrix} 2 & 2 & 2 \\ 2 & 2+x & 2 \\ 2 & 2 & 2+y \end{vmatrix} = 2xy$$

Hint: C_2: $C_2 - C_1$ and C_3: $C_3 - C_1$

17. Prove that

$$\begin{vmatrix} x+\pi & 2x & 2x \\ 2x & x+\pi & 2x \\ 2x & 2x & x+\pi \end{vmatrix} = (5x+\pi)(x-\pi)^2$$

Hint: Apply C_1: $C_1 + C_2 + C_3$

Answers

1. – 15 2. – 195 3. 0

CHAPTER 11

Theory of Matrices-II

Definition

Singular Matrix

A square matrix say A is said to be singular matrix if its determinant $|A| = 0$

For Example

If $A = \begin{vmatrix} 6 & -12 \\ 2 & -4 \end{vmatrix}$ then $|A| = \begin{vmatrix} 6 & -12 \\ 2 & -4 \end{vmatrix} = -24 + 24 = 0$

Therefore A is singular matrix of order 2

Non–singular Matrix

A square matrix A is said to be non singular matrix if its determinant $|A| \neq 0$

For Example

If $A = \begin{vmatrix} 7 & 4 \\ 3 & 2 \end{vmatrix}$ then $|A| = \begin{vmatrix} 7 & 4 \\ 3 & 2 \end{vmatrix} = 14 - 12 = 2; = 0$

Therefore A is non singular matrix of order 2

Adjoint of a Square Matrix

Assume A is square matrix such that $A = [a_{ij}]_{n \times n}$ i.e. A is matrix of order n and $|A|$ is the determinant formed by the corresponding elements of A and if A_{ij} is the cofactor of elements a_{ij}, then the matrix obtained by transposing the cofactor matrix of elements of A is called the adjugate or adjoint matrix denoted by AdjA.

Thus if, $A = \begin{pmatrix} a_{11} & a_{12} & a_{13} \\ a_{21} & a_{22} & a_{23} \\ a_{31} & a_{32} & a_{33} \end{pmatrix}$

And $|A| = \begin{vmatrix} a_{11} & a_{12} & a_{13} \\ a_{21} & a_{22} & a_{23} \\ a_{31} & a_{32} & a_{33} \end{vmatrix}$

The Cofactor of A is given by $\begin{vmatrix} A_{11} & A_{12} & A_{13} \\ A_{21} & A_{22} & A_{23} \\ A_{31} & A_{32} & A_{33} \end{vmatrix}$

Then adjoint of A is given by

$$\text{Adj } A = (\text{Cofactor})^T = \begin{vmatrix} A_{11} & A_{12} & A_{13} \\ A_{21} & A_{22} & A_{23} \\ A_{31} & A_{32} & A_{33} \end{vmatrix}$$

Properties of Adjoint of a Matrix

(i) If A = [aij] is a square matrix of order n and I is the identity matrix **then A(Adj A) = (Adj A) A = |A| I**

Proof:

Let $A = \begin{pmatrix} a_{11} & a_{12} & a_{13} \\ a_{21} & a_{22} & a_{23} \\ a_{31} & a_{32} & a_{33} \end{pmatrix}$; and $|A| = \begin{vmatrix} a_{11} & a_{12} & a_{13} \\ a_{21} & a_{22} & a_{23} \\ a_{31} & a_{32} & a_{33} \end{vmatrix}$

and $\text{Adj } A = \begin{pmatrix} A_{11} & A_{21} & A_{31} \\ A_{12} & A_{22} & A_{32} \\ A_{13} & A_{23} & A_{33} \end{pmatrix}$

Where A_{ij} are the cofactor of a_{ij} in A, Then,

$$A(\text{Adj}A)=\begin{pmatrix} a_{11}A_{11}+a_{12}A_{12}+a_{13}A_{13} & a_{11}A_{11}+a_{12}A_{22}+a_{13}A_{23} & a_{11}A_{11}+a_{12}A_{32}+a_{13}A_{33} \\ a_{21}A_{11}+a_{22}A_{12}+a_{23}A_{13} & a_{21}A_{21}+a_{22}A_{22}+a_{23}A_{23} & a_{21}A_{11}+a_{22}A_{32}+a_{23}A_{33} \\ a_{31}A_{11}+a_{32}A_{12}+a_{33}A_{13} & a_{31}A_{21}+a_{32}A_{22}+a_{33}A_{23} & a_{31}A_{11}+a_{32}A_{32}+a_{33}A_{33} \end{pmatrix}$$

$$= \begin{pmatrix} |A| & 0 & 0 \\ 0 & |A| & 0 \\ 0 & 0 & |A| \end{pmatrix} = \Rightarrow$$

Similarly, we can prove that

$$(\text{Adj } A)\, A = |A|\, I$$

And hence $A(\text{Adj } A) = (\text{Adj } A)A = |\, A\, |\, I$

(ii) **Adj (A^T) = (Adj A)^T**

Proof:

Let

$$A = \begin{pmatrix} a_{11} & a_{12} & a_{13} \\ a_{21} & a_{22} & a_{23} \\ a_{31} & a_{32} & a_{33} \end{pmatrix} \text{ and Adj } A = \begin{pmatrix} A_{11} & A_{21} & A_{31} \\ A_{12} & A_{22} & A_{32} \\ A_{13} & A_{23} & A_{33} \end{pmatrix}$$

Then $A^T = \begin{pmatrix} a_{11} & a_{21} & a_{31} \\ a_{12} & a_{22} & a_{32} \\ a_{13} & a_{23} & a_{33} \end{pmatrix}$

And

$$\text{Adj }(A^T) = \begin{pmatrix} A11 & A12 & A13 \\ A21 & A22 & A23 \\ A31 & A32 & A33 \end{pmatrix}$$

Since

$$(\text{Adj } A) = \begin{pmatrix} A_{11} & A_{21} & A_{31} \\ A_{12} & A_{22} & A_{32} \\ A_{13} & A_{23} & A_{33} \end{pmatrix}$$

$$(\text{Adj } A)^T = \begin{pmatrix} A_{11} & A_{21} & A_{31} \\ A_{12} & A_{22} & A_{32} \\ A_{13} & A_{23} & A_{33} \end{pmatrix}$$

Hence proved

(iii) If A and B are the two square matrix then Adj (A B) = Adj A × Adj B.

The proof is to left to students to work out

(iv) |Adj A| = |A|$^{n-1}$

Illustration No. 1

$$\text{If } A = \begin{pmatrix} 2 & 1 & 0 \\ 0 & 2 & 0 \\ 3 & 1 & 4 \end{pmatrix}$$

Then find the adjoint of the matrix, and also verify that A (Adj A) = (Adj A) A = |A| = I?

Solution:

$$\text{Given the matrix} \quad A = \begin{pmatrix} 2 & 1 & 0 \\ 0 & 2 & 0 \\ 3 & 1 & 4 \end{pmatrix}$$

$$\text{Then} \quad |A| = \begin{vmatrix} 2 & 1 & 0 \\ 0 & 2 & 0 \\ 3 & 1 & 4 \end{vmatrix} = 2\begin{vmatrix} 2 & 0 \\ 1 & 4 \end{vmatrix} - 1\begin{vmatrix} 0 & 0 \\ 3 & 4 \end{vmatrix} + 0\begin{vmatrix} 0 & 2 \\ 3 & 1 \end{vmatrix}$$

$$= 2\,(8 - 0) - 1\,(0 - 0) + 0\,(0 - 6) = 16$$

$$\text{and Adj} \quad A = \begin{pmatrix} A_{11} & A_{21} & A_{31} \\ A_{12} & A_{22} & A_{32} \\ A_{13} & A_{23} & A_{33} \end{pmatrix}$$

Where

$$A_{11} = (-1)^{1+1}\begin{vmatrix} 2 & 0 \\ 1 & 4 \end{vmatrix} = (1)\,(8 - 0) = 8$$

$$A_{12} = (-1)^{1+2} = \begin{vmatrix} 0 & 0 \\ 3 & 4 \end{vmatrix} (-1)\,(0 - 0) = 0$$

$$A_{13} = (-1)^{1+3}\begin{vmatrix} 0 & 2 \\ 3 & 1 \end{vmatrix} = (1)\,(0-6) = -6$$

$$A_{21} = (-1)^{2+1}\begin{vmatrix} 1 & 0 \\ 1 & 4 \end{vmatrix} = (-1)(4-0) = -4$$

$$A_{22} = (-1)^{2+2}\begin{vmatrix} 2 & 0 \\ 3 & 4 \end{vmatrix} = (1)\,(8-0) = 8$$

$$A_{23} = (-1)^{2+3}\begin{vmatrix} 2 & 1 \\ 3 & 1 \end{vmatrix} = (-1)\,(2-3) = 1$$

$$A_{31} = (-1)^{3+1}\begin{vmatrix} 1 & 0 \\ 2 & 0 \end{vmatrix} = (1)\,(0-0) = 0$$

$$A_{31} = (-1)^{3+2}\begin{vmatrix} 2 & 0 \\ 0 & 0 \end{vmatrix} = (-1)\,(0-0) = 0$$

$$A_{31} = (-1)^{3+3}\begin{vmatrix} 2 & 1 \\ 0 & 2 \end{vmatrix} = (1)\,(4-0) = 4$$

Therefore,

$$\text{Adj } A = \begin{pmatrix} 8 & -4 & 0 \\ 0 & 8 & 0 \\ -6 & 1 & 4 \end{pmatrix}$$

As for the 2nd part i.e. A (Adj A)

$$= \begin{pmatrix} 2 & 1 & 0 \\ 0 & 2 & 0 \\ 3 & 1 & 4 \end{pmatrix}\begin{pmatrix} 8 & -4 & 0 \\ 0 & 8 & 0 \\ -6 & 1 & 4 \end{pmatrix}$$

$$= \begin{pmatrix} 16+0+0 & -8+8+0 & 0+0+0 \\ 0+0+0 & 0+16+0 & 0+0+0 \\ 24+0-24 & -12+8+4 & 0+0+16 \end{pmatrix}$$

$$= \begin{pmatrix} 16 & 0 & 0 \\ 0 & 16 & 0 \\ 0 & 0 & 16 \end{pmatrix}$$

$$= 16 \begin{pmatrix} 1 & 0 & 0 \\ 0 & 1 & 0 \\ 0 & 0 & 1 \end{pmatrix}$$

$$= |A|\ I$$

Inverse of a Matrix

If A and B and two square matrix of the same under such that AB = BA = I, where I is the identity matrix, then either A is the inverse of B or B is the inverse of A. Then in first case we write $A = B^{-1}$ and in the 2nd case it is $B = A^{-1}$.

So, $AA^{-1} = A^{-1}A = I$ and $B\,B^{-1} = B^{-1}B = I$

The Expression of inverse of matrix since we know that

$$A\,(\text{Adj } A) = (\text{Adj } A)\,A = |A|\ I$$

Or

$$A\,\frac{\text{Adj } A}{|A|} = \frac{\text{Adj } A}{|A|}\,A = I \qquad \text{provided } |A| \neq 0$$

$$\frac{\text{Adj } A}{|A|} = A^{-1}\ I \qquad \text{where } |A| \neq 0$$

Since

$$A^{-1} = \frac{\text{Adj } A}{|A|} = \frac{1}{|A|}\begin{pmatrix} A_{11} & A_{21} & A_{31} \\ A_{12} & A_{22} & A_{32} \\ A_{13} & A_{23} & A_{33} \end{pmatrix}$$

Properties of Inverse of a Matrix

Property (i)

If the inverse of a square matrix exits then it is unique i.e. there cannot be two different inverse for the same matrix.

Proof:

Let A be square matrix with the inverse as A^{-1} then

$$A A^{-1} = A^{-1}A = I \qquad .....(1)$$

Now, we assume that B be another inverse of the square matrix A. Then,

$$AB = BA = I \qquad .....(2)$$

From (1) and (2), we have,

$$A A^{-1} = A B$$

Therefore $A^{-1} (AA^{-1}) = A^{-1} (AB)$

Or $\quad (A^{-1} A) A^{-1} = (A^{-1} A) B \quad$ *(By Associative Property)*

Or $\quad I A^{-1} = I B \Rightarrow A^{-1} = B$

Thus it shows that the square matrix A has only one inverse.

Property (ii)

If A^{-1} exists then

(a) $(A^{-1})^{-1} = A$

(b) $(A^T) = (A^{-1})^T$

Proof:

(a)Let us assume A^{-1} is the inverse of matrix A, then we know that $(AA^{-1}) = (AA^{-1}) = I$

As we know that A^{-1} unique to A

Therefore as $(AA^{-1}) = I$

$$A = (A^{-1})^{-1} I$$

$$A = (A^{-1})^{-1}$$

Hence proved

(b) As we know $(AA^{-1})^T = (A^{-1})^T A^T \quad$ [Since $(A B)^T = B^T A^T$].

Or $\quad I^T = (A^{-1})^T A^T$

Or $\quad I = (A^{-1})^T A^T \quad$ [Since $AA^{-1} = I$] $\qquad .....(1)$

Similarly,

$$A^T (A^{-1}) = (A^{-1} A)^T = (I^T) = I \qquad .....(2)$$

From (1) and (2) we have both result (equal to I)

Therefore,

$$(A^T)(A^{-1})^T = I$$

Or

$$(A^{-1})^T = (A^T)^{-1} I$$

$$= (A^T)^{-1}$$

Hence proved

Property (iii)

If A and B are two non singular matrix of the same order then

$(AB)^{-1} = B^{1} A^{-1}$, where A^{-1}, B^{-1} are the respective inverse of A and B

Proof:

As A^{-1} and B^{-1} are respectively inverse A and B, we have

$AA^{-1} = A^{-1}A = I$ and $BB^{-1} = B^{-1} B = I$

Now $(AB)(B^{-1} A^{-1}) = A(BB^{-1}) A^{-1} = A(I) A^{-1} = (AI) A^{-1} = AA^{-1} = I \qquad .....(1)$

And $(B^{-1} A^{-1})(AB) = B^{-1}(A^{-1}A) B = B^{-1}(I) B = (BI) B^{-1} = BB^{-1} = I \qquad .....(2)$

From (1) and (2), we get

$$(AB)(B^{-1} A^{-1}) = (B^{-1} A^{-1}) = (B^{-1} A^{-1})(AB) = I$$

Or $\qquad (AB)^{-1} = B^{-1} A^{-1}$

Illustration No. 2

Find the inverse of the matrix

$$\begin{pmatrix} 2 & -1 & 0 \\ 3 & 1 & -2 \\ 0 & 1 & 2 \end{pmatrix}$$

Solution:

Let $$A = \begin{pmatrix} 2 & -1 & 0 \\ 3 & 1 & -2 \\ 0 & 1 & 2 \end{pmatrix}$$

Then $$|A| = \begin{vmatrix} 2 & -1 & 0 \\ 3 & 1 & -2 \\ 0 & 1 & 2 \end{vmatrix} = 2\begin{vmatrix} 1 & -2 \\ 1 & 2 \end{vmatrix} - 1\begin{vmatrix} 3 & -2 \\ 0 & 2 \end{vmatrix} + 0\begin{vmatrix} 3 & 1 \\ 0 & 1 \end{vmatrix}$$

$$= 2(2 + 2) - 1(6 - 0) + 0\,(3 + 0) = 2$$

and $$\text{Adj}\,A = \begin{pmatrix} A_{11} & A_{21} & A_{31} \\ A_{12} & A_{22} & A_{32} \\ A_{13} & A_{23} & A_{33} \end{pmatrix}$$

where

$$A_{11} = (-1)^{1+1}\begin{vmatrix} 1 & -2 \\ 1 & 2 \end{vmatrix} = (1)\,(2 + 2) = 4$$

$$A_{12} = (-1)^{1+2}\begin{vmatrix} 3 & -2 \\ 0 & 2 \end{vmatrix} = (-1)\,(6 - 0) = -6$$

$$A_{13} = (-1)^{1+3}\begin{vmatrix} 3 & 1 \\ 0 & 1 \end{vmatrix} = (1)\,(3 - 0) = 3$$

$$A_{21} = (-1)^{2+1}\begin{vmatrix} -1 & 0 \\ 1 & 2 \end{vmatrix} = (-1)(-2 - 0) = 2$$

$$A_{22} = (-1)^{2+2}\begin{vmatrix} 2 & 0 \\ 0 & 2 \end{vmatrix} = (1)\,(4 - 0) = 4$$

$$A_{23} = (-1)^{2+3}\begin{vmatrix} 2 & -1 \\ 0 & 1 \end{vmatrix} = (-1)\ (2-0) = -2$$

$$A_{31} = (-1)^{3+1}\begin{vmatrix} -1 & 0 \\ 1 & -2 \end{vmatrix} = (1)\ (2-0) = 2$$

$$A_{31} = (-1)^{3+2}\begin{vmatrix} 2 & 0 \\ 3 & -2 \end{vmatrix} = (-1)\ (-4-0) = 4$$

$$A^{31} = (-1)^{3+3}\begin{vmatrix} 2 & -1 \\ 3 & 1 \end{vmatrix} = (1)\ (2+3) = 5$$

Therefore,

$$\text{Adj A} = \begin{pmatrix} 4 & 2 & 2 \\ -6 & 4 & 4 \\ 3 & -2 & 5 \end{pmatrix}$$

And hence,

$$A^{-1} = \frac{\text{Adj A}}{|A|} = \frac{1}{2}\begin{pmatrix} 4 & 2 & 2 \\ -6 & 4 & 4 \\ 3 & -2 & 5 \end{pmatrix}$$

$$= \begin{pmatrix} 2 & 1 & 1 \\ -3 & 2 & 2 \\ \frac{3}{2} & -1 & \frac{5}{2} \end{pmatrix}$$

Illustration No. 3

Find the inverse of the matrix given by

$$\begin{pmatrix} 1 & 2 & 1 \\ 1 & 3 & -1 \\ 2 & 3 & 5 \end{pmatrix}$$

Solution:

Let $$A = \begin{pmatrix} 1 & 2 & 1 \\ 1 & 3 & -1 \\ 2 & 3 & 5 \end{pmatrix}$$

Then $$|A| = 1\begin{vmatrix} 3 & -1 \\ 3 & 5 \end{vmatrix} - 2\begin{vmatrix} 1 & -1 \\ 2 & 5 \end{vmatrix} + 1\begin{vmatrix} 1 & 3 \\ 2 & 3 \end{vmatrix}$$

$$= 1\ (15 + 3) - 2\ (5 + 2) + 1\ (3 - 6)$$

$$= 18 - 14 - 3$$

$$= 18 - 17 = 1 \neq 0$$

Therefore A^{-1} exists.

Now finding the cofactor Matrix of A i.e.

$$A_{11} = (-1)^{1+1}\begin{vmatrix} 3 & -1 \\ 3 & 5 \end{vmatrix} = 18$$

$$A_{12} = (-1)^{1+2}\begin{vmatrix} 1 & -1 \\ 2 & 5 \end{vmatrix} = -7$$

$$A_{13} = (-1)^{1+3}\begin{vmatrix} 1 & 3 \\ 2 & 3 \end{vmatrix} = -3$$

$$A_{21} = (-1)^{2+1}\begin{vmatrix} 2 & 1 \\ 3 & 5 \end{vmatrix} = -7$$

$$A_{22} = (-1)^{2+2}\begin{vmatrix} 1 & 1 \\ 2 & 5 \end{vmatrix} = 3$$

$$A_{23} = (-1)^{2+3} \begin{vmatrix} 1 & 2 \\ 2 & 3 \end{vmatrix} = 1$$

$$A_{31} = (-1)^{3+1} \begin{vmatrix} 2 & 1 \\ 3 & -1 \end{vmatrix} = -5$$

$$A_{32} = (-1)^{3+2} \begin{vmatrix} 1 & 1 \\ 1 & -1 \end{vmatrix} = 2$$

$$A_{33} = (-1)^{3+3} \begin{vmatrix} 1 & 2 \\ 1 & 3 \end{vmatrix} = 1$$

Therefore the cofactor matrix of A is given by

$$\text{Cofactor} \quad A = \begin{pmatrix} 18 & -7 & -3 \\ -7 & 3 & 1 \\ -5 & 2 & 1 \end{pmatrix}$$

But,

$$\text{Adj } A = (\text{Cofactor } A)^T$$

And hence,

$$\text{Adj } A = \begin{pmatrix} 18 & -7 & -5 \\ -7 & 3 & 2 \\ -3 & 1 & 1 \end{pmatrix}$$

Since,

$$A^{-1} = \frac{\text{Adj } A}{|A|} = \frac{1}{1} \begin{pmatrix} 18 & -7 & -5 \\ -7 & 3 & 2 \\ -3 & 1 & 1 \end{pmatrix}$$

$$= \begin{pmatrix} 18 & -7 & -5 \\ -7 & 3 & 2 \\ -3 & 1 & 1 \end{pmatrix}$$

Orthogonal Matrix

A square matrix A of any order is said to be orthogonal if $AA^T = A^TA = I$ where A^T is the transpose of the matrix A and I is the identity matrix of the same order.

Properties of Orthogonal Matrix

If A is an orthogonal matrix, then A^T is the inverse of the matrix A i.e. $A^{-1} = A^T$

{Since $AA^T = A^TA = I$}

Note

For a matrix A to be orthogonal we have, $AA^T = I$

Therefore,

$$|A|^2 = 1 \qquad \{\text{Since } |A^T| = |A|\}$$

Therefore,

$$|A| = \pm 1$$

If $|A| = 1$ then A is called as proper orthogonal matrix and if $|A| = -1$ then A is called as improper orthogonal matrix.

Solving Simultaneous Linear Equations Using Matrix Method

The solution of simultaneous linear equations can be obtained by using the following methods

(i) Cramer's Rule

(ii) Matrix Inversion Method

(iii) Gauss–Elimination Method

We will elaborate on each of the following techniques

Cramer's Rule

Let's understand solving linear simultaneous equations using Cramer's Rule

Assume the given linear equation in the following form

$$a_1x + b_1 y = c_1 \qquad \text{.....(i)}$$

$$a_2x + b_2y = c_2 \qquad \text{.....(ii)}$$

We solve the above two equations by elimination method i.e. we eliminate y from the both the above equation by multiplying (i) by b_2 and (ii) by b_1.

We get

$$a_1 b_2 x + b_1 b_2 y = c_1 b_2$$

$$a_2 b_1 x + b_1 b_2 y = c_2 b_2$$

Subtracting the above equations we get

$$(a_1 b_2 - a_2 b_1)\ x = c_1 b_2 - c_2 b_2 \qquad \text{.....(iii)}$$

Similarly we eliminate x by multiplying (i) by a_2 and (ii) by a_1

Therefore we get

$$a_1 a_2 x + b_1 a_2 y = c_1 a_2$$

$$a_1 a_2 x + a_1 b_2 y = c_2 a_1$$

Subtracting the above equations we get

$$(b_1 a_2 - a_1 b_2) y = c_1 a_2 - c_2 a_1 \qquad \text{....(iv)}$$

The values of x and y are derived by

$$\frac{x}{c_1 b_2 - c_2 b_2} = \frac{y}{a_1 c_2 - a_2 c_1} = \frac{1}{a_1 b_2 - a_2 b_1}$$

the above solution can be expressed in determinant form

$$\frac{x}{\begin{vmatrix} c_1 & b_1 \\ c_2 & b_2 \end{vmatrix}} = \frac{y}{\begin{vmatrix} a_1 & c_1 \\ a_2 & c_2 \end{vmatrix}} = \frac{1}{\begin{vmatrix} a_1 & b_1 \\ a_2 & b_2 \end{vmatrix}}$$

$$x = \frac{\begin{vmatrix} c_1 & b_1 \\ c_2 & b_2 \end{vmatrix}}{\begin{vmatrix} a_1 & b_1 \\ a_2 & b_2 \end{vmatrix}} \text{ and } y = \frac{\begin{vmatrix} a_1 & c_1 \\ a_2 & c_2 \end{vmatrix}}{\begin{vmatrix} a_1 & b_1 \\ a_2 & b_2 \end{vmatrix}}$$

Let

$$D_x = \begin{vmatrix} c_1 & b_1 \\ c_2 & b_2 \end{vmatrix},\ D_y = \begin{vmatrix} a_1 & c_1 \\ a_2 & c_2 \end{vmatrix} \text{ and } D = \begin{vmatrix} a_1 & b_1 \\ a_2 & b_2 \end{vmatrix}$$

Then

$$x = \frac{Dx}{D} \text{ and } y = \frac{Dx}{D}$$

Similarly the same can be extended to solve 3 linear equation of the form

$$a_1x + b_1 y + c_1z = d_1$$
$$a_2x + b_2y + c_2 z = d_2$$
$$a_3x + b_3y + c_3z = d_3$$

Illustrations No.4

Solve the linear equation by Cramer's Rule

$$3x + 4y = 12$$
$$2x + 5y = 10$$

Solution:

Let us express the above equation in Matrix form

$$A = \begin{pmatrix} 3 & 4 \\ 2 & 5 \end{pmatrix}, X = \begin{pmatrix} x \\ y \end{pmatrix} \text{ and } B = \begin{pmatrix} 12 \\ 10 \end{pmatrix}$$

And calculating the determinant of the matrix A,

We get,

$$\text{Det } A = \begin{pmatrix} 3 & 4 \\ 2 & 5 \end{pmatrix} = (15 - 8) = 7 \neq 0$$

The system has a unique solution

Therefore the value of x and y is given by

$$x = \begin{vmatrix} 12 & 4 \\ 10 & 5 \end{vmatrix} / 7 = \frac{20}{7}$$

$$y = \begin{vmatrix} 3 & 12 \\ 2 & 10 \end{vmatrix} / 7 = \frac{6}{7}$$

Hence the solutions are $x = \frac{20}{7}$ and $y = \frac{6}{7}$

Illustration No.5

Solve the following linear equations by Cramer's Rule

$$x + 2y + 3y = 5$$
$$4x - y + z = 4$$
$$2x + 2y + z = 2 \ ?$$

Solution:

The above equations can be presented in Matrix form as AX = B where

$$A = \begin{pmatrix} 1 & 2 & 3 \\ 4 & -1 & 1 \\ 2 & 2 & 1 \end{pmatrix} \quad x = \begin{pmatrix} x \\ y \\ z \end{pmatrix} \text{ and } B = \begin{pmatrix} 5 \\ 4 \\ 2 \end{pmatrix}$$

Therefore the determinant of the matrix A is given by |D|

$$|D| = 1 \begin{vmatrix} -1 & 1 \\ 2 & 1 \end{vmatrix} - 2 \begin{vmatrix} 4 & 1 \\ 2 & 1 \end{vmatrix} + 3 \begin{vmatrix} 4 & -1 \\ 2 & 2 \end{vmatrix}$$

$$= 1\,(-1-2) - 2\,(4-2) + 3\,(8+2)$$

$$= -3 - 4 + 30 = 23, \text{ hence } |D| \neq 0, \text{ now}$$

$$x = \frac{Dx}{|D|} = \frac{\begin{vmatrix} 5 & 2 & 3 \\ 4 & -1 & 1 \\ 2 & 2 & 1 \end{vmatrix}}{23}$$

D_x is obtained by replacing the x – column by the constant matrix B

$$= \frac{5 \begin{vmatrix} -1 & 1 \\ 2 & 1 \end{vmatrix} -2 \begin{vmatrix} 4 & 1 \\ 2 & 1 \end{vmatrix} +3 \begin{vmatrix} 4 & -1 \\ 2 & 2 \end{vmatrix}}{23}$$

$$= \frac{5(-3) - 2(2) + 3(10)}{23} = \frac{11}{23}$$

Similarly,

$$y = \frac{Dy}{|D|} = \frac{\begin{vmatrix} 1 & 5 & 3 \\ 4 & 4 & 1 \\ 2 & 2 & 1 \end{vmatrix}}{23}$$

D_y is obtained by replacing the y – column by the constant matrix B

$$= \frac{1\begin{vmatrix} 4 & 1 \\ 2 & 1 \end{vmatrix} - 5\begin{vmatrix} 4 & 1 \\ 2 & 1 \end{vmatrix} + 3\begin{vmatrix} 4 & 4 \\ 2 & 2 \end{vmatrix}}{23}$$

$$= \frac{1(2) - 5(2) + 3(0)}{23}$$

$$= -\frac{8}{23}$$

Similarly,

$$z = \frac{Dz}{|D|} = \frac{\begin{vmatrix} 1 & 2 & 5 \\ 4 & -1 & 4 \\ 2 & 2 & 2 \end{vmatrix}}{23}$$

D_z is obtained by replacing the z – column by the constant matrix B

$$= \frac{1\begin{vmatrix} -1 & 4 \\ 2 & 2 \end{vmatrix} - 2\begin{vmatrix} 4 & 4 \\ 2 & 2 \end{vmatrix} + 5\begin{vmatrix} 4 & -1 \\ 2 & 2 \end{vmatrix}}{23}$$

$$= \frac{1(-10) - 2(0) + 5(10)}{23}$$

$$= \frac{40}{23}$$

Therefore the solution of the system of equations are

$$x = \frac{11}{23},\ y = -\frac{8}{23} \text{ and } z = \frac{40}{23}$$

Illustration No.6

Solve the following linear equation by Cramer's Rule?

$x + y + z = 10$

$x - y = 8$

$x + 2y - z = 4$

Solution:

The above equations can be presented in Matrix form as AX = B where

$$A = \begin{pmatrix} 1 & 1 & 1 \\ 1 & 1 & 0 \\ 1 & 2 & -1 \end{pmatrix} \quad x = \begin{pmatrix} x \\ y \\ z \end{pmatrix} \text{ and } B = \begin{pmatrix} 10 \\ 0 \\ 4 \end{pmatrix}$$

Therefore the determinant of the matrix A is given by |D|

$$|D| = 1\begin{vmatrix} -1 & 0 \\ 2 & -1 \end{vmatrix} - 1\begin{vmatrix} 1 & 0 \\ 1 & -1 \end{vmatrix} + 1\begin{vmatrix} 1 & -1 \\ 1 & 2 \end{vmatrix}$$

$$= 1\,(1 - 0) - 1\,(-1 - 0) + 1(2 + 1)$$

$$= 1 + 1 + 3 = 5$$

And hence |D| ≠ 0, now

$$x = \frac{Dx}{|D|} = \frac{\begin{vmatrix} 10 & 1 & 1 \\ 8 & -1 & 0 \\ 4 & 2 & -1 \end{vmatrix}}{5}$$

D_x is obtained by replacing the x – column by the constant matrix B

$$= \frac{10\begin{vmatrix} -1 & 0 \\ 2 & -1 \end{vmatrix} - 1\begin{vmatrix} 8 & 0 \\ 4 & -1 \end{vmatrix} + 1\begin{vmatrix} 8 & -1 \\ 4 & 2 \end{vmatrix}}{5}$$

$$= \frac{10(1) - 1(-8) + 1(20)}{5} = \frac{38}{5}$$

Similarly,

$$y = \frac{Dy}{|D|} = \frac{\begin{vmatrix} 1 & 10 & 1 \\ 1 & 8 & 0 \\ 1 & 4 & -1 \end{vmatrix}}{5}$$

D_y is obtained by replacing the y – column by the constant matrix B

$$= \frac{1\begin{vmatrix} 8 & 0 \\ 4 & -1 \end{vmatrix} - 10\begin{vmatrix} 1 & 0 \\ 1 & -1 \end{vmatrix} + 1\begin{vmatrix} 1 & 8 \\ 1 & 4 \end{vmatrix}}{5}$$

$$= \frac{1(-8) - 10(-1) + 1(-4)}{5}$$

$$= -\frac{2}{5}$$

Similarly,

$$z = \frac{Dz}{|D|} = \frac{\begin{vmatrix} 1 & 1 & 10 \\ 1 & -1 & 8 \\ 1 & 2 & 4 \end{vmatrix}}{5}$$

D_z is obtained by replacing the z – column by the constant matrix B

$$= \frac{1\begin{vmatrix} -1 & 8 \\ 2 & 4 \end{vmatrix} - 1\begin{vmatrix} 1 & 8 \\ 1 & 4 \end{vmatrix} + 10\begin{vmatrix} 1 & -1 \\ 1 & 2 \end{vmatrix}}{5}$$

$$= \frac{1(-20) - 1(-4) + 10(3)}{5}$$

$$= \frac{14}{5}$$

Therefore the solution of the system of equations are $x = \frac{38}{5}$, $y = -\frac{2}{5}$ and $z = \frac{14}{5}$

Illustration No.7

Solve the following linear equation by Cramer's Rule

$x + 2y + 2z = 6$

$3x - y + z = 4$

$2x + 4y - z = 10$

Solution:

The above equations can be presented in Matrix form as AX = B where

$$A = \begin{pmatrix} 1 & 2 & 2 \\ 3 & -1 & 1 \\ 2 & 4 & -1 \end{pmatrix} \quad X = \begin{pmatrix} x \\ y \\ z \end{pmatrix} \text{ and } \begin{pmatrix} 6 \\ 4 \\ 10 \end{pmatrix}$$

Therefore the determinant of the matrix A is given by |D|

$$|D| = 1\begin{vmatrix} -1 & 1 \\ 4 & -1 \end{vmatrix} - 2\begin{vmatrix} 3 & 1 \\ 2 & -1 \end{vmatrix} + 2\begin{vmatrix} 3 & -1 \\ 2 & 4 \end{vmatrix}$$

$$= 1\,(1 - 4) - 2\,(-3 - 2) + 2\,(12 + 2)$$

$$= 35$$

And hence $|D| \neq 0$, now

$$x = \frac{Dx}{|D|} = \frac{\begin{vmatrix} 6 & 2 & 2 \\ 4 & -1 & 1 \\ 10 & 4 & -1 \end{vmatrix}}{35}$$

D_x is obtained by replacing the x – column by the constant matrix B

$$= \frac{6\begin{vmatrix} -1 & 1 \\ 4 & -1 \end{vmatrix} - 2\begin{vmatrix} 4 & 1 \\ 10 & -1 \end{vmatrix} + 2\begin{vmatrix} 4 & -1 \\ 10 & 4 \end{vmatrix}}{35}$$

$$= \frac{6(1-4) - 2(-14) + 2(26)}{35} = \frac{62}{35}$$

Similarly,

$$y = \frac{Dy}{|D|} = \frac{\begin{vmatrix} 1 & 6 & 2 \\ 3 & 4 & 1 \\ 2 & 10 & -1 \end{vmatrix}}{35}$$

D_y is obtained by replacing the y – column by the constant matrix B

$$= \frac{1\begin{vmatrix} 4 & 1 \\ 10 & -1 \end{vmatrix} - 6\begin{vmatrix} 3 & 1 \\ 2 & -1 \end{vmatrix} + 2\begin{vmatrix} 3 & 4 \\ 2 & 10 \end{vmatrix}}{35}$$

$$= \frac{1(-14)-6(-5)+2(22)}{35}$$

$$= \frac{60}{35} \text{ or } \frac{12}{7}$$

Similarly,

$$z = \frac{Dz}{|D|} = \frac{\begin{vmatrix} 1 & 2 & 6 \\ 3 & -1 & 4 \\ 2 & 4 & 10 \end{vmatrix}}{35}$$

D_z is obtained by replacing the z – column by the constant matrix B

$$= \frac{1\begin{vmatrix} -1 & 4 \\ 4 & 10 \end{vmatrix} - 2\begin{vmatrix} 3 & 4 \\ 2 & 10 \end{vmatrix} + 6\begin{vmatrix} 3 & -1 \\ 2 & 4 \end{vmatrix}}{35}$$

$$= \frac{1(-26)-2(22)+6(14)}{35}$$

$$= \frac{14}{35}$$

Therefore the solution of the system of equations are $x = \frac{62}{35}$, $y = \frac{12}{7}$ and

$z = \frac{14}{35}$

Illustration No.8

Solve the following linear equation by Cramer's Rule

$$4x + 5y + z = 4$$

$$6x + 9y + 3z = 18$$

$$x + y + z = 10$$

Solution:

The above equations can be presented in Matrix form as AX = B where

$$A = \begin{pmatrix} 4 & 5 & 1 \\ 6 & 9 & 3 \\ 1 & 1 & 1 \end{pmatrix} \quad X = \begin{pmatrix} x \\ y \\ z \end{pmatrix} \text{ and } \begin{pmatrix} 8 \\ 18 \\ 10 \end{pmatrix}$$

Therefore the determinant of the matrix A is given by $|D|$

$$|D| = 4\begin{vmatrix} 9 & 3 \\ 1 & 1 \end{vmatrix} - 5\begin{vmatrix} 6 & 3 \\ 1 & 1 \end{vmatrix} + 1\begin{vmatrix} 6 & 9 \\ 1 & 1 \end{vmatrix}$$

$$= 4(9 - 3) - 5(6 - 3) + 1(6 - 9)$$

$$= 6$$

And hence $|D| \neq 0$, now

$$x = \frac{Dx}{|D|} = \frac{\begin{vmatrix} 8 & 5 & 1 \\ 18 & 9 & 3 \\ 10 & 1 & 1 \end{vmatrix}}{6}$$

D_x is obtained by replacing the x – column by the constant matrix B

$$= \frac{8\begin{vmatrix} 9 & 3 \\ 1 & 1 \end{vmatrix} - 5\begin{vmatrix} 18 & 3 \\ 10 & 1 \end{vmatrix} + 1\begin{vmatrix} 18 & 9 \\ 10 & 1 \end{vmatrix}}{6}$$

$$= \frac{8(9-3) - 5(18-30) + 1(18-90)}{6} = \frac{36}{6} = 6$$

Similarly,

$$y = \frac{Dy}{|D|} = \frac{\begin{vmatrix} 4 & 8 & 1 \\ 6 & 18 & 3 \\ 1 & 10 & 1 \end{vmatrix}}{6}$$

D_y is obtained by replacing the y – column by the constant matrix B

$$= \frac{4\begin{vmatrix}18 & 3\\ 10 & 1\end{vmatrix} - 8\begin{vmatrix}6 & 3\\ 1 & 1\end{vmatrix} + 1\begin{vmatrix}6 & 18\\ 1 & 10\end{vmatrix}}{6}$$

$$= \frac{1(18-30)-8(6-3)+1(60-18)}{35} = -\frac{30}{6} \text{ or } -5$$

Similarly,

$$z = \frac{Dz}{|D|} = \frac{\begin{vmatrix}4 & 5 & 8\\ 6 & 9 & 18\\ 1 & 1 & 10\end{vmatrix}}{6}$$

D_z is obtained by replacing the z – column by the constant matrix B

$$= \frac{1\begin{vmatrix}9 & 18\\ 1 & 10\end{vmatrix} - 5\begin{vmatrix}6 & 18\\ 1 & 10\end{vmatrix} + 8\begin{vmatrix}6 & 9\\ 1 & 1\end{vmatrix}}{6}$$

$$= \frac{4(90-18)-5(60-18)+8(6-9)}{6}$$

$$= \frac{54}{6} = 9$$

Therefore the solution of the system of equations are x = 6, y = –5 and z = 9.

Matrix Inversion Method

Assume the given linear equation in the following form

$$a_1x + b_1 y = c_1 \qquad \text{.....(i)}$$

$$a_2x + b_2y = c_2 \qquad \text{.....(ii)}$$

such that,

$$A = \begin{pmatrix} a_1 & b_1 \\ a_1 & b_2 \end{pmatrix}, X = \begin{pmatrix} x \\ y \end{pmatrix} \text{ and } B = \begin{pmatrix} c_1 \\ c_2 \end{pmatrix}$$

Then i and ii can be written in matrix form as

AX = B

Assuming that Det A ≠ 0, then A^{-1} exists, then we have

$$A^{-1}(AX) = A^{-1}B$$

$$(A^{-1} A) X = A^{-1}B$$

$$I X = A^{-1}B \text{ or, } X = A^{-1}B$$

Illustration No.9

Solve the following equations using matrix inversion method

$$4x - 6y = 2$$

$$2x + 10y = 14$$

Solution:

Let,

$$A = \begin{pmatrix} 4 & -6 \\ 2 & 10 \end{pmatrix}, X = \begin{pmatrix} x \\ y \end{pmatrix} \text{ and } B = \begin{pmatrix} 2 \\ 14 \end{pmatrix}$$

Then

$$\text{Det } A = |A| = \begin{pmatrix} 4 & -6 \\ 2 & 10 \end{pmatrix} = 40 + 12 = 52 \neq 0$$

Therefore, A^{-1} exists

Rearranging the above set of equation in matrix form i.e.

$$AX = B \text{ or } X = A^{-1} B$$

But $$A^{-1} = \frac{1}{|A|} \begin{pmatrix} A_{11} & A_{21} \\ A_{12} & A_{22} \end{pmatrix}$$

$$= \frac{1}{52} \begin{pmatrix} 10 & 6 \\ -2 & 4 \end{pmatrix}$$

Since

A_{11} = Cofactor of 4 = $(-1)^{1+1}$ 10 = 10

A_{12} = Cofactor of – 6 = $(-1)^{1+2}$ 2 = –2

A_{21} = Cofactor of 2 = $(-1)^{2+1}$ (– 6) = 6

A_{22} = Cofactor of 10 = $(-1)^{2+2}$ 4 = 4

Since $$X = A^{-1} B = \frac{1}{52}\begin{pmatrix} 10 & 6 \\ -2 & 4 \end{pmatrix}\begin{pmatrix} 2 \\ 14 \end{pmatrix}$$

$$= \frac{1}{52}\begin{pmatrix} 20 + 84 \\ -4 + 56 \end{pmatrix}$$

$$= \frac{1}{52}\begin{pmatrix} 104 \\ 52 \end{pmatrix} = \begin{pmatrix} 2 \\ 1 \end{pmatrix}$$

And hence x = 2 and y = 1

Illustration No.10

Solve the following equations using matrix inversion method

$2x + 3y = 1$

$x + 4y = 6$

Solution:

Let

$$A = \begin{pmatrix} 2 & 3 \\ 1 & 4 \end{pmatrix}, X = \begin{pmatrix} x \\ y \end{pmatrix} \text{ and } B = \begin{pmatrix} 1 \\ 6 \end{pmatrix}$$

But $$|A| = \begin{vmatrix} 2 & 3 \\ 1 & 4 \end{vmatrix} = 8 - 3 = 5 \neq 0$$

Since determinant of the matrix i.e. $|A| \neq 0$, therefore A^{-1} exists

The above linear equations can be expressed as in matrix form as AX = B

Or $$X = A^{-1}B$$

But $$A^{-1} = \frac{\text{Adj } A}{|A|} = \frac{1}{|A|}\begin{pmatrix} A_{11} & A_{21} \\ A_{12} & A_{22} \end{pmatrix}$$

Where A_{11}, A_{21}, A_{12} and A_{22} are respective cofactor's of elements a_{11}, a_{21}, a_{12} and a_{22}.

$$= \frac{1}{5}\begin{pmatrix} 4 & -3 \\ -1 & 2 \end{pmatrix}$$

Since A_{11} = Cofactor of a_{11} = 4

A_{12} = Cofactor of a_{12} = – 1

A_{21} = Cofactor of a_{21} = – 3

A_{22} = Cofactor of a_{22} = 2

$$X = A^{-1}B = \frac{1}{5}\begin{pmatrix} 4 & -3 \\ -1 & 2 \end{pmatrix}\begin{pmatrix} 1 \\ 6 \end{pmatrix}$$

$$= \frac{1}{5}\begin{pmatrix} 4\times 1 + (-3)\times 6 \\ (-1)\times 1 + 2\times 6 \end{pmatrix}$$

$$= \frac{1}{5}\begin{pmatrix} 4 + (-18) \\ (-1) + 12 \end{pmatrix}$$

$$= \frac{1}{5}\begin{pmatrix} -14 \\ 11 \end{pmatrix} = \begin{pmatrix} -\frac{14}{5} \\ \frac{11}{5} \end{pmatrix}$$

Hence $$x = -\frac{14}{5} \text{ and } y = \frac{11}{5}$$

Illustration No.11

Using matrix inversion method solve the following linear equations

$$x - y = 3$$

$$2x + 3y + 4z = 17$$

$$y + 2z = 7$$

Solution:

Let the above equations are represented as

$$A = \begin{pmatrix} 1 & -1 & 0 \\ 2 & 3 & 4 \\ 0 & 1 & 2 \end{pmatrix}, X = \begin{pmatrix} x \\ y \\ z \end{pmatrix} \text{ and } B = \begin{pmatrix} 3 \\ 17 \\ 7 \end{pmatrix}$$

Now, the given equation can be expressed as

$$AX = B \text{ or } X = A^{-1} B$$

But $$A^{-1} = \frac{\text{Adj } A}{|A|}$$

Therefore, $$|A| = \begin{vmatrix} 1 & -1 & 0 \\ 2 & 3 & 4 \\ 0 & 1 & 2 \end{vmatrix} = 1\begin{vmatrix} 3 & 4 \\ 1 & 2 \end{vmatrix} + 1\begin{vmatrix} 2 & 4 \\ 0 & 2 \end{vmatrix} + 0\begin{vmatrix} 2 & 3 \\ 0 & 1 \end{vmatrix}$$

$$= (6 - 4) + (4 - 0) + 0$$

$$= 2 + 4 = 6$$

But, (Adj A) = (Cofactor A)^T

Hence,

$$A_{11} = (-1)^{1+1} \begin{vmatrix} 3 & 4 \\ 1 & 2 \end{vmatrix} = 2$$

$$A_{12} = (-1)^{1+2} \begin{vmatrix} 2 & 4 \\ 0 & 2 \end{vmatrix} = -4$$

$$A_{13} = (-1)^{1+3} \begin{vmatrix} 2 & 3 \\ 0 & 1 \end{vmatrix} = 2$$

$$A_{21} = (-1)^{2+1} \begin{vmatrix} -1 & 0 \\ 1 & 2 \end{vmatrix} = 2$$

$$A_{22} = (-1)^{2+2} \begin{vmatrix} 1 & 0 \\ 0 & 2 \end{vmatrix} = 2$$

$$A_{23} = (-1)^{2+3} \begin{vmatrix} 1 & -1 \\ 0 & 1 \end{vmatrix} = -1$$

$$A_{31} = (-1)^{3+1} \begin{vmatrix} -1 & 0 \\ 3 & 4 \end{vmatrix} = -4$$

$$A_{32} = (-1)^{3+2} \begin{vmatrix} 1 & 0 \\ 2 & 4 \end{vmatrix} = -4$$

$$A_{33} = (-1)^{3+3} \begin{vmatrix} 1 & -1 \\ 2 & 3 \end{vmatrix} = 5$$

Since Cofactor $A = \begin{pmatrix} 2 & -4 & 2 \\ 2 & 2 & -1 \\ -4 & -4 & 5 \end{pmatrix}$

$$\text{Adj } A = \begin{pmatrix} 2 & -4 & -4 \\ -4 & 2 & -4 \\ 3 & -1 & 5 \end{pmatrix}$$

$$A^{-1} = \frac{1}{6}\begin{pmatrix} 2 & 2 & -4 \\ -4 & 2 & -4 \\ 2 & -1 & 5 \end{pmatrix}$$

Hence,

$$X = A^{-1} B = \frac{1}{6}\begin{pmatrix} 2 & 2 & -4 \\ -4 & 2 & -4 \\ 2 & -1 & 5 \end{pmatrix}\begin{pmatrix} 3 \\ 17 \\ 7 \end{pmatrix} = \frac{1}{6}\begin{pmatrix} 6 + 34 - 28 \\ -12 + 34 - 28 \\ 6 - 17 + 35 \end{pmatrix} = \begin{pmatrix} 2 \\ -1 \\ 4 \end{pmatrix}$$

Therefore $x = 2$, $y = -1$ and $z = 4$

Illustration No.12

Solve by matrix inversion Method

$$2x + 2y + 2z = 12$$

$$2x + 4z = 14$$

$$6x + 2y + 2z = 24$$

Solution:

Let,

$$A = \begin{pmatrix} 2 & 2 & 2 \\ 2 & 0 & 4 \\ 6 & 2 & 2 \end{pmatrix}, X = \begin{pmatrix} x \\ y \\ z \end{pmatrix} \text{ and } B = \begin{pmatrix} 12 \\ 14 \\ 24 \end{pmatrix}$$

Now calculating Det A, we get

$$\text{Det } A = \begin{vmatrix} 2 & 2 & 2 \\ 2 & 0 & 4 \\ 6 & 2 & 2 \end{vmatrix}$$

$$= 2\begin{vmatrix} 0 & 4 \\ 2 & 2 \end{vmatrix} - 2\begin{vmatrix} 2 & 4 \\ 6 & 2 \end{vmatrix} + 2\begin{vmatrix} 2 & 0 \\ 6 & 2 \end{vmatrix}$$

$$= 2(0 - 8) - 2(4 - 24) + 2(4 - 0)$$

$$= 2(-8) - 2(-20) + 8$$

$$= -16 + 40 + 8$$

$$= 32 \neq 0$$

Hence the inverse of A exists,

Now, the given equation can be expressed as $AX = B$ or $X = A^{-1}B$

But $$A^{-1} = \frac{\text{Adj } A}{|A|}$$

Now finding the Adjoint matrix of the matrix A is given by

$$(\text{Adj } A) = (\text{Cofactor})^T$$

Therefore the cofactor matrix of A are

$$\text{Cofactor} \quad A = \begin{pmatrix} A_{11} & A_{12} & A_{13} \\ A_{21} & A_{22} & A_{23} \\ A_{31} & A_{32} & A_{33} \end{pmatrix}$$

$$A_{11} = (-1)^{1+1}\begin{vmatrix} 0 & 4 \\ 2 & 2 \end{vmatrix} = 8$$

$$A_{12} = (-1)^{1+2}\begin{vmatrix} 2 & 4 \\ 6 & 2 \end{vmatrix} = 20$$

$$A_{13} = (-1)^{1+3}\begin{vmatrix} 2 & 0 \\ 6 & 2 \end{vmatrix} = 4$$

$$A_{21} = (-1)^{2+1}\begin{vmatrix} 2 & 2 \\ 2 & 2 \end{vmatrix} = 0$$

$$A_{22} = (-1)^{2+}\begin{vmatrix} 2 & 2 \\ 6 & 2 \end{vmatrix} = -8$$

$$A_{23} = (-1)^{2+3}\begin{vmatrix} 2 & 2 \\ 6 & 2 \end{vmatrix} = 8$$

$$A_{31} = (-1)^{3+1}\begin{vmatrix} 2 & 2 \\ 0 & 4 \end{vmatrix} = 8$$

$$A_{32} = (-1)^{3+2}\begin{vmatrix} 2 & 2 \\ 2 & 4 \end{vmatrix} = -4$$

$$A_{33} = (-1)^{3+3}\begin{vmatrix} 2 & 2 \\ 2 & 0 \end{vmatrix} = -4$$

$$\text{Cofactor } A = \begin{pmatrix} -8 & 20 & 4 \\ 0 & -8 & 8 \\ 8 & -4 & -4 \end{pmatrix}$$

$$\text{Therefore } \text{Adj}A = \begin{pmatrix} -8 & 0 & 8 \\ 20 & -8 & -4 \\ 4 & 8 & -4 \end{pmatrix}$$

And hence

$$A^{-1} = \frac{\text{Adj } A}{|A|} = \frac{1}{32}\begin{pmatrix} 8 & 0 & 8 \\ 20 & -8 & -4 \\ 4 & 8 & -4 \end{pmatrix}$$

Since $X = A^{-1} B$

$$x = \frac{1}{32}\begin{pmatrix} -8 & 0 & 8 \\ 20 & -8 & -4 \\ 4 & 8 & -4 \end{pmatrix}\begin{pmatrix} 12 \\ 14 \\ 24 \end{pmatrix}$$

$$= \frac{1}{32}\begin{pmatrix} -96+0+192 \\ 240-112-96 \\ 48+112-96 \end{pmatrix}$$

$$= \frac{1}{32}\begin{pmatrix} 96 \\ 32 \\ 64 \end{pmatrix}$$

$$= \begin{pmatrix} 3 \\ 1 \\ 2 \end{pmatrix}$$

Therefore the values of $x = 3$, $y = 1$ and $z = 2$

Gauss–Jordan Elimination Method

This method is used to solve linear equations by making use of elementary row operations and hence termed as matrix reduction method. Following are the procedural steps in solving a linear equation using Gauss–Jordan Elimination method.

1. Write the system of equations in matrix form i.e. $AX = O$ in case of homogenous equations or $AX = B$ in case of non homogenous equations.

2. Construct the augmented matrix $[A \mid B]$ by introducing the vector B to the right of the coefficient matrix of A.
3. Then apply row operations to transform the matrix into its Echelon form.

The following example will illustrate the working steps.

Illustration No. 13

Solve the system of linear equations using Gauss–Jordan Method

$$2x + 2y + 2z = 12$$
$$2x - 2y + 2z = 4$$
$$4x + 2y - 2z = 2$$

Solution:

Write down the system of equations in an augmented form

$$\left(\begin{array}{ccc|c} 2 & 2 & 2 & 12 \\ 2 & -2 & 2 & 4 \\ 4 & 2 & -2 & 2 \end{array}\right)$$

$$\left(\begin{array}{ccc|c} 2 & 2 & 2 & 12 \\ 0 & -4 & 0 & -8 \\ 0 & -2 & -6 & -22 \end{array}\right)$$

- By applying $R_2: R_2 - R_1$ and $R_3: R_3 - 2R_1$

$$\left(\begin{array}{ccc|c} 2 & 2 & 2 & 12 \\ 0 & 1 & 0 & 2 \\ 0 & -2 & -6 & -22 \end{array}\right)$$

- By multiplying R_2 by $(-\frac{1}{4})$

$$\left(\begin{array}{ccc|c} 1 & 1 & 1 & 6 \\ 0 & 1 & 0 & 2 \\ 0 & -1 & -3 & -11 \end{array}\right)$$

- By multiplying R_1 and R_3 by $(\frac{1}{2})$

$$\left(\begin{array}{ccc|c} 1 & 0 & 1 & 4 \\ 0 & 1 & 0 & 2 \\ 0 & 0 & -3 & -9 \end{array}\right)$$

- By applying $R_1 : R_1 - R_2$ and $R_3 : R_3 + R_2$

$$\left(\begin{array}{ccc:c} 1 & 0 & 1 & 4 \\ 0 & 1 & 0 & 2 \\ 0 & 0 & 1 & 3 \end{array}\right)$$

- By multiplying R_3 by $(-1/3)$

$$\left(\begin{array}{ccc:c} 1 & 0 & 0 & 1 \\ 0 & 1 & 0 & 2 \\ 0 & 0 & 1 & 3 \end{array}\right)$$

- By applying $R_1 : R_1 - R_3$

Hence the solution are $x = 1$, $y = 2$ and $z = 3$

Rank of Matrix

If A is non zero matrix of any order then the rank of a matrix is the highest order non singular matrix contained in A. Denoted by $\rho(A)$.

Illustration No.14

If $A = \begin{pmatrix} 2 & 4 & 6 \\ -8 & 0 & 10 \end{pmatrix}$ find the rank of the matrix A?

Solution:

The Rank of the matrix A is 2 as the highest order non singular matrix which is contained in A is

$$\begin{pmatrix} 2 & 4 \\ -8 & 0 \end{pmatrix}_{2 \times 2}$$

Such that the determinant $\neq 0$

Illustration No. 15

If $A = \begin{pmatrix} 6 & 6 & -4 \\ -4 & 0 & 10 \end{pmatrix}$ find the rank of the matrix A?

Solution:

The Rank of the matrix A is 2 as the matrix A contains at least one non-singular matrix whose order is 2 × 2

For Example

$$\begin{pmatrix} 6 & -4 \\ -4 & 10 \end{pmatrix}_{2 \times 2}$$

Such that the determinant ≠ 0

Illustration No. 16

If $A = \begin{pmatrix} 21 & -3 & 0 \\ 3 & 3 & 12 \\ 39 & -9 & -12 \end{pmatrix}$ find the rank of the matrix A?

Solution:

The Rank of the matrix A is 2 as the matrix A contains at least one non-singular matrix whose order is 2 × 2

For Example

$$\begin{pmatrix} 3 & 12 \\ -9 & -12 \end{pmatrix}_{2 \times 2}$$

Such that the determinant ≠ 0

Application of Theory of Matrix to Business and Economics

Illustration No. 17

Producer of Panther Brand of cars has availed the services of fast track a leading transport co. to transfer cars from the manufacturing unit to showroom. Fast track has used three different variety of load capacity trailer T_1, T_2 and T_3 to transport the V-Series, X-Series of panther cars. The load capacities of each type of trailer in term of 3 different models of cars are furnished below.

Cars / Trailer	T_1	T_2	T_3
V – Series	4	3	5
X – Series	5	2	3
Z – Series	2	2	3

(i) Find the number of trailer of each type utilized to 131 units of V–Series, 107 units of X–Series and 77 units of Z–Series cars?

(ii) Due to heavy demand of cars additional load capacity was utilized to meet the demand and no. of trailers was uncharged to 20 for T_1, 25 for T_2 and 20 for T_3. Find with the addition how many cars can be made available to the showroom?

Solution:

Let x, y, and z be the no of trailers T_1, T_2 and T_3 used by fast track to transport Panther brand cars to show room. Thus we can express the data as

$$\begin{pmatrix} 4x + 3y + 5z \\ 5x + 2y + 3z \\ 2x + 2y + 3z \end{pmatrix} \begin{matrix} = 131 \\ = 107 \\ = 77 \end{matrix}$$

In matrix from we have

$$\begin{pmatrix} 4 & 3 & 5 \\ 5 & 2 & 3 \\ 2 & 2 & 3 \end{pmatrix} \begin{pmatrix} x \\ y \\ z \end{pmatrix} \begin{pmatrix} 131 \\ 107 \\ 77 \end{pmatrix}$$

i.e.

$$A X = B$$

$$\Rightarrow \quad X = A^{-1} B$$

i.e $$\begin{pmatrix} x \\ y \\ z \end{pmatrix} \begin{pmatrix} 4 & 3 & 5 \\ 5 & 2 & 3 \\ 2 & 2 & 3 \end{pmatrix}^{-1} = \begin{pmatrix} 131 \\ 107 \\ 77 \end{pmatrix}$$

But $\quad A^{-1} = \dfrac{AdjA}{|A|}$

Since $\quad |A| = 4(6 - 6) - 3(15 - 6) + 5(10 - 4) = 3$

Let A_{ij} be the cofactor for the element a_{ij}

Therefore, we have

$$A_{11} = (-1)^{1+1} \begin{pmatrix} 2 & 3 \\ 2 & 3 \end{pmatrix} = 0$$

$$A_{12} = (-1)^{1+2} \begin{pmatrix} 5 & 3 \\ 2 & 3 \end{pmatrix} = -9$$

$$A_{13} = (-1)^{1+3} \begin{pmatrix} 5 & 2 \\ 2 & 2 \end{pmatrix} = 6$$

$$A_{21} = (-1)^{2+1} \begin{pmatrix} 3 & 5 \\ 2 & 3 \end{pmatrix} = 1$$

$$A_{22} = (-1)^{2+2} \begin{pmatrix} 4 & 5 \\ 2 & 3 \end{pmatrix} = 2$$

$$A_{23} = (-1)^{2+3} \begin{pmatrix} 4 & 3 \\ 2 & 2 \end{pmatrix} = -2$$

$$A_{31} = (-1)^{3+1} \begin{pmatrix} 3 & 5 \\ 2 & 3 \end{pmatrix} = -1$$

$$A_{32} = (-1)^{3+2} \begin{pmatrix} 4 & 5 \\ 5 & 3 \end{pmatrix} = -13$$

$$A_{31} = (-1)^{3+1} \begin{pmatrix} 4 & 3 \\ 5 & 2 \end{pmatrix} = -7$$

The cofactor elements of matrix A is given by

Cofactor $$A = \begin{pmatrix} A_{11} & A_{12} & A_{13} \\ A_{21} & A_{22} & A_{23} \\ A_{31} & A_{32} & A_{33} \end{pmatrix}$$

$$= \begin{pmatrix} 0 & -9 & 6 \\ 1 & 2 & -2 \\ -1 & 13 & -7 \end{pmatrix}$$

But,

$$Adj\ A = (\text{Cofactor } A)^T$$

$$= \begin{pmatrix} 0 & -9 & 6 \\ 1 & 2 & -2 \\ -1 & 13 & -7 \end{pmatrix}^T$$

$$= \begin{pmatrix} 0 & 1 & -1 \\ -9 & 2 & 13 \\ 6 & -2 & -7 \end{pmatrix}^T$$

As, we know that, $A^{-1} = \frac{Adj A}{|A|}$ therefore,

$$A^{-1} = \frac{1}{3}\begin{pmatrix} 0 & 1 & -1 \\ -9 & 2 & 13 \\ 6 & -2 & -7 \end{pmatrix}$$

But $X = A^{-1} B$ and hence,

$$x = \frac{1}{3}\begin{pmatrix} 0 & 1 & -1 \\ -9 & 2 & 13 \\ 6 & -2 & -7 \end{pmatrix}\begin{pmatrix} 131 \\ 107 \\ 77 \end{pmatrix}$$

$$= \frac{1}{3}\begin{pmatrix} 0 \times 131 + 1 \times 107 + (-1)77 \\ (-9) \times 131 + 2 \times 107 + 13 \times 77 \\ 6 \times 131 + (-2)(107) + (-7)77 \end{pmatrix}$$

$$= \begin{pmatrix} 30/3 \\ 36/3 \\ 33/3 \end{pmatrix} = \begin{pmatrix} 10 \\ 12 \\ 11 \end{pmatrix}$$

No. of trailer used of type T_1, T_2 and T_3 are 10, 12, 11

(ii) As the demand has exceed the expectation more no. of trailer were put to use. The number of trailer of $T_1 = 20$ $T_2 = 25$ and $T_3 = 20$ made available to meet the extra demand.

Hence the no. of cars which can be transfer to show room

$$\begin{pmatrix} V - \text{Series} \\ X - \text{Series} \\ Z - \text{Series} \end{pmatrix} \begin{pmatrix} 4 & 3 & 5 \\ 5 & 2 & 3 \\ 2 & 2 & 3 \end{pmatrix} \begin{pmatrix} 20 \\ 25 \\ 20 \end{pmatrix}$$

$$= \begin{pmatrix} 4\times 20 + 3\times 25 + 5\times 20 \\ 5\times 20 + 2\times 25 + 3\times 20 \\ 2\times 20 + 2\times 25 + 3\times 20 \end{pmatrix}$$

$$= \begin{pmatrix} 225 \\ 210 \\ 150 \end{pmatrix}$$

Hence the number of vehicles which can be made available are 225, 210 and 150.

Illustration No. 18

Three friends Abhishek, Vimal and Devesh are short term investors. They analyzed the performance of a few IT – Co's and decided to invest in three Co's namely Magnatech, Zizatech and Megatech each listed in the stock market. Each friend decided to purchase contain no. of shares as per available resources which are below.

Purchase of Stock

Name of the investor	Magnatech	Zizatech	Megatech	Funds Available
Abhishek	5	6	10	1430
Vimal	4	8	12	1580
Devash	6	10	6	1670

using matrix inversion method. Find the price of share at which the friends have purchased them?

Solution:

Let x, y and z be the price of share purchased by Abhishek, Vimal and Devesh. From the above available information we obtain the following set of simultaneous equation.

$$5x + 6y + 10z = 1430$$

$$4x + 8y + 12z = 1580$$

$$6x + 10y + 6z = 1670$$

The above system of equation is represented in Matrix from

$$\begin{pmatrix} 5 & 6 & 10 \\ 4 & 8 & 12 \\ 6 & 10 & 6 \end{pmatrix} \begin{pmatrix} x \\ y \\ z \end{pmatrix} \begin{pmatrix} 1430 \\ 1530 \\ 1670 \end{pmatrix}$$

i.e $AX = B$

$\Rightarrow X = A^{-1}B$

Therefore,

$$\begin{pmatrix} x \\ y \\ z \end{pmatrix} = \begin{pmatrix} 5 & 6 & 10 \\ 4 & 8 & 12 \\ 6 & 10 & 6 \end{pmatrix}^{-1} \begin{pmatrix} 1430 \\ 1580 \\ 1670 \end{pmatrix}$$

But $A^{-1} = \dfrac{\text{Adj}A}{|A|}$

Since determinant of A i.e $|A| = 5(48 - 120) - 6(24 - 72) + 10(40 - 48)$

Since $|A| = -360 + 288 - 80$

$= -152$

Let A_{ij} be the cofactor for the elements a_{ij} of Matrix A

Therefore we have

$$A_{11} = (-1)^{1+1}\begin{pmatrix} 8 & 12 \\ 10 & 6 \end{pmatrix} = -72$$

$$A_{12} = (-1)^{1+2}\begin{pmatrix} 4 & 12 \\ 6 & 6 \end{pmatrix} = 48$$

$$A_{13} = (-1)^{1+3}\begin{pmatrix} 4 & 8 \\ 6 & 10 \end{pmatrix} = -8$$

$$A_{21} = (-1)^{2+1}\begin{pmatrix} 6 & 10 \\ 10 & 6 \end{pmatrix} = 64$$

$$A_{22} = (-1)^{2+2}\begin{pmatrix} 5 & 10 \\ 6 & -10 \end{pmatrix} = -30$$

$$A_{23} = (-1)^{2+3}\begin{pmatrix} 5 & 6 \\ 6 & -10 \end{pmatrix} = -14$$

$$A_{31} = (-1)^{3+1}\begin{pmatrix} 6 & 10 \\ 8 & 12 \end{pmatrix} = -8$$

$$A_{32} = (-1)^{3+2}\begin{pmatrix} 5 & 10 \\ 4 & 12 \end{pmatrix} = -20$$

$$A_{33} = (-1)^{3+3}\begin{pmatrix} 5 & 6 \\ 4 & 8 \end{pmatrix} = 16$$

The cofactor Matrix A is given by

$$\text{Cofactor } A = \begin{pmatrix} -72 & 48 & -8 \\ 64 & -30 & -14 \\ -8 & -20 & 16 \end{pmatrix}$$

Therefore,

$$[\text{Adj } A] = [\text{Cofactor } A]^T$$

$$= \begin{pmatrix} -72 & 64 & -8 \\ 48 & -30 & -20 \\ -8 & -14 & 16 \end{pmatrix}$$

But $$A^{-1} = \frac{\text{Adj}A}{|A|}$$

$$= -\frac{1}{152}\begin{pmatrix} -72 & 64 & -8 \\ 48 & -30 & -20 \\ -8 & -14 & 16 \end{pmatrix}$$

As, $$X = A^{-1} B$$

$$= -\frac{1}{152}\begin{pmatrix} -72 & 64 & -8 \\ 48 & -30 & -20 \\ -8 & -14 & 16 \end{pmatrix}\begin{pmatrix} 1430 \\ 1580 \\ 1670 \end{pmatrix}$$

$$= -\frac{1}{152}\begin{pmatrix} -102960 + 101120 - 13360 \\ 68640 - 47400 - 33400 \\ -11440 - 22120 + 26720 \end{pmatrix}$$

$$= -\frac{1}{152}\begin{pmatrix} -15200 \\ -12160 \\ -6840 \end{pmatrix} = \begin{pmatrix} 100 \\ 80 \\ 45 \end{pmatrix}$$

Hence the price of share are Rs. 100, Rs. 80 and Rs. 45.

Illustration No. 19

Riga tyre Ltd, manufacture of radial tyres for motor cycle uses three different qualities of processed rubber R_1, R_3, and R_3 for producing three different type of tyre namely born tough, lifelong and anti-skid. The processed rubber are mixed in different portions in (kg) to produce born tough, lifelong & anti-skid which are furnished below

Type of Tyres

Type of Rubber	Born tough	Lifelong	Anti-skid
R_1	2	3	1
R_2	3	2	2
R_3	1	3	3

Using matrix inversion method find the number of types all the three variety that can be manufactured if available qualities of produced rubber R_1 is 3500 kg, R_2 is 4100 kg and R_3 is 4400 kg.

Solution:

Let x, y and z be the number of units of born tough, lifelong and anti-skid be produced

From available data we frame the following simultaneous equation

$$2x + 3y + z = 3500$$

$$3x + 2y + 2z = 4100$$

$$x + 3y + 3z = 4400$$

Expressing in matrix form – AX = B

$$\begin{pmatrix} 2 & 3 & 1 \\ 3 & 2 & 2 \\ 1 & 3 & 3 \end{pmatrix} \begin{pmatrix} x \\ y \\ z \end{pmatrix} = \begin{pmatrix} 3500 \\ 4100 \\ 4400 \end{pmatrix}$$

where

$$A = \begin{pmatrix} 2 & 3 & 1 \\ 3 & 2 & 2 \\ 1 & 3 & 3 \end{pmatrix} \quad X = \begin{pmatrix} x \\ y \\ z \end{pmatrix}$$

And

$$B = \begin{pmatrix} 3500 \\ 4100 \\ 4400 \end{pmatrix}$$

As $X = A^{-1} B$, but $A^{-1} = \frac{AdjA}{|A|}$

Calculating the determinant of the matrix A, we get

$$|A| = 2(6-6) - 3(9-2) + 1(9-2)$$

$$= 0 - 21 + 7 = -14$$

Let A_{ij} be the Cofactor of elements a_{ij} of matrix A

Then,

$$A_{11} = (-1)^{1+1} \begin{vmatrix} 2 & 2 \\ 3 & 3 \end{vmatrix} = 0$$

$$A_{12} = (-1)^{1+2} \begin{vmatrix} 3 & 2 \\ 1 & 3 \end{vmatrix} = -7$$

$$A_{13} = (-1)^{1+3} \begin{vmatrix} 3 & 2 \\ 1 & 3 \end{vmatrix} = 7$$

$$A_{21} = (-1)^{2+1} \begin{pmatrix} 3 & 1 \\ 3 & 3 \end{pmatrix} = -6$$

$$A_{22} = (-1)^{2+2} \begin{pmatrix} 2 & 1 \\ 1 & 3 \end{pmatrix} = 5$$

$$A_{23} = (-1)^{2+3} \begin{pmatrix} 2 & 3 \\ 1 & 3 \end{pmatrix} = -3$$

$$A_{31} = (-1)^{3+1} \begin{pmatrix} 3 & 1 \\ 2 & 2 \end{pmatrix} = 4$$

$$A_{32} = (-1)^{3+2} \begin{pmatrix} 2 & 1 \\ 3 & 2 \end{pmatrix} = -1$$

$$A_{33} = (-1)^{3+3} \begin{pmatrix} 2 & 3 \\ 3 & 2 \end{pmatrix} = -5$$

The cofactor of matrix A is given by

$$\text{Cofactor} \quad A = \begin{pmatrix} 0 & -7 & 7 \\ -6 & 5 & -3 \\ -4 & -1 & -5 \end{pmatrix}$$

Therefore,

$$\text{Adj } A = [\text{Cofactor } A]^T = \begin{pmatrix} 0 & -6 & 4 \\ -7 & 5 & -1 \\ 7 & -3 & -5 \end{pmatrix}$$

As,

$$A^{-1} = \frac{\text{Adj}A}{|A|} = -\frac{1}{14}\begin{pmatrix} 0 & -6 & 4 \\ 7 & 5 & -1 \\ 7 & -3 & -5 \end{pmatrix}$$

Therefore $X = A^{-1} B$

$$= -\frac{1}{14} \times \begin{pmatrix} 0 & -6 & 4 \\ -7 & 5 & -1 \\ 7 & -3 & -5 \end{pmatrix} \begin{pmatrix} 3500 \\ 4100 \\ 4400 \end{pmatrix}$$

$$= -\frac{1}{14} \begin{pmatrix} 0 - 24600 + 17600 \\ -24500 + 20500 - 4400 \\ 24500 - 12300 - 22000 \end{pmatrix}$$

$$= -\frac{1}{14} \begin{pmatrix} -7000 \\ -8400 \\ -9800 \end{pmatrix} = \begin{pmatrix} 500 \\ 600 \\ 700 \end{pmatrix}$$

Hence the numbers of tyres produced in all the three varieties are 500, 600 and 700.

Illustration No.20

A dietician has prescribed health supplements for better stamina among athletes, they are mega boost, endura and ultimate. Each of the supplements contain certain proportion of calcium, protein and vitamins as for worked below in (units)

	Mega Boost	**Endura**	**Utimate**
Calcium	4	5	7
Protein	8	6	12
Vitamins	10	10	9

If an average athlete's needs 49 unit's of calcium and 82 unit's proteins and 98 unit's of vitamins as daily intake. In what proportion the supplements should be mixed to fulfill desired requirement? Solve by matrix inversion method?

Solution:

From the information, we get,

$$4x + 5y + 7z = 49$$

$$8x + 6y + 12z = 82$$

$$10x + 10y + 9z = 98$$

$$\begin{pmatrix} 4 & 5 & 7 \\ 8 & 6 & 12 \\ 10 & 10 & 9 \end{pmatrix} \begin{pmatrix} x \\ y \\ z \end{pmatrix} = \begin{pmatrix} 49 \\ 82 \\ 98 \end{pmatrix}$$

$$A X = B$$

$$X = A^{-1} B$$

Where,

$$A = \begin{pmatrix} 4 & 5 & 7 \\ 8 & 6 & 12 \\ 10 & 10 & 9 \end{pmatrix} \quad X = \begin{pmatrix} x \\ y \\ z \end{pmatrix} \quad B = \begin{pmatrix} 49 \\ 82 \\ 98 \end{pmatrix}$$

But $$A^{-1} = \frac{\text{Adj } A}{|A|}$$

Determinant of $$A = |A| = 4(-66) - 5(-48) + 7(20)$$

$$= -264 + 240 + 140$$

$$= 116$$

$$A_{11} = (-1)^{1+1} \begin{pmatrix} 6 & 12 \\ 10 & 9 \end{pmatrix} = -66$$

$$A_{12} = (-1)^{1+2} \begin{pmatrix} 8 & 12 \\ 10 & 9 \end{pmatrix} = 48$$

$$A_{13} = (-1)^{1+3} \begin{pmatrix} 8 & 6 \\ 10 & 10 \end{pmatrix} = 20$$

$$A_{21} = (-1)^{2+1} \begin{pmatrix} 5 & 7 \\ 10 & 9 \end{pmatrix} = 25$$

$$A_{22} = (-1)^{2+2} \begin{pmatrix} 4 & 7 \\ 10 & 9 \end{pmatrix} = -34$$

$$A_{23} = (-1)^{2+3} \begin{pmatrix} 4 & 5 \\ 10 & 10 \end{pmatrix} = 10$$

$$A_{31} = (-1)^{3+1} \begin{pmatrix} 5 & 7 \\ 6 & 12 \end{pmatrix} = 18$$

$$A_{32} = (-1)^{3+2} \begin{pmatrix} 4 & 7 \\ 8 & 12 \end{pmatrix} = 8$$

$$A_{33} = (-1)^{3+3} \begin{pmatrix} 4 & 5 \\ 8 & 6 \end{pmatrix} = -16$$

Cofactor $A = \begin{pmatrix} -66 & 48 & 20 \\ 25 & -34 & 10 \\ 18 & 8 & -16 \end{pmatrix}$

Adj A = [Cofactor $A]^T = \begin{pmatrix} -66 & 25 & 18 \\ 48 & -34 & 8 \\ 20 & 10 & -16 \end{pmatrix}$

Since,

$$A^{-1} = \frac{1}{116} \begin{pmatrix} -66 & 25 & 18 \\ 48 & -34 & 8 \\ 20 & 10 & -16 \end{pmatrix}$$

But $X = A^{-1} B$

$$= \frac{1}{116} \begin{pmatrix} -66 & 25 & 18 \\ 48 & -34 & 8 \\ 20 & 10 & -16 \end{pmatrix} \begin{pmatrix} 49 \\ 82 \\ 98 \end{pmatrix}$$

$$= \frac{1}{116} \begin{pmatrix} -3234 + 2050 + 1764 \\ 2352 - 2788 + 784 \\ 980 + 820 - 1568 \end{pmatrix} = \begin{pmatrix} 5 \\ 3 \\ 2 \end{pmatrix}$$

Hence the required quantity are 5 units of mega boost, 3 units of endura and 2 units of ultimate

Illustration No. 21

Shyam has invested a sum of Rs. 12,000 in three different investments at the rate 8%, 10% and 12% per annum. At the end of holding period of one year his interest earning was Rs 1240. If the sum of the first and third investable sum exceeds the second by 4000, then find by matrix inversion method what are the different amounts invested?

Solution:

Let x, y and z be the amount invested. Therefore from the data we get

$$x + y + z = 12000$$

$$0.08x + 0.10y + 0.12y = 1240$$

or

$$8x + 10y + 12y = 124000$$

$$x - y + z = 4000$$

Expressing in matrix form i.e

$$\begin{pmatrix} 1 & 1 & 1 \\ 8 & 10 & 12 \\ 1 & -1 & 1 \end{pmatrix} \begin{pmatrix} x \\ y \\ z \end{pmatrix} = \begin{pmatrix} 12000 \\ 124000 \\ 4000 \end{pmatrix}$$

i.e. $AX = B$

Where

$$A = \begin{pmatrix} 1 & 1 & 1 \\ 8 & 10 & 12 \\ 1 & -1 & 1 \end{pmatrix}, X = \begin{pmatrix} x \\ y \\ z \end{pmatrix} \text{ and } B = \begin{pmatrix} 12000 \\ 124000 \\ 4000 \end{pmatrix}$$

$$X = A^{-1} B$$

But $A^{-1} = \dfrac{\text{Adj } A}{|A|}$

Since $|A| = 1\ (10 + 12) - 1\ (8 - 12) + 1\ (-8 - 10)$

$$= 22 + 4 - 18 = 8$$

Let A_{ij} be the cofactor of elements of the matrix A i.e. a_{ij}

$$A_{11} = (-1)^{1+1} \begin{vmatrix} 10 & 12 \\ -1 & 1 \end{vmatrix} = 22$$

$$A_{12} = (-1)^{1+2} \begin{pmatrix} 8 & 12 \\ 1 & 1 \end{pmatrix} = 4$$

$$A_{13} = (-1)^{1+3} \begin{pmatrix} 8 & 10 \\ 1 & -1 \end{pmatrix} = -18$$

$$A_{21} = (-1)^{2+1} \begin{pmatrix} 1 & 1 \\ -1 & 1 \end{pmatrix} = -2$$

$$A_{22} = (-1)^{2+2} \begin{pmatrix} 1 & 1 \\ 1 & 1 \end{pmatrix} = 0$$

$$A_{23} = (-1)^{2+3} \begin{pmatrix} 1 & 1 \\ 1 & -1 \end{pmatrix} = 2$$

$$A_{31} = (-1)^{3+1} \begin{pmatrix} 1 & 1 \\ 10 & 12 \end{pmatrix} = 2$$

$$A_{32} = (-1)^{3+2} \begin{pmatrix} 1 & 1 \\ 8 & 12 \end{pmatrix} = -4$$

$$A_{33} = (-1)^{3+3} \begin{pmatrix} 1 & 1 \\ 8 & 10 \end{pmatrix} = 2$$

therefore,

Cofactor $$A = \begin{pmatrix} 22 & 4 & -18 \\ -2 & 0 & 2 \\ 2 & -4 & 2 \end{pmatrix}$$

$$\text{Adj } A = [\text{Cofactor } A]^T = \begin{pmatrix} 22 & -2 & 2 \\ 4 & 0 & -4 \\ -18 & 2 & 2 \end{pmatrix}$$

$$A^{-1} = (1/8)\begin{pmatrix} 22 & -2 & 2 \\ 4 & 0 & -4 \\ -18 & 2 & 2 \end{pmatrix}$$

But, $X = A^{-1} B$

$$= 1/8 \begin{pmatrix} 22 & -2 & 2 \\ 4 & 0 & -4 \\ -18 & 2 & 2 \end{pmatrix}\begin{pmatrix} 12000 \\ 124000 \\ 4000 \end{pmatrix}$$

$$= 1/8 \begin{pmatrix} 264000 - 248000 + 8000 \\ 48000 + 0 - 16000 \\ -216000 + 248000 + 8000 \end{pmatrix} = \begin{pmatrix} 3000 \\ 4000 \\ 5000 \end{pmatrix}$$

Illustration No. 22

A company earns profit before tax of Rs 50000. As a social responsibility initiative the firm has decided to denote 20% of after tax profit. If the central government levies corporate tax at the rate of 40% of profit after deducting charitable contribution and local taxes which is payed at the rate of 10% of its profit after deducting of charitable donation. Calculate how much the company has paid for corporate tax, local tax and donation to charitable institution using Cramer's rule?

Solution:

Let us assume that the amount donated to charity be 'x', corporate tax paid is y and local tax be 'z'. From the above information we get

$$x = 20\% \text{ of } (50{,}000 - y - z)$$

$$x = \frac{20}{100} \times (50{,}000 - y - z)$$

$$5x = 50000 - y - z$$

$$5x + y + z = 50{,}000 \qquad \text{.....(i)}$$

Similarly,

$$y = 40\% \ (50000 - x - z) = 40/100 \ (50{,}000 - x - z)$$

$$= 2/5 \ (50{,}000 - x - z)$$

$$2x + 5y + 2z = 100000$$

Similarly,

$$z = 10\% \ (50000 - x)$$

$$x + 10z = 50000$$

Therefore the three Linear equation are

$$5x + y + z = 50000$$

$$2x + 5y + 2z = 100000$$

$$x + 0.y + 10z = 50000$$

AX = B (form)

$$A = \begin{pmatrix} 5 & 1 & 1 \\ 2 & 5 & 2 \\ 1 & 0 & 10 \end{pmatrix}, X = \begin{pmatrix} x \\ y \\ z \end{pmatrix} \quad B = \begin{pmatrix} 50000 \\ 100000 \\ 50000 \end{pmatrix}$$

$$D = \begin{pmatrix} 5 & 1 & 1 \\ 2 & 5 & 2 \\ 1 & 0 & 10 \end{pmatrix}$$

$$= 5\ (50 - 2) - 1\ (20 - 2) + 1\ (0 - 5)$$

$$= 5\ (48) - 1\ (18) - 5 = 240 - 23 = 217 \neq 0$$

$$D_x = \begin{pmatrix} 50000 & 1 & 1 \\ 100000 & 5 & 2 \\ 50000 & 0 & 10 \end{pmatrix}$$

$$= 50000\ (48) - 1\ (900000) + 1\ (-250000)$$

$$= 2400000 - 1150000 = 1250000$$

$$D_y = \begin{pmatrix} 5 & 50000 & 1 \\ 2 & 100000 & 2 \\ 1 & 50000 & 10 \end{pmatrix}$$

$$= 5(900000) - 50000\ (18) + 0$$

$$= 45,00,000 - 9,00,000$$

$$= 36,\ 00,\ 000$$

$$D_z = \begin{pmatrix} 5 & 1 & 50000 \\ 2 & 5 & 100000 \\ 1 & 0 & 50000 \end{pmatrix}$$

$$= 5(250000) - 1\ (0) + 50000(-5)$$

$$= 1250000 - 250000$$

$$= 1000000$$

Therefore,

$$x = \frac{D_x}{|D|} = \frac{1250000}{217} = 5760$$

$$y = \frac{D_y}{|D|} = \frac{3600000}{217} = 16589.8$$

$$Z = \frac{D_z}{|D|} = \frac{1000000}{217} = 4608.29$$

Illustration No. 23

A chocolates manufacturer produces three different brand of chocolates C_1, C_2, C_3. Which he sells in three different markets M_1, M_2 and M_3, the monthly sales & sales price are given below

Selling price

$$\begin{matrix} & C_1 & C_2 & C_3 \\ M_1 \\ M_2 \\ M_3 \end{matrix} \begin{pmatrix} 500 & 600 & 800 \\ 1000 & 1400 & 1500 \\ 1400 & 1700 & 2000 \end{pmatrix} \begin{matrix} S_1 & S_2 & S_3 \end{matrix} \begin{pmatrix} 3 & 4 & 5 \\ 4 & 5 & 7 \\ 5 & 7 & 8 \end{pmatrix}$$

If the cost of manufacture of C_1, C_2 and C_3 is Rs. 2, Rs. 3 and Rs. 4, then find profit earned by the manufacture by matrix algebra method.

Solution:

The selling price and cost price per unit of three different chocolates brands in three different markets can be written as

Selling price per unit

$$\begin{pmatrix} 3 & 4 & 5 \\ 4 & 5 & 7 \\ 5 & 7 & 8 \end{pmatrix}$$

Cost per unit

$$\begin{pmatrix} 2 & 3 & 4 \\ 2 & 3 & 4 \\ 2 & 3 & 4 \end{pmatrix}$$

Therefore profit earned per unit

$$(\pi) = \begin{pmatrix} 3 & 4 & 5 \\ 4 & 5 & 7 \\ 5 & 7 & 8 \end{pmatrix} - \begin{pmatrix} 2 & 3 & 4 \\ 2 & 3 & 4 \\ 2 & 3 & 4 \end{pmatrix}$$

$$\begin{pmatrix} 1 & 1 & 1 \\ 2 & 2 & 3 \\ 3 & 4 & 4 \end{pmatrix}$$

And then the total profit earned in three market

=Profit in market M_1 +Profit earned in market M_2 + Profit market M_3

$$= [500 \quad 600 \quad 800] \begin{pmatrix} 1 \\ 1 \\ 1 \end{pmatrix} + [1000 \quad 1400 \quad 1000] \begin{pmatrix} 2 \\ 2 \\ 3 \end{pmatrix} +$$

$$[1400 \quad 1700 \quad 2000] \begin{pmatrix} 3 \\ 4 \\ 4 \end{pmatrix}$$

$$= [500 + 600 + 800] + [2000 + 2800 + 4500]$$
$$+ [4200 + 6800 + 8000]$$
$$= 1900 + 9300 + 19000$$
$$= 30200$$

Illustration No. 24

A crop infested by pest required pest control treatment by three different chemicals A, B and C in different proportion that is 58, 26 and 32 units. Dinesh a farmer whose farm is affected by pest purchased three different types of spray Meganol containing 4 units A, 6 units of B and 8 units of C, Hexanol containing 2 units of A, 2 units B and 4 unit C and Dexanol containg 6 units of A, 4 units of B and 2 units C. Find how many barrels of each type of spray will be used to fulfill the specific requirement? (Solve by Cramer's Rule)

Solution:

Let x units Meganol, y units of Hexanol and 7 units of Dexanol are been utilized. Then the above problem can be represented in equation from

$$4x + 6y + 8z = 58$$
$$2x + 2y + 4z = 26$$
$$6x + 4y + 2z = 32$$

The above system of equation can be written as AX = B

Where $A = \begin{pmatrix} 4 & 6 & 8 \\ 2 & 2 & 4 \\ 6 & 4 & 2 \end{pmatrix}$ $X = \begin{pmatrix} x \\ y \\ z \end{pmatrix}$ and $B = \begin{pmatrix} 58 \\ 26 \\ 32 \end{pmatrix}$

Calculating determinant of the matrix A, we get

Det $A = 4 (4 - 16) - 6 (4 - 24) + 8 (8 - 12) = 40$

But

$$D_x = \begin{pmatrix} 58 & 6 & 8 \\ 26 & 2 & 4 \\ 32 & 4 & 2 \end{pmatrix}$$

$$= 58 [4 - 16] - 6 [52 - 128] + 8 [104 - 64] = 80$$

$$D_y = \begin{pmatrix} 4 & 58 & 8 \\ 2 & 26 & 4 \\ 6 & 32 & 2 \end{pmatrix}$$

$$= 4 (52 - 128) - 58 (4 - 24) + 8 (64 - 156) = 120$$

$$D_z = \begin{pmatrix} 4 & 6 & 58 \\ 2 & 2 & 26 \\ 6 & 4 & 32 \end{pmatrix}$$

$$= 4 (64 - 104) - 6 (64 - 156) + 58 (8 - 12) = 160$$

$$x = \frac{D_x}{|D|} = \frac{80}{40} = 2$$

$$y = \frac{D_y}{|D|} = \frac{120}{40} = 3$$

$$z = \frac{D_z}{|D|} = \frac{160}{40} = 4$$

Illustration No. 25

Paramount Automobiles a leading scooter manufacturer produced three different varieties S_1, S_2 and S_3, that needs working on three different machinery M_1, M_2 and M_3 to assemble. The number days of machinery required by each scooter are given below

	M_1	M_2	M_3
S_1	3	6	3
S_2	6	9	6
S_3	9	6	6

If the total available time in each machine is 72, 126 and 117 days, find the number of scooter S_1, S_2 & S_3 that can be produced with the available time? (Solve by Cramer's Rule)

Solution:

Let x units of S_1, y units of S_2 and z units of S_3 are produced with the available time

Then,

$$3x + 6y + 3z = 72$$

$$6x + 9y + 6z = 126$$

$$9x + 6y + 6z = 117$$

Then above system can be written in Matrix form as AX = B

$$|D| = \begin{pmatrix} 3 & 6 & 3 \\ 6 & 9 & 6 \\ 9 & 6 & 6 \end{pmatrix}$$

$$= 3\,(18) - 6\,(-18) + 3\,(-45)$$

$$= 54 + 108 - 135$$

$$= 27$$

$$D_x = \begin{pmatrix} 72 & 6 & 3 \\ 126 & 9 & 6 \\ 117 & 6 & 6 \end{pmatrix}$$

$= 72\,(54 - 36) - 6\,(756 - 702) + 3\,(756 - 1053)$

$= 81$

$$D_y = \begin{pmatrix} 3 & 72 & 3 \\ 6 & 126 & 6 \\ 9 & 117 & 6 \end{pmatrix}$$

$= 3\,(756 - 702) - 72\,(36 - 54) + 3\,(702 - 1134)$

$= 162$

$$Dz = \begin{pmatrix} 3 & 6 & 72 \\ 6 & 9 & 126 \\ 9 & 6 & 117 \end{pmatrix}$$

$= 3\,(1053 - 756) - 6\,(702 - 1134) + 72\,(36 - 81)$

$= 243$

Since the number units of x, y and z

$$x = \frac{|D_x|}{|D|} = \frac{81}{27} = 3$$

$$y = \frac{|D_y|}{|D|} = \frac{162}{27} = 6$$

$$z = \frac{|D_z|}{|D|} = \frac{243}{27} = 9$$

Summary

- A square matrix say A is said to be singular matrix if its determinant $|A| = 0$
- Assume A is square matrix such that $A = [a_{ij}]_{n \times n}$ i.e. A is matrix of order n and $|A|$ is the determinant formed by the corresponding elements of A and if A_{ij} is the cofactor of elements a_{ij}, then the matrix obtained by transposing the cofactor matrix of elements of A is called the adjugate or adjoint matrix denoted by Adj A.

- A square matrix A of any order is said to be orthogonal if $AA^T = A^TA = I$ where A^T is the transpose of the matrix A and I is the identity matrix of the same order.
- If A is non zero matrix of any order then the rank of a matrix is the highest order non singular matrix contained in A, denoted by $\rho(A)$.

Self Practice Exercises

Solve the following by Cramer's Rule

1. $5x - 6y + 4z = 8$
   $3x + y + 2z = 17$
   $-x + 2y + 3z = 13$

2. $2x + 2y + z = 8$
   $x + 5y + 6z = 31$
   $-5x + y + 4z = 7$

3. $x + y + z = 6$
   $-3x + 4y + 2z = 11$
   $-x - y + 4z = 9.$

4. $2x + 2y + 3z = 10$
   $-3x + 5y + z = 4$
   $x - 3y + 4z = 6$

Solve the following by Matrix Inversion Method

5. $x + 3y + 4z = 25$
   $x - y + 2z = 5$
   $5x + 5y + 3z = 27$

6. $2x + 8y + 5z = 19$
   $x + y + z = 2$
   $x + 2y - z = 2$

7. $3x + 4y + 2z = 9$
   $2x + y + z = 4$
   $x + 5y + 3z = 9$

Solve the following by Gauss Elimination Method

8. $-2x + y + 3z = 15$
   $-x - y + 2z = 11$
   $x + 2y + z = 0$

9. $3x - 5y + 6z = 6$
   $x + y + 3z = 20$
   $-2x + 2y - z = -6$

10. $4x - 2y + 3z = 19$
    $x + y - 2z = -5$
    $2x + y + z = 6$

11. Gammon Transporters uses three different variety of trailers A, B and C to transport three different types of vehicles X, Y and Z. If the

capacity of each trailer in terms of different variety of vehicles are as follows

	X	Y	Z
A	2	3	1
B	1	1	3
C	3	2	4

Using matrix inversion method , find the number of trailers that will be required to transport 46, 52 and 94 vehicles of type X, Y and Z respectively?

12. Two companies X and Y are holding stake in each other company, while X is holding 30% stake in Y whereas Y in turn is holding 20% stake in X. If both the companies have separately earned profits of Rs.1,26,000 and Rs.50,000, then find the total profit of each company using matrix inversion method?

13. To control certain crop disease it is necessary to spray 11 units of chemical P, 9 units of chemical Q and 14 units of chemical R. If one drum of spray X contains 1 unit of P, 2 units of Q and 3 units of R and one drum of spray Y contains 3 units of P, 1 unit of Q and 2 units of R, whereas one drum of spray Z contains 4 units of P, 3 units of Q and 2 units of R. Find how many drums of spray X, Y and Z will be required to meet the desired requirement? Use Cramer's Rule to solve the problem?

14. A mix is made from 3 foods A, B and C. The three foods contains certain level of nutrients P, Q and R which is furnished below

	P	Q	R
A	2	4	10
B	6	2	2
C	8	4	2

Find how much of each of A, B and C are required to fulfill 16 units of P, 10 units of Q and 14 units of R? Use Cramer's Rule to solve the problem?

15. A company manufactures three different types of bearings B_1, B_2 and B_3. The sum of the production of all the three varieties is 200

units. If it is found that the sum of the first – two types of bearing exceeds the third type by 40 units and sum of the second and third bearing produced is three times that of the first then determine the production level of each of kind of bearing using Gaussian/Matrix inversion Method.

16. An amount of Rs. 20,000 is put into three different investments earning 8%, 10% and 12% rate of interest annually. The combined income earned from the investment is Rs. 2120 whereas the combined income of the first and second investment is 280 less than the third investment, then find the investment made in each of the part using Cramer's Rule?

Answers

1. x = 2, y = 3 and z = 4
2. x = – 1, y = 4 and z = 2
3. x = 1, y = 2 and z = 3
4. x = 1, y = 1 and z = 2
5. x = 0, y = 3 and z = 4
6. x = –1, y = 2 and z = 1
7. x = 1, y = 1 and z = 1
8. x = – 2, y = –1 and z = 4
9. x = 8, y = 6 and z = 2
10. x = 2, y = – 1 and z = 3
11. X = 10, Y = 12 and Z = 10
12. X = 1,50,000 and Y = 80,000
13. X = 1, Y = 2 and Z = 2
14. A = 1, B = 1 and C = 1
15. B_1 = 50, B_2 = 70 and B_3 = 80
16. x = 4000, y = 6000 and z = 10,000

CHAPTER 12

Input Output Analysis

Brief History

Input – Output Model was largely developed and modified by Wassiley. W Leontief for which he won the Nobel Prize in Economics in 1973. The initial origins of these work can be traced to the inter-industry analysis contained in the 'tableau economique' of Francois Quesnay and Marx's 'scheme of reproduction of capital'. The hypothesis on which IO model is mainly centered around is the idea of structural inter-industry dependence and inter-relationship in an economy. The model propounds that each industries use the products of other industries to produce their own products i.e. the output of one industry is the input for another. For example – when an automobile producers use steel, glass, rubber, and plastic products to produce automobiles. And when you are buying a car, you are affecting the demand for glass, plastic and steel etc.

Basically classified into two types:

(i) The '***Closed type***', in which whatever is the productional output it is consumed or absorbed within the industry, only benefiting the participants. For, instances assume three firms X, Y & Z whose produce is absorbed among themselves in various proportions depending upon the requirement. To elaborate let X consume 40% of its own produce and rest is shared by Y & Z in varied proportion say 25% and 35% depending upon capacity requirement. Similarly the same can be envisaged about Y & Z. The consumption pattern is presented in a tabular form.

Producer Industry	Consumer Industry (%)		
	X	Y	Z
X	40	25	35
Y	30	30	40
Z	25	50	25

In contrast,

(ii) the ***'open type'***, in which a part of the output is consumed outside the industry by the consumers from the general population (external demand) and rest is placed within the industry (internal demand). For example, when a tyre manufacturing company like MRF Ltd produces tyres and tubes in the perspective of not only for fulfilling the demand of the intermediate users like automobile industry (For e.g., Suzuki India Ltd) but also for the individual consumers who are in the possession of either a 2-wheeler or a 4-wheeler who needs a replacement because of wear and tear happening with usage. In this, chapter we will elaborate on the open type modeling.

Let us consider an economy operating with three firms X, Y and Z each producing a single product and the output of the firms are x_1, x_2 and x_3 respectively. In order to ascertain the inter-dependency assume that a_{ij} is input contribution of one firm to the another in order to produce one unit of final output i.e. i^{th} firm contributing to the production of one unit of the final output of the jth firm. In this case it can be any of the firms X, Y and Z. In matrix terminology a_{ij} is called as technical coefficient.

Apart from the industry demand there exist a final demand d_1, d_2 and d_3 outside the industry serving the requirements from the population load. The information is represented in a tabular form

Firms	**Production**	**The Technical Coefficient**			**Final Demand**
X	x_1	a_{11}	a_{12}	a_{13}	d_1
Y	x_2	a_{21}	a_{22}	a_{23}	d_2
Z	x_3	a_{31}	a_{32}	a_{33}	d_3

From the table, the column represents different firms and the row highlights the firms inputs as a commodity into the column industry. It is clear that out of x_1 units produced by the firm X, a_{11} x_1 is self consumed, a_{12} x_2 is consumed by Y as one unit x_2 of firm Y requires a_{12}

units of x_1 of firm X for production. Similarly firm Z will consume $a_{13}x_3$ for its requirements apart from this an added final demand of d_1 and similarly the same can be thought for the other 2 firms.

The Alternate Explanation

(i) The firm X uses a_{11} unit of its own product for self consumption for producing one more unit of x_1. Uses a_{21} number of units from Y to produce one more unit of x_1 and atlast uses a_{31} number of units from Z to produce one more unit of x_1.

(ii) The firm Y uses a_{21} unit of firm X for producing one more unit of x_2. Uses a_{22} number of units of own product for self consumption to produce one more unit of x_2 and atlast uses a_{23} number of units from Z to produce one more unit of x_2.

(iii) The firm Z uses a_{31} unit of firm X for producing one more unit of x_3. Uses a_{32} number of units from Y to produce one more unit of x_3 and atlast uses a_{33} number of units of own product for self production to produce one more unit of x_3.

The above tabular figures are expressed in linear form

Hence we can write

$$x_1 = a_{11}x_1 + a_{12}x_2 + a_{13}x_3 + d_1$$

In general form, the equation can be rewritten as

$$x_1 = \sum_{j=1}^{3} a_{ij} + d_i$$

= Need of each firm (including itself) + Final demand (Population Load).

Similarly x_2 and x_3 rewritten in the general form as expressed for x_1.

i.e.

$$x_2 = a_{21}x_1 + a_{22}x_2 + a_{23}x_3 + d_2,$$

$$x_3 = a_{31}x_1 + a_{32}x_2 + a_{33}x_3 + d_3.$$

The above equations can be written in matrix notation as

$$\begin{vmatrix} x_1 \\ x_2 \\ x_3 \end{vmatrix} = \begin{vmatrix} a_{11} & a_{12} & a_{13} \\ a_{21} & a_{22} & a_{23} \\ a_{31} & a_{32} & a_{33} \end{vmatrix} \begin{vmatrix} x_1 \\ x_2 \\ x_3 \end{vmatrix} + \begin{vmatrix} d_1 \\ d_2 \\ d_3 \end{vmatrix}$$

So the matrix form can represented as

$$X = AX + D$$

where

$$X = \begin{vmatrix} x_1 \\ x_2 \\ x_3 \end{vmatrix} \quad A = \begin{vmatrix} a_{11} & a_{12} & a_{13} \\ a_{21} & a_{22} & a_{23} \\ a_{31} & a_{32} & a_{33} \end{vmatrix} \text{ and } D = \begin{vmatrix} d_1 \\ d_2 \\ d_3 \end{vmatrix}$$

or $X - AX = D$,

i.e. $X(I - A) = D$, where I = Identity Matrix, since $IX = I$.

Hence $X = D(I - A)^{-1}$ where $(I - A)^{-1}$ is Leontief Inverse Matrix.

Here this is a case when the economy is functioning with only 3 firms. But this model can be extended to 'n' numbers of firms operating in the economy and the technique is similarly model based on above illustration.

Illustration No.1

A 2 sector economy has the following input – output coefficient matrix A and a final demand D

$$A = \begin{vmatrix} 0.3 & 0.2 \\ 0.2 & 0.5 \end{vmatrix} \quad D = \begin{vmatrix} 80 \\ 30 \end{vmatrix}$$

Find the gross output of each industry to meet the final demand?

Solution:

Let x_1 and x_2 be the gross out of both the sectors to meet the final demand.

The matrix notation for the 2 – sector is

$$AX + D = X$$

$$X(I - A) = D \leftrightarrow X = (I - A)^{-1} D$$

$$I - A = \begin{vmatrix} 1 & 0 \\ 0 & 1 \end{vmatrix} - \begin{vmatrix} 0.3 & 0.2 \\ 0.2 & 0.5 \end{vmatrix} = \begin{vmatrix} 1-0.3 & 0-0.2 \\ 0-0.2 & 1-0.5 \end{vmatrix}$$

$$= \begin{vmatrix} 0.7 & -0.2 \\ -0.2 & 0.5 \end{vmatrix}$$

Since the valve of (I – A) is calculated let's calculate the Leontief' s Inverse Matrix ..

i.e. $(I - A)^{-1}$

Since the adjoint (I – A) $= \begin{vmatrix} 0.5 & 0.2 \\ 0.2 & 0.7 \end{vmatrix}$

i.e. Adj (I – A) $= \begin{vmatrix} 0.5 & 0.2 \\ 0.2 & 0.7 \end{vmatrix}$

And the determinant of

$$(I - A) = \begin{vmatrix} 0.7 & -0.2 \\ -0.2 & 0.5 \end{vmatrix} = (0.7) \times (0.5) - (-0.2) \times (-0.2)$$

$$= 0.35 - 0.04 = 0.31$$

Therefore Det (I – A) = 0.31

Since $- A) -1 = \dfrac{\text{Adj}(I-A)}{\text{Det}(I-A)}$

$$= \frac{1}{0.31} \begin{vmatrix} 0.5 & 0.5 \\ 0.2 & 0.7 \end{vmatrix}$$

But $X = (I - A)^{-1} D$

$$\begin{vmatrix} x_1 \\ x_2 \end{vmatrix} = \frac{1}{0.31} \begin{vmatrix} 0.5 & 0.2 \\ 0.2 & 0.7 \end{vmatrix} \times \begin{vmatrix} 80 \\ 30 \end{vmatrix} = \frac{1}{0.31} \begin{vmatrix} 0.5 \times 80 & + & 0.2 \times 30 \\ 0.2 \times 80 & + & 0.7 \times 30 \end{vmatrix}$$

$$= \frac{1}{0.31} \begin{vmatrix} 40 & + & 6 \\ 16 & + & 21 \end{vmatrix} = \frac{1}{0.31} \begin{vmatrix} 46 \\ 37 \end{vmatrix} = \begin{vmatrix} 46/0.31 \\ 37/0.31 \end{vmatrix}$$

$$\begin{vmatrix} x_1 \\ x_2 \end{vmatrix} = \begin{vmatrix} 148.3 \\ 119.3 \end{vmatrix}$$

Hence the 2 - sector produces approximately 148.3 and 119.3 units to fulfill the requirement of final demand.

Illustration No.2

Note: This illustration will exhibit how to calculate the technical coefficients Matrix.

Given the following transaction matrix, frame the technical coefficient matrix and also find the gross output to meet the final demand of 300 units of Chemical industry and 400 units of Paint industry

Production	**Chemical Industry**	**Paint Industry**	**Final Demand**
Chemical Industry	200	300	200
Paint Industry	300	400	300

Solution:

The gross output of Chemical Industry = 200 + 300 + 200 = 700.

Similarly the gross output of Paint industry = 300 + 400 + 300 = 1000

(***Note:*** Total demand = Industry Demand + Final Demand)

Production	**Chemical Industry**	**Paint Industry**	**Final Demand**
Chemical Industry	200	300	200
Paint Industry	300	400	300

** **Doubled head Arrow points self consumption, Single arrow refers the contribution to other industry**

Let A be the technical Coefficient matrix represented as

$$= \begin{vmatrix} a_{11} & a_{12} \\ a_{21} & a_{22} \end{vmatrix}$$

For Chemical Industry,

we have

a_{11} = 200/700 = 0.28.

(Chemical Industry's output being utilized for self consumption)

a_{12} = 300/1000 = 0.30.

(Chemical industry's output being utilized by paint industry as input for its output i.e. the contribution of chemical industry share in production of a unit of output in paint industry)

Similarly

For

a_{21} = 300/700 = 0.42.

(Paint industry's output being utilized by chemical industry as input for its output i.e. the contribution of Paint industry share in production of a unit of output in Chemical industry).

a_{22} = 400/1000 = 0.40

(paint Industry's output being utilized for self consumption).

$$\text{Therefore the technical matrix} = \begin{vmatrix} 0.28 & 0.30 \\ 0.42 & 0.40 \end{vmatrix}$$

Let's assume that x_1 and x_2 be the output of Chemical and paint industry

The matrix equation for both industry are

$$AX + D = X$$

$$X (I - A) = D \text{ or } X = (I - A)^{-1} D$$

Calculating Leontief Inverse Matrix

i.e. $$(I - A)^{-1} = \frac{Adj(I - A)}{Det(I - A)}$$

$$\text{Therefore } I - A = \begin{vmatrix} 1 & 0 \\ 0 & 1 \end{vmatrix} - \begin{vmatrix} 0.28 & 0.30 \\ 0.42 & 0.40 \end{vmatrix} = \begin{vmatrix} 0.72 & -0.30 \\ -0.42 & 0.60 \end{vmatrix}$$

$$Adj\ (I - A) = \begin{vmatrix} 0.60 & 0.30 \\ 0.42 & 0.72 \end{vmatrix}$$

$$Det\ (I - A) = \begin{vmatrix} 0.72 & -0.30 \\ -0.42 & 0.60 \end{vmatrix} = 0.72 \times 0.60 - (-0.30) \times (-0.42)$$

$$= 0.432 - 0.126 = 0.306.$$

Therefore the Leontief Inverse Matrix

$$(I - A)^{-1} = \frac{Adj(I - A)}{Det(I - A)} = 1/0.306 \begin{vmatrix} 0.60 & 0.30 \\ 0.42 & 0.72 \end{vmatrix}$$

The gross output of both the industry that needs to fulfill the final demand Of 300 and 400 units are

$$\begin{vmatrix} x_1 \\ x_2 \end{vmatrix} = \frac{1}{0.306}\begin{vmatrix} 0.60 & 0.30 \\ 0.42 & 0.72 \end{vmatrix}\begin{vmatrix} 300 \\ 400 \end{vmatrix} = \frac{1}{0.306}\begin{vmatrix} 180 & + & 120 \\ 126 & + & 288 \end{vmatrix}$$

$$= \begin{vmatrix} 300/0.306 \\ 414/0.306 \end{vmatrix} = \begin{vmatrix} 980.39 \\ 1352.94 \end{vmatrix}$$

Hence chemical and paints industry needs to produce an output equivalent to 980.39 units and 1352.94 units to satisfy the final demand.

Simon–Hawkins Conditions

In order to make input–output model viable two essential conditions needs to be satisfied, they are

(i) The determinant of (I – A) matrix cannot be zero i.e. Det (I – A) > 0, a positive number. Since a zero determinant will convert the model into an indeterminate form. As we need to calculate the Leontief Inverse Matrix, if determinant found zero then the inverse matrix cannot be obtained (calculated). Further the elements of the inverse matrix cannot be negative because increase in final demand will reduce output which is technically illogical.

(ii) The diagonal elements $1- a_{11}$, $1- a_{22}$, $1-a_{33}$ ....$1-a_{mn}$ of the (I – A) matrix needs to be positive. The logic behind is that an industry cannot use more of its own output as an input for further production generation. So the range of $1- a_{ij}$ is given by $0 < 1- a_{ij} < 1$.

Illustration No. 3

An economy produces two types of commodities x and y serving as an intermediate inputs for each other's production with 0.3 units of x and 0.6 units of y are required to produce an unit of x. Similarly 0.2 units of x and 0.7 unit of y are required to produce an unit of y, find the viability of the system. And if the economy needs 80 units of x and 40 units of y, calculate the gross output of the two commodities.

Solution:

Lets frame the input–output coefficient table.

Output	Input	
	x	**y**
X	0.3	0.2
Y	0.6	0.7

From the above table the coefficient matrix

$$A = \begin{vmatrix} 0.3 & 0.2 \\ 0.6 & 0.7 \end{vmatrix}$$

$$\text{The Leontief matrix } (I - A) = \begin{vmatrix} 1 & 0 \\ 0 & 1 \end{vmatrix} - \begin{vmatrix} 0.3 & 0.2 \\ 0.6 & 0.7 \end{vmatrix} = \begin{vmatrix} 0.7 & -0.2 \\ -0.6 & 0.3 \end{vmatrix}$$

In order to prove the model viable it needs to prove both the Simon Hawkins Condition

1. Det (I – A) > 0.
2. $1 - a_{11}$, $1 - a_{22}$ both be positive.

Therefore

Det (I – A) = 0.7 × 0.3 – (– 0.2) × (– 0.6) = 0.09 a +ve number hence 1st condition is satisfied.

Then both diagonal elements are positive and less than 1 hence even the 2nd condition is satisified.

And hence the system is viable.

With the available data we construct the system equation

Let x_1 and x_2 gross output of both commodities then

We have

$$0.3x_1 + 0.2\,x_2 + 80 = x_1$$

$$0.6x_1 + 0.7x_2 + 40 = x_2$$

$$\begin{vmatrix} 0.3 & 0.2 \\ 0.6 & 0.7 \end{vmatrix} \begin{vmatrix} x_1 \\ x_2 \end{vmatrix} + \begin{vmatrix} 80 \\ 40 \end{vmatrix} = \begin{vmatrix} x_1 \\ x_2 \end{vmatrix}$$

i.e.

AX + D = X

where $A = \begin{vmatrix} 0.3 & 0.2 \\ 0.6 & 0.7 \end{vmatrix}$ $D = \begin{vmatrix} 80 \\ 40 \end{vmatrix}$ and $X = \begin{vmatrix} x_1 \\ x_2 \end{vmatrix}$

Calculating the Leontief Inverse Matrix

$$(I - A)^{-1} = \frac{Adj(I - A)}{Det(I - A)}$$

$$(I - A) = \begin{vmatrix} 0.3 & 0.2 \\ 0.6 & 0.7 \end{vmatrix}$$

And the adjoint of I – A matrix is given by

Adj $(I - A) = \begin{vmatrix} 0.3 & 0.2 \\ 0.6 & 0.7 \end{vmatrix}$ and the Det (I – A) = 0.09.

∴ the Leontief Inverse matrix = $\frac{1}{0.09}\begin{vmatrix} 0.3 & 0.2 \\ 0.6 & 0.7 \end{vmatrix}$

Since

$$X = (I - A)^{-1} D = \frac{1}{0.09}\begin{vmatrix} 0.3 & 0.2 \\ 0.6 & 0.7 \end{vmatrix}\begin{vmatrix} 80 \\ 40 \end{vmatrix}$$

$$= \frac{1}{0.09}\begin{vmatrix} 32 \\ 76 \end{vmatrix} \begin{vmatrix} 32/0.09 \\ 76/0.09 \end{vmatrix} = \begin{vmatrix} 355.5 \\ 844.4 \end{vmatrix}$$

Hence the total output x_1 = 355.5 and x_2 = 844.4.

Illustration No. 4

The inter–industry relationship between two industries are P and Q are given below

Industries	P	Q	Final Demand
P	60	80	60
Q	70	90	40

If the gross output is raised to 300 and 400 find the demand which can satisfied. Also check the viability of the system.

Solution:

The gross output of P = 60 + 80 + 60 = 200.

Similarly the gross output of Q = 70 + 90 + 40 = 200

(*Note: Total demand = Industry Demand + Final Demand*)

Let A be the technical Coefficient matrix represented as = $\begin{vmatrix} a_{11} & a_{12} \\ a_{21} & a_{22} \end{vmatrix}$

For Industry P, we have

$a_{11} = 60/200 = 0.30.$

$a_{12} = 80/200 = 0.40$

Similarly for Q, we have

$a_{21} = 70/200 = 0.35.$

$a_{22} = 90/200 = 0.45$

Therefore the technical matrix = $\begin{vmatrix} 0.30 & 0.40 \\ 0.35 & 0.45 \end{vmatrix}$

Let's assume that d_1 and d_2 be the final demand that needs to be satisfied when the gross output increases to 300 and 400.

The matrix equation for both industry are

$$AX + D = X$$

$$X (I - A) = D$$

Therefore $$I - A = \begin{vmatrix} 1 & 0 \\ 0 & 1 \end{vmatrix} - \begin{vmatrix} 0.30 & 0.40 \\ 0.35 & 0.45 \end{vmatrix} = \begin{vmatrix} 0.70 & -0.40 \\ -0.35 & 0.55 \end{vmatrix}$$

Hence

$$\begin{vmatrix} d_1 \\ d_2 \end{vmatrix} = \begin{vmatrix} 0.70 & -0.40 \\ -0.35 & 0.55 \end{vmatrix} \begin{vmatrix} 300 \\ 400 \end{vmatrix} = \begin{vmatrix} 50 \\ 115 \end{vmatrix}$$

Since $$I - A = \begin{vmatrix} 0.70 & -0.40 \\ -0.35 & 0.55 \end{vmatrix}$$

But to test the viability of the system we need to check Simon Hawkins Condition.

Therefore the determinant of $(I - A) = 0.70 \times 0.55 - (-0.40) \times (-0.35)$

$= 0.245\ (\neq 0).$

And the diagonal elements are positive and hence the both the Simon Hawkins condition are satisfied.

Solution for a 2 – Sector Model with Primary Inputs

Following are the details are furnished in the table

Sector	X	Y	Final Demand
X	a_{11}	a_{12}	d_1
Y	a_{21}	a_{22}	d_2
Primary Input	l_1	l_2	

Where l_1 and l_2 are the labour requirements for producing one unit of X and Y respectively.

Since

$$X = (I - A)^{-1} D$$

Therefore

$$X = \begin{vmatrix} (1-a_{11}) & -a_{21} \\ -a_{12} & (1-a_{22}) \end{vmatrix}^{-1} \begin{vmatrix} d_1 \\ d_2 \end{vmatrix}$$

$$= \frac{1}{D} \begin{vmatrix} A_{11} & A_{12} \\ A_{21} & A_{22} \end{vmatrix} \begin{vmatrix} d_1 \\ d_2 \end{vmatrix}$$

Where D is the determinant of (I – A) matrix and A_{11}, A_{12}..... are the cofactors of the matrix (I – A)

Hence the total labour requirement is

$$L X = (l_1 \; l_2) \begin{vmatrix} X_1 \\ X_2 \end{vmatrix} = l_1X_1 + l_2X_2$$

In Case of 2 Primary Inputs Utilized

In this case instead of one primary input two inputs are utilized then the primary input coefficient can be arranged in the following way

Sector	X	Y
Input I_1	l_{11}	l_{12}
Input I_2	l_{21}	l_{22}

Hence the requirement will be given as

$$\begin{vmatrix} I_1 \\ I_2 \end{vmatrix} = \begin{vmatrix} l_{11} & l_{12} \\ l_{21} & l_{22} \end{vmatrix} (X) \qquad \text{where} \quad X = (I - A)^{-1} D$$

Let us understand the above topic by an illustration

Illustration No. 5

An economy produces two types of commodities Copper and Iron serving as an intermediate inputs for each other's production with 0.4 units of Copper and 0.5 units of Iron are required to produce an unit of copper. Similarly 0.3 units of copper and 0.6 unit of iron are required to produce an unit of iron. Find the viability of the system. If 3 and 4 labor days are required to produce an unit of x and y and if the economy needs 100 units of copper and 60 units of iron, calculate the gross output of the two commodities and the total labour requirement ?

Solution:

Let's frame the input–output coefficient table.

Output	**Input**	
	Copper	**Iron**
Copper	0.4	0.5
Iron	0.3	0.6

From the above table the coefficient matrix

1. $A = \begin{vmatrix} 0.4 & 0.5 \\ 0.3 & 0.6 \end{vmatrix}$

2. The Leontief matrix $(I - A) = \begin{vmatrix} 1 & 0 \\ 0 & 1 \end{vmatrix} - \begin{vmatrix} 0.4 & 0.5 \\ 0.3 & 0.6 \end{vmatrix} = \begin{vmatrix} 0.6 & -0.5 \\ -0.3 & 0.4 \end{vmatrix}$

   In order to prove the model viable it needs to prove both the Simon Hawkins Condition

3. $\text{Det}\,(I - A) > 0$.

4. $1 - a_{11}$, $1 - a_{22}$ both be positive.

Therefore

$\text{Det}\,(I - A) = 0.6 \times 0.4 - (-0.5) \times (-0.3) = 0.09$ a +ve number hence 1st condition is satisfied.

Then both diagonal elements are positive and less than 1 hence even the 2nd condition is satisfied.

And hence the system is viable.

With the available data we construct the system equation

Let x_1 and x_2 gross output of both commodities then

We have

$$0.4x_1 + 0.5x_2 + 100 = x_1$$

$$0.3x_1 + 0.6x_2 + 60 = x_2$$

$$\begin{vmatrix} 0.4 & 0.5 \\ 0.3 & 0.6 \end{vmatrix} \begin{vmatrix} x_1 \\ x_2 \end{vmatrix} + \begin{vmatrix} 100 \\ 60 \end{vmatrix} = \begin{vmatrix} x_1 \\ x_2 \end{vmatrix}$$

i.e.

$$AX + D = X$$

where

$$A = \begin{vmatrix} 0.4 & 0.5 \\ 0.3 & 0.6 \end{vmatrix} \quad D = \begin{vmatrix} 100 \\ 60 \end{vmatrix} \quad \text{and} \quad X = \begin{vmatrix} x_1 \\ x_2 \end{vmatrix}$$

Calculating the Leontief Inverse Matrix

$$(I - A)^{-1} = \frac{\text{Adj } (I - A)}{\text{Det } (I - A)}$$

$$(I - A) = \begin{vmatrix} 0.6 & -0.5 \\ -0.3 & 0.4 \end{vmatrix}$$

And the adjoint of I – A matrix is given by

$$\text{Adj } (I - A) = \begin{vmatrix} 0.4 & 0.5 \\ 0.3 & 0.6 \end{vmatrix} \quad \text{and the Det } (I - A) = 0.09.$$

Therefore the Leontief Inverse matrix $= \frac{1}{0.09}\begin{vmatrix} 0.4 & 0.5 \\ 0.3 & 0.6 \end{vmatrix}$

Since

$$X = (I - A)^{-1} D = \frac{1}{0.09}\begin{vmatrix} 0.4 & 0.5 \\ 0.3 & 0.6 \end{vmatrix} \begin{vmatrix} 100 \\ 60 \end{vmatrix}$$

$$\frac{1}{0.09}\begin{vmatrix} 70 \\ 60 \end{vmatrix} = \begin{vmatrix} 70 / 0.09 \\ 66 / 0.09 \end{vmatrix} = \begin{vmatrix} 777.78 \\ 733.34 \end{vmatrix}$$

Hence the total output $x_1 = 777.78$ and $x_2 = 733.34$

But to produce 1 unit of Copper requires 3 labour days and 1 unit of iron requires 4 labour days then the total amount of labour required

$$\begin{vmatrix} 3 & 4 \end{vmatrix} \begin{vmatrix} 777.78 \\ 733.34 \end{vmatrix} = 2333.34 + 2933.33 = 5266.67 \text{ labour days.}$$

Determination of Equilibrium Prices (2 – Sector Model)

Let us assume a 2–sector model with two commodities X and Y priced at p_1 and p_2 and the primary factor priced at w respectively.

The technology or transaction matrix can be expressed in value forms as

Sector	Commodity X	Commodity Y	Final Demand
1	$a_{11}x_1p_1$	$a_{12}x_2p_1$	d_1p_1
2	$a_{21}x_1p_2$	$a_{22}x_2p_2$	d_2p_2
Primary Input	l_1x_1w	l_2x_2w	
Total Cost	$a_{11}x_1p_1 + a_{21}x_1p_2 + l_1x_1w$	$a_{12}x_2p_1 + a_{22}x_2p_2 + l_2x_2w$	

In case of pure competition revenue matches cost i.e. a case of zero profit.

1st Industry

Revenue = Output × Price.

= $x_1 \times p_1$.

Total Cost = $a_{11}x_1p_1 + a_{21}x_1p_2 + l_1x_1w$

Since Revenue = Total Cost

$$x_1 p_1 = a_{11}x_1p_1 + a_{21}x_1p_2 + l_1x_1w$$

Similarly for the 2nd Industry

Revenue = Total Cost

$$x_2 p_2 = a_{12}x_2p_1 + a_{22}x_2p_2 + l_2x_2w$$

Rearranging the above two equations we have,

$$p_1 - a_{11}p_1 - a_{21}p_2 = l_1w$$

$$p_2 - a_{12}p_1 - a_{22}p_2 = l_2w$$

or

$$(1 - a_{11})\, p_1 - a_{21}p_2 = l_1w$$

$$- a_{12}\, p_1 + (1 - a_{22})\, p_2 = l_2w$$

The above equations can be expressed in Matrix Form as

$$\begin{vmatrix} (1-a_{11}) & -a_{21} \\ -a_{12} & (1-a_{22}) \end{vmatrix} \begin{vmatrix} p_1 \\ p_2 \end{vmatrix} = \begin{vmatrix} l_1 w \\ l_2 w \end{vmatrix}$$

Or

$(I - A)^T P = L$ where L stands value added coefficients.

Or

$P = [(I - A)^T]^{-1} L$

i.e.

$$\begin{vmatrix} p_1 \\ p_2 \end{vmatrix} = \begin{vmatrix} (1 - a_{11}) & - a_{21} \\ - a_{12} & (1 - a_{22}) \end{vmatrix}$$

Illustration No. 6

An economy consists two sectors manufacturing and agriculture intermediate inputs for each other's production are 0.6 units of manufacturing and 0.20 units of agriculture are required to produce an unit of manufacturing output. Similarly 0.4 units of manufacturing and 0.55 unit of agriculture are required to produce a unit of agriculture, find the viability of the system. If 7 and 5 labor days are required to produce an unit of manufacturing and agriculture unit and if the total requirements are 200 units of manufacture and 150 units of agriculture, calculate the gross output of the two sectors and the total labour requirement? If the wage rate is Rs 50 per day find the equilibrium prices of the both the sectors?

Solution:

Lets frame the input–output coefficient table.

Output	Input	
	Manufacturing	**Agriculture**
Manufacturing	0.60	0.20
Agriculture	0.40	0.55

From the above table the coefficient matrix

$$A = \begin{vmatrix} 0.60 & 0.20 \\ 0.40 & 0.55 \end{vmatrix}$$

$$\text{The Leontief matrix } (I - A) = \begin{vmatrix} 1 & 0 \\ 0 & 1 \end{vmatrix} - \begin{vmatrix} 0.60 & 0.20 \\ .040 & 0.55 \end{vmatrix} = \begin{vmatrix} 0.4 & -0.2 \\ -0.4 & 0.45 \end{vmatrix}$$

In order to prove the model viable it needs to prove both the Simon Hawkins Condition

Det (I – A) > 0, and

1– a_{11}, 1-a_{22} both be positive.

Therefore

Det (I – A) = 0.4 × 0.45 – (– 0.2) × (– 0.4) = 0.10, a +ve number hence 1st condition is satisfied.

Then both diagonal elements are positive and less than 1 hence even the 2nd condition is satisfied. And hence the system is viable.

With the available data we construct the system equation

Let x_1 and x_2 gross output of both commodities then,

We have

$$0.60x_1 + 0.20\,x_2 + 200 = x_1$$

$$0.40x_1 + 0.55x_2 + 150 = x_2$$

$$\begin{vmatrix} 0.60 & 0.2 \\ 0.40 & 0.55 \end{vmatrix} \begin{vmatrix} x_1 \\ x_2 \end{vmatrix} + \begin{vmatrix} 200 \\ 150 \end{vmatrix} = \begin{vmatrix} x_1 \\ x_2 \end{vmatrix}$$

i.e.

$$AX + D = X$$

where

$$A = \begin{vmatrix} 0.60 & 0.2 \\ 0.40 & 0.55 \end{vmatrix} \quad D = \begin{vmatrix} 200 \\ 150 \end{vmatrix} \quad \text{and } X = \begin{vmatrix} x_1 \\ x_2 \end{vmatrix}$$

Calculating the Leontief Inverse Matrix

$$(I - A)^{-1} = \frac{Adj(I-A)}{Det(I-A)}$$

$$(I - A) = \begin{vmatrix} 0.4 & -0.2 \\ -0.6 & 0.45 \end{vmatrix}$$

And the adjoint of I – A matrix is given by

$$Adj\ (I - A) = \begin{vmatrix} 0.4 & 0.2 \\ 0.6 & 0.45 \end{vmatrix} \text{ and the Det } (I - A) = 0.10.$$

Therefore the Leontief Inverse matrix = $\frac{1}{0.10}\begin{vmatrix} 0.45 & 0.2 \\ 0.6 & 0.4 \end{vmatrix}$

Since

$$X = (I - A)^{-1} D = \frac{1}{0.10}\begin{vmatrix} 0.45 & 0.2 \\ 0.6 & 0.4 \end{vmatrix}\begin{vmatrix} 200 \\ 150 \end{vmatrix}$$

$$\frac{1}{0.10}\begin{vmatrix} 120 \\ 180 \end{vmatrix}\begin{vmatrix} 120/0.10 \\ 180/0.10 \end{vmatrix} = \begin{vmatrix} 1200 \\ 1800 \end{vmatrix}$$

Hence the total output x_1 = 1200 and x_2 = 1800

But to produce 1 unit of x requires 7 labour days and 1 unit of y requires 5 labor days then the total output will require

$$\begin{vmatrix} 7 & 5 \end{vmatrix}\begin{vmatrix} 1200 \\ 1800 \end{vmatrix} = 8400 + 9000 = 17400 \text{ labour days.}$$

The equilibrium prices when the wage rate Rs 50 per day is given by

$$\begin{vmatrix} p_1 \\ p_2 \end{vmatrix} = \begin{vmatrix} (1 - a_{11} & -a_{21}) \\ -a_{12} & (1 - a_{22}) \end{vmatrix}^{-1}\begin{vmatrix} l_1 w \\ l_2 w \end{vmatrix}$$

$$= \frac{1}{0.10}\begin{vmatrix} 0.45 & 0.2 \\ 0.6 & 0.4 \end{vmatrix}\begin{vmatrix} 350 \\ 250 \end{vmatrix}$$

$$= \frac{1}{0.10}\begin{vmatrix} 157.50 + 50 \\ 210 + 100 \end{vmatrix} = \begin{vmatrix} 2075 \\ 3100 \end{vmatrix}$$

The equilibrium prices of both the sector are p_1 = 2075 and p_2 = 3100.

Importance of Input–Output Analysis

- The ultimate goal of the Input-Output Analysis technique is to generate a Total Requirements Table that shows the flows of between industries in the production of output for a given sector.

- To arrive at this final result, IO Analysis requires two earlier steps:
  1. *Transactions table*: Contains basic data on the flows of goods and services among suppliers and purchasers during a study year.
  2. *Direct requirements table*: Derived from the transactions table, this shows the inputs required directly from different suppliers by each intermediate purchaser for each unit of output that purchaser produces.
- "Input output analysis can be thought of as documenting and exploring the precise systems of inter-industry exchange through which different components of products become different components of income."
- Input output analysis is a good forecasting technique in order to understand and estimate future requirements.

Practical Issues

- Data needs and complexity: IO models are tremendously complex and data hungry. This typically places these models in the hands of experts.

Theoretical Issues

- *Time/Data issues*: Usually a single year's data are used to develop the Total Requirements Table. But 1. purchases may actually reflect a longer term investment and 2. short term trends may impact the data.
- *Stability of the technical coefficients*: Technology changes, prices change, and demand changes, all affecting the coefficients in the Total Requirement Table. This can impact the results if the coefficients are "out of date".
- IO assumes a linear relationship between increasing demand for inputs and outputs: This assumes 1) externalities and 2) increasing/decreasing returns to scale.
- *Industrial categorization*: IO models still assume that each industry (1) has a single, homogeneous production function and (2) each produces one product. These assumptions do not reflect the real economy very well.

Finally despite all these problems IO analysis stands tremendously popular and powerful analytical tool.

Review Questions

1. Write about Lieontief input-output model and its significance?
2. Write short notes on
   (a) Closed type Model.
   (b) Open type Model.
3. Derive Lientiof Open type mathematical model?
4. What are the significant assumptions in input-output modeling?
5. Write about the advantages and disadvantages of using input-output modeling?
6. Derive the dynamic input–output model?

Self Practice Exercises

1. For a 2 sector economy the following transaction matrix is furnished below

Sector	Sales		Final Demand	Gross Output
	X	Y		
X	5	4	11	20
Y	8	6	9	24

   (i) Find the technical coefficient matrix?
   (ii) Also find the change in gross output if the final demand increases to 22 and 18 units?

2. The following table gives the inter industry flow of two industries

Industries	Consumption		Domestic Demand	Gross Output
	I	II		
I	300	500	200	1000
II	400	1000	600	2000

   Calculate the technological coefficient matrix and if the domestic demand changes to 400 and 1200 calculate the gross output?

3. An 2–sector economy has the following input–output coefficient matrix A and final demand D as

$$A = \begin{pmatrix} 0.4 & 0.3 \\ 0.6 & 0.1 \end{pmatrix} \text{ and } D = \begin{pmatrix} 110 \\ 80 \end{pmatrix}$$

   Check for the viability of the system and also calculate the gross output of each sector to meet the final demand?

4. Two industries X and Y input–output relationship is furnished below along with final demand

Sectors	X	Y	Final Demand
X	100	120	80
Y	80	100	70

If the gross output of X and Y increases to 600 and 450 respectively then determine the final demand which can be satisfied. And also check the Simon Hawkins condition?

5. Given below is the transaction matrix of a 3–sector economy.

Sectors	A	B	C	Final Demand
A	40	90	30	50
B	60	80	40	70
C	30	60	50	80

(i) Construct the technological matrix and Check for the viability of the system?

6. Two commodities A and B are interdependent on each for their production. A unit of A requires 0.3 units of A and 0.8 units of B whereas a unit of B requires 0.2 of A and 0.7 units of B. If there is external demand of 50 units of A and 25 units B then, calculate the gross output of the two commodities and if the labor hrs required for A and B are 4 and 3 then find the total labor hrs required?

7. An economy has two industries Agriculture and Manufacturing with following input–out table

Sectors	Agriculture	Manufacturing	Final Demand
Agriculture	0.4	0.4	70
Manufacturing	0.6	0.3	60

(i) Check for the viability of the system?

(ii) If 6 and 4 labour hrs are required to produced an unit of Agriculture and manufacturing output then find the total labour requirement?

(iii) Also calculate the gross output and if the wage rate is Rs. 8 per hour find the equilibrium prices?

8. An economy has only two sector producing only coal and steel. Both the commodities produce serve as the intermediate inputs in each other production. To produce a unit of steel requires 0.6 units of steel and 0.3 units of coal. Similarly to produce a unit of coal requires 0.4 units of steel and 0.5 units of coal. And if the domestic demand for both the commodity are 100 and 120 units. Then find the gross output of coal and steel and if 2 and 3 labour days are required to an unit of coal and steel then find the total labour days required for the production?

Answers

1. $\begin{pmatrix} 0.30 & 0.25 \\ 0.40 & 0.5 \end{pmatrix}$, Gross Output of X = 139 units (approx.) and Y = 98 Units (approx.)
2. $\begin{pmatrix} 0.30 & 0.25 \\ 0.40 & 0.5 \end{pmatrix}$, Gross Output are 2000 and 4000 units.
3. System is viable and the gross output are 273 and 179 units.
4. Simon Hawkins condition is valid and the gross output to meet the changed demand conditions are 2078 units and 1645.6 units.
5. $\begin{pmatrix} 0.19 & 0.36 & 0.13 \\ 0.28 & 0.32 & 0.18 \\ 0.14 & 0.24 & 0.22 \end{pmatrix}$ and Simon Hawkins condition is valid.
6. Gross Output of A : 400 and B : 1150 and Total Labour (hrs) = 6100.
7. Gross Output Ag. = 288.8 and Manu. = 433.3, Total Labor (hrs) = 3466 and equilibrium prices are Rs. 257.7 and Rs. 266.6.
8. Gross Output Steel: 1225 and Coal: 975 and total lab. Days = 5375.

CHAPTER 13

Theory of Equation and Inequation

Introduction

Any two algebraic expressions which are connected by an equality sign (=) are termed as equation. For example, $ax + b = 0$ is a linear equation or an equation of degree one where a and b are constants and x is unknown variable.

Some Useful Definitions

Polynomial

Let a_0, a_1, a_2 ... a_n be real numbers and x is a variable then the expression $f(x) = a_0 + a_1x + a_2x^2 + ... + a_n x^n$ is called as a polynomial of real variable x with real coefficients.

Degree of an Equation

The highest power of the variable x of a non zero coefficient of a polynomial or an equation is the degree of that polynomial or equation.

If, $f(x) = a_0 + a_1x + a_2x^2 + ...+ a_nx^n$ where $a_n \neq 0$ is called as the nth degree equation.

For example

$3x^2 + 4x + 7$ is an equation of degree 2 and $x^3 + 3x^2 + 4x + 5$ is an equation of degree 3.

Solution of an Equation

The value or values of x which satisfies a given equation is called as the solution of that equation otherwise termed as the roots of the equation. The set of values of the variable x are the solution set of the equation.

For example

Let $x - 3 = 0$ is a equation of degree one. Now by substituting $x = 3$ in the equation both the sides turns equal. Thus it shows that the equation has a single solution ($x = 3$) and no other value except 3 will satisfy the equation. Hence the equation will have a single root.

Linear Equation

An equation whose unknown variable x with non–zero coefficient having the highest power as one are termed as linear equations or equations whose standard form can be expressed as $ax \pm b = 0$ are linear equations.

For example

$3x + 7$ is a linear equation.

Quadratic Equation

An equation of the form $ax^2 + bx + c = 0$, a, b, c $\in$ R, $a \neq 0$ is called as quadratic equation or an equation of second degree. And if $b = 0$ then the equation gets reduced to $ax^2 + c = 0$ which is called as pure equation.

For example

$2x^2 + 5x + 8$ is a quadratic equation or equation of degree 2.

Solution of a Quadratic Equation

Some of the common methods of solving quadratic equations are

(i) Factorization Method

In this method the quadratic equation is split into linear factors say $(ax + b)$ and $(cx + d)$ and equate them to zero separately to find the values of the unknown variable x

(ii) Method of Computing Perfect Square

In this method the equation $ax^2 + bx + c = 0$ is rewritten as $ax^2 + bx = - c$

Dividing both side by 'a 'we get,

$$x^2 + \frac{b}{a}x = -\frac{c}{a}$$

Adding both side by $\frac{b^2}{4a^2}$, we get

$$x^2 + \frac{b}{a}x + \frac{b^2}{4a^2} = -\frac{c}{a} + \frac{b^2}{4a^2}$$

$$\left(x + \frac{b}{2a}\right)^2 = \frac{b^2 - 4ac}{4a^2}$$

Therefore $$x + \frac{b}{2a} = \frac{\pm\sqrt{b^2 - 4ac}}{2a}$$

Or $$x = \frac{-b \pm \sqrt{b^2 - 4ac}}{2a}$$

Or $$x = \frac{-b + \sqrt{b^2 - 4ac}}{2a}, \frac{-b - \sqrt{b^2 - 4ac}}{2a}$$

are the required values of x. The values of x are called as the roots of the equation $ax^2 + bx + c = 0$.

Illustration No.1

Solve

$$x^2 - 16 = 0$$

Solution:

$$x^2 - 16 = 0$$

$$x^2 - 4^2 = 0$$

$$(x - 4)(x + 4) = 0 \qquad \text{Since } a^2 - b^2 = (a - b)(a + b)$$

Or $$(x - 4) = 0 \text{ or } (x + 4) = 0$$

That implies x = 4 or – 4

Therefore the solution set of the equation is {4, –4}

Illustration No. 2

Solve $x^2 - 10x + 16 = 0$

Solution:

$$x^2 - 10x + 16 = 0$$

$$x^2 - 8x - 2x + 16 = 0$$

$$x(x - 8) - 2(x - 8) = 0$$

or $$(x - 2)(x - 8) = 0$$

that implies $$x = 2 \text{ or } 8$$

Alternatively

From the equation $x^2 - 10x + 16 = 0$.

We have,

$a = 1$, $b = -10$ and $c = 16$

Since we have,

$$x = \frac{-b \pm \sqrt{b^2 - 4ac}}{2a}$$

$$= \frac{-(-10) \pm \sqrt{(-10)^2 - 4.1.16}}{2.1}$$

$$= \frac{-(-10) \pm \sqrt{(-10)^2 - 4.1.16}}{2}$$

$$= \frac{10 \pm \sqrt{36}}{2}$$

$$= \frac{10 \pm 6}{2}$$

therefore,

$$x = 8, 2$$

Illustration No. 3

Solve $3x^2 - 2x + \frac{1}{3} = 0$

Solution:

$$3x^2 - 2x + \frac{1}{3} = 0$$

$$9x^2 - 6x + 1 = 0$$

$$9x^2 - 3x - 3x + 1 = 0$$

$$3x(3x - 1) - 1(3x - 1) = 0$$

$$(3x - 1)(3x - 1) = 0$$

That implies the required solution is $x = \frac{1}{3}$

Therefore the solution set is $x = \frac{1}{3}, \frac{1}{3}$

Illustration No. 4

Solve $\frac{3}{(x+2)}+\frac{1}{(x+3)}=\frac{5}{(x+5)}$

Solution:

$$\frac{3}{(x+2)}+\frac{1}{(x+3)}=\frac{5}{(x+5)}$$

Multiplying with the common factors we have

$$\frac{3(x+3)+(x+2)}{(x+2)(x+3)}=\frac{5}{(x+5)}$$

$$(x+5)\{3(x+3)+(x+2)\} = 5(x+2)(x+3)$$

$$3(x+5)(x+3)+(x+5)(x+2) = 5(x+2)(x+3)$$

$$3(x^2 + 8x + 15) + (x^2 + 7x + 10) = 5(x^2 + 5x + 6)$$

$$(3x^2 + 24x + 45) + (x^2 + 7x + 10) = (5x^2 + 25x + 30)$$

$$4x^2 + 31x + 55 = 5x^2 + 25x + 30$$

$$x^2 - 6x - 25 = 0$$

But

$$x = \frac{-b \pm \sqrt{b^2-4ac}}{2a}$$

Since a = 1, b = – 6 and c = – 25

$$= \frac{-(-6) \pm \sqrt{(-6)^2 - 4.1.(-25)}}{2.1}$$

$$= \frac{(6) \pm \sqrt{36+100}}{2}$$

$$= \frac{(6) \pm \sqrt{136}}{2}$$

$$= \frac{6 \pm 2\sqrt{34}}{2}$$

$$= 3 \pm \sqrt{34}$$

Therefore $x = 3+\sqrt{34}$ or $3-\sqrt{34}$

Illustration No. 5

Solve $\sqrt{\frac{x}{1-x}} + \sqrt{\frac{1-x}{x}} = 4\frac{1}{4}$

Solution:

Let us assume $\sqrt{\frac{x}{1-x}} = y$

$$y + \frac{1}{y} = \frac{17}{4}$$

$$4y^2 + 4 = 17y$$

$$4y^2 - 17y + 4 = 0$$

$$4y^2 - 16y - y + 4 = 0$$

$$4y(y-4) - 1(y-4) = 0$$

$$(4y-1)(y-4) = 0$$

$$y = 4 \text{ or } \frac{1}{4}$$

But $\sqrt{\frac{x}{1-x}} = y$

Therefore, when y = 4

$$\sqrt{\frac{x}{1-x}} = 4$$

Squaring both side we have,

Or $\frac{x}{1-x} = 16$

$$x = 16 - 16x$$

or $x = \frac{16}{17}$

And when $x = \frac{1}{4}$, we have,

$$\sqrt{\frac{x}{1-x}} = \frac{1}{4}$$

Squaring both side we have,

Or $$\frac{x}{1-x} = \frac{1}{16}$$

$$16x = 1 - x$$

$$17x = 1, x = \frac{1}{17}$$

Illustration No. 6

Solve $x^{2/3} - 10x^{1/3} + 9 = 0$

Solution:

Given $$x^{2/3} - 10x^{1/3} + 9 = 0$$

Let $$x^{1/3} = y$$

$$y^2 - 10y + 9 = 0$$

$$y^2 - 9y - y + 9 = 0$$

$$y(y - 9) - 1(y - 9) = 0$$

$$(y - 1)\ (y - 9) = 0$$

$$y = 1 \text{ or } 9$$

Since $x^{1/3} = y$ then substituting the value of $y = 1$ or 9, we get,

When $y = 1$, we have,

$$x^{1/3} = 1,$$

Now cubing both sides, we get,

$$x = 1$$

And when $y = 9$, we get,

$$x^{1/3} = 9$$

Now cubing both sides, we get,

$$x = 729$$

Therefore the solution set is {1, 729}

Illustration No. 7

Solve $\dfrac{\sqrt{5+x} + \sqrt{5-x}}{\sqrt{5+x} - \sqrt{5-x}} = \dfrac{5}{x}$ where $x \neq 0$

Solution:

Given

$$\frac{\sqrt{5+x}+\sqrt{5-x}}{\sqrt{5+x}-\sqrt{5-x}}=\frac{5}{x}$$

Now using componendo and dividendo method, we get,

$$\frac{\sqrt{5+x}+\sqrt{5-x}+\sqrt{5+x}-\sqrt{5-x}}{\sqrt{5+x}-\sqrt{5-x}-\sqrt{5+x}-\sqrt{5-x}}=\frac{(5+x)}{(5-x)}$$

$$\frac{2\sqrt{5+x}}{2\sqrt{5-x}}=\frac{5+x}{5-x}$$

$$\frac{\sqrt{5+x}}{\sqrt{5-x}}=\frac{5+x}{5-x}$$

Squaring both side

$$\left(\frac{5+x}{5-x}\right)=\left(\frac{5+x}{5-x}\right)^2$$

Applying componendo and dividendo method, once again both side we get,

$$\frac{5+x+5-x}{5-x-5-x}=\frac{(5+x)^2+(5-x)^2}{(5+x)^2-(5-x)^2}$$

Now, we have,

$$\frac{10}{-2x}=\frac{50+2x^2}{-20x}$$

$$100 = 50 + 2x^2$$

Or $2x^2 = 50$

Or $x^2 = 25$

Or $x = \pm 5$

Alternatively

The given equation is

$$\frac{\sqrt{5+x}+\sqrt{5-x}}{\sqrt{5+x}-\sqrt{5-x}}=\frac{5}{x}$$

Now rationalize the denominator by the rationalizing factor, we have

$$\frac{\sqrt{5+x} + \sqrt{5-x}}{\sqrt{5+x} - \sqrt{5-x}} \times \frac{\sqrt{5+x} + \sqrt{5-x}}{\sqrt{5+x} + \sqrt{5-x}} = \frac{5}{x}$$

$$\frac{\left(\sqrt{5+x} + \sqrt{5-x}\right)^2}{5 + x - 5 + x} = \frac{5}{x}$$

$$10 + 2\sqrt{25-x^2} = 10$$

$$\sqrt{25-x^2} = 0$$

Or

$$x = \pm 5$$

Illustration No. 8

Solve $5\left(x - \frac{1}{x}\right)^2 - 6\left(x + \frac{1}{x}\right) + 12 = 0$

Solution:

Given $5\left(x - \frac{1}{x}\right)^2 - 6\left(x + \frac{1}{x}\right) + 12 = 0$

$$5\left(x^2 + \frac{1}{x^2} - 2\right) - 6\left(x + \frac{1}{x}\right) + 12 = 0$$

$$5\left(x^2 + \frac{1}{x^2}\right) - 6\left(x + \frac{1}{x}\right) + 2 = 0$$

Let us assume that $x + \frac{1}{x} = y$

Now squaring both sides, we get,

$$\left(x^2 + \frac{1}{x^2}\right) = y^2 - 2$$

$$5(y^2 - 2) - 6y + 2 = 0$$

$$5y^2 - 6y - 8 = 0$$

$$5y^2 - 10y + 4y - 8 = 0$$

$$5y(y - 2) + 4(y - 2) = 0$$

$$(5y + 4)(y - 2) = 0$$

Or $$y = -\frac{4}{5} \text{ or } 2$$

Now when y = 2, we have,

$$x + \frac{1}{x} = 2$$

$$x^2 + 1 = 2x$$

$$x^2 - 2x + 1 = 0$$

or $$(x - 1)^2 = 0 \text{ or } x = 1$$

Similarly when $$y = -\frac{4}{5}$$

$$x + \frac{1}{x} = -\frac{4}{5}$$

$$5x^2 + 5 = -4x$$

$$5x^2 + 4x + 5 = 0$$

But

$$x = \frac{-b \pm \sqrt{b^2 - 4ac}}{2a}$$

Since a = 5, b = 4 and c = 5

$$= \frac{-4 \pm \sqrt{(4)^2 - 4.5(5)}}{2.1}$$

$$= \frac{-4 \pm \sqrt{16 - 100}}{2}$$

$$= \frac{-4 \pm \sqrt{-84}}{2}$$

$$= \frac{-4 \pm 2\sqrt{-21}}{2}$$

$$= -2 \pm \sqrt{-21} \qquad \text{as } \sqrt{-1} = i$$

We get

$$= -2 \pm \sqrt{21 \times (-1)}$$

$$x = -2 \pm \sqrt{21}\, i$$

Illustration No.9

Solve $\sqrt{2x+3} + \sqrt{3x+2} = \sqrt{5x+5}$

Solution:

Given

$$\sqrt{2x+3} + \sqrt{3x+2} = \sqrt{5x+5}$$

Squaring both sides, we get,

$$(2x+3) + (3x+2) + 2\left(\sqrt{2x+3} + \sqrt{3x+2}\right) = 5x+5$$

Therefore, we have,

$$2\left(\sqrt{2x+3}\right)\times\left(\sqrt{3x+2}\right) = 0$$

Or

$$\left(\sqrt{2x+3}\right)\times\left(\sqrt{3x+2}\right) = 0$$

Squaring both sides, we get,

$$(2x+3)(3x+2) = 0$$

Or

$$x = -\frac{3}{2} \text{ or } -\frac{2}{3}$$

Illustration No.10

Solve $\sqrt{8-4x} + \sqrt{2x-19} = 3$

Solution:

Given

$$\sqrt{8-4x} + \sqrt{2x-19} = 3$$

Now rearranging the above equation we get,

$$\sqrt{8-4x} = \left(3 - \sqrt{2x-19}\right)$$

Squaring both sides we get,

$$8 - 4x = 9 + 2x - 19 - 6\sqrt{2x-19}$$

$$8 - 4x = 2x - 10 - 6\sqrt{2x-19}$$

$$18 - 6x = -6\sqrt{2x - 19}$$

Or

$$-6(x - 3) = -6\sqrt{2x - 19}$$

Or

$$(x - 3) = \sqrt{2x - 19}$$

Now squaring both sides, we get,

$$(x - 3)^2 = (2x - 19)$$

$$x^2 - 6x + 9 = 2x - 19$$

$$x^2 - 8x + 28 = 0$$

But

$$x = \frac{-b \pm \sqrt{b^2 - 4ac}}{2a}$$

Since a = 1, b = –8 and c = 28

$$= \frac{-(-8) \pm \sqrt{(-8)^2 - 4.1.(28)}}{2.1}$$

$$= \frac{8 \pm \sqrt{64 - 112}}{2}$$

$$= \frac{8 \pm \sqrt{-48}}{2}$$

$$= \frac{8 \pm 2\sqrt{12i}}{2}$$

$$= 4 \pm \sqrt{12\,i} \text{ where } i = \sqrt{-1}$$

Illustration No. 11

Solve $\frac{x}{3} + \frac{3}{x} = \frac{5}{3} + \frac{3}{5}$

Solution:

Given $\frac{x}{3} + \frac{3}{x} = \frac{5}{3} + \frac{3}{5}$

Rearranging the equations we get,

$$\frac{x}{3} - \frac{5}{3} = \frac{3}{5} - \frac{3}{x}$$

$$\frac{x-5}{3} = \left(\frac{3}{5} - \frac{3}{x}\right)$$

$$\frac{x-5}{3} = 3\left(\frac{1}{5} - \frac{1}{x}\right)$$

$$\frac{x-5}{3} = \frac{3(x-5)}{5x}$$

$$5x = 9 \text{ or } x = \frac{9}{5}$$

Illustration No. 12

Solve $x = \sqrt{5 + \sqrt{5 + \sqrt{5 + \sqrt{5 \ldots \infty}}}}$

Solution:

Given

$$x = \sqrt{5 + \sqrt{5 + \sqrt{5 + \sqrt{5 \ldots \infty}}}}$$

Let us assume that $x = \sqrt{5 + \sqrt{5 + \sqrt{5 \ldots \infty}}}$

Then, we have,

$$x = \sqrt{5 + x}$$

Now squaring both sides we get,

$$x^2 = 5 + x$$

$$x^2 - x - 5 = 0$$

But

$$x = \frac{-b \pm \sqrt{b^2 - 4ac}}{2a}$$

Since a = 1, b = –1 and c = – 5

$$= \frac{-(-1) \pm \sqrt{(-1)^2 - 4.1(-5)}}{2.1}$$

$$= \frac{1 \pm \sqrt{1+20}}{2}$$

$$= \frac{1 \pm \sqrt{21}}{2}$$

$$x = \frac{1+\sqrt{21}}{2} \text{ or } \frac{1-\sqrt{21}}{2}$$

Illustration No.13

Solve $(3 + x)^{2/3} + (3 - x)^{2/3} = 4(9 - x^2)^{1/3}$

Solution:

Given

$$(3 + x)^{2/3} + (3 - x)^{2/3} = 4(9 - x^2)^{1/3}$$

Cubing both sides, we get

$\{(3 + x)^{2/3}\}^3 + \{(3 - x)^{2/3}\}^3 + 3(3 + x)^{2/3} (3 - x)^{2/3}\{(3 + x)^{2/3} + (3 - x)^{2/3}\} = 64(9 - x^2)$

As $(a + b)^3 = a^3 + b^3 + 3ab(a + b)$

$$(3 + x)^2 + (3 - x)^2 + 3(9 - x^2)^{2/3} \{4 (9 - x^2)^{1/3}\} = 64(9 - x^2)$$

$$18 + 2x^2 - 468 + 52x^2 = 0$$

$$54x^2 = 450$$

$$x = \pm\sqrt{\frac{450}{54}}$$

The Nature of the Roots of the Quadratic Equation

$$ax^2 + bx + c = 0$$

The nature of the roots of the quadratic equation $ax^2 + bx + c = 0$ depends upon the value of $\left(b^2 - 4ac\right)$ called the discriminant and represented by D or Δ.

Assuming a, b and c are real and rational we can have the following properties.

1. When $\Delta = \left(b^2 - 4ac\right) = 0$ then the roots are real and equal.

2. When $\Delta = (b^2 - 4ac) > 0$ and perfect square then the roots are real, rational and unequal.
3. When $\Delta = (b^2 - 4ac) > 0$ and not perfect square then the roots are real, irrational and unequal.
4. When $\Delta = (b^2 - 4ac) < 0$ and not perfect square then the roots are imaginary and both the roots are complex and unequal.

Illustration No. 14

Examine the nature of the roots of the following equations

(i) $2x^2 - 8x + 7 = 0$

(ii) $3x^2 + 4x + 8 = 0$

Solution:

(i) The given quadratic equation is

$2x^2 - 8x + 7 = 0$ where $a = 2$, $b = -8$ and $c = 7$

Therefore $\Delta = (b^2 - 4ac) = (-8)^2 - 4 . 2 . 7$

$= 64 - 56 = 8 > 0$

Therefore as the discriminant of the equation is greater than zero hence the roots of the equation are real and unequal.

(ii) The given quadratic equation is

$3x^2 + 4x + 8 = 0$ where $a = 3$, $b = 4$ and $c = 8$

Therefore $\Delta = (b^2 - 4ac) = (4)^2 - 4 . 3 . 8$

$= 16 - 96 = -80 < 0$

Therefore as the discriminant of the equation is less than zero hence the roots of the equation are imaginary and unequal.

Illustration No.15

For what values of k will the equation $2x^2 - 2(2 + 6k)x + 7(6 + 4k) = 0$ will have equal roots?

Solution:

The given quadratic equation is $2x^2 - 2(2 + 6k)x + 7(6 + 4k) = 0$ where $a = 2$, $b = -2(2 + 6k)$ and $c = 7(6 + 4k)$

Therefore $\Delta = (b^2 - 4ac) = [-2(2 + 6k)]^2 - 4. 2. 7(6 + 4k)$

For roots to be equal, the discriminant is zero,

Therefore, $4(2 + 6k)^2 - 56\ (6+ 4k) = 0$

Or $(2 + 6k)^2 - 14\ (6 + 4k) = 0$

Or, $4 + 36k^2 + 24k - 84 - 56k = 0$

$$36k^2 - 32k - 80 = 0$$

$$9k^2 - 8k - 20 = 0$$

Therefore,

$$k = \frac{-(-8) \pm \sqrt{(-8)^2 - 4.9.(-20)}}{2.9}$$

Since a = 9, b = – 8 and c = – 20

$$= \frac{(8) \pm \sqrt{64 + 720}}{18}$$

$$= \frac{(8) \pm 28}{18}$$

$$= \frac{(8) + 28}{18}, \frac{8 - 28}{18}$$

$$= 2, \frac{-10}{9}$$

Relation between Roots and Coefficients of a Quadratic Equation

The quadratic equation $ax^2 + bx + c = 0$ $(a \neq 0)$ and the roots are given as

$$\frac{-b + \sqrt{b^2 - 4ac}}{2a}, \frac{-b - \sqrt{b^2 - 4ac}}{2a}$$

Now let α and β are the two roots of the equation such that,

$$\alpha = \frac{-b + \sqrt{b^2 - 4ac}}{2a},$$

$$\beta = \frac{-b - \sqrt{b^2 - 4ac}}{2a}$$

Adding both α and β, we get

$$\alpha + \beta = -\frac{b}{2a}$$

$$= -\frac{\text{coeff of } x}{\text{coeff of } x^2}$$

By multiplying α and β, we get

$$\alpha\beta = \frac{-b + \sqrt{b^2 - 4ac}}{2a} \times \frac{-b - \sqrt{b^2 - 4ac}}{2a}$$

$$= \frac{b^2 - \left(b^2 - 4ac\right)}{4a^2}$$

$$= \frac{4ac}{4a^2}$$

$$= \frac{c}{a}$$

$$= \frac{\text{cons tan t Term}}{\text{coeff. of } x^2}$$

Thus the sum and the product of the roots can be expressed in terms of the coefficients of the quadratic equation.

Formulation of an Equation using given Roots

Let us assume that α and β be the roots of an equation $ax^2 + bx + c = 0$ where $(a \neq 0)$.

Then, sum of the roots, $\alpha + \beta = -\frac{b}{a}$ and product of two roots, $\alpha\ \beta = \frac{c}{a}$

Now dividing the equation $ax^2 + bx + c = 0$ by a, we get,

$$x^2 + \frac{b}{a}x + \frac{c}{a} = 0 \text{ or } x^2 - \left(-\frac{b}{a}\right)x + \frac{c}{a} = 0$$

Now substituting the value of $-\frac{b}{a}$ and $\frac{c}{a}$ in terms α and β, we get,

$$x^2 - (\alpha + \beta) + \alpha\ \beta = 0$$

$$x^2 - (\text{Sum of the roots}) + (\text{Product of the roots}) = 0$$

Illustration No. 16

Find the quadratic equation whose roots are

(i) 5, 4

(ii) $2+\sqrt{3}, 2-\sqrt{3}$

Solution:

(i) The equation whose roots are 5, 4 is given by

x^2 – (Sum of the roots) + (Product of the roots) = 0

Or $x^2 - (\alpha + \beta) + \alpha\,\beta = 0$ where $\alpha = 5, \beta = 4$

$$x^2 - (5 + 4)\,x + 4\,.\,5 = 0$$

$$x^2 - 9x + 20 = 0$$

(ii) The equation whose roots are $2+\sqrt{3}, 2-\sqrt{3}$ is given by

x^2 – (Sum of the roots) + (Product of the roots) = 0

Or $x^2 - (\alpha + \beta) + \alpha\,\beta = 0$ where $\alpha = 2+\sqrt{3}$, $\beta = 2-\sqrt{3}$

$$x^2 - \left(2+\sqrt{3}+2-\sqrt{3}\right) x + \left(2+\sqrt{3}\right)\left(2-\sqrt{3}\right) = 0$$

$$x^2 - 4x + 1 = 0$$

Illustration No. 17

If α and β are the roots of the equation $x^2 - 5x + 4 = 0$ then find the value of

(i) $\left(\frac{\alpha}{\beta} + \frac{\beta}{\alpha}\right)$?

(ii) $\left(\frac{\alpha}{\beta} - \frac{\beta}{\alpha}\right)^2$?

Solution:

As α and β are the roots of the equation $x^2 - 5x + 4 = 0$

where a = 1 b = – 5 and c = 4, then,

Sum of the roots $(\alpha + \beta) = 5$

Product of the roots $(\alpha\,\beta) = 4$

Therefore,

$$\left(\frac{\alpha}{\beta}+\frac{\beta}{\alpha}\right)=\frac{\alpha^2+\beta^2}{\alpha\beta}=\frac{(\alpha+\beta)^2-2\alpha\beta}{\alpha\beta}=\frac{5^2-2.4}{4}=\frac{25-8}{4}=\frac{17}{4}$$

(ii) $$\left(\frac{\alpha}{\beta}-\frac{\beta}{\alpha}\right)^2=\left(\frac{\alpha^2-\beta^2}{\alpha\beta}\right)^2=\frac{\{(\alpha+\beta)\ (\alpha-\beta)\}^2}{\alpha\beta}$$

$$\frac{(\alpha+\beta)^2(\alpha-\beta)^2}{\alpha\beta}=\frac{(\alpha+\beta)^2\{(\alpha+\beta)^2-4\alpha\beta\}}{\alpha\beta}$$

$$=\frac{(5)^2\{(5)^2-4.4\}}{4.4}$$

$$=\frac{25.9}{16}=\frac{225}{16}$$

Illustration No. 18

If α and β are the roots of the equation $ax^2 + bx + c = 0$ then find the equations whose roots are

(i) $\frac{1}{\alpha^2}$ and $\frac{1}{\beta^2}$

(ii) α^2 and β^2

Solution:

Since α and β are the roots of the equation $ax^2 + bx + c = 0$, then we have,

Sum of the roots $(\alpha + \beta) = -\frac{b}{a}$

Product of the roots $(\alpha\beta) = \frac{c}{a}$

(i) The equation whose roots are given by $\frac{1}{a^2}$ and $\frac{1}{\beta^2}$, then

Sum of the roots

$$\left(\frac{1}{\alpha^2}+\frac{1}{\beta^2}\right)=\frac{\alpha^2+\beta^2}{(\alpha\beta)^2}=\frac{(\alpha+\beta)^2-2\alpha\beta}{(\alpha\beta)^2}=\frac{\left(\frac{b}{a}\right)^2-2\left(\frac{c}{a}\right)}{\left(\frac{c}{a}\right)^2}$$

$$=\frac{b^2-2ac}{c^2}$$

Products of the roots $\left(\frac{1}{\alpha^2}.\frac{1}{\beta^2}\right)=\frac{1}{(\alpha\beta)^2}=\frac{a^2}{c^2}$

Therefore the equation with roots as $\frac{1}{\alpha^2}$ and $\frac{1}{\beta^2}$

x^2 – (Sum of the roots) x + (Product of the roots) = 0

$$x^2-\left(\frac{b^2-2ac}{c^2}\right)x+\frac{a^2}{c^2}=0$$

$$c^2x^2-\left(b^2-2ac\right)x+a^2=0$$

(ii) The equation whose roots are given by α^2 and β^2, then

Then sum of the roots $=\alpha^2+\beta^2=(\alpha+\beta)^2-2\alpha\beta=\left(\frac{b}{a}\right)^2-\frac{2c}{a}$

$$=\frac{b^2-2ac}{a^2}$$

Similarly the product of the roots $=\alpha^2\beta^2=(\alpha\beta)^2=\left(\frac{c}{a}\right)^2$

Therefore the equation with roots as α^2 and β^2

x^2 – (Sum of the roots)x + (Product of the roots)= 0

$$x^2-\left(\frac{b^2-2ac}{a^2}\right)x+\left(\frac{c}{a}\right)^2=0$$

$$a^2x^2-\left(b^2-2ac\right)x+c^2=0$$

Illustration No. 19

If α and β are the roots of the equation $x^2 + 3x + 4 = 0$ then find the equations whose roots are

(i) $\dfrac{\beta}{\alpha+3}, \dfrac{\alpha}{\beta+3}$

(ii) $\dfrac{\alpha}{\beta}, \dfrac{\beta}{\alpha}$

Solution:

Since α and β are the roots of the equation $x^2 + 3x + 4 = 0$, then we have,

Sum of the roots $(\alpha + \beta) = -3$

Product of the roots $(\alpha\beta) = 4$

(i) The equation whose roots are given by $\dfrac{\beta}{\alpha+3}$ and $\dfrac{\alpha}{\beta+3}$

then sum of the roots $= \dfrac{\beta}{\alpha+3}$ and $\dfrac{\alpha}{\beta+3} = \dfrac{\beta(\beta+3)+\alpha(\alpha+3)}{(\alpha+3)(\beta+3)}$

$$\frac{\alpha^2+\beta^2+3(\alpha+\beta)}{(\alpha+3)(\beta+3)}$$

$$= \frac{(\alpha+\beta)^2-2(\alpha\beta)+3(\alpha+\beta)}{3(\alpha+\beta)+\alpha\beta+9}$$

$$\frac{(\alpha+\beta)^2-3(\alpha+\beta)-2(\alpha\beta)}{3(\alpha+\beta)+\alpha\beta+9}$$

$$\frac{9+3(-3)-2.4}{3(-3)+4+9} = -\frac{8}{4} = -2$$

And product of the roots $\left(\dfrac{\beta}{\alpha+3}\right) \times \left(\dfrac{\alpha}{\beta+3}\right) = \dfrac{\alpha\beta}{3(\alpha+\beta)+\alpha\beta+9}$

$$\frac{4}{3(-3)+4+9} = \frac{4}{4} = 1$$

Therefore the equation with roots as $\frac{\beta}{\alpha+3}$ and $\frac{\alpha}{\beta+3}$

x^2 – (Sum of the roots)x + (Product of the roots) = 0

$x^2 - 2x + 1 = 0$

(ii) The equation whose roots are given by $\frac{\alpha}{\beta}$ and $\frac{\beta}{\alpha}$

Sum of the roots

$$= \frac{\alpha}{\beta} + \frac{\beta}{\alpha} = \frac{\alpha^2+\beta^2}{\alpha\beta} = \frac{(\alpha+\beta)^2 - 2(\alpha+\beta)}{\alpha\beta} = \frac{9-2\times 4}{4} = \frac{1}{4}$$

Product of the roots = $\frac{\alpha}{\beta} \times \frac{\beta}{\alpha} = 1$

Hence the equation is given by

x^2 – (Sum of the roots)x + (Product of the roots) = 0

$$x^2 - \left(\frac{1}{4}\right)x + 1 = 0$$

$4x^2 - x + 4 = 0$

Illustration No. 20

If α and β are the roots of the equation $p^2 + qx + r = 0$ then find the equations whose roots are $\alpha^3\beta$ and $\beta^3\alpha$?

Solution:

Since α and β are the roots of the equation $px^2 + qx + r = 0$, then we have,

Sum of the roots $(\alpha + \beta) = -\frac{q}{p}$

Product of the roots $(\alpha\beta) = \frac{r}{p}$

The equation whose roots are given by $\alpha^3\beta$ and $\beta^3\alpha$

Sum of the roots $= \alpha^3\beta + \beta^3\alpha$

$$= \alpha\beta\left(\alpha^2 + \beta^2\right) = \alpha\beta\ (\alpha + \beta)^2 - 2(\alpha\beta)$$

$$= \frac{r}{p}\left(\frac{q}{p}\right)^2 - 2\left(\frac{r}{p}\right)$$

$$= \frac{r}{p}\left(\frac{q^2 - 2rp}{p^2}\right)$$

And the product of the roots are = $\alpha^3\beta \times \beta^3\alpha = (\alpha\beta)^4 = \left(\frac{r}{p}\right)^4$

Therefore the equation is given by

x^2 – (Sum of the roots)x + (Product of the roots) = 0

$$x^2 - \left\{\frac{r}{p}\left(\frac{q^2 - 2rp}{p^2}\right)\right\}x + \left(\frac{r}{p}\right)^4 = 0$$

$$p^4x^2 - rp\left(q^2 - 2rp\right)x + (r)^4 = 0$$

Illustration No. 21

Find the condition that the roots of the equation $x^2 + bx + c = 0$ are in the ratio 2 : 3?

Solution:

Let 2α and 3α be the roots of the equation $x^2 + bx + c = 0$ then

Sum of the roots = $2\alpha + 3\alpha = -b$ i.e. $5\alpha = -b$ or $\alpha = -\frac{b}{5}$

And product of the roots = $(2\alpha \times 3\alpha) = c$

$$6\alpha^2 = c \text{ or } \alpha^2 = \frac{c}{6}$$

But $\alpha = -\frac{b}{5}$ therefore $\left(-\frac{b}{5}\right)^2 = \frac{c}{6}$

And hence, $6b^2 = 25c$

Solving Simultaneously Linear Equation Having Two Variables

A simultaneous equation characterized by two or more variables they are generally represented as

$$a_1x + b_1y + c_1 = 0$$

$$a_2x + b_2y + c_2 = 0$$

By solution of the system of equations, we mean the values of x and y satisfying both equation. The system of equation can have

(i) Unique Solution when $a_1b_2 - a_2b_1 \neq 0$ or $\frac{a_1}{a_2} \neq \frac{b_1}{b_2}$

(ii) Infinite Solution, when $a_1b_2 - a_2b_1 = 0$, $a_1c_2 - a_2c_1 = 0$ or $c_1b_2 - c_2b_1 = 0$.

(iii) No solution when $a_1b_2 - a_2b_1 = 0$ and at least one of $a_1c_2 - a_2c_1$ or $c_1b_2 - c_2b_1$ is not equal to zero.

Basically they are three different methods for solving linear simultaneous equation

(i) Method of Substitution.

(ii) Method of Elimination.

(iii) Method of Cross Multiplication.

Let us consider two simultaneous equation of the form

$$a_1x + b_1y + c_1 = 0 \quad .....(1)$$

$$a_2x + b_2y + c_2 = 0 \quad .....(2)$$

then,

(i) Method of Substitution

let us rearrange the equation 1

i.e. $a_1x = -(b_1y + c_1)$

or $$x = -\frac{b_1y + c_1}{a_1}$$

Now we substitute the value of x in equation 2, we get

$$a_2\left(-\frac{b_1y + c_1}{a_1}\right) + b_2y + c_2 = 0$$

$$-a_2b_1y - a_2c_1 + a_1b_2y + a_1c_2 = 0$$

or $$y = -\frac{a_1c_2 - a_2c_1}{a_1b_2 - a_2b_1}$$

Now substituting the value of y in any of the equations 1 or 2, we get,

$$x = \frac{b_1c_2 - b_2c_1}{a_1b_2 - a_2b_1}$$

(ii) Method of Elimination

In this case we eliminate x from equations 1 or 2 by multiplying 1 by a_2 and 2 by a_1, then we get,

$$a_1 a_2 x + b_1 a_2 y + c_1 a_2 = 0 \quad .....(3)$$

$$a_2 a_1 x + b_2 a_1 y + c_2 a_1 = 0 \quad .....(4)$$

Now subtracting 3 and 4, we get,

$$(a_1b_2 - a_2b_1)y = -(a_1c_2 - a_2c_1)$$

Or
$$y = -\frac{a_1c_2 - a_2c_1}{a_1b_2 - a_2b_1}$$

Substitute the value of y in any of the above equations, we get,

$$x = \frac{b_1c_2 - b_2c_1}{a_1b_2 - a_2b_1}$$

(iii) Method of Cross Multiplication

By multiplying 1 by a_2 and 2 by a_1, then we get,

$$a_1 a_2 x + b_1 a_2 y + c_1 a_2 = 0 \quad .....(5)$$

$$a_2 a_1 x + b_2 a_1 y + c_2 a_1 = 0 \quad .....(6)$$

Now subtracting 5 and 6, we get,

$$(a_1b_2 - a_2b_1)\, y = -(a_1c_2 - a_2c_1)$$

Or
$$(a_1b_2 - a_2b_1)\, y = a_2c_1 - a_1c_2$$

$$\frac{y}{a_2c_1 - a_1c_2} = \frac{1}{(a_1b_2 - a_2b_1)} \quad .....(7)$$

Similarly by multiplying 1 by b_2 and 2 by b_1, we get

$$a_1 b_2 x + b_2 a_2 y + c_1 b_2 = 0 \quad .....(8)$$

$$a_2 b_1 x + b_1 a_1 y + c_2 b_1 = 0 \quad .....(9)$$

Now subtracting 8 and 9, we get

$$(a_1b_2 - a_2b_1)\, x = b_2c_1 - b_1c_2$$

or

$$\frac{x}{b_2c_1 - b_1c_2} = \frac{1}{(a_1b_2 - a_2b_1)} \quad .....(10)$$

Therefore, from equation 7 and 10, we have,

$$\frac{x}{b_2c_1 - b_1c_2} = \frac{y}{a_2c_1 - a_1c_2} = \frac{1}{(a_1b_2 - a_2b_1)}$$

Illustrative No. 22

Solve the equations 3x + 4y = 12 and 2x + 4y = 8?

Solution:

Given equations are 3x + 4y =12 and 2x + 4y = 8. But from the second equation, we get $x = 4 - 2y$. Substituting the value x in the first equation, we get,

$$3(4 - 2y) + 4y = 12$$

$$12 - 6y + 4y = 12 \text{ or } - 2y = 0 \text{ or } y = 0$$

Now, substituting the value of y in $x = 4 - 2y$, we get,

$$x = 4 - 2 \times 0 = 4$$

Hence the required solutions are x = 4 and y = 0

Illustrative No. 23

Solve the equations 5x + 4y = 18 and x + y = 4?

Solution:

The equations are rearranged in the standard form as

$$5x + 4y - 18 = 0$$

$$x + y - 4 = 0$$

Now by the method of cross multiplication, we have,

$$\frac{x}{4(-4)-1(-18)} = \frac{y}{1(-18)-5(-4)} = \frac{1}{5(1)-4(1)}$$

$$\frac{x}{-16+18} = \frac{y}{-18+20} = \frac{1}{5-4}$$

$$\frac{x}{2} = \frac{y}{2} = 1$$

Therefore, the required solutions are x = 2 and y = 2

Illustration No. 24

Solve the following equations $x + 2y = 12$ and $2x + y = 15$ using elimination method?

Solution:

The given equations are $x + 2y = 12$ and $2x + y = 15$. In order to eliminate x from both the equations multiply the 1st equation with coefficient of x in the 2nd equation and vice versa, so we get,

$$2x + 4y = 24 \quad .....(1)$$

$$2x + y = 15 \quad .....(2)$$

Subtracting 1 and 2, we get,

$$3y = 9 \Rightarrow y = 3$$

Substituting the value of y in any one of the equations we get,

$$x + 6 = 12$$

$$\Rightarrow \quad x = 6$$

Hence the required solutions are $x = 6$ and $y = 3$

Illustrative No. 25

Solve $x^2 + y^2 = 20$ and $x + y = 2$?

Solution:

The given equations are

$$x^2 + y^2 = 20 \quad .....(1)$$

and $$x + y = 2 \quad .....(2)$$

But from 2, we get $y = 2 - x$

Substituting 2 in 1, we get,

$$x^2 + (2 - x)^2 = 20$$

$$x^2 + 4 - 4x + x^2 = 20$$

$$2x^2 - 4x - 16 = 0$$

$$x^2 - 2x - 8 = 0$$

$$x^2 - 4x + 2x - 8 = 0$$

$$x(x - 4) + 2(x - 4) = 0$$

$$(x - 4)(x + 2) = 0$$

Or $$x = 4 \text{ or } -2$$

Substituting the values of x in 2 we get y = – 2 at x = 4 and y = 4 when x = – 2.

Hence the required solution x = 4 and y = – 2, x = – 2 and y = 4

Illustrative No. 26

Solve $\frac{1}{x^2} + \frac{1}{y^2} = 10$ and $\frac{1}{x} + \frac{1}{y} = 2$?

Solution:

Given equations are $\frac{1}{x^2} + \frac{1}{y^2} = 10$ and $\frac{1}{x} + \frac{1}{y} = 2$

Substituting $\frac{1}{x} = a$ and $\frac{1}{y} = b$ then the equations changes to

$$a^2 + b^2 = 10 \qquad .....(1)$$

and $$a + b = 2 \qquad .....(2)$$

since $$a + b = 2 \text{ or } b = 2 - a$$

substituting the value of b in 1 we get,

$$a^2 + (2 - a)^2 = 10$$

$$a^2 + 4 - 4a + a^2 = 10$$

$$2a^2 - 4a - 6 = 0$$

$$a^2 - 2a - 3 = 0$$

$$a^2 - 3a + a - 3 = 0$$

$$a(a - 3) + 1\,(a - 3) = 0$$

$$(a + 1)\,(a - 3) = 0$$

$$a = -1 \text{ or } 3$$

Substituting the value of 'a' we get the b = –1 and 1

∴ We get the values of x and y against the values of a and b are

$$x = \frac{1}{3} \text{ and } -1,\ y = -1 \text{ and } 1$$

And hence the roots are $x = \frac{1}{3}$, $y = -1$ or $x = -1$, $y = 1$

Cubic and Biquadratic Equations

An equation in which the highest power of the variable is three is termed as a cubic equation and the general form of the cubic equation is $ax^3 + bx^2 + cx + d$ where $a \neq 0$ and a, b, c and d ϵ R.

Relation between Roots and Coefficients of Cubic Equations

Let us take a general cubic equation of the form

$$ax^3 + bx^2 + cx + d \text{ where } a \neq 0 \quad .....(1)$$

Now divide 1 by a, we get,

$$x^3 + \frac{b}{a}x^2 + \frac{c}{a}x + \frac{d}{a} = 0 \quad .....(2)$$

An equation whose roots are α, β and ν can be represented as

$$(x - \alpha)(x - \beta)(x - \nu) = 0$$

$$\Rightarrow \quad x^3 - (\alpha + \beta + \nu)x^2 + (\alpha\beta + \beta\nu + \alpha\nu)x - \alpha\beta\nu = 0 \quad .....(3)$$

Since equation 2 and 3 represents the same equation therefore comparing the coefficients, we get,

Sum of the Roots $= \alpha + \beta + \nu = -\frac{b}{a}$

Sum of the product of roots taken two at a time

$$= \alpha\beta + \beta\nu + \alpha\nu = +\frac{c}{a} = +\frac{(\text{coeff. of } x)}{\text{coeff. of } x^3}$$

Product of the roots $= \alpha\beta\nu = -\frac{d}{a} = -\frac{\text{constant Term}}{\text{coeff. of } x^3}$

Biquadratic Equation

An equation in which the highest power of the variable is four is termed as a biquadratic equation and the general form of the biquadratic equation is $ax^4 + bx^3 + cx^2 + dx + e$ where $a \neq 0$ and a, b, c, d and e ϵ R.

Then,

Sum of the Roots $= \alpha + \beta + \nu + \delta = -\frac{b}{a}$

Sum of the product of roots taken two at a time

$$= \alpha\beta + \beta\nu + \nu\delta + \delta\alpha = +\frac{c}{a}$$

Sum of the product of roots taken three at a time

$$= \alpha\beta\nu + \beta\nu\delta + \alpha\nu\delta + \beta\nu\delta = -\frac{d}{a}$$

Product of the roots $= \alpha\beta\nu\delta = \frac{e}{a}$

Illustration No. 27

For the cubic equation $x^3 + bx^2 + cx + d = 0$, find the value of the following

(i) $\sum\alpha^2$, (ii) $\sum\alpha^2\beta^2$ (iii) $\sum\alpha^2\beta\nu$

Solution:

The given cubic equation is $x^3 + bx^2 + cx + d = 0$

Therefore,

$$\alpha + \beta + \nu = -b$$

$$\alpha\beta + \beta\nu + \alpha\nu = +c$$

$$\alpha\beta\nu = -d$$

For (i),

$$\sum\alpha^2 = \alpha^2 + \beta^2 + \gamma^2 = (\alpha + \beta + \gamma)^2 - 2(\alpha\beta + \beta\nu + \alpha\nu)$$

$$= b^2 - 2c$$

For (ii)

$$\left(\sum\alpha\beta\right)^2 = \sum\alpha^2\beta^2 + 2\,\alpha\beta\gamma\left(\sum\alpha\right)$$

$$\sum\alpha^2\beta^2 = \left(\sum\alpha\beta\right)^2 - 2\,\alpha\beta\gamma\left(\sum\alpha\right)$$

$$= c^2 - 2(-d)(-b)$$

$$= c^2 - 2bd.$$

For (iii)

$$\sum\alpha^2\beta\nu = \alpha\beta\gamma(\alpha + \beta + \gamma) = (-d)(-b) = bd$$

Illustration No. 28

Solve the equation $x^3 - 3x^2 - 4x + 12 = 0$, given that the product of the two roots is 6?

Solution:

Let α, β and ν are the roots of the cubic equation $x^3 - 3x^2 - 4x + 12 = 0$ and let $\alpha\beta = 6$.

Since, sum of the roots $\alpha + \beta + \gamma = 3$ and product of the roots $\alpha\beta\gamma = -12$

But $\alpha\beta = 6$, therefore $6\gamma = -12$ or $\gamma = -2$

Since $\gamma = -2$ and hence $\alpha + \beta = 5$

As $\alpha + \beta = 5$ and $\alpha\beta = 6$ then the quadratic equation with sum and product of the roots is given by

$$x^2 - (\alpha + \beta)x + \alpha\beta = 0 \text{ or } x^2 - 5x + 6 = 0$$

or
$$(x - 3)(x - 2) = 0 \text{ or } x = 3 \text{ or } 2$$

therefore $\alpha = 3$ and $\beta = 2$. Hence the required roots are 3, 2 and – 2.

Illustration No. 29

Solve the equation $x^4 - 8x^3 + 17x^2 - 2x - 14 = 0$ which has $3 + \sqrt{2}$ as one of the roots?

Solution:

Since one of the roots of the equation is $3 + \sqrt{2}$; the other root will be $3 - \sqrt{2}$. Let α and β be the other two roots of the equation then,

Sum of the roots $= \alpha + \beta + 3 + \sqrt{2} + 3 - \sqrt{2} = 8$ or $\alpha + \beta = 2$

Similarly product of the roots $\alpha\beta(3 + \sqrt{2})(3 - \sqrt{2}) = -14$

Therefore $\alpha\beta = -2$

If we assume that α and β are the roots of a quadratic equation then

$$t^2 - (\alpha + \beta)t + \alpha\beta = 0,$$

then,

$$t^2 - 2t - 2 = 0$$

i.e.
$$t = \frac{2 \pm \sqrt{12}}{2} \text{ or } 1 \pm \sqrt{3}$$

Hence the required roots are $1 \pm \sqrt{3}$, $3 \pm \sqrt{2}$.

Inequalities

If a real number say 'a' can be either be positive, negative or zero, when it is positive it can be written as $a > 0$, when negative it can be written as $a < 0$ otherwise we can write $a = 0$ which means a is neither positive nor negative.

Two numbers a and b are not equal i.e. $a \neq b$ then either a is greater than b or a is lesser than b, which means

(i) if $a > b$ then $a - b > 0$.

(ii) if $a < b$ then $a - b < 0$.

Definition

If two real numbers or two algebraic expressions are related by the symbol > (Greater than) or < (Lesser than) and also the signs ≥ or ≤ (greater than equal or lesser than equal) form an inequality. An inequality with (< or >) sign are called as strict inequality while an inequality having (≥ or ≤) sign are called as slack inequality. The direction in which the inequalities points is called as sense of the inequality. When two or more inequalities contains the same sign either < or > then they are inequalities of the same sense otherwise opposite in sense.

For example

If a and b are two real numbers then $a > b$, $a < b$ are plain inequalities and if $a \geq b$, $a \leq b$ are called as inequations.

Understanding Slack Inequalities

(i) The inequality $a \geq b$ means that $a > b$ or $a = b$ but a is not less than b.

(ii) The inequality $a \leq b$ means that $a < b$ or $a = b$ but a is not greater than b.

Some Examples of Inequalities

$$2x + 3 < 0,$$

$$5x + 4 > 0$$

$$3x + 4 \leq 0 \text{ and}$$

$$5x + 3 \geq 0.$$

Properties of Inequalities

(i) if $a > b$ then $b < a$ and vice versa.

(ii) if $a > b$ and $b > c$ then $a > c$ (transitivity property)

(iii) An inequality remains true if the same number is added on both sides of the inequality i.e. if $a > b$ then for any c, we have $a + c > b + c$.

(iv) Both sides of an inequality can be multiplied or divided by the same positive number without changing the sense of the inequality i.e.

if $a > b$ for any c , we have $a c > b c$ or $\frac{a}{c} > \frac{b}{c}$

(v) Both sides of the inequality may be multiplied or divided by the same negative number but then the sense of the inequality gets reversed i.e.

if $a > b$ and $c < 0$, then $ac < bc$.

(vi) Two inequalities of the same sense can be added term wise i.e. if $a > b$ and $c > d$ then $a + c > b + d$

(vii) Two inequalities of the same sense can be mulitiplied term wise and the resulting inequality will have the same sense as the multiplied inequality i.e. if $a > b$ and $c > d$ then $a c > b d$.

(viii) Two inequalities of the opposite sense can be subtracted retaining the sense of the inequality from which other is subtracted i.e. if $a > b$ and $c < d$ then $a - c > b - d$.

(ix) For any number the side can be changed but the sign of the number gets reversed maintaining the sense of the inequality.

Illustration No. 1

Solve $25x < 200$ when x is natural number and x is real number

Solution:

Given $25x < 200 \Rightarrow x < 8$.

When x is a natural number the following of the values of x holds good i.e. 1, 2, 3, 4, 5, 6, 7.

And when x is an real number $x \in (-\infty, 8)$

Illustration No.2

Solve $6x - 4 < 4x + 2$ when x is an integer?

Solution:

Given $6x - 4 < 4x + 2$

or $6x - 4x < 4 + 2$

or $2x < 6$

or $x < 3$

Thus when x is an integer it takes values ... – 3, – 2, – 1, 0, 1, 2

Illustration No. 3

Solve the inequation $\frac{4x-6}{4} + 18 \geq 6 + \frac{8x}{3}$?

Solution:

The given inequation is $\frac{4x-6}{4} + 18 \geq 6 + \frac{8x}{3}$

$$\left(\frac{4x-6}{4}\right) - \left(\frac{8x}{3}\right) \geq 6 - 18$$

$$\frac{3(4x-6) - 4(8x)}{12} \geq -12$$

$$\frac{12x - 18 - 32x}{12} \geq -12$$

$$12x - 18 - 32x \geq -144$$

$$-20x - 18 \geq -144$$

$$-20x \geq -126$$

Or $20x \leq 126$

Or $x \leq \frac{63}{10}$

Therefore the solution set of the inequation is $x \in (-\infty, \frac{63}{10})$

Illustration No. 4

Solve the inequation $|2x - 3| \leq \frac{1}{3}$?

Solution:

The given inequation is $|2x - 3| \leq \frac{1}{3}$

But,

$$|x-a| \le b \leftrightarrow a-b \le x \le a+b$$

$$\therefore \quad |2x-3| \le \frac{1}{3} \leftrightarrow 3-\frac{1}{3} \le 2x \le 3+\frac{1}{3}$$

$$\leftrightarrow \frac{8}{3} \le 2x \le \frac{10}{3}$$

$$\leftrightarrow \frac{4}{3} \le x \le \frac{5}{3}$$

Therefore $\quad x \in \left[\frac{4}{3}, \frac{5}{3}\right]$

Illustration No.5

Solve the inequation $|x-3| \ge 4$?

Solution:

The given inequation $|x-3| \ge 4$

As we know that $\quad |x-a| \ge b \leftrightarrow x \le a-b$ or $x \ge a+b$

Therefore,

$$|x-3| \ge 4 \leftrightarrow x \le 3-4 \text{ or } x \ge 3+4$$

And hence $\quad x \le -1$ or $x \ge 7$

Therefore the solution set of the inequation $x \in (-\infty, -1] \cup [7, \infty)$

Illustration No.6

Solve $|x-2| + |x-3| \ge 6$?

Solution:

The inequality in the left hand side has two components both containing modulus. Now by equating both the components to zero we get, x = 2, 3 as critical points as these points divide the real into three parts viz. $(-\infty, +2]$, $[2, 3]$ and $[3, \infty)$. Now, we consider the following three cases

Case I

When x lies between $-\infty$ and 2 i.e. $-\infty < x < +2$

In this case we have $|x-2| = -(x-2)$ and $|x-3| = -(x-3)$

Therefore,

$$|x-2| + |x-3| \geq 6$$

$$\Rightarrow \quad -(x-2)-(x-3) \geq 6$$

$$\Rightarrow \quad -x+2-x+3 \geq 6$$

$$\Rightarrow \quad -2x+5 \geq 6$$

$$\Rightarrow \quad -2x \geq 1$$

$$x \leq -\frac{1}{2}$$

But, x lies between $-\infty < x < +2$ and hence in this case the solution set of the given inequation is $(-\infty, -\frac{1}{2})$ ..... (1)

Case II

When $2 \leq x \leq 3$, in this case, we have $|x-2| = (x-2)$ and $|x-3| = -(x-3)$

Then

$$|x-2| + |x-3| \geq 6$$

Therefore

$$\Rightarrow \quad (x-2)-(x-3) \geq 6$$

$$\Rightarrow \quad x-2-x+3 \geq 6$$

$$\Rightarrow \quad 1 \geq 6,$$

which is an absurd result and hence the inequation has no solution for $x \in [2, 3)$

Case III

When $x \geq 3$, in this case, we have $|x-2| = (x-1)$ and $|x-3| = (x-3)$

Therefore,

$$|x-2| + |x-3| \geq 6$$

And hence

$$(x-2)+(x-3) \geq 6$$

$$x-2+x-3 \geq 6$$

$$2x - 5 \geq 6$$

$$x \geq \frac{11}{2}$$

As x ≥ 3, in this case the solution set of the given inequation is

$$\left(\frac{11}{2}, \infty\right) \qquad .....(ii)$$

Combining the results of (i) and (ii), we get,

$$\left(-\infty, -\frac{1}{2}\right] \cup \left[\frac{11}{2}, \infty\right)$$

Illustration No. 7

Solve the following inequation

$$\frac{|x + 4| + x}{x + 3} > 1$$

Solution:

The given inequation is

$$\frac{|x + 4| + x}{x + 3} > 1$$

Now,

$$\frac{|x + 4| + x}{x + 3} - 1 > 0$$

$$\frac{|x + 4| + x - x - 3}{x + 3} > 0$$

$$\frac{|x + 4| - 3}{x + 3} > 0$$

Now there exist two cases for the modulus case

Case II

When $|x + 4| \geq 0$ i.e. $x \geq -4$

Then $|x + 4| = x + 4$,

Therefore

$$\frac{|x+4|-3}{x+3} > 0$$

$$\frac{|x+4|-3}{x+3} > 0$$

$$\frac{x+1}{x+3} > 0$$

i.e. $x \in (-\infty, -3) \cup (-1, \infty)$

But $x \geq -4$, therefore the solution set of the given inequation in this case is

$$[-4, -3) \cup (-3, \infty) \qquad \text{.....(i)}$$

Case III

When $|x+4| < 0$ i.e. $x < -4$

In this case $|x+4| = -(x+4)$

Therefore,

$$\frac{|x+4|-3}{x+3} > 0$$

$$\frac{-(x+4)-3}{x+3} > 0$$

$$-\frac{(x+7)}{x+3} > 0$$

$$\frac{(x+7)}{x+3} > 0$$

And hence $x \in (-7, -3)$

But $x < -4$, therefore, the solution set of the inequation in this case is

$$(-7, -4) \qquad \text{.....(ii)}$$

From (i) and (ii), we obtain the solution set of the inequation as $[-4, -3) \cup (-3, \infty) \cup (-7, -4) = (-7, -3) \cup (-3, \infty)$.

Graphs of Linear Inequations or Inequalities in Two variables

To plot the graph of an inequation or inequality of the form

$$ax + by + c \le 0 \text{ or } ax + by + c \ge 0$$

In the first step we convert the given inequalities in equation form i.e. $ax + by + c = 0$ and plot the straight line with reference to the coordinate axis.

The line $ax + by + c = 0$ divides the plane into two half's and one of the half's will have the desired feasible solution of

$$ax + by + c \ge 0 \text{ or } ax + by + c \le 0.$$

In order to determine the solution set we take any point not lying on $ax + by + c = 0$ and substitute the coordinates in the given inequation. If the point satisfies the inequation then the solution set is the region in which the point is contained, otherwise if the inequation is not satisfied then the solution set is the region not containing the point.

For example the coordinates of origin (0, 0) not lying on $ax + by + c = 0$ and is substituted in the given inequation and found to satisfy the inequation the region containing origin will the solution set. Otherwise the non–origin side of the line will be the solution set.

Illustration No. 8

Plot the graphs of the following inequations (inequalities)

(i) $5x + 2y < 10$

(ii) $3x + 4y > 12$

Solution

(i) we have the given inequation as $5x + 2y < 10$ .....(1)

Now the corresponding equation in reference to the above inequation (1) is

$$5x + 2y = 10 \text{ or } \frac{x}{2} + \frac{y}{5} = 1 \quad .....(2)$$

Plotting (2) with reference to the coordinate axis XOX′ and YOY′, we get, the following figure,

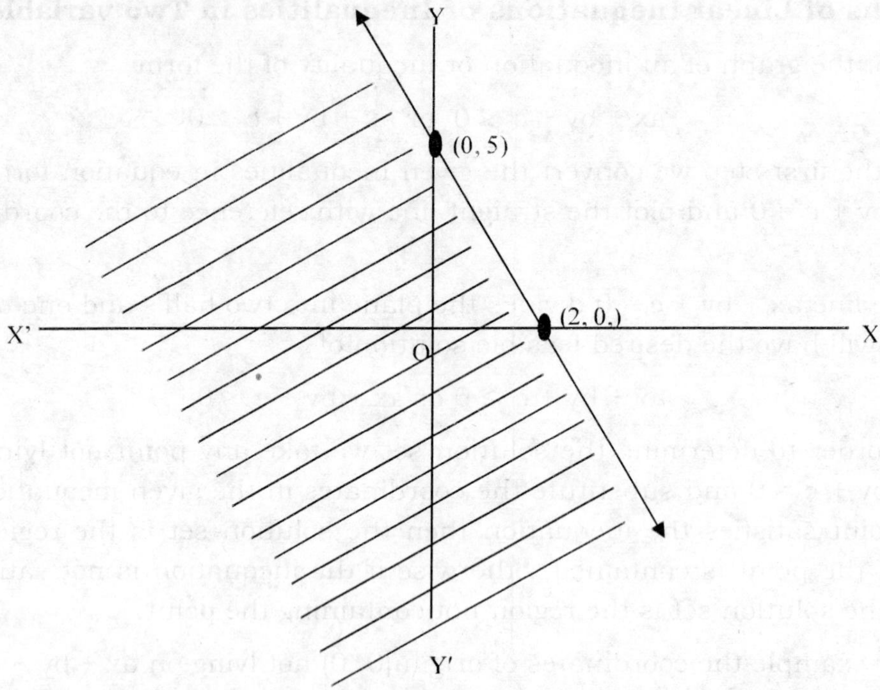

The line divides the plane into two half's with one half being the feasible solution. Now let us take a point (0, 0) not lying on the line. Therefore substituting x = 0 and y = 0, we get 5.0 + 2.0 < 10, which is true and hence the point satisfy the inequation 5x + 2y < 10 and the solution set lie on the side of the origin of the line i.e. below the line which contains the origin not including those on the line since the inequation being less than type.

(ii) we have the given inequation as 3x + 4y > 12 .....(1)

Now the corresponding equation in reference to the above inequation (1) is

$$3x + 4y = 12 \text{ or } \frac{x}{4} + \frac{y}{3} = 1 \qquad .....(2)$$

Plotting (2) with reference to the coordinate axis XOX' and YOY', we get, the following figure,

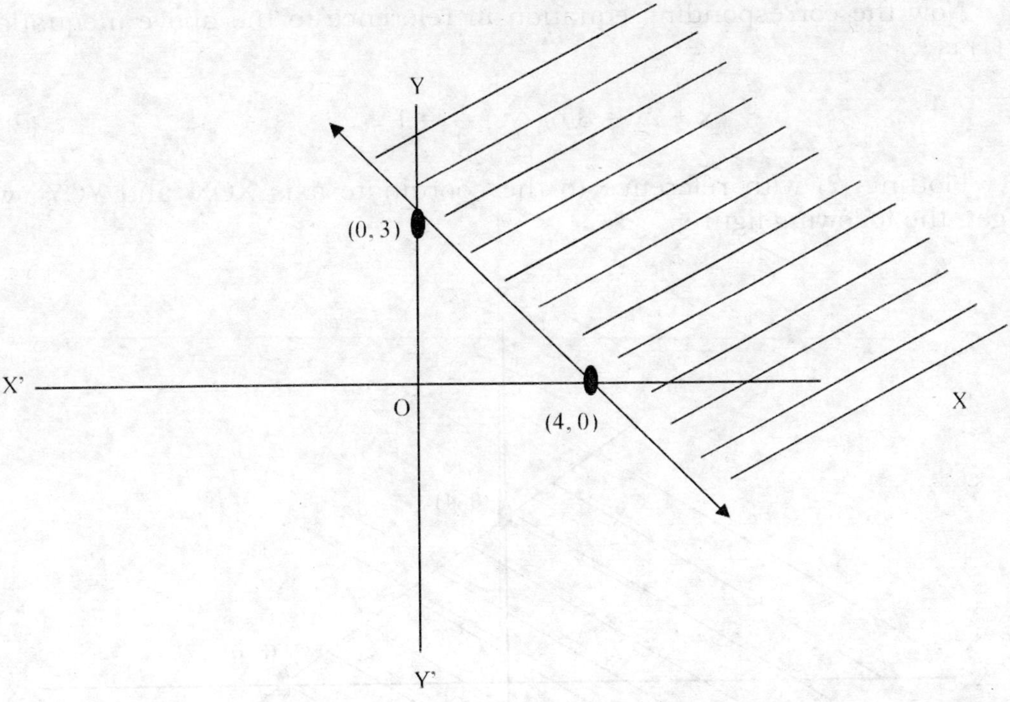

The line divides the plane into two half's with one half being the feasible solution. Now let us take a point (0, 0) not lying on the line. Therefore substituting x = 0 and y = 0, we get 3.0 + 4.0 > 12, which is not true and hence the point doesn't satisfy the inequation 3x + 4y >12 and hence the solution set lie on the non-origin side of the line i.e. above the line not including those solution which are on the line since the inequation being greater than type.

Illustration No. 9

Plot the graphs of the following inequations (inequalities)

(i) $x + 2y \leq 8$

(ii) $2x + 3y \geq 12$

Solution:

(i) we have the given inequation as $x + 2y \leq 8$ .....(1)

Now the corresponding equation in reference to the above inequation (1) is

$$x + 2y = 8 \text{ or } \frac{x}{8} + \frac{y}{4} = 1 \qquad .....(2)$$

Plotting (2) with reference to the coordinate axis XOX′ and YOY′, we get, the following figure,

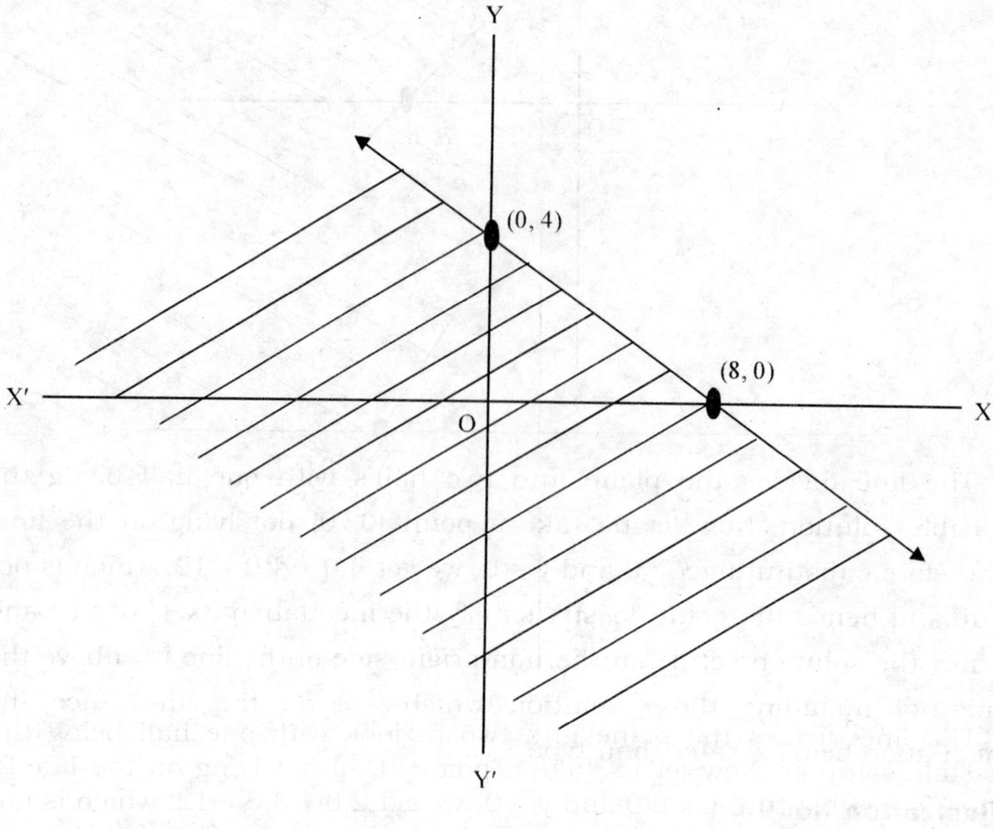

The line divides the plane into two half's with one half being the feasible region satisfying the inequation. Now let us take a point (0, 0) not lying on the line on 2. Therefore substituting x = 0 and y = 0, we get 0 + 2.0 < 8, which is true and hence the solution set lie on the side of the origin of the line i.e. below the line which contains the origin including those on the line since the inequation being lesser than equal to type.

(ii) we have the given inequation as $2x + 3y \geq 12$ .....(1)

Now the corresponding equation in reference to the above inequation (1) is

$$2x + 3y = 12 \text{ or } \frac{x}{6} + \frac{y}{4} = 1 \qquad .....(2)$$

Plotting (2) with reference to the coordinate axis XOX' and YOY', we get, the following figure,

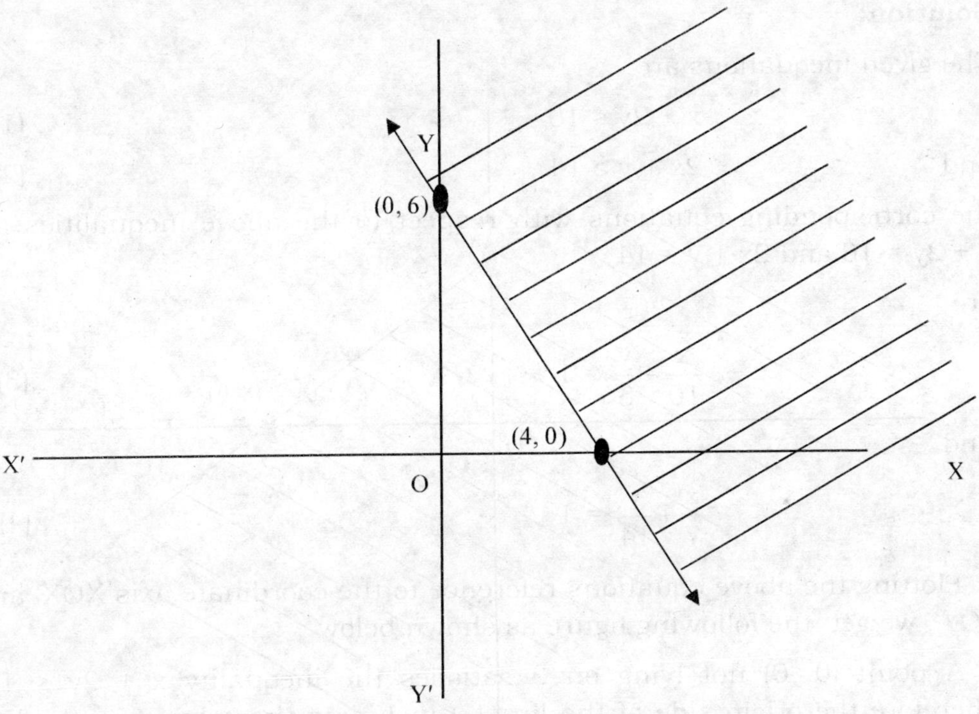

The line divides the plane into two regions with one half being the feasible solution. Now let us take a point (0, 0) not lying on the line 2. Therefore substituting x = 0 and y = 0, we get 2.0 + 3.0 > 12, which is not true and hence the point doesn't satisfy the inequation 2x + 3y >12 and hence the solution set lie on the non-origin side of the line i.e. above the line including those solution which are on the line since the inequation being greater than equal to type.

Graphs of System of Linear Equations

Any two or more linear inequalities or inequations having common solution are called as simultaneous inequations or inequalities. The solution set can be obtained by plotting both the system of inequations having common area. The illustration below will help to understand the concept.

Illustration No. 10

Plot the following system of equations and find the solution set of $x + 2y \leq 10$ and $2x + y \geq 14$?

Solution:

The given inequations are

$$x + 2y \leq 10 \quad .....(1)$$

and

$$2x + y \geq 14 \quad .....(2)$$

the corresponding equations with respect to the above inequalities are $x + 2y = 10$ and $2x + y = 14$

or

$$\frac{x}{10} + \frac{y}{5} = 1 \quad .....(3)$$

And

$$\frac{x}{7} + \frac{y}{14} = 1 \quad .....(4)$$

Plotting the above equations reference to the coordinate axis XOX' and YOY', we get, the following figure as shown below

A point (0, 0) not lying on 3 satisfies the inequality $x + 2y \leq 10$. Therefore the origin side of the line (3) including those lying on the line are the solution set of $x + 2y \leq 10$.

Similarly, the point (0, 0) not lying on the (4) doesn't satisfy the inequality $2x + y \geq 14$. Therefore the solution set lies on the non origin side of the line including those on the line as the inequality is of greater than equal to type.

And hence the common solution set is given by the area in the shaded region ABC.

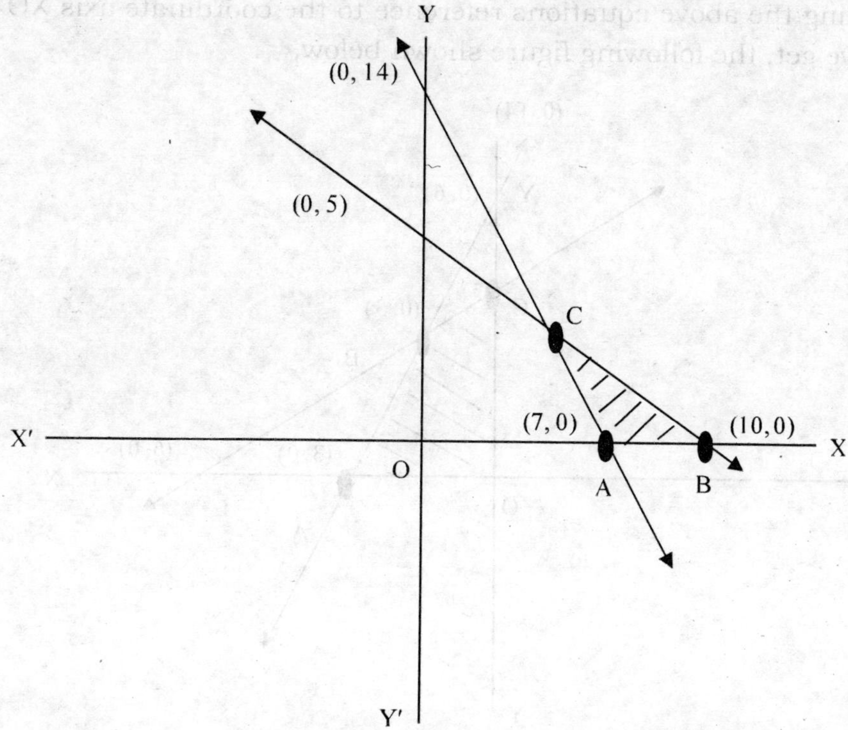

Illustration No. 11

Plot the following system of equations and find the solution set of

$$x + 2y \le 6 \text{ and } 2x + y \le 6 ?$$

Solution:

The given inequations are

$$x + 2y \le 6 \quad .....(1)$$

and $$2x + y \le 6 \quad .....(2)$$

the corresponding equations with respect to the above inequalities are $x + 2y = 6$ and $2x + y = 6$

or

$$\frac{x}{6} + \frac{y}{3} = 1 \quad .....(3)$$

And $$\frac{x}{3} + \frac{y}{6} = 1 \quad .....(4)$$

Plotting the above equations reference to the coordinate axis XOX' and YOY', we get, the following figure shown below,

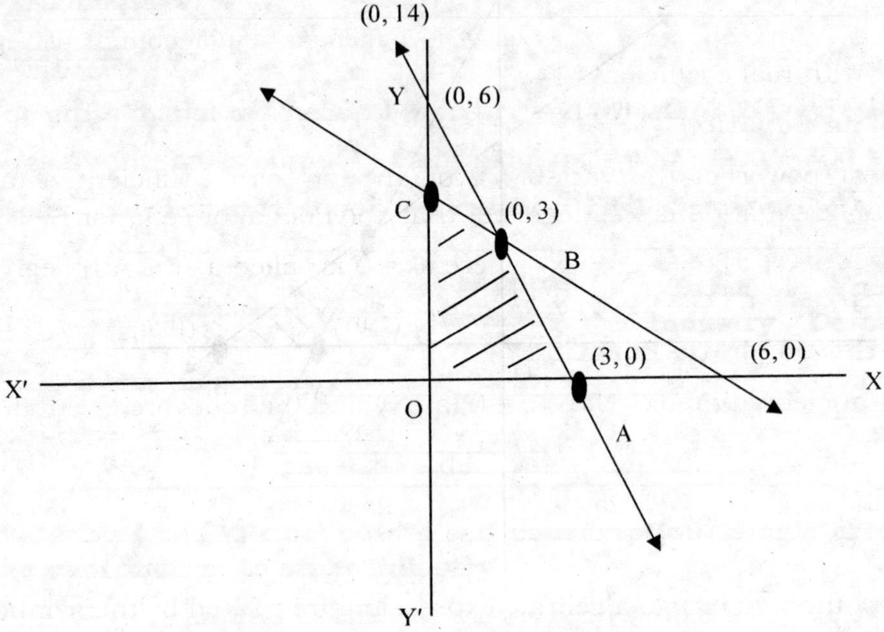

A point (0, 0) not lying on 3 satisfies the inequality x + 2y ≤ 6. Therefore the origin side of the line (3) including those lying on the line are the solution set of x + 2y ≤ 0.

Similarly, the point (0, 0) not lying on the (4) also satisfies the inequality 2x + y ≤ 6. Therefore the solution set also lies on the origin side of the line including those on the line as the inequality are all lesser than equal to type the solution set will lie in the bounded area by the lines and below it and move towards origin.

And hence the common solution set is given by the area bounded by the shaded region ABC.

Summary

Equation

Any two algebraic expressions which are connected by equality sign (=) are termed as equation.

Polynomial

Let a_0, a_1, a_2 ... a_n be real numbers and x is a variable then the expression

$f(x) = a_0 + a_1x + a_2x^2 + ... + a_nx^n$ is called as a polynomial of real variable x with real coefficients.

Degree of an Equation

The highest power of the variable x of an non zero coefficient of an polynomial or equation is the degree of that polynomial or equation.

If, $f(x) = a_0 + a_1x + a_2x^2 + ... + a_nx^n$ where $a_n \neq 0$ is called as the nth degree equation.

Roots of the Quadratic Equation

The quadratic equation $ax^2 + bx + c = 0$ $(a \neq 0)$ and the roots are given as

$$\frac{-b+\sqrt{b^2-4ac}}{2a}, \frac{-b-\sqrt{b^2-4ac}}{2a}$$

Inequality

If two real numbers or two algebraic expression are related by the symbol > (Greater than) or < (Lesser than) and also the signs ≥ or ≤ (greater than equal or lesser than equal) form an inequality. An inequality with (< or >) sign are called as strict inequality while an inequality having (≥ or ≤) sign are called as slack inequality.

Self Practice Exercises

1. Solve $\frac{\sqrt{7+x}+\sqrt{7-x}}{\sqrt{7+x}-\sqrt{7-x}} = \frac{7}{x}$ where $x \neq 0$

2. Solve $x = \sqrt{6+\sqrt{6+\sqrt{6+\sqrt{6 ... \infty}}}}$

3. If the product of the roots of the equation $px^2 + 8x + (4p - 1) = 0$ is -2 then find the value of p?

4. Find the value of p satisfying the equation $\log_2(p^2 + 4p + 8) = 2$?

5. Solve the quadratic equation $x^2 - \left(3\sqrt{2} - 2i\right)x - 6\sqrt{2i} = 0$ and find the root of the equation?

6. Examine the nature of the roots of the equation

   (i) $4x^2 - 9x + 7 = 0$.

   (ii) $9x^2 - 6\sqrt{6}\,x + 6 = 0$

7. If α and β are the roots of the equation $x^2 + bx + c = 0$ then find the equation whose roots are $\frac{1}{a+b}$ and $\frac{1}{\beta+b}$?

8. Show that the roots of the equation $(x - p)(x - q) + (x - q)(x - r) + (x - r)(x - p) = 0$ are always real and they cannot be equal until and unless $p = q = r$?

9. If α and β are the roots of the equation $2x^2 + 4x + 6 = 0$ then find the equations whose roots are

   (i) $\alpha + 4$ and $\beta + 4$

   (ii) $\frac{\alpha}{\beta}$ and $\frac{\beta}{\alpha}$

10. Solve the quadratic equation $2x^2 - (3 + 7i)x - (3 - 9i) = 0$ and find the roots of the equation?

11. Solve $\frac{4x+4}{2x-4} < 5$?

12. Solve $|x - 4| \geq 7$?

13. Solve $2 \leq |x - 1| \leq 3$?

14. Solve $|x^2 - 2x| > x$?

15. $\frac{|x-1|}{x+2} < 1$

16. Solve the system of equations $|x - 3| \leq 6$ and $|x| \geq 3$?

17. Solve $x^2 - |5x + 8| > 0$

18. Plot the system of equations and find the solution set of $6x + 8y \geq 24$, $y \geq 1$ and $x \geq 0$?

19. Plot the following system of equations and find the solution set of $3x + 2y \leq 12$, $x + y \geq 1$, $x \geq 1$, $y \leq 3$ where $x, y \geq 0$?

20. Plot the following system of equations and find the solution set of $4x + 3y \leq 12$, $3x + 4y \geq 12$ where $x, y \geq 0$?

Answers

1. $x = -7$
2. $x = -2, 3$
3. $p = 1/6$
4. $p = -2$
5. $\alpha = 3\sqrt{2}, \beta = -2i$
6. (i) Roots are imaginary and unequal

   (ii) Roots are real and equal.
7. $cx^2 - bx + 1 = 0$
9. (i) $x^2 - 10x + 27 = 0$ and (ii) $3x^2 + 2x + 3$
10. $\alpha = \frac{3}{2} + \frac{1}{2} i$ and $\beta = 3i$
11. $x > 4$
12. $x \in (-\infty, -3] \cup [11, \infty)$
13. $x \in [-2, -1] \cup [3, 4]$
14. $1 < x < 3$
15. $(-\infty, -2) \cup (-½, \infty)$
16. $(-3, -2] \cup [3, 9]$
17. $x < \frac{5 - \sqrt{57}}{2}$ and $x < \frac{5 + \sqrt{57}}{2}$

CHAPTER 14

Progression

Arithmetic Progression

A sequence of number are said to be in arithmetic progression if the difference of the term and the proceeding term is always same i.e., $a_{n+1} - a_n$ = constant (=d) for all $n \in N$

For Example

1 + 3 + 5 + 7 ... here the first term is a = 1 and the common difference is d = 2

General Form of an A.P

Let a be the first term and d the common different of an arithmetic progression then the general form of the A.P is given by

$$a, (a + d), (a + 2d), (a + 3d), \ldots$$

And the corresponding arithmetic series is

$$a + (a + d) + (a + 2d) + \ldots$$

The General Term of an A.P

If a is the first term and d is the common different of an A.P then its nth term is given by a + (n – 1) d i.e. $a_n = a + (n - 1) d$

Proof:

Let $a_1, a_2, a_3, \ldots a_n \ldots$ given A.P.

Then a_1 = first term

$$= a + (1 - 1) d = a$$

Similarly
$$a_2 - a_1 = d = a_2 = d + a_1 = d + a$$
$$= a + (2 - 1)d$$
$$a_3 - a_2 = d = a_3 = a_2 + d = (a + d) + d$$
$$= a + 2d$$
$$= a + (3 - 1) d$$

Continuing the process, we get

$$a_n - a_{n-1} = d = a_n = a_{n-1} + d$$
$$= a + (n - 2 + 1) d$$
$$= a + (n - 1) d$$

Sum of n–terms of an A.P

Sum of n term of an A.P whose term is 'a' and common different 'd' is given by

$$S_n = \frac{n}{2}[2a + (n-1)d] \quad \text{or} \quad S_n = \frac{n}{2}[a + l]$$

where l = last term = $a + (n - 1)d$

Proof:

Let $a_1, a_2, a_3 \ldots$ are in A.P with first term a and common difference d, then $a_1 = a$, $a_2 = a + d$, $a_3 = a + 2d \ldots$ and $a_n = a + (n - 1)d$

Now, $\quad S_n = a_1 + a_2 + a_3 \ldots + a_{n-1} + a_n$

$$= a + (a + d) + \ldots a + (n - 1) d \qquad \ldots..(1)$$

Now write 1 in reserve order we get,

$$S_n = a + (n - 1) d + a + (n - 2)d + \ldots + (a + d) + a \qquad \ldots..(2)$$

Adding 1 and 2, we get

$$2S_n = [2a + (n - 1)d] + [2a + (n - 1)d] + \ldots + [2a + (n - 1)d]$$

{Since 2a + (n – 1) d repeated n times}

Therefore,

$$2S_n = n [2a + (n - 1) d]$$

And hence,

$$S_n = \frac{n}{2}[2a + (n-1)d]$$

As l is the last term where $l = a + (n - 1) d$

Therefore,

$$S_n = \frac{n}{2}[2a + (n-1)d]$$
$$= \frac{n}{2}[a + \{a + (n-1)d\}]$$
$$= \frac{n}{2}[a + l]$$

Hence proved

Properties of an Arithmetic Progression

(i) If a constant is added or subtracted from each term of an A.P. then the sequence is also an A.P. with same difference.

Proof:

Let the given A.P be a, a + d, a + 2d ...

Adding each term by a nonzero constant k, then the sequence turns into

$$a + k, (a + d + k), (a + 2d + k) \ldots$$

which is also an A.P series with first term a + k and common difference d. Similarly the sequence obtained by subtracting each term by k we get,

$$a - k, (a + d - k), (a + 2d - k)\ldots$$

which is also an A.P series with first term a − k and common difference d

(ii) If each term of an A.P. is multiplied or divided by the same constant then the sequence is also in A.P. with common difference kd or d/k where d is the common different in A.P.

Proof:

Let the given A.P be a, a + d, a + 2d ...

Multiplying each term by a nonzero constant k, then the sequence turns into

$$ak, (a + d)k, (a + 2d)k \ldots$$

$$ak, (ak + dk), (ak + 2dk)\ldots$$

which is clearly a A.P with common difference dk and first term ak

When every term is divided by k, then

$$\frac{a}{k}, \frac{a+d}{k}, \frac{a+2d}{k}, \ldots$$

which is also in A.P with first term $\frac{a}{k}$ and common difference $\frac{d}{k}$

(iii) If the corresponding terms of two A.P's are added or subtracted, then the resulting sequence is also in A.P.

Proof:

Let the two A.P series of n terms be

$a_1, (a_1 + d_1), (a_1 + 2d_1) \ldots$

$a_2, (a_2 + d_2), (a_2 + 2d_2) \ldots$

then the sequence obtained by adding the corresponding terms is

$(a_1 + a_2), [(a_1 + d_1) + (a_2 + d_2)], [(a_1 + 2d_1) + (a_2 + 2d_2)] \ldots$

$(a_1 + a_2), [(a_1 + a_2) + (d_1 + d_2)], [(a_1 + a_2) + 2(d_1 + d_2)] \ldots$

which is clearly an A.P with the first term $(a_1 + a_2)$ and common difference $(d_1 + d_2)$

(iv) In a finite A.P. the sum of the terms equidistant from the beginning and end is equal is always same and is equal to the sum of first and last term i.e. $a_k + a_{n-(k-1)} = a_1 + a_n$ for all k = 1, 2, 3, n – 1

Proof:

Let $a_1, a_2, a_3, \ldots a_n$ be an A.P series, with common difference d, then the kth term from the beginning is $a_k = a_1 + (k - 1)d$ and k^{th} term from the end is (n – k + 1) term from the beginning then $a_{n-k+1} = a_1 + (n - k + 1 - 1) = a_1 + (n - k)d$. Therefore, the sum of the kth term from the beginning and k^{th} term from the end

$$a_k + a_{n-k+1} = a_1 + (k - 1)d + a_1 + (n - k)d$$
$$= a_1 + a_1 + (n - 1)d$$
$$= a_1 + a_n$$

(v) Three numbers a, b, c in A.P if and only if 2b = a + c.

(Left to students to proof)

Illustration No.1

Which term of the series, 14 + 11 + ... is equal to (i) – 45 (ii) –105, if the series is in A.P. with first term as 14 and common different as (–3) ?

Solution:

Since $t_n = a + (n - 1)\, d$

$= 12 + (n - 1)\,(-3)$

$= 12 + 3 - 3n = 15 - 3n$

But the nth term is – 45 then

$15 - 3n = -45$

Or, $3n = 60$ or $n = 20$

Similarly,

$$-105 = 15 - 3n \text{ or } 3n = 120 \text{ or } n = 40$$

Illustration No. 2

Find an A.P. whose third term is 9 and eighth term is 29 ?

Solution:

Since,

$$t_3 = 3^{rd} \text{ term} = a + (3 - 1)\, d$$
$$= a + 2d$$

And,

$$t_8 = 8^{th} \text{ term} = a + (8 - 1)\, d$$
$$= a + 7d$$

But given that $t_3 = 9$ and $t_8 = 29$, thus we have

$$a + 2d = 9 \qquad \text{..... (i)}$$

$$a + 7d = 29 \qquad \text{.....(ii)}$$

Solving (i) and (ii) we get, a = 1 and d = 4

Hence the series is 1, 5, 9, 13, ...

Illustration No. 3

Find the total number of all divisible of 5 in between 50 to 300?

Solution:

The first term which is divisible by 5 is 50 and the last term which is divisible by 5 is 300

Therefore of number of terms in A.P with the series 50 + 55 + 60 ... 300 is given by

$$t_n = a + (n - 1)\, d$$

$$300 = 50 + (n - 1)\, 5$$

$$\left(\frac{300 - 50}{5}\right) + 1 = n \text{ or } n = 51$$

Illustration No. 4

Find the value of expression 2 – 7 + 3 – 8 + 4 – 9 + ... to 100?

Solution:

The series can be rewritten as

(2 + 3 + 4 ... to 50 terms) – (7 + 8 + 9 + ... 50 terms)

Both these are in AP with values of a and d as a = 2, n = 50 and d = 1 and a = 7, n = 50 and d = 1 respectively

Therefore,

$$= \frac{50}{2}[2\times 2+(50-1)1]-\frac{50}{2}[2\times 7+(50-1)1]$$

$$= 25(4+49)-25(14+49)$$

$$= 25(53-63)$$

$$= 25(-10) = -250$$

Illustration No.5

Find the three numbers in A.P such that the sum of the numbers is 30 and there product is 960?

Solution:

Let the three numbers in A.P are a – d, a, a + d,

Now we have that sum of the numbers is 30

i.e.

$$a + d + a + a - d = 30$$

or $$3a = 30 \rightarrow a = 10$$

But the products of the numbers is 960,

Therefore,

$$(a - d) \times a \times (a + d) = 960$$

Substituting the value of a = 10 in the equation, we have

$$(10 - d) \times 10 \times (10 + d) = 960$$

$$(100 - d^2) = 96$$

$$d^2 = 4 \text{ or } d = \pm 2$$

Now if a = 10 and d = 2 then the required number are 8, 10, 12 and if a = 10 and d = – 2 then the required numbers are 12, 10, 8, the same numbers in reserve order.

Arithmetic Mean (A. M)

If in between two given quantities a and b, we insert n quantities A_1, A_2 ... A_n such that a, A_1, A_2 ... A_n b form an A.P., then we can say that A_1, A_2... A_n are arithmetic means between a and b.

Let A_1, A_2 ... A_n be n – arithmetic means between two quantities a and b. Then, a, A_1, A_2 ... A_n b are in A.P.

Let d be the common different of the A.P series, then

$$b = (n + 2)^{th} \text{ term} = a + (n + 1)\, d$$

therefore,

$$d = \frac{b-a}{n-1}$$

Now,

$$A_1 = a + d = a + \left(\frac{b-a}{n-1}\right)$$

$$A_2 = a + 2d = a + 2\left(\frac{b-a}{n-1}\right)$$

.

.

.

$$A_n = a + n\,d = a + n\left(\frac{b-a}{n-1}\right)$$

These are the required arithmetic means a and b.

Illustration No. 6

Between 1 and 40 are inserted n arithmetic means so that the ratio of the 6th and $(n - 2)^{th}$ means is 4 : 7, then find the value of n?

Solution:

Let A_1, A_2, ... A_n be n arithmetic means in between 1 and 40. Then 1 A_1, A_2, ... A_n , 40 is an A.P with common difference (d)

Therefore

$$d = \left(\frac{40-1}{n+1}\right) = \frac{39}{n+1}$$

But,

$$A_6 = 1 + 5d$$

$$= 1 + 5\left(\frac{39}{n+1}\right) = \frac{(n+1)+195}{n+1}$$

and, $$A_{n-2} = 1 + (n-3)d = 1 + (n-3)\left(\frac{39}{n+1}\right)$$

$$= \frac{(n+1) + (n-3)39}{n+1} = \frac{40n - 116}{n+1}$$

It is given that $$\frac{A_6}{A_{n-2}} = \frac{4}{7}$$

Therefore,

$$\frac{\frac{n+196}{n+1}}{\frac{40n-116}{n+1}} = \frac{4}{7}$$

$$7n + 1372 = 160n - 464$$

$$153n = 1836$$

$$n = 12.$$

Illustration No. 7

Prove that the sum of n arithmetic means between two numbers is n times the single A.M between them?

Solution:

Let $A_1, A_2, \ldots A_n$ be n arithmetic means in between a and b. Then a $A_1, A_2, \ldots A_n$, b is an A.P with common difference (d)

Therefore

$$d = \frac{b-a}{n-1}$$

Now,

$$A_1 + A_2 + \ldots + A_n = \frac{n}{2}(A_1 + A_n)$$

$$= \frac{n}{2}(a+b)$$

That implies, $$A_1 + A_2 + \ldots + A_n = n\left(\frac{a+b}{2}\right)$$

$$= n \text{ (A.M between a and b)}$$

Illustration No. 8

The sum of the two numbers is 34. An even number of arithmetic means are being inserted between them such that their sum exceeds their number by 192, then find the number of means inserted.

Solution:

Let a and b be the two numbers such that a + b = 34,

Assume A_1 , A_2, ... A_{2n} be 2n arithmetic means between a and b

Then,

$$A_1 + A_2 + \dots + A_{2n} = 2n\left(\frac{a+b}{2}\right)$$

$$A_1 + A_2 + \dots + A_{2n} = n\,(a + b) = 34n$$

It is given that $A_1 + A_2 + \dots + A_{2n} = 2n + 192$

$$34n = 2n + 192$$

$$32n = 192$$

Or $n = 6$

Representation of Terms in A.P

The convenient representation of terms of an A.P are as follows

1. 3 terms : a – d, a, a + d
2. 4 terms : a – 3d, a – d, a + d, a + 3d
3. 5 terms : a – 2d, a – d, a, a + d, a + 2d

Note: In case of odd number of terms the middle terms is a whereas in case of even number of terms the middle terms are a – d and a + d and the common difference is 2d.

Illustration No. 9

Find the four numbers in A.P such that the sum of number is 24 and the difference between product of means and extremes is 8?

Solution:

Let the numbers in A.P are a – 3d, a – d, a + d, a + 3d

Now sum of the numbers are,

$$a - 3d + a - d + a + d + a + 3d = 24$$

$$4a = 24 \text{ or } a = 6$$

But the difference between the product of means and extremes is equal to 8

Therefore,

$$(a - d)(a + d) - (a - 3d)(a + 3d) = 8$$

$$(a^2 - d^2) - (a^2 - 9d^2) = 8$$

$$a^2 - d^2 - a^2 + 9d^2 = 8,$$

$$8d^2 = 8 \text{ or } d = \pm 1$$

And hence the numbers are when a = 6 and d = 1 are 3, 5, 7, 9 and when a = 6 and d = –1 we have the numbers as 9, 7, 5, 3 i.e. the same numbers in reserve order.

Illustration No.10

If the sum of 6th and 9th term of an A.P are in the ratio of 36 : 81, then show that the ratio of the 6th term and 9th term of A.P 11 : 17?

Solution:

Let the given A.P be a, a + d, a + 2d + ...

then we have

$$\frac{S_6}{S_9} = \frac{\frac{6}{2}(2a + 5d)}{\frac{9}{2}(2a + 8d)} = \frac{36}{81}$$

$$\frac{(2a + 5d)}{(2a + 8d)} = \frac{2}{3}$$

$$6a + 15d = 4a + 16d$$

Or $d = 2a$

But,

$$\frac{t_6}{t_9} = \frac{(a + 5d)}{(a + 8d)}$$

Therefore, substituting the value of d = 2a , we get

$$\frac{t_6}{t_9} = \frac{a + 5(2a)}{a + 8(2a)} = \frac{11a}{17a} = \frac{11}{17}$$

Hence Proved

Illustration No.11

A polygon with 15 sides whose length follows an A.P series. If the sum of all the sides is equal to 240 meters and the length of the largest side is 15 times that of the smallest side, then find the smallest and the largest sides of the polygon.

Solution:

Let 'a' be the smallest side of the polygon and 'd' be the common difference since sum of all sides is 240 meters

Therefore,

$$240 = \frac{n}{2}[2a + (n-1)d]$$

Here n = 15

And hence,

$$240 = \frac{15}{2}[2a + (15-1)d]$$

$$32 = 2a + 14d$$

$$16 = a + 7d \qquad \text{..... (i)}$$

But as the largest side is 15 times the smallest side

Therefore

$$a + (15-1)d = 15a$$

$$a + 14d = 15a$$

or $$a = d$$

Substituting the value of a in (i) we get

$$16 = d + 7d$$

$$16 = 8d \text{ or } d = 2$$

As $d = 2$ and $d = a$

Therefore $a = 2$

And hence the smallest side of the polygon is a = 2 meters and the length of the largest side is 15 × 2 = 30 meters.

Illustration No. 12

The two persons X and Y have joined Master Consultant with X started with a salary of Rs 5000 and that of Y is Rs 6000 with annual increment of Rs.1000 and Rs. 900. In which year will X overtake Y in salary?

Solution:

The current difference between Mr. X and Mr. Y is Rs 1000. Now, the annual difference is Rs 100 per year. At this rate Mr. X will take 10 years to equal the salary with Mr. Y and hence in the 11th year Mr. X will earn more than Mr. Y.

Illustration No.13

'A' number 28 is divided into four parts that are in A.P such that the product of second and third is in the ratio of 5 : 6. Find the second and fourth parts.

Solution:

Let the numbers be a – 3d, a – d, a + d and a + 3d

As sum of the numbers a – 3d + a – d + a + d + a + 3d = 28

Therefore,

$$4a = 28 \text{ or } a = 7$$

But product of 1st and 4th part and 2nd and 3rd part are in the ratio of 5:6

Therefore,

$$(a - 3d)\ (a + 3d) : (a - d)\ (a + d): 5:6$$

$$(a^2 - 9d^2): (a^2 - d^2) : 5:6$$

$$\frac{49 - 9d^2}{49 - d^2} = \frac{5}{6}$$

$$49 = 49d^2$$

$$\therefore \quad d^2 = \pm 1$$

Thus the numbers are 4, 6, 8 and 10.

Illustration No. 14

If p^2, q^2 and r^2 are in A.P then prove that $\frac{1}{p+q}, \frac{1}{q+r}$ and $\frac{1}{p+r}$ are also in A.P?

Solution:

Since p^2, q^2 and r^2 are in A.P

Therefore,

$$p^2 + (pq + pr + p),\ q^2 + (pq + qr + rp) \text{ and } r^2 + (pq + qr + rp)$$

Since by adding (pq + qr + rp) to each of p^2, q^2 and r^2 makes no difference as their difference is again p^2, q^2 and r^2

The above equation can be rewritten as

$$p(p + q) + r(q + p),\ q(q + p) + r(q + p),\ r(p + r) + q(p + r)$$

$$(p + q)(p + r),\ (q + p)(q + r),\ (q + r)(p + r)$$

Now divide each by (p + q) (q + v) (p + r)

We get,

$$\frac{1}{p+q}, \frac{1}{q+r}, \frac{1}{p+r} \text{ which is also in A.P}$$

Illustration No.15

A man agrees to repay a debt of Rs. 7200 in number of installments where each next installment exceeding the former by Rs 4. If 1st installment is Rs. 2, then find out in how many installments the loan will be completely repaid?

Solution:

Given the 1st installment is Rs 2 i.e. a = 2 and d = 4 and S_n = 7200

$$S_n = \frac{n}{2}[2a + (n-1)d]$$

$$7200 = \frac{n}{2}[4 + (n-1)4]$$

$$14400 = n(4 + 4n - 4)$$

$$4n^2 = 14400$$

$$n^2 = 3600$$

hence n = ± 60, as n cannot be negative, therefore n = 60. So the loan will be repaid in 60 installments

Illustration No. 16

The sum of digits of a 3 digit number is 15, the digits in the number form an A.P series. If the digits are reversed then number exceeds the original number by 198. Find the number?

Solution:

Let the digits be a + d, a, a – d

Therefore,

$$a + d + a + a - d = 15$$

or $3a = 15 \rightarrow a = 5$

But the 3 digit number can expressed as

$$100(a - d) + 10(a) + a + d = 100\,a - 100d + 10\,a + a + d$$

$$= 111a - 99d$$

And when the number is reversed i.e.

$$100(a + d) + 10a + (a - d) = 111a + 99d$$

But it is given that reversed number exceed the former by 198

Therefore,

$$111a + 99d = 111a - 99d + 198$$

$$198d = 198$$

That implies d = 1

So the numbers are 4, 5 and 6 and the required number is 456.

Geometric Progression

A sequence of numbers are said to form a geometric progression if the ratio of the term and its preceding term bears constant quantity. Otherwise a sequence is in G.P if the terms in the sequence either increase or decreases by a constant quantity. The constant quantity in G.P is called as common ratio.

For example

3, 6, 12, 24 ... is an infinite G.P profession with first sum as 3 and common ratio is 2.

Similarly, 4, 16, 64 ... is a G.P whose 1st term is 4 and common ratio is 4

General Form of a G.P

If 'a' is the first term and r the common ratio of the sequence. Then the general terms of G.P is of the form a, ar, ar^2 ...

N^{th} term of a G.P

Theorem

Prove that the nth term of an G.P with first term a and common ratio r is given by $a_n = a\, r^{n-1}$

Proof:

Let $a_1, a_2, a_3 \ldots a_n \ldots$ be a given G.P then

$$a_1 = a = a_1 = a \times r^{1-1}$$

since $a_1, a_2, a_3 \ldots$ are G.P with a common ratio r,

Therefore,

$$\frac{a_2}{a_1} = r = a_2 = a\, r = a\, r^{2-1}$$

Similarly,

$$\frac{a_3}{a_2} = r = a_3 = a_2 r = (ar)\, r = ar^2$$

$$= ar^{3-1}$$

Proceeding further

$$\frac{a_4}{a_3} = r = a_4 = a_3 r = (ar^2) \times r$$

$$= ar^3$$

$$= ar^{4-1}$$

Continuous this process we have, $a_n = ar^{n-1}$

Hence proved

Sum of n terms of a G.P

The sum of n terms of a G.P with first terms a and common ratio r is given $S_n = \frac{a(r^n - 1)}{r - 1}$ for r > 1

$$= \frac{a(1 - r^n)}{1 - r} \text{ for } r < 1$$

Proof:

Let S_n denoted the sum of n terms of the G.P with first term as a and common ratio r.

Therefore S_n can written as $S_n = a + ar\ ar^2 + ... + ar^{n-1}$ .....(i)

Now multiplying both side by 'r' we have,

$$r\,S_n = ar + ar^2 + ar^3 + ... + ar^n \quad \text{..... (ii)}$$

Subtracting (i) and (ii) we get $S_n - rS_n = a - ar^n$

$$S_n(1 - r) = a\,(1 - r^n)$$

i.e. $$Sn = \frac{a(1 - r^n)}{1 - r} \text{ where } r \neq 1$$

Now changing the signs of numerator and denominator we have

$$S_n = \frac{a(r^n - 1)}{r - 1}$$

Note: The two formulas of calculating G.P are identical

Sum of an Infinite G.P

The sum of an infinite G.P with first term a and common ratio where $|r| < 1$ is

$$S_n = \frac{a}{r - 1}$$

Proof:

Consider an infinite G.P with first term a and common ratio r where $|r| < 1$ i.e. $-1 < r < 1$

But the sum of n terms of G.P is given by

$$S_n = \frac{a(1 - r^n)}{1 - r}$$

$$= \frac{a}{r - 1} - \frac{ar^n}{1 - r} \quad \text{.....(i)}$$

As r lies between –1 and 1 i.e. $-1 < r < 1$

Therefore r^n decrease as n increases and r^n tends to zero as n trends to infinity i.e. $r^n \to 0$ as $n \to \infty$

Hence substituting the condition in '1' we get,

The sum of the an infinite G.P is

$$S_\infty = \lim_{n\to\infty} S_n = \lim_{n\to\infty}\left(\frac{a}{r-1} - \frac{ar^n}{1-r}\right)$$

$$= \frac{a}{r-1} \text{ for } |r|<1$$

Properties of Geometric Progression

(i) If all the terms of a G.P be multiplied or divided by the same number then it remains a G.P with same common ratio

(ii) The reciprocals of the terms of a G.P also forms a G.P

Brief Proof:

Let $a_1, a_2 \ldots a_n \ldots$ be G.P with common ratio r

Then $$\frac{a_{n+1}}{a_n} = r \text{ for all } n \in N$$

Now the sequence formed by the reciprocals of the terms of the given G.P is

$$\frac{1}{a_1}, \frac{1}{a_2} \ldots \frac{1}{a_n}, \ldots$$

We have $$\frac{\frac{1}{a_{n+1}}}{\frac{1}{a_n}} = \frac{a_n}{a_{n+1}} = \frac{1}{r}$$

So, the new series is with common ratio as $1/r$

(iii) Three non zero number a, b, c are in G.P if and only if $b^2 = ac$

(iv) In a infinite G.P the product of the terms equidistant from the beginning and the end is always the same and is equal to the product of the first and the last term

Proof:

Let $a_1, a_2 \ldots a_n$ be infinite G.P with common ratio r then the kth term from the beginning a_1r^{k-1} and k^{th} term from the end = $(n - k + 1)^{th}$ term from the beginning = $a_{n-k+1} = a_1 r^{n-k}$

And hence the product of kth term from beginning and k^{th} term from the end

$$= a_k a_{n-k+1}$$

$$= a_1 r^{k-1} a_1 r^{n-k}$$

$$= a_1 a_1 r^{n-1}$$

$$= a_1 a_n \text{ for all } k = 2, 3 \ldots n - 1$$

(v) If $a_1, a_2, a_3 \ldots$ are G.P then $\log a_1, \log a_2, \ldots \log a_n$ are also in A.P and vice versa.

Illustration No. 17

If 0.005 + 0.02 + 0.08 + 0.32 + ... 81.92 are in G.P. Then find which term of the series is 81.92?

Solution:

Hence the ratio $\frac{t_2}{t_1} = \frac{0.02}{0.05} = 4$ and $\frac{t_3}{t_2} = \frac{0.08}{0.02} = 4$

Since the series in G.P with common ratio as 4 and the 1st term a = 0.005

Let 81.92 be the nth term. Then $t_n = ar^{n-1} = 0.005 \times 4^{n-1}$.

But $t_n = 81.92$,

Therefore $81.92 = 0.005 \times 4^{n-1}$

And hence $\frac{81.92}{0.005} = 4^{n-1}$

$16384 = 4^{n-1}$

or $4^7 = 4^{n-1}$

i.e. $7 = n - 1$ or $n = 8$

Hence the 8th term of the series = 81.92

Illustration No.18

The second term of G.P is cube of the first term and the 4th term is 128. Then find the series?

Solution:

Let a be the first terms and r the common ratio then $t_n = ar^{n-1}$

Therefore,

$$t_2 = ar^{2-1} = ar \text{ and } t_4 = ar^{4-1} = ar^3$$

But $ar = a^3$ .....(i)

and $ar^3 = 128$ .....(ii)

Therefore $r = a^2$,

Now substituting $r = a^2$ in (ii) we get,

$$a \times r^3 = a \times (a^2)^3 = 128$$

or $a^7 = 2^7$

therefore $a = 2$ since $a = 2$ and hence $r = 4$;

Hence the series is 2, 8, 32 ...

Illustration No. 19

Three numbers whose sum is 21 are in A.P. If 2, 5 and 8 are added to them then the numbers are in G.P. Then find the numbers?

Solution:

Let the three number in A.P are $a + d$, a, $a - d$

Such that,

$$a - d + a + a + d = 21 \text{ or } 3a = 21$$

or, $a = 7$

Also $(a + d + 2)$, $(a + 5)$ and $(a + d + 8)$ are in G.P

so that $(a - d + 2)(a + d + 8) = (a + 5)^2$

i.e. $(7 - d + 2)(7 + d + 8) = (12)^2$

$$(9 - d)(15 + d) = 144$$

$$135 + 9d - 15d - d^2 = 144$$

$$d^2 + 6d + 9 = 0$$

$$(d + 3)^2 = 0$$

Or

$$d = -3$$

Therefore the numbers are 4, 7 and 10

Illustration No. 20

Find the sum to n[th] term of the series

(i) 5 + 55 +555 + ...

(ii) .4 + .14 + .444 + ...

Solution:

$$S_n = 5 + 55 + 555 \ldots \text{to n terms}$$

$$= 5(1 + 11 + 111 + \ldots \text{to n terms}$$

$$= \frac{5}{9}(9 + 99 + 999 + \ldots \text{to n terms})$$

$$= \frac{5}{9}[(10-1) + (10^2 - 1) + (10^3 - 1) + \ldots \text{to n terms}]$$

$$= \frac{5}{9}(10 + 10^2 + 10^3 \ldots \text{to n terms}) - (1 + 1 + \ldots \text{n terms})$$

$$= \frac{5}{9}\left[10\left(\frac{10^n - 1}{10 - 1}\right) - n\right] = \frac{5}{9}\left[\left(\frac{10^{n+1} - 10}{9}\right) - n\right]$$

$$= \left[\frac{5}{81}(10^{n+1} - 10) - \frac{5}{9}n\right]$$

(ii) $S_n = 0.4 + 0.44 + 0.444 + \ldots \text{to n terms}$

$$= 4(0.1 + 0.11 + 0.111 + \ldots \text{to n terms}$$

$$= \frac{4}{9}(0.9 + 0.99 + 0.999 + \ldots \text{to n terms})$$

$$= \frac{4}{9}[(1 - \frac{1}{10}) + (1 - \frac{1}{10^2}) + (1 - \frac{1}{10^3}) + \ldots \text{to n terms}]$$

$$= \frac{4}{9}[(1 + 1 + \ldots \text{n terms}) - (\frac{1}{10} + \frac{1}{10^2} + \frac{1}{10^3} \ldots \text{to n terms})$$

$$= \frac{4}{9}\left[n - \frac{1}{10}\left(\frac{1 - \left(\frac{1}{10}\right)^n}{1 - \frac{1}{10}}\right)\right] = \frac{4}{9}\left[n - \frac{1}{9}\left(1 - \frac{1}{10^n}\right)\right]$$

Illustration No. 21

Find the sum of the following series

$$\frac{1}{3}+\frac{1}{4^2}+\frac{1}{3^3}+\frac{1}{4^4}+\frac{1}{3^5}+\frac{1}{4^6}+\ldots\infty$$

Solution:

$$S_\infty=\left(\frac{1}{3}+\frac{1}{3^3}+\frac{1}{3^5}+\ldots\infty\right)+\left(\frac{1}{4^2}+\frac{1}{4^4}+\frac{1}{4^6}+\ldots\infty\right)$$

$$= \text{(An infinite G.P with } a=\frac{1}{3} \text{ and } r=\frac{1}{9})$$

$$+ \text{(An infinite G.P with } a=\frac{1}{16} \text{ and } r=\frac{1}{16})$$

$$=\frac{\frac{1}{3}}{1-\frac{1}{9}}+\frac{\frac{1}{16}}{1-\frac{1}{16}}=\frac{3}{8}+\frac{1}{15}=\frac{53}{120}$$

Illustration No. 22

The sum of an infinite G.P is p and the sum of the squares is q then find the first term and the common ratio of the series?

Solution:

Let the infinite G.P be a, ar, ar^2 ... where a is the 1st term r is the common ratio

Then $p = a + ar + ar^2 + \ldots\infty$

$$p=\frac{a}{1-r} \quad \ldots..(1)$$

Similarly $q = a^2 + a^2r^2 + a^2r^4 + \ldots$

$$=\frac{a^2}{1-r^2} \quad \ldots..(2)$$

Now, squaring 1, we get,

$$p^2=\left(\frac{a}{1-r}\right)^2=\frac{a^2}{(1-r)^2} \quad \ldots..(3)$$

Therefore dividing 3 and 2, we get

$$\frac{p^2}{q} = \frac{a^2}{(1-r)^2} \times \frac{1-r^2}{a^2} = \frac{(1-r)(1+r)}{(1-r)(1-r)}$$

$$= \frac{(1+r)}{(1-r)}$$

$$q(1 + r) = p^2(1 - r)$$

$$q + qr = p^2 - p^2r$$

$$(p^2 + q)\, r = p^2 - q$$

Therefore,

$$r = \frac{p^2 - q}{p^2 + q}$$

Now substituting the value of r in (1) we get,

$$p = \frac{a}{1 - \left(\frac{p^2 - q}{p^2 + q}\right)}$$

$$= \frac{a(p^2 + q)}{2q}$$

or

$$a = \frac{2pq}{(p^2 + q)}$$

Geometric Means

Let a and b are two numbers. If n numbers G_1, G_2 ... G_n are inserted in between a and b such that the sequence a_1, G_1, G_2 ... G_n b are in G.P. Then the numbers G_1, G_2 ... are known as geometric means (GM) between a and b

Geometric Mean

Let G a single geometric mean inserted between a and b, then a, G, b will be in G.P

Therefore,

$$\frac{G}{a} = \frac{b}{G}$$

Or $\quad G^2 = ab$

Or $G = \sqrt{ab}$

Insertion of Geometric Means

Let $G_1, G_2 \ldots G_n$ be the n geometric means between a and b then a, $G_1, G_2, \ldots G_n$, b are in G.P.

Let r be the common ratio including the terms are a and b where the total number of terms are (n + 2).

But,

$$b = (n + 2)^{th} \text{ term in G.P} = ar^{(n+2)-1}$$
$$= ar^{n+1}$$

Hence, $\frac{b}{a} = r^{n+1}$

or $r = \left(\frac{b}{a}\right)^{\frac{1}{n+1}}$

Since $G_1 = ar = a\left(\frac{b}{a}\right)^{\frac{1}{n+1}}$

$G_2 = ar^2 = a\left(\frac{b}{a}\right)^{\frac{2}{n+1}}$

.

.

$G_n = ar^n = a\left(\frac{b}{a}\right)^{\frac{n}{n+1}}$ respectively

Product of n Geometric Means

If $G_1, G_2 \ldots G_n$ are n geometric means

Then,

$$G_1, G_2 \ldots G_n = a^n\left(\frac{b}{a}\right)^{\frac{1}{n+1}} \times \left(\frac{b}{a}\right)^{\frac{2}{n+1}} \ldots \left(\frac{b}{a}\right)^{\frac{n}{n+1}} = a^n\left(\frac{b}{a}\right)^{\frac{1+2+3+\ldots n}{n+1}}$$

$$= a^n\left(\frac{b}{a}\right)^{\frac{n(n+1)}{2(n+1)}}$$

$$= a^n\left(\frac{b}{a}\right)^{\frac{n}{2}}$$

Representation of terms in G.P

The convenient representation of the terms of G.P is

1. 3 terms: $\frac{a}{r}$, a, ar

2. 4 terms: $\frac{a}{r^3}, \frac{a}{r}$, ar, ar^3 , and

3. 5 terms: $\frac{a}{r^2}, \frac{a}{r}$, a, ar, ar^2 etc.

Illustration No. 23

A square is drawn by joining the mid points of the sides of a square. Similarly a third square is drawn inside the second square in the same manner and the process is the square is 4 m. Find the sum of the area of the squares

Solution:

Let ABCD be the 1st square who's each side is equal to 4 m and EFGH is the 2nd square formed by joining the midpoint of ABCD

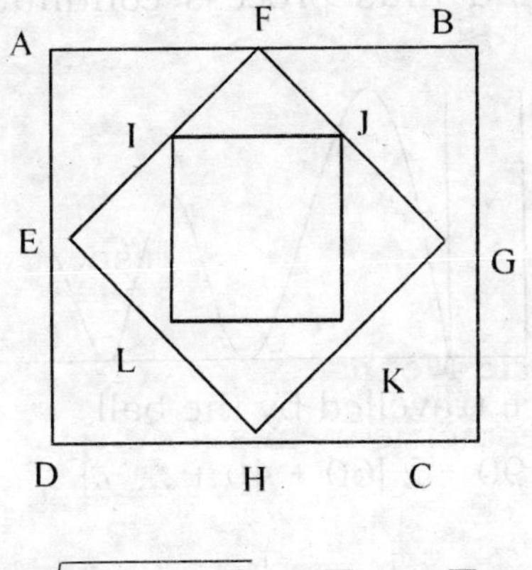

Then $$EF = \sqrt{(2)^2 + (2)^2} = \sqrt{8} = 2\sqrt{2}\,m$$

Now let IJKL be the mid points of the side EF, FG, GH and EH. Then

$$IJ = \sqrt{\left(\sqrt{2}\right)^2 + \left(\sqrt{2}\right)^2} = \sqrt{4} = 2m$$

Similarly proceeding further we get the side of the fourth square is $\sqrt{2}$ m and so on. Therefore the sum of the areas of the square formed is

$$= [4^2 + \left(2\sqrt{2}\right)^2 + 2^2 + \left(\sqrt{2}\right)^2 + \ldots \infty$$

$$= 16 + 8 + 4 + 2 + \ldots \infty$$

$$= \frac{16}{1 - \frac{1}{2}} = 32$$

Illustration No. 24

After striking the floor a ball dropped from certain height rebounds $(2/3)^{rd}$ of the height. Find the total distance travelled before the ball coming to complete rest assuming that the ball has been dropped from a height of 90 meters?

Solution:

As the ball falls from the height of 90 meters. Once the ball strikes the floor the ball rebounds to the $(2/3)^{rd}$ of initial height. Now it falls from a height (2/3) of 90 meters i.e. 60 meters. Now after striking the floor the ball rises to $(2/3)^{rd}$ height. Now the ball falls from a height of $(2/3)^{rd}$ of 60 m i.e. 40 meter. And thus process continues till the ball comes to complete stop.

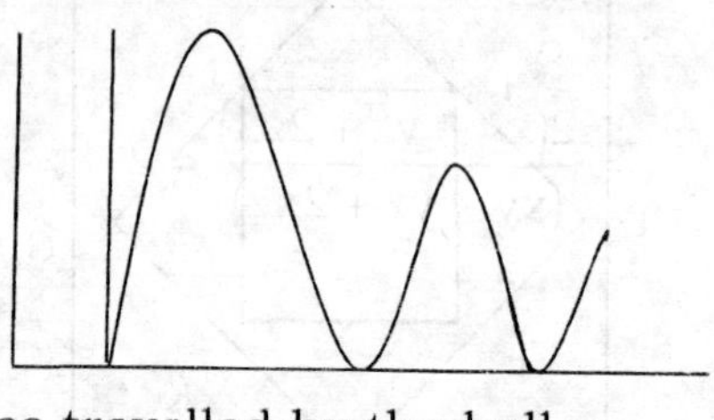

Hence the total distance travelled by the ball

$$= 90 + 2\ [60 + 40 + \ldots \infty]$$

$$= 90 + 2\left(\frac{60}{1 - \frac{2}{3}}\right)$$

$$= 90 + 360$$

$$= 450 \text{ m}$$

Illustration No. 25

If x, y, z are in G.P and p, q are the A.M of x, y and y, z separately, then prove that

(i) $\frac{x}{p} + \frac{z}{q} = 2$

(ii) $\frac{1}{p} + \frac{1}{q} = \frac{2}{y}$

Solution:

(i) If x, y and z are in G.P then $y^2 = x z$

And since p is the A.M of x and y therefore $2p = x + y$

Similarly $2q = y + z$

$$q = \frac{y+z}{2}$$

Now, $$\frac{x}{p} + \frac{z}{q} = \frac{2x}{x+y} + \frac{2z}{y+z}$$

$$= \frac{2x(y+z) + 2z(x+y)}{(x+y)(y+z)}$$

$$= \frac{2(xy + yz + 2zx)}{xy + yz + xz + y^2}$$

$$= \frac{2(xy + yz + 2zx)}{(xy + yz + 2xz)} = 2$$

(ii) $\frac{1}{p} + \frac{1}{q} = \frac{2}{y}$

$$= \frac{2(y+z) + 2(x+y)}{(x+y)(y+z)}$$

$$= \frac{2(y+z) + 2(x+y)}{(x+y)(y+z)}$$

$$= \frac{2(x+z+2y)}{(xy + yz + 2y^2)}$$

$$= \frac{2(x+z+2y)}{y(x+z+2y)} = \frac{2}{y}$$

Illustration No. 26

If $x^2 + y^2$, $xy + yz$ and $y^2 + z^2$ are in G.P then proved that x, y and z are also in G.P

Solution:

Given $x^2 + y^2$, $xy + yz$ and $y^2 + z^2$ are in G.P

Therefore, $(xy + yz)^2 = (x^2 + y^2)(y^2 + z^2)$

$$x^2 y^2 + y^2 z^2 + 2xzy^2$$

$$= x^2 y^2 + x^2 z^2 + y^4 + y^2 z^2$$

$$2xzy^2 = x^2 z^2 + y^4$$

$$x^2 z^2 + y^2 - 2xzy^2 = 0$$

$$(xz - y^2)^2 = 0$$

$$(xy - y^2) = 0$$

$$y^2 = xz$$

Hence x y and z are in G.P

Illustration No. 27

If p, q, r are in G.P and $p^{\frac{1}{x}} = q^{\frac{1}{y}} = c^{\frac{1}{z}}$ then prove that x , y and z are in A.P

Solution:

Since $$p^{\frac{1}{x}} = q^{\frac{1}{y}} = c^{\frac{1}{z}} = \lambda \quad \text{(say)}$$

Therefore,

$$p = \lambda^x,\ q = \lambda^y \text{ and } r = \lambda^z$$

Now x, y and z are in G.P

And hence

$$y^2 = x\,z$$

therefore,

$$(\lambda^y)^2 = \lambda^x \times \lambda^z$$

$$\lambda^{2y} = \lambda^{x+z}$$

comparing both side, We get, $2y = x + z$

Therefore x, y and z are in A.P

Illustration No. 28

Find two positive numbers whose difference is 24 and who's A.M exceeds G.M by 8

Solution:

Let the two numbers be a and b

But, a – b = 24

As A.M – G.M = 8

Therefore,

$$\frac{a+b}{2} - \sqrt{ab} = 8$$

$$a + b - 2\sqrt{ab} = 16$$

$$\left(\sqrt{a} - \sqrt{b}\right)^2 = 16 \text{ or } \sqrt{a} - \sqrt{b} = \pm 4 \quad .....(i)$$

But a – b = 24

$$\left(\sqrt{a} - \sqrt{b}\right)\left(\sqrt{a} + \sqrt{b}\right) = 24$$

$$\left(\sqrt{a} + \sqrt{b}\right) = \pm 6$$

Since $\left(\sqrt{a} + \sqrt{b}\right) = \pm 6$ ..... (ii)

Solving (i) and (ii) we get

$$2\sqrt{a} = \pm 10$$

or $\sqrt{a} = \pm 5$

or $a = 25$ and $b = 1$.

the required numbers

Illustration No. 29

The $(p + q)^{th}$ term and $(p - q)^{th}$ term of a G.P are U and V respectively.

Then show that the pth and q^{th} terms are $\sqrt{uv}$ and $u\left(\frac{u}{v}\right)^{\frac{p}{2q}}$

Solution:

Let a be the first term of and r common ratio then

$$t_{p+q} = u \text{ and } t_{p-q} = v$$

i.e. $ar^{p+q-1} = u$ ... (i) and $ar^{p-q-1} = v$ ... (ii)

Dividing (i) and (ii) we get

$$\frac{ar^{p+q-1}}{ar^{p-q-1}} = \frac{u}{v}$$

$$r^{2q} = \frac{u}{v} \text{ or } r = \left(\frac{u}{v}\right)^{\frac{1}{2q}}$$

Therefore $$\frac{1}{r} = \left(\frac{v}{u}\right)^{\frac{1}{2q}}$$

But $t_p = a\,r^{p-1}$

$$t_p = ar^{p+q-1}\left(\frac{1}{r}\right)^q$$

$$= t_{p+q}\left(\frac{1}{r}\right)^q$$

Since $t_{p+q} = ar^{p+q-1}$

But,

$$t_p = u\left(\frac{v}{u}\right)^{\frac{q}{2q}} == u\left(\frac{v}{u}\right)^{\frac{1}{2}} \sqrt{uv}$$

And $t_q = ar^{q-1}$

$$= ar^{p+q-1}\left(\frac{1}{r}\right)^p$$

$$= a\,t_{p+q}\left(\frac{1}{r}\right)^p$$

$$= u\left(\frac{v}{u}\right)^{\frac{p}{2q}} \qquad \left\{\text{Since } t_q = u\left(\frac{v}{u}\right)^{\frac{p}{2q}}\right\}$$

Illustration No. 30

The sum of 3 numbers is 28 and their product is 512 then find the numbers?

Solution:

Let $\frac{a}{r}$, a, ar be the three numbers

Then $$\frac{a}{r} \times a \times ar = 512 \Leftrightarrow a^3 = 512 \Rightarrow a = 8$$

But the sum of the number is

$$\frac{a}{r} + a + ar = 28 \Leftrightarrow 8\left(\frac{1}{r} + 1 + r\right) = 28$$

$$2(1 + r + r^2) = 7r$$

$$2 + 2r + r^2 = 7r$$

Or $$2r^2 - 5r + 2 = 0$$

$$2r(r - 2) - 1(r - 2) = 0 \text{ or } r = 2 \text{ or } \frac{1}{2}$$

Relationship between A.M and G.M

Let A and G be the arithmetic mean and geometric mean of two given positive real numbers a and b then,

Arithmetic Mean = $\frac{a+b}{2}$ and Geometric mean = $\sqrt{ab}$

And hence,

$$A - G = \frac{a+b}{2} - \sqrt{ab}$$

$$= \frac{a + b - \sqrt{ab}}{2}$$

$$= \frac{\left(\sqrt{a} - \sqrt{b}\right)^2}{2} \geq 0 \qquad .....(1)$$

Hence from 1, we have,

$$A \geq G$$

Illustration No. 31

If the A.M and the G.M of two positive numbers a and b are 10 and 8 then find the numbers?

Solution:

Given that $\quad A.M = \frac{a+b}{2} = 10 \quad$ .....(1)

And $\quad G.M = \sqrt{ab} = 8 \quad$ .....(2)

From 1 and 2, we have,

$$a + b = 20 \quad .....(3)$$

$$ab = 64$$

But

$$(a - b)^2 = (a + b)^2 - 4ab$$

$$= 400 - 256$$

$$= 144$$

Therefore,

$$a - b = \pm 12 \quad .....(4)$$

Solving 3 and 4, we get,

$a = 4$, $b = 16$ or $a = 16$, $b = 4$,

thus the numbers are a = 4, 16 and b = 16, 4.

Harmonic Progression

A set of numbers $a_1, a_2, a_3 \ldots a_n$ are said to be in H.P then their reciprocals $\frac{1}{a_1}, \frac{1}{a_2}, \frac{1}{a_3} \ldots \frac{1}{a_n}$ are in A.P and vice–versa.

The nth term of H.P

The nth term of a H.P is the reciprocal of the nth term of the corresponding A.P. i.e. if the nth term of A.P is $a + (n - 1)d$ then the nth term of H.P will be $\frac{1}{a + (n-1)d}$

Harmonic Mean between Two Quantities

Let a and b be two quantities, then their harmonic mean, 1/a . 1/H, 1/b are in A.P therefore

$$\frac{1}{H}-\frac{1}{a}=\frac{1}{b}-\frac{1}{H}$$

$$\frac{2}{H}=\frac{1}{a}+\frac{1}{b} \text{ or } \frac{2ab}{a+b}$$

Illustration No. 32

If the harmonic mean between two quantities is to their geometric means as $2\sqrt{6}$ to 5, then prove that the quantities are in the ratio of 2 : 3?

Solution:

Let x and y are two numbers then

$$\text{H.M} = \frac{2xy}{x+y} \text{ and G.M} = \sqrt{xy}$$

But

$$\frac{2xy}{x+y} : \sqrt{xy} : 2\sqrt{6} : 5$$

Therefore

$$\frac{10xy}{x+y} = 2\sqrt{6}\sqrt{xy}$$

$$\frac{5\sqrt{xy}}{x+y} = \sqrt{6}$$

Squaring both sides, we get,

$$25\,xy = 6\,(x+y)^2$$

$$6x^2 + 12xy + 6y^2 = 25xy$$

$$6x^2 - 13\,xy + 6y^2 = 0$$

$$(3x-2y)\,(2x-3y) = 0$$

Therefore x : y : 2: 3

Summary

- A sequence of number are said to be in arithmetic progression if the difference of the term and the proceeding term is always same i.e. $a_{n+1} - a_n$ = constant (=d) for all $n \in N$.
- If a is the first term and d is the common difference of an A.P then its nth term is given by a + (n – 1) d i.e. $a_n = a + (n-1)\,d$

- In a finite A.P. the sum of the terms equidistant from the beginning and end is equal is always same and is equal to the sum of first and last term i.e. $a_k + a_{n-(k-1)} = a_1 + a_n$ for all k = 1, 2, 3, n – 1
- A sequence of numbers are said to form a geometric progression if the ratio of the term and its preceding term bears constant quantity. Otherwise a sequence is in G.P if the terms in the sequence either increase or decreases by a constant quantity. The constant quantity in G.P is called as common ratio.
- If all the terms of a G.P be multiplied or divided by the same number then it remains a G.P with same common ratio
- A set of numbers a_1, a_2, a_3 ... a_n are said to be in H.P then their reciprocals $\frac{1}{a_1}, \frac{1}{a_2} \dots \frac{1}{a_n}$ are in A.P and vice–versa.

Self Practice Exercises

1. Find the three numbers in A.P whose sum is 35 and the product is 3135?
2. Find the three numbers in A.P such that their sum is 12 and the sum of their cubes is 288?
3. If p, q, and r are in A.P then show that $p^2(q + r)$, $q^2(p + r)$ and $r^2(p + q)$ are also in A.P?
4. Find the four numbers in A.P such that their sum is 16 and the sum of their squares is 84?
5. The sum of two numbers is 13/6 and an even number of arithmetic means are inserted between them such that their sum exceeds their number by 1. Find the number of means inserted?
6. Show that the sum of $(p + q)^{th}$ term and $(p - q)^{th}$ term of an A.P is equal to twice the pth term?
7. A man has saved altogether Rs. 20,000 in 11 years. In each year after the first he had saved Rs. 500 more than he did in the preceding year. Find how much he had saved in the first year.
8. If p^{th} term of an A.P is equal to q and the qth term is p then show that $(p + q)^{th}$ term of the A.P is zero?

9. A man agrees to repay his debt of Rs. 5,000 in number of installments. If each installment exceeds the preceding one by Rs. 200 and the first installment being Rs. 100 then find how many installments will be required to square off the debt completely?

10. The sum of n, 2n and 3n terms of an A.P are respectively X, Y and Z, then prove that Z = 3 (Y – X)?

11. The ratio of the sums of p and q terms of an A.P is $p^2 : q^2$, then show that the ratio of the pth and q^{th} term is (2p – 1) : (2q – 1) ?

12. Find the sum of n terms of the series 2 + 4 + 6 + ... n terms ?

13. If p, q and r are in A.P then show that $\frac{1}{qr}, \frac{1}{rp}$ and $\frac{1}{pq}$ are also in A.P?

14. Find the sum of the 8 terms of the series 1 + 2 + 4 + ... ?

15. Find the sum of n terms of the series 11 + 85 + 735 + ... n terms ?

16. Find the sum to n terms of the series 4 + 44 + 444 + ... n terms ?

17. Find the sum of the infinite series $\frac{1}{4} + \frac{1}{5^2} + \frac{1}{4^3} + \frac{1}{5^3} \ldots \infty$?

18. Find the sum of the n terms of the series 0.3 + 0.33 + 0.333 + ... n terms?

19. If p, q and r are in G.P then prove that $p(q^2 + r^2) = r(p^2 + q^2)$?

20. Find the three numbers in G.P such that their sum is 84 and the sum of their squares is 4368?

21. Insert 7 geometric mean between 3 and 768?

22. Find the two positive numbers whose difference is 12 and A.M exceeds the G.M by 2?

23. If t_1, t_2 and t_3 be respectively the sum of n, 2n and 3n terms of a G.P, then prove that $t_1((t_3 - t_2) = (t_2 - t_1)^2$.

24. If a, b, c in G.P then $\frac{1}{1+\log_{10} a}, \frac{1}{1+\log_{10} b}$ and $\frac{1}{1+\log_{10} c}$ are in H.P

Answers

1.	(11, 15, 19)	2.	(2, 4, 6)
4.	(1, 3, 5, 7)	5.	n = 6
7.	Rs. 15,000	9.	n = 10
12.	n(n + 1)	14.	255
15.	$\frac{9}{8}\left[9^n - 1\right] + n(n+1)$	16.	$\frac{4}{81}\left[10^{n+1} - 10\right] - \frac{4}{9}n$
17.	$\frac{111}{360}$	18.	$\frac{3}{9}\left[n - \frac{1}{9}\left(1 - \frac{1}{10^n}\right)\right]$
20.	4, 16, 64, ...	21.	6, 12, 24, ...
22.	a = 16 and b = 4		

CHAPTER 15

Permutation and Combination

A Brief Historical Background

The roots of permutation and combination can be traced in ancient India with the credit being placed on Jains (one who practices Jainism) and especially Mahavira for giving the first general formulae for permutation and combination. Other contributors are Sushruta in his work "Sushruta Samhita", Sanskrit scholar Pingala in his work "Chhanda Sutra" and Bhaskaracharya treated the subject in his work "Lilavati" named as "Anka Pasha". Outside India the Greeks, the Latins, the Chinese, the Arabs and the Hebrews have contributed at different levels but the Arabs and Hebrew writers are known to have utilized the knowledge of the subject in studying astronomy, for instance Rabi Ben Ezra calculated the number of combination of known planets taken two, three at a time. Another Hebrew writer Levi Ben Gerson gave the general formulae of ^nC_r. However, it was Swiss mathematician, James Bernoulli, who was the first to give a thorough treatment to the subject in his book titled "Arts Conjectandi" what we know as permutation and combination today in its modern form.

The Fundamental Principles of Counting

Let us understand the above concept by a simple illustration.

Abhisek a newly passed out management graduate is supposed to attend an interview with Glaxo Smithline India Ltd. He has altogether 2 pants, 3 shirts and 2 ties to choose, in how many ways can he dress up for the interview. Since he has 2 pants there are 2 ways in which a pant can be chosen. Similarly a shirt can be chosen in 3 ways and hence there are $2 \times 3 = 6$ ways to wear pants and shirts after which he needs to choose among two ties which can be chosen in 2 ways. Therefore in altogether he has $6 \times 2 = 12$ different ways in which he can dress up for the interview.

Multiplication Rule

If an event can occur in m ways, another event can occur in 'n' ways, and then both of the events can occur in m × n ways. The same can be extended to any finite number of operations.

For example

How many 2–digit numbers can be formed from the digits 2, 3, 4 and 5?

Since a two digit number will have a unit's and a ten's place and out of the four digits any one of the digits can take the units place.

After filling the unit's place we are left with 3 digits to filled in the ten's place.

Therefore by the application of fundamental rule of multiplication the number of 2–digit numbers that can be formed are 4 × 3 = 12.

Addition rule

If an experiment can be performed in 'n' ways, another experiment can be performed in 'm' ways then either of the two experiments can be performed in (m + n) ways. This rule can be extended to any finite number of experiments.

Permutation

Defination

Permutation is an arrangement of a number of objects in a definite order taken some or all at a time.

For example

When two coins are tossed we can have four different outcomes. They are HT, TH, HH and TT all different set of permutations.

Theorem:

The number of permutation of n different things taken r at time where $r \leq n$ is given by ${}^{n}P_{r} = \dfrac{n!}{(n-r)!}$

The number of permutation of n distinct objects taken r at a time means number of ways of filling in r vacant places by the n distinct objects.

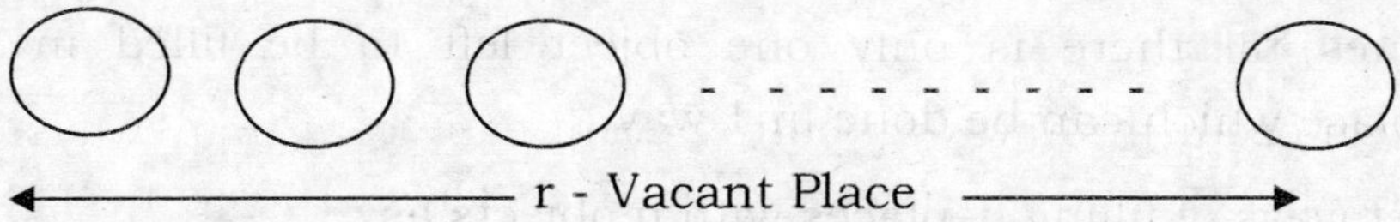

The first place can be filled in by any of the n objects in n ways.

Following which we have (n – 1) objects which can filled in (n – 1) ways in the second place.

And hence the first two places can be filled in n (n – 1) ways

Now the for third place since we are left with (n – 2) objects it can be filled in by (n – 2) ways.

Therefore the first three places can be filled in n(n –1) (n – 2) ways. Proceeding further the number of ways of filling r^{th} place can be in {n – (r – 1)} ways.

Thus we have seen that in each step the number of places to be filled in and every factor is one less than the proceeding factor.

Therefore the number of ways of filling in r vacant places in succession is n(n – 1) (n – 2) . . . . {n – (r – 1)}

or n(n – 1) (n – 2) . . . (n – r + 1)

By multiplying and dividing by (n – r) (n – r – 1)

we get,

$$= \frac{n(n-1)(n-2)...............(n-r+1)(n-r)(n-r-1)....3.2.1}{(n-r)(n-r-1).........3.2.1}$$

$$= \frac{n!}{(n-r)!} \qquad .....(15.1)$$

Hence Proved

Corollary

The number of permutation of n things taken all at a time is n! i.e. ${}^{n}P_{n}$ = n!.

In this case we need to fill n places with n different objects

We knew that the first place can be filled by n ways (n objects and n places). And the second place with (n – 1) objects can be filled in (n – 1) place in (n – 1) ways. Similarly the third place can be filled in (n – 3) ways this continues till there is only one object left to be filled in last remaining place which can be done in 1 way.

Hence we get ways of filling n–places with n objects as,

$$n(n-1)(n-2)\ldots\ldots\ldots\ldots 3.2.1 = n! = {}^{n}P_{n}$$

Alternatively,

$${}^{n}P_{n} = \frac{n!}{(n-n)!} = \frac{n!}{(0)!} = \frac{n!}{1} = n!.$$

(Note: 0! = 1).

Illustration No. 1

Prove that ${}^{n-1}P_r + r.\,{}^{n-1}P_r = {}^{n}P_r$

Solution:

$${}^{n-1}P_r + r.\,{}^{n-1}P_r = \frac{(n-1)!}{(n-1-r)!} + \frac{r.(n-1)!}{[n-1-(r-1)]!}$$

$$= \frac{(n-1)!}{(n-1-r)!} + \frac{r.(n-1)!}{(n-r)!}$$

Since $(n-r)! = (n-r)(n-r-1)!$

$$= \frac{(n-1)!}{(n-1-r)!}\left[1 + \frac{r}{(n-r)}\right]$$

$$= \frac{(n-1)!}{(n-1-r)!}\left[\frac{n-r+r}{(n-r)}\right]$$

$$= \frac{(n-1)!}{(n-1-r)!}\left[\frac{n}{(n-r)}\right]$$

$$= \frac{n!}{(n-r)!}$$

$$= {}^{n}P_r$$

Hence proved

Illustration No. 2

Calculate ${}^{8}P_{5}$=?

Solution:

$$\frac{8!}{(8-5)!} = 6720$$

Illustration No. 3

Find the number of permutated of 7 things taken 5 at a time?

Solution:

The number of permutation of 7 things taken 5 at a time is given by

$$= {}^7P_5 = 2520$$

Illustration No. 4

In how many can the letter A, B, C, D, E can be arranged so that

(i) AD are always together?

(ii) AD are never together?

Solution:

Had there was no restriction in arrangement the letter A, B, C, D, E could have been arranged in 5! = 120 ways.

But now since restriction is placed that A and D should go together. Assume that A and D as single (unit) denoted by AD.

Now it is like arranging (5 – 2 + 1) letters i.e. = 4 letters since AD are considered as a single unit which can be done in 4! ways.

However the solution is incomplete since the internal arrangement of A and D is not counted.

Since A and D can be internally arranged in 2! ways either (AD or DA) therefore the net arrangements is given by 4! × 2! = 48 ways.

(ii) In the second case we have to arrange A and D in such a fashion that A and D can never be together.

But without restriction the number arrangement 5! and from above we know that if A and D are always together then the number of arrangement will be 4! × 2! ways.

Hence number of arrangement when A and D are never together is given by (5! - 4! 2!) = 120 – 48 = 72 ways.

Illustration No. 5

In how many 6 books of different languages English, Hindi, Tamil, Telugu, Urdu and Kanad can be arranged so that English and Urdu books are never together?

Solution:

There are 6 books and 6 books can be arranged in 6! ways i.e. $^6P_6$ ways = 6! ways.

Assume that English and Urdu books are always together so the arrangement can be done in 5! × 2! ways.

Hence the required number of permutation when both English and Urdu books are never together is 6! – 5! × 2! ways.

$$= 6 \times 5! - 5! \times 2!$$

$$= 5!\,(6 - 2)$$

$$= 4 \times 5! = 480.$$

Permutations when some of the objects are Similar

Theorem

The number of permutations of n objects of which p are of one kind and (n – p) of another kind taken all at a time is $\frac{n!}{p!(n-p)!}$

Proof:

Let the number of requisite permutation be 'x'. Consider a particular one of these 'x' permutations in which p objects of one kind are all together followed by (n – p) objects of the other kind.

$$\text{i.e } \underbrace{1\ 2\ 3\ 4\ 5\ \ldots\ldots p}_{p \text{ similar objects}}\ \ \underbrace{p+1,\ p+2 \ldots\ldots n}_{(n-p) \text{ similar objects.}}$$

Let us assume that all p objects are different from all (n – p) objects.

Therefore number of permutation of p objects taken all at a time is

$$^pP_p = p!$$

The number of permutation of (n – p) objects taken all at a time is

$$^{n-p}P_{n-p} = (n-p)!$$

By the application of Fundamental principle of counting, the cases of x permutation will give rise p! (n – p)! permutations by arranging the objects of a particular kind within the places occupied by them.

Therefore the x permutations will give rise to x × p! (n – p)! permutations and if all the n objects are distinct, then the number of permutations $= {}^nP_n = n!$.

$$\Leftrightarrow \quad x \times p!\,.\,(n-p)! = n!$$

Therefore $$x = \frac{n!}{p!(n-p)!}$$

Corollary: Generalization

The number of different permutation of n objects of which p_1 are similar of one kind , p_2 are similar of second kind and so on until p_k are similar of kth kind is given where $p_1 + p_2 + p_3 + \ldots\ldots\ldots\ldots\ldots p_k = n$

Illustration No. 6

How many new words can be made from the letters of the word PERMUTATION?

Solution:

There are 11 words 9 are all different (P, E, R, M, U, A, I, O, N) and 2 are similar [T].

Hence the required number of arrangement are $\frac{11!}{2!}$

But we need to devise new words other than the word permutation which is included in above arrangement.

Hence the required numbers of new words are $\left(\frac{11!}{2!}\right) - 1$

[Since deleting the WORD permutation from the arrangement].

Illustration No. 7

In how many ways can the letters of the following words can be arranged

(a) ECONOMICS?

(b) MATHEMATICS?

Solution:

The word ECONOMICS contains 9 letters out of which 2 are C's , 2 are O's and rest are all different letters.

Hence the '9' letters can be arranged in 9!. But with 2 number of C's and O's

The number of arrangement is given by $\frac{9!}{2!\,2!}$ ways.

(b) The word MATHEMATICS contains 11 letters out of which 2 are A's, 2 are M's and 2 are T's and the rest are all different.

So, the '11' letters can be arranged in (11!) ways and with 3 letters appearing twice in the word. Hence the required numbers of arrangement are $\frac{11!}{2!\,2!\,2!}$ ways.

Illustration No. 8

Find number of different signals that can be made from 3 Saffron, 4 white and 2 green flags by arranging all of them vertically on a flag staff?

Solution:

Here the total number of flags = 3 + 4+ 2 = 9.

And the number of Saffron Flags = 3.

the number of White Flags = 4.

the number of Green Flags = 2.

Hence the required number of signals that can be made are

$$\frac{9!}{2!\,3!\,4!} = 1260$$

Theorem

Number of permutations of n-things, taken 'r' at a time when each thing can be repeated r-times is given by = n^r.

Proof

Let us consider that r places needs to be filled in from n objects when the repetition is allowed.

The number of ways of filling-up first place is n

Since repetition is allowed, so

The number of ways of filling-up second-place = n

Similarly, number of ways of filling-up third-place = n.

Therefore the number of ways of filling in the 1st three places = $n \times n \times n = n^3$.

Continuing the process, we can conclude that

Number of ways of filling-up rth place = n

Hence total number of ways in which first, second ....rth, places can be filled up in

$$= n \times n \times n \text{ ...... r factors.}$$

$$= n^r$$

Illustration No. 9

A child has 4 pocket and 5 coins. In how many ways can he put the coins in his pocket?

Solution

The first coin can be put in 4 ways, similarly second, third, fourth and fifth coins also can be put in 4 ways.

So total number of ways = $4 \times 4 \times 4 \times 4 \times 4 = 4^5 = 1024$.

Illustration No 10.

A circuit network contains 10 switches such that each switch may have 2 possible positions. In how many different ways switching are possible?

Solution:

Since each of the 10 switches has 2 possible positions

And hence the total number of switchings = $2 \times 2 \times 2 \times$ .... 10 times.

$$= 2^{10} = 1024$$

Circular Permutation

Arrangement of objects along the circumference of a circle are circular permutation.

As in circular permutation it is important to note that there is neither a beginning nor any end; in such arrangements we only consider the relative position of the different objects.

For example

Let's consider the circular permutation of five objects a, b, c, d and e. If arranged in linear form they all stand different but in case of circular permutation a, b, c, d and e or e, a, b, c and d are not different but one and the same.

Illustration along two circular

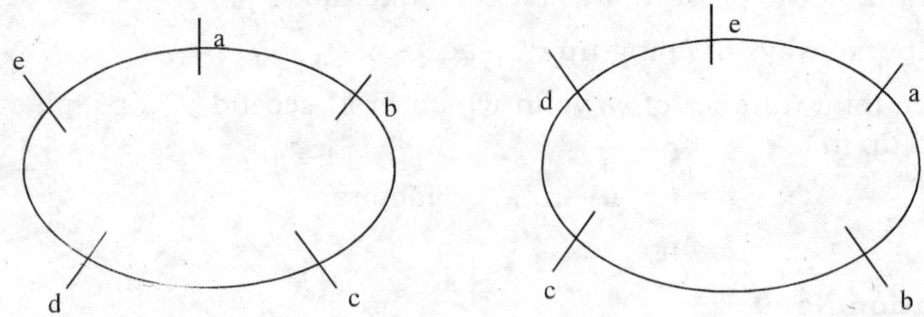

As circular permutation depends on relative positioning of the objects we fix the position of one of the object and arrange the remaining.

If there are n objects after fixing one object we are remaining with (n – 1) objects which can be arranged (n – 1)! way

And hence the number of circular permutation of n different objects

$$= (n - 1)!$$

Note

When clock-wise and anti-clock wise arrangements are not different, then the arrangements can be made from both sides, and this will be the same.

Here two permutations will be counted as one, so the total permutations will be half, hence in this case circular–permutations

$$= \frac{(n-1)!}{2}$$

Illustration No. 11

In how many ways can 4 Hindus and 4 Muslims can be made to seat in a circular arrangement such that no two Muslims are in adjacent position?

Solution:

First we can place the 4 Hindus in a circular arrangement which can be done in 3! ways.

Once 4 Hindus are placed we have to place 4 Muslims in the same circular arrangement with no two Muslims are in adjacent positions.

Now since we have already placed Hindus, the Muslims can be placed in 4! Ways

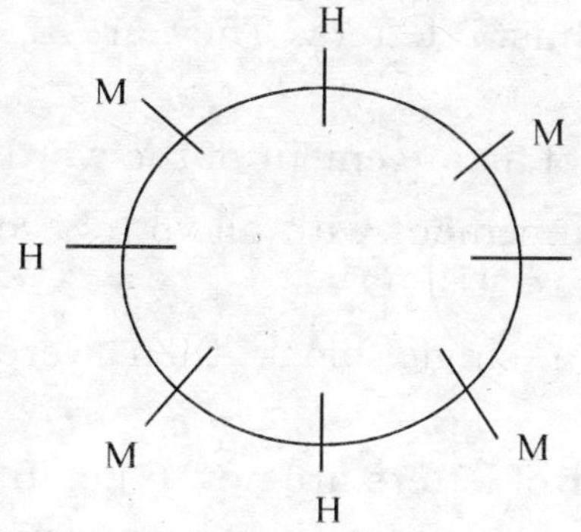

Hence the total number of arrangements are given by 4! × 3! = 144 ways.

Illustration No. 11

In how many can 6 paper can be arranged with the best and the worst paper coming always together?

Solution:

Since there are 6 papers without any restriction it can be arranged in 6! ways.

But we need an arrangement in which the best and the worst paper are always together.

Consider the best and worst paper as a single unit [Best + Worst] = 1.

Now we have remaining with 5 sets of paper i.e. 4 + 1 (Best + Worst).

We can now have 5! way of arrangement of these 5 papers which is 120 ways (5! = 120). But we need to bring in the internal arrangement of both papers i.e. either (B and W) or (W and B) which can be done in 2! ways.

Hence the required arrangement of 6 papers with the Best and the Worst coming together is 5! × 2! = 240 ways.

Illustration No. 12

Find the ways in which the letters of the word LUCKNOW can be arranged? Also find the number of ways in which the vowels are always together and vowels are never together?

Solution:

The word "LUCKNOW" has 7 letters. They are L, U, C, K, N, O, W all different.

The required number of arrangement of the word = 7! ways.

In order to find the arrangement with all vowels coming together. Let us identify the vowels they are "U", "O".

Consider (U, O) as a single block then were left with remaining 5 letters {L, C, K, N, W}.

Thus the total number of letters are now 6 i.e. (5 + 1 unit of vowels).

Therefore 6 letters can be arranged in 6! Ways. But the block of vowels contains 2 letters which can be arranged in among themselves is 2! ways.

Hence the required number of arrangement in which vowels are always together = 6! × 2! = 1440 ways.

Part 2

We also need to find an arrangement in which the vowels do not come together. Since the word LUCKNOW can be arranged 7! (*Note*: Without any discrimination between vowels and the consonanats).

And we have an arrangement in which vowels are coming together i.e. 6! × 2! ways.

Since we need to find the formations in which vowel do not come together is given by 7! – 6! × 2! = (7 – 2) 6!. = 5.6! = 3600 ways.

Illustration No. 13

How many 6 books of management of subjects QT, OR, FM, MM, ME, OB such that the books of QT and OR are always together?

Solution:

There are 6 books (QT, OR, FM, MM, ME, OB) these 6 books can be arranged in ${}^{6}P_{6} = 6!$ ways

But we need to find an arrangement in which the books of the subject of QT and O R should be always together.

So consider the books of the subject of QT and OR as 1 unit and these two can be arranged among themselves in 2! ways.

Since we are left 5 units of books (4 + 1 combined unit of OR, QT) which can be arranged in 5! Ways.

Since the required arrangement with OR and QT book always coming together is = 5! × 2! Ways = 240 ways.

Illustration No. 14

How many 5 digits number can be formed using the digits from 0 to 9. If the repetition of digits is not allowed?

Solution:

Since there are 10 digits from 0 to 9 and we need to form a 5 digit numbers from the given numbers.

For example 53412 or 12314.

Hence the number of 5 digits numbers which can be formed out of 10 digits is ${}^{10}P_5$ ways. But in this ${}^{10}P_5$ ways includes a case in which zero can come in the first place that's needs to be eliminated.

Thus numbers with zero in the first place are ${}^{9}P_4$ [for e.g. 01234. Etc]

Hence the required number of 5 digits number are $({}^{10}P_5 - {}^{9}P_4)$ ways.

Illustration No. 15

How many numbers lying in between 100 and 1000 that can be formed from the digits 0, 1, 2, 3, 4, 5, 6 if the digits are not repeated?

Solution:

Since all numbers lying between 100 and 1000 are 3 digits numbers which can be formed out of {0, 1, 2, 3, 4, 5, 6} i.e. 7 digits is ${}^{7}P_3$ ways.

But in this ${}^{7}P_3$ ways there are numbers with zero in the first place for e.g. 084........... etc which is essentially a 2 digit number. Hence such numbers needs to be eliminated from the set of ${}^{7}P_3$ to get the required set of number.

For such cases '0' fixed in the first placed and the number that can be formed with zero in 1st place i.e. (100 place) ${}^{6}P_2$.

Hence the required number of arrangement are ${}^{7}P_3 - {}^{6}P_2 = 180$ ways.

Illustration No. 16

In how many can 3 red balls, 4 green balls and 5 black balls can be arranged in a row if the balls of same color is indistinguishable?

Solution:

The total number of balls are = 12 balls i.e. (3 red in color, 4 green in color, 5 black in colour). These 12 balls can be arranged in row in 12! ways. But out of which we have 3 different colored ball all indistinguishable between them. Hence the required number of arrangements is $\frac{12!}{3!\,4!\,5!}$ ways.

Illustration No. 17

For a final university examination 6 examination papers has to be divided in a particular order without disclosures? It was found that one of the order has leaked out. Then find in how many ways the order can be changed?

Solution:

Since we can arrange 6 examination papers in $^6P_6 = 6! = 720$ ways. Therefore out of 720, one has been leaked.

So we will be left with (720 – 1) = 719 more ways in which the order can be changed.

Illustration No. 18

A secret code has 5 digits, consisting of two distinct English alphabets followed by 3 distinct numbers between 1 to 9. In how many ways such code can be generated. And how many such code end with odd integers?

Solution:

Since there are all together 26 alphabets in English Nomenclature. And any two distinct alphabets can be arranged in $^{26}P_2$ ways.

Out of 9 numbers (i.e. 1 to 9) any three number can be arranged in $^9P_3$ ways.

Therefore the total numbers of code that can be generated are $^{26}P_2 \times {}^9P_3$ ways.

In the second case the code word should end with an odd integer (i.e. 1, 3, 5, 7, 9).

The code's unit place can only filled by any of the five odd integers in $^5P_1 = 5$ ways and the ten's place filled any of the remaining digits including the rest of odd integers in $^8P_1$ ways. Similarly the hundredth place can be filled in $^7P_1$ ways.

Hence the total number of code words ending in odd integer are $^{26}P_2 \times {^5P_1} \times {^8P_1} \times {^7P_1}$ ways.

Illustration No 19

In how many 5 digits number can be formed out of the digits 1,2,3,4,5,6,7,8,9 such that the number do not include digits 3, 4?

Solution:

Since the number formed is devoid of the digits 3, 4. Therefore we are left with 7 digits and hence we can form a 5 digits number from 7 digits in $^7P_5 = 2520$ ways.

Illustration No. 20

In how many 5 digits number can be formed out of the digits 1, 2, 3, 4, 5, 6, 7, 8, 9 Such that the number of ways includes the digits 3, 4?

Solution:

The 5 digits number formed should always include the digits (3, 4), therefore out of 5 places 2 places are filled by (3, 4) which can be done in $^5P_2$ ways. After filling of the 2 places, are left with 3 places that needs to filled up from 7 digits which is done in $^7P_3$ ways.

Thus the total number of arrangements will be $^7P_3 \times {^5P_2}$ ways.

Illustration No. 21

In how many 3 medals can be awarded to 5 students if any student is likely to win all of them?

Solution:

Since there is a likelihood that each student can either win 1 medal or 2 medal or all the medals. Therefore the first medal can be awarded to any of the 5 boys after the distribution of 1st medal, 2nd medal can also be given to any of the students as each are equal likely to receive the same.

Thus both medal can be awarded in $5 \times 5 = 25$ ways

Once again the third medal can be given in 5 ways (as each are equal likely) therefore the required number of arrangements are $5^3 = 5 \times 5 \times 5 = 125$ ways.

Illustration No. 22

A panel of 8 jury are made to seat in a circular formation to discuss a important case. Find in how many ways two particular judge A and B (a) are always together and (b) never together?

Solution:

(a) As 8 jury members can be arranged in circular fashion in

$(8 - 1) = 7!$ ways.

Since two particular judges A and B are supposed be always together consider them as a single unit such that were left with 7 jury members. (Rest 6 + 1 unit of 2 judges).

Therefore 7 judge member can be arranged in circular fashion in 6! ways. However the single unit of 2 judge members can arranged among themselves 2! Hence the required members of arrangements are $6! \times 2! = 1440$ ways

(b) In second case we found that judge members can be arranged in order to find two particular judge members are never together. We subtract the arrangement were these both judge members always together.

i.e. (Total arrangement) – (Arrangement with two particular judge member are always together) = $7! - 6!\,2! = 3600$ ways.

Illustration No. 23

How many different words can be formed from the letters of the word?

(a) ECONOMETRICS?

(b) CONSTANTINE?

Solution:

(a) The word ECONOMETRICS has 12 letters in which there are 2 E's, 2 O's, 2 C's and the rest of all are different i.e. N, M, T, R, I, S.

Therefore the number of arrangements are $\dfrac{12!}{2!\,2!\,2!}$ ways

(b) The word CONSTANTINE has 11 word out of them there are 3 N, 2 T and the rest are all different i.e. O, S, C, T, E, I.

Hence the number of arrangements are = $\frac{11!}{2!\,3!}$ ways.

Combination

A Combination is a basically a selection of object without considering the order of the objects from a given set of objects - Mathematically expressed the number of combination of n objects taken 'r' at time is given by ^nC_r.

Therefore: The number of combinations of n distinct objects taken as at a time is given by $^nC_r = \frac{n!}{r!(n-r)!}$

where $1 \le r \le n$

Proof:

Let's assume ^nC_r denote the number of combination of n different things taken r at a time.

But as we know in this case that the consideration of order is not significant and hence for every combination of r objects there is corresponding r! arrangements.

Hence ^nC_r combinations with in all give $^nC_r \times r!$ permutation.

Hence $$^nC_r \times r! = \frac{n!}{(n-r)!} = {}^nP_r$$

or $$^nC_r = \frac{n!}{r!(n-r)!}$$

Theorem

Prove that $^nC_r = {}^nC_{n-r}$ where n and r are natural numbers and $n \ge r$

Proof:

Since we knew $$^nC_r = \frac{n!}{r!(n-r)!}$$

Substitute r = n – r we get

$$^nC_{n-r} = \frac{n!}{[n-(n-r)]!(n-r)!}$$

$$= \frac{n!}{(n-n+r)!(n-r)!}$$

$$= \frac{n!}{r!(n-r)!} = {}^nC_r$$

Hence Proved.

Illustration No. 1

Prove that ${}^nC_r + {}^nC_{r-1} = {}^{n+1}C_r$ where n and r natural numbers and $n \geq r$?

Solution:

$${}^nC_r + {}^nC_{r-1} = \frac{n!}{r!(n-r)!} + \frac{n!}{[n-(r-1)]!(r-1)!}$$

$$= \frac{n!}{(n-r)!(r-1)!}\left[\frac{1}{r} + \frac{1}{n-r+1}\right]$$

Since $r! = r(r-1)!$ and $(n-r+1)! = (n-r+1)(n-r)!$

$$= \frac{n!}{(n-r)!(r-1)!}\left[\frac{n-r+1+r}{r(n-r+1)}\right]$$

$$= \frac{n!}{(n-r)!(r-1)!}\left[\frac{n+1}{r(n-r+1)}\right]$$

$$= \frac{(n+1)\,n!}{(n-r+1)(n-r)!\,r\,(r-1)!}$$

$$= \frac{(n+1)!}{(n-r+1)!\,r!} = {}^{n+1}C_r$$

Note:

1. $(n+1)n! = (n+1)!$
2. $r\,(r-1)! = r!$
3. $(n-r+1)\,(n-r)! = (n-r+1)!$

Illustration No. 2

Prove that ${}^nC_r + {}^nC_{r+1} = {}^{n+1}C_{r+1}$ where r and n are natural number and $r \le n$

Solution:

$$ {}^nC_r + {}^nC_{r+1} = \frac{n!}{r!(n-r)!} + \frac{n!}{[n-(r+1)]!(r+1)!} $$

$$ = \frac{n!}{r!(n-r)(n-r-1)!} + \frac{n!}{[n-r-1]!(r+1)\,r!} $$

$$ = \frac{n!}{(n-r-1)!\,r!}\left[\frac{1}{n-r} + \frac{1}{r+1}\right] $$

$$ = \frac{n!}{(n-r-1)!\,r!}\left[\frac{r+1+n-r}{(r+1)(n-r)}\right] $$

$$ = \frac{(n+1)\,n!}{(n-r+1)(n-r)!\ (r+1)\,r!} $$

$$ = \frac{(n+1)\,!}{(n-r)!\,(r+1)!} = {}^{n+1}C_{r+1} $$

Hence proved.

Illustration No. 3

Prove that the product of r consecutive positive integers is divisible by r! ?

Solution:

Let the 'r' consecutive positive integers be

(n + 1) (n + 2) (n +3)- - - - - - - - - (n + r - 1) (n +r).

And thus product is (n + 1) (n + 2) (n + 3) - - - - - - - - (n + r – 1) (n + r)

By multiplying and dividing by n!

We have,

$$ = \left[\frac{n! \times (n+1)(n+2)............(n+r-1)(n+r)\}}{n!}\right] $$

Since $\quad n! = 1 \times 2 \times 3 \times - - - - - \times n$

Therefore,

$$= \left[\frac{\{1.2.3 \ldots\ldots\ldots\ldots (n-1)\,n\} \times (n+1)(n+2) \ldots\ldots\ldots\ldots (n+r-1)(n+r)\}}{n!}\right]$$

$$= \left[\frac{(n+r)!}{n!}\right]$$

Now multiplying numerator and denominator by r!

We get

$$= \left[\frac{(n+r)!}{n!}\right]\left[\frac{r!}{r!}\right]$$

$$= r!\left[\frac{(n+r)!}{r!\,n!}\right] = r!\,{}^{n+r}C_r$$

Since we have seen that there is a factor r! in the product.

Therefore, $\dfrac{{}^{n+r}C_r\, r!}{r!} = {}^{n+r}C_r,$

Hence proved.

Illustration No. 4

For all $1 \le r \le n$, prove that $n\,{}^{n-1}C_r = (n-r+1)\,{}^{n}C_{r+1}$

Solution:

$n\,{}^{n-1}C_r$ can be written as

$$= n \times \frac{(n-1)!}{[(n-1)-(r-1)]!(r-1)!}$$

$$= \frac{n(n-1)!}{(n-r)!(r-1)!}$$

$$= \frac{n!}{(n-r)!\,(r-1)!}$$

Multiplying and dividing the numerator and denominator by (n – r + 1)

$$= \frac{n!}{(n-r)!\,(r-1)!}\left[\frac{n-r+1}{n-r+1}\right]$$

$$= \frac{(n-r+1)\,n!}{(n-r)(n-r-1)!\ (r-1)!}$$

$$= (n-r+1)^nC_{r-1}$$

Hence proved.

Illustration No. 5

Prove that for all $r \le n$, $n\left[{}^{n-1}C_{r-1}\right] = r^nC_r$

Solution:

$$n\left[{}^{n-1}C_{r-1}\right] = n \times \frac{(n-1)!}{[(n-1)-(r-1)]!(r-1)!}$$

$$= \frac{n(n-1)!}{(n-r)!(r-1)!}$$

$$= \frac{n!}{(n-r)!\,(r-1)!}$$

Multiplying and Divide numerator and denominator by 'r'

Then we get,

$$= \frac{n!}{(n-r)!\,(r-1)!}\left[\frac{r}{r}\right]$$

$$= \frac{r \times n!}{(n-r)!\ r(r-1)!}$$

$$= \frac{r \times n!}{(n-r)!\ r!} = r^nC_r$$

Hence proved.

Illustration No. 6

A Company has 4 chartered accountants, 5 MBA's, 3 Company Secretary working in the firm, a core audit team needs to be selected containing 2 member from each disciple. In how many team can formed ?

Solution:

Since a core audit team needs to be selected containing 2 members from each discipline i.e. chartered accountants, MBA's and Company Secretary.

Therefore a 2 member chartered accountants can be selected out of 4 chartered accountants can be done in $^4C_2$ ways.

→ 2 member can be selected out of 5 MBA's in $^5C_2$ ways and finally 2 member can be selected out of 3 Company Secretary in $^3C_2$ ways.

Hence,

The required no. of combination using fundamental rule of counting

$$= {}^4C_2 \times {}^5C_2 \times {}^3C_2 \text{ ways}$$

Illustration No. 7

Find the number of triangle that can be formed using the vertices of an octagon?

Solution:

Octagon is a eight sided figure having 8 vertices and in order to form a triangle we need any three vertices.

Now forming a triangle will need selection of any 3 vertices out of 8 vertices which can be done in $^8C_3$ ways.

Thus the number of triangle that can be formed is $^8C_3 = \dfrac{8!}{5!3!} = 56$ ways.

Illustration No. 8

A question paper has 2 parts. Part I and Part II each containing 8 question each. If a student has to choose 5 questions from Part I and 4 question from Part II, in how many ways the student can choose the questions?

Solution:

Since the question paper has 2 parts I and II and in totality 9 questions needs to be answered 5 from part I and 4 from part II.

A student can select a set of 5 questions from part I in $^8C_5$ ways.

Similarly a student can select a set of 4 questions from part II in $^8C_4$ ways.

Hence the required no. of selection using fundamental rule is given by

$$= {}^8C_4 \times {}^8C_5 = 3920 \text{ ways}$$

Illustration No. 9

In how many a football team can be selected consisting of 11 players out of 18 players. Also find in how many teams can select to always include the captain and the goal keeper?

Solution:

Since the total number of players available are 18 and the number of players need to be selected is 11.

Hence the requirement number of selection = ${}^{18}C_{11}$ ways.

We need to select a team which should always include the team captain and the goal keeper.

Therefore we exclude the 2 players (captain + goal keeper) from 18 players i.e. = 18 – 2 = 16.

Now we have 16 players available for selection and we need 9 out of them.

Hence the selection of team to include (captain + goal keeper) can be done is ${}^{16}C_9$ ways.

Illustration No. 10

A committee of 5 members needs to be formed out of 5 men and 3 women. Find in how ways the selection can be done to include at least 2 women?

Solution:

Since a committee of 5 needs to be formed out of 5 men and 3 women to include at least 2 women in the committee. The combinations are,

(i) 3 Men and 2 Women

The number of ways of selection of 1st combination = ${}^{5}C_3 \times {}^{4}C_2$.

(ii) 2 Men and 3 Women.

The number of ways of selection of 2nd combination = ${}^{5}C_2 \times {}^{3}C_3$

Hence the required number of selection are

$$= {}^{5}C_3 \times {}^{4}C_2 + {}^{5}C_2 \times {}^{3}C_3 \text{ ways}$$

Illustration No. 11

Out of 8 toy's manufactured 3 were found to be defective without knowing which one is defective. In how many ways 3 toys can be selected? And also find in how many of these selected will include at least 2 defective toys?

Solution:

Since out of 8 toys any 3 toys can selected in $^8C_3$ ways

{Note without discrimination between defective and non defective}

Now we need to make a selection to at least 2 defective

The combinations are:

(i) 1 non defective and 2 defective

The number of ways of selection of 1st combination = $^5C_1 \times {}^3C_2$

(ii) All 3 defective.

The number of ways of selection of 2nd combination = $^3C_3$.

Hence the required a number of selection is

$$= {}^5C_1 \times {}^3C_2 + {}^3C_3$$

$$= 5 \times 3 + 1 = 16$$

Illustration No. 12

A bag contains 5 green balls, 3 black balls, 4 red balls. In many ways the selection of 3 balls can be made so that to include:

1. To include one from each color?
2. 2 Red and 1 Black?
3. None is Black in color?

Solution:

Since the bag contain in all 12 balls (5 green balls + 3 black balls + 4 Red balls)

Case I: We need to be selected one ball from each color and the total number of balls are 3. Therefore the number of ways of selection of a ball from each color is $^5C_1 \times {}^4C_1 \times {}^3C_1 = 5 \times 4 \times 3 = 60$ ways

Case II: In second case the selection needs to only include 2 Red balls and 1 Black ball and the selection can be done = $^4C_2 \times {}^3C_1 = 6 \times 3 = 18$ ways.

Case III: We need to make a selection of 3 balls in which none of the black color ball will be included. Now after eliminating all black balls we are left with 12 – 3 = 9 balls and hence 3 balls can be selected out of 9 balls in ${}^9C_3$ ways.

Illustration No. 13

Find the number ways a person can invite one or more of his 5 friend in a party?

Solution:

A person can invite either 1- friend or 2 – friends, or 3 – friends or 4 - friends or 5 friends in the following ways.

He can invite one friend in ${}^5C_1$ ways.

He can invite two friends in ${}^5C_2$ ways.

He can invite three friends in ${}^5C_3$ ways.

He can invite four friends in ${}^5C_4$ ways.

He can invite five friends in ${}^5C_5$ ways.

Hence the total number of ways of inviting one or more friends.

$${}^5C_1 + {}^5C_2 + {}^5C_3 + {}^5C_4 + {}^5C_5 = 5 + 10 + 10 + 5 + 1 = 31 \text{ ways.}$$

Alternatively:

There exist only 2 possibilities either he can invite a friend or may not invite. So there exits 5 such possibilities i.e. $2 \times 2 \times 2 \times 2 \times 2 = 2^5 = 32$. But this includes the combinations were none of the friend are invited. Therefore the required no ways a person can invite his friends are $= 2^5 - 1 = 32 - 1 = 31$ ways.

Illustration No. 14

If 6 parallel lines are intersected by other 4 parallel lines then find the number of rectangle that can be formed?

Solution:

A rectangle is a four sided figure with each two sides are parallel to one another.

Therefore any two parallel lines can be selected from first 6 parallel lines in ${}^6C_2$ ways.

Similarly another set of two parallel lines can be selected from 4 parallel lines in ${}^4C_2$ ways.

Hence the number of selections of 4 lines i.e. 2 from the 1st set and 2 from the 2nd set can be done in = ${}^6C_2 \times {}^4C_2$ ways.

But each such selection of 4 lines gives a rectangle. Therefore number of rectangle that can be formed are ${}^6C_2 \times {}^4C_2 = 90$.

Illustration No. 15

Five cross marks 'x' need to be place in given below formation such that each row should at least contain one cross marks. Find the number of ways, this can be done?

Solution:

The combination's are

	1st Row	2nd Row	3rd Row
1st	1	1	3
2nd	2	1	2
3rd	2	2	1
4th	1	2	2
5th	1	3	1
6th	1	1	3

Hence the required ways are

$$= {}^3C_1 \times {}^3C_1 \times {}^3C_3 + {}^3C_2 \times {}^3C_1 \times {}^3C_2 + {}^3C_2 \times {}^3C_2 \times {}^3C_1 + {}^3C_1 \times {}^3C_2 \times {}^3C_2$$

$$+ {}^3C_1 \times {}^3C_3 \times {}^3C_1 + {}^3C_1 \times {}^3C_1 \times {}^3C_3$$

$= 9 + 27 + 27 + 27 + 9 + 9$

$= 108$ ways

Illustration No. 16

A cricket team needs to be selected out of 15 players. It includes 3 wicket keepers' and 4 bowler. In how many ways a team of eleven be selected if we have to compulsorily have 1 wicket keeper and at least 3 bowler?

Solution:

We need to select a 11 player which should include 1 wicket keeper and at least 3 bowlers i.e. 1 wicket keeper + 3 bowlers or 1 wicket keeper + 4 bowlers.

Therefore required number ways of selected a team is given by

$$= {}^{3}C_{1} \times {}^{4}C_{3} \times {}^{8}C_{7} + {}^{3}C_{1} \times {}^{4}C_{4} \times {}^{8}C_{6}$$

$$= 3 \times 4 \times 8 + 3 \times 1 \times 28 = 180 .$$

Illustration No. 17

A hockey team of 11 players needs to be selected from a team of 21 members. Find in how many ways the team can be selected?

Solution:

As a team of 11 players needs to be selected from a total of 21member it can be done ${}^{21}C_{11}$ ways.

Illustration No.18

Find the number of diagonals that can be formed from the vertices of a hexagon?

Solution:

A hexagon contains 6 vertices and a line can be formed by joining any two vertices of a hexagon?

The number of lines that can be formed by joining the two vertices is ${}^{6}C_{2}$

However in this combination also includes the sides of the hexagon that needs to be subtracted from the above combination as the number stands at 6 since a hexagon has 6 sides.

Thus the number of diagonals that can be in a hexagon is ${}^{6}C_{2} - 6 = 9$.

Illustration No. 19

Find how many lines can be formed out of 10 points in a plane with 3 of them being collinear?

Solution:

Since to form a line we need two points to be connected and the plane contains 10 points and hence the number of lines that can be formed is $^{10}C_2$.

However in this case there are 3 points are collinear (means lying in a straight line) And the above combination also needs to be eliminated because of double counting.

The number of such lines which needs to be eliminated are $^3C_2$.

But we need to consider all these collinear points and count them as a single line.

Thus we need to adjust by reducing the collinear points by $^3C_2 - 1$.

Hence the number of lines that can be formed is $^{10}C_2 - (^3C_2 - 1) = 43$.

Illustration No. 20

In a Quantitative Techniques paper contains 12 questions divided into two sections. Section A and Section B of 6 question each. How many ways an examinee can answer 8 questions with at least two questions from section A and at least 3 question from section B?

Solution:

The examinee has following combinations

1. 2 Section A + 6 Section B.
2. 3 Section A + 5 Section B.
3. 4 Section A + 4 Section B.
4. 5 Section A + 3 Section B.

The number of ways of selection of 1st combination $= {^6C_2} \times {^6C_2}$

The number of ways of selection of 2nd combination $= {^6C_3} \times {^6C_5}$

The number of ways of selection of 3rd combination $= {^6C_4} \times {^6C_4}$

The number of ways of selection of 4th combination $= {^6C_5} \times {^6C_3}$

Therefore the total number ways of answering the question paper is

$$= {^6C_2} \times {^6C_2} + {^6C_3} \times {^6C_5} + {^6C_4} \times {^6C_4} + {^6C_5} \times {^6C_3} = 690 \text{ ways}$$

Illustration No. 21

Each team have participated against each other in kabadi tournament if the total number of games played is 15. Find the number of teams participated in the tournament?

Solution:

Let 'n' be the number of teams participating against each other. Since for a game need 2 teams to play against each other.

Thus, the number of game played by n teams = $^nC_2$

But the total games played is 15

Since $^nC_2 = 15$

or $n(n-1) = 15$

or $n^2 - n - 15 = 0.$

or $n^2 - 6n + 5n - 15 = 0$

or $(n - 6)(n + 5) = 0$

or $n = 6$ as $n \neq$ -ve number i.e. -5

Therefore the total number of teams participating in the tournament is 6

Illustration No. 22

Find the number of ways of selecting 4 cards from a pack of well shuffled card such that.

1. It includes a card belonging to each suit?
2. Two are red in color and two are black in color?

Solution:

Since a pack of card contains 52 cards with 4 different suits they are

1. Diamonds
2. Hearts
3. Clubs
4. Spade

So they are 13 cards of diamond, 13 cards of hearts, 13 of clubs and 13 of spade.

Therefore 1 diamond card can be chosen out of 13 diamond card in $^{13}C_1$ ways.

(i) Hearts card can be chosen out of 13 cards (Hearts)in $^{13}C_1$ ways.

1 – Clubs card can be chosen out of 13 cards(Clubs) in $^{13}C_1$ ways and 1- Spade card can be chosen out of 13 cards(Spade) in $^{13}C_1$ ways.

Hence the required number of ways of selecting 4 card by using multi theorem is

$$= {}^{13}C_1 \times {}^{13}C_1 \times {}^{13}C_1 \times {}^{13}C_1 = 13^4$$

(ii) Since there are 26 cards Red in color and 26 in Black in color. Therefore 2 Red cards can be chosen out of 26 Red cards in $^{26}C_2$ ways.

Similarly 2 Black cards can be chosen out of 26 Black cards in $^{26}C_2$ ways.

Therefore the required number of ways selecting 4 cards 2 of Red and other 2 Black is

$$= {}^{26}C_2 \times {}^{26}C_2 \text{ ways.}$$

Illustrations No. 23

A core team of 6 members needs to be selected out of 5 officers and 8 clerks. Find how many ways the team can be selected if the team should include at least 3 clerks?

Solution:

The combination are:

(i) 3 clerks, 3 officers

(ii) 4 clerks, 2 officers

(iii) 5 clerks, 1 officer

The number of ways of selecting 3 clerks and 3 officers can be done in $^6C_3 \times {}^8C_3$ ways.

The number of ways of selecting 4 clerks and 2 officers can be done in $^6C_2 \times {}^8C_4$ ways.

The number of ways of selecting 5 clerks and 1 officer can be done $^6C_1 \times {}^8C_5$ ways.

Therefore the required number of ways selecting the team is $= {}^6C_3 \times {}^8C_3 + {}^6C_2 \times {}^8C_4 + {}^6C_1 \times {}^8C_5$ ways.

Illustration No. 24

In a business meeting each and every member is the meeting shakes hand with every other person. If there are all together 55 handshakes, find the number of member attended the meeting?

Solution:

Let 'n' be the number of people attended the meeting. Since in a hand shake two people's needs to come together.

Therefore the number of hand shake which can happen with 'n' number of people attending the business meet is ${}^nC_2$.

But the total numbers of hand shake are 55.

Therefore $\quad {}^nC_2 = 55$

$$\frac{n(n-1)}{2} = 55$$

$$n^2 - n - 110 = 0$$

$$n^2 - 11n + 10n - 110 = 0$$

$$(n-11)(n+10) = 0$$

$$n = 11 \text{ or } -10$$

Since n cannot be negative.

Hence $\quad n = 11$

Illustration No. 24

5 balls of different colors needs to be placed in 4 boxes of different sizes such that no box remains empty. Find the number of possible arrangements?

Solution:

The number of combination of placing 5 balls in 4 boxes

(i) 1, 1, 1, 2

(ii) 1, 1, 2, 1

(iii) 1, 2, 1, 1

(iv) 2, 1, 1, 1

In the first case the arrangement are ${}^5C_1 \times {}^4C_1 \times {}^3C_1 \times {}^2C_2 = 60$ ways.

In the second case the arrangement are ${}^5C_1 \times {}^4C_1 \times {}^3C_2 \times {}^1C_1 = 60$ ways.

In the third case the arrangement are ${}^5C_1 \times {}^4C_2 \times {}^2C_1 \times {}^1C_1 = 60$ ways.

In the fourth case the arrangement are ${}^5C_2 \times {}^3C_1 \times {}^2C_1 \times {}^1C_1 = 60$ ways.

Hence the required number of arrangement are = 60 + 60 + 60 + 60 = 240 ways.

Illustration No. 25

If a polygon has 35 diagonals find the number of sides of the polygon?

Solution:

Let the polygon be with n sided. In order to form a diagonal we need to join to vertices. As such we can form in all with 'n' vertices ${}^nC_2$ number of lines which also includes the sides of the 'n' sided polygon.

Hence the required number of arrangement of forming diagonals one obtained by deducting 'n' straight line from the formation of ${}^nC_2 - n$.

But number of diagonals are 35 therefore ${}^nC_2 - n = 35$.

$$\frac{n(n-1)}{2} - n = 35$$

$$n^2 - n - 2n = 70$$

$$n^2 - 3n - 70 = 0$$

$$n^2 - 10n + 3n - 70 = 0$$

$$n(n-10) + 3(n-10) = 0$$

$$(n-10)(n+3) = 0 \text{ as } n \neq \text{- ve}$$

$$n = 10.$$

Brief Summary

1. **Multiplication Rule:** If an event can occur in m ways, another event can occur in 'n' ways, then both of the event can occur in m x n ways. It can be extended to any finite number of operations.

2. Permutation is an arrangement of a number of objects in a definite order taken some or all at a time.

3. The number of different permutation of n objects of which p_1 are similar of one kind , p_2 are similar of second kind and so on until p_k are similar of kth kind is given by n! / p_1! p_2 ! p_3 ! .... n_k! where p_1+ p_2+ p_3+................ p_k = n.

4. Number of permutations of n-things, taken 'r' at a time when each thing can be repeated r-times is given by = n^r.

5. The number of circular permutation of n objects = (n – 1) !.

6. A Combination is a basically a selection of object without considering the order of the objects from a given set of objects - Mathematically expressed the number of combination of n objects taken 'r' at time is given by ^nC_r .

Self Practice Questions

1. Find the number of triangles that can be formed in a hexagon? {*Hint*: to form a triangle 3 value of the 6 vertices needs to be selected.}

2. Find the how many the word HYDERABAD can be arranged with no vowels coming together?

3. How many new words can be formed out of the word PERCEPTION?

4. A drama troupe of 5 members needs to be selected out of 3 girls and 6 boys. In how many the troupe can formed if the troupe contains

   (i) At least 2 girls

   (ii) No girls

5. From the digits 1,2,3,4,5,6,7 how many 3 digits even numbers can be formed when

   (i) repetition of digits is not allowed ?

   (ii) repetition of digits is allowed?

6. How many signal of different type can be made from 3 white flags, 5 blue flags and 2 red flags?

7. In how many ways 4 cards can be chosen from a well shuffled pack of cards such that all the four cards are of the same suit?

8. An examination paper consists of 8 questions in two parts. Part I 4 question and Part II 4 questions. An examinee who is taking the

examination needs to attempt 5 questions from both the selection with the condition that he needs to attempt at least 2 question from each part. Find the number of ways to answering the question?

9. In a social gathering every member greets each other. If there are 93 greetings all together, find the number of people attending the social gathering?

10. In the implementation of a child welfare program a village is selected with 50 families of which 32 families have at most 3 children. As per the schedule of the program 25 families needs to be selected of which at least 22 families have at most 3 children. Then in how many ways the choice can be made?

11. 4 balls of different color are to be placed in 3 boxes of different sizes. In how many different ways can we place the balls so that no box remains empty?

12. There are 8 points in a plane out of which 3 are collinear. Find the number of straight lines formed by forming them

13. In how many ways 8 books can be arranged so that the best and the worst book are always come together and in the second case they never come together?

14. If 5 Indians, 4 Bangladeshi and 6 Singhalese are made to seat in a row in a particular order such that all persons from the same nationality are made to sit together. In how many ways the sitting arrangement can be made?

15. If ${}^{n}C_8 = {}^{n}C_3$ then find 'n'?

16. In how many ways can 8 identical toys be distributed among 5 chilldren so that each child gets at least one gift?

17. If there are 3 routes from Delhi to Lucknow and 4 routes from Lucknow to Kolkata then how many routes are possible for a travelers to reach Kolkata from Delhi via Lucknow?

18. How many distinct words can be formed from out of the word "TRACKER" such that the word starts with A and ends with E?

19. Several teams have taken part in the IPL Cricket Tournament Season 2 if each team has to play one game with all other teams then how many has participated in the tournament if the total number of games played is 28?

20. There are 6 different types of cricket bats and 5 different types of cricket balls are available in a sport shop. How many purchases of cricket bat and ball can be made?

21. How many new words can be made out of the word "MEENAKSHI"?

22. In how many ways can we make a 5 letter word having 3 consonants and 2 vowels if there are 6 consonants and 4 vowels to choose from a word jumble?

23. Vimal passed out engineering graduate received a call letter for an interview with Prism Cement. He has altogether 4 shirts, 5 trousers and 3 Ties to choose to dress for the interview. In how many ways he can dress himself with a combination of all three?

24. If there are 5 letters and 5 envelopes, in how many ways the wrong choices can be made?

    Hint: Total number of choices are 5! Out of which only one is correct.

25. In how many ways 5 boys and 5 girls are made to sit in round table such that no 2 boys nor 2 girls sitting next to each other?

26. How many diagonals are there in 'n' sided polygon?

27. In how many ways 10 oranges can be divided into two sets each having 5 oranges?

28. In how many ways can 6 letters can be posted in 4 post boxes, if any number of letters can be posted in all the four post boxes?

29. A four digit number is formed from the digits 1, 2, 3, 4, 5, 6, 7. In how many ways the arrangement can be done so that the digits 3, 7 are always excluded?

30. In how many ways 6 balls can be arranged such that the red ball and the black ball are never together?

Answers

1. $^6C_3$

2. $\dfrac{9!-7!3!}{2!2!}$

3. $\dfrac{10!}{2!2!}-1$

4. (i) $^3C_2 \times {}^6C_3 + {}^3C_3 \times {}^6C_2$, (ii) $^6C_5$

5. (i) 120, (ii) 196

6. $\dfrac{10!}{5!3!2!}$
7. 2860
8. 48
9. n = 13
10. ${}^{32}C_{22} \times {}^{18}C_3 + {}^{32}C_{23} \times {}^{18}C_2 + {}^{32}C_{24} \times {}^{18}C_1 + {}^{32}C_{25}$
11. 36
12. ${}^{8}C_2 - {}^{3}C_2 + 1$
13. (i) 7! 2!, (ii) 8! – 7! 2!
14. 3! 5! 4! 6!
15. 11
16. 5^8
17. 12
18. $\dfrac{5!}{2!}$
19. 8
20. 30
21. $\dfrac{9!}{2!} - 1$
22. 1440
23. 60
24. 5! -1
25. 4! 5!
26. ${}^{n}C_2 - n$
27. $\dfrac{{}^{10}C_5 \times {}^{5}C_5}{2!}$
28. 4^6
29. ${}^{5}P_4$
30. 6! – 5!2!

CHAPTER 16

Theory of Probability-I

Introduction

The first and foremost question which comes to mind is what is probability? Let us begin with an illustration, consider you have decided to visit a mall on a sunny day, what is the likeliness that it might rain and spoil your plan. A second instance, imagine that you have purchased a lottery ticket with a bonanza of Rs. 1 Crore. What is the chance of your winning the jackpot? In both the statements you might have noticed that the statements are not precise which means that one cannot form a definite idea about the outcome of the event but can only guess a basic proposition that a particular event may or may not occur. In such events where there is a chance of happening of an event either going in favor or going against is the probability associated with that event.

A Brief Historical Background

The theory of probability's origins can be traced to the works of Italian Mathematician Jerome Cardan (1501-1576) entitled "Liber de Ludo Aleae" meaning "Book on Games of Chance" published in 1663 after his death. But the scientific development to the theory of probability was laid by Blaise Pascal (1623-62) and Pierre de Fermat (1601-1665) when approached by a French gambler Chevalier de Mere in solving the sharing the benefits of a incomplete gambling match. The lengthy deliberation between the mathematicians resulted what we know now as the modern theory of probability. Other stalwarts who have contributed immensely are James Bernoulli (1654-1705) his work "Arts Conjectandi", De Moivre (1667-1705) work "Doctrine of Chances", Thomas Bayes (1702-1761), Laplace(1749-1827), Fisher, Von Mises and Russian Mathematician like Chebychev (1821-1894), Markov (1856-1922) and

Kolmogorov who is also credited with the development of axiomatic theory.

"Probability can be defined as the chance of an event to happen".

Now let us understand the concept. Imagine you have tossed an unbiased coin; the outcome may be either head or tail. Then what is the chance of head appearing in a single toss? Answer is $1/2$ how?

Explanation

Since a coin has two sides a head and a tail. We can write the sample space to contain S = {Head, Tail}.

Now number of events in favour i.e. head appearing is 1 and against i.e. tail appearing is also 1

But the total number of events are 2 (favourable + unfavourable)

Therefore the probability is number of favourable events to total number of possible events.

$$P(\text{Head Appearing}) = \frac{\text{Number of Favourable Events}}{\text{Total Number of Possible Events}} = \frac{1}{2}$$

(By Classical Approach)

The above example only illustrates a different notion of defining probability. We proceed further as it is needed to discuss certain key definitions and concepts relating to the theory of probability before entering into main theme. Some of them are:

Experiment

Any process or operation that results in the happening of an outcome whether single or more than one is an experiment

Random Experiment

Imagine a die being rolled any of the following numbers from 1 to 6 may appear on the face of the die. So appearance of one of the six numbers' is an outcome of an activity. But we are definitely not certain which among them may appear i.e. the resulting outcome is not unique but one among the possible alternatives available; may be 3 or 6. Such activities where the resulting outcome is not unique but one among the several possible options available is called as random experiment. Generally random experiments are performed under homogenous conditions where the result of the experiment is not unique but one among the probables is a random experiment.

For example

Drawing of a card from a well shuffled pack of cards, tossing of an unbiased coin etc.

Trial and Event

Performing a random experiment is trial and the outcome is event.

For example

Tossing of an unbiased coin is a trial and the appearance of either a head or tail is two different events.

Remember that events can be simple or composite. Tossing a coin and getting a head or tail is a simple or elementary event where as rolling 2 die's and getting a sum of 10 is a composite event. Here the pairing can be (4, 6),(5, 5) or (6, 4).

Favorable and Unfavorable Events

The outcome of a random experiment which results in the happening of an event going in favour then the events are termed favourable otherwise unfavourable.

Equally Likely Events

Events are equally likely taking all factors into consideration there is no reason that any one of the event will be preferred over the other.

For example

When a card is drawn at random from a well shuffled pack of cards may result in drawing any card among the following 4 suits. Or when a coin is tossed both the outcomes head and tail are equally likely to occur and none is preferred over the other.

Mutually Exclusive Events

When one event prevents the occurrence or happening of the other i.e. two events cannot happen simultaneously.

For example

When a coin is tossed if head appears prevents the occurrence of tail and vice–versa.

Complementary Events

The complementary events are the number of unfavorable outcomes of an experiment.

Independent and Dependent Events

Events are said to independent when the happening of an event is not influenced by the happening of the other then the events are termed independent otherwise termed dependent.

For example

If a coin is tossed the appearance of tail in the first throw will no way influence the appearance of either head or tail in the second throw.

Sample Space or Exhaustive Events

The total number of results or outcomes i.e. to include all possible cases of a random experiment is a sample space or exhaustive outcomes or events.

For example

A coin is tossed we can either get a head or a tail but the total number of possibilities are 2(Head + Tail).

Similarly a die contains 6 faces numbered {1, 2, 3, 4, 5, 6}. Therefore the total exhaustive cases are 6.

Different Approaches to Probability

Some of the approaches to probability theory are:

1. The Classical Approach.
2. The Empirical Approach.
3. The Subjective Approach.
4. The Axiomatic Approach.

The Classical Approach

According to Classical Approach, probability is defined as the ratio of favorable events to total number of all possible events

$$\text{Probability} = \frac{\text{Favourable Events}}{\text{Total Number of all Possible Events}}$$

If there are 'q' mutually exclusive and equally likely events out of which 'p' are favorable denoted by event A, then the happening of A

$$P(A) = \frac{p}{q} = \frac{\text{Favourable Events}}{\text{Total Number of all Possible Events}}$$

Since 'p' are in favour then 'q – p' are unfavorable i.e. the event will not happen say event B, then the happening of the event B

$$P(B) = \frac{q-p}{q} = \frac{\text{Unfavourable Events}}{\text{Total Number of all Possible Events}}$$

$$P(B) = 1 - \frac{p}{q}$$

$$= 1 - P(A) \text{ or } P(A) + P(B) = 1$$

Note:

Probability is clearly a number ranging from 0 to 1. When the happening of the event is absolute certain i.e. p = q then the probability P(A) = 1 and when the occurrence is definitely impossible i.e. p = 0 then the probability of the impossible event to happen P(A) = 0.

But $0 \le p \le q$

Now divide both sides by 'q'.

$$0 \le \frac{p}{q} \le 1 \leftrightarrow 0 \le P(A) \le 1 \qquad \left(\text{since } P(A) = \frac{p}{q}\right)$$

Assumptions of Classical Approach

(i) All the outcomes are equally likely.

(ii) Total number of exhaustive events is finite. If there is any violation of any one of the above assumptions then classical approach fails.

For example

An unbiased die is rolled what is the probability for 4 to appear?

Explanation

A unbiased die has 6 faces numbered {1, 2, 3, 4, 5, 6}, therefore the total no. outcomes = 6.

When the die is rolled any of the numbers 1 or 2 or 3 or 4 or 5 or 6 can appear (since events are equally likely).

Now '4' can appear in only one way. Therefore the number of chance in favour for 4 appearing is 1.

By Classical Approach

$$P(\text{Favourable}) = \frac{\text{Number of favourable Event}}{\text{Total Number of Events}} = \frac{1}{6}$$

Illustration No. 1

A card is drawn from a well shuffled pack of cards. What is the probability that the card will be an ace card?

Solution:

A pack of cards contain 52 cards divided into 4 suits (Spade, Diamonds, Clubs, Hearts) each suit containing 13 cards. Numbered {A, 2, 3, 4, 5, 6, 7, 8, 9, 10, J, Q and K}.

When we are speaking about aces each suit will contribute an ace i.e. ace of Heart, ace of Diamond, ace of Spade, ace of Clubs.

Total number of aces are 4 and selection of an ace (random selection) can be done in ${}^4C_1$ ways.

(Since selection of r things out of n things can be done in ${}^nC_r$ ways)

Selection of a card from 52 cards can be done in ${}^{52}C_1$ ways.

$$\text{Hence the probability of drawing an ace card} = \frac{{}^4C_1}{{}^{52}C_1} = \frac{4}{52} = \frac{1}{13}$$

Emphirical or Statistical Approach (Von Mises)

If a random experiment is performed essentially and repetitively under homogenous conditions, then the limiting value of the ratio of the number of times the event happens to the total number of trial, as the number of trials becomes indefinitely large, is called the probability of the occurrence of that event, based on the assumption that the limit is finite and definite number.

If an event A occurs m times out of n trials, then the probability P of the happening of the event A is given by

$$P(A) = \lim_{n \to \infty} \frac{m}{n}$$

in the anticipation that the limit exists and is finite.

Conditions for Application Empirical Approach: If events are not equally probable and the exhaustive cases in a trial is infinite then classical approach stands limited to address such issues . In cases empirical approach is applied which is based on the relative frequency $\frac{m}{n}$ as $n \to \infty$ with the assumption that the experiment has been performed essentially under homogeneous conditions.

In short as conditions cannot remain identical when the experiment is conducted with large number repetitions and the relative frequency $\frac{m}{n}$ may not attain a unique limiting value as $n \to \infty$.

Subjective Approach or Personalistic Probability

In this approach the probability of success is based on personal assessment.

It can be defined as the degree of credence or confidence placed on the happening of an event by an individual with available empirical evidence which is either qualitative or quantitative in nature. Since the happening of an event is based on personal perception it is envisaged differently by different individuals. Subjective approach is finding its way in forecasting and in decision making in situations where both classical and empirical approach fails.

For example

Estimating that the price of onions can surpass Rs 50 a kilo. Or Indian economy achieving a GDP growth of 10% are all cases of subjective probability.

Axiomatic Approach

(Developed by A.N Kolmogorov, Russian Mathematician)

Axiomatic approach had drawn largely from theoretical aspects of set theory and introduced probability as a set function.

It can be understood as the probability of occurrence of an event A, given a sample space S such that A subset of S then we define probability function P(A) on S if the following axioms are satisfied

(i) $P(A) \geq 0$.

(ii) $P(S) = 1$ and

(iii) If A_1, A_2,......are sequentially mutually exclusive events in S, then $P(A_1 \cup A_2 \cup ....) = P(A_1) + P(A_2) + ...$

Illustration 2

A bag contains 4 red balls and 6 black balls. A ball is drawn at random. Find the probability that the ball drawn is black in color?

Solution:

Total number of Balls are 10 {4 Red + 6 Black}

The number of ways of selecting a ball out of 10 balls can be done in $^{10}C_1 = 10$ ways

Selection of a black ball out of 6 black balls can be done in $^{6}C_1 = 6$ ways

Therefore the probability that the ball is black

$$P(\text{Black}) = \frac{\text{Number of Favourable Events}}{\text{Total Number of Events}} = \frac{6}{10} = \frac{3}{5}$$

Illustration 3

From a bag 3 balls are drawn at random containing 4 red balls, 5 green balls and 3 blue balls. What is the probability that the balls drawn are

(i) All Blue

(ii) 2 Green and 1 Red

Solution:

Total no. of Balls are 4 + 5 + 3 = 12 balls. {4 red balls, 5 green balls and 3 blue balls}

Hence 3 balls can be drawn from 12 balls $^{12}C_3 = 220$ ways.

In case (i) since all balls are selected are blue, so their selection can be done in $^{3}C_3 = 1$ way. {Number of blue balls are 3}.

The probability of selecting 3 blue balls $= \frac{1}{220}$.

In case (ii) 2 green and 1 red is selected and the no. of favorable events are $^{5}C_2 \times {}^{4}C_1 = 40$ ways.

But total number ways selection of 3 balls = 220.

Therefore the probability calculated for the event $= \frac{40}{220} = \frac{2}{11}$

Illustration No. 4

What is the probability that a leap year contains 53 Sundays?

Solution:

A leap contains 366 days i.e. 52 weeks and 2 additional days.

Now the last 2 days can be a combination of either {(Saturday, Sunday), (Sunday, Monday), (Monday, Tuesday), (Tuesday, Wednesday), (Wednesday, Thursday), (Thursday, Friday), (Friday, Saturday)}.

Number of times Sunday is getting repeated are the combination

= {(Saturday, Sunday), (Sunday, Monday)} = 2

Out of 7 possible outcomes 2 are in favour

Therefore the probability a leap will contain 53 Sundays P(S) = $\frac{2}{7}$.

Illustration No. 5

In a game of die, if Vikram throws 3, 4 or 7 he losses the game, whereas if he throws 5 or 10 he wins. Find the ratio of the probability of his winning to losing the game?

Solution:

Let us assume

A : the event that he throws 3, 4 or 7

B : the event that he throws 5 or 10.

The combination of scoring 3, 4 or 7 are {(1, 2), (2, 1) (1, 3), (2, 2), (3, 1), (1, 6), (2, 5), (3, 4), (4, 3), (5, 2), (6, 1)} i.e. A = 11.

The combination of scoring 5 or 10 are {(1, 4), (2, 3), (3, 2), (4, 1), (4, 6), (5, 5), (6, 4)} i.e. B = 7.

But total no. of outcomes when two die's are thrown = $6^2 = 36$

Therefore the probability of losing P(A) = $\frac{11}{36}$

And the probability of winning P(B) = $\frac{7}{36}$

Hence the ratio of winning to losing $= \frac{11}{36} : \frac{7}{36} = 11 : 7$

Illustration No. 6

A number is selected at random from a set of 20 numbers numbered from 1 to 20. What is the probability that the number appearing is a multiple of 5?

Solution:

Sample space is numbered S = {1, 2, 3, 4, 5, ...20}

The numbers which are multiple of 5 are {5, 10, 15, 20}.

Hence the no. of favourable cases = 4

Therefore the required probability of selecting a no. which is a multiple of 5 i.e $P(5) = \frac{4}{20} = \frac{1}{5}$

Illustration No. 7

Two unbiased coins are tossed. Find what is the probability that

(i) At least one head appears.

(ii) 1 tail and 1 head appear.

Solution:

When two coins are tossed the number of possible outcomes are {HH, TH, HT, TT}

And we have 3 cases where either one or maximum 2 heads appear. They are {HH, TH, HT}.

So, the number of favourable cases are 3

And the total number of outcomes are 4

Hence the required probability were at least 1 Head appears

$$P(H \geq 1) = \frac{3}{4}$$

{Note: At least 1 head means it may range from 1 head appearing to maximum possible number which can appear is 2}

In case (ii) we have 2 events that favours the appearance of a head or a tail i.e. {HT, TH}

Therefore the number of favourable events are 2

Hence the required probability i.e

$$P(HT \text{ or } TH) = \frac{2}{4} = \frac{1}{2}$$

Illustration No. 8

Vipul and Vinay throw a two unbiased die. If Vipul throws 9 find the probability of Vinay throwing a higher number?

Solution:

Let us assume that

A : the event that Vipul throws 9.

B : the event that Vinay throws a no. higher than 9.

The no.'s are {10, 11, 12}.

But when the dies are rolled the total number of outcomes = 36.

The events in favour of A are { (3, 6), (4, 5), (5, 4), (6, 3)} = 4

But the events in favour of Vinay scoring over Vipul are 10, 11 and 12 the combinations are {(4, 6), (5, 5), (6, 4), (5, 6), (6, 5), (6, 6)} = 6

Hence required probability that Vinay scoring over Vipul i.e.

$$P(B) = \frac{6}{36} = \frac{1}{6}$$

Illustration No. 9

If 8 people sit in a circular fashion in a business meet, find the probability that

(i) 2 friends Mr. X and Mr. Y are always sit together and

(ii) They are never together.

Solution:

When 8 people are sitting in a business meet in circular form the no. ways they can arranged (8 – 1)! = 7! ways.

{*Note*: n people can be arranged in circular way in (n – 1)! ways}

Case i

In order to find the number of ways in which Mr X and Mr.Y are always together consider Mr X and Y as 1 unit and now the number of persons are reduced to 7 and these 7 men can be arranged in 6! ways in circular form. However the number of ways in which the two particular person can be arranged among themselves is 2! ways

Therefore the no. of favorable ways = 6! × 2!

Hence the required probability $= \frac{6! \times 2!}{7!}$

Case ii

In the second case the required probability that the friends are never together is

$$1 - \frac{6! \times 2!}{7!}$$

Illustration No. 10

A committee of 3 persons needs to be formed from 4 men and 3 women. Find what is the probability that the committee will have

(i) All men

(ii) 2 men and 1 women.

Solution:

Total number of persons available for selection = 7 {4 men + 3 women}

Out of 7 persons, 3 can selected in ${}^7C_3$ ways.

To select a committee to be formed with all men in the committee i.e 3 men.

Therefore selection of 3 men from a group of 4 men can be done in ${}^4C_3$ ways.

Hence the probability associated with the selection procedure $= \frac{{}^4C_3}{{}^7C_3}$

In case (ii) 2 men can be selected from 4 men in ${}^4C_2$ ways

And 1 women can be selected out of a group of 3 in ${}^3C_1$ ways

Together they can be selected in ${}^4C_2 \times {}^3C_1$ ways

Therefore the probability of selecting 2 Men and 1 Women in the committee $= \frac{{}^4C_2 \times {}^3C_1}{{}^7C_3}$

Addition Theorems

Theorem 1

If A and B are two events and they are not necessarily mutually exclusive, then the probability of occurrence of at least one of the two events A and B i.e. $P(A \cup B)$ is given by

$$P(A \cup B) = P(A) + P(B) - P(A \cap B)$$

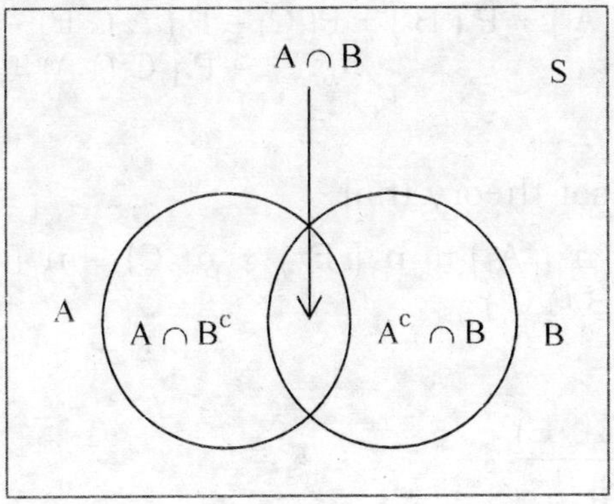

Let S be a sample space and A & B any two events of S

Since $$P(A \cup B) = \frac{n(A \cup B)}{n(S)}$$

where $n(A \cup B)$ is the number of sample points (Outcomes) in favor of $(A \cup B)$

$n(S)$ = Total no. of all possible sample points in S

But from the Figure

$$n(A \cup B) = n(A \cap B^c) + (A \cap B) + n\,(A^c \cap B)$$

But $$n(A \cap B^c) = n(A) - n(A \cap B)$$

Similarly $n(A^c \cap B) = n(B) - n(A \cap B)$

Therefore $n(A \cup B) = n(A) - n(A \cap B) + n(A \cap B) + n(B) - n(A \cap B)$

Cancelling $n(A \cup B)$,

We get $n(A \cup B) = n(A) + n(B) - n(A \cap B)$

Therefore,

$$P(A \cup B) = \frac{n(A \cup B)}{n(S)} = \frac{n(A) + n(B) - n(A \cap B)}{n(S)}$$

$$= \frac{n(A)}{n(S)} + \frac{n(B)}{n(S)} - \frac{n(A \cap B)}{n(S)}$$

$$= P(A) + P(B) - P(A \cap B)$$

Hence proved

Generalisation: For any three events A, B and C the probability of occurrence at least either one of them i.e. $P(A \cup B \cup C)$ is given by

$$P(A \cup B \cup C) = P(A) + P(B) + P(C) - P(A \cap B) - P(B \cap C) - P(C \cap A) + P(A \cap B \cap C)$$

Proof:

Since we know from set theory that

$$n(A \cup B \cup C) = n(A) + n(B) + n(C) - n(A \cap B) - n(B \cap C) - n(C \cap A) + n(A \cap B \cap C)$$

Therefore,

$$P(A \cup B \cup C) = \frac{n(A \cup B \cup C)}{n(S)}$$

$$= \frac{n(A) + n(B) + n(C) - n(A \cap B) - n(B \cap C) - n(C \cap A) + n(A \cap B \cap C)}{n(S)}$$

$$= \frac{n(A)}{n(S)} + \frac{n(B)}{n(S)} + \frac{n(C)}{n(S)} - \frac{n(A \cap B)}{n(S)} - \frac{n(B \cap C)}{n(S)} - \frac{n(C \cap A)}{n(S)} + \frac{n(A \cap B \cap C)}{n(S)}$$

$$= P(A) + P(B) + P(C) - P(A \cap B) - P(B \cap C) - P(C \cap A) + P(A \cap B \cap C).$$

Hence Proved.

Theorem 2

If A and B are two mutually exclusive events i.e. $A \cap B = \{\Phi\}$ then the probability that either A or B to occur is the sum of the individual probabilities of the events A and B.

Proof:

Let S be the total number of possible events out of which 'm' are favorable to ' A ' and ' n ' are favorable to 'B', then the number of events in favour to either A or B is 'm + n' (since A and B are exclusive events).

Hence

$$P(A \cup B) = \frac{\text{The Probability that either A or B to occur}}{\text{Number of Favourable Events}}$$

$$= \frac{m+n}{S} = \frac{m}{S} + \frac{n}{S}$$

$$= P(A) + P(B) \qquad \text{Since } P(A) = \frac{m}{S} \text{ and } P(B) = \frac{n}{S}$$

Corollary

If A and B are mutually exclusive events then $A \cap B = \{\Phi\}$ by addition theorem of probability we have

$$P(A \cup B) = P(A) + P(B) - P(A \cap B)$$

But A and B being mutually exclusive therefore $P(A \cap B) = \{0\}$

Hence we have

$$P(A \cup B) = P(A) + P(B) - 0$$

$$P(A \cup B) = P(A) + P(B)$$

Theorem 3

If A and B are two non mutually exclusive events then the probability of occurrence of event A i.e P(A) is given by

$$P(A) = P(A \cap B^c) + P(A \cap B)$$

Proof:

Since $P(A) = \dfrac{n(A)}{n(S)}$

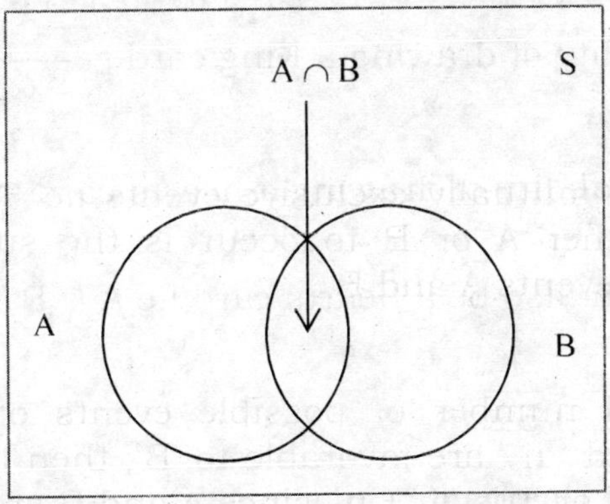

But from the above diagram

We get,

$$n(A) = n(A \cap B^c) + n(A \cap B)$$

Replacing the value of n(A)

$$P(A) = \frac{n(A \cap B^c) + n(A \cap B)}{n(S)}$$

$$= \frac{n(A \cap B^c)}{n(S)} + \frac{n(A \cap B)}{n(S)}$$

Therefore,

$$P(A) = P(A \cap B^c) + P(A \cap B)$$

Hence proved

Solved Exercises

Illustration No. 1

A card is drawn from a well shuffled pack of cards. Find the probability that the card drawn is either a king or a hearts?

Solution:

Let A be the event of drawing an Ace card and B a diamonds card.

But in a pack of cards there are 4 suits (Diamonds, Hearts, Spade, Clubs)

Each suit will contribute a king i.e. King – Hearts, King – Diamonds, King – Spade, King – Clubs.

Thus, the probability of drawing a King card = $\frac{{}^{4}C_1}{{}^{52}C_1} = \frac{4}{52} = \frac{1}{13}$ ways.

Similarly the probability of drawing a hearts card = $\frac{{}^{13}C_1}{{}^{52}C_1} = \frac{13}{52} = \frac{1}{4}$

But the king card can also be a hearts card i.e A ∩ B

Hence $P(A \cap B) = \frac{{}^{1}C_1}{{}^{52}C_1} = \frac{1}{52}$

Therefore by addition theorem

$$P(A \cup B) = P(A) + P(B) - P(A \cap B)$$

$$= \frac{1}{13} + \frac{1}{4} - \frac{1}{52} = \frac{4+13-1}{52} = \frac{16}{52}$$

Illustration No. 2

If A and B are two events in an experiment for which P(A) = 0.70, P(B) = 0.65 and P(A ∪ B) = 0.80 find P(A ∩ B) = ?

Solution:

Using Addition theorem we have

$$P(A \cup B) = P(A) + P(B) - P(A \cap B)$$

$$0.80 = 0.70 + 0.65 - P(A \cap B)$$

$$P(A \cap B) = 1.35 - 0.80 = 0.55$$

Illustration No. 3

If three unbiased die are rolled. Find the probability that either a triplet or a sum of 8 will appear?

Solution:

When two unbiased dies are rolled the total number of outcomes are $6^3 = 216 \ (6 \times 6 \times 6)$

Let us assume that

A : the event that a triplet Appears.

B : the event that the sum appearing is 8.

Since the combination of triplet are {(1, 1, 1), (2, 2, 2), (3, 3, 3), (4, 4, 4), (5, 5, 5) and (6, 6, 6)}

Total Combination = 6

$$P(A) = \frac{6}{216} = \frac{1}{36}$$

And the combination for the sum to be 8 are {(1, 1, 6), (1, 6, 1), (1, 2, 5), (1, 5, 2), (1, 3, 4), (1, 4, 3), (2, 1, 5), (2, 5, 1), (2, 3, 3), (2, 2, 4), (2, 4, 2), (3, 1, 4), (3, 4, 1), (3, 2, 3), (3, 3, 2), (4, 1, 3), (4, 3, 1), (4, 2. 2), (5, 1, 2), (5, 2, 1), (6, 1, 1)}

Total Combination = 21

$$P(B) = \frac{21}{216} = \frac{7}{72}$$

Since the events are mutually exclusive therefore A ∩ B = Φ

i.e $\quad P(A \cap B) = 0$

By Addition theorem

$$P(A \cup B) = P(A) + P(B)$$

$$= \frac{1}{36} + \frac{7}{72} = \frac{9}{72} = \frac{1}{8}$$

Illustration No. 4

If A and B are 2 events not mutually exclusive if $P(A) = \frac{2}{5}$, $P(B) = \frac{1}{4}$ and $P(A \cup B) = \frac{1}{2}$ then find the probability of

(i) $P(A \cap B^c)$

(ii) $P(A^c \cap B^c)$

Solution:

Since we know

$$P(A) = P(A \cap B) + P(A \cap B^c)$$

But from Addition theorem

$$P(A \cap B) = P(A) + P(B) - P(A \cap B)$$

$$P(A \cap B) = \frac{2}{5} + \frac{1}{4} - \frac{1}{2}$$

$$= \frac{5+8-10}{20} = \frac{3}{20}$$

$$P(A) = P(A \cap B) + P(A \cap B^c)$$

$$P(A \cap B^c) = P(A) - P(A \cap B)$$

$$= \frac{2}{5} - \frac{3}{20} = \frac{8-3}{20} = \frac{1}{4}$$

(ii) By De Morgans Law

we have $(A \cap B)^c = A^c \cup B^c$

$P(A^c \cup B^c) = P(A \cap B)^c = 1 - [P(A \cap B)]$

$$= 1 - \frac{3}{20} = \frac{17}{20}$$

Illustration No 5

A bag contains balls numbered from 1 to 15. Find the probability that a ball drawn at random is a multiple of 3 or 5?

Solution:

In this case the sample size 15

Let A be the event that the ball drawn is multiple of 3 (i.e 3, 6, 9, 12, 15) and B the ball drawn is a multiple of 5 (i.e. 5 , 10, 15)

Therefore,

The probability of drawing a ball which is 3 or its multiple $P(A) = \frac{^5C_1}{^{15}C_1} = \frac{5}{15} = \frac{1}{3}$, and that of 5 or its multiple $P(B) = \frac{^3C_1}{^{15}C_1} = \frac{3}{15} = \frac{1}{5}$

But there is a common multiple 15 among A and B i.e. $A \cap B = 1$

Therefore $P(A \cap B) = \frac{1}{15}$

Hence the required probability by addition theorem,

$$P(A \cup B) = P(A) + P(B) - P(A \cap B)$$

$$= \frac{1}{3} + \frac{1}{5} - \frac{1}{15} = \frac{7}{15}$$

Illustration No. 6

If A, B and C are 3 mutually exclusive and exhaustive events then find P(A) if $\frac{2}{3} P(C) = \frac{3}{4} P(A)$ and $\frac{2}{5} P(B) = \frac{5}{7} P(A)$?

Solution:

Since A, B and C are mutually exclusive and exhaustive events

We have

$$P(A \cup B \cup C) = P(A) + P(B) + P(C)$$

Therefore

$$1 = P(A) + \frac{25}{14} P(A) + \frac{9}{8} P(A) \quad \{\text{since } P(A \cup B \cup C) = 1\}$$

$$1 = \frac{56P(A) + 100P(A) + 63P(A)}{56}$$

$$P(A) = \frac{56}{219}$$

Illustration No. 7

If three horses A, B and C are competing in a race and B has twice likely to win as C and A has thrice likely to win over B , then find their individual probabilities of winning the race?

Solution:

Let 'p' be the probability that C wins the race i.e. P (C) = p

Then the associated probability of B winning the race is twice that of C

Therefore, P(B) = 2p

But A has thrice likely to winning over B thus the probability of A winning the race i.e. P(A) = 3 × 2p = 6p

As either of A or B or C can win the race

Therefore

$$P(A \cup B \cup C) = P(A) + P(B) + P(C)$$

Since $P(A \cup B \cup C) = 1$

$$1 = 6p + 2p + p = 9p \text{ or } p = \frac{1}{9}$$

The probability of A winning i.e $P(A) = \frac{6}{9} = \frac{2}{3}$

The probability of B winning i.e $P(B) = \frac{2}{9}$

The probability of C winning i.e $P(C) = \frac{1}{9}$

Illustration No. 8

The probability that Vivek will pass his Mathematics paper is $3/4$ and that of Micro economics is $1/3$. And that the probability he will pass atleast one of the examination is $\frac{4}{15}$. Find the probability that he will pass both the examinations?

Solution:

Let us define A as the event that Vivek passes the Mathematics paper and B the event that he passes the Microeconomics.

But $P(A) = \frac{3}{4}$, $P(B) = \frac{1}{3}$ and $P(A \cup B) = \frac{4}{15}$

Using Addition theorem

$$P(A \cup B) = P(A) + P(B) - P(A \cap B)$$

$$\frac{4}{15} = \frac{3}{4} + \frac{1}{3} - P(A \cap B)$$

$$P(A \cap B) = \frac{49}{60}$$

Illustration No. 9

Mayank an investment manager predicts that the price of the share of a company will rise and the odds against is 2 : 3 that the price will remain constant is 1: 2. Find the probability that the price will fall?

Solution:

Let us define

A : the event that the price of the stock will rise

B : the event that the price of the stock will remain the same

And $P(A) = \frac{3}{5}$ and $P(B) = \frac{1}{3}$

As the events are mutually exclusive $P(A \cap B) = \{0\}$

Therefore the probability that the price of the share will fall i.e $P(A^c \cap B^c)$ is given by

$$P(A^c \cap B^c) = 1 - P(A \cup B) = 1 - \{P(A) + P(B)\}$$

$$= 1 - \{\frac{3}{5} + \frac{1}{3}\}$$

$$= \frac{1}{15}$$

Illustration No. 10

From a production team of 100, one will be selected to undergo special training in Total Quality Management Programme. Out of 100, 60 are engineers, 70 are MBA's in Production system and two-thirds of engineers are also MBA degree holder. Find the probability that a person chosen at random will neither be engineer nor MBA degree holder?

Solution:

Let us assume

A : the event that an engineer is selected

B : the event that a MBA is selected

Therefore the associated probability of selection

$$P(A) = \frac{60}{100} = \frac{3}{5} \text{ and } P(B) = \frac{70}{100} = \frac{7}{10}$$

But if a person is selected at random who is engineer and MBA i.e.

$$A \cap B = \frac{2}{3} \times 60 = 40$$

Therefore $$P(A \cap B) = \frac{40}{100} = \frac{2}{5}$$

By addition rule the probability of selection of either an engineer or a MBA is given by

$$P(A \cup B) = P(A) + P(B) - P(A \cap B)$$

$$= \frac{3}{5} + \frac{7}{10} - \frac{2}{5} = \frac{9}{10}$$

Hence the required probability that a person selected is neither Engineer nor MBA is $P(A^c \cap B^c)$

$$P(A^c \cap B^c) = 1 - P(A \cup B) = 1 - \frac{9}{10} = \frac{1}{10}$$

Compound Probability

The probability of the simultaneous occurrence of two or more events is called as compound probability and for any two events A and B it is denoted by $(A \cap B)$

Conditional Probability

Let A and B be two events such that $P(B) > 0$, then the probability of the occurrence of the event B assuming that A has already occurred is called the conditional probability of B and is denoted by $P\left(\frac{B}{A}\right)$ or P(A, B) and read as the probability of B given A.

Multiplication Theorem

The probability of simultaneous occurrence of two events A and B is given by the product of the unconditional probability of the event A by the conditional probability of B, with the assumption that A has already occurred

Symbolically

$$P(A \cap B) = P(A)\, P\left(\frac{B}{A}\right) \text{ or } P(B \cap A) = P(B)\, P\left(\frac{A}{B}\right)$$

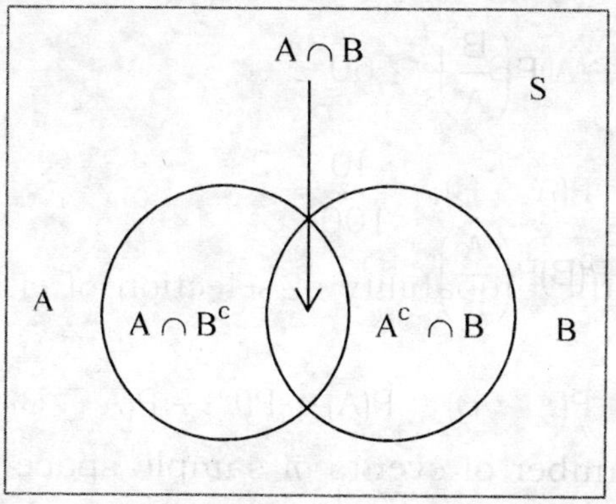

Let A be a subset of the sample space S such that n (A) ≠ 0 i.e. event A occurs.

Together with n(A) consider the event n(A ∩ B) which is a favourable to both events A and B.

Since n(A) is the number of possible events for the conditional event $\frac{B}{A}$ to happen on the assumption that A has already occurred out of which n(A ∩ B) are favourable to $\left(\frac{B}{A}\right)$.

In this case A being reduced sample thereby restricting the occurrence of B to A ∩ B

The conditional probability of B assuming A has already occurred

$$P\left(\frac{B}{A}\right) = \frac{n(A \cap B)}{n(A)}$$

Divide both numerator and denominator with n(S)

We get,

$$P\left(\frac{B}{A}\right) = \frac{\frac{n(A \cap B)}{n(S)}}{\frac{n(A)}{n(S)}}$$

$$= \frac{P(A \cap B)}{P(A)}$$

$$P(A \cap B) = P(A)P\left(\frac{B}{A}\right)$$

Similarly

$$P(B \cap A) = P(B)P\left(\frac{A}{B}\right)$$

Alternate Proof

Let n be the total number of events of sample space S which are equally likely out of which m are in favour of the event A. While the cases in favour to both A and B i.e. $A \cap B$ is say k where $k \leq m$. Then the probability that both A and B will occur is given by

$$P(A \cap B) = \frac{\text{The Number of Outcomes in favour of } A \cap B}{\text{Total Number of Events}} = \frac{k}{n} \quad \text{......(1)}$$

$$P(A) = \frac{\text{The Number of Outcomes in favour of } A}{\text{Total Number of Events}} = \frac{m}{n} \quad \text{.....(2)}$$

In order to compute $P\left(\frac{B}{A}\right)$ i.e. the probability of the event B to have occurred knowing that A has already occurred, now we will restrict our discussion to reduced sample space of A. Thus with respect to new sample space A the number of events in favour of B consists of those in favour to both A and B i.e. $A \cap B$.

Thus,

$$P\left(\frac{B}{A}\right) = \frac{\text{The Number of Outcomes in favour of } B \cap A}{\text{Total Number of Events}} = \frac{k}{n} \quad \text{......(3)}$$

Note:

Since there are only m equally likely events are left for B as the remaining (n – m) are unfavourable to B and as k events are in favour to B. Therefore the ratio $\frac{k}{m}$ represents conditional probability $P\left(\frac{B}{A}\right)$ assuming A has already occurred.

Now multiply and divide equation 3 by m, we get,

$$P(A \cap B) = \frac{k}{n} \times \frac{m}{m}$$

$$= \frac{k}{m} \times \frac{m}{n}$$

$$= P(A)P\left(\frac{B}{A}\right)$$

Hence proved

Generalisation

Theorem 2

For 3 events A, B and C

$$P(A \cap B \cap C) = P(A)P\left(\frac{B}{A}\right)P\left(\frac{C}{A \cap B}\right)$$

Proof:

$$P(A \cap B \cap C) = P[(A \cap B) \cap C] = P(A \cap B)P\left(\frac{C}{A \cap B}\right)$$

$$= P(A \cap B)P\left(\frac{B}{A}\right)P\left(\frac{C}{A \cap B}\right)$$

$$\left\{\text{Since } P(A \cap B) = P(A)P\left(\frac{B}{A}\right)\right\}$$

Corollary

If two events A and B are independent then $P(A \cap B) = P(A)\, P(B)$

The same can be extended to n mutually exclusive events $A_1, A_2, A_3, \ldots\ldots A_n$ i.e. $P(A_1 \cap A_2 \cap A_3 \cap \ldots\ldots \cap A_n) = P(A_1)\, P(A_2) \ldots\ldots P(A_n)$

Theorem 3:

If two events A and B are independent then their complements are also independent.

Proof:

Since we knew,

$$P(A \cup B) + P(A^c \cap B^c) = 1 \qquad \text{(Demorgan's law)}$$

$$\begin{aligned} P(A^c \cap B^c) &= 1 - P(A \cup B) \\ &= 1 - \{P(A) + P(B) - P(A \cap B)\} \\ &= 1 - P(A) - P(B) + P(A \cap B) \\ &= 1 - P(A) - P(B) + P(A) \times P(B) \\ &= 1 - P(A\,) - P(B)\,\{1 - P(A)\} \\ &= \{(1 - P(A)\,(1 - P(B)\} \\ &= P(A^c) \times P(B^c) \end{aligned}$$

Hence proved

Illustrations 1

A bag contains 9 red and 5 black balls, 2 balls are drawn at random one after the other with replacement. What is the probability that the 2nd ball drawn is black, knowing that the first ball drawn is also black?

Solution:

Let A = {the event that the 1st ball drawn is Black}

B = {The event that the 2nd ball drawn is black}

Since total number of balls are = 9 + 5 = 14

Therefore, the number of ways 2 balls can be drawn from a bag containing 14 balls is $^{14}C_2$ ways.

The number of ways of drawing 2 black balls out of 5 Black balls is $^{5}C_2$ ways.

Therefore,

$$P(A \cap B) = \frac{^5C_2}{^{14}C_2} = \frac{10}{91}$$

But the probability of drawing the 1st black ball is $P(A) = \frac{^5C_1}{^{14}C_1} = \frac{5}{14}$

From conditional probability

We knew,

$$P(A \cap B) = P(A)P\left(\frac{B}{A}\right)$$

$$P\left(\frac{B}{A}\right) = \frac{P(A \cap B)}{P(A)} = \frac{\frac{10}{91}}{\frac{5}{14}} = \frac{28}{91}$$

Illustration No. 2

Two fair dices are rolled, find the probability of getting a total of 10 when it is known that the second dice will show a smaller number than the 1st die?

Solution:

Let A be the event that the sum be 10 in 2 dice the combinations are {(4, 6), (5, 5) and (6, 4)}

And B be the event where the 1st die scores over the 2nd

i.e. (2, 1), (3, 1), (4, 1), (5, 1), (6, 1), (3, 2), (4, 2), (5, 2), (6, 2), (4, 3), (5, 3), (6, 3), (5, 4), (6, 4), (6, 5).

So, there are 15 number of events in favour i.e. 1st scoring over the 2nd.

Again the points are in common P(A ∩ B) =(6, 4) = 1 point

$$P(A \cap B) = \frac{1}{36}$$ (Since total number of events are 6^2 = 36)

And $$P(B) = \frac{15}{36}$$

Therefore $$P\left(\frac{A}{B}\right) = \frac{P(A \cap B)}{P(B)} = \frac{\frac{1}{36}}{\frac{15}{36}} = \frac{1}{15}$$

Illustration No. 3

In a lot containing 12 bags it was found that 4 are defective. If 4 bags are drawn one after the other without replacement and then find the probability that all the four are defective?

Solution:

Let us assume that

A : the event that the 1st drawn bag is a defective.

B : the event that the 2nd drawn bag is a defective item.

C : the event even the 3rd drawn bag is also defective.

D : the event that the 4th drawn bag is also defective.

Since the bags are drawn one after the another without replacement and hence all four events are not independent.

Hence by conditional probability

$$P(A \cap B \cap C \cap D) = P(A)\, P\left(\frac{B}{A}\right) P\left(\frac{C}{A \cap B}\right) P\left(\frac{D}{A \cap B \cap C}\right)$$

$$= \frac{4}{12} \times \frac{3}{11} \times \frac{2}{10} \times \frac{1}{9} = \frac{1}{495}$$

Illustration No. 4

Two groups A and B are competing against each to win the local body election. The probability that A will win is 0.7 and B's is 0.3. As per the promise made by both the groups that if elected to power will restructure the functioning of the local body. The probability that group A can introduce restructuring process is 0.7 and that of the competing group B is 0.40. Find what is the probability that the restructuring process will be introduced?

Solution:

Let us assume that

A : the event that group A is elected to power.

B : the event that the group B is elected to power.

C : the event that the restructuring is introduced.

The event C can only happen when one of the mutually exclusive event $A \cap C$ and $B \cap C$ happens.

Therefore

$$P(C) = (A \cap C) + (B \cap C)$$

But $P(A) = 0.7$, $P(B) = 0.3$ and $P\left(\frac{C}{A}\right) = 0.7$ and $P\left(\frac{C}{B}\right) = 0.4$

Since $P(A \cap C) = P(A)\,P\left(\frac{C}{A}\right)$

$$P(B \cap C) = P(B)\,P\left(\frac{C}{B}\right)$$

Therefore

$$P(C) = P(A)\,P\left(\frac{C}{A}\right) + P(B)\,P\left(\frac{C}{B}\right)$$

$$= 0.7 \times 0.7 + 0.3 \times 0.4$$

$$= 0.49 + 0.12 = 0.61$$

Illustration No. 5

Abhishek has received call letters from 2 firms X and Y each having one opening. But for X's opening 3 candidates were short listed and for Y's opening 2 were shortlisted. Find what is the probability that Abhishek at least getting selected in one post?

Solution:

Let us assume that

A : the event that Abhishek's selection for X's opening.

B : the event that Abhishek's selection for Y's opening.

Since the selection of each candidate is equally likely, therefore,

$$P(A) = \frac{1}{3} \Rightarrow P(A^c) = \frac{2}{3}$$

And $P(B) = \frac{1}{2} \Rightarrow P(B^c) = \frac{2}{3}$

Hence Abhishek's selection for atleast one job is given by

$$P(A \cup B) = 1 - P(A^c \cap B^c)$$
$$= 1 - P(A^c)\, P(B^c)$$

Since A, B are independent events

Therefore, $$P(A \cup B) = 1 - \frac{2}{3} \times \frac{1}{2} = \frac{2}{3}$$

Illustration No. 6

The probability that a management trainee will stay with an organization is 0.70 and that the probability of an employee working for the firm will earn Rs 5,000 p.m is 0.5. And the employee, who is also a management trainee and will continue with the firm is 0.8. Find the probability that the employee earns Rs 5,000 p.m and also a management trainee?

Solution:

Let us define each events

A : the event that the management trainee stays with the firm.

B : The event that the employee will earn Rs 5,000.

Therefore the probability concerned of $P(A) = 0.70$ and $P(B) = 0.5$ and the probability of employee a management trainee continued with the firm i.e $P(A \cup B) = 0.8$

Since

$$P\left(\frac{B}{A}\right) = \frac{P(A \cap B)}{P(A)}$$

But from Addition theorem we know

$$P(A \cup B) = P(A) + P(B) - P(A \cap B)$$

Or

$$P(A \cap B) = P(A) + P(B) - P(A \cup B)$$
$$= 0.7 + 0.5 - 0.8 = 0.4$$

Therefore,

$$P\left(\frac{B}{A}\right) = \frac{P(A \cap B)}{P(A)}$$

$$= \frac{0.40}{0.70} = 0.57$$

Illustration No. 7

In a game X, Y and Z toss a coin successively, whoever throws head first wins the game. If X starts throwing first then Y and Z follow. Then find their respective probability of winning the game?

Solution:

Let

A : the event that X throws head in a toss.

B : the event that Y throws a head in a toss.

C : the event that Z throws a head in a toss.

Since all the 3 events are independent, therefore

$$P(A) = \frac{1}{2}, \quad P(B) = \frac{1}{2} \quad \text{and} \quad P(C) = \frac{1}{2}$$

But if A starts the game he can in the following mutually exclusive way i.e. he will win in the 1st throw or his second chance appears only if Y and Z fails to throw head in their respective chances and the pattern continues till A throws head

i.e. $A, A^c \cap B^c \cap C^c \cap A, A^c \cap B^c \cap C^c \cap A^c \cap B^c \cap C^c \cap A$.......

By Addition theorem, the probability that X wins is given by

$$P(A) + P(A^c \cap B^c \cap C^c \cap A) + P(A^c \cap B^c \cap C^c \cap A^c \cap B^c \cap C^c \cap A)....$$

as $P(A^c) = \frac{1}{2}, \quad P(B^c) = \frac{1}{2}$ and $P(C^c) = \frac{1}{2}$

$$= \frac{1}{2} + \left(\frac{1}{2}\right)^3 \frac{1}{2} + \left(\frac{1}{2}\right)^6 \frac{1}{2} ....$$

$$= \frac{1}{2}\left\{1 + \left(\frac{1}{2}\right)^3 + \left(\frac{1}{2}\right)^6 ....\right\}$$

$$= \left(\frac{1}{2}\right)\left\{\frac{1}{1 - \left(\frac{1}{2}\right)^3}\right\} = \frac{4}{7}$$

Similarly the winning chances of Y are

i.e $A^c \cap B,\ A^c \cap B^c \cap C^c \cap A^c \cap B$

$$= P(A^c \cap B) + P(A^c \cap B^c \cap C^c \cap A^c \cap B) +$$

$$= P(A^c) \times P(B) + P(A^c) \times P(B^c) \times P(C^c) \times P(A^c) \times P(B)$$

$$= \left\{\frac{1}{2} \times \frac{1}{2} + \left(\frac{1}{2}\right)^4 \frac{1}{2} + \ldots\ldots\ldots\ldots\right\}$$

$$= \left(\frac{1}{2}\right)^2 \left\{1 + \left(\frac{1}{2}\right)^3 + \ldots\ldots\ldots\ldots\right\}$$

$$= \left(\frac{1}{4}\right)\left\{\frac{1}{1-\left(\frac{1}{2}\right)^3}\right\} = \frac{2}{7}$$

Similarly Z's winning chances are $= 1 - \left(\frac{4}{7} + \frac{2}{7}\right) = \frac{1}{7}$

Illustration No. 8

The odds in favour of Manish solving a problem in probability is 7: 5 and the odds against Ankur is 5 : 4. If both of them attempt the problem independently then,

(i) Find the probability that the problem will be solved

(ii) Neither of them could solve the problem.

Solution:

Let us define

A : The event that Manish solves the problem.

B : The event that Ankur solves the problem.

But the odds in favour of Manish = 7 : 5 and odds against Ankur = 5:4

Note: {Odds in favour = No. of favourable events: No. of unfavourable events,

Odds in against = No. of unfavourable events: No. of favourable events}.

Therefore $P(A) = \frac{7}{12},\ P(B) = \frac{4}{9}$

(i) the required probability that the problem will be solved if they try independently of each other is the event P(A U B)

By addition theorem

$$P(A \cup B) = P(A) + P(B) - P(A \cap B)$$

But $P(A \cap B) = P(A) \times P(B)$ (Since the events are independent)

$$= \frac{7}{12} \times \frac{4}{9} = 0.26$$

Therefore

$$P(A \cup B) = \frac{7}{12} + \frac{4}{9} - \frac{7}{27} = 0.73 \text{ (Approx).}$$

(ii) The required probability that neither solves the problem = 1 – {Probability that at least one of Manish or Ankur Solves the problem}

$$= 1 - \{ P(A \cup B) \} = 1 - 0.77 = 0.23.$$

Illustration No. 9

A bank locker can be opened only when the combination key X and Y function properly. If the probability that the key X fails is 0.20 and that of Y is 0.10. What is the probability that the locker could not be opened because of failed key operation?

Solution:

Let us define

A : the event that key X fails.

B : the event that key Y fails.

Since the events A and B are independent and

$$P(A) = 0.20 \text{ and } P(B) = 0.10$$

$$P(A^c) = 0.80 \text{ and } P(B^c) = 0.90$$

The probability that the locker will fail to open because of failed operation of key X and Y is given by

$$P(A \cup B) = 1 - P(A^c \cap B^c) = 1 - P(A^c) \times P(B^c) = 1 - 0.80 \times 0.90 = 0.28$$

Illustration No.10

In a final semester examination it was observed that 35 % has failed in accounts paper and 45% has failed in managerial economics and the those who have failed in at least one of the papers are 60%.If a student is selected at random.

(i) Find what is the probability that the student has failed in Managerial economics paper?

(ii) And that he has passed Accounts paper knowing that the he failed in Managerial economics paper?

Solution:

Let us define

A : the event that the student has failed in Accounts Paper.

B : the event that the students has failed in managerial economics Paper.

Therefore $P(A) = 0.35$, $P(B) = 0.45$ and $P(A \cup B) = 0.60$

But we know by addition rule

$$P(A \cup B) = P(A) + P(B) - P(A \cap B)$$

$$0.60 = 0.35 + 0.45 - P(A \cap B)$$

Therefore

$$P(A \cap B) = 0.80 - 0.60 = 0.20$$

Hence the probability that the selected student has failed in Managerial Economics

$$P(A^c \cap B) = P(B) - P(A \cap B) = 0.45 - 0.20 = 0.25$$

In the second case the required probability of finding the conditional probability of passing accounts paper knowing that failing in Managerial economics paper i.e.

$$P\left(\frac{A^c}{B}\right) = \frac{P(A^c \cap B)}{P(B)} = \frac{0.25}{0.45} = 0.55$$

Bayes' Theorem

(Developed by Rev. Thomas Bayes)

Statement

Let B_i's were i = 1, 2, 3, ...n be n mutually exclusive and disjoint events with the probability of $P(B_i$' s) $\neq 0$ then for any arbitrary event A which is a subset of $B_1 \cup B_2 \cup B_3$...............$\cup B_n$ such that $P(A) > 0$ then we have

$$P\left(\frac{B_i}{A}\right) = \frac{P(B_i)P\left(\frac{A}{B_i}\right)}{\sum_{i=1}^{n} P(B_i)P\left(\frac{A}{B_i}\right)}$$

Proof:

Let B_1, B_2, B_3..............B_n are mutually exclusive and exhaustive events such that A occurs only when one of B_1, B_2, B_3..............B_n occurs i.e A occurs in combination with any one of the mutually exclusive event

$(A \cap B_1)$, $(A \cap B_2)$,...... $(A \cap B_n)$ where all $(A \cap B_1)$, $(A \cap B_2)$,...... $(A \cap B_n)$ are disjoint.

So, we can write,

$$A = (A \cap B_1) \cup (A \cap B_2) \cup \ldots\ldots \cup (A \cap B_n)$$

By applying Additional Rule of Probability, we get

$$P(A) = P(A \cap B_1) + P(A \cap B_2) + \ldots\ldots.. + P(A \cap B_n) \quad \ldots\ldots(i)$$

Since $P(A \cap B_i) = P(B_i \cap A)$

Therefore eq. (i) can be rewritten as

$$P(A) = P(B_1 \cap A) + P(B_2 \cap A) + \ldots\ldots.. + P(B_n \cap A) \quad \ldots\ldots(ii)$$

Again by Conditional Probability Rule

$$P(B_i \cap A) = P(B_i)P\left(\frac{A}{B_i}\right)$$

Therefore,

$$P(A) = P(B_1)P\left(\frac{A}{B_1}\right) + P(B_2)P\left(\frac{A}{B_2}\right) + P(B_3)P\left(\frac{A}{B_3}\right) + \ldots\ldots\ldots\ldots$$

$$P(A) = \sum_{i=1}^{n} P(B_i)P\left(\frac{A}{B_i}\right) \quad \ldots\ldots(iii)$$

But

$$P(A \cap B_i) = P(A)P\left(\frac{B_i}{A}\right)$$

$$P(B_i \cap A) = P(A)P\left(\frac{B_i}{A}\right)$$

(Since $P(A \cap B_i) = P(B_i \cap A)$)

$$P(B_i)P\left(\frac{A}{B_i}\right) = P(A)P\left(\frac{B_i}{A}\right)$$

$$\Rightarrow \quad P\left(\frac{B_i}{A}\right) = \frac{P(B_i)P\left(\frac{A}{B_i}\right)}{P(A)}$$

Writing the value of P(A) from eq. (iii)

$$\Rightarrow \quad P\left(\frac{B_i}{A}\right) = \frac{P(B_i)P\left(\frac{A}{B_i}\right)}{\sum_{i=1}^{n} P(B_i)P\left(\frac{A}{B_i}\right)}$$

Hence proved.

Simplified Version

An event A can only occur if and only if one of the mutually exclusive and exhaustive event B_1, B_2, B_3.........B_n occur where the individual probability of $P(B_1)$, $P(B_2)$........$P(B_n)$ are known so does the conditional probability of $P(A/B_i)$ where i = 1,2,3.......n

Then the conditional probability of $P(B_i/A)$ where A has said to have occurred is given by

$$\Rightarrow \quad P\left(\frac{B_i}{A}\right) = \frac{P(B_i)P\left(\frac{A}{B_i}\right)}{\sum_{i=1}^{n} P(B_i)P\left(\frac{A}{B_i}\right)}$$

Solved Exercises

Illustration No. 1

Out of three bags each containing 4 white and 4 black balls, 5 white and 3 black balls and 6 white and 7 black balls, one bag is chosen at random and a ball is drawn and found to be white. What is the probability that the ball drawn is from 1st bag?

Solution:

Let A be the event of drawing a white ball and B_i = {i^{th} bag is chosen} where i = 1, 2 and 3

Therefore, the individual probabilities of selecting the bags are

$$P(B_1) = \frac{1}{3},\ P(B_2) = \frac{1}{3},\ \text{and } P(B_3) = \frac{1}{3}$$

But, the conditional probability of drawing a white ball i.e. P(A/B_i) from each of the bags is

$$P\left(\frac{A}{B_1}\right) = \frac{^4C_1}{^8C_1} = \frac{1}{2}$$

$$P\left(\frac{A}{B_2}\right) = \frac{^5C_1}{^8C_1} = \frac{5}{8}$$

and

$$P\left(\frac{A}{B_3}\right) = \frac{^6C_1}{^{13}C_1} = \frac{6}{13}$$

Therefore, the probability of drawing a white from the 1st bag i.e. the conditional probability $P\left(\frac{B_1}{A}\right)$,

Using Baye's Rule

$$P\left(\frac{B_1}{A}\right) = \frac{P(B_1)P\left(\frac{A}{B_1}\right)}{P(B_1)P\left(\frac{A}{B_1}\right) + P(B_2)P\left(\frac{A}{B_2}\right) + P(B_3)P\left(\frac{A}{B_3}\right)}$$

$$= \frac{\frac{1}{3}\times\frac{1}{2}}{\frac{1}{3}\times\frac{1}{2}+\frac{1}{3}\times\frac{5}{8}+\frac{1}{3}\times\frac{6}{13}} = 0.307$$

Illustration No. 2

A class consists of 30 students and the ratio of boys to girls is 2:1. It was also found that among the students 4 boys and 3 girls are rank holders.

A student is selected at a random was found to be a rank holder. What is the probability that the student selected is a boy?

Solution:

Let A_1 = The event that the chosen student is a boy

A_2 = The event that the chosen student is a girl and

B = The event that the student selected is a rank holder.

Since the total number of students in the class is 30 and bifurcated in the ratio of 2:1.

Therefore, the number of boys $= 30 \times \frac{2}{3} = 20$,

the number of girls $= 30 \times \frac{1}{3} = 10$

The probability of selecting a boy student $= P(A_1) = \frac{20}{30} = \frac{2}{3}$,

The probability of selecting a girl student $= P(A_2) = \frac{10}{30} = \frac{1}{3}$,

And the probability of selecting a rank holder and boy student by the conditional probability rule

$$P\left(\frac{B}{A_1}\right) = \frac{4}{20} = \frac{1}{5}$$

(Since the number of rank holder boys are 4 in a total of 20)

The probability of selecting a rank holder and girl student $P\left(\frac{B}{A_2}\right) = \frac{3}{10}$

(Since the number of rank holder girls are 3 in a total of 10)

Therefore the conditional probability of choosing a student a boy and a rank holder $P\left(\frac{A_1}{B}\right)$

By Baye's Rule

$$P\left(\frac{A_1}{B}\right) = \frac{P(A_1)P\left(\frac{B}{A_1}\right)}{P(A_1)P\left(\frac{B}{A_1}\right) + P(A_1)P\left(\frac{B}{A_2}\right)}$$

$$= \frac{\frac{2}{3} \times \frac{1}{5}}{\frac{2}{3} \times \frac{1}{5} + \frac{1}{3} \times \frac{3}{10}} = \frac{4}{7}$$

Illustration No. 3

A manufacturing firm uses two different machines M_1 and M_2 for producing an article. The level of output of M_1 is 40% and that of M_2 is 60% of the gross output. It was also observed that 6% and 4% of the output produced by the machines are defective. An article selected at random from the lot was found defective. What is the probability that the defective has come from the M_2?

Solution:

Let A_1 = The event that the article is selected from M_1

A_2 = The event that the article is selected from M_2

B = The event that the article is defective.

The probability of selecting an article from M_1 = $P(A_1) = \frac{40}{100} = 0.40$

The probability of selecting an article from M_2 = $P(A_2) = \frac{60}{100} = 0.60$

And the conditional probability of selecting a defective article from M_1

$$P\left(\frac{B}{A_1}\right) = \frac{6}{10} = 0.06$$

And the conditional probability of selecting a defective article from M_2

$$P\left(\frac{B}{A_2}\right) = \frac{4}{10} = 0.04$$

Therefore the probability that of selecting an article from machine M_2 and defective

The required probability using Bayes Rule is

$$P\left(\frac{A_2}{B}\right) = \frac{P(A_2)P\left(\frac{B}{A_2}\right)}{P(A_1)P\left(\frac{B}{A_1}\right) + P(A_2)P\left(\frac{B}{A_2}\right)}$$

$$= \frac{0.60 \times 0.04}{0.60 \times 0.04 + 0.40 \times 0.06} = 0.50$$

Illustration No. 4

A corporate is decided to avail the services either of management consultant X, Y and Z for their upcoming project and their respective probability of selection are 0.3, 0.4 and 0.3. But it was noticed that the disaster management capabilities of the consultants are 0.7, 0.6 and 0.4 and the corporate has executed the project and faced a bottleneck in the project after initializing the process, then find the probability that X's SELECTION for disaster management?

Solution:

Let A_1 = The event that X is chosen as consultant

A_2 = The event that Y is chosen as consultant

A_3 = The event that Z is chosen as consultant

B = The event expressing the disaster management abilities.

The individual probability of selection

$$P(A_1) = 0.3,\ P(A_2) = 0.4 \text{ and } P(A_3) = 0.3$$

And the probability of disaster management abilities of X, Y and Z are

$$P\left(\frac{B}{A_1}\right) = 0.70,\ P\left(\frac{B}{A_2}\right) = 0.60 \text{ and } P\left(\frac{B}{A_3}\right) = 0.40$$

Therefore, the conditional probability of X's selection for disaster management is given by

Using Baye's Rule

$$P\left(\frac{A_1}{B}\right) = \frac{P(A_1)P\left(\frac{B}{A_1}\right)}{P(A_1)P\left(\frac{B}{A_1}\right) + P(A_2)P\left(\frac{B}{A_2}\right) + P(A_3)P\left(\frac{B}{A_3}\right)}$$

$$= \frac{0.30 \times 0.70}{0.30 \times 0.70 + 0.40 \times 0.60 + 0.30 \times 0.40} = 0.36$$

Illustration No. 5

A newly installed machine has a chance of breakdown because of faulty design and the chance that the fault exists is 12%. The chance that the machine may breakdown with faulty design is 80% and otherwise 60%. It was observed that the machine experienced breakdown. Find the probability that the breakdown is due to faculty design?

Solution:

Let A_1 = The event that the machine design is faulty

A_2 = The event that the machine design is faultless

B = The event that the machine has experienced breakdown.

Therefore the probability that the machine design is faulty

$= P(A_1) = 0.12$

And that of faultless design = $P(A_2) = 0.88$

And the associated conditional probability for both events are

$$P\left(\frac{B}{A_1}\right) = 0.80 \text{ and } P\left(\frac{B}{A_2}\right) = 0.60$$

Therefore the probability of design fault causing breakdown is given by

$$P\left(\frac{A_1}{B}\right) = \frac{P(A_1)\,P\left(\frac{B}{A_1}\right)}{P(A_1)\,P\left(\frac{B}{A_1}\right) + P(A_2)\,P\left(\frac{B}{A_2}\right)}$$

$$= \frac{0.12 \times 0.80}{0.12 \times 0.80 + 0.88 \times 0.60} = 0.153$$

Illustration No. 6

A book publisher has launched a book on Management found that the book sales are not exciting. On analyzing found that the quality of print and poor presentation of materials are the reasons for the sales not picking up and their respective probability of discontent with both the factors are 0.60 and 0.40 respectively. It was also evaluated that the probability of students going away on account of these factors are 0.70

and 0.60. The publisher wants to know if a customer is discontented, what is the probability that is due to poor presentation of materials?

Solution:

Let A_1 = The event that the quality of print contributing to falling sales.

A_2 = The event that the poor presentation contributing to falling sales.

B = The event that the students moving away having experienced discontent.

Therefore the individual probability

$= P(A_1) = 0.60$ and that of $P(A_2) = 0.40$.

And the associated conditional probability for both events i.e students moving away because of poor quality is

$P\left(\frac{B}{A_1}\right) = 0.70$ and that of poor materials is $P\left(\frac{B}{A_2}\right) = 0.60$

Therefore the probability of poor material causing student's discontent is given by

$$P\left(\frac{A_2}{B}\right) = \frac{P(A_2)P\left(\frac{B}{A_2}\right)}{P(A_1)P\left(\frac{B}{A_1}\right) + P(A_2)P\left(\frac{B}{A_2}\right)}$$

$$= \frac{0.40 \times 0.60}{0.40 \times 0.60 + 0.60 \times 0.70} = 0.36$$

Illustration No.7

A General Insurance Co. has insured 500 people with Medical Insurance policy from a residential society with Male to Female ratio being 3:2. If the probability that a male falling sick is 0.10 and that of female is 0.11, find the probability that if an insured person falls sick is from female gender?

Solution:

Let us assume

A_1 = The event of choosing an insured male

A_2 = The event of choosing an insured female

B = The event that an person falls sick

Since the ratio Male to female population is 3:2,

therefore male population = 300 and that of female = 200

Hence the associated probabilities are $P(A_1) = \frac{3}{5}$, $P(A_2) = \frac{2}{5}$.

But $$P\left(\frac{B}{A_1}\right) = 0.10 \text{ and } P\left(\frac{B}{A_2}\right) = 0.11,$$

Using Bayesian Theorem

$$P\left(\frac{A_2}{B}\right) = \frac{P(A_2)P\left(\frac{B}{A_2}\right)}{P(A_1)P\left(\frac{B}{A_1}\right) + P(A_2)P\left(\frac{B}{A_2}\right)}$$

$$= \frac{0.40 \times 0.11}{0.60 \times 0.10 + 0.40 \times 0.11} = 0.423$$

Illustration No. 8

Animesh speaks truth 3 out of 4 counts. On a certain occasion he has reported in a friend circle that seeing Aamir Khan shooting for his upcoming film Peepli Live and the probability that he actually saw Aamir shooting and telling truth is 0.60 and lying was 0.40. Find the probability that he was actually speaking the truth?

Solution:

A_1 = the event that Animesh speaks truth.

A_2 = The event of that Animesh lies.

B = The event that he seen Aamir Khan shooting.

Since he speaks 3 out 4 times, hence the associated probabilities are

$$P(A_1) = \frac{3}{4} = 0.75,\ P(A_2) = \frac{1}{4} = 0.25$$

But the probability that watching Aamir shooting for Peepli live and Animesh also telling truth $= P\left(\frac{B}{A_1}\right) = 0.60$,

And Animesh lying watching Aamir shoot = $P\left(\frac{B}{A_2}\right) = 0.40$

There we need to find that Animesh is actually speaking truth that he has indeed saw Aamir shoot i.e. $P\left(\frac{A_1}{B}\right)$

Using Bayesian Theorem

$$P\left(\frac{A_1}{B}\right) = \frac{P(A_1)\,P\left(\frac{B}{A_1}\right)}{P(A_1)\,P\left(\frac{B}{A_1}\right) + P(A_2)\,P\left(\frac{B}{A_2}\right)}$$

$$= \frac{0.75 \times 0.60}{0.75 \times 0.60 + 0.25 \times 0.40} = 0.818$$

Illustration No. 9

In a mega event stampede has caused heavy injuries among the visitors and the investigating team set has reported that the result of stampede is heavy influx and lack proper infrastructure during the event. The contribution of heavy influx is 60% and that of infrastructure deficiencies is 40% for stampede and the associated probabilities for causing injuries by heavy influx is 0.30 and that of because poor infrastructure is 0.20. Find the probability that if the injury has resulted then it is because of heavy influx?

Solution:

A_1 = stampede resulted because of heavy influx.

A_2 = stampede resulted because of Poor Infrastructure.

B = Visitor received injuries.

Since the associated probabilities are $P(A_1) = 0.60$, $P(A_2) = 0.4$

But the conditional probabilities of injuries resulting from of the reasons as mentioned

$$P\left(\frac{B}{A_1}\right) = 0.30 \quad \text{and} \quad P\left(\frac{B}{A_1}\right) = 0.20$$

Therefore the probability of heavy influx causing stampede $P\left(\frac{A_1}{B}\right)$

Using Bayesian Theorem

$$P\left(\frac{A_1}{B}\right) = \frac{P(A_1)\,P\left(\frac{B}{A_1}\right)}{P(A_1)\,P\left(\frac{B}{A_1}\right) + P(A_2)\,P\left(\frac{B}{A_2}\right)}$$

$$= \frac{0.60 \times 0.30}{0.60 \times 0.30 + 0.40 \times 0.20} = 0.69$$

Illustration No. 10

In a book exhibition, it was reported that 40% of the visitors have actually made purchases. It came to light that non students' community has made the major purchases accounting to 60% among the purchasers. Also, that the visitors who made no purchase in them non students were 30%. If a visitor who visits the exhibition found to be non student, find the probability that he will actually purchase a book?

Solution:

Let

A_1 = the event where the visitor actually purchases the book

A_2 = The event where the visitor does not purchases the book

B = The visitor is a non student.

Since

$$P(A_1) = 40/100 = 0.4,\ P(A_2) = 1 - 0.40 = 0.60$$

But

$$P\left(\frac{B}{A_1}\right) = 0.60 \Leftrightarrow P\left(\frac{B^c}{A_1}\right) = 0.40$$

$$P\left(\frac{B}{A_2}\right) = 0.30 \Leftrightarrow P\left(\frac{B^c}{A_2}\right) = 0.70$$

The probability of finding a non student actually purchased book is $P(A_1/B)$

Using Bayesian Theorem

$$P\left(\frac{A_1}{B}\right) = \frac{P(A_1)\,P\left(\frac{B}{A_1}\right)}{P(A_1)\,P\left(\frac{B}{A_1}\right) + P(A_2)\,P\left(\frac{B}{A_2}\right)}$$

$$= \frac{0.40 \times 0.60}{0.40 \times 0.60 + 0.60 \times 0.30} = 0.57$$

Brief Summary

1. "Probability can be defined as the chance of an event to happen".
2. A experiment is a random experiment if performed under similar conditions and the outcome of the experiment is not unique but one among the several alternatives available.
3. **Equally Likely Events**: Events are equally likely taking all factors into consideration there is no reason that any one of the event will be preferred over the other.
4. **Mutually Exclusive Events**: when one event prevents the occurrence or happening of the other i.e. two events cannot happen simultaneously.
5. **The Classical Approach**

   According to Classical Approach probability is defined as the ratio of favorable events to total number of all possible events

$$\text{Probability} = \frac{\text{Favourable Event}}{\text{Total Number of all Possible Events}}$$

6. **Axiomatic Approach:**

   (Developed by A.N Kolmogorov, Russian mathematician)

In Axiomatic approach had drawn largely from theoretical aspects of set theory and introduced probability as a set function.

It can be understood as the probability of occurrence of an event A, given a sample space S such that A subset of S then we define probability function P (A) on S if the following axioms are satisfied

(i) $P(A) \geq 0$.

(ii) P(S) = 1 and

(iii) If A_1, A_2,......are sequentially mutually exclusive events in S, then $P(A_1 \cup A_2 \cup \ldots\ldots\ldots\ldots) = P(A_1) + P(A_2) + \ldots\ldots\ldots\ldots$

7. Addition Rule

If A and B are two events and they are not necessarily mutually exclusive, then the probability of occurrence of atleast one of the two events A and B i.e., P(A U B) is given by

$$P(A \cup B) = P(A) + P(B) - P(A \cap B)$$

8. Conditional Probability

The probability of simultaneous occurrence of two events A and B is given by the product of the unconditional probability of the event A by the conditional probability of B, with the assumption that A has already occurred

Symbolically

$$P(A \cap B) = P(A)\, P\left(\frac{B}{A}\right) \text{ or } P(B \cap A) = P(B)\, P\left(\frac{A}{B}\right)$$

9. Bayes' Theorem

(Developed by Rev. Thomas Bayes)

Statement: Let B_i's where I = 1, 2, 3, ...n be n mutually exclusive and disjoint events with the probability of $P(B_i\text{'s}) \neq 0$ then for any arbitrary event A which is a subset of $B_1 \cup B_2 \cup B_3 \ldots\ldots\ldots\ldots \cup B_n$ s.t $P(A)>0$ then we have

$$\Rightarrow \qquad P\left(\frac{B_i}{A}\right) = \frac{P(B_i)P\left(\frac{A}{B_i}\right)}{\sum_{i=1}^{n} P(B_i)P\left(\frac{A}{B_i}\right)}$$

Simplified Version

An event A can only occur if and only if one of the mutually exclusive and exhaustive event B_1, B_2, $B_3 \ldots\ldots\ldots B_n$ occur where the individual probability of $P(B_1)$, $P(B_2)$........$P(B_n)$ are known so does the conditional probability of $P(A/B_i)$ where i= 1, 2, 3.......n

Then the conditional probability of $P(B_i/A)$ where A has said to have occurred is given by

$$\Rightarrow \qquad P\left(\frac{B_i}{A}\right) = \frac{P(B_i)P\left(\frac{A}{B_i}\right)}{\sum_{i=1}^{n} P(B_i)P\left(\frac{A}{B_i}\right)}$$

Review Questions

1. Define the following terms
   (a) Exhaustive Event
   (b) Favourable Event
   (c) Dependent and Independent Event
   (d) Mutually exclusive Event.
2. State and prove Additional theorem of probability?
3. Define the term probability and explain the significance using an example?
4. Write the difference between
   (a) Simple event and Compound event
   (b) Trail and Event
5. Write on how probability is explained in terms of Classical and empirical approach? Which stands superior over the other. Explain?
6. State and prove Conditional theory of probability?
7. State and explain Axiomatic Approach of Probability?
8. State and Prove Bayesian Theorem? And explain its significance?

Self Practice Exercises

1. If 3 coins are tossed, what is the probability of getting

(i) 2 Heads and 1 tail?

(ii) neither 3 Heads nor 3 tails?

2. If 2 fair dice are thrown, what is the probability that the sum of the no's appearing is a prime number?

3. A bag contains 4 green balls and 6 white balls 2 balls are drawn at random what is the probability that

(i) Both are of different colors?

(ii) Both are green.

4. A card is drawn from a well shuffled pack of cards. Find the probability that

(i) It is a card of hearts?

(ii) It is not a spade?

5. What is the probability that a non leap year will contain 53 Sundays?

6. The odds in favour of solving a particular problem by A is 3 : 4 and odds against B is 2 : 3. What is the probability that if both have attempted the problem, then the problem is solved?

7. A committee of 4 needs to be formed among 2 Mathematician, 4 Economists and 3 Statistician. Find the probability that the committee consists of

(i) 2 Economists?

(ii) Atleast 1 Mathematician?

(iii) One from each field?

8. In a particular business venture A is to make a profit of Rs. 500 with the probability of 0.7 and loss of Rs.200 with a probability of 0.3. Find A's expectation?

9. In rolling 2 dices - find the probability that

(i) There is at least a six? 8?

(ii) Sum is greater than equal to 8?

10. A company with a staff strength of 20, out of which 6 are professionally qualified and rest are none. What is the probability that if 4 are hand picked?

(i) They are all professionals?

(ii) At least 2 are professional?

11. A problem is given to 3 students X, Y and Z with their respective probability of solving are 2/3, $^3/_4$ and 2/5. What is the probability that if all the 3 attempt the question, the problem gets solved?

12. If P(A) = 2/3, P(B) = $^1/_4$ and P(A ∩ B) = 3/5 find P(A^c U B^c) = ?

13. There are 7 persons sitting in a circular formation. What is the probability that 2 particular persons are always together?

14. Out of 1 to 100 no's – one is selected at random. What is the probability that the number is either divisible by 5 or 7?

15. In a game of 2 dice the probability that A will lose if his first throw is either 2, 6 or 10 and win if throws 5 or 7. Find the ratio between his probability of losing and winning in the 1st throw?

16. In a lot of 20 articles – 6 are found to be defective. If 5 articles are drawn at random, what is the probability that exactly 3 of the articles are defective?

17. The odds against 3 students Jyothi, Abhishek and Vimal appearing in an examination standing first are 1: 2, 2:3 and 1:5 respectively. What is the probability that either one of them will stand first in the examination?

18. A bag contains 20 balls – marked 1 to 20, a ball is drawn, find the probability that the ball drawn is a multiple of 3 or 5?

19. From a well shuffled pack of cards 3 cards are drawn at random. What is the probability that they are a king, a queen and an ace?

20. If Subbah and Aishwarya seeking admission into MBA program with their respective probabilities are 0.6 and 0.5. Find the probability that both of them gets selected?

    (*Hint:* Use Addition Rule)

21. What is the probability that 4 A's come consecutively in the word ALLAHABAD?

22. From a group of 8 men and 5 women a committee needs to 7 is formed. What is the probability that the committee has

    (i) Exact 4 men and 3 women?

    (ii) At least 4 women?

23. Mr X speaks truth in 2 out of 3 times and tossing a coin he has reported that head has appeared. Find the probability he is actually speaking a truth?

Hint: (Apply Bayesian theorem)

24. Find the probability that a leap year selected at random will contain either 53 Saturdays or 53 Sundays? (Apply Addition theorem):

25. A restaurant advertises about its service in cable network and local news paper estimating that there is a 15% chance that an individual will view the advertisement in television and a 20% chance that one will notice the Ad in the newspaper. If an individual is selected at random then what is the probability that he or she is exposed to at least one method of advertisement?

26. An Financial analyst predicts that the odds against that ITC's share will rise in the stock market is 2:5 and the odds in favour to SAIL steel to rise 2:3? What is the probability that the price of both the stock will rise? (Apply Addition theorem):

27. A bank is experiencing displeasure among its customers. On probing it was found that the service quality, product packaging and ambience are the contributing factor. Further analysis has revealed that the probabilities of displeasure with the three factors are 0.35, 0.40 and 0.25 respectively. On conducting customer meet, it was found that the probabilities of a customer going away on account of these factors are 0.5, 0.80 and 0.60. The service manager if wants to know if a customer is dissatisfied, what is the probability that it is due to service quality?

Hint: (Apply Bayesian theorem)

28. Three magazines X, Y and Z are published in Lucknow. In a survey the readership for the following magazines among the population is 25% read magazine X, 18% read magazine Y and 25% read magazine Z, 7% read both magazine X and Y, 6% read magazine both Y and Z , 5% read magazine both X and Z and the readership all the three are 2%. Find the probability what percentage of the population read non of the magazines?

Hint: *(Apply Set theory and Addition theorem)*

29. An investor predicts that the price of the stock will rise is 2/3 and to remain constant is 1/4. Find what is the probability that the price of the stock will go down? (*Apply Addition theorem*)?

30. 5 boys and 5 girls sit in a randomly in a row. Find the probability that all the girls sit together?

31. A bag contains 10 dolls of various colors of which 4 are red, 3 are blue and 3 yellow. If 4 dolls are drawn at random, find what is the probability that
    (i) 2 are Red & 2 are Yellow
    (ii) All red.
    (iii) One of each color

32. 10 people sit in a circular fashion. Find what is the probability that
    (i) Two particular persons always sit together?
    (ii) They are never sit together?

Answers

1. (i) $\frac{3}{8}$ (ii) $\frac{3}{4}$ 2. $\frac{5}{12}$

3. (i) $\frac{8}{15}$ (ii) $\frac{2}{15}$ 4. (i) $\frac{1}{4}$ (ii) $\frac{3}{14}$

5. $\frac{1}{7}$ 6. $\frac{27}{35}$

7. (i) $\frac{{}^4C_2 \times {}^7C_2}{{}^9C_4}$ (ii) $\frac{{}^2C_1 \times {}^8C_3 + {}^2C_2 \times {}^7C_2}{{}^9C_4}$ (iii) $\frac{{}^2C_1 \times {}^4C_1 \times {}^3C_1 \times {}^6C_1}{{}^9C_4}$

8. 290 9. (i) $\frac{11}{36}$ (ii) $\frac{5}{12}$

10. (i) $\frac{{}^6C_4}{{}^{20}C_4}$ (ii) $\frac{{}^6C_2 \times {}^{14}C_2 + {}^6C_3 \times {}^{14}C_1 + {}^6C_6}{{}^{20}C_4}$

11. $\frac{19}{20}$ 12. $\frac{41}{60}$

13. $\frac{1}{3}$ 14. 0.33

15. 9 : 10 16. $\frac{{}^{6}C_3 \times {}^{14}C_2}{{}^{20}C_5}$

17. $\frac{17}{90}$ 18. $\frac{9}{20}$

19. 0.002 20. 0.10

21. $\frac{1}{21}$

22. (i) $\frac{{}^{8}C_4 \times {}^{5}C_3}{{}^{13}C_7}$ (ii) $\frac{{}^{8}C_3 \times {}^{5}C_4 + {}^{8}C_2 \times {}^{5}C_5}{{}^{13}C_7}$

23. $\frac{2}{3}$ 24. $\frac{3}{7}$

25. 32% 26. $\frac{8}{21}$

27. 0.271 28. 48%

29. $\frac{1}{12}$ 30. $\frac{1}{42}$

31. (i) $\frac{{}^{4}C_2 \times {}^{3}C_2}{{}^{10}C_4}$ (ii) $\frac{{}^{4}C_4}{{}^{10}C_4}$ (iii) $\frac{{}^{4}C_1 \times {}^{3}C_1 \times {}^{3}C_1 \times {}^{7}C_1}{{}^{10}C_4}$

32. (i) $\frac{8!2!}{9!}$ (ii) $1-\left(\frac{8!2!}{9!}\right)$

CHAPTER 17

Theory of Probability - II

Random Variable

Imagine a coin is tossed or a die is rolled in both the cases the outcome of the experiment is one of the possible alternatives i.e. when a coin is tossed either a head or tail can appear. Each having associated probability of $\frac{1}{2}$. Similarly rolling a die with the sample space S {1, 2, 3, 4, 5, 6} each face of the die have equal chance of appearance i.e. $\frac{1}{6}$. Experiments whose outcomes are result of chance are random experiment and a variable whose value is a result of a random experiment is called a chance variable or a random variable.

Further a variable assuming discrete set of values i.e. tossing of a coin or casting of a die, then the variable is called as discrete random variable. On the contrary if a random variable assumes all real values within a given interval are the examples of continuous discrete variable.

Probability Function (or Density Function) of a Random Variable

If 'X' is a discrete random variable assuming a set of values x_1, x_2, x_3, ... x_n

Then the function defined by

$$f(x_i) = P(X = x_i) = p_i$$

i.e. the probability that 'X' takes the value $x_i = p_i$ is called the probability function or density function.

This mathematical function gives the corresponding probability values for the different values of $x_1, x_2, x_3, \ldots x_n$ of the variable 'X' and satisfies two essential conditions.

(i) $f(x_i) \geq 0$ for all i = 1, 2, 3, ... n

(ii) $\sum_{i=1}^{n} f(x_i) = 1$

Discrete Probability Distribution

If a discrete random variable can assume values of $x_1, x_2, x_3, \ldots x_n$ with the corresponding probabilities $p_1, p_2, p_3, \ldots p_n$ such that $\sum_{i=1}^{n} p_i = 1$ then the occurrence of the values x_i (i = 1, 2, ... n) with the probabilities p_i (i = 1, 2, ... n) is called the discrete probability distribution.

For example

Suppose when a die is rolled any one of the values between 1 to 6 can appear with the individual probabilities being $\frac{1}{6}$.

Thus sum of the probabilities $\sum_{i=1}^{n} p_i = 1 = \frac{1}{6} + \frac{1}{6} + \frac{1}{6} + \frac{1}{6} + \frac{1}{6} + \frac{1}{6} = 1$

Therefore discrete probability distribution is the values of x with their corresponding probabilities.

x :	**1**	**2**	**3**	**4**	**5**	**6**	**Total**
p:	$\frac{1}{6}$	$\frac{1}{6}$	$\frac{1}{6}$	$\frac{1}{6}$	$\frac{1}{6}$	$\frac{1}{6}$	1

Illustration No. 1

If X is discrete random variable having the following probability distribution

X	0	1	2	3	4	5
P(X= xi)	0	k	2k	3k	4k	5k

Determine

1. The value of k?
2. Find p(x < 3)?

3. What will be $p(0 < x < 4)$?
4. And calculate $p(x \geq 4)$?

Solution:

As per the probability function

1. $f(x_i) = P(X = x_i)$ should satisfy two conditions

   (i) $f(x_i) \geq 0$ for all i = 1, 2, 3, ... n.

   (ii) $\sum_{i=1}^{n} f(x_i) = 1$

   therefore from (ii) we have,

   $0 + k + 2k + 4k + 6k + 8k = 1$ or $21k = 1$

   Or $k = \frac{1}{21}$

2. $p(x < 3) = f(0) + f(1) + f(2)$

   $= 0 + \frac{1}{21} + \frac{2}{21} = \frac{3}{21}$

   $= \frac{1}{7}$

3. $p(0 < x < 4) = f(1) + f(2) + f(3)$

   $= \frac{1}{21} + \frac{2}{21} + \frac{3}{21} = \frac{6}{21}$

4. $p(x \geq 4) = f(4) + f(5)$

   $= \frac{4}{21} + \frac{5}{21} = \frac{9}{21}$

Illustration No. 2

If X is discrete random variable having the following probability distribution

x	0	1	2	3	4	5	6
$P(X = x_i)$	0	k	$6k^2$	2k	3k	$4k^2$	3k

1. Find the value of k?
2. Find $p(x \geq 3)$?
3. Find $p(1 \leq x \leq 4)$?
4. And also calculate $p(x < 2)$?

Solution:

1. As per the probability function

$f(x_i) = P(X = x_i)$ should satisfy two conditions

(i) $f(x_i) \geq 0$ for all i = 1, 2, 3, ... n

(ii) $\sum_{i=1}^{n} f(x_i) = 1$

therefore from (ii) we have,

$$0 + k + 6k^2 + 2k + 3k + 4k^2 + 3k = 1$$

$$10k^2 + 9k = 1$$

Or $\quad 10k^2 + 9k - 1 = 0$

Or $\quad 10k(k + 1) - 1(k + 1) = 0$

Therefore

$$(k + 1)(10k - 1) = 0$$

Or $\quad k = -1$ or $\frac{1}{10}$

Since from (i) $f(x_i) > 0$

Therefore k with negative value cannot be accepted and hence $k = \frac{1}{10}$.

2. $p(x \geq 3) = f(3) + f(4) + f(5) + f(6)$

$$= \frac{2}{10} + \frac{3}{10} + \frac{4}{100} + \frac{3}{10}$$

$$= \frac{84}{100}$$

$$= \frac{22}{25}$$

3. $p(1 \leq x \leq 5) = f(1) + f(2) + f(3) + f(3) + f(4) + f(5)$

$$= \frac{1}{10} + \frac{6}{100} + \frac{2}{10} + \frac{3}{10} + \frac{4}{100}$$

$$= \frac{70}{100} = \frac{7}{10}$$

4. $p(x < 2) = f(0) + f(1)$

$$= 0 + \frac{1}{10} = \frac{1}{10}$$

Illustration No. 3

For what value of 'x', f(x) will be probability density function of discrete random variable 'X' where $f(x) = n^2(x + 1)$ for x_i where i = 1, 2, 3, ... 10?

Solution:

X	1	2	3	4	5	6	7	8	9	10
f(x)	$2n^2$	$3n^2$	$4n^2$	$5n^2$	$6n^2$	$7n^2$	$8n^2$	$9n^2$	$10n^2$	$11n^2$

Since f(x) is probability distribution function therefore needs to satisfy the following conditions

(i) $f(x_i) \geq 0$ for all i = 1, 2, 3, ... n.

(ii) $\sum_{i=1}^{n} f(x_i) = 1$

From the 2nd condition we have,

$$2n^2 + 3n^2 + 4n^2 + 5n^2 + 6n^2 + 7n^2 + 8n^2 + 9n^2 + 10n^2 + 11n^2 = 1$$

i.e. $$65n^2 = 1, n^2 = \frac{1}{65}$$

$$n = \pm\frac{1}{\sqrt{65}} \text{ or } n = \frac{1}{\sqrt{65}}$$

Illustration No. 4

For what value 'a', f(x) = ax where x_i = 1, 2, ... n will be a discrete random variable?

Solution:

Since f(x) needs to be p.d.f has to satisfy two essential conditions

(i) $f(x_i) \geq 0$ for all i = 1, 2, 3, ... n.

(ii) $\sum_{i=1}^{n} f(x_i) = 1$

but f(x) = ax

$$\sum_{i=1}^{n} f(x_i) = \left[a\sum_{i=1}^{n} x_i\right] = 1 = a\left(\frac{n(n+1)}{2}\right),$$

Therefore $$a = \frac{2}{n(n+1)}$$

Illustration No. 5

Find the probability distribution of the number of tails appearing when two coins are tossed?

Solution:

Let 'X' denote the number of tails then X – is a random variable which can assume value 0, 1 and 2

Therefore

$$P(X = 0) = P(\text{No Tails}) = P(HH) = \frac{1}{4}$$

$$P(X = 1) = P(\text{Exactly 1 Tail}) = P(HT \text{ or } TH) = \frac{2}{4} = \frac{1}{2}$$

$$P(X = 2) = P(\text{Two Tails}) = P(TT) = \frac{1}{4}$$

The probability distribution of X is given

X	0	1	2
P(X = x)	$\frac{1}{4}$	$\frac{1}{2}$	$\frac{1}{4}$

Illustration No. 6

If 3 balls are drawn from a bag containing 3 green balls and 5 white balls without replacement. And if the event 'X' is a random experiment which denotes the number of white ball drawn. Construct a probability distribution of X?

Solution:

Since the event is a random experiment i.e. drawing white balls. Let X denote the number of white balls drawn and it can assume values from 0, 1, 2, and 3 with associated probability as shown below

$$P(X = 0) = P(\text{No White Ball}) = \frac{{}^3C_3}{{}^8C_3}$$

$$P(X = 1) = P(\text{One White Ball}) = \frac{{}^3C_2 \times {}^5C_1}{{}^8C_3}$$

$$P(X = 2) = P(\text{Two White Ball}) = \frac{{}^3C_1 \times {}^5C_2}{{}^8C_3}$$

$$P(X = 3) = P(\text{Three White Ball}) = \frac{{}^5C_3}{{}^8C_3}$$

X	0	1	2	3
P(x)	$\frac{{}^3C_3}{{}^8C_3}$	$\frac{{}^3C_2 \times {}^5C_1}{{}^8C_3}$	$\frac{{}^3C_1 \times {}^5C_2}{{}^8C_3}$	$\frac{{}^5C_3}{{}^8C_3}$

Expectation of Random Variable

If a discrete random variable x assumes values $x_1, x_2, x_3, \ldots x_n$ with corresponding probabilities $p_1, p_2, p_3, \ldots p_n$ such that $\sum_{i=1}^{n} p_i = 1$ i.e. $p_1 + p_2, + p_3, \ldots + p_n = 1$ then the expectation or mathematical expectation is

$$E(x) = \sum_{i=1}^{n} p_i x_i = p_1 x_1 + p_2 x_2 + \ldots + p_n x_n$$

i.e. sum of the product of different values of x and their respective probabilities.

Similarly if the probabilities are replaced by relative frequencies $\frac{f_i}{N}$ (Empirical method of probability) where $N = \sum f_i$ then $E(x) = \sum \frac{fx}{N}$ the mean of the probability distribution of x as N becomes larger and larger the relative frequencies $\frac{f_i}{N} \to p_i$ (i.e. $\frac{f_i}{N}$ approaches p_i).

The expected value of x^2 is defined by $E(x^2) = \sum_{i=1}^{n} p_i x_i^2$

Variance and Standard Deviation of a Random Variable

Variance is used to measure the dispersion or variation of x

Since variance = $E\left(x-\overline{x}\right)^2$ were $\overline{x}$ = mean

Therefore $\cdot$ $Var(x) = E(x^2 - 2x\overline{x} + \overline{x}^2)$

$$= E(x^2) - 2E(x)\overline{x} + E(\overline{x}^2)$$

$$= E(x^2) - 2\overline{x}\overline{x} + \overline{x}^2$$

$$= E(x^2) - \overline{x}^2$$

$$= E(x^2) - \{E(\overline{x})\}^2$$

Standard Deviation (S.D) = $\sqrt{Variance} = \sqrt{E(x^2) - \{E(\overline{x})\}^2}$

Illustration No. 7

A Random variable 'x' has the following probability distribution.

X	0	1	2	3
P(x)	0.2	0.4	0.3	0.1

Calculate the mean, variance and S.D?

Solution:

Since

$$E(x) = \sum_{i=1}^{n} p_i x_i = p_1 x_1 + p_2 x_2 + \ldots + p_n x_n$$

$$= 0 \times 0.2 + 1 \times 0.4 + 2 \times 0.3 + 3 \times 0.1$$

$$= 1.3$$

Therefore calculating

$$E(x^2) = \sum_{i=1}^{n} p_i x_i^2 = 0 \times (0.2) + 1 \times (0.4) + 4 \times (0.3) + 9$$

$$(0.1)$$

$$= 2.5$$

Hence $Var(x) = E(x^2) - \{E(\overline{x})\}^2$

$$= 2.5 - (1.3)^2$$

$$= 2.5 - 1.69 = 0.81$$

Standard Deviation (S.D) = $\sqrt{\text{Variance}}$

$= \sqrt{0.81}$

$= 0.9$

Illustration No. 8

If $f(x) = k,\ x = 1, 2, 3, 4$ and 5

$= 0$ elsewhere

is a probability distribution function. Find the value of 'k' and also find the mean and standard deviation?

Solution:

Since we knew that

$$\sum_{i=1}^{n} f(x_i) = 1$$

Or $f(1) + f(2) + f(3) + f(4) + f(5)$

Or $k + k + k + k + k = 1$

Or $5k = 1$ or $k = \frac{1}{5}$.

The probability distribution function

X	1	2	3	4	5
P	$\frac{1}{5}$	$\frac{1}{5}$	$\frac{1}{5}$	$\frac{1}{5}$	$\frac{1}{5}$

Therefore mean $E(x) = \sum_{i=1}^{n} p_i x_i$

$= \frac{1}{5}(1 + 2 + 3 + 4 + 5)$

$= 3$

$\text{Var}(x) = E(x^2) - \{E(x)\}^2$

$= \frac{1}{5}(1^2 + 2^2 + 3^2 + 4^2 + 5^2) - (3)^2.$

$= 2$

Standard Deviation (S.D) = $\sqrt{\text{Variance}} = \sqrt{2} = 1.414.$

Illustration No. 9

The following table gives the probability distribution of a random variable

x	3	4	5	6
p	0.3	0.3	0.2	0.2

Find the expectation and variance of the random variable?

Solution:

Expectation $= E(x) = \sum_{i=1}^{n} p_i x_i$

$= 3 \times 0.3 + 4 \times 0.3 + 5 \times 0.2 + 6 \times 0.2.$

$= 4.3$

$Var(x) = E(x^2) - \{E(\bar{x})\}^2$

$= 3^2 \times 0.3 + 4^2 \times 0.3 + 5^2 \times 0.2 + 6^2 \times 0.2 - (4.3)^2$

$= 1.21$

Illustration No. 10

A box containing 10 bulbs of which 4 are found to be defective. If 4 bulbs are drawn at random, find the number of defective drawn and construct a probability distribution function?

Solution:

Since the bulbs are drawn at random, the random variable can assume values from zero defective to all defective i.e. 0, 1, 2, 3 & 4 with associated probability as shown below

$$P(X = 0) = P(\text{No Defective Bulb}) = \frac{{}^6C_4}{{}^{10}C_4}$$

$$P(X = 1) = P(\text{One Defective Bulb}) = \frac{{}^4C_1 \times {}^6C_3}{{}^{10}C_4}$$

$$P(X = 2) = P(\text{Two Defective Bulb}) = \frac{{}^4C_2 \times {}^6C_2}{{}^{10}C_4}$$

$$P(X = 3) = P(\text{Three Def. Bulb}) = \frac{{}^4C_3 \times {}^6C_1}{{}^{10}C_4}$$

$$P(X = 4) = P(\text{Four Def. Bulb}) = \frac{{}^4C_4}{{}^{10}C_4}$$

The probability distribution function is given by

X	0	1	2	3	4
P(x)	$\frac{^6C_4}{^{10}C_4}$	$\frac{^4C_1 \times {}^6C_3}{^{10}C_4}$	$\frac{^4C_2 \times {}^6C_2}{^{10}C_4}$	$\frac{^4C_3 \times {}^6C_1}{^{10}C_4}$	$\frac{^4C_4}{^{10}C_4}$

Illustration No. 11

A player tosses two unbiased coins, if two head appears he wins Rs. 6, if one head appears he wins Rs. 3 and if no head appears – he wins nothing. Construct the probability distribution function and also calculate his expected gain?

Solution:

Let X is the random variable showing the players gain in rupees.

Then X may assume values from X = 6, 3, 0

With probability distribution is given by

$$P(X = 6) = P(HH) = \frac{1}{4}$$

$$P(X = 3) = P(HT \text{ or } TH) = \frac{2}{4} = \frac{1}{2}$$

$$P(X = 0) = P(TT) = \frac{1}{4}$$

The probability distribution table is

X	6	3	0
P	$\frac{1}{4}$	$\frac{1}{2}$	$\frac{1}{4}$

The expected gain of the player = $E(X) = 6 \times \frac{1}{4} + 3 \times \frac{1}{2} + 0 \times \frac{1}{4} = \text{Rs.}3$

Binomial Distribution

Developed by Swiss Mathematician James Bernoulli also called as Bernoulli's distribution is utilized under the following conditions where the trials are performed under homogenous conditions say 'n' finite number of times and the outcome of the trials has discrete outcomes viz. success or failure which are independent i.e. result of one trail has no bearing on the result of the preceding or succeeding trails and the probability of success or failure remains constant even after repeated number of trails.

For example

Tossing of a coin may result in either Head or Tail whether performed once or n number of times and the outcome of each trial practically remain independent of the outcome of the other.

Binomial Distribution is the most fundamental discrete probability distribution and is defined by the probability function

$$f(x) = {}^nC_x p^x q^{n-x} \qquad (x = 0, 1, 2, 3 \ldots n)$$

where 'x' gives the probability of success in n trial ($x \le n$) with p being the probability of success in a single trial and

q the probability of failure = 1 – p.

Theorem

To prove the probability of x success in a series of trials of n independent trials where($x \le n$) is $f(x) = {}^nC_x p^x q^{n-x}$ where p is the probability of success and q the probability of failure (q = 1 – p.) in a single trial.

Proof:

Let there be n independent trials and a random variable 'x' denoting the number of success in n trials. Consider 'p' be the probability of a success in each trial and 'q' the probability of failure in each trial i.e. q = 1 – p.

Further we denote 'S' = Success and 'F' = Failure

Now lets' find the probability that first 'x' trials are success and the rest (n – x) are failures i.e.

$$\underbrace{S\,S\,S \ldots S}_{x \text{ Success}} \qquad \underbrace{F\;F\;F \ldots F}_{(n-x) \text{ failures}}$$

Therefore the probability of 'x' successes is

$$\underbrace{p \times p \times p \times p \ldots \times p}_{x \text{ times}} = p^x$$

and the probability of failures

$$\underbrace{q \times q \times q \times q \ldots \times q}_{(n-x) \text{ times}} = q^{n-x}$$

And hence the probability of success of 'x' trials and the rest (n – x) failure trials is $p^x q^{n-x}$ where q = 1 – p. Similarly the probability of obtaining x successes and n – x failures in any other order is also $p^x q^{n-x}$.

But we are more keen to find out that in n independent trials any x are successful and therefore the rest (n – x) are failures. Since x trails can be selected out n trials in ^nC_x mutually exclusive ways of which the probability of x successes is $p^x q^{n-x}$.

Therefore by the theorem of total probability, the probability f(x) of x success without considering the order in a series of n independent trails is given by

$$f(x) = {}^nC_x p^x q^{n-x} \qquad \text{where } q = 1 - p$$

Hence proved

Mean (or expectation) and Variance of the Binomial Distribution

Let the random variable 'x' assumes values of $x_1, x_2, \ldots x_n$ with the associated probabilities $p_1, p_2, \ldots p_n$ then the mean of the random variable is given by

$$E(x) = \sum_{i=1}^{n} p_i x_i .$$

Also called as the average or the expected value of the probability distribution of x.

Since we knew that for a binomial distribution

$$f(x) = {}^nC_x p^x q^{n-x} \qquad (x = 0, 1, 2, 3\ldots n)$$

and hence

$$E(x) = \sum_{i=1}^{n} p_i x_i = \sum_{i=1}^{n} x_i\, {}^nC_x p^x q^{n-x}$$

$$= 0\,{}^nC_0 p^0 q^n + 1.{}^nC_1 p^1 q^{n-1} + 2\,{}^nC_2 p^2 q^{n-2} + \ldots + n\,{}^nC_n p^n q^{n-n}$$

$$= 0 + 1.{}^nC_1 p^1 q^{n-1} + 2\,{}^nC_2 p^2 q^{n-2} + \ldots + n\,{}^nC_n p^n q^0$$

$$= 1.\, np^1 q^{n-1} + 2\frac{n(n-1)}{2!} p^2 q^{n-2} + \ldots + n\,{}^nC_n p^n$$

$$= np(q^{n-1} + (n-1)p^1 q^{n-2} + \ldots + n\,{}^nC_n p^{n-1})$$

$$= np(q+p)^{n-1}$$

$$= np \qquad \text{Since } p + q = 1$$

Hence mean of the binomial distribution is np

Variance (σ²)

$$\sigma^2 = \sum_{x=0}^{n} x^2 f(x) - m^2 \quad \text{where m = mean}$$

$$= \sum_{x=0}^{n} x^2 f(x) - n^2p^2$$

$$= \sum_{x=0}^{n} x^2\, {}^nC_x p^x q^{n-x} - n^2p^2$$

$$= \sum_{x=0}^{n} [x + x(x-1)]\, {}^nC_x p^x q^{n-x} - n^2p^2$$

$$= \sum_{x=0}^{n} x\, {}^nC_x p^x q^{n-x} + \sum_{x=0}^{n} x(x-1)\, {}^nC_x p^x q^{n-x} - n^2p^2$$

$$= np + \sum_{x=2}^{n} x(x-1)\, {}^nC_x p^x q^{n-x} - n^2p^2$$

As $\sum_{x=0}^{n} x\, {}^n C_x p^x q^{n-x}$

= np and the second term value is zero for x = 0 and x = 1

$$= np + 2.1.{}^nC_2 p^2 q^{n-2} + 3.2\,{}^nC_3 p^3 q^{n-3} + \ldots + n(n-1)\,{}^nC_n p^n q^{n-n} - n^2p^2$$

$$= np + n(n-1)p^2q^{n-2}\left(1+\frac{p}{q}\right)^{n-2} - n^2p^2$$

$$= np + n(n-1)p^2q^{n-2}\left(\frac{1}{q}\right)^{n-2} - n^2p^2 \qquad \text{(Since p + q = 1)}$$

$$= np + n(n-1)\,p^2 - n^2p^2$$

$$= np(1-p)$$

$$= npq$$

Hence the variance of the binomial distribution is npq

And since Standard Deviation (σ) = $\sqrt{\sigma^2} = \sqrt{npq}$

Similarly the moment of skewness and Kurtosis is given by

$$\text{Skewness} = \frac{q-p}{\sqrt{npq}}$$

$$\text{Kurtosis} = \frac{1-6pq}{npq}$$

Illustration No. 1

Five coins are tossed simultaneously. What is the probability of getting?

(i) 3 heads,

(ii) Atleast 2 Heads

Solution:

When a coin is tossed the probability of getting a head is considered as a successful event and be represented by 'p'.

So p = the probability of success of getting a Head $= \frac{1}{2}$

Whereas the probability of failure be 'q'

i.e q = the probability of failure $= 1 - p = 1 - \frac{1}{2} = \frac{1}{2}$

Here n = 5

The probability function

$$f(x) = {}^nC_x p^x q^{n-x}$$

$$= {}^5C_x \left(\frac{1}{2}\right)^x \left(\frac{1}{2}\right)^{5-x}$$

Therefore the probability of getting 3 – heads i.e. putting x = 3,

Hence

$$f(3) = {}^5C_3 \left(\frac{1}{2}\right)^3 \left(\frac{1}{2}\right)^2$$

$$= {}^5C_3 \left(\frac{1}{2}\right)^5$$

(ii) In the second case getting at least 2 heads

$$= f(2) + f(3) + f(4) + f(5)$$

$$= {}^5C_2 \left(\frac{1}{2}\right)^2 \left(\frac{1}{2}\right)^{5-2} + {}^5C_3 \left(\frac{1}{2}\right)^3 \left(\frac{1}{2}\right)^{5-3} + {}^5C_4 \left(\frac{1}{2}\right)^4 \left(\frac{1}{2}\right)^{5-4} + {}^5C_5 \left(\frac{1}{2}\right)^5 \left(\frac{1}{2}\right)^{5-5}$$

$$= \left(\frac{1}{2}\right)^5 ({}^5C_2 + {}^5C_3 + {}^5C_4 + {}^5C_5)$$

Illustration No. 2

Determine the Binomial distribution for which the mean = 8 and variance = 6?

Solution:

Since the mean np = 8 and the variance npq = 6

Therefore substituting the value of np = 8 in variance npq = 6

We get,

$$8 \times q = 6$$

$$q = \frac{6}{8} = \frac{3}{4}$$

And hence the value of $p = 1 - q = 1 - \frac{3}{4} = \frac{1}{4}$

Substituting the value of 'p' in the mean equation,

We get $n \times \frac{1}{4} = 4$ or $n = 16$

Hence the required probability distribution distribution is

$$f(x) = {}^{n}C_{x} p^{x} q^{n-x}$$

$$= {}^{16}C_{x} \left(\frac{1}{4}\right)^{x} \left(\frac{1}{4}\right)^{16-x}$$

Illustration No. 3

Triveni Electricals manufacturer of electrical bulbs found that 10% of the bulbs manufactured in bulk are defective. Find what is the probability that in a sample of 15 bulbs will not include more than 1 defective bulb?

Solution:

Let

p = probability that the bulb being defective.

q = 1- p probability that the bulb being non-defective.

Here

$$p = \frac{10}{100} = \frac{1}{10}$$

$$q = 1 - \frac{1}{10} = \frac{9}{10}$$

Since the probability distribution

$$f(x) = {}^{n}C_{x} p^{x} q^{n-x}$$

Therefore not more than 1 defect is the case {f(0) + f(1)}

Hence,

$$f(0) + f(1) = {}^{15}C_0 \left(\frac{1}{10}\right)^0 \left(\frac{9}{10}\right)^{15-0} + {}^{15}C_1 \left(\frac{1}{10}\right)^1 \left(\frac{9}{10}\right)^{15-1}$$

$$= {}^{15}C_0 \left(\frac{9}{10}\right)^{15} + {}^{15}C_1 \left(\frac{1}{10}\right)\left(\frac{9}{10}\right)^{14}$$

Illustration No. 4

In a MAT exam the overall percentage of failures are 30%. Then find out the probability that out of a group of 5 students only 2 have passed the examination?

Solution:

Let us consider the student passing the examination as a successful event.

i.e. p = Probability of success.

$q = 1 - p$ = probability of failure.

But $q = 30\ \%$ (given), $q = \frac{30}{100} = \frac{3}{10}$

Therefore

$$p = 1 - q = 1 - \frac{3}{10} = \frac{7}{10}$$

Here $n = 5$ i.e. the number of students in the group.

Therefore the probability function is given by

$$f(x) = {}^{n}C_{x} p^{x} q^{n-x}$$

$$= {}^{5}C_{x} \left(\frac{7}{10}\right)^{x} \left(\frac{3}{10}\right)^{5-x}$$

Here the required probability = the probability of 2 success in 5 trials

$$= {}^5C_2\left(\frac{7}{10}\right)^2\left(\frac{3}{10}\right)^{5-2}$$

$$= {}^5C_2\left(\frac{7}{10}\right)^2\left(\frac{3}{10}\right)^3$$

Illustration No. 5

With the outbreak of swine flu in the city has reported that 10 people have tested positive and were isolated for treatment. If the probability that a patient undergoing treatment recovers is 60% then find the probability that At least 8 patients have fully recovered?

Solution:

Let 'x' be the number of patients recovered then,

$$p = \text{the probability of recovery} = \frac{60}{100} = \frac{3}{5}$$

q = the probability of non–recovery

$$q = 1 - p = 1 - \frac{3}{5} = \frac{2}{5}$$

The probability function is given by

$$f(x) = {}^nC_x p^x q^{n-x}$$

To find the probability that at least 8 recovers out of 10 patients

$$= f(8) + f(9\ ) + f(10)$$

$$= {}^{10}C_8\left(\frac{3}{5}\right)^8\left(\frac{2}{5}\right)^{10-8} + {}^{10}C_9\left(\frac{3}{5}\right)^9\left(\frac{2}{5}\right)^{10-9} + {}^{10}C_{10}\left(\frac{3}{5}\right)^{10}\left(\frac{2}{5}\right)^{10-10}$$

$$= {}^{10}C_8\left(\frac{3}{5}\right)^8\left(\frac{2}{5}\right)^2 + {}^{10}C_9\left(\frac{3}{5}\right)^9\left(\frac{2}{5}\right)^1 + {}^{10}C_{10}\left(\frac{3}{5}\right)^{10}$$

Illustration No. 6

If the probability of hitting a target is 1 out of 8 and if 12 shots are fired, then find the probability that the target being hit

(i) Exactly four times?

(ii) At least thrice?

Solution:

Let us consider hitting the target is a successful event

Then

$$p = \text{the probability of hitting the target} = \frac{1}{8} \text{ (given)}$$

$$q = \text{the probability of not hitting the target} = 1 - p = 1 - \frac{1}{8} = \frac{7}{8}$$

Here number of shots fired are 12 i.e. n = 12

Therefore the binomial distribution is given by

$$f(x) = {}^{n}C_{x} p^{x} q^{n-x}$$

(i) In the first case we need to find the probability of exact four times the target being hit

$$\text{i.e.} \quad f(4) = {}^{12}C_{4}\left(\frac{1}{8}\right)^{4}\left(\frac{7}{8}\right)^{12-4} = {}^{12}C_{4}\left(\frac{1}{8}\right)^{4}\left(\frac{7}{8}\right)^{8}$$

(ii) In the second case the probability the target being at least being hit thrice is given by

$$f(x \geq 3) = 1 - f(x < 3) = 1 - \{f(0) + f(1) + f(2)\}$$

$$= 1 - \left[{}^{12}C_{0}\left(\frac{7}{8}\right)^{12} + {}^{12}C_{1}\left(\frac{1}{8}\right)^{1}\left(\frac{7}{8}\right)^{12-1} + {}^{12}C_{2}\left(\frac{1}{8}\right)^{2}\left(\frac{7}{8}\right)^{12-2}\right]$$

$$= 1 - \left[{}^{12}C_{0}\left(\frac{7}{8}\right)^{12} + {}^{12}C_{1}\left(\frac{1}{8}\right)^{1}\left(\frac{7}{8}\right)^{11} + {}^{12}C_{2}\left(\frac{1}{8}\right)^{2}\left(\frac{7}{8}\right)^{10}\right]$$

Illustration No. 7

A bag contains 3 Red balls, 5 Black Balls and 7 Blue Balls. If 3 balls are drawn with replacement find what is the probability that

(i) All black in color?

(ii) None are black?

Solution:

Let

$$p = \text{the probability of drawing a black ball} = \frac{5}{15} = \frac{1}{3}$$

Therefore $q = 1 - p = 1 - \frac{1}{3} = \frac{2}{3}$

the binomial distribution is given by

$$f(x) = {}^{n}C_{x} p^{x} q^{n-x}$$

(i) We need to find the probability of all black balls drawn i.e. f(3)

therefore, $f(3) = {}^{3}C_{3}\left(\frac{1}{3}\right)^{3}$

(ii) In the second case the probability of drawing zero black balls i.e f(0).

$$f(0) = {}^{3}C_{0}\left(\frac{1}{3}\right)^{0}\left(\frac{2}{3}\right)^{3-0} = {}^{3}C_{0}\left(\frac{2}{3}\right)^{3}$$

Illustration No. 8

Green–Leaf Ltd. a tea manufacturer is planning to launch a new variety of CTC tea in the market. In order to promote the product a team was formed in order to oversee the complete marketing activities. The team formed has devised 5 – different strategies in order to make the promotional campaign a success. If the team past success rate was 70% find the probability of 3 strategies being successful?

Solution:

Let us consider that.

p = Probability of success of a strategy

q = 1 – p = probability of failure of a strategy.

But $p = 70\ \%\ \text{(given)} = \frac{70}{100} = \frac{7}{10}$

Therefore

$$q = 1 - p = 1 - \frac{7}{10} = \frac{3}{10}$$

Here n = 5

Therefore the probability function is given by

$$f(x) = {}^{n}C_{x} p^{x} q^{n-x}$$

$$= {}^{5}C_{x}\left(\frac{7}{10}\right)^{x}\left(\frac{3}{10}\right)^{5-x}$$

Here the required probability = the probability of 3 success in 5 trials

$$= {}^5C_3\left(\frac{7}{10}\right)^3\left(\frac{3}{10}\right)^{5-3}$$

$$= {}^5C_3\left(\frac{7}{10}\right)^3\left(\frac{3}{10}\right)^2$$

Fitting of a Binomial Distribution

Assume a random experiment having n trials satisfying the binomial distribution conditions where the experiment is repeated N times. Then the frequency of x successes is given by

$$N\,f(x) = N\,{}^nC_x p^x q^{n-x} \qquad (x = 0, 1, 2, \dots n)$$

By substituting the values of x = 0, 1, 2, ... n we get the requisite expected or theoretical frequencies of the Binomial distribution as shown in the table.

No. of Successes (x)	Expected Frequencies N × f(x)
0	Nq^n
1	$N\,{}^nC_1p^1q^{n-1}$
2	$N\,{}^nC_2p^2q^{n-2}$
3	$N\,{}^nC_3p^3q^{n-3}$
⋮	⋮
N	$N\,{}^nC_np^nq^{n-n}$

If the probability of success 'p' is known then the expected frequencies can be calculated as shown in the table but if 'p' is not known, we first calculate the mean of the given frequency distribution by using the formula $\overline{X} = \frac{\sum fx}{\sum f}$ and equate to np, the mean of the binomial distribution.

i.e. $\overline{X} = np \leftrightarrow \frac{\overline{X}}{n} = p$, where q = 1 − p with the values of p and q; the expected or theatrical values can then be calculated using the above mentioned formula.

Illustration No. 9

Fit a binomial distribution for the following data

X	F
0	50
1	40
2	30
3	20
4	10

Solution

Here we have n = 4 and $N = \sum f = 150$

The mean of the given distribution is

$$\overline{X} = \frac{\sum fx}{\sum f} = \frac{0 \times 50 + 1 \times 40 + 2 \times 30 + 3 \times 20 + 4 \times 10}{150} = \frac{200}{150} = \frac{4}{3}$$

If 'p' the parameter then mean of the Binomial distribution is

$$np = \text{Mean of the distribution} = \frac{4}{3}$$

$$4p = \frac{4}{3} \leftrightarrow p = \frac{1}{3} \text{ and } q = 1 - p = \frac{2}{3}$$

Hence from the formula of Binomial distribution, the expected frequencies can be calculated by N × f(x).

$= N\, {}^nC_x p^x q^{n-x}$

$= 150\, {}^4C_x p^x q^{4-x}$ where x = 0, 1, 2, 3, 4.

X	**Expected Binomial Frequency**
0	$150\, {}^4C_0 \left(\frac{2}{3}\right)^4 = 30$
1	$150\, {}^4C_1 \left(\frac{1}{3}\right)\left(\frac{2}{3}\right)^3 = 59$
2	$150\, {}^4C_2 \left(\frac{1}{3}\right)^2\left(\frac{2}{3}\right)^2 = 44$
3	$150\, {}^4C_3 \left(\frac{1}{3}\right)^3\left(\frac{2}{3}\right) = 15$
4	$150\, {}^4C_4 \left(\frac{1}{3}\right)^4 = 2$

Hence the fitted Binomial distribution

X	F
0	30
1	59
2	44
3	15
4	2

Poissons Distribution

Developed by French mathematician Simeon D. Poisson may be understood as a limiting case of binomial distribution under the following restraint

(i) the number of trials is indefinitely large i.e. $n \to \infty$.

(ii) the probability of success in each trial is indefinitely small i.e. $p \to 0$.

(iii) $np = m$ is finite

Under these conditions the binomial distribution tends to the probability function of the Poisson distribution as given below

$$f(x) = \frac{e^{-m} m^x}{x!} \qquad \text{where} \quad x = 0, 1, 2, 3, \ldots \infty.$$

where x represents the number of successes

Poisson Distribution as a Limit of the Binomial Distribution

As we have discussed Binomial distribution where the number of trials n is extremely large and the probability of success 'p' in a trial is very small where $np = m$ remains constant i.e. if $n \to \infty$, $p \to 0$, where $np = m$ = constant then the binomial distribution tends towards Poisson distribution.

In Binomial distribution the probability P(x) of x successes in a series of n independent trials is given by

$$P(x) = {}^nC_x p^x q^{n-x}, \qquad \text{where } p+q = 1$$

$$= \frac{n(n-1)(n-2)\ldots(n-x+1)}{x!} p^x q^{n-x}$$

Let np = m or $p = \frac{m}{n}$ and $q = 1 - p = 1 - \frac{m}{n}$

Therefore P(x) can be rewritten as

$$P(x) = \frac{n(n-1)(n-2)\ldots(n-x+1)}{x!}\left(\frac{m}{n}\right)^x\left(1-\frac{m}{n}\right)^{n-x}$$

$$= \frac{1.\left(1-\frac{1}{n}\right).\left(1-\frac{2}{n}\right)\ldots\left[1-\left(x-\frac{1}{n}\right)\right]}{x!}(m)^x \frac{\left(1-\frac{m}{n}\right)^n}{\left(1-\frac{m}{n}\right)^x}$$

As, $n \to \infty$ then $1/n \to 0$ where x, m are fixed

But,

$$\left(1-\frac{1}{n}\right),\left(1-\frac{2}{n}\right),\ldots\left[1-\left(x-\frac{1}{n}\right)\right],\left(1-\frac{m}{n}\right)^x \text{ all} \to 1$$

And $$\lim_{n\to\infty}\left[1-\left(\frac{m}{n}\right)^n\right] = \lim_{N\to-\infty}\left[\left(1+\frac{1}{N}\right)^{-mN}\right] = \left[\left\{\lim_{N\to-\infty}\left(1+\frac{1}{N}\right)^{-m}\right\}^N\right] = e^{-m}$$

$$-\frac{m}{n} = \frac{1}{N} \text{ or } n = -mN \text{ where } N \to -\infty \text{ as } n \to \infty$$

Hence when $n \to \infty$

We get $$\lim_{n\to\infty} P(x) = \frac{e^{-m}m^x}{x!}$$

Therefore $$f(x) = \frac{e^{-m}m^x}{x!} \qquad x = 0, 1, 2, \ldots$$

where $\lim_{n\to\infty} P(x)$ replaced f(x)

Hence proved.

Mean and Variance of Poisson's Distribution

The probability function for a Poisson distribution is given by

$$f(x) = \frac{e^{-m}m^x}{x!} \qquad \text{where } x = 0, 1, 2, 3, \ldots \infty.$$

$$\text{Mean} = E(x) = \sum_0^{\infty} x.f(x)$$

$$= \sum_0^{\infty} x \,.\, \frac{e^{-m} m^x}{x!}$$

$$= e^{-m}\left[0 + 1.m + 2.\frac{m^2}{2!} + \ldots + r.\frac{m^r}{r!} + \ldots\right]$$

$$= e^{-m}\left[m + \frac{m^2}{1!} + \ldots + \frac{m^r}{(r-1)!} + \ldots\right]$$

$$= me^{-m}\left[1 + \frac{m}{1!} + \frac{m^2}{2!} \ldots + \frac{m^{r-1}}{(r-1)!} + \ldots\right]$$

$$= me^{-m}e^{m} = m$$

$$\text{Variance} = E(x^2) - \{E(x)\}^2 .$$

$$= \sum_0^{\infty} x^2\, f(x) - m^2$$

$$= \sum_0^{\infty} [x(x-1) + x]\, f(x) - m^2 .$$

$$= \sum_0^{\infty} x(x-1)f(x) + \sum_0^{\infty} x\, f(x) - m^2$$

$$= \sum_0^{\infty} x(x-1)f(x) + m - m^2$$

$$= \sum_0^{\infty} x(x-1)\left[e^{-m}\frac{m^x}{x!}\right] + m - m^2$$

$$= e^{-m}\sum_0^{\infty} x(x-1)\left[\frac{m^2 m^{x-2}}{x(x-1)(x-2)!}\right] + m - m^2$$

$$= e^{-m}m^2\sum_0^{\infty}\left[\frac{m^{x-2}}{(x-2)!}\right] + m - m^2$$

$$= m^2e^{-m}\left[\frac{m^0}{0!} + \frac{m^1}{1!} + \frac{m^2}{2!} + \ldots\right] + m - m^2$$

$$= m^2e^{-m}\left[1+\frac{m}{1!}+\frac{m^2}{2!}...\right]+m-m^2$$

$$= m^2e^{-m}e^m + m - m^2$$

$$= m^2 + m - m^2$$

$$= m$$

Illustration No. 1

Wizard Electronics company manufacturer of solar cells found that the 4% of the output manufactured are defective. In a lot 100 units, find what is the probability that there are exactly 2 defective cells?

Solution:

Let us assume that the occurrence of defective solar cells is taken as success

Since $p = \frac{4}{100} = 0.04$

And $n = 100$

Therefore the mean (m) = np = 0.04 × 100 = 4

The probability function of a Poisson distribution is given by

$$f(x) = \frac{e^{-m}m^x}{x!}$$

thus the Poisson distribution with parameter 4 is

$$f(4) = \frac{e^{-4}(4)^x}{x!}$$

the probability of getting exact 2 defective i.e. x = 2.

$$f(2) = \frac{e^{-4}(4)^2}{2!} = \frac{0.0183 \times 16}{2} = 0.146$$

Illustration No. 2

In a national highway as reported by the traffic officials that during a month the number of road accidents follow Poisson distribution with mean 2. Find then in a certain month the number of accidents reported

(i) Is not more than 1?

(ii) Is between 2 and 3?

Solution:

Since The probability function of a Poisson distribution is given by

$$f(x) = \frac{e^{-m} m^x}{x!}$$

In this case the mean (m) = np = 2.

(i) In the first case we need to find the probability f(0) and f(1)

$$f(0) = \frac{e^{-2}(2)^0}{0!} = 0.1353 \times 1 = 0.1353$$

$$f(0) = \frac{e^{-2}(2)^1}{1!} = 0.1353 \times 2 = 0.2706$$

Therefore the probability of reporting not more than 1 accidents is

$$f(0) + f(1\ ) = 0.1353 + 0.2706$$

$$= 0.4059$$

(ii) In the second case we need to find the number accidents in between 2 and 3 i.e. f(2) and f(3)

$$f(2) = \frac{e^{-2}(2)^2}{2!} = \frac{0.1353 \times 4}{2} = 0.2706$$

$$f(3) = \frac{e^{-2}(2)^3}{3!} = \frac{0.1353 \times 8}{6} = 0.1804$$

Therefore the required probability is

$$f(2) + f(3) = 0.451$$

Illustration No. 3

With the outbreak of Bird–flu in the city a new type of vaccination is being administered in order to safe-guard against any further spread. Out of a sample of 1000 that were administered with the vaccination reported signs of reaction with the associated probability being 0.002 then show that exactly three suffered from reaction?

Solution:

Here the sample size n = 1000 and p = 0.002

Therefore the mean = np = 1000 × 0.002 = 2

We know that the probability function of a Poisson distribution is given by

$$f(x) = \frac{e^{-m} m^x}{x!}$$

But we need to find the probability of exact three suffered reaction i.e f(3)

And hence substituting the values

$$f(3) = \frac{e^{-2}(2)^3}{3!} = \frac{0.1353 \times 8}{6} = 0.1804$$

Hence the probability that exact three suffered reaction is 0.1804.

Illustration No. 4

The number of complains received by a branch of Bharat Nirman Bank is a random variable having a Poisson distribution with mean (m) = 3. Find the probability that the branch will receive

(i) Only 4 complaints per day?

(ii) At most 3 complaints per day?

Solution:

In this case m = 3 and we know that the probability function of a Poisson distribution is given by

$$f(x) = \frac{e^{-m} m^x}{x!}$$

Therefore substituting the values

$$f(4) = \frac{e^{-3} 3^4}{4!} = \frac{0.050 \times 81}{24} = 0.168 \quad (\text{Since } e^{-3} = 0.050)$$

Thus the probability of the branch receiving 4 complains per day is 0.1687.

(ii) In second case we need to find $f(x \leq 3)$ i.e. f(0) + f(1) + f(2) + f(3).

Therefore

$$f(0) = \frac{e^{-3} 3^0}{0!} = \frac{0.050 \times 1}{1} = 0.050$$

$$f(1) = \frac{e^{-3} 3^1}{1!} = \frac{0.050 \times 3}{1} = 0.15$$

$$f(2) = \frac{e^{-3}3^2}{2!} = \frac{0.050 \times 9}{2} = 0.225$$

$$f(3) = \frac{e^{-3}3^3}{3!} = \frac{0.050 \times 27}{6} = 0.225$$

Hence f(0) + f(1) + f(2) + f(3) = 0.050 + 0.15 + 0.225 + 0.225 = 0.65.

The probability of receiving at most 3 complains per day is 0.65.

Illustration No. 5

At a busy traffic junction the probability that the car of an individual will break down is p = 0.0002. In peak hours the number of vehicles passing the junction is 2000. Under this situation what is the probability that exactly 2 cars have experienced break down?

Solution:

In this particular problem we have n = 2000. And p = 0.0002.

Since n is sufficiently large we can calculate the requisite probability using Poisson distribution.

Since the probability function of a Poisson distribution is given by

$$f(x) = \frac{e^{-m}m^x}{x!}$$

We need to find exact probability to 2 break downs i.e. f(2)

Since n = 2000 and p =0.0002 therefore mean (m) = 0.4

Hence

$$f(2) = \frac{e^{-0.4}(0.4)^2}{2!} = 0.0536$$

Hence the required probability is 0.0536.

Illustration No. 6

In a group of people it was found that 10% of the lot have hearing problem. Then from a random sample of 35 what is the probability that at least 2 people are suffering from hearing problem

Solution:

Here it is furnished that

$$p = \frac{10}{100} = \frac{1}{10}$$

the value of n = 35

therefore mean (m) = np = 35 × 0.1 = 3.5

we need find the probability that atleast 2 – people are suffering hearing problem i.e $f(x \geq 2)$

$$f(x \geq 2) = 1 - f(x \leq 1)$$

$$= 1 - [f(0) + f(1)]$$

But,

$$f(0) = \frac{e^{-3.5}(3.5)^0}{0!} = \frac{0.030 \times 1}{1} = 0.030$$

$$f(1) = \frac{e^{-3.5}(3.5)^1}{1!} = \frac{0.030 \times 3.5}{1} = 0.105$$

Therefore

$$f(x \geq 2) = 1 - [f(0) + f(1)]$$

$$= 1 - (0.135)$$

$$= 0.865.$$

Illustration No. 7

The probability that a man aged 40 years will expire before reaching 45 years is 0.02. In a lot of 300 men in the age bracket of 40 years, find what is the approximate probability that exactly three men will die within the next 5 years?

Solution:

We have n = 300 and p be the probability that the man aged 40 years will die within the coming 5 years = 0.02

Therefore Mean = m = np = 300 × 0.02 = 6

As we know that the probability function of a Poisson distribution is given by

$$f(x) = \frac{e^{-m} m^x}{x!}$$

We need to find that exactly 3 men will die before reaching the age of 45 years

$$f(6) = \frac{e^{-6}(6)^3}{3!} = \frac{0.0024 \times 216}{6} = 0.0864$$

And hence the probability that three will die before reaching the age of 45 years is 0.0864.

Illustration No. 8

A typist was given an assignment to type 200 pages of a project report. It was found that the report contains 30 typing errors and the errors are randomly distributed. If 10 pages are selected at random, then find

(i) there are zero errors?

(ii) between 1 and 2 errors?

Solution:

Since there are 30 errors in 200 pages, therefore we have

$$p = 30/200 = 0.15$$

Here $n = 10$

So mean = np = 10 × 0.15 = 1.5

As we know that the probability function of a Poisson distribution is given by

$$f(x) = \frac{e^{-m} m^x}{x!}$$

We need to find that exact zero typing error in random selection 10 pages

$$f(0) = \frac{e^{-1.5}(1.5)^0}{0!} = \frac{0.223 \times 1}{1} = 0.223$$

In the second case we need to find between 1 and 2 typing errors i.e. f(1) + f(2)

$$f(1) = \frac{e^{-1.5}(1.5)^1}{1!} = \frac{0.223 \times 1.5}{1} = 0.334$$

$$f(2) = \frac{e^{-1.5}(1.5)^2}{2!} = \frac{0.223 \times 2.25}{2} = 0.250$$

Therefore the number of typing errors between 1 and 2 errors is

$$f(1) + f(2) = 0.334 + 0.250.$$

$$= 0.584.$$

Fitting of a Poisson Distribution

Fitting a Poisson distribution to a given frequency distribution is by computing the mean $(\overline{X})$ of the given distribution and equate it to the mean of the Poisson distribution (m) i.e. $\overline{X} = m$. Once m is calculated then the various probabilities of the Poisson's distribution can be calculated using the formula

$$f(x) = \frac{e^{-m} m^x}{x!} \quad \text{where } x = 0, 1, 2, \ldots$$

If 'N' is the total observed frequency, then the expected frequencies of the Poisson's distribution are given by = N × f(x).

The expected frequencies of the Poisson's distribution can be computed as given

Expected Poisson Frequencies

Variables (x)	Probability f(x)	Expected or theoretical frequencies = N × f(x)
0	$f(0) = e^{-m}$	$= N \times e^{-m}$
1	$f(1) = \frac{e^{-m} m^1}{1!}$	$= N \times \frac{e^{-m} m^1}{1!}$
2	$f(2) = \frac{e^{-m} m^2}{2!}$	$= N \times \frac{e^{-m} m^2}{2!}$
3	$f(3) = \frac{e^{-m} m^3}{3!}$	$= N \times \frac{e^{-m} m^3}{3!}$
4	$f(4) = \frac{e^{-m} m^4}{4!}$	$= N \times \frac{e^{-m} m^4}{4!}$

Illustration No. 9

Fit a Poisson distribution to the following data and calculate the Expected frequencies?

X	F
0	50
1	20
2	15
3	10
4	5

Solution:

Fitting of the Poisson's Distribution

X	F	f X
0	50	0
1	20	20
2	15	30
3	10	30
4	5	0
	$N = \sum f = 100$	$\sum fX = 80$

Therefore the mean (m)= $\bar{X} = \frac{\sum fX}{\sum f} = \frac{80}{100} = 0.80$

Thus the mean (m) of the expected distribution = 0.80

By Poisson's distribution law, the expected frequencies are given by

$$= N.f(x) = 100.\frac{e^{-m}m^{x}}{x!}$$

Computation of Expected Frequencies

X	N × f (x)
0	$100 \times e^{-0.80} = 45$
1	$100 \times \frac{e^{-0.80}(0.80)^{1}}{1!} = 36$
2	$100 \times \frac{e^{-0.80}(0.80)^{2}}{2!} = 14.4$
3	$100 \times \frac{e^{-0.80}(0.80)^{3}}{3!} = 3.84$
4	$100 \times \frac{e^{-0.80}(0.80)^{4}}{4!} = 0.786$

(Note: $e^{-0.80} = 0.450$)

Illustration No. 10

If 200 mobiles phone go through regular inspection check as they are assembled in the production line and the number of defects per set is enlisted below?

X	F
0	155
1	25
2	15
3	5
4	0

Solution:

Fitting of the Poisson's Distribution

X	F	f X
0	155	0
1	25	25
2	15	30
3	5	15
4	0	0
	N = ∑ f = 200	∑ f X = 70

Therefore the mean $= \overline{X} = \frac{\sum fX}{\sum f} = \frac{70}{200} = 0.35$

Thus the mean (m) of the expected distribution = 0.35

By Poisson's distribution law, the expected frequencies are given by

$$= N.f(x) = 200.\frac{e^{-m}m^x}{x!}$$

Computation of Expected Frequencies

X	N × f (x)
0	$200.e^{-0.35} = 140$
1	$200 \times \frac{e^{-0.35}(0.35)^1}{1!} = 49$
2	$200 \times \frac{e^{-0.35}(0.35)^2}{2!} = 8.575$
3	$200 \times \frac{e^{-0.35}(0.35)^3}{3!} = 1.00$
4	$200 \times \frac{e^{-0.35}(0.35)^4}{4!} = 0.0875$

(Note: $e^{-0.35} = 0.70$)

Normal Distribution

A Normal Distribution is a continuous probability distribution defined by probability density function

$$f(x) = \frac{1}{\sigma\sqrt{2\pi}} e^{\frac{-(x-m)^2}{2\sigma^2}}$$

$$-\infty < x < \infty$$

where m = mean, σ = Standard Deviation (S.D)

$$\pi = 3.1416 \text{ or } \frac{22}{7}, \; e = 2.71823 \text{ (approximately)}$$

The curve representing normal distribution is called normal curve.

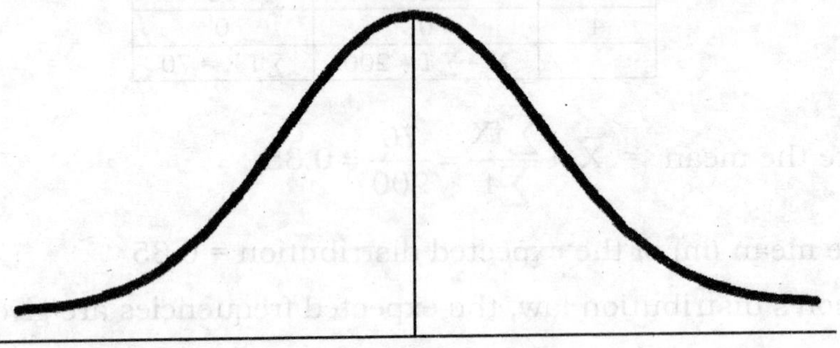

The area under the curve between the ordinate x = a and x = b where (b > a) represents the probability that x lies between a and b i.e P(a < x < b)

Thus the area under the normal curve between the lines x = a and x = b

i.e. $$P(a < x < b) = \int_a^b f(x)dx$$

The curve is symmetrical to mean and bell shaped and both the tails to the right and the left are asymptotic in nature i.e. they extend to infinity in both the direction.

Properties of Normal Distribution

1. It's a continous distribution having two parameters mean (m) and Standard deviation (σ).
2. Normal curve is bell shaped and perfectly symmetrical about the mean i.e. the two halves of the curve coincide with each other.

3. Mean = m, S.D = σ, Skewness = 0, Kurtosis = 0 and Mean = Median = Mode = m.
4. It is unimodal i.e. having only one mode.
5. The quartiles Q_1 and Q_3 are equidistant from the mean

   Q_1= m – 0.67 σ (approximately)

   Q_3= m + 0.67 σ (approximately)

   Q.D = 0.67 σ.

   M.D = 0.80 σ.

6. The maximum ordinate is at x = m and its value is $\frac{1}{\sigma\sqrt{2\pi}}$
7. The area under the curve is in between the curve and the horizontal axis and is equal to the number of frequencies. The standard deviation (σ) distributes the area in the following way.

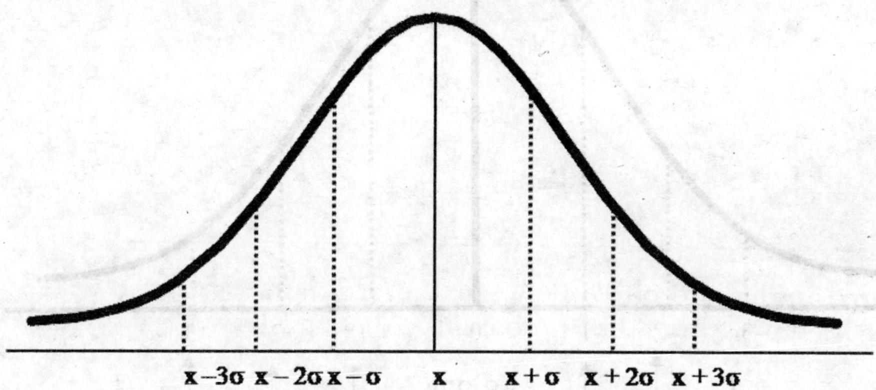

(i) 68.27% of area lies between x = ± σ

(ii) 95.45% of area lies between x = ± 2σ

(iii) 99.73 % of area lies between x = ± 3σ

Importance of Normal Distribution

1. Binomial, Poisson can be approximated to normal distribution.
2. It is used to find the confidence levels of the population parameters.
3. It is used to find the control limits in statistical quality control.

Standard Normal Distribution

If we take $\frac{x-m}{\sigma} = z$ we get the standard normal distribution with mean 'm = 0' and standard deviation (σ) = 1 defined by the density function

$$f(x) = \frac{1}{\sigma\sqrt{2\Pi}} e^{\frac{-z^2}{2\sigma^2}} \qquad (-\infty < z < \infty)$$

Here z is called the standard normal variable with mean '0' and standard deviation '1'. In short it can be written as p(z) is N(0, 1).

Area under the Curve

The standard normal curve y = p(z) with percentage distribution of area under the curve is depicted in the figure.

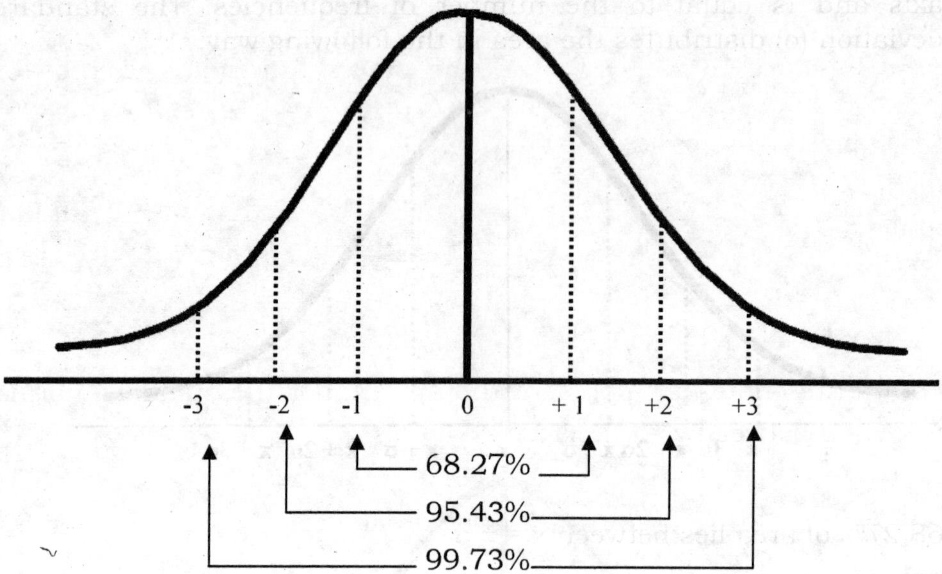

The area include between z= ± 1 = 68.27%.

The area include between z= ± 2 = 95.45%.

The area include between z= ± 3 = 99.73%.

And the total area is '1'.

Let us assume that Φ(z) is the area under the standard normal curve to the left of the ordinate i.e.

$$z = \int_{-\infty}^{z} p(z)dz \text{ or } z = \int_{-\infty}^{z} \frac{1}{\sigma\sqrt{2\pi}} e^{\frac{-z^2}{2\sigma^2}} dz$$

Then Φ(–z) the area under the standard normal curve to the left of the ordinate (–z) or the area under the standard normal to the right of the ordinate (z) i.e.,

$$(-z) = \int_{-\infty}^{-z} p(z)dz \text{ or } z = \int_{-\infty}^{-z} \frac{1}{\sigma\sqrt{2\pi}} e^{\frac{-z^2}{2\sigma^2}} dz$$

$$= 1 - \Phi(z) = 1 - \int_{-\infty}^{z} p(z)dz$$

Therefore the sum of the area under the standard normal curve to the left and to the right of the ordinate z is 1.

For example

Since the standard normal curve with the density function is given by

$$p(z) = \frac{1}{\sqrt{2\pi}} \int e^{-\frac{z^2}{2}} dz$$

And for z> 0 we define

$$p(z) = \frac{1}{\sqrt{2\pi}} \int_{0}^{z} e^{-\frac{z^2}{2}} dz$$

Then the function p(z) represent the area under the standard normal curve ranging between z = 0 and z = z

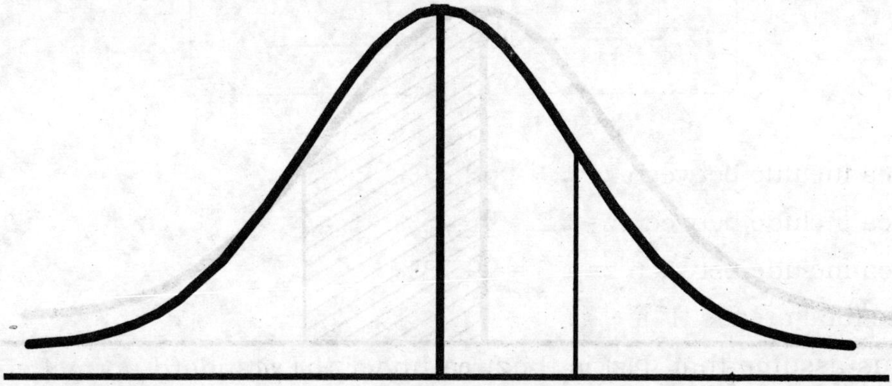

As the area under the normal curve represent probability. Therefore the function p(z) represent the probability when the variable varies from 0 to z.

$$p(0 < Z < z) \quad = \quad p(z) = \frac{1}{\sqrt{2\pi}} \int_0^z e^{-\frac{z^2}{2}} dz$$

To illustrate let's find the probability where z lies between 0 and 1.86 or P(0 < z < 1.86). First locate the value of z equal 1.8 in the first column and then moving across the row to the column labeled 0.06 and the value so obtained can be read as 0.4686. Thus P(0 < z < 1.86) = 0.4686.

Note: Normal curve being symmetric therefore the negative value of z is identical to the corresponding positive value.

Thus $\quad P(-1.86 < z < 0) = 0.4686$

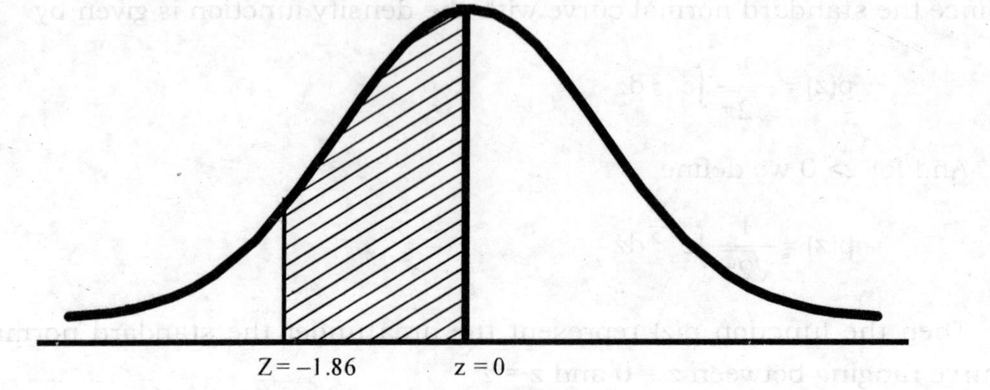

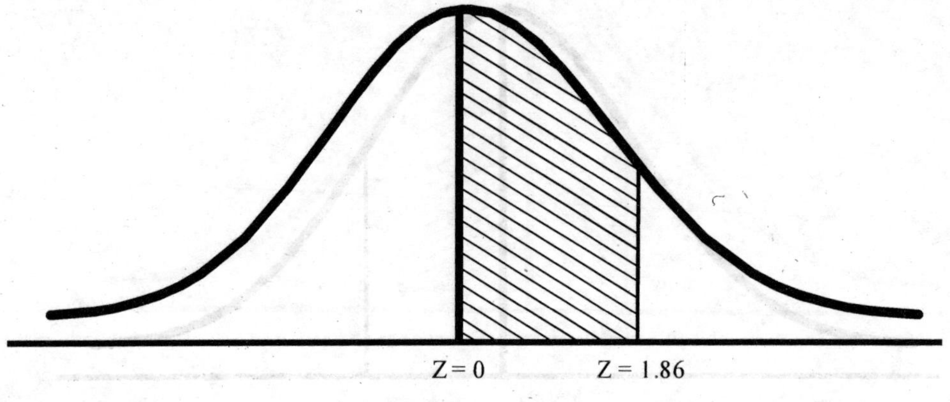

Solved Exercises

Illustration No. 1

Find the area under the standard normal curve between –1.46 and z = 0?

Solution:

The required Area = (Area between z = –1.46 and z = 0)

+ (Area between z = 0 and z = 1.46) ... (By Symmetry)

To find the area between z = 0 and z = 1.46. From the table the value is 0.4279, therefore the value is 0.4279 represent area lying between –1.46 and 0

i.e. $P(-1.46 \leq z \leq 0)$

Illustration No. 2

Find the area between z = – 0.56 and z = 1.73?

Solution:

The required area = (Area in between z = – 0.56 and z = 0) + (Area in between z = 0 and z = 1.73).

i.e. (Area in between z = 0 and z = 0.56) + (Area in between z = 0 and z =1.73). (By symmetry)

From table the values are

$$= 0.2123 + 0.4582$$

$$= 0.6705$$

Clearly $P(-0.56 \leq z \leq 1.73) = 0.6705$

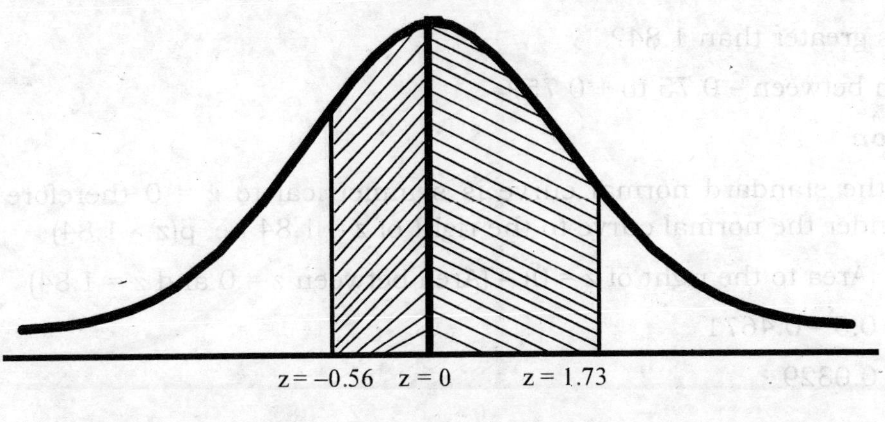

Illustration No. 3

Find the probability that the standard normal variate is less than – 0.72?

Solution:

As standard normal curve is symmetrical about z = 0

Therefore the area under the normal curve to the left of z = 0

= We need to find the required probability P(z < – 0.72)

= Area under the curve to the left of z = – 0.72

= (Area to the left of Z = 0) – (Area between z = – 0.72 and z = 0)

= 0.5 – 0.2642

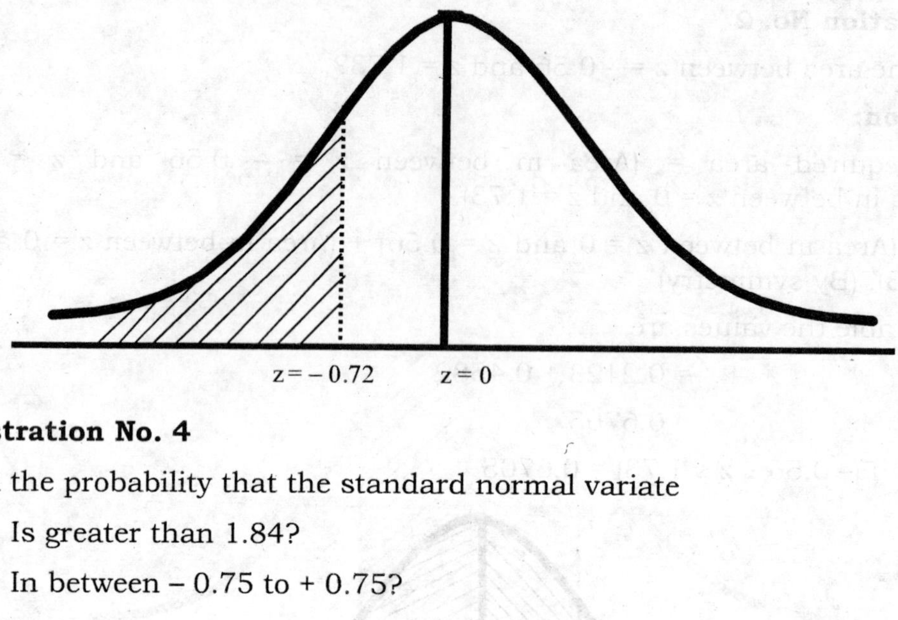

Illustration No. 4

Find the probability that the standard normal variate

(i) Is greater than 1.84?

(ii) In between – 0.75 to + 0.75?

Solution

Since the standard normal curve is symmetrical to z = 0 therefore the area under the normal curve to the right of z = 1.84 i.e. p(z > 1.84)

= (Area to the right of z = 0) – (Area between z = 0 and z = 1.84)

= 0.5 – 0.4671

= 0.0329

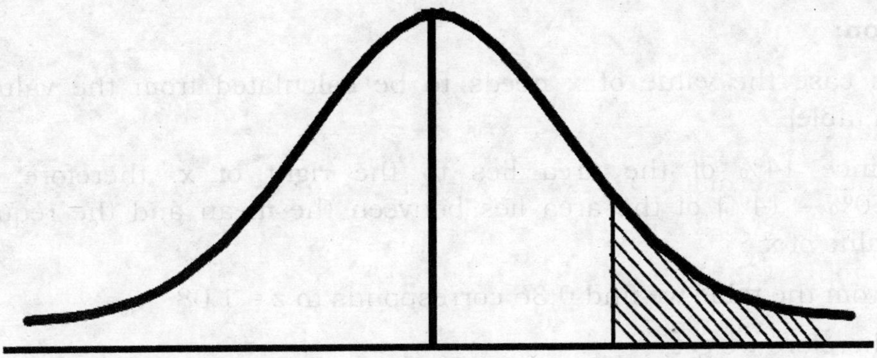

(ii) The required probability p(– 0.75 < z < + 0.75)

= the area under the standard normal curve

= (Area between z = – 0.75 and z = 0) + (Area between z = 0 and z = + 0.75)

= 2(Area between z = 0 and z = + 0.75) (By symmetry)

= 2 × 0.2734

= 0.5468

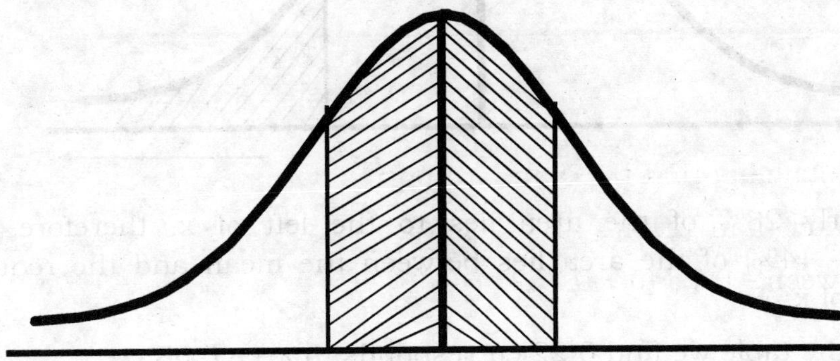

Illustration No. 5

For a normal distribution with mean = 45 and S.D = 8, find the value of that has

(i) 14 % of area to its right?

(ii) 28 % area to its left?

Solution:

In this case the value of x needs to be calculated from the value of z (from table)

(i) Since 14% of the area lies to the right of x. therefore 36% (50% – 14%) of the area lies between the mean and the required value of x.

From the table we find 0.36 corresponds to z = 1.08

Therefore we get,

$$1.08 = \frac{x-45}{8}$$

Or

$$x = 45 + 8.64 = 53.64$$

(ii) Similarly 28% of the area lies to the left of x. therefore 22% (50% – 14%) of the area lies between the mean and the required value of x.

From the table we find 0.22 corresponds to z = – 0.58

Therefore we get,

$$-0.58 = \frac{x-45}{8}$$

Or

$$x = 45 - 4.64 = 40.36$$

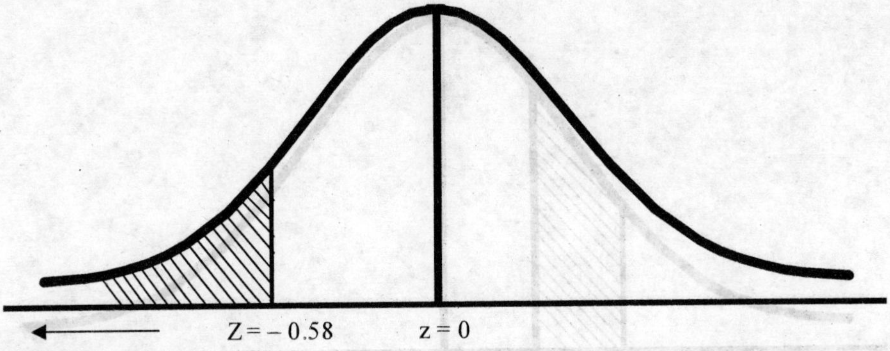

Illustration No. 6

Te average weight of 200 students in a certain school is 60 Kg. and the standard deviation is 10 kg. Assuming the weights are normally distributed find

(i) How many students weight in between 40 to 50 Kg?

(ii) More than 65 kg?

Solution:

Since the mean m = 60 Kg and S.D = 5 Kg

Therefore we have $z = \frac{x - m}{\sigma}$

Where the value of x is in standard units

(i) 40 Kg in standard units

$$= \frac{40 - 60}{10} = -2$$

And 50 Kg in standard units

$$= \frac{50 - 60}{10} = -1$$

Therefore the number of students lying between 40 and 50 kg is the area under the standard normal curve between z = – 2.0 and z = – 1.0

= (Area between z = – 2.0 and z = 0) – (Area between z = – 1.0 and z = 0)

(By Symmetry)

= 0.4772 – 0.3413

= 0.1359

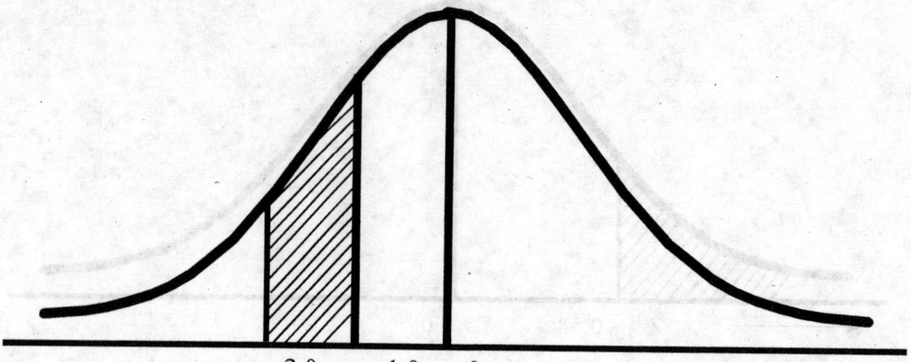

Hence the required number of students lying between 40 and 50 kg are 200 × 0.1359 = 27.18 i.e. 27 students rounded to the nearest decimal.

(ii) 65 kg in standard units = $\frac{65-60}{10} = 0.5$

Therefore the proportion of students whose weights exceeds 65 kg

= (Area under the normal curve to the right of z = 0.5)

= (Area to the right z = 0) – (Area between z = 0 and z = 0.5)

= 0.5 – 0.1915

= 0.3085

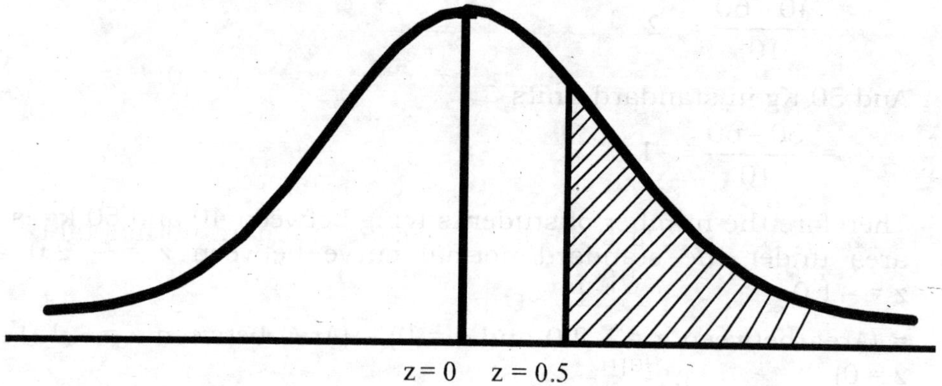

thus the required number of students are 200 × 0.3085 = 61.7 i.e 62 approximately rounded to nearest decimal.

Illustration No. 7

In a normal distribution 25% of the items are under 40 and 10% of the items are over 55. Find the mean and S.D of the distribution?

Solution:

We are given that

$$P(x < 40) = 0.25.$$

And

$$P(x > 55) = 0.10.$$

Let 'm' and 'σ' be the mean and S.D of x. Converting into standard form

i.e.,

$$z = \frac{x - m}{\sigma}$$

therefore for $x_1 = 40\ ,\ z_1 = \frac{40 - m}{\sigma}.$

and for $x_2 = 55\ ,\ z_2 = \frac{55 - m}{\sigma}$

Thus the area under the standard normal curve lies to the left of z_1 and since the area is less than 0.5, therefore z_1 should be negative.

Hence $p(x < z_1) = 0.25$

$$= 0.5 - p(z_1 < x < 0) = 0.50 - 0.25 = 0.25$$

Or

$$P(0 < x < -z_1) = 0.25 \text{ (By symmetry)}$$

$$-z_1 = 0.68 \qquad \text{(Using standard normal table)}$$

i.e. $(-)\frac{40 - m}{\sigma} = 0.68$

Or $40 - m = -0.68\,\sigma \Rightarrow m - 0.68\,\sigma = 40$ ..... (1)

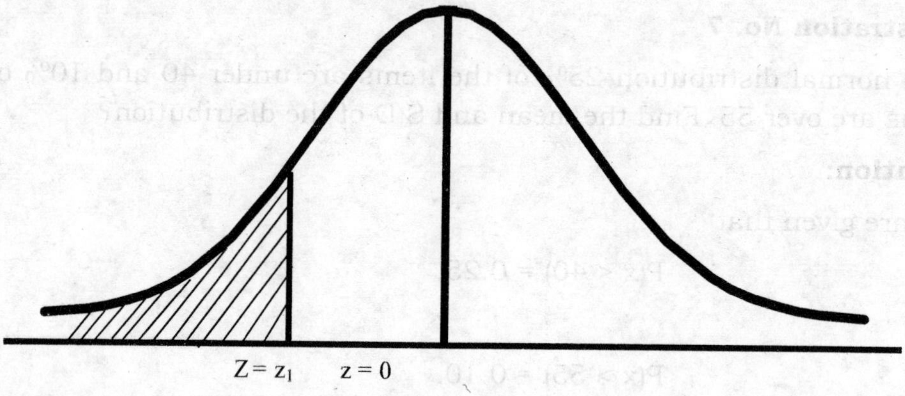

Similarly

$$P(x > 55) = 0.10$$

$$= 0.50 - P(0 < x > z_2) = 0.10$$

Or $\quad P(0 < x < z_2) = 0.40$

Or $\quad z_2 = 1.28$

Therefore $\quad \frac{55 - m}{\sigma} = 1.28 \Rightarrow m + 1.28\ \sigma = 55$ .....(2)

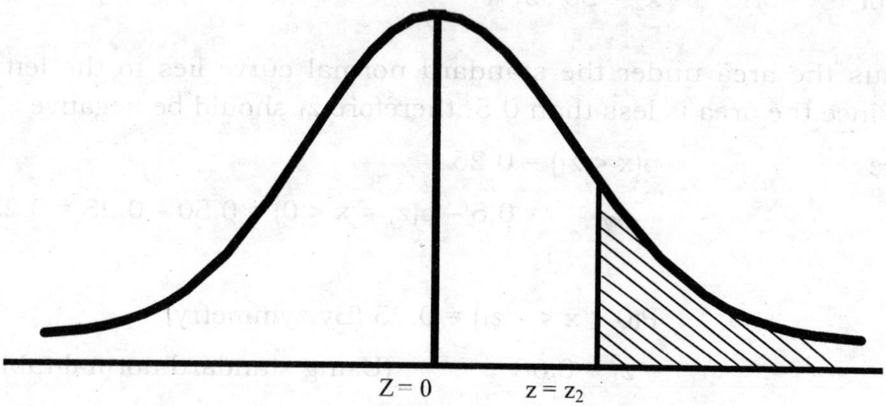

Solving the above two simultaneous equations we get

Mean (m) = 45.2, S.D (σ) = 7.65

Illustration No. 8

In a godown 500 floursent bulbs have been installed with mean life of 150 days. If the length of the life of the bulbs is normally distributed with standard deviation of 75 days, find

(i) How many bulbs will expire in less than 100 days?

(ii) If at all the bulbs are simultaneously replaced at what interval should be allowed if not more than 20% expires before replacement?

Solution:

As we know that the standardized value of x is given by

$$z = \frac{x - m}{\sigma}$$

Here mean (m) = 150 and S.D (σ) = 50.

Therefore

$$z = \frac{100 - 150}{75} = -0.66$$

Thus the probability of the bulbs to expire less than 100 Hrs.

= Area under the normal curve to the left of z = – 2.0

= Area to the right of z = + 2.0 (*By Symmetry*)

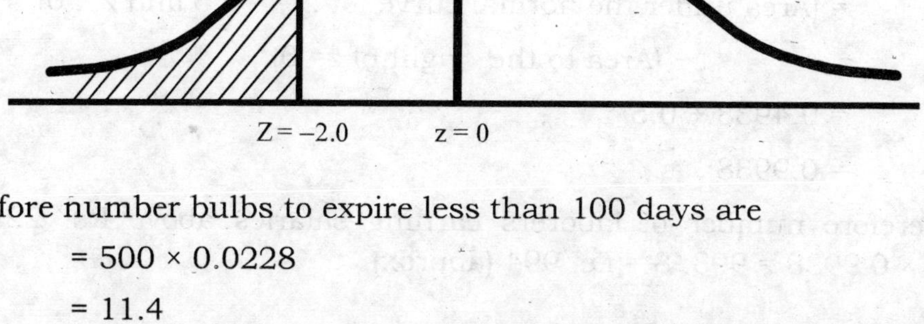

Therefore number bulbs to expire less than 100 days are

= 500 × 0.0228

= 11.4

approximately 11 bulbs rounded to the nearest decimal.

(ii) The value of Z corresponding to an area = 0.3 (0.5 – 0.20) under the normal curve is given by 0.84

Since

$$z = \frac{x - m}{\sigma}$$

therefore $\frac{x - 150}{75} = 0.84$

or $x = 213$ days

Hence the need to be replaced before 213 days

Illustration No. 9

In a factory employing 1000 laborers' whose salaries are normally distributed with mean salary Rs. 2500 and the standard deviation Rs. 100. Show that 99% of laborers have salaries exceeding Rs. 2250 and 6.65% having salaries exceeding 2650?

Solution:

As we know that the standardized value of x is given by

$$z = \frac{x - m}{\sigma}$$

Here mean (m) = 2500 and S.D (σ) = 2250

Therefore

$$z = \frac{2250 - 2500}{100} = -2.5$$

And hence the area to the right of the ordinate at z = -2.5

= (Area under the normal curve for z = – 2.5 and z = 0)

+ (Area to the right of z= 0).

= 0.4938 + 0.5

= 0.9938

Therefore number of laborers earning salaries above Rs. 2250 are 1000 × 0.9938 = 993. 8 = i.e. 994 (approx).

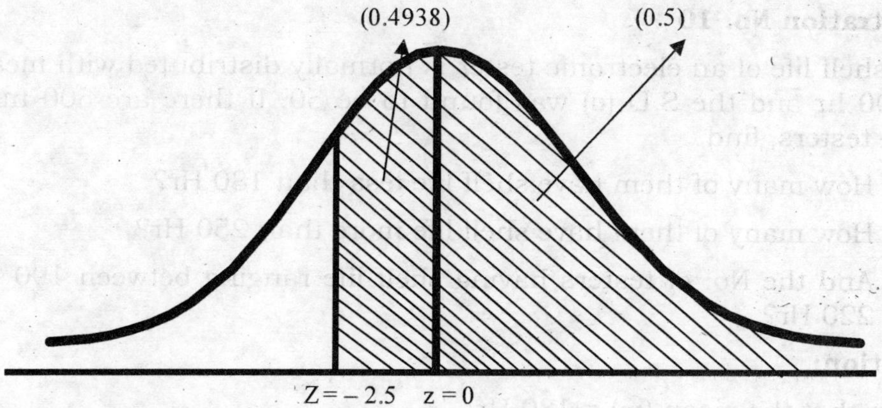

Which is nearly equally to a figure close to 99%

Whereas in the second case the standard normal variate corresponding to x = 2650 is

$$z = \frac{2650 - 2500}{100} = 1.5$$

therefore the area to the right to z = 1.5

= (Area to the right of z = 0) – (Area between z = 0 and z = 1.5)

= 0.5 – 0.4332

= 0.0668

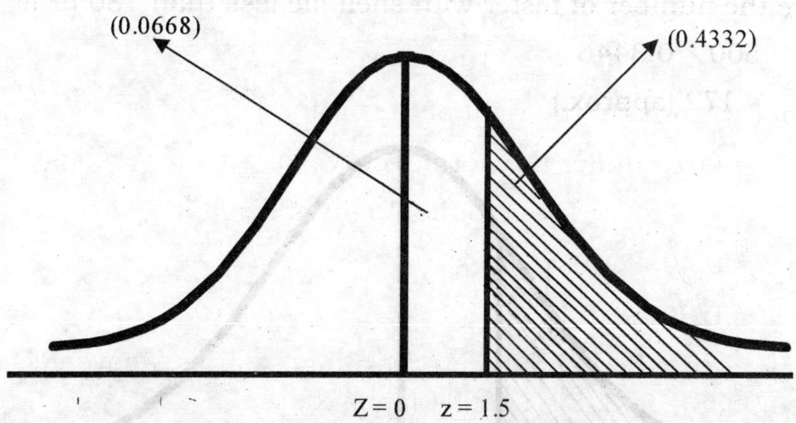

Hence the number of laborers earning salaries above Rs. 2650 are 1000 × 0.0668 = 66.8 i.e. 67 (approx), which is a figure very close 6.65%.

Illustration No. 10

The shelf life of an electronic tester is normally distributed with mean life of 200 hr and the S.D (σ) was found to be 50. If there are 500 number such testers, find

(i) How many of them have shelf life less than 180 Hr?

(ii) How many of them have shelf life more than 250 Hr?

(iii) And the No. of testers having shelf life ranging between 190 Hr. to 220 Hr?

Solution:

Given that the mean (m) = 180 Hr

S.D (σ) = 50 Hr

As,
$$z = \frac{x - m}{\sigma}$$
$$= \frac{180 - 200}{50} = -0.4$$

Thus the probability of the tester with shelf life less than 180 Hrs

= Area under the normal curve to the left of z = – 0.4

= Area to the right of z = + 0.4 (*By Symmetry*)

= 0.5 – 0.1554

= 0.3446

Therefore the number of tester with shelf life less than 180 hr is

500 × 0.3446

≈ 172 (approx.)

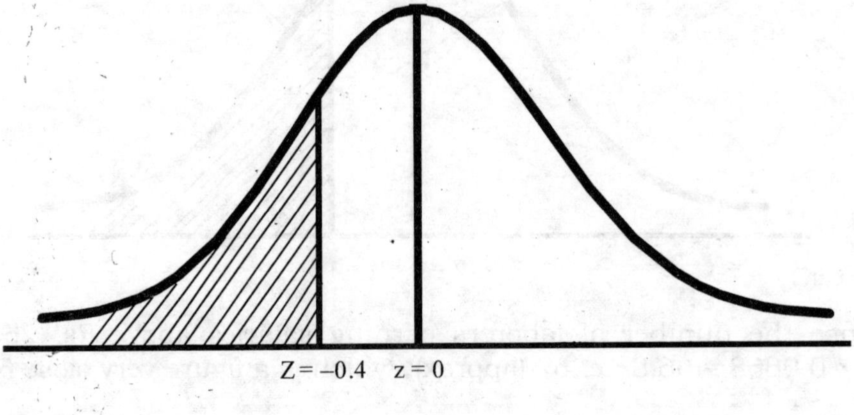

(ii) In order to determine the number of testers with shelf life more than 250 Hr

Since, $$z = \frac{x - m}{\sigma}$$

$$= \frac{250 - 200}{50} = 1.0$$

Thus the probability of the tester with shelf life more than 250 Hr

= Area under the normal curve to the right of z = 1.0.

= (Area to the right of z = 0) – (Area between z = 0 and z = 1)

= 0.5 – 0.3413

= 0.1587

Therefore the number of tester with shelf life more than 250 hr is

500×0.1587

$\approx$ 79 (approx.)

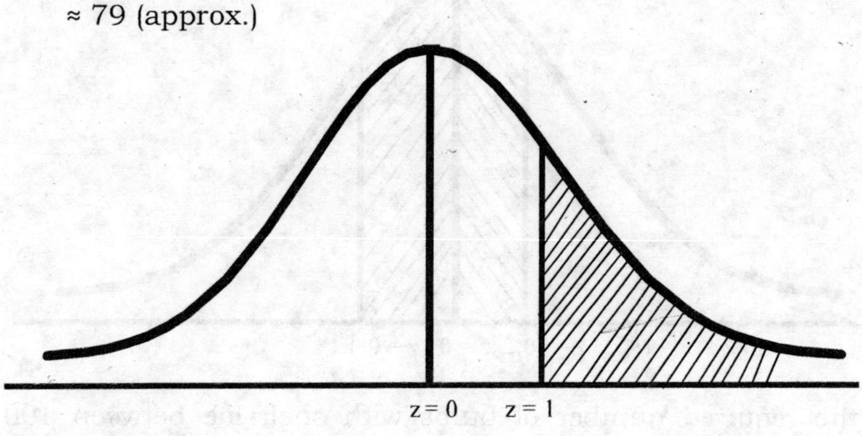

(iii) The standardized value of 190 is

$$z = \frac{x - m}{\sigma}$$

$$= \frac{190 - 200}{50} = -0.20$$

Similarly,

The standardized value of 220 is

$$z = \frac{x - m}{\sigma}$$

$$= \frac{220 - 200}{50} = 0.40$$

Therefore,

The probability that the tester's shelf life is lying between 190 and 220 Hr.

= the area under the normal curve between z = –0.2 and z = 0.4

= (Area between z = 0 and z = 0.2) (By symmetry) + (Area between z = 0 and z 0.4)

= 0.0793 + 0.1554

= 0.2347

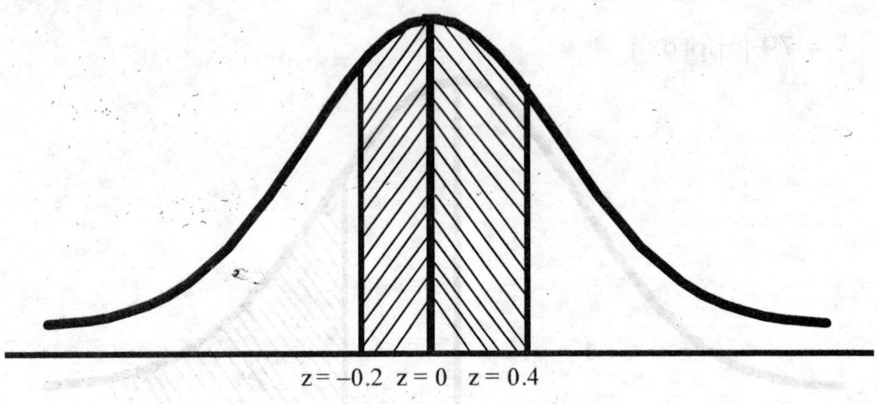

Hence the required number of bulbs with shelf life between 190 and 220 hr is

Equal to 500 × 0.2347

= 117 (approx.)

Normal Approximation to Binomial Distribution

When n is sufficiently large and neither p or q are not too small, then binomial distribution is approximation to normal distribution.

i.e if x is a binomial random variable with mean (m) = np and standard deviation (σ) = $\sqrt{npq}$, then the limiting form of the distribution is

$$z = \frac{x - np}{\sqrt{npq}}$$

As n is sufficiently large ($n \to \infty$), is the standard normal distribution.

Illustration No.11

Students appearing for CAT – 2010 examination in totality a multiple choice questionnaire having 150 questions with 4 alternatives to each question out of which 1 is correct answer. Find what is the probability that if students make intelligent guess work will yield 20 to 25 correct answers from 60 questions out of 150 about which the students have no clue?

Solution:

Since

$$n = 60 \text{ and } p = \frac{1}{4} \text{ and } q = 1 - p = \frac{3}{4}$$

(note: p being the probability of getting a right answer which 1 out of 4)

So, calculating the mean (m) = np = $60 \times \frac{1}{4} = 15$

And the standard deviation (σ) = $\sqrt{npq} = \sqrt{15 \times \frac{3}{4}} = 3.35$

Now we need to find $f(20 \le x \le 25)$

Therefore for x = 20 we have,

$$z = \frac{x - np}{\sqrt{npq}} = \frac{20 - 15}{3.35} = 1.49$$

And for x = 25 we have,

$$z = \frac{x - np}{\sqrt{npq}} = \frac{25 - 15}{3.35} = 2.98$$

Therefore

$$f(20 \le x \le 25) = f(1.49 \le z \le 2.98) = 0.4986 - 0.4319$$
$$= 0.0667$$

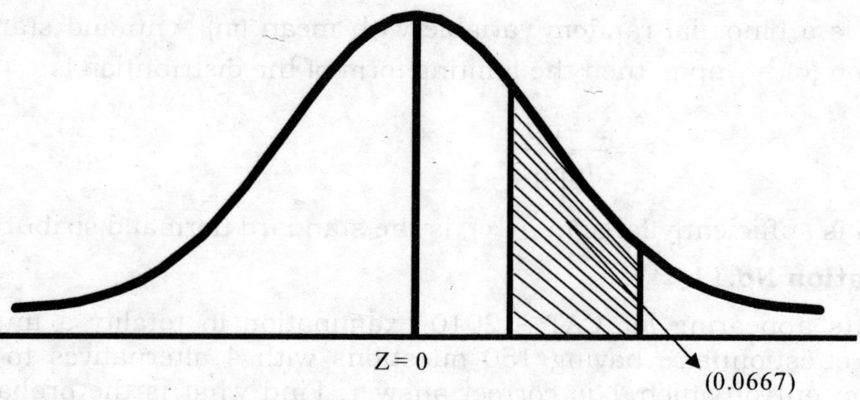

Normal Approximation to Poisson Distribution

As binomial distribution is approximated to normal distribution similarly when the mean(m) gets larger and larger then the normal distribution can be used as approximation to Poisson distribution.

i.e if x is a Poisson random variable with mean = m and standard deviation (σ) = $\sqrt{m}$, then the limiting form of the distribution of

$$z = \frac{x - m}{\sqrt{m}}$$

As mean is sufficiently large (i.e m → ∞), is the standard normal distribution.

Illustration No. 12

The average number of customers visiting a mall is 200 per day. Calculate the probability that in a particular day the number of customers visiting the mall has exceeded 220 per day?

Solution:

Since the mean (m) = 200

And we need to find the probability of f(x > 220). Since m is sufficiently large we can estimate z(x > 220) with mean = 200 and Standard Deviation (σ) = $\sqrt{200}$.

Converting into standard normal variate

We know that

$$z = \frac{x - m}{\sqrt{m}} = \frac{220 - 200}{\sqrt{200}} = 1.414$$

Therefore,

$$f(x > 220) = f(\ z\ > 1.414)$$
$$= 0.5 - 0.4192.$$
$$= 0.0808$$

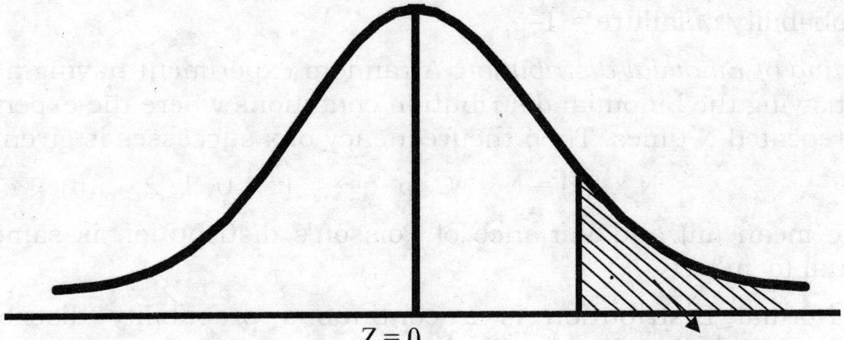

Brief Summary

1. If 'X' is a discrete random variable assuming a set of values $x_1, x_2, x_3, \ldots x_n$

   The function defined by

   $$f(x_i) = P(X = x_i) = p_i$$

   i.e. the probability that 'X' the value $x_i = p_i$ (say) is called the probability function or density function.

2. If a discrete random variable can assume values of $x_1, x_2, x_3, \ldots x_n$ with the corresponding probabilities $p_1, p_2, p_3, \ldots p_n$ such that $\sum p_i = 1$ then the occurrence of the values x_i ($i = 1, 2, \ldots n$) with the probabilities p_i ($i = 1, 2, \ldots n$) is called the discrete probability distribution.

3. If a discrete random variable x assumes values $x_1, x_2, x_3, \ldots x_n$ with corresponding probabilities $p_1, p_2, p_3, \ldots p_n$ such that $\sum p_i = 1$ i.e. $p_1, + p_2, + p_3, \ldots + p_n = 1$ then the expectation or mathematical expectation is

   $$E(x) = \sum_{i=1}^{n} p_i x_i = p_1x_1 + p_2x_2 + \ldots p_nx_n.$$

4. Variance of a Random Variable = $E(x^2) - \{E(x)\}^2$.

5. Binomial Distribution is the most fundamental discrete probability distribution and is defined by the probability function

$$f(x) = {}^nC_x\, p^x\, q^{n-x} \qquad (x = 0, 1, 2, \ldots n)$$

where 'x' gives the probability of success in n trial ($x \le n$) with p being the probability of success in a single trial and q the probability of failure = 1–p.

6. *Fitting of Binomial distribution*: A random experiment having n trials satisfying the binomial distribution conditions where the experiment is repeated N times. Then the frequency of x successes is given by

$$N \times f(x) = N \times {}^nC_x\, p^x\, q^{n-x} \qquad (x = 0, 1, 2, \ldots n)$$

7. The mean (m) and Variance of Poisson's distribution is same and equal to 'm'

8. A Normal Distribution is a continuous probability distribution defined by probability density function

$$f(x) = \frac{1}{\sigma\sqrt{2\Pi}} e^{\frac{-(x-m)^2}{2\sigma^2}}$$

$$-\infty < x < \infty$$

where m = mean, σ = Standard Deviation (S.D).

Review Questions

1. What do you understand by random variable and its mathematical expectation?
2. Explain the concept of mathematical expectation and its application?
3. Derive the mean and variance of a random variable?
4. What do you understand by Bernoulli's trials and explain with an example?
5. Prove that the probability of r success in a series of n independent trial is ${}^nC_x\, p^x\, q^{n-x}$ where p is the probability of success in a single trial and that of failure q = 1 – p?
6. What do you understand by the fitting of a Binomial distribution and explain by an example?
7. Define Binomial distribution? Derive the mean and variance of a binomial distribution?

8. Write short notes on Poisson's distribution and under what conditions Poisson's is applicable?
9. Prove that the mean and the variance of a Poisson's distribution are equal?
10. Prove that Poisson's distribution is a limiting case of Binomial distribution?
11. What do you understand fitting of a Poisson's distribution? Explain?
12. Define Normal distribution with its characteristics?
13. Write down on the different properties of Normal distribution?
14. Write under what condition Normal distribution is an approximation to Poisson's and Binomial distribution?
15. Write down the probability density function of a standard normal variate?

Self Practice Questions

1. A random variable x has the following probability distribution

X	0	1	2	3
P	$\frac{1}{5}$	$\frac{2}{5}$	$\frac{1}{5}$	$\frac{2}{5}$

Find the mean and hence calculate the variance of the distribution?

2. A random variable x has the following probability distribution

X	2	4	6	8
P	0.4	0.3	0.2	0.1

Find the mean and the standard deviation?

3. If an unbiased die is tossed and 'x' represents the number of points on a die then calculate the expectation.
4. A and B play against each in a poker's game. The rule of the game that the player who firsts throws '6' wins the game. A gets the first opportunity to cast the die and if he fails then B throws and if he fails again A throws and the process continues. Find their respective probabilities?

5. A bag contains 4 white balls and 5 black balls and 4 balls are drawn at random without replacement. If x denotes the random experiment of drawing black balls, construct a probability distribution of x?

6. The probability mass function of a random variable (X) is given by

$$f(x) = \begin{cases} 3p & \text{for } x = 1 \\ 2p & \text{for } x = 2 \\ p & \text{for } x = 3 \\ 0 & \text{otherwise} \end{cases}$$

Calculate the value of p and E(X). Also calculate the standard deviation?

7. If x is discrete random variable having the following probability distribution

X	1	2	3	4	5
P	k	4k	k	5k	2k

1. Find the value of k?
2. Find $p(x \geq 3)$?

8. A coin is tossed until tail appears what is the expected number of tosses?

9. A random variable x has the following probability distribution

X	0	1	2	3	4	5
P	0	K	3k	5k	7k	9k

Find

(i) The value of k?

(ii) $p(2 < x < 4)$?

(iii) $p(x > 3)$?

10. Three unbiased coin is tossed. If x denotes the number of heads. Construct the probability distribution of 'x' by writing all the possible outcomes and hence calculate the expected value of standard deviation of x?

Binomial Distribution

1. A die is thrown 4 times if getting an odd number is a successful event then what is the probability of
   (i) Exactly 3 scuccess?
   (ii) At least 3 Success?
2. If 10 coins are tossed simultaneously find the probability of getting
   (i) 5 heads?
   (ii) At least 8 heads?
3. Animesh Infrastructure has published in national newspaper regarding to job opening in the company. If 10 people has gone through the selection process with the probability of success being 80%. Then show that only 3 candidates have received the call letters?
4. If the probability of defective bulb is $\frac{1}{5}$, find the following for the binomial distribution of defective bulb in a total size of 200
   (i) Mean?
   (ii) Variance?
   (iii) Moment of Skewness.
   (iv) Moment of Kurtosis?
5. If the average number of defective reported in a particular gift item is 1 in 10. Then find the probability of getting exactly 4 defectives in a packet of 20 articles selected at random?
6. Kaun Banega Crorepati a highly successful television-show has reported 90% viewership of the programme. Find what is the probability that at least 90% of the viewers in a random sample of 10 watch the programme?
7. A lock manufacturer has reported that the probability of defective locks is 0.3. Find the mean and variance of the defective locks in a lot size of 1000.
8. Odds in favour of Pramita winning a game of chess against Vidhushi is 5:2. Find the probability out of 8 games the probability of Vidhushi winning 2 games?

9. The probability associated with the case of cars stolen and recovered in Lucknow city is $\frac{1}{2}$. Find the probability that at least one out of 10 cars stolen in a particular day will be recovered?

10. The probability that a bomb dropped hitting a target is $\frac{1}{4}$. If two bombs are enough to destroy the target and if 5 bombs are dropped at the target, what is the probability that the target is destroyed?

Poisson Distribution

1. The probability that a certain car will have flat tyre while travelling on an over - bridge is given by p = 0.0001. If 10,000 vehicles are driven over the bridge then find the probability that not more than one car will have a flat type?

2. Zorrik Blades manufacturer of razor blades found that one fifth percent of the blades turn out to be defective. The blades are supplied in packets of 10. Use Poisson distribution to calculate the approximate number of packets with single defective blades in a consignment of 10,000?

3. The probability that a man aged 35 years will expire before reaching 40 years is 0.03. In a lot of 200 men in the age bracket of 30 years, find what is the approximate probability that exact two men will die within the next 5 years?

4. In corporate office between 10 and 11 A.M the average number of incoming phone calls per minute is 3. Then find the probability that during a particular minute there will be

   (i) No phone calls?

   (ii) Exact 3 Phone calls?

5. The number of cases of accidents reported on a highway during a month follows a Poisson distribution with mean 8, find the probability that in a particular month the number of accidents will be

   (i) Not more than 1

   (ii) exact 2 accidents?

6. Air Fly a low cost airlines has reported 3 air crashes in last 30 years of services offered. The company wants to make the coming decade of operation free from any such incidence. Presuming similar trend to continue find what is the probability that Air fly meets the target?
7. In a coal mine the incidence of accident cases found to be on the average 2 per year. Find the probability that in a particular year there will be at the most 1 accident?
8. Baush and Comb a manufacturer of eye lenses has received complain regarding lens with outside chance of $\frac{1}{1000}$. If the lenses are supplied in packet of 10, find the probability that there exactly
   (i) Zero defect?
   (ii) Exactly 2 defects in a lot size of 10,000?
9. Assuming typing error per page committed by a typist following a Poisson distribution, find the expected frequencies of the following distribution of typing mistakes

No. of Errors per page.	No. of pages.
0	30
1	25
2	15
3	10

10. 200 audio players are inspected as they come off the production process and the number of defects per set as recorded are furnished below

No. of defects.	No. of Sets.
0	140
1	30
2	20
3	7
4	3

Normal Distribution

1. Find the probability that the standard normal variate is
   (i) less than 1.26?
   (ii) greater than 1.46?

2. In a normal distribution 30% of the items are under 50 and 12% over 72. Find the mean and standard deviation of the normal distribution?

3. A job opening in Krishna Textiles has witnessed 100 female candidates appearing for the interview. The mean age of the group was found to be 25 yrs and standard deviation 4 years. If the age of the candidates are normally distributed, find how many range between 20 to 28 years?

4. If the mean of a normal distribution is 100 and 10% of the values are greater than 110, then find the standard deviation?

5. Find the area under the standard normal curve

   (i) Between – 0.56 to + 0.56?

   (ii) Between – 2.16 to – 1.16?

6. For a normal distribution with mean = 70 and S.D = 20, find the value of that has

   (i) 25% of area to its right?

   (ii) 15% area to its left?

7. An infrastructure Co. taken a project to build a runway for an airport project and the time taken to construct the runway is normal variate with mean as 450 labor days and standard deviation 150 labour days. If the Company promises to construct the runway in 500 labor days or agrees to pay a penalty for the delay in excess of 500 at the rate of Rs. 5000 for each labor day, find that the company lands up paying Rs. 100000 as penalty?

   (Hint : Calculate the additional labor days using the penalty paid i.e. Extra days = Penalty / Penalty per day)

8. A sample of 100 car batteries are tested to find the length of life has produced the following result

   (a) The mean of the sample is 24 months?

   (b) The Standard deviation of the sample is 4 months?

   Assuming the data exhibited normal distribution, find what percentage of car batteries are expected to have

   (i) life less than 20 months?

   (ii) life more than 28 months?

9. Surya Bulbs Ltd manufacturer of electric bulbs found that the 5% of the bulbs manufactured are defective then find what is the probability that out of 300 bulbs selected less than 10 are defective?

10. The average number of customer visiting Venky's Super Mart is 576 per week. Find the probability that there being more than 525 customers per week?

Answers

Random Variable

1. 2, 0.8
2. 4, 2
3. $\frac{17}{2}$
4. $\frac{6}{11}, \frac{5}{11}$
5.

x:	0	1	2	3	4
p:	$\frac{^4C_4}{^9C_4}$	$\frac{^4C_3 \times {^5C_1}}{^9C_4}$	$\frac{^4C_2 \times {^5C_2}}{^9C_4}$	$\frac{^4C_1 \times {^5C_3}}{^9C_4}$	$\frac{^5C_4}{^9C_4}$

6. $\frac{1}{10}$, 1, 1
7. $\frac{1}{13}$, 0.53
8. 2.
9. $\frac{1}{25}, \frac{1}{5}, \frac{16}{25}$
10.

x:	0	1	2	3
p:	$\frac{1}{8}$	$\frac{3}{8}$	$\frac{3}{8}$	$\frac{1}{8}$

S.D = 0.866

Binomial Distribution

1. $(i)\,{}^4C_3\left(\frac{1}{2}\right)^4, (ii)\left(\frac{1}{2}\right)^4\left({}^4C_3 + {}^4C_4\right)$

2. $((i)\,{}^{10}C_5\left(\frac{1}{2}\right)^{10}, (ii)\left(\frac{1}{2}\right)^{10}\left[{}^{10}C_8 + {}^{10}C_9 + {}^{10}C_{10}\right]$

3. ${}^{10}C_3\left(\frac{4}{5}\right)^3\left(\frac{1}{5}\right)^3$

4. $40, 32, \frac{3}{20\sqrt{2}}, \frac{1}{800}$

5. ${}^{20}C_4\,(0.1)^4\,(0.9)^{16}$

6. ${}^{10}C_9\left(\frac{9}{10}\right)^9\left(\frac{1}{10}\right)^1 + {}^{10}C_{10}\left(\frac{9}{2}\right)^{10}$

7. 300, 63

8. ${}^8C_2\left(\frac{2}{7}\right)^2\left(\frac{5}{7}\right)^6$

9. $1-\left(\frac{1}{2}\right)^{10}$

10. $\frac{47}{128}$

Poisson's Distribution

1. (0.734)
2. 196
3. (0.0432)
4. (i) 0.050, (ii) 0.225
5. (i) 0.0297, (ii) 0.1056
6. 0.905
7. 0.27

8. (i) $e^{-0.01}$, (ii) $\frac{e^{-0.01}}{100}$

9. 29.44, 31.02, 16.44, 5.8

10. 120.1, 61.85, 15.92, 0.35

Normal Distribution

1. (i) (0.1038) (ii) (0.0721)
2. Mean = 57 and S.D (σ) = 13
3. (0.6678)
4. (7.81)
5. (i) (0.4246), (ii) (0.1076)
6. (i) (83.4), (ii) (49.2)
7. (0.3228)
8. (i) (31 approx.), (ii) (22 approx.)
9. 0.4066.
10. 0.9830.

Appendix

Exponential Distribution

The probability density function for the exponential distribution is given by

$$f(x) = \begin{cases} \alpha\, e^{-\alpha x}, & \text{for } 0 < x < \infty \\ 0 & \text{otherwise,} \end{cases}$$

where α is a given positive parameter. The distribution is also known as negative exponential distribution.

The exponential density function is characterized by exponential decreasing function with maximum at x = 0 and the maximum value α. And the mean and the variance of the distribution are

$$\text{Mean} = \frac{1}{\alpha}$$

$$\text{Variance} = \frac{1}{\alpha^2}$$

Exponential distribution and Poisson's distribution have close relation. For instance if Poisson's distribution represent number of failures per unit time in contrast exponential distribution represents the time between two successive failures. And find application in waiting line theory or Queuing theory.

Note :

The cumulative density function of the exponential distribution is

$$f(x) = \int_0^x \alpha e^{-\alpha x} dx = 1 - e^{-\alpha x}$$

Beta Distribution

The probability density function for a beta distribution is given by

$$f(x) = \frac{x^{m-1}(1-x)^{n-1}}{\beta(m,n)}$$

where $0 \le x \le 1$; $m , n > 0$ and

$$\beta(m,n) = \int_0^1 x^{m-1}(1-x)^{n-1} dx \text{ is the beta function.}$$

The mean and the variance of the random variable x in beta distribution is given by

$$\text{Mean} = \frac{m}{m+n}$$

$$\text{Variance} = \frac{mn}{(m+n)^2(m+n+1)}$$

This distribution finds application PERT technique where activity times are estimated under a specific range.

CHAPTER 18

Linear Programming Formulation and Graphical Method

Brief History

Linear programming was developed by George B Dantzing also referred as the father of linear programming; his work was primarily focused on finding methods to support logistical problems of US Air force during World War II. His research received further support from fellow scholars like Neumann, Hurwitz and Koopmans who were also working on the same line during that period. Initially the technique was originally called as "*Programming of interdependent activities in a linear structure*" which was shortened to "*Linear Programming*". Further Charnes. A and Cooper W.W played a vital role in refinement and introducing the concept to industrial problems.

Introduction

The term linear programming consists of two words one linear which suggests that the relationship between the variables in the objective function and the constraints are linear in nature and that the word programming suggest the process of determination of a course of action or an action plan. Thus linear programming is a mathematical technique to find the most optimal solution of a problem by the allocation of limited resources subjected to a set of constraints. The function which needs to be optimized is called the objective function and the parameters which set a bounded limit to resources are called constraint functions.

According to Samuelson, Dorfman and Solow linear programming is "The analysis of problems in which linear function of a number of variables is to be maximized or minimized when those variables are subjected to a number of restraints in the form of inequalities".

Whereas in the words of David Smith "Linear Programming is an optimization technique for finding an optimal (maximum or minimum) value of a function, called objective function of several independent variables, the variables being subjected to various constraints (or restrictions) expressed as equations or inequalities. The term linear indicates that the function to be maximized or minimized is linear in nature and the corresponding constraints represented by a system of linear inequalities or linear equations involving the variables"

Basic Requirement of Linear Programming

(i) *Decision Variables*

Decision variable refers to whose variables which are competing against one another for allocation of limited resources, usually inter related in terms of resource utilization and required synchronized solutions with linear relationship existing among them.

(ii) *Clearly defined objective function*

The objectives needs to be clearly identified either to maximize contribution in terms of profit or to minimize cost within the limited restriction placed on resources. The objective function must be able to be expressed in mathematical form as,

Optimize (Maximize or Minimize) $Z = c_1x_1 + c_2x_2 + c_3x_3 + \dots\dots\dots + c_nx_n$

where Z is the measure of performance variable, which is the function of $x_1, x_2, x_3, \dots x_n$ and $c_1, c_2, c_3, \dots c_n$ represents the contribution of unit of the respective variable $x_1, x_2, x_3, \dots x_n$ to the measure of performance of Z.

(iii) *Constraints*

As the resources are not available in abundance there should be some limitation placed on the usage of resources which are allocated among the various competing activities. Limitation in terms of production capacity, manpower, space, raw material, market potential and time etc which places a restriction on the degree to which an objective can be achieved. These constraints must be able to be expressed as linear equalities or inequalities in terms of decision variables.

(iv) *Non-negative Restrictions*

All decision variables must assume non-negative values as the existence of negative values of the decision variables in regular course will imply an undesirable condition in real life situation. If x and y are the decision variables, their non-negative restrictions can be expressed as $x \geq 0$ and $y \geq 0$.

(v) *Quantitative Measurability*

It is essential to have each element being able to be measured in quantitative terms. Numerical values needs to be furnished so that the relationship among the variables can be calculated.

(vi) *Linear Relationship*

Linear programming essentially deals with problems in which the objective and the restraints can be expressed in linear mathematical form (as equations or inequalities). Basically linear relationship has two properties viz. first proportionality which means the relationships are directly proportional and second divisibility which means the variables are not restricted only to integer values; even fractional values are well accepted.

(vii) *Alternative Line of Action*

There must be alternative course of action to choose from. For instance different permutations and combinations should be made available in order to make selection among productive factors in terms of machines, times and raw materials.

(viii) *Feasible Solution*

A set of values of decision variables which equally satisfies all the given constraints and non-negativity condition is called as feasible solution. Generally a linear programming problem has a large number of feasible solutions to offer.

(ix) *Optimality in Solution*

As mentioned above a feasible solution which helps in reaching desired objectives is called as optimal solution. It is the best solution among the feasible solution yielding the highest value in maximization case or lowest in case of minimization case of the objective function.

Assumptions of Linear Programming Problems

In order to solve Linear Programming (LP) problems the following assumptions are made

(i) *Proportionality*

The amount of resources utilized and its contribution to the objective function must be in proportion to the value of decision variable. For example for a doing a particular unit of job if requires 3 man hours (resource) then to finish 3 units of the same job will require 9 man hours (resource).

(ii) *Certainty*

The availability and the consumption of resources per unit utilized must be known with certainty.

(iii) *Additivity*

The total resources utilization by different activities must be collectively equal to the sum of the individual resource requirements.

(iv) *Divisibility*

In this assumption divisibility means the variables are not restricted only to integer values as such, even fractional values are permitted and where divisibility of variables are not permitted there must be assigned integer values. Cases where the variable can only assume integer or discrete number of values, LP model is not applicable but an integer programming model can be applied.

(v) *Finiteness*

The number of activities which are included in linear programming problem must be finite so does the constraints.

Advantages of Linear Programming

(i) Linear programming helps in optimal utilization of scarce resources by effective and efficient allocation of these recourses.

(ii) Linear programming problem is suitable and flexible in analyzing multi dimensional tasks.

(iii) Linear programming provides quality in decision making by offering feasible and practical solution to problems by replacing intuition by objective analysis.

(iv) It provides a platform for offering solutions to changing conditions by re evaluation process. For example it is found that a particular plan partly executed found ineffective, in such cases the reminder of the plan can be modified in order to get best results.

(v) The major significance of linear programming model is to highlight the bottlenecks in situations which can arise due to variable environmental and economical factors.

Disadvantages of Linear Programming Problem

(i) Linear programming problems are applicable under certainty conditions but in real life situation of business it may not be possible to state all the conditions with absolute certainty.

(ii) Linear programming problem assumes linear relationship among decision variables however in reality neither the objective function nor the constraints are linearly related.

(iii) Linear programming model does not take into account the effect of time and uncertainty as such LP model should be defined in a manner to incorporate any internal or external change.

(iv) As the parameters appearing in the model are assumed to be constant but in reality the situation is exactly opposite i.e. parameters are unknown and uncertain.

(v) Linear programming deals with single objective whereas in real life situation they may be multi dimensional objectives.

(vi) Linear programming assumes that all inputs are quantifiable which are not so in reality.

(vii) Linear programming allows variables to take fractional values but in cases a fractional unit cannot be produced nor procured. In this situation rounding off to the nearest integer value may not yield optimal solution.

Business Application of Linear Programming Problem

(i) *Industrial Application*

Linear programming finds extensive usage in dealing with industrial problems. Some of the key areas are as follows:

- *Production planning*: The major objective is to reduce the cost of production operation keeping in view of the criticalities in terms of inventory position, manpower efficiency and constraints on production capacity.
- *Product mix*: A firm produces assortment of products with varying cost and selling pricing using limited resources. In such

cases it needs to be determined the marginal contribution of each product so as to maximize the total contribution or profitability.

- *Trim loss:* In manufacturing facilities the essential products are manufactured in standard size and shapes but the orders are placed for different sizes, quantities and shapes. The problem is to find the combination of requirements that should be produced from materials in order to reduce trim loss.
- *Blending problem:* In blending problem the goal is to determine the blending of ingredients in such a way that will achieve desired quality objectives at least cost combination.
- *Shipping at least Cost*: The transportation models of linear programming deals with finding optimal cost for transportation to various destined points for catering to market demands subject to various constraints.

(ii) *Managerial Application*

- *Financial Angle:* Linear programming is used in financial management essentially to create wealth and maximize profit from investments on capital goods and services. It also deals with the right blending of portfolio mix in order to reduce the risk and cost of investment.
- *Marketing Angle*: Selection of right media mix, physical distribution and routing of salesman are the some of the problems which can be dealt with linear programming models subject to various sales budget constraints.
- *Human Resource Angle*: LPP helps in manpower planning so as to bring efficiency in working, hiring and deploying coupled with fitting suitable individuals according to the job requirements.

(iii) *Miscellaneous Applications*

- *Agriculture*: Linear programming problem can be used in farm land planning to allocate scarce resources to derive maximum gain and reduce cost of cultivation.
- *Military*: Linear programming has found extensive usage in military applications in terms of logistical support extended to frontline soldiers who needs to be steadfast against any hostility or aggression from the enemy end.

General formulation of Linear Programming Model

Consider the following Linear programming Problem with n decision variables and m constraints as stated below

$$\text{Maximize } (Z) = c_1x_1 + c_2x_2 + c_3x_3 + \dots + c_nx_n$$

Subject to

$$\left.\begin{array}{l} a_{11}x_1 + a_{12}x_2 + a_{13}x_3 + \dots a_{1n}x_n \le b_1 \\ a_{21}x_1 + a_{22}x_2 + a_{23}x_3 + \dots a_{2n}x_n \le b_2 \\ \quad \vdots \qquad\qquad \vdots \qquad\qquad \vdots \\ a_{m1}x_1 + a_{m2}x_2 + a_{m3}x_3 + \dots a_{mn}x_n \le b_m \\ \text{and} \quad x_1, x_2, \dots x_n \ge 0 \end{array}\right\} \text{Constraints}$$

the above formulation can be expressed in concise form as

$$\text{Optimize (Max. or Min) } Z = \sum_{j=1}^{n} c_jx_j \quad \text{...Objective Function}$$

Subject to constraints

$$\sum_{j=1}^{n} a_{ij}x_j (\le, =, \ge); \quad (i = 1, 2, 3 \dots m) \quad \text{...Constraints}$$

And $x_j \ge 0$; $j = 1, 2, 3, \dots n$ ... (Non-negativity Conditions)

Where c_js are the coefficient per unit contribution of decision variables x_j in the objective function. The a_{ij}s are input-output coefficients or technological coefficients and represents the amount of resource consumed per unit variable x_j and a_{ij}s can take values which are positive, negative or zero and b_i is the total availability of the ith resource and it is assumed to have value greater than zero i.e. $b_i \ge 0$ for all i. If b_i found negative, in such case both side of the constraint is multiplied with – 1 to make $b_i > 0$ i.e. reverse the inequality.

Stages of Linear Programming Model

Basically LP problem contains three stages which includes:

(i) ***Problem Identification***

A stage in which the problem is identified which calls for enumeration of all available information with respect to problem and also establish the relationship among variables and specification of constraints.

(ii) *Problem Formulation*

A stage of developing a mathematical model from the available information. It calls for identifying the decision variables, specifying the objectives, setting the mathematical equations or inequations for the constraints and the objective function in a comprehensive format which is the most critical stage in LP formulation.

(iii) *Problem Solving*

In this stage appropriate method is selected for obtaining feasible solution of the problem and test for the optimality.

Guidelines for Formulation of Linear Programming Problem

(i) Study the available information and identify the decision variables.

(ii) Assigning proper symbols to the decision variables for e.g. (x_1, x_2, ...x_n).

(iii) Stating the feasible alternatives in terms of variables. Basically feasible alternatives are non negative constraints expressed as

$$x_1 > 0, x_2 > 0......$$

(iv) Identify the constraints or restriction and express them as linear equations or inequations in terms of decision varibles.

(v) Identify the objective function which needs to optimized and present it as a linear function of decision variables multiplied by their unit contributions.

(vi) At last express the objective function, the constraints and the non-negative conditions in a linear programming format.

Some illustrations of Linear Programming Formulation

Illustration No. 1

A firm manufacturing two products X and Y. In order to produce a unit of X requires 2 hours of machining and 1 hour of craftsmanship whereas Y requires 3 hours of machining and 2 hours of craftsmanship. If the total available machine hours per week is 100 hours and craftsmanship is 80 hours, then what should be the output produced of X and Y so that the profit is maximized ? Knowing that by selling each unit of X, Y generates profit at the rate of Rs. 12 and Rs. 10 respectively. Formulate the LP problem? (Production type model)

Solution:

Step 1: let us assume that

x: Number of units of X produced.

y: Number of units of Y produced.

where x and y are the decision variables.

Data Summary

Decision Variables	**Products**	**Machining (Hrs)**	**Craftsmanship (Hrs)**	**Profit per unit**
x	X	2	1	12
y	Y	3	2	10
Total Availability		**100**	**80**	

Step 2: The objective of the firm is to maximize profit (P), Hence the objective function can be expressed as

$$\text{Maximize Profit (P)} = 12\,x + 10\,y$$

Step 3: But the constraints are the available time for machining and craftsmanship and the time needs to be utilized so that the output level produce will maximize profit.

Therefore, the constraint can be expressed as mathematically

(i) Machining hours consumed by x units of X + Machining hours consumed by y units of Y ≤ Total available machine hours.

i.e. $2\,x + 3\,y \leq 100$

(ii) Craftsmanship hours consumed by x units of X + Craftsmanship hours consumed by y units of Y ≤ Total available craftsmanship hours.

i.e. $1\,x + 2\,y \leq 80$

Step 4: As both x and y are non-negative restrictions

We have $x, y \geq 0$

Hence the LP problem can be stated as

> Maximize $P = 12x + 10y$ (Objective function)
>
> Subject to
>
> $\left.\begin{array}{l} 2x + 3y \leq 100, \\ 1x + 2y \leq 80 \end{array}\right\} \longrightarrow$ Constraints
>
> with $x, y \geq 0$

Illustration No. 2

Animesh Chemicals Ltd. produces 3 brands of Soap Lily, Dew and Snow-White. And to produce Lily, Dew and Snow-White three different chemicals C_1, C_2 and C_3 are used in different proportions. One unit of Lily requires 3 units of C_1, 2 units of C_2 and 5 units of C_3; One unit of Dew requires 2 units of C_1, 5 units of C_2 and 4 units of C_3 and One unit of Snow-White requires 1 unit of C_1, 3 units of C_2 and 2 units of C_3. The available stock of the raw materials with the company are 24 units of C_1, 40 units of C_2 and 30 units of C_3. The profit earned by selling Lily, Dew and Snow-White are Rs. 3, Rs. 4 and Rs. 6 respectively. Formulate the LP problem for profit maximizing condition?

Solution:

Step 1

Let us assume that

x: Number of units of Lily produced.

y: Number of units of Dew produced.

z: Number of units of Snow-White produced

where x, y and z are decision variables.

Data Summary

Decision Variables	Products	Raw Materials			Profit per unit
		C_1	C_2	C_3	
x	Lily	3	2	5	3
y	Dew	2	5	4	4
z	Snow-White	1	3	2	6
Total Availability		**24**	**40**	**30**	

Step 2

The objective of the firm is to maximize profit (P), Hence the objective function can be expressed as

$$\text{Maximum (P)} = 3x + 4y + 6z \text{ (Objective function)}$$

Step 3

But the profit maximization output is hindered by the constraints in available raw materials.

Therefore, the constraint can be expressed as mathematically

(i) Units of C_1 used in Lily + Units of C_1 used in Dew + Units of C_1 used in Snow-White $\leq$ Total Availability of C_1

i.e. $3x + 2y + z \leq 24$,

(ii) Units of C_2 used in Lily + Units of C_2 used in Dew + Units of C_2 used in Snow-White $\leq$ Total Availability of C_2

i.e. $2x + 5y + 3z \leq 40$,

(iii) Units of C_3 used in Lily + Units of C_3 used in Dew + Units of C_3 used in Snow-White $\leq$ Total Available of C_3

i.e. $5x + 4y + 2z \leq 30$

Step 4

We cannot have negative units of x , y and z as x , y and z ≥ 0

Hence the LP problem can be stated as

Maximize $P = 3x + 4y + 6z$ (Objective function)

Subject to

$$\left.\begin{aligned} 3x + 2y + z &\leq 24, \\ 2x + 5y + 3z &\leq 40 \\ 5x + 4y + 2z &\leq 30 \end{aligned}\right\} \rightarrow \text{Constraints}$$

With x, y and z ≥ 0.

Illustration No. 3

A growing child as prescribed by a child nutritionist should regularly intake Vitamins A and B in definite proportions for healthy grow. In the market natural food supplements F_1 and F_2 are available; both are enriched sources of vitamins A and B with F_1 containing 2 units of Vitamin A and 4 units of Vitamin B whereas F_2 containing 4 units of Vitamin A and 2 units of Vitamin B. If F_1 and F_2 are priced at Rs. 4 and Rs. 5 per unit respectively and the minimum dietary requirement is 80 and 70 units of Vitamin A and B. Neglecting any excess intake as harmless, find out the optimizing mixture of F_1 and F_2 to minimize cost and also meet the daily dietary requirements? Formulate the LP problem (Dietary type problem)

Solution:

Step 1 :

Let us assume that

x: Number of units of F_1 consumed.

y: Number of units of F_2 consumed.

where x and y are decision variables

Data Summary

Decision Variables	Foods Consumed	Contents of Vitamins in Food Supplements		Cost per unit
		A	B	
x	F_1	2	4	4
y	F_2	4	2	5
Min. Vitamin Req.		**80**	**70**	

Step 2:

The objective is to minimize cost (Z), Hence the objective function can be expressed as

$$\text{Minimize (C)} = 4x + 5y$$

Step 3 :

But the minimum dietary requirements Vitamins A and B are 80 and 70 units respectively. Therefore the above constraints can be expressed mathematically as

$$2x + 4y \geq 80$$

$$4x + 2y \geq 70$$

(Note: Follow the matching techniques as shown in earlier examples)

Step 4:

As both x and y are non negative restrictions,

we have $x, y \geq 0$

Hence the LP problem can be stated as

Minimize (C) = $4x + 5y$ (Objective function)

Subject to

$2x + 4y \geq 80$,
$4x + 2y \geq 70$ } → Constraints

With $x, y \geq 0$

Illustration No. 4

Vinod who has just retired from his service wants to invest a portion of his retirement fund into bond market. His investment consultant has recommended in investing in two types of bonds B_A and B_B with AAA rating and A+ rating. The bond B_A yields a 10% return against which the bond B_B yields 12% return on the amount invested. If Vinod invest in all Rs. 1,00,000 with due consideration to invest at least Rs. 20,000 in bond B_A and not more than Rs. 60,000 in bond B_B and he wants to invest in bond B_A as in bond B_B. What should be the consultant's advice in order to maximize his return on investment keeping in view with the risk involved in investment? Formulate the linear programming problem? (Investment type problem)

Solution:

Step 1 :

Let us assume that

x: The amount invested in bond B_A.

y: The amount invested in bond B_B.

where x and y are decision variables

Step 2:

The objective of the Vinod is to maximize his return on investment (R). Hence the objective function can be expressed as

Maximize Return on Investment (ROI) = $0.10x + 0.12y$

Step 3 :

Subject to constraints are

$x + y \leq 1{,}00{,}000$ (Investable Sum).

$x \geq 20{,}000$ (Constraint in the investment in bond B_A).

$y \leq 60{,}000$ (Constraint in the investment in bond B_B).

Step 4:

As both x and y are non negative restrictions as investment cannot be negative therefore we have $x, y \geq 0$

Hence the LP problem can be stated as

Maximize (R) = $0.10x + 0.12y$ (Objective function)

Subject to

$$\left.\begin{array}{l} x + y \leq 1{,}00{,}000, \\ x \geq 20{,}000, \\ y \leq 60{,}000 \end{array}\right\} \longrightarrow \text{Constraints}$$

with $x, y \geq 0$

Illustration No. 5

Kamkhya furniture a wholesale dealer in furniture deals in only two items viz. office tables and chairs. The firm has invested Rs.1,00,000 for acquisition and a showroom space to store at most 90 pieces of tables and chairs. If a table cost Rs.1500 and that a chair Rs. 1,000 and that the firm earns Rs. 250 as profit by selling a table and that of Rs. 150 by selling a chair. Assuming that the firm sells all the items that have been acquired, find the product mix such that the furniture dealer's profit is maximized. Formulate the above LPP? (Space Constraint Type)

Solution:

Step 1:

Let us assume that

x: The number of tables acquired and sold.

y: The number of chairs acquired and sold.

where x and y decision variables.

Step 2:

The objective of the furniture dealer is to maximize his profit. Hence the objective function can be expressed as

$$\text{Maximum (P)} = 250\,x + 150\,y$$

Step 3 :

Since the cost of each table is Rs. 1500 and that of chair is Rs. 1000 therefore the total cost of x units of tables and y units of chairs is

$1500x + 1000\,y \leq 1{,}00{,}000$ (Maximum fund available for acquisition)

And, the maximum number of pieces of tables and chairs that can be stored is

$x + y \leq 90$ (Constraint of storage space)

Step 4:

As both x and y are non negative restrictions therefore we have $x, y \geq 0$. Hence the LP problem can be stated as

Maximize $P = 250\,x + 150\,y$ (Objective function)

Subject to

$$\left.\begin{array}{l} 1500x + 1000\,y \leq 1{,}00{,}000, \\ x + y \leq 90 \end{array}\right\} \longrightarrow \text{Constraints}$$

$x, y \geq 0$

Illustration No. 6

A manufacturer producing two different types of bearings B_1 and B_2 requires the use of three different machineries M_1, M_2 and M_3. The time required for each bearing in days and the total time available in days in each machine are as follows

Machine	B_1	B_2	Available days
M_1	4	2	140
M_2	2	2	80
M_3	2	6	180

If the profit earned in bearing B_1 is Rs. 120 and that of bearing B_2 is Rs. 80, find how many units of each of the bearing needs to be manufactured so as to maximize the profit of the manufacturer. Formulate the above linear programming problem?

Solution:

Step 1

Let us assume that

x: The number of bearing B_1 manufactured

y: The number of bearing B_2 manufactured

x and y are decision variables

Data Summary

Machine	B_1	B_2	Available days
M_1	4	2	140
M_2	2	2	80
M_3	2	6	180

Step 2

The objective of the furniture dealer is to maximizing his profit. Hence the objective function can be expressed as

$$\text{Maximum (P)} = 120\,x + 80\,y$$

Step 3

But the time limitation of the three machines can represented as

$$4x + 2y \leq 140 \text{ (Maximum available time in } M_1)$$

And

$$2x + 2y \leq 80 \text{ (Maximum available time in } M_2)$$

Similarly

$$2x + 6y \leq 180 \text{ (Maximum available time in } M_3)$$

Step 4

As both x and y are non-negative restrictions therefore we have $x, y \geq 0$.

Hence the LP problem can be stated as

$$\text{Maximum } (P) = 120x + 80y$$

Subject to

$$\left.\begin{aligned} 4x + 2y &\le 140 \\ 2x + 2y &\le 80 \\ 2x + 6y &\le 180 \end{aligned}\right\} \longrightarrow \text{Constraints}$$

and $x, y \ge 0$

Illustration No. 7

A mobile manufacturer has introduced a new type of touch screen mobile into the market. In order to promote the product the firm has taken the assistance of an advertising company for media campaign and in the process is ready to spend Rs. 10,00,000 for the promotional work. The advertisement firm after brainstorming has reduced to using three different platforms for promotional work with respect to exposure and visual impact to attract the customers; they are the bill boards, the news paper and the television. The agency's own research data on cost effectiveness of the advertising media has the following information

Media	Cost per unit space(Rs.)	Estimated Hits
Bill Boards	30,000	1,00,000
News Paper	60,000	4,00,000
Television	1,00,000	8,00,000

The mobile phone company wants to maximize exposure to customers and at the same time restricting maximum expenditure on bill board advertisement to Rs. 1,50,000 and Rs. 3,00,000 for newspaper and at the same time wants to target at least 10,00,000 prospective customer by the TV commercial. Formulate the LPP which maximizes the mobile manufacturer's goal? (Advertizing Type)

Solution:

Step 1 :

Let us assume that

x: The number of Bill Boards advertising units space needs to be purchased.

y: The number of News Paper advertising units space needs to be purchased.

z: The number of Television advertising units space that needs to be purchased.

x, y and z are decision variables

Step 2:

The objective is to maximize the number of exposures to customers of the advertisement through all of the following three mediums, therefore the objective function can be expressed as

$$\text{Maximum } (Z) = 1,00,000\,x + 4,00,000y + 8,00,000z$$

Step 3:

But following are the constraints,

(i) The maximum total expense on advertisement cannot exceed Rs. 10,00,000

Therefore,

$30,000x + 60,000y + 1,00,000z \le 10,00,000$

(Advertizing Budget Constraint)

(ii) The expenditure incurred on Bill Board advertisement cannot exceed Rs. 1,50,000 i.e.

$30,000\,x \le 1,50,000$

(iii) The expenditure on news paper not exceed Rs. 3,00,000

$60,000\,y \le 3,00,000$

(iv) At least 10,00,000 prospective customers i.e.

$8,00,000\,z \ge 10,00,000$

Step 4

As both x, y and z are non-negative restrictions therefore we have x, y and z ≥ 0. Hence the LP problem can be stated as

$$\text{Maximum } (Z) = 1,00,000\,x + 4,00,000y + 8,00,000z$$

Subject to

$30,000x + 60,000y + 1,00,000z \le 10,00,000$ (Ad. Budget Constraint)

$$\left.\begin{aligned} 30,000\,x &\le 1,50,000 \\ 60,000\,y &\le 3,00,000 \\ 8,00,000\,z &\ge 10,00,000 \end{aligned}\right\} \longrightarrow \text{Constraints}$$

where we have x , y and z ≥ 0

Illustration No. 8

A fertilizer manufacturer is selling bio fertilizer into the market essentially produced from two types of ingredients X and Y. The major compositions of the ingredients are as follows

Ingredients	Nitrogen	Phosphorous	Calcium
X	30%	40%	30%
Y	20%	60%	20%

The company decided that the fertilizer should contain

(i) At least 45% Phosphorous.

(ii) At least 25% Nitrogen.

(iii) At least 20% Calcium.

(iv) And must be sold in bags of 10 kg.

The cost of each ingredient i.e. X is Rs. 100 per kg and that of Y is Rs. 90 per kg. Find the manner in which the ingredients must be mixed in order to minimize the cost for the fertilizer manufacturer. Formulate the above LPP? (Mixing Type Problem)

Solution:

Step 1

Let us assume that

x: The amount of ingredients X in a fertilizer bag of 10 kg

y: The amount of ingredients Y in a fertilizer bag of 10 kg

Step 2

The objective is to minimize the cost (C) of the ingredients i.e.

$$\text{Minimize Cost (C)} = 100x + 90y$$

Step 3

Now each bag of fertilizer must contain 25% Nitrogen in a bag of 10 kgs therefore, the weight must be at least 2.5 kg

$$0.30x + 0.20y \geq 2.5$$

Similarly the net contain of Phosphorus and Calcium in the bag must exceed 4.5 Kg and 2 Kg respectively, therefore, we have

$$0.40x + 0.60y \geq 4.5 \text{ and } 0.30x + 0.20y \geq 2$$

Step 4

As both x and y are non-negative restrictions therefore we have x and $y \geq 0$. Hence the LP problem can be stated as

Minimize Cost (C) = 100x + 90 y

Subject to

$$\left.\begin{array}{l} 0.30\,x + 0.20\,y \geq 2.5 \\ 0.40x + 0.60y \geq 4.5 \\ 0.30x + 0.20\,y \geq 2, \end{array}\right\} \longrightarrow \text{Constraints}$$

where $x, y \geq 0$

Graphical Method of Solving Linear Programming Problem

Graphical method is one of the most sort methods to solve linear programming problem with two variables. In this chapter we will discuss on two graphical solution approaches – "The Corner or Extreme Point Method" and then "The Iso-Profit or Iso-Cost method".

Some of the key terminologies

(i) ***Constraints***

A constraint sets the upper limit for the available resource or the lower limit for desired level to achieve.

(ii) ***Corner Points or Vertices***

Corner points are the intersection of two or more boundaries of the feasible region.

(iii) ***Feasible Region***

The region defined by the constraints and the non-negative constraints of a linear programming problem

(iv) ***Solution***

The set of values of decision variables which satisfies the constraints of the linear programming problem is the solution of the LP problem.

(v) *Feasible Solution*

The set of values of decision variable satisfying all the constraints and non-negative conditions of the LP problem simultaneously are termed as feasible solutions

(vi) *Infeasible Solution*

The set of values of decision variable which does not satisfy all the constraints and non-negative conditions of the LP problem simultaneously are termed as infeasible solutions

(vii) *Basic Solution*

For a given set of m simultaneous equations in n variables where n > m, a solution obtained by setting (n – m) variables equal to zero and solving the rest m equations in m variables is called as basic solution. The (n – m) whose values have not appeared in this solution are termed as non basic variables and the rest m variables are called basic variables.

For example

Consider 4 unknowns a, b, c and d and only two equations. In such case you can set any two variables say a and b to zero i.e. a = b = 0 and now solve the equations to derive the value of c and d. Solution found in this manner is called as basic solution

(viii) *Basic Feasible Solution*

A feasible solution which is also a basic solution of a linear programming problem is a basic feasible solution.

(ix) *Optimal Solution*

Any feasible solution which optimizes the objective function whether maximization or minimization case is called optimal solution.

(x) *Unbounded Solution*

A LP problem which is permitted to have infinitely large solution can be termed as unbounded solution.

(xi) *Degenerate Solution*

A basic feasible solution is termed as degenerate if one or more of the basic variables becomes zero.

(xii) *Non-degenerate Solution*

A basic feasible solution is termed as non-degenerate if one or more of the basic variables becomes non-zero and positive.

Some Important Theorems

While solving the linear programming problem for optimal solution following statements of some of the important theorems are utilized

(i) The feasible solution region comprises of a convex set.

(ii) Feasible solution corresponding to the corner or extreme points within the feasible region constitutes the feasible solution.

(iii) Within the feasible region the number of basic feasible solutions are finite.

(iv) If there exists a optimum feasible solution then it will occur at one or more of the extreme points of the basic feasible solution.

Note

Convex Set is a polygon and by convex we mean if any two points in the polygon when randomly selected then the straight line joining the two points will completely lie within the polygon. The extreme points of the convex set are the basic feasible solutions of the linear programming problem.

The Corner or Extreme Point Method

The following are the steps involved in this method

(i) First develop the linear programming model.

(ii) Plot the constraints by converting the inequalities into equalities and identify the feasible region.

(iii) Find the coordinates of each corner or extreme point of the feasible region by solving the simultaneous equations.

(iv) Substitute the coordinates in the objective function to calculate the profit or cost.

(v) Identify the corner or extreme point offering the optimal solution i.e. highest profit (maximization case) or least cost (minimization case).

Illustration No. 1

Maximize $Z = 30x + 20y$

Subject to constraints

$$5x + 2y \le 150$$

$$x + y \le 60$$

where $x, y \ge 0$.

Solution:

Since the variable x, y are non- negative in nature i.e. x, y ≥ 0 thus the solution set lie on the first quadrant of the xy-plane.

In order to plot the constraints we convert the inequalities into equalities

Therefore, we get,

$$5x + 2y = 150.$$

$$x + y = 60.$$

Since the constraints represent linear functions, they are represented as straight lines which requires two joining points to plot them and the points for each line can be identified by finding the intercept for the both axes (x-axis and y-axis). The intercept for x-axis is obtained by putting y = 0 in the constraint and the intercept of y-axis is obtained by putting x = 0 in the constraint.

Therefore substituting y = 0 in 5x + 2y = 150, we get,

$$5x = 150 \text{ or } x = 30$$

Similarly substituting x = 0 in 5x + 2y = 150, we get,

$$2y = 150 \text{ or } y = 75$$

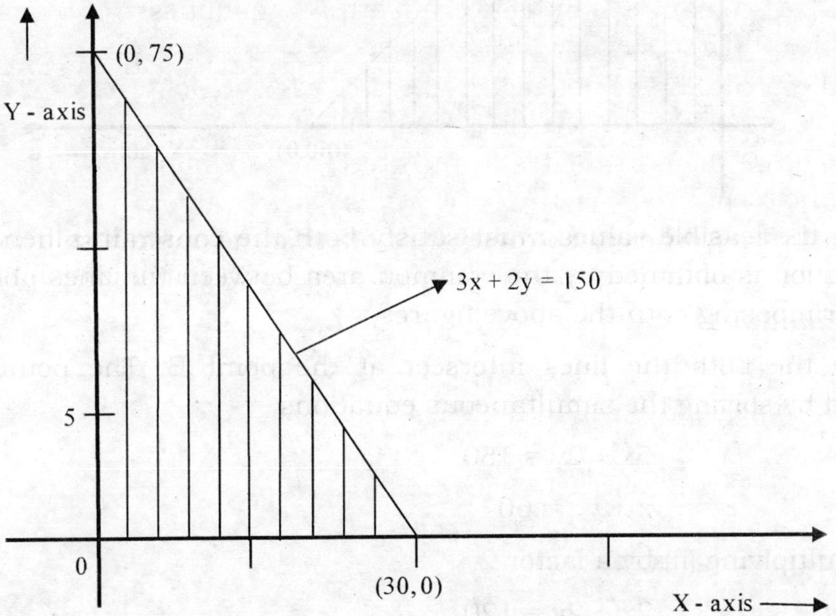

Therefore plotting of the straight line 5x + 2y = 150 can be done by joining the x – intercept 30 with the y- intercept 75 i.e. the straight line joining the point (30, 0) on x-axis with (0, 75) on y-axis. Since the constraint is (≤) type the feasible region will lie on or below the constraint.

Similarly we can plot the line x + y= 60 by substituting y = 0 in x + y = 60, we get x = 60 and substituting x = 0, we get y = 60. On plotting the line by the joining the points (60, 0) on x – axis with (0,60) on y-axis (≤) constraint type, we get the following figure as shown below

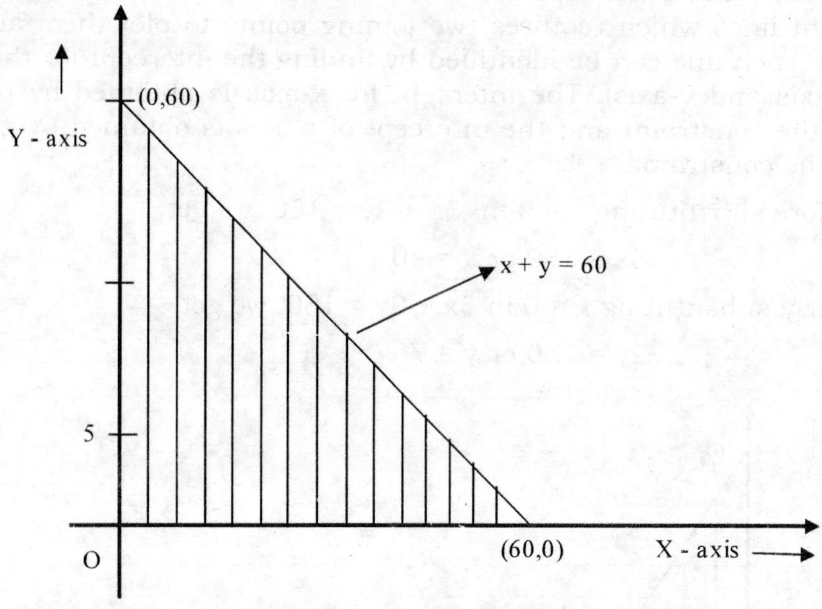

Since the feasible values must satisfy both the constraints hence the combination is obtained by the common area between the lines obtained by superimposing both the above figures.

Since the both the lines intersect at the point B. The point B is obtained by solving the simultaneous equations

$$5x + 2y = 150. \quad \text{.....(i)}$$

$$x + y = 60. \quad \text{.....(ii)}$$

By multiplying (ii) by a factor 2,

we get $$2x + 2y = 120 \quad \text{.....(iii)}$$

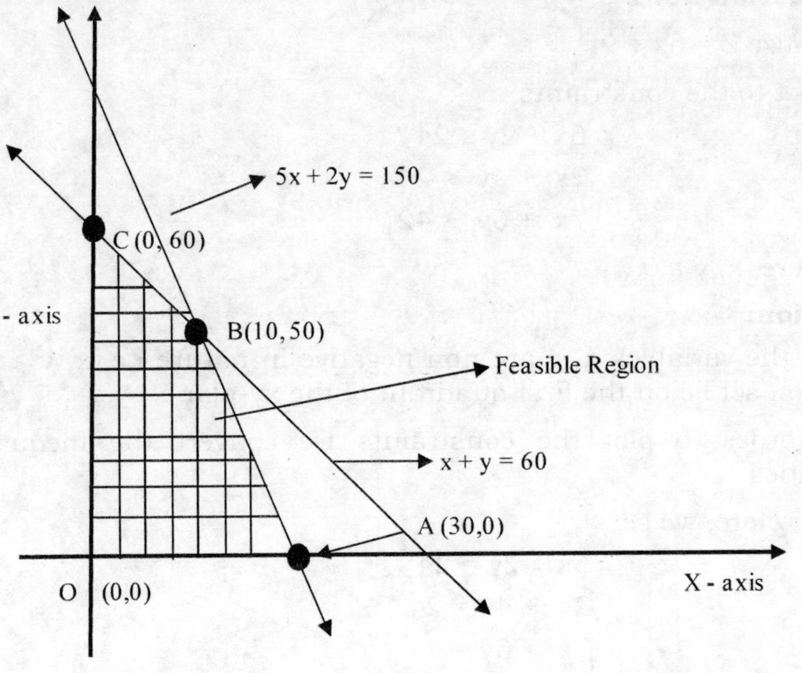

Subtracting (i) and (iii)

We get x = 10 and the value of y can be obtained from (ii). By substituting the value of x we get y = 50

Therefore the polygon OABC represents the region of values of x and y satisfying the constraints as well as the non-negative constraints called as feasible solution of LPP.

The corners of the feasible region are O(0,0), A (30,0), B(10, 50) and C (0,60)

The value of the objective function at these points are

$$Z_O = 30 \times 0 + 20 \times 0 = 0$$

$$Z_A = 30 \times 30 + 20 \times 0 = 900$$

$$Z_B = 30 \times 10 + 20 \times 50 = 1300$$

$$Z_C = 30 \times 0 + 20 \times 60 = 1200$$

Since the objective of the function is to maximize Z and Z is maximized at B (10, 50) and thus the optimal solution is x = 10 and y = 50 and the optimal value is 1300.

Illustration No. 2

Maximize Z = 6x + 9y

Subject to the constraints

$$6x + 2y \leq 24$$
$$2x + 4y \leq 24$$
$$7x + 6y \leq 42$$

where x, y ≥ 0.

Solution:

Given the variables x, y are non-negative in nature i.e. x, y ≥ 0 thus the solution set lie on the first quadrant of the xy-plane.

In order to plot the constraints we convert the inequalities into equalities

Therefore, we get,

$$6x + 2y = 24$$
$$2x + 4y = 24$$
$$7x + 6y = 42$$

Since the constraints represent linear functions, they are represented as straight lines which requires two joining points to plot them and the points for each line can be identified by finding the intercept for the both axes (x-axis and y-axis). The intercept for x-axis is obtained by putting y = 0 in the constraint and the intercept of y-axis obtained by putting x = 0 in the constraint. Thus the above functions can be represented as

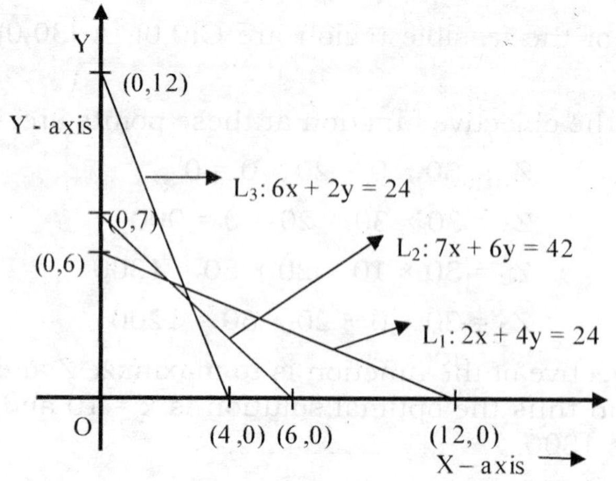

As we have identified each line as L_1, L_2 and L_3 solving each of three equations we get the point of intersection of L_1 and L_2, L_2 and L_3 and L_1 and L_3. Let us represent the boundaries in the graph as A, B, C and D.

Since all the lines are all less than equal type (≤) therefore the feasible region is represented by the lines must satisfy all the constraints hence the combination is obtained by the common area lying inside within the lines.

Now by solving the above set of equations we get A = (4, 0), $B = \left(\frac{30}{11}, \frac{42}{11}\right)$ (i.e. by the combination of lines L_3 and L_2), $C = \left(\frac{3}{2}, \frac{21}{4}\right)$ (i.e. by the combination of lines L_1 and L_2) and D = (0, 6). Therefore the polygon OABCD represents the region of values of x and y satisfying the constraints as well the non negative constraints called as feasible solution of LPP. And thus represented in the graph

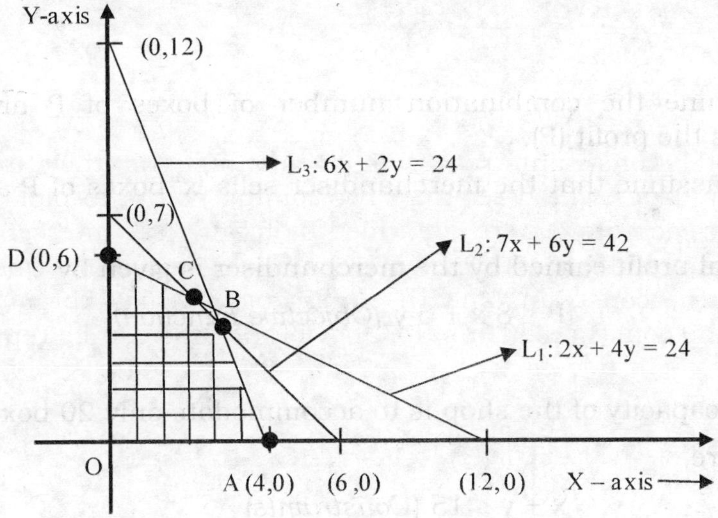

The value of the objective function at these points are

$$Z_O = 6 \times 0 + 9 \times 0 = 0$$

$$Z_A = 6 \times 4 + 9 \times 0 = 24$$

$$Z_B = 6 \times \frac{30}{11} + 9 \times \frac{42}{11} = 50.72$$

$$Z_C = 6 \times \frac{3}{2} + 9 \times \frac{21}{4} = 56.25$$

$$Z_D = 6 \times 0 + 9 \times 6 = 54$$

Since the objective of the function is to maximize Z and Z is maximized at $C\left(\frac{3}{2}, \frac{21}{4}\right)$ and the optimal solution is $x = \frac{3}{2}$ and $y = \frac{21}{4}$ and the optimal value is 56.25.

Illustration No. 3

A merchandiser selling two different articles P and Q in packed boxes and thus invested Rs. 600 for acquiring them keeping in view of the capacity load of his shop which is to accommodate a maximum of 15 boxes. He sells article P at a profit of Rs.8 per box and that of Q at Rs.5 per box. In the anticipation that the merchandiser has sold all his capacity and if his acquiring cost of article P and Q are Rs. 50 and Rs. 25 per box, find the maximum no. of boxes of P and Q he should purchase in order to maximize his profit? Also calculate the maximum profit? (Using Graphical Method)?

Solution

Step I:

To determine the combination number of boxes of P and Q which maximizes the profit (P).

Let us assume that the merchandiser sells 'x' boxes of P and 'y' boxes of Q.

The total profit earned by the merchandiser is given by

$$P = 8x + 5y \text{ (Objective Function)} \qquad .....(1)$$

Step II :

Since the capacity of the shop is to accommodate only 20 boxes

Therefore,

$$x + y \leq 15 \text{ (Constraints)} \qquad .....(2)$$

And taking into account cost per boxes and the availability of funds, we get

$$50x + 25y \leq 600$$

or

$$2x + y \leq 24 \text{ (Constraints)} \qquad .....(3)$$

But clearly $x, y \geq 0$

In order to plot the constraints we convert the inequalities into equalities.

Therefore, we get,

$$x + y = 15 \text{ and } 2x + y = 24$$

Since the constraints represent linear functions, they are represented as straight lines which requires two joining points to plot them and the points for each line can be identified by finding the intercept for the both axes (x-axis and y-axis). The intercept for x-axis is obtained by putting y = 0 in the constraint and the intercept of y-axis obtained by putting x = 0 in the constraint. Thus the above functions can be represented as

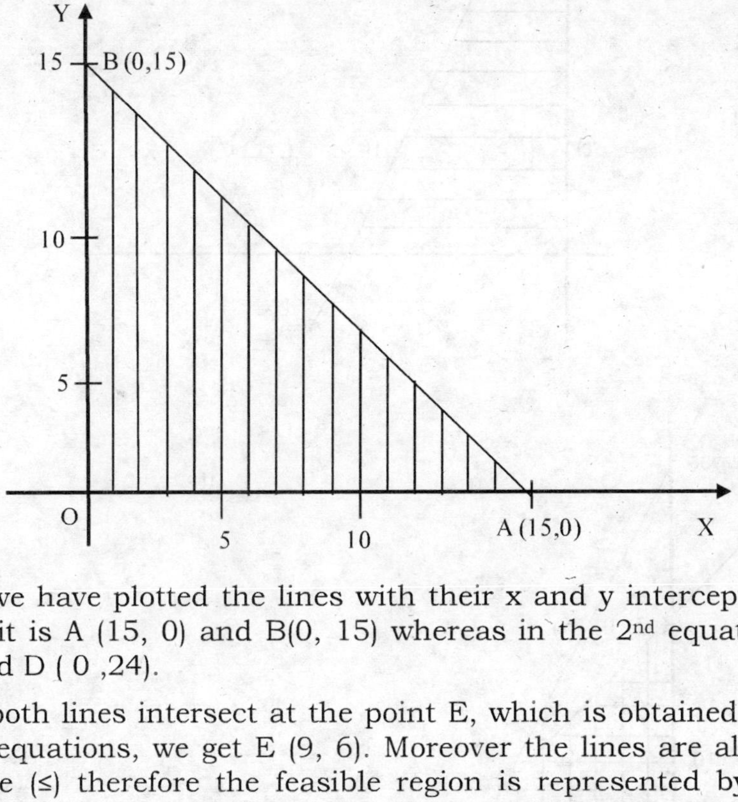

Since we have plotted the lines with their x and y intercept in the 1st equation it is A (15, 0) and B(0, 15) whereas in the 2nd equation it is C (12, 0) and D (0 ,24).

Since both lines intersect at the point E, which is obtained by solving both the equations, we get E (9, 6). Moreover the lines are all less than equal type (≤) therefore the feasible region is represented by the lines must satisfy all the constraints hence the combination is obtained by the common area lying inside within the lines.

Therefore the polygon OCED represents the region of values of x and y satisfying the constraints as well as the non-negative constraints called the feasible solution of the LPP.

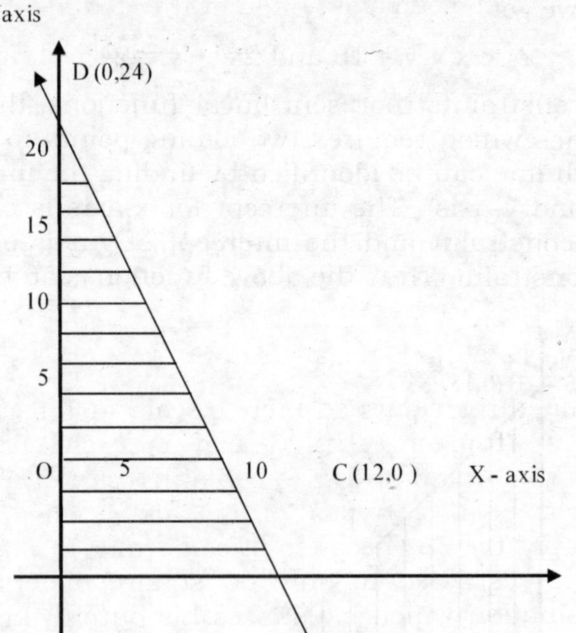
Y- axis
D (0,24)
20
15
10
5
O
5
10
C (12,0)
X - axis

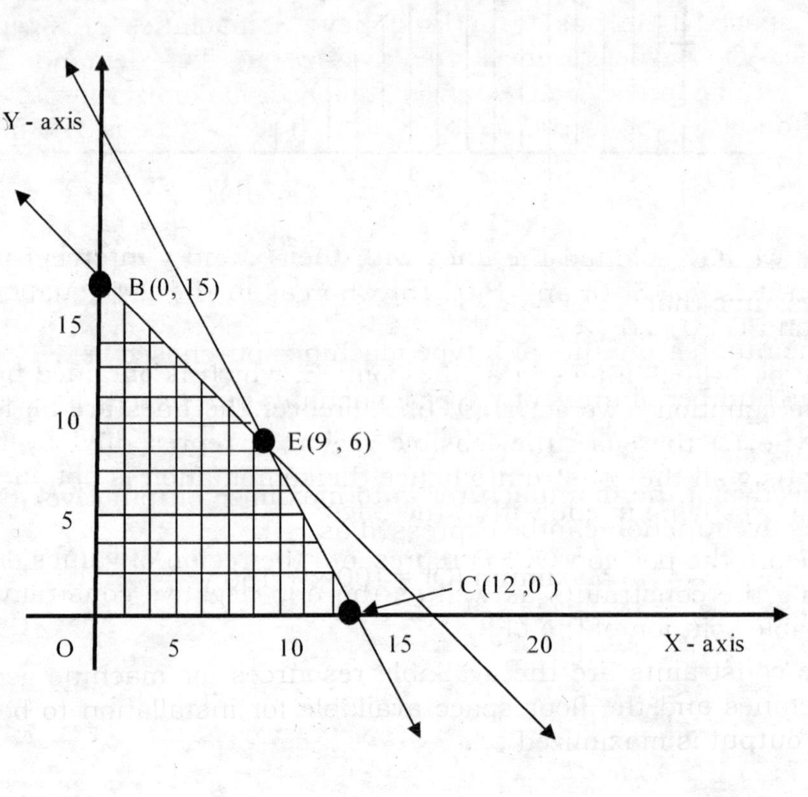
Y - axis
B (0, 15)
15
10
E (9 , 6)
5
C (12 ,0)
O
5
10
15
20
X - axis

The value of the objective function at these points are

$$Z_O = 8 \times 0 + 5 \times 0 = 0$$

$$Z_D = 8 \times 12 + 5 \times 0 = 96$$

$$Z_B = 8 \times 9 + 5 \times 6 = 102$$

$$Z_C = 8 \times 0 + 5 \times 15 = 75$$

Since the objective function is to maximize Z therefore Z is maximized at E (9, 6) and the optimal solution is x = 9 and y = 6 and the optimal value is 102.

Illustration No. 4

A wire manufacturer wants to increase his install capacity by erecting two new types of machineries X and Y, each costing Rs 4000 and Rs. 8000. He has a plan outlay to invest not more than Rs 1, 60,000 for the acquisition of both types of machines which will enhance his production capacity. But because of constraint in floor area in his work shop which is restricted to only 60 square meters he purchases an optimal combination which maximizes his output, knowing that each of machine X type requires 4 square meters and Y-type requires 2 square meters space. If he has to at least have 3 machines of X-type and 4 machines of Y-type to meet the average market demand. Find the number of machines needs to be installed to maximize output if output generation of X-type is 100 meters and that of Y - type is 150 meters per hour?

Solution:

Step 1:

let us assume that

x: The number of units of X type machines purchased

y: The number of units of Y types machines purchased

Step 2:

The objective of the manufacturer is to maximize output level (Q), Hence the objective function can be expressed as

$$\text{Maximum (Q)} = 100\,x + 150\,y$$

Step 3 :

But the constraints are the available resources for machine acquisition for machines and the floor space available for installation to be utilized so that output is maximized.

Therefore, the constraint can be expressed mathematically

(i) $4000\,x + 8000\,y \le 160000$ (Budget Constraint)

i.e $x + 2y \le 40$

(ii) $4\,x + 2\,y \le 40$ (Floor Space Constraint)

i.e $2x + y \le 20$

Step 4:

Using the demand restriction that he needs to have certain quantities of both X and Y type of machines in order to meet sales demand.

Both x and y are non-negative restrictions and can be expressed as

$$x \ge 3,$$

$$y \ge 4.$$

Step 5:

Since the variable x, y are non-negative in nature i.e. $x, y \ge 0$ thus the solution set lie on the first quadrant of the xy-plane.

In order to plot the constraints we convert the inequalities into equalities

Therefore, we get,

$$x + 2y = 40.$$

$$2x + y = 20.$$

Using substitution method to determine feasible points to plot the constraints as shown in earlier examples.

Therefore by substituting $y = 0$ in $x + 2y = 40$, we get

$$x = 40$$

Similarly substituting $x = 0$ in $x + 2y = 40$

we get $2y = 40$ or $y = 20$.

Therefore plotting of the straight line $x + 2y = 40$ can be done by joining the x-intercept 40 with the y-intercept 20 i.e. the straight line joining the point (40, 0) on x-axis with (0, 20) on y-axis. Since the constraint is (≤) type the feasible region will lie on or below the constraint.

Similarly we can plot the line $2x + y = 20$ by substituting,

$y = 0$ in $2x = 20$

We get x = 10 and substituting x = 0 we get y = 20. On plotting the line by joining the points (10, 0) on x-axis with (0, 20) on y-axis (≤) constraint type.

Further using the other two constraints x = 3, y = 4 to plot the feasible zone.

Since the feasible values must satisfy all the constraints hence the combination is obtained by the common area between the lines in the plane and the shaded region highlights the feasible zone. The corners of the feasible region are A (3, 4), B (8, 4) and C (3, 14).

(i) At the point A (3, 4) :

Total No. of Machines Purchased are : 3 + 4 = 7 and the output (Q) = 100 × 3 + 150 × 4 = 900.

(ii) At the point B (8, 4) :

Total No. of Machines Purchased are : 8 + 4 = 12 and the output (Q) = 100 × 8 + 150 × 4 = 1400.

(iii) At the point C(3, 4) :

Total No. of Machines Purchased are : 3 + 14 = 17 and the output (Q) = 100 × 3 + 150 × 14 = 2400.

Thus the maximum number of machines he can purchase is 17 with budget constraint and the maximum output generated in this process 2400 meters of wire.

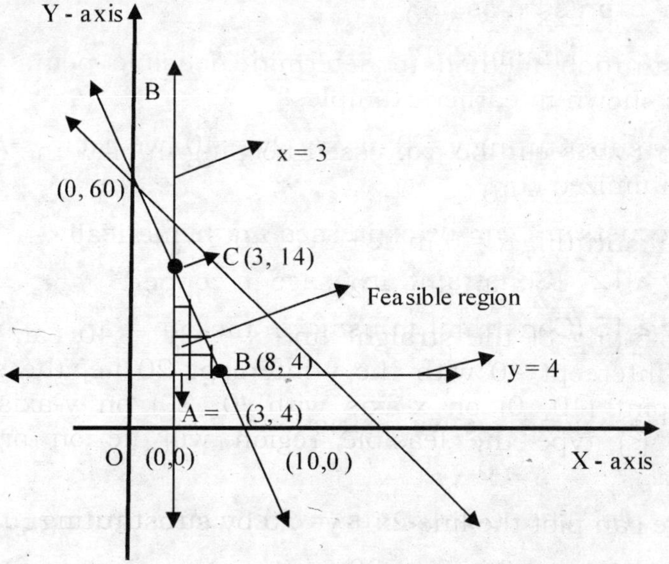

Illustration No. 5

A wire manufacturer has developed a new type of wire mixing two different grades of copper- nickel combination i.e. Grade I and Grade II in which the proportion of copper and nickel are furnished in the table below

Grade	Copper	Nickel
Grade I	1	2
Grade II	2	1

If the pricing of Grade I and Grade II are Rs. 60 per unit and Rs. 90 per unit and at least 12 units of copper and 15 units of nickel have gone in the production of the wire. Find the combination of Grade I and Grade II that needs to be utilized to produce the wire at the minimized cost. Formulate the LP problem and solve using graphical method?

Solution:

Step 1:

let us assume that

x: The number of units of Grade I to be utilized

y: The number of units of Grade II to be utilized

Step 2:

The objective of the firm is minimize cost (C), Hence the objective function can be expressed as

$$\text{Minimize (P)} = 60\,x + 90\,y$$

Step 3 :

The above constraints can be expressed keeping in view the output level produced at minimized cost.

Therefore, the constraint can be expressed mathematically

(i) $x + 2y \geq 12$ (Constraint in usage in copper)

(ii) $2x + y \geq 15$ (Constraint in usage in nickel)

Step 4:

As both x and y are non-negative restrictions,

we have $x, y \geq 0$

In order to plot the constraints we convert the inequalities into equalities

Therefore, we get,

$$x + 2y = 12.$$

$$2x + y = 15.$$

Using substitution method to determine feasible points to plot the constraints as shown in earlier examples.

Therefore by substituting y = 0 in x + 2y = 12,

we get x = 12

Similarly substituting x = 0 in x + 2y = 12

we get 2y = 12 or y = 6.

Therefore plotting of the straight line x + 2y = 12 can be done by joining the x-intercept 12 with the y-intercept 6 i.e. the straight line joining the point (12, 0) on x-axis with (0, 6) on y-axis. Since the constraint is (≥) type the feasible region will lie on or outward from origin.

Similarly we can plot the line 2x + y= 15 by substituting y = 0

in 2x = 15, we get x = 7.5 and substituting x = 0 we get y = 15. On plotting the line by joining the points (7.5, 0) on x-axis with (0, 15) on y-axis (≥) constraint type.

Further solving the two equations we get x = 6, y = 3 we plot the feasible zone. Since the feasible values must satisfy all the constraints hence the combination is obtained by the common area between the lines in the plane and the shaded region highlights the feasible zone. The corners of the feasible region are B (12, 0), C (6, 3) and D (0, 15).

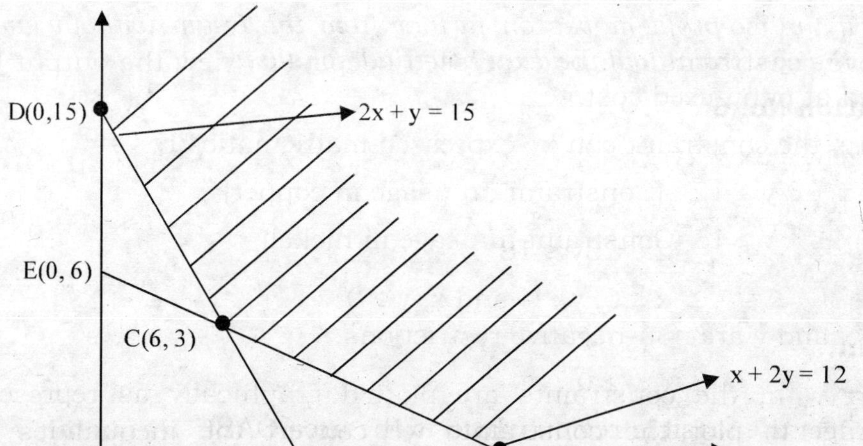

Now the value of the objective function at these points are

$$C_B = 60 \times 12 + 90 \times 0 = 720$$

$$C_C = 60 \times 6 + 90 \times 3 = 810$$

$$C_D = 60 \times 0 + 90 \times 15 = 1350$$

Since the objective of the function is to minimize Z and Z is minimized at B (12, 0) and the optimal solution is x = 12 and y = 0 and the optimal value is 720.

Iso-Profit or Iso-Cost Method

In this method the optimal solution is determined by utilizing the slope of the objective function equation. An iso-profit or cost line is a locus of points which give the same value of the objective function and assigning different values of Z we determine different profit or cost lines and plot graphically many such lines each parallel to the other. The following the steps are involved:

(i) Determine the feasible region and the corner points of the feasible region.

(ii) Iso-profit or cost lines are plotted within feasible region by choosing a specific profit or cost figure.

(iii) Then move the iso-profit or cost lines in parallel direction of increasing or decreasing objective function until the line intersect with one specific corner point yielding the single optimal solution.

(iv) Compute the optimal profit or the cost.

Note : *If the iso-profit moves on further from the origin without limit as against the constraint will yield un bounded solutions.*

Illustration No. 6

Maximize Z = 3x + 2y

Subject to

$$2x + 3y \leq 18$$

$$x \leq 6,\ y \leq 4 \text{ and } x, y \geq 0$$

Solution:

To start with, the constraints are plotted graphically as represented below to get the feasible region which is the area OABC.

Thus a family of lines at various levels of objective function as represented in the figure are called as iso-profit lines which are drawn with arbitrarily chosen values of Z (say Z = 12) thus the iso-profit function takes an equation form 2x + 3y = 18 and then is plotted in the same way as the constraints are plotted. The line is moved away from the origin in the upward direction until it intersects at an extreme point in the feasible region; in this case it is B (6, 2) which is obtained by solving the two linear equations (x = 6, 2x + 3y = 18).

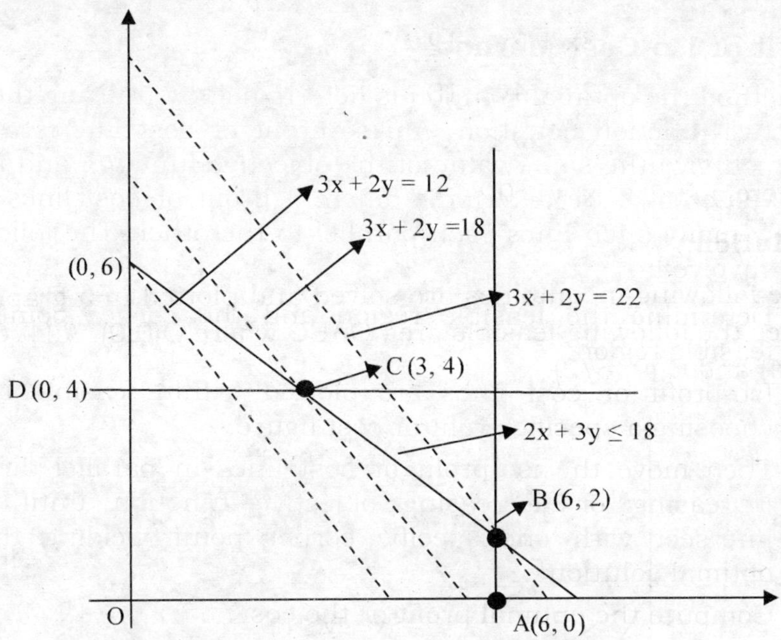

And thus by substituting the value of coordinates of B in the objective function we get the value of Z = 22 which is the maximum in the present case.

Special Cases of Linear Programming

They are certain typical problems faced by a linear programming problem i.e. all LP problem may not give an unique optimal solution. It is found that in some special cases there might be problems which may erupt because of the following reasons

(i) ***Redundant Constraint***

Each constraint defines certain restriction or boundary in LP problem as it sets the area for the solution set, however a constraint which does not define any boundaries for the feasible solution is a redundant constraint as redundant constraint is not prerequisite for the solution set and hence cannot be omitted at the time LP formulation.

Illustrative Example

Maximize 3x + 4 y

Subject to

$$3x + 2y \leq 12$$

$$2x + 4y \leq 10$$

$$x + 2y \leq 12$$

where $x, y \geq 0.$

Solution:

The following inequalities are solved and plotted in a graphical form gives the following feasible area OABC where O(0,0), A (4, 0), B (7/2, 3/4) and C (0, 5/2).

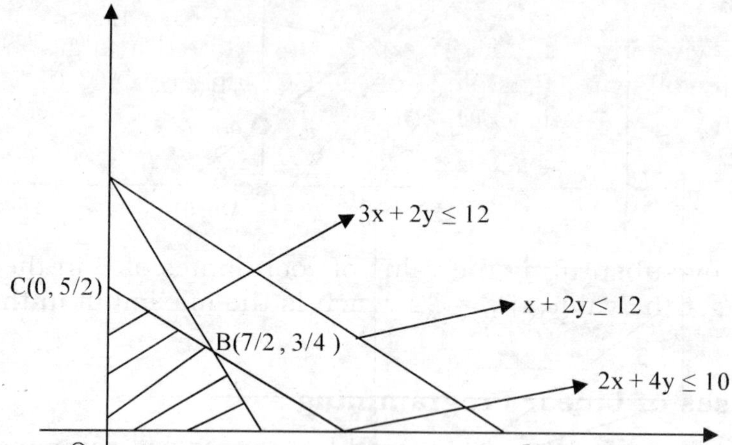

The constraint x + 2y ≤ 12 stands redundant for finding the feasible solution as the feasible solution can be easily determined by the other two constraints.

(ii) ***Alternative Optimal Solution***

In certain LP problem cases there may exist multiple optimal solutions yielding the same optimal value of the objective function this may result because of the slope of the objective function being equal to that of the constraint forming the boundary of the feasible solution region and the other condition that the constraints are binding or active i.e. left hand side of a constraint equals the right hand side. An active constraint is one that passes through one of the extreme feasible solution space.

Illustrative Example

Maximize 2x + 3 y

Subject to

$$2x + 3y \leq 36$$
$$2x + 4y \leq 40$$
$$x \leq 15,$$
$$y \leq 10$$

where x, y ≥ 0.

Solution:

The following inequalities are solved and plotted in a graphical form gives the following feasible area OABCD where O(0, 0), A (15, 0), B(15, 2), C (12, 4) and D (0, 10).

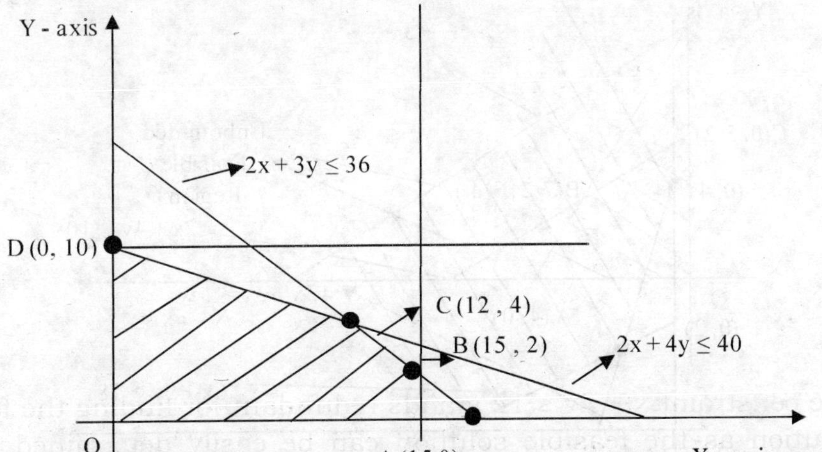

Now substituting the values of the feasible solution in the objective function we get at the point B the value of the objective function is

36, similarly the value of the objective function at the point C is again 36.

i.e. the two solution are B (15, 2) and C (12, 4).

(iii) ***Unbounded Solution***

In cases where the LP problem may not have a finite solution but is permitted to have infinitely large solution then it is known as unbounded solution. Unbounded solution results in maximization case if all the constraints are greater than or equal to type or in case of minimization case all the constraints are lesser than or equal to type.

Illustrative Example

Maximize 5x + 10 y

Subject to

$$6x + 9y \geq 18$$

$$12x + 3y \geq 12,$$

where x, y ≥ 0.

Solution:

When the above two inequalities are plotted graphically i.e. 6x + 9 y ≥ 18 and 12x + 3y ≥ 12 the resulting feasible area OABC which is shown in the figure below,

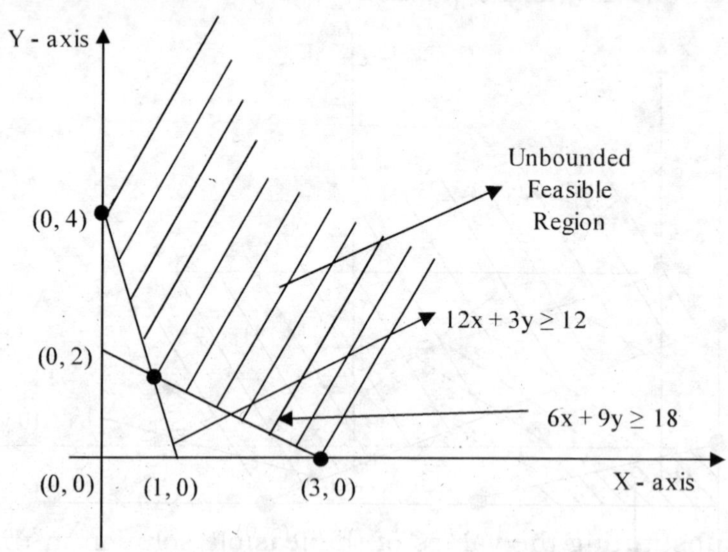

As the objective is to maximization of the objective function and as we move from the origin there exists an infinite number of solutions in the convex sets which increase the value of the objective function without violating the feasibility conditions i.e. both the decision variables and the objective function can be increased infinitely and hence unbounded solution.

(iv) ***Infeasible Solution***

If a linear programming problem has no solution satisfying the constraints then it is said to infeasible which is generally cause by faulty formulation of the problem.

Illustrative Example

Maximize $4x + 3y$

Subject to

$$6x + 10y \geq 30$$
$$4x + 6y \leq 12$$

where $x, y \geq 0$.

Solution:

The objective is to maximize $4x + 3y$ but when the above two inequalities are plotted graphically i.e. $6x + 10y \geq 30$ and $2x + 6y \leq 6$ the resulting areas does interest and have no common point which shown in the figure below,

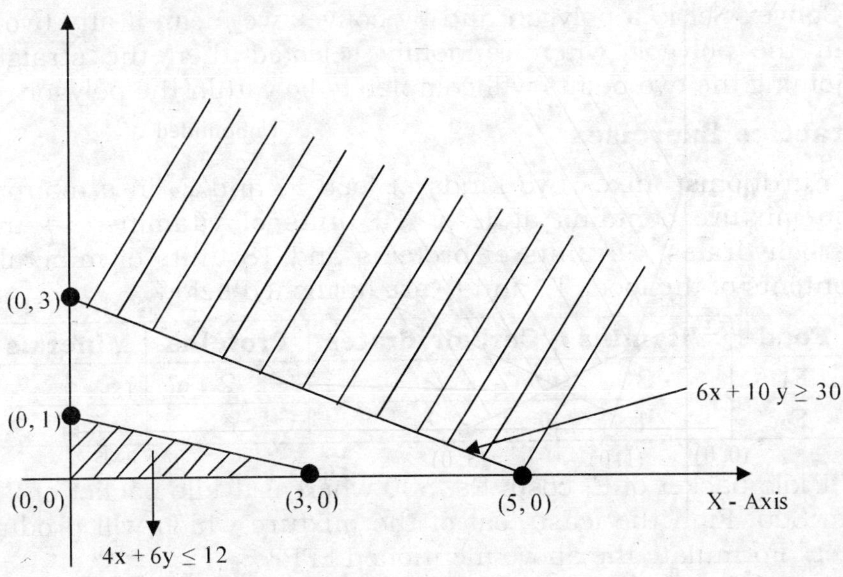

Since there is no point which can lie in both regions and hence there is no points that satisfies all the available constraints. Therefore, there exists no feasible solution.

Summary

- Linear Programming is an optimization technique for finding an optimal (maximum or minimum) value of a function, called objective function of several independent variables, the variables being subjected to various constraints (or restrictions) expressed as equations or inequalities.
- Decision variable refers to whose variables which are competing against one another for allocation of limited resources, usually inter related in terms of resource utilization and required synchronized solutions with linear relationship existing among them.
- A set of values of decision variables which equally satisfies all the given constraints and non-negativity condition is called as feasible solution.
- For a given set of m simultaneous equations in n variables where n > m, a solution obtained by setting (n – m) variables equal to zero and solving the rest m equations in m variables is called as basic solution. The (n – m) whose values have not appeared in this solution are termed as non-basic variables and the rest m variables are called basic variables.
- Convex Set is a polygon and by convex we mean if any two points in the polygon when randomly selected then the straight line joining the two points will completely lie within the polygon.

Self Practice Exercises

1. A nutritionist mixes two kinds of food F_1 and F_2 in a manner that the mixture contains at least 12 units of vitamins, 14 units of carbohydrates, 22 units of proteins and 18 units of minerals. The contents of the foods F_1 and F_2 are furnished below.

Food	Vitamins	Carbohydrates	Proteins	Minerals
F_1	3	2	2	4
F_2	4	2	6	2

A 1 kilo packet of F_1 costs Rs. 500 whereas 1 kilo packet of F_2 costs Rs. 800. Find the least cost of the mixture which will produce the diet. Formulate the above mentioned LPP?

2. Sanchit an investor wants to invest certain investable amount of Rs. 50,000 in different proportions in various financial instruments. On analysis, he decided to invest at least Rs. 25,000 in Govt. of India bonds yielding a return of 12% and not more than 15,000 in Company Fixed Deposit yielding a return of 15%. What should be the suitable mix of both the financial instrument that will maximize his return on his investment?

3. A electrical appliances manufacturer produces two types , electrical heaters H_1 and H_2. Each of these product requires 3 different operation viz.. Assembling, machining and Testing. Each of the heaters require certain hrs of operation which is furnished below

Operation	**H_1**	**H_2**
Assembling	2	4
Machining	6	2
Testing	8	6

If the total available capacity of each of the operation I hours are 60 for machining, 120 for assembling and 300 for testing, then find the number of heaters that needs to be manufactured in order to maximize the profit of the manufacturer knowing that each unit of H_1 earns Rs 50 and H_2 earns Rs. 30 as profit per unit. Formulate the above LPP?

4. A production manager of an oil refinery company has to decide on the optimal mix of two possible blending process of which inputs and outputs are furnished below

Process	**Input**		**Output**	
	Grade 1	Grade 2	Gasoline A	Gasoline B
1	4	3	8	10
2	6	4	5	5

The maximum amount of Grade 1 and Grade 2 crudes are 400 and 300 units and the market demand stands at least 250 units of gasoline A and 200 units of gasoline B. If the profits per production run for both the process i.e. P_1 and P_2 are Rs. 100 and Rs. 150 respectively, then formulate the above LPP of blending problem?

5. Greenway Biotech a fertilizer firm in order to meet a specific order needs to manufacter 500 kgs. of bio degradable fertilizer. The ingredients which have gone into produce the fertilizer are X, Y and Z. The ingredient X cost Rs. 9 per kg, Y cost Rs. 10 per kg. and Z

cost Rs. 11 per kg. But in order to produce the fertilizer not more than 150 kg of X, at least 75 kg of Y and 100 kg of Z are required. Determine the amount each of the ingredients that needs to be used so that the cost is minimized? Formulate the above LPP?

6. A publisher of management books has introduced a "Business Mathematics" book in both in hard bound and paper back form. It was found that the market demand for the hard bound book will not exceed 5000 copies and that of paper back will exceed 3000 copies. If the introductory cost of hard bound and paper back are Rs. 50 and Rs. 40 and it is further known that the time required for binding the books are 20 and 15 min with total available time of 1200 hr. Find the number of books that needs to be bounded from time available to maximize profit. Formulate the LPP?

7. Zig Zag Advertizing wants to reach two types of consumers type A and type B. The firm has short listed two medium for advertising in order to reach the consumers – one the television and the second being the radio. If the annual advertizing budget of the company is Rs. 3,00,000 and the contractual requirement for TV advertisement being at least two advertisements each costing Rs. 60,000 and that of radio of being limited to 5 costing Rs. 30,000 each. Then find the media mix so as to maximize the total reach, if it is known that the a single TV advertisement is set reach 3,00,000 consumers of type A and 1,00,000 of type B whereas the radio reaches 1,00,000 in type A and 2,00,000 in type B. Formulate the above LPP?

8. A drug manufacturing company claims to have found treatment for common cough and cold consisting of two drugs X and Y. Human trials have indicated that in order the treatment to be effective a minimum daily dosage of X and Y needs to be administered. The dosages are 12 mg. of drug X and 8 mg of Y. The ingredients that have gone in to produce X and Y is available in chemical A and B, where each unit A contains 5 mg of X and 3 mg of Y whereas B contains 3 mg of drug X and 4 mg of drug Y. If each unit of A cost Rs. 4 per unit and B cost Rs. 6 per unit find the least cost combination of the two chemicals that will yield the drug designed to meet the minimum daily recommended course? Formulate the LPP?

9. Solve by Graphical Method the following LPP

   Maximize $Z = 20x + 50y$

Subject to Constraints

$$2x + 4y \leq 1000.$$

$$5x + 4y \leq 1300.$$

where $x, y \geq 0$.

10. Solve by Graphical Method the following LPP

Maximize $Z = 16x + 24y$

Subject to Constraints

$$3x + 3y \leq 27.$$

$$6x + 15y \leq 108.$$

$$0 \leq x \leq 4.$$

$$0 \leq y \leq 6.$$

where $x, y \geq 0$.

11. Solve by Graphical Method the following LPP

Maximize $Z = 12x + 8y$

Subject to Constraints

$$-4x + 2y \leq 4.$$

$$2x - 2y \leq 4.$$

$$6x + 4y \leq 18.$$

where $x, y \geq 0$.

12. Solve by Graphical Method the following LPP

Minimize $Z = 6x + 4y$

Subject to Constraints

$$10x + 2y \geq 20.$$

$$x + y = 6.$$

$$2x + 8y \geq 24.$$

where $x, y \geq 0$.

13. Solve by Graphical Method the following LPP

Minimize $Z = 10x + 5y$

Subject to Constraints

$$2x + 4y \leq 40.$$

$$3x + y \geq 30.$$
$$8x + 6y \geq 60.$$

where $x, y \geq 0$.

14. Solve by Graphical Method the following LPP

Maximize $Z = 4x + 3y$

Subject to Constraints

$$5x + 2y \leq 39.$$
$$3x + 6y \leq 57.$$
$$x \leq 6.$$

where $x, y \geq 0$

15. Solve by Graphical Method the following LPP

Minimize $Z = 2x + 7y$

Subject to Constraints

$$3x + 8y \leq 25.$$
$$2x + 4y \geq 14.$$
$$0 \leq x \leq 8.$$

where $x, y \geq 0$.

16. Solve by Graphical Method the following LPP

Maximize $Z = 5x + 3y$

Subject to Constraints

$$4x + 6y \leq 44.$$
$$0 \leq x \leq 5.$$
$$0 \leq y \leq 6.$$

where $x, y \geq 0$

17. Solve by Graphical Method the following LPP

Minimize $Z = 20x + 10y$

Subject to Constraints

$$4x + 4y \geq 42.$$
$$-2x + 3y \geq 12.$$
$$0 \leq x \leq 6.$$

where $x, y \geq 0$.

18. Solve by Graphical Method the following LPP

Minimize Z = 7x + 3y

Subject to Constraints

$$2x + 6y \geq 46.$$

$$0 \leq x \leq 8.$$

$$0 \leq y \leq 6.$$

where x, y ≥ 0.

Answers

1. Minimize C = 500x + 800y *(Objective Function)*

Subject to Constraints

$$3x + 4y \geq 12.$$

$$2x + 2y \geq 14.$$

$$2x + 6y \geq 22.$$

$$4x + 2y \geq 18.$$

where x, y ≥ 0.

2. Maximize Z = 0.12 x + 0.15 y *(Objective Function)*

Subject to Constraints

$$x + y \leq 50000$$

$$x \geq 25000$$

$$y \leq 15000$$

and x, y ≥ 0.

3. Maximize Profit (P) = 50 x + 30y *(Objective Function)*

Subject to Constraints

$$2x + 4y \leq 60$$

$$6x + 2y \leq 120$$

$$8x + 6y \leq 300$$

and x, y ≥ 0.

4. Maximize Profit (P) = 100 x + 150y *(Objective Function)*

   Subject to Constraints

   $$4x + 6y \leq 400$$
   $$3x + 4y \leq 300$$
   $$8x + 6y \geq 250$$
   $$10x + 5y \geq 250$$

   and x, y ≥ 0.

5. Minimize Z = 9x + 10 y + 11z

   Subjected to

   $$x + y + z = 500.$$
   $$x \leq 150.$$
   $$y \geq 75.$$
   $$z \geq 100, \text{ where } x, y, \text{ and } z \geq 0$$

6. Maximize Z = 50x + 40y

   Subjected to

   $$20x + 15y \leq 72{,}000.$$
   $$x \leq 5000.$$
   $$y \geq 3000.$$

   where x, y ≥ 0.

7. Maximize Z = 4,00,000 x + 3,00,000y

   Subjected to

   $$60{,}000\,x + 30{,}000\,y \leq 3{,}00{,}000.$$
   $$x \geq 2$$
   $$y \leq 5$$

   where x, y ≥ 0.

8. Minimize Z = 4x + 6y

   Subjected to

   $$5x + 3y = 12$$
   $$3x + 4y = 8$$

   where x, y ≥ 0.

9. $x = 0$, $y = 250$ and $Z = 12{,}500$.
10. $x = 3$, $y = 6$ and $Z = 192$.
11. $x = 5$, $y = 3$ and $Z = 84$.
12. $x = \frac{28}{19}, y = \frac{50}{19}$ and $Z = 19.30$
13. $x = 10$, $y = 0$ and $Z = 100$.
14. $x = 5$, $y = 7$ and $Z = 41$.
15. $x = 7$, $y = 0$ and $Z = 14$.
16. $x = 5$, $y = 4$ and $Z = 37$.
17. $x = \frac{18}{5}, y = \frac{32}{5}$ and $Z = 136$.
18. $x = 5$, $y = 6$ and $Z = 53$.

CHAPTER 19

The Simplex Method and Duality

Brief Introduction

Developed by George B. Dantzig (1947) to solve linear programming problems involving multiple decision variables as against graphical method which stands unviable to handle multiple variables in linear programming problems. As in real life situation we deal with many variables simplex technique stands a feasible option to reach optimum solution using a systematic procedure involving fixed computational rules. Though having a close proximity to graphical method, simplex method is an iterative process which examines extreme points of the feasible solution space to search for optimum solution in a logical way by repeating a series of same set of steps which can be applied effectively to solve large linear programming problem by formulating a linear objective function subject to constraints with no restriction placed on the number decision variables or constraints in a problem.

Simplex Method for Maximization Case (all ≤ constraint type)

Consider the following linear programming problem as stated below

Maximize $(Z) = c_1x_1 + c_2x_2 + c_3x_3 + \ldots + c_nx_n$

Subject to

$$\left.\begin{array}{l} a_{11}x_1 + a_{12}x_2 + a_{13}x_3 + \ldots a_{1n}x_n \le b_1 \\ a_{21}x_1 + a_{22}x_2 + a_{23}x_3 + \ldots a_{2n}x_n \le b_2 \\ \vdots \qquad\qquad\qquad\qquad\qquad \vdots \\ a_{m1}x_1 + a_{m2}x_2 + a_{m3}x_3 + \ldots \; a_{mn}x_n \le b_m \end{array}\right\} \text{Constraints}$$

and $x_1, x_2, \ldots x_n \ge 0$

Converting inequalities to equalities by the introduction of slack variables.

Objective is a desired result that needs to be achieved in form of profit maximization or cost minimization etc.

Constraints represents limitation in resource capacity in the form of Machine Hours, Labour Hours, Space constraint and Investment constraint.

Slack Variables represents unutilized resource capacity with zero profit contribution and hence the slack variables introduced into the objective function with zero coefficient value.

The modified form of constraint are

$$a_{11}x_1 + a_{12}x_2 + a_{13}x_3 + \dots a_{1n}x_n + s_1 = b_1$$

$$a_{21}x_1 + a_{22}x_2 + a_{23}x_3 + \dots a_{2n}x_n + s_2 = b_2$$

$$\dots$$

$$a_{m1}x_1 + a_{m2}x_2 + a_{m3}x_3 + \dots a_{mn}x_n + s_m = b_m$$

and $x_1, x_2, \dots x_n \geq 0, s_1, s_2, \dots s_m \geq 0.$

Initial basic solution is obtained by setting

$$x_1 = x_2 = \dots = x_n = 0.$$

Thus we get,

$$s_1 = b_1, s_2 = b_2, s_3 = b_3 \dots s_n = b_n$$

Therefore the above Linear Programming Problem can be expressed as

Maximize $Z = c_1x_1 + c_2x_2 + \dots + c_nx_n + 0.s_1 + 0.s_2 + \dots + 0.s_n.$

(Objective Function)

Subject to

$$\left.\begin{array}{l} a_{11}x_1 + a_{12}x_2 + a_{13}x_3 + \dots a_{1n}x_n + s_1 = b_1 \\ a_{21}x_1 + a_{22}x_2 + a_{23}x_3 + \dots a_{2n}x_n + s_2 = b_2 \\ \dots \\ a_{m1}x_1 + a_{m2}x_2 + a_{m3}x_3 + \dots a_{mn}x_n + s_m = b_m \\ \text{and} \quad x_1, x_2, \dots x_n \geq 0, s_1, s_2, \dots s_m \geq 0, \end{array}\right\} \text{Constraints}$$

The expression can be summarized in a tabular form

Contribution C_j			c_1	$c_2 \ldots$	c_n	0	$0 \ldots 0$
C_B	Basic Variables	Solution Variables	x_1	x_2	x_n	s_1	s_2 s_m
0	s_1	b_1	a_{11}	a_{12}	a_{1n}	1	$0 \ldots 0$
0	s_2	b_2	a_{21}	a_{22}	a_{2n}	0	$1 \ldots 0$
$\vdots$	$\vdots$	$\vdots$	$\vdots$		$\vdots$		$\vdots$
0	s_m	b_n	a_{m1}	a_{m2}	a_{mn}	0	$0 \ldots 1$
Contribution Loss per unit (Z_j)			0	$0 \ldots 0$		0	$0 \ldots 0$
Net Evaluation Row ($C_j - Z_j$)			c_1	$c_2 \ldots c_n$		0	$0 \ldots 0$

Body Matrix   Identity (Basis) Matrix

The first row of the simplex table also called as objective row containing the coefficients of the objective function. The second row contains the major column headings for the simplex table and remains unchanged in the succeeding tables.

The first column (C_B) is called the objective column contains the coefficients of the basic variables i.e. the slack variables. The second column the product mix column singles out the basic variables (*Slack Variables*) in the initial simplex table. And the third column (*Solution Variables*) provides the information about the resources or the values of the corresponding basic variables.

Each simplex table contains an identity matrix which represents the coefficient of slack variables which are added to the constraints to convert them to equalities which lies exactly beneath the basic variables and the body matrix contains the coefficients of non-basic variables as shown in the figure above. The numbers a_{ij} called the exchange coefficients which represent the rate at which resources i (i = 1, 2, ... m) are utilized by each unit of an activity j (j = 1, 2, ... m) can take either a positive, negative or zero values.

Two more rows the Z_j row and $C_j - Z_j$ row are added to the simplex table where Z_j represents the amount by which the value of objective function would increase (decrease) if one unit of a given variable in j^{th} column is brought into the solution or it may be interpreted as the contribution of profit (loss) per unit by the introduction of one unit of non basic variable replacing a basic variable in the solution i.e. Z_j is calculated by summing the products of the elements of jth column by the corresponding elements in the objective column (C_B)

$$Z_j = \sum \begin{pmatrix} \text{Cofficients of basic elements } (C_B) \times \text{corresponding} \\ \text{element in the } j^{th} \text{ column} \end{pmatrix}$$

And the last row ($C_j - Z_j$) also called as net evaluation row represents the amount by which the objective function will increase (decrease) with the introduction of one unit of the variable represented x_j in the column head i.e. net profit (loss).

$C_j - Z_j$ (net effect) = (C_j) incoming unit of profit/loss – (Z_j) Outgoing unit of profit/cost.

Now calculate the value of $C_j - Z_j$, if all the elements are found negative, then the current solution is the optimal solution or else if there exists some positive numbers then the current solution can be further improved upon by removing one basic variable with a non basic variable and the process terminates if the solution is found optimal.

The Simplex Method (Maximization Case)

Step I: Formulation stage of a Linear Programming Model

Construct a LP – Model for the real life problem into a standard linear programming form i.e. mathematical representation of Objective and Constraint function.

Step II: The mathematical expression is converted into standard form by introducing slack variables meant to convert inequalities into equality. An inequality arises out of unlikely full utilization of available resources.

Step III: Set up of Initial Simplex Table

1. The first row of the table is the objective row (C_j) contains the coefficients of the variables in the objective function which are directly obtained from the objective function and the values of (C_j) would remain the same in the succeeding table.
2. The second row provides the main column headings of the simplex table.
3. (i) The first column – The objective column C_B – lists the coefficients of the current basic variables in the objective function. In this case the basic variables is the slack variables.

(ii) The Second Column – The basic variable column or the product mix column. In the initial simplex table the basic variables are Slack Variables.

(iii) The Third Column – Called the solution variables or quantity column indicating the resources or the solution values of the basic variables.

(iv) To the right of the solution variables is the column of decision variables/slack variables containing the coefficient of variable constraints? The column of decision variables forms the body matrix and the slack forms the identity matrix in the simplex tableau.

Note: The simplex method starts with zero solution and progresses through a series of iterations to determine the optimal solution. Commonly termed as initial feasible solution. To obtain initial feasible solution the decision variable are taken as zero.

Decision Variables are variables competing against one another to grab a share of the limited resources

Test for the optimality, examine the index row if all the elements are negative then the current solution is optimal otherwise if there exist some positive numbers then the current solution can be further improved upon by removing one basic variable with a non basic variable. And the process terminates if the solution is found to be optimal.

Step IV : Determination of Entering variable

Identify the column with highest positive index number from the table $(C_j - Z_j)$ row which is identified as the key (pivot column). The identification of key – column suggest that a non basic variable will replace a basic solution in the product mix column in order to improve profit.

Determine the Departing Variable

The next is to identify the departing variable which will be replaced by an entering variable; this is done by the identification of the key or pivot row. The selection procedure is divide each element in the solution variable by the corresponding element in the key column and the row identified with the smallest quotient is the key row.

We compute the ratio $\{b_1/a_{1j}, b_2/a_{2j}, b_3/a_{3j}, \dots b_m/a_{mj},\}$ the row corresponding to minimum quotient is the key row. Let $\{b_k/a_{kj}\}$ be identified with least ratio then Sk will be the departing variable in the next table.

Step V: Incorporating the necessary changes in the Simplex Table and developing the second Simplex table

Having identified the Key Row and Column a revised simplex table is developed incorporating the necessary changes as furnished below.

Effecting Changes in the Key Row

(i) The departing variable is replaced by the entering variable with its contribution.

(ii) Each element in the key row is divided by the key element to obtain the new row

$$\text{New Element} = \frac{\text{Corresponding Elements in the key row}}{\text{Key Element}}$$

Now the new values for the remaining rows are calculated using the following formula

$$\text{New Row} = (\text{Elements in the key row}) - (\text{Intersecting Element}) \times \left\{\begin{pmatrix}\text{Corresponding Element} \\ \text{in new Row}\end{pmatrix}\right\}$$

The new entries in the C_B column and the solution variables are entered into the new table and now compute the Z_j row and $C_j - Z_j$ row as shown earlier and if all the elements in the index row are zero or negative then the optimal solution reached.

Step VI: Otherwise the procedural step are continued until an optimum solution is reached.

Let's understand Simplex Method by an example

Illustration No. 1

A firm is manufacturing two different variety of electrical switches A and B which needs to undergo certain hours of fabrication and testing time as furnished below.

Switches Type	Fabrication (Hr)	Testing (Hr)
A	4	2
B	3	1

If the total fabrication time available is 240 hr and that of testing is 100 hr, find the profit maximizing output of both the switches, if by selling each unit of A and B generates profit of Rs. 3 and Rs. 2. Formulate the LPP and solve by Simplex Method?

Step I: Formulation of a Linear Programming Model

Construct a LP – Model for the real life problem into a standard linear programming form i.e. Mathematical representation of Objective and Constraint function, the procedure is as follows:

As per the above problem the aim of the firm is to find the output mix which will result in profit maximization (Z).

Let us assume that

x : The quantity of switch A be produced.

y : The quantity of switch B be produced.

where x, y are decision variables

Data Summary

Decision Variables	**Switches Type**	**Fabrication (Hr)**	**Testing (Hr)**	**Profit per unit**
X	A	4	2	3
Y	B	3	1	2
Available (Hr)		240	100	

The objective i.e. Profit (Z) – Maximization

Maximize Profit (Z) = 3x + 2y (Objective function)

As resources are limited with respect to fabrication time (hr) and testing time (hr). In the above case it is 240 hr and 100 hr respectively representing resource constraints.

Therefore the constraint equations can be written as

$$\left.\begin{array}{l} 4x + 3y \le 240 \\ 2x + y \le 100 \end{array}\right\} \text{Constraints for Fabrication and Testing time}$$

Thus the above LP problem can be expressed as

Maximum Profit (Z) = 3 x + 2 y (Objective function)

Subject to

$$\left.\begin{array}{l} 4x + 3y \le 240 \\ 2x + y \le 100 \end{array}\right\} \text{Constraint}$$

where $x, y \ge 0$ (non-negative constraints)

Step II : The mathematical expression is converted into standard form by introducing slack variables meant to convert inequalities into equality. An inequality arises out of unlikely full utilization of available resources.

Let s_1 and s_2 be the slack variables introduced into constraint equations. Therefore the constraint equations changes into

$$x + 2y + 1 \times s_1 + 0 \times s_2 = 90$$

$$2x + y + 0 \times s_1 + 1 \times s_2 = 60$$

The objective function in current form does not meet the simplex requirement, as objective function contains every variable even slack variable. Since slack variables represents unutilized capacity and the profit contribution of slack variables is zero, therefore the coefficients assigned to slack variables will also be zero in the objective function.

The rewritten form of the objective function is

$$(Z) = 3x + 2y + 0 \times s_1 + 0 \times s_2$$

Step III : Set up of Initial Simplex Table

1. The first row of the table is the objective row(C_j) contains the coefficients of the variables in the objective function which are directly obtained from the objective function and the values of (C_j) would remain the same in the succeeding table.

2. The second row provides the main column headings of the simplex table.

3. (i) The first column – The objective column C_B – lists the coefficients of the current basic variables in the objective function. In this case the basic variables are slack variables

   (ii) The Second Column – The basic variable column or the product mix column. In the initial simplex table the basic variables are Slack Variables.

   (iii) The Third Column – Called the solution variables or quantity column indicating the resources or the solution values of the basic variables.

   (iv) To the right of the solution variables is the column of decision variables/Slack variables containing the coefficient of variable constraints. The column of decision variables forms the body matrix and the slack forms the identity matrix in the simplex tableau.

Note: The simplex method starts with zero solution (zero production level) and progresses through a series of iterations to determine the optimal solution. Commonly termed as initial feasible solution. To obtain initial feasible solution the decision variable x, y are taken as zero so the constraint reduces to $s_1 = 240$ and $s_2 = 100$ exhibiting existing capacity is idle capacity. It is observed that the initial basic solution has two non zero solutions (Slack Variables) called as basic variables and two decision variables x, y with zero values are called as non basic variables.

Basic Simplex Table with Initial feasible Solutions

Contribution (C_j)			3	2	0	0
C_B	Basic Variables	Solution Variables	x	y	s_1	s_2
0	s_1	240	4	3	1	0
0	s_2	100	2	1	0	1

Body Matrix (x, y columns) Identity Matrix (s_1, s_2 columns)

Note: What does the value of x and y column convey? Say x coefficient $\begin{bmatrix} 4 \\ 2 \end{bmatrix}$ and y $\begin{bmatrix} 3 \\ 1 \end{bmatrix}$. It means each unit of switch A produced 4 units of s_1 and 2 units of s_2 from the solution quantity, as to produce 1 unit of switch A will require 4 units of fabrication time and 2 hours of testing time. These coefficients can be also thought as substitution rates.

4. Two more rows the Z_j row and $Z_j - C_j$ row are added to the simplex table were Zj represents the amount by which the value of objective function would decrease if one unit of a given variable in j^{th} column is brought into the solution. Z_j is calculated by summing the products of the elements of jth column by the corresponding elements in the objective column (C_B)

$Z_j = \sum$ (Coefficients of basic elements (C_B) × Corresponding element in the jth column)

For e.g. The values are calculated as

$$Z_j \text{ element for the x - Column} = 0 \times 4 + 0 \times 3 = 0$$

$$Z_j \text{ element for the y - Column} = 0 \times 2 + 0 \times 1 = 0$$

$$Z_j \text{ element for the } s_1 \text{ - Column} = 0 \times 1 + 0 \times 0 = 0$$

$$Z_j \text{ element for the } s_2 \text{ - Column} = 0 \times 0 + 0 \times 1 = 0.$$

And the last row ($C_j - Z_j$) also called as net evaluation row represents the amount by which the objective function will increase with the introduction of one unit of the variable represented by the column head.

($C_j - Z_j$) value for the x - Column = 3 – 0 = 3

($C_j - Z_j$) value for the y - Column = 2 – 0 = 2

($C_j - Z_j$) value for the s_1 - Column = 0 – 0 = 0

($C_j - Z_j$) value for the s_2 - Column = 0 – 0 = 0

Simplex Table with Z_j and $C_j - Z_j$ Values

Contribution (C_j)			3	2	0	0
C_B	Basic Variables	Solution Variables	x	y	s_1	s_2
0	s_1	240	4	3	1	0
0	s_2	100	2	1	0	1
Contribution loss per unit (Z_j)			0	0	0	0
($C_j - Z_j$) index or net evaluation			3	2	0	0

Examining the values of the $C_j - Z_j$, we can interpret that the total profit can be increased by Rs. 3 for switch A and by Rs.2 for switch B for each unit added to the solution. Now the value of the objective function is calculated in the current solution by multiplying C_B column with its corresponding elements in the solution variables column

$$Z = 0 \times 90 + 0 \times 60 = 0$$

Note: Test for optimality

Examine the index row if all the elements are negative then the current solution is optimal otherwise if there exist some positive numbers then the current solution can be further improved upon by removing one basic variable with a non basic variable.

The positive values ($C_j - Z_j$) row indicates scope for improvement as positive number reflects foregone profit by not having a particular value in the solution and so there exist a better solution over the initial feasible solution. And the optmal solution is reached when $C_j - Z_j$ row contains non positive number.

Step IV : Determination of Entering variable

Identify the column with highest positive index number from the table it's the 'x' column which is identified as the key (pivot column). The identification of key – column suggest that a non basic variable x will replace a basic solution in the product mix column in order to improve profit. Thus the variable 'x' is called as entering variable.

Determine the Departing Variable

The next is to identify the departing variable which will be replaced by an entering variabl, this is done by the identification of the key or pivot row. The selection procedure is divide each element in the solution variable by the corresponding element in the key column and the row identified with the smallest quotient is the key row. For the present problem

Quotient of the $s_1 = 240/4 = 60$, Quotient of the $s_2 = 100/2 = 50$.

Since the smallest non-negative quotient is 50 the second row become the key row and the slack variable 's_2' becomes the departing variable and 'x' the entering variable. The intersection of the key row and key column is the key element in this case it is '2' and element other than key element is the intersecting element.

Intersecting Element

Simplex table with key column and key row identified

Contribution (C_j)			3	2	0	0	Min. Ratio
C_B	Basic Variables	Solution Variables	x	y	s_1	s_2	
0	s_1	240	4	3	1	0	60
0	s_2	100	**2**	1	0	1	50
Contribution loss per unit (Z_j)			0	0	0	0	Min = 50
($C_j - Z_j$) index or net evaluation			3	2	0	0	

Key Element

Key Column ***Key Row***

Step V: *Incorporating the necessary changes in the Simplex Table and developing the second Simplex table*

Having identified the Key Row and Column a revised simplex table is developed incorporating the necessary changes as furnished below.

Effecting Changes in the Key Row

(i) The departing variable is replaced by the entering variable with its contribution.

(ii) Each element in the key row is divided by the key element to obtain the new row

The calculation of new key row is as follows

$$\frac{100}{2}=50,\ \frac{2}{2}=1,\ \frac{1}{2}=\frac{1}{2},\ \frac{0}{2}=0,\ \frac{1}{2}=\frac{1}{2}$$

The New Row

3	X	50	**1**	$\frac{1}{2}$	0	$\frac{1}{2}$

Now the new values for the remaining rows are calculated using the following formula

New Row

$$=(\text{Elements in the old row})-(\text{Intersecting Element})\times\left\{\begin{pmatrix}\text{Corresponding Element}\\ \text{in New Row}\end{pmatrix}\right\}$$

Using the above formula to calculate the S_1 row

Elements in the old s_1 row (1)	Intersecting Element of the old s_1 row (2)	Corresponding elements of the new row (3)	New s_1 row = (1) – (2) × (3)
240	4	50	40
4	4	1	0
3	4	1 / 2	1
1	4	0	1
0	4	1 / 2	– 2

Therefore the new s_1 row

0	s_1	60	**0**	3/2	1	– 2

The revised table after the incorporation of necessary changes

Second Simplex table with key column and key row identified

Contribution (C_j)			3	2	0	0
C_B	Basic Variables	Solution Variables	x	y	s_1	s_2
0	s_1	40	0	1	1	– 2
3	x	50	**1**	½	0	1/2
Contribution loss per unit (Z_j)			3	1	0	– 3/2
(C_j – Z_j) index or net evaluation			0	½	0	– 3/2

As discussed earlier Calculation of Z_j Values

Zj element for the x - Column = $0 \times 0 + 1 \times 3 = 3$

Zj element for the y - Column = $0 \times 1 + 3 \times 1/2 = 3/2$

Zj element for the s_1 - Column = $0 \times 1 + 3 \times 0 = 0$

Zj element for the s_2 - Column = $0 \times -2 + 3 \times 1/2 = 3/2$.

Calculation of $C_j - Z_j$ Values

$(C_j - Z_j)$ value for the x - Column = $3 - 3 = 0$

$(C_j - Z_j)$ value for the y - Column = $2 - 1/2 = 3/2$

$(C_j - Z_j)$ value for the s_1 - Column = $0 - 0 = 0$

$(C_j - Z_j)$ value for the s_1 - Column = $0 - 3/2 = -3/2$.

And the value of the objective function is obtained by substituting the values of the variables in the objective function

The improved solution $Z = 0 \times 40 + 3 \times 50 = 150$

Now the current solution yielded a profit of Rs.150 producing 50 units of switches A and 0 units of switch B.

The solution is not yet optimized as the presence of a positive number in the index row indicates further scope of improvement.

Step VI : The column with the positive number is identified as the key column and to select the key row as discussed earlier the quotients are calculated and the least of the non negative quotients is identified which becomes the key row.

Simplex table with improved solution identified

Contribution (C_j)			3	2	0	0	Min Ratio.
C_B	Basic Variables	Solution Variables	x	y	s_1	s_2	
0	s_1	40	0	**1**	1	– 2	= **40** ←
3	x	50	**1**	1/2	0	½	100
Contribution loss per unit (Z_j)			3	3/2	0	3/2	Min = 40
($C_j - Z_j$) index or net evaluation			0	1/2	0	– 3/2	

Proceeding the way as mentioned earlier that y will replace S_1 in the solution along with its contribution. The next is to calculate the new row as shown earlier

The New Key Row

2	y	40	**0**	1	1	– 2

The New Intersecting Element Row

3	x	30	**1**	0	– 1/2	3/2

Simplex table with 2nd improved solution identified

Contribution (C_j)			3	2	0	0
C_B	Basic Variables	Solution Variables	x	y	S_1	S_2
2	y	40	0	**1**	1	– 2
3	x	30	**1**	0	– 1/2	3/2
Contribution loss per unit (Z_j)			3	2	1/2	1/2
($C_j – Z_j$) index or net evaluation			0	0	– 1/2	– 1/2

Calculating the values of Z_j and $C_j – Z_j$ as discussed earlier and all the entries in the index row are found to be negative which indicates no further improvement and the optimal solution is reached and identified as x = 30, y = 40 and the max (Z) = Rs 170.

Hence the solution can be interpreted as the firm maximizes its profit at level of output of 30 units of A type switches and 40 units of B type switches.

Illustration No. 2

An icecream manufacturing firm producing two different varieties of icecream Fruit-Nut and Chocó-bites. Each of Fruit–Nut variety requires 10 minutes of processing and 40 minutes of freezing time whereas Chocó-bites requires 20 minutes of processing and 30 minutes of freezing time. If the total processing time available is 400 minutes and that of freezing is 1200 minutes find the profit maximizing output, if by selling a Fruit–Nut generates Rs 4 and Choco–bites Rs.5 as profit per unit, Formulate the LPP and solve by simplex method?

Solution:

Step I: Formulation of a Linear Programming Model

Construct a LP – Model for the problem into a standard linear programming form i.e. Mathematical representation of Objective and Constraint function, the procedure is as follows:

As per the above problem the aim of the firm is to find the output mix which will result in profit maximization (Z).

Let us assume that

x : The quantity of Fruit–Nut produced

y : The quantity of Choco–bites produced

Were x, y are decision variables

The objective i.e. Profit (Z) - Maximization

Maximum Profit (Z) = 4 x + 5 y (Objective function)

As resources are limited with respect to processing time (min) and freezing time (min). In the above case it is 400 min and 1200 min respectively representing resource constraints.

Therefore the constraint equations can be written as

$$\left.\begin{array}{l} 10x + 20y \le 400 \\ 40x + 30y \le 1200 \end{array}\right\} \text{Constraints for Processing and freezing time}$$

Thus the above LP problem can be expressed as

Maximum Profit (Z) = 4 x + 5 y. *(Objective function)*

Subject to

$$\left.\begin{array}{l} 10x + 20y \le 400 \\ 40x + 30y \le 1200 \end{array}\right\} \text{Constraint}$$

where x, y ≥ 0 (non–negative constraints)

Step II:

The mathematical expression is converted into standard form by introducing slack variables meant to convert inequalities into equality.

Let s_1 and s_2 be the slack variables introduced into constraint equations. Therefore the constraint equations changes into

$$10x + 20y + 1 \times s_1 + 0 \times s_2 = 400$$

$$40x + 30y + 0 \times s_1 + 1 \times s_2 = 1200$$

Since slack variables represents unutilized capacity and profit contribution of slack variables is zero, therefore the coefficients assigned to slack variables will also be zero in the objective function.

Therefore the rewritten form of the objective function is

$$(Z) = 4x + 5y + 0 \times s_1 + 0 \times s_2$$

Step III : Set up of Initial Simplex Table

Basic Simplex Table with Initial feasible Solutions

Contribution (C_j)			4	5	0	0
C_B	Basic Variables	Solution Variables	x	y	s_1	s_2
0	s_1	400	10	20	1	0
0	s_2	1200	40	30	0	1

Body Matrix Identity Matrix

Z_j row is calculated as

Z_j element for the x - Column = $0 \times 10 + 0 \times 40 = 0$

Z_j element for the y - Column = $0 \times 20 + 0 \times 30 = 0$

Z_j element for the s_1 - Column = $0 \times 1 + 0 \times 0 = 0$

Z_j element for the s_2 - Column = $0 \times 0 + 0 \times 1 = 0$.

And the last row $(C_j - Z_j)$

$(C_j - Z_j)$ value for the x - Column = $4 - 0 = 4$

$(C_j - Z_j)$ value for the y - Column = $5 - 0 = 5$

$(C_j - Z_j)$ value for the s_1 - Column = $0 - 0 = 0$

$(C_j - Z_j)$ value for the s_2 - Column = $0 - 0 = 0$

Simplex Table with Z_j and $C_j - Z_j$ Values

Contribution (C_j)			4	5	0	0
C_B	Basic Variables	Solution Variables	x	y	s_1	s_2
0	s_1	400	10	20	1	0
0	s_2	1200	40	30	0	1
Contribution loss per unit (Z_j)			0	0	0	0
($C_j - Z_j$) index or net evaluation			4	5	0	0

The value of the objective function is calculated in the current solution by multiplying C_B column with its corresponding elements in the solution variables column

$$Z = 0 \times 400 + 0 \times 1200 = 0$$

The positive values ($C_j - Z_j$) row indicates scope for improvement as positive number reflects foregone profit by not having a particular value in the solution and so there exist a better solution over the initial feasible solution.

Step IV:

Determination of Entering variable

The column identified with highest positive index number from the table it is the 'y' column which is identified as the key (pivot column). The identification of key – column suggest that a non basic variable y will replace a basic solution in the product mix column in order to improve profit. Thus the variable 'y' is called as entering variable.

Determine the Departing variable

The next is to identify the departing variable which will be replaced by an entering variable, this is done by the identification of the key or pivot row. The selection procedure is divide each element in the solution variable by the corresponding element in the key column and the row identified with the smallest quotient is the key row. For the present problem

Quotient of the s_1 = 400/20 = **20**, Quotient of the s_2 = 1200/30 = 40.

Since the smallest non-negative quotient is 20 the 1st row become the key row and the slack variable 'S_1' becomes the departing variable and 'y' the entering variable. The intersection of the key row and key column is the key element in this case it is '30' and element other than key element is the intersecting element.

Key Element

Simplex table with key column and key row identified

Contribution (C_j)			3	2	0	0	Min. Ratio
C_B	Basic Variables	Solution Variables	x	y	s_1	s_2	
0	s_1	400	10	**20**	1	0	400 /20 = **20**
0	s_2	1200	40	30	0	1	1200 /30 = 40
Contribution loss per unit (Zj)			0	0	0	0	Min = 30
(Cj – Zj) index or net evaluation			3	2	0	0	

Intersecting Element

Key Column Key Row

Step V: *Incorporating the necessary changes in the Simplex Table and developing the second Simplex table*

Having identified the Key Row and Column a revised simplex table is developed incorporating the necessary changes.

The calculation of new key row is shown below

The New Row

5	y	20	1/2	1	1/20	0

Now the new values for the remaining rows are calculated as shown in the table

Elements in the old s_2 row (1)	Intersecting Element of the old s_1 row (2)	Corresponding elements of the new row(3)	New s_2 row = (1) – (2) × (3)
1200	30	20	600
40	30	½	25
30	30	1	0
0	30	1/20	–3/2
1	30	0	1

Therefore the new s_2 row

0	s_2	600	25	0	– 3/2	1

The revised table after the incorporation of necessary changes

Revised Simplex table

Contribution (C_j)			4	5	0	0
C_B	Basic Variables	Solution Variables	X	Y	s_1	s_2
5	y	20	½	1	1/20	0
0	s_2	600	25	0	– 3/2	1
Contribution loss per unit (Z_j)			5/2	5	1/4	0
(C_j – Zj) index or net evaluation			3/2	0	– 1/4	0

As discussed earlier Calculation of Z_j Values

Z_j element for the x - Column = 5 × 1/2 + 0 × 25 = 5/2

Z_j element for the y - Column = 5 × 1 + 0 × 0 = 5

Z_j element for the s_1 - Column = 5 × 1/20 + 0 × –3/2 = 1/4

Z_j element for the s_2 - Column = 5 × 0 + 0 × 1 = 0

Calculation of C_j – Z_j Values

(C_j – Z_j) value for the x - Column = 4 – 5/2 = 3/2

(C_j – Z_j) value for the y - Column = 5 – 5 = 0.

(C_j – Z_j) value for the S_1 - Column = 0 – 1/4 = – 1/4

(C_j – Z_j) value for the S_1 - Column = 0 – 0 = 0

And the value of the objective function is obtained by substituting the values of the variables in the objective function

The improved solution Z = 5 × 20 + 0 × 600 = 100

The presence of a positive number in the index row indicates further scope of improvement and the optimal solution not yet obtained.

Step VI : The column with the positive number is identified as the Key Column and to select the key row as discussed earlier the quotients are calculated and the least of the non negative quotients is identified which becomes the key row.

Second Simplex table with key column and key row identified

Contribution (C_j)			4	5	0	0	Min Ratio
C_B	Basic Variables	Solution Variables	x	y	s_1	s_2	
5	y	20	1/2	1	1/20	0	20/(1/2) = 40
0	S_2	600	25	0	– 3/2	1	600/25 = 24 ←
Contribution loss per unit (Zj)			5/2	5	1/4	0	
(C_j – Z_j) index or net evaluation			3/2 ↑	0	– 1/4	0	

Proceeding the way as mentioned earlier that x will replace S_2 in the solution along with its contribution. The next is calculate the new row as performed earlier

The New Key Row

4	X	24	**1**	0	– 3/50	1/25

The New Intersecting Element Row

5	Y	8	**0**	1	1/50	– 1/50

Simplex table with improved solution identified

Contribution (C_j)			4	5	0	0
C_B	Basic Variables	Solution Variables	x	Y	s_1	s_2
5	Y	8	0	1	1/50	–1/50
4	X	24	1	0	–3/50	1/25
Contribution loss per unit (Z_j)			4	5	–7/50	3/50
(C_j – Z_j) index or net evaluation			0	0	–1/4	–3/50

Calculating the values of Z_j and C_j – Z_j as discussed earlier and all the entries in the index row are found to be negative which indicates no further improvement and the optimal solution is reached and identified as x = 24, y = 8 and the max (Z) = Rs 136.

Hence the solution can be interpreted as the ice-cream manufacturer maximizes its profit at level of output of 24 units of fruit - nut and 8 units of Choco-bites.

Illustration No. 3

Maximize $z = 2x + 5y + 7z$

Subject to

$$4x + 2y + 6z \le 72,$$

$$2x + 4y + 2z \le 84$$

where $x, y, z \ge 0$.

Solution:

Maximize (Z) = 2x + 5 y + 7z *(Objective function)*

Subject to

$$\left.\begin{array}{l} 4x + 2y + 6z \le 72, \\ 2x + 4y + 2z \le 84 \end{array}\right\} \text{Constraint}$$

where x, y and $z \ge 0$ (non-negative constraints)

The mathematical expression is converted into standard form by introducing slack variables meant to convert inequalities into equality. Let s_1 and s_2 be the slack variables introduced into constraint equations. Therefore the constraint equations changes into

$$4x + 2y + 6z + 1 \times s_1 + 0 \times s_2 = 72$$

$$2x + 4y + 2z + 0 \times s_1 + 1 \times s_2 = 84$$

The rewritten form of the objective function is

$$(Z) = 2x + 5y + 7z + 0 \times s_1 + 0 \times s_2.$$

Set up of Initial Simplex Table

Simplex Table with Initial feasible Solutions

Contribution (C_j)			2	5	7	0	0
C_B	Basic Variables	Solution Variables	x	y	z	s_1	s_2
0	s_1	72	4	2	6	1	0
0	s_2	84	2	4	2	0	1

Introducing the Z_j and $C_j - Z_j$ row in the initial simplex table

Simplex Table with Z_j and $C_j - Z_j$ row Initial feasible Solutions

Contribution (C_j)			2	5	7	0	0
C_B	Basic Variables	Solution Variables	x	Y	z	s_1	s_2
0	s_1	72	4	2	6	1	0
0	s_2	84	2	4	2	0	1
	Z_j		0	0	0	0	0
	$C_j - Z_j$		2	5	7	0	0

Note: The values of Z_j and $C_j - Z_j$ are calculated as shown in the earlier example.

The positive values ($C_j - Z_j$) row indicates scope for improvement so there exist a better solution over the initial feasible solution.

Determination of Entering and Departing variable

z column is identified with highest positive number therefore z column is the (pivot column) and next is to identify the departing variable which will be replaced by an entering variable, this is done by the identification of the key or pivot row.

For the present problem

Quotient of the s_1 = 72/6 = 12

Quotient of the s_2 = 84/2 = 42

Since the smallest non-negative quotient is 12 the first row is the key row.

Simplex table with key column and key row identified

Contribution (C_j)			2	5	7	0	0	Min Ratio
C_B	Basic Variables	Solution Variables	x	y	z	s_1	s_2	
0	s_1	72	4	2	**6**	1	0	72/ 6 = 12
0	s_2	84	2	4	2	0	1	84/ 2 =42
	Z_j		0	0	0	0	0	
	$C_j - Z_j$		2	5	**7**	0	0	

Key Element Key Column Key Row

The calculation of new values key row is given by

The New Row

7	Z	12	**2/3**	**1/ 3**	1	1/6	0

And the values of second row are

0	s_2	60	**2/3**	**10/ 3**	0	– 1/3	1

Replacing the values in the simplex table we get,

Revised Simplex table

Contribution (C_j)			2	5	7	0	0
C_B	Basic Variables	Solution Variables	x	y	z	s_1	s_2
7	z	12	2/3	1/3	**1**	1/6	0
0	S_2	60	2/3	10/3	0	– 1/3	1
Z_j			14/3	7/3	7	7/6	0
$C_j - Z_j$			– 8/3	8/3	**0**	– 7/6	0

Since there exists a positive value in the net evaluation row there is further scope for improvement in solution.

Revised Simplex table

Contribution (C_j)			2	5	7	0	0	Min ratio
C_B	Basic Variables	Solution Variables	x	y	z	s_1	s_2	
7	z	12	2/3	1/3	**1**	1/6	0	12/(1/3) = 36
0	s_2	60	2/3	10/3	0	– 1/3	1	60/(10/3) = 18
Z_j			14/3	7/3	0	7/6	0	
$C_j - Z_j$			– 8/3	8/3	**0**	– 7/6	0	

Key column key row

Revised Rows are

5	y	18	1/5	1	0	– 1/10	3/10

And the values of second row are

7	z	6	3/5	0	1	1/5	– 1/10

Replacing the values in the simplex table we get,

Revised Simplex table

Contribution (C_j)			2	5	7	0	0
C_B	Basic Variables	Solution Variables	x	y	z	s_1	s_2
7	Z	6	3/5	0	**1**	1/5	– 1/10
5	Y	18	1/5	1	0	– 1/10	3/10
Z_j			26/5	5	7	9/10	8/10
$C_j - Z_j$			– 16/5	0	**0**	– 9/10	– 8/10

Since all the numbers in the index row is either zero or negative hence the optimal solution is reached at x = 0, y = 18 and z = 6 and the maximum value (Z) is 132.

Simplex Method for Minimization Case (all ≥ constraint type) Using Charne's Big – M Method

In the previous section we have elaborated on how to apply simplex method to linear programming problems where the main objective is to maximize the profit with less than or equal to constraints. There might exist many situations in which objective may not be profit maximization but minimization cost or time with greater than equal to type constraints. In such cases simplex needs certain modification to obtain optimal policy which is mentioned below.

Formulate proper mathematical structure for the Linear Programming Problem i.e.

Minimize $Z = c_1x_1 + c_2x_2 + c_3x_3 + \dots + c_nx_n$

Subject to

$$\left.\begin{array}{l} a_{11}x_1 + a_{12}x_2 + a_{13}x_3 + \dots a_{1n}x_n \ge b_1 \\ a_{21}x_1 + a_{22}x_2 + a_{23}x_3 + \dots a_{2n}x_n \ge b_2 \\ \dots \\ a_{m1}x_1 + a_{m2}x_2 + a_{m3}x_3 + \dots a_{mn}x_n \ge b_m \\ x_1, x_2, \dots x_n \ge 0 \end{array}\right\} \text{Constraints}$$

and (applies to the last line above)

Converting inequalities to equalities

Introducing surplus variable (negative slack variables) to convert inequalities to equalities

$$a_{11}x_1 + a_{12}x_2 + a_{13}x_3 + \dots a_{1n}x_n - s_1 = b_1$$

$$a_{21}x_1 + a_{22}x_2 + a_{23}x_3 + \dots a_{2n}x_n - s_2 = b_2$$

$$\dots$$

$$a_{m1}x_1 + a_{m2}x_2 + a_{m3}x_3 + \dots a_{mn}x_n - s_m = b_m$$

and $x_1, x_2, \dots x_n \geq 0, \; s_1, s_2, \dots s_m \geq 0.$

Any initial basic solution is obtained by setting

$$x_1 = x_2 = \dots = x_n = 0$$

Thus we get,

$$-s_1 = b_1 \quad \text{or} \quad s_1 = -b_1$$

$$-s_2 = b_2 \quad \text{or} \quad s_2 = -b_2$$

$$\dots\dots\dots \quad \text{or} \quad \dots\dots\dots$$

$$-s_m = b_m. \quad \text{or} \quad s_m = -b_m$$

Which is not feasible because it violates the non-negative stipulation, (i.e $s_i \geq 0$). Thereby we make a slight change by introducing artificial variables, the set of constraint can be rewritten as

$$a_{11}x_1 + a_{12}x_2 + a_{13}x_3 + \dots a_{1n}x_n - s_1 + A_1 = b_1$$

$$a_{21}x_1 + a_{22}x_2 + a_{23}x_3 + \dots a_{2n}x_n - s_2 + A_2 = b_2$$

$$\dots$$

$$a_{m1}x_1 + a_{m2}x_2 + a_{m3}x_3 + \dots a_{mn}x_n - s_m + A_m = b_m$$

and $x_1, x_2, \dots x_n \geq 0, \; s_1, s_2, \dots s_m \geq 0,$

$A_1, A_2, \dots A_n \geq 0$

Now, an initial basic feasible can be reached by setting all the decision and surplus variables to zero. Thus initial basic solution of the LPP is

$$A_1 = b_1, \quad A_2 = b_2, \quad A_3 = b_3 \dots A_m = b_m.$$

Now in order to obtain an optimal solution we have to derive the artificial variables by assigning a large value (M) to each artificial variable and zero values to surplus variables in the objective function. Therefore LPP can be rewritten as

Minimize Z

$$= c_1x_1 + c_2x_2 + \ldots + c_nx_n + 0.s_1 + 0.s_2 + \ldots + 0.s_n + M.A_1 + M.A_2 + \ldots + MA_m.$$

Subject to

$$a_{11}x_1 + a_{12}x_2 + a_{13}x_3 + \ldots a_{1n}x_n - s_1 + A_1 = b_1$$

$$a_{21}x_1 + a_{22}x_2 + a_{23}x_3 + \ldots a_{2n}x_n - s_2 + A_2 = b_2$$

$$\ldots$$

$$a_{m1}x_1 + a_{m2}x_2 + a_{m3}x_3 + \ldots a_{mn}x_n - s_m + A_m = b_m$$

and $x_1, x_2, \ldots x_n \geq 0,\ s_1, s_2, \ldots s_m \geq 0,$

$A_1, A_2, \ldots A_n \geq 0$

The modification to simplex method for solving LPP with a very large penalty cost M assigned to artificial variable is called as Charnes Big M Method or Penalty Method.

- Artificial variables are fictitious in nature and used only for computational purpose with very large unit of penalty assigned in the objective function.
- Constraints with ≥ type require both surplus and artificial variables and if one of the constraints is a strict equality only an artificial variable is added to get to initial solution without neither slack nor surplus variable.
- If all the artificial variable cannot be driven out there exist no feasible solution and the infeasible is caused by the presence of inconsistent constraint in the formulation of the LPP.

Steps involved in Simplex Algorithm in Minimization Case

Step I : Formulate the appropriate mathematical problem for the LPP at hand with added artificial and surplus variable in the constraints and objective function.

Step II : Set up the initial simplex table and the initial basic solution is obtained by assigning zero value to decision variables i.e. $x_1 = x_2 = \ldots = x_n = 0$, so that $A_1 = b1$, $A_2 = b_2$, $A_3 = b3 \ldots A_m = b_m$. The initial table is shown below.

Contribution C_j			c_1 c_2 ...... c_n	0 0 0	M M .... M
C_B	Basic Variables	Solution Variables	X_1 x_2 x_n	s_1 s_2 s_m	A_1 A_2. A_m
M	A_1	b_1	a_{11} a_{12} a_{1n}	– 1 0 0	M 0 0
M	A_2	b_2	a_{21} a_{22} a_{22}	0 – 1 0	0 M 0
.	.	...	....	....	....
.	.	...			
M	A_m	b_n	a_{m1} a_{m2} a_{mn}	0 0 .... – 1	0 0 ..... M
Z_j			0 0 0	0 0...... 0	0 0 0
Net Evaluation Row($C_j - Z_j$)			c_1 c_2 ...... c_n	0 0 ... 0	M M .... M

Step III : Test the optimality if (Cj – Zj) has all positive values then the solution is optimal otherwise if atleast one value (Cj – Zj) row is found negative then choose the largest negative number in the net evaluation row as it determines the pivot column and the incoming variable in the solution basis.

Step IV : With the identification of Pivot Column, Row and the pivot element or key element continue the iterative process as discussed in the maximization case till the optimal solution is determined.

- The significant difference between Maximization and minimization case the treatment of the index or net evaluation row. Since the objective is cost or time minimization case the entering variable to enter the solution will be the one identified with the largest negative number and the optimality is reached once all elements in the index row is either zero or positive.
- Second a very large value is assigned to artificial variables in the objective function.
- All variables except artificial variable can re enter the solution once driven out in an iteration but not artificial variable because of the large M associated with it in the objective function.
- If there exist at least one artificial variable in the basis at zero level i.e the corresponding entry in solution variable is zero and also the coefficient of M in net evaluation row is non-negative then there is no solution to the problem and the current basic solution is degenerate.
- If there exist atleast one artificial variable in the basis at non-zero level i.e the corresponding entry in solution variable is non - zero and also the coefficient of M in net evaluation row is non – negative then the LPP does not have basic feasible solution.

Illustration No. 6

Minimize Cost (Z) = 5 x + 7 y (Objective function)

Subject to

$$\left.\begin{array}{l} x + 2y \geq 140 \\ 2x + 2y \geq 160 \end{array}\right\} \longrightarrow \text{Constraints}$$

where $x, y \geq 0$ (non-negative constraints)

The mathematical expression is converted into standard form by introducing surplus variables meant to convert inequalities into equality. The main idea remains same with maximization problem the difference lies in maximization case we begin the initial solution with no profit whereas we begin initial solution with a very high cost.

Let s_1 and s_2 be the surplus variables introduced into constraint equations. Therefore the constraint equations changes form

i.e.

$$x + 2y - 1 \times s_1 = 140$$

$$2x + 2y + - 1 \times s_2 = 160$$

Since surplus variables represents extra units in the lowest cost mixture. Let x and y be zero in the initial solution thus we have

$$- s_1 = 140 \quad \text{or} \quad s_1 = - 140$$

$$- s_2 = 160 \quad \text{or} \quad s_2 = - 160$$

Setting the values of x and y has resulted the values of surplus variables to be negative which is not possible. In order to prevent the surplus variable appearing in the initial solution we set the values of surplus variables (s_1, s_2) to zero. We introduce artificial variables A_1 and A_2 at high cost M per unit

$$x + 2y - 1 \times s_1 + A_1 = 140$$

$$2x + 2y - 1 \times s_2 + A_2 = 160$$

Setting $x, y, s_1, s_2 = 0$, we get $A_1 = 140$ and $A_2 = 160$

Now the standard form of LPP

Minimize (Z) = $5x + 7y + 0.\ s_1 + 0\ s_2 + M.\ A_1 + M.\ A_2$

Subject to the constraints

$$x + 2y - 1 \times s_1 + A_1 = 140.$$

$$2x + 2y - 1 \times s_2 + A_2 = 160.$$

where $x, y, s_1, s_2, A_1, A_2 \geq 0.$

Set up of Initial Simplex Table. The initial basic feasible is obtained by setting x, y, s_1, s_2 = 0 and A_1 = 140, A_2 = 160.

Basic Simplex Table with Initial feasible Solutions

Contribution (C_j)			5	7	0	0	M	M
C_B	Basic Variables	Solution Variables	x	y	s_1	s_2	A_1	A_2
M	A_1	140	1	2	− 1	0	1	0
M	A_2	160	2	2	0	− 1	0	1

Z_j row is calculated as

Z_j element for the x - Column = M × 1 + M × 2 = 3 M

Z_j element for the y - Column = M × 2 + M × 2 = 4M

Z_j element for the S_1 - Column = M × − 1 + M × 0 = − M

Z_j element for the S_2 - Column = M × 0 + M × − 1 = − M.

Z_j element for the A_1 - Column = M × 1 + M × 0 = M

Z_j element for the A_2 - Column = M × 0 + M × 1 = M.

And the last row ($C_j - Z_j$)

($C_j - Z_j$) value for the x - Column = 5 − 3M

($C_j - Z_j$) value for the y - Column = 7 − 4M

($C_j - Z_j$) value for the S_1 - Column = M

($C_j - Z_j$) value for the S_1 - Column = M

($C_j - Z_j$) value for the S_1 - Column = 0

($C_j - Z_j$) value for the S_1 - Column = 0

Simplex Table with Z_j and $C_j - Z_j$ Values

Contribution (C_j)			16	12	0	0	M	M
C_B	Basic Variables	Solution Variables	x	y	s_1	s_2	A_1	A_2
M	A_1	140	1	2	− 1	0	1	0
M	A_2	160	2	2	0	− 1	0	1
(Z_j)			3M	4M	− M	− M	M	M
($C_j - Z_j$) index or net evaluation			5-3M	7- 4M	M	M	0	0

Determination of Entering variable

Identify the column with highest negative index number from the table it is the 'y' column which is identified as the key (pivot column). The identification of key – column suggest that a non basic variable y will replace a basic solution in the product mix column in order to reduce cost. Thus the variable 'y' is called as entering variable.

Determine the Departing variable

The next is to identify the departing variable which will be replaced by an entering variable, this is done by the identification of the key or pivot row. The selection procedure is divide each element in the solution variable by the corresponding element in the key column and the row identified with the smallest quotient is the key row. For the present problem

Quotient of the A_1 = 140/2 = **70**, Quotient of the A_2 = 160/2 = 80.

Since the smallest non-negative quotient is 70 the 1st row become the key row and the artificial variable 'A_1' becomes the departing variable and 'y' the entering variable.

Simplex table with key column and key row identified

Contribution (C_j)			16	12	0	0	M	M	Min Ratio
C_B	Basic Variables	Solution Variables	x	y	s_1	s_2	A_1	A_2	
M	A_1	140	1	2	– 1	0	1	0	140/2 = 70 ←
M	A_2	160	2	2	0	– 1	0	1	160/2 = 80
	(Z_j)		3M	4M	– M	– M	M	M	
(C_j – Z_j) index or net evaluation			5-3M	7- 4M ↑	M	M	0	0	

Incorporating the necessary changes in the Simplex Table and developing the second Simplex table

The New Row

7	y	70	1/2	1	– 1/2	0	–	0

Now the new values for the remaining row is calculated and the values of the other new row is

M	A_2	20	1	0	1	– 1	–	1

The revised table after the incorporation of necessary changes

Revised Simplex table

Contribution (C_j)			5	7	0	0	M	M
C_B	Basic Variables	Solution Variables	x	y	s_1	s_2	A_1	A_2
7	Y	70	1/2	1	– 1/2	0	–	0
M	A_2	20	1	0	1	– 1	–	1
(Z_j)			7/2 + M	7	7/2 – M	– M		M
(C_j – Z_j) index or net evaluation			3/2 – M	0	7/2 – M	M		0

The presence of a negative number in the index row indicates further scope of improvement and the optimal solution not yet obtained.

The column with the highest negative number is identified as the Key Column and to select the key row as discussed earlier the quotients are calculated and the least of the non negative quotients is identified which becomes the key row.

Second Simplex table with key column and key row identified

Contribution (C_j)			5	7	0	0	M	M	Min Ratio
C_B	Basic Variables	Solution Variables	x	y	s_1	s_2	A_1	A_2	
7	y	70	1/2	1	– 1/2	0	–	0	140
M	A_2	20	1	0	1	– 1	–	1	20 ←
(Z_j)			7/2 + M	7	7/2 – M	– M		M	
(C_j – Z_j) index or net evaluation			3/2 – M ↑	0	7/2 – M	M		0	

Proceeding the way as mentioned earlier that x will replace A_2 in the solution along with its contribution. The next is calculate the new row as shown earlier

The New Key Row

5	x	20	1	**0**	**1**	- 1

The New Intersecting Element Row

7	y	60	**0**	1	–1	½

Simplex table with improved solution identified

	Contribution (C_j)		5	7	0	0	M	M
C_B	Basic Variables	Solution Variables	x	y	s_1	s_2	A_1	A_2
7	Y	60	**0**	1	– 1	1/2		
5	X	20	1	**0**	**1**	– 1		
	(Z_j)		5	7	– 2	– 3/2		
	(C_j – Z_j) index or net evaluation		0	0	2	3/2		

Calculating the values of Z_j and $C_j - Z_j$ as discussed earlier and all the entries in the index row are found to be positive which indicates no further improvement and the optimal solution is reached and identified as x = 20, y = 60 and the Min (Z) = Rs 520.

Some Typical Problems Encounter in Simplex method

Some of the typical problems encountered while using simplex method are

(i) ***Degeneracy*****:** Degeneracy occurs when there is tie between the minimum ratio's leading to the possibility of selecting more than one key row i.e. existence of selecting more than one departing variable.

(ii) ***Tie for Key Column*****:** This type of problem occurs when the two or more columns have the same maximum positive index number in the index row thereby giving the possibility of selecting more than one column.

(iii) ***Unbounded Solution*****:** This type of problem occurs when there is no constraint on the solution such that one or more of the decision variables can increase indefinitely without violating any of the constraints, mainly occurs in maximization case.

(iv) ***Multiple Optimal Solutions*****:** It is case where there exists multiple number of solutions for a given problem.

(v) ***Infeasible Solution*****:** Infeasible solution occurs when there exist no solution that satisfies all the constraints of LP problem.

Duality

In cases simplex method is utilized it starts with an initial basic feasible solution to reach the optimal solution i.e. all non negative constants be ≥0 and once optimality is reached the simplex method gets terminated.

But it is found in certain cases there exists some negative constants even though the optimality is reached in such cases simple method cannot be utilized instead dual is utilized to reach optimal solution. In this section the formulation of dual largely elaborated.

Formulation of Dual Linear Programming Problem

Definition of Primal Problem

A general linear programming problem is in primal form is given by

Maximize $(Z_x) = c_1x_1 + c_2x_2 + c_3x_3 + \dots + c_nx_n$

Subject to

$$\left.\begin{array}{ll} & a_{11}x_1 + a_{12}x_2 + a_{13}x_3 + \dots a_{1n}x_n \le b_1 \\ & a_{21}x_1 + a_{22}x_2 + a_{23}x_3 + \dots a_{2n}x_n \le b_2 \\ & \dots \\ & a_{m1}x_1 + a_{m2}x_2 + a_{m3}x_3 + \dots a_{mn}\, x_n \le b_m \\ \text{and} & x_1, x_2, \dots x_n \ge 0 \end{array}\right\} \text{Constraints}$$

Then the corresponding dual LP problem is stated as

Definition of Dual Problem

Minimize $(Z_y) = b_1y_1 + b_2y_2 + b_3y_3 + \dots + b_my_m$

Subject to

$$\left.\begin{array}{ll} & a_{11}y_1 + a_{21}y_2 + a_{31}y_3 + \dots a_{m1}y_m \ge c_1 \\ & a_{12}y_1 + a_{22}y_2 + a_{32}y_3 + \dots a_{m2}y_m \ge c_2 \\ & \dots \\ & a_{1n}x_1 + a_{2n}x_2 + a_{3n}x_3 + \dots a_{mn}\, x_m \ge c_n \\ \text{and} & y_1, y_2, \dots y_m \ge 0 \end{array}\right\} \text{Constraints}$$

We now place the matrix formulation of the primal and the dual problem side by side

Primal	Dual
Maximize $(Z_x) = cx$	Minimize $(Zy) = b^T y$
$Ax \le b$	$A^Ty \ge c^T$
$x \ge 0$	$y \ge 0$

The above pair of primal and dual problems can presented in general form as

Primal	Dual
$\text{Max}(Z_x) = \sum_{j=1}^{n} c_j x_j$	$\text{Min}(Z_y) = \sum_{j=1}^{n} b_i y_i$
Subject to the constraints	Subject to the constraints
$\sum_{j=1}^{n} a_{ij} x_j \le b_i,\ i = 1, 2, \ldots m$	$\sum_{i=1}^{m} a_{ji} y_i \ge c_j,\ j = 1, 2, \ldots n$
and	and
$x_j \ge 0;\ j = 1, 2 \ldots n$	$y_i \ge 0;\ i = 1, 2, \ldots m$

From the above we find the following relationships

(i) The primal problem is a maximization problem whereas the dual problem is a minimization one.

(ii) The maximization problem has less than (≤) type of inequalities whereas the dual problem has greater than (≥) type inequalities in the constraints.

(iii) Number of variables in dual is equal to the number of constraints in the primal and vice versa.

(iv) The cost vector in the objective function of the primal is the right hand side vector in the set of constraints of the dual and vice versa.

(vi) The ith variable in the dual variable is unconstrained in sign if and only if i^{th} the constraint in the original is an equality.

(vii) The coefficient matrix of the constraints of the primal is transpose of the coefficient matrix of the constraints of the dual and vice versa.

Illustration No. 1

Formulate the dual of the following L.P.P

Maximize $Z = 4x_1 + 10x_2 + 12x_3$

Subject to the constraints

$$6x_1 + 12x_2 - 2x_3 \le 8$$

$$-4x_1 + 2x_2 + 6x_3 \le 6$$

$$2x_1 - 8x_2 + 4x_3 \le 4$$

$$10x_1 + 6x_2 - 14x_3 \le 10$$

$$x_1, x_2, x_3 \ge 0$$

Solution:

The problem can be rewritten as

Maximize $(Z_x) = cx$

Subject $Ax \le b, x \ge 0$

Here $x = [x_1, x_2, x_3]$ and $c = [4, 10, 12]$

$$A = \begin{bmatrix} 6 & 12 & -2 \\ -4 & 2 & 6 \\ 2 & -8 & 4 \\ 10 & 6 & -14 \end{bmatrix} \text{ and } b = \begin{bmatrix} 8 \\ 6 \\ 4 \\ 10 \end{bmatrix}$$

Since there are four constraints in the primal therefore the number of dual variables are y_1, y_2, y_3 and y_4.

Now the dual problem is to minimize $(Z_y) = (8, 6, 4, 10)\ [y_1, y_2, y_3, y_4]$

i.e $8y_1 + 6y_2 + 4y_3 + 10\ y_4$.

$(Z_y) = 8y_1 + 6y_2 + 4y_3 + 10\ y_4$.

Subject to the constraints $A^T y \ge c^T$, $y \ge 0$

i.e

$$\begin{bmatrix} 6 & -4 & 2 & 10 \\ 12 & 2 & -8 & 6 \\ 2 & 6 & 4 & -14 \end{bmatrix} \begin{bmatrix} y_1 \\ y_2 \\ y_3 \\ y_4 \end{bmatrix} \ge \begin{bmatrix} 4 \\ 10 \\ 12 \end{bmatrix}$$

i.e.

$$6\ y_1 - 4y_2 + 2y_3 + 10y_4 \ge 4$$

$$12y_1 + 2y_2 - 8y_3 + 6y_4 \ge 10$$

$$-2y_1 + 6y_2 + 4y_3 - 14\ y_4 \ge 12$$

$$y_1, y_2, y_3, y_4 \ge 0$$

Illustration No. 2

Maximize $Z = 3x_1 + 4x_2 + x_3$

Subject to the constraints

$$6x_1 + 3x_2 - 3x_3 \le 6$$

$$-12x_1 + 6x_2 + 8x_3 \le -12$$

$4x_1 + 5x_2 + 3x_3 \leq 5$

$x_1, x_2, x_3 \geq 0$

Solution:

The problem can be rewritten as

Maximize $(Z_x) = cx$

Subject $Ax \leq b, x \geq 0$

Here $x = [x_1, x_2, x_3]$ and $c = [3, 4, 1]$

$$A = \begin{bmatrix} 6 & 3 & -3 \\ -12 & 6 & 8 \\ 4 & 5 & 3 \end{bmatrix} \text{ and } b = \begin{bmatrix} 6 \\ -12 \\ 5 \end{bmatrix}$$

Since there are three constraints in the primal therefore the number of dual variables are y_1, y_2 and y_3

Now the dual problem is to minimize $(Z_y) = (6, -10, 5)\ [y_1, y_2, y_3]$

i.e $6y_1 - 10y_2 + 5y_3$

$(Z_y) = 6y_1 - 10y_2 + 5y_3$

Subject to the constraints $A^T y \geq c^T$, $y \geq 0$

$$\begin{bmatrix} 6 & -12 & 4 \\ 3 & 6 & 5 \\ -3 & 8 & 3 \end{bmatrix} \begin{bmatrix} y_1 \\ y_2 \\ y_3 \end{bmatrix} \geq \begin{bmatrix} 3 \\ 4 \\ 1 \end{bmatrix}$$

i.e.

$6y_1 - 12y_2 + 4y_3 \geq 3$

$3y_1 + 6y_2 + 5y_3 \geq 4$

$-3y_1 + 8y_2 + 3y_3 \geq 1$

$y_1, y_2, y_3 \geq 0$

Illustration No. 3

Maximize $Z = x_1 + 2x_2 + 3x_3$

Subject to the constraints

$8x_1 + 3x_2 + x_3 \leq 11$

$-5x_1 + 6x_2 + 2x_3 \leq 13$

$2x_1 + 2x_2 + 4x_3 \leq 7$

$x_1, x_2, x_3 \geq 0$

Solution:

The problem can be rewritten as

Maximize $(Z_x) = cx$

Subject $Ax \le b, x \ge 0$

Here $x = [x_1, x_2, x_3]$ and $c = [1, 2, 3]$

$$A = \begin{bmatrix} 8 & 3 & 1 \\ -5 & 6 & 2 \\ 2 & 2 & 4 \end{bmatrix} \text{ and } b = \begin{bmatrix} 11 \\ 13 \\ 7 \end{bmatrix}$$

Since there are three constraints in the primal therefore the number of dual variables are y_1, y_2 and y_3.

Now the dual problem is to minimize $(Z_y) = (11, 13, 7)\ [y_1, y_2, y_3]$

i.e $11y_1 + 13y_2 + 7y_3.$

$(Z_y) = 11y_1 + 13y_2 + 7y_3.$

Subject to the constraints $A^T y \ge c^T$, $y \ge 0$.

$$\begin{bmatrix} 8 & -5 & 2 \\ 3 & 6 & 2 \\ 1 & 2 & 4 \end{bmatrix} \begin{bmatrix} y_1 \\ y_2 \\ y_3 \end{bmatrix} \ge \begin{bmatrix} 1 \\ 2 \\ 3 \end{bmatrix}$$

i.e.

$8y_1 - 5y_2 + 2y_3 \ge 1$

$3y_1 + 6y_2 + 2y_3 \ge 2$

$y_1 + 2y_2 + 4y_3 \ge 3$

$y_1, y_2, y_3 \ge 0$

Illustration No. 4

Formulate the dual of the following L.P.P

Minimize $(Z_x) = 2x_1 + 4x_2 + 3x_3$

Subject to the constraints

$6x_1 + 10x_2 + 8x_3 \ge 14$

$12x_1 + 2x_2 + 6x_3 \ge 8$

$8x_1 + 14x_2 - 4x_3 \ge 4$

$x_1, x_2, x_3 \ge 0$

Solution:

The problem can be rewritten as a primal problem in matrix notation

Minimize $(Z_x) = cx$

Subject $Ax \geq b,\ x \geq 0$

Here $x = [x_1, x_2, x_3]$ and $c = [2, 4, 3]$

$$A = \begin{bmatrix} 6 & 10 & 8 \\ 12 & 2 & 6 \\ 14 & 4 & 10 \\ 8 & 14 & -4 \end{bmatrix} \text{ and } b = \begin{bmatrix} 14 \\ 8 \\ 6 \\ 4 \end{bmatrix}$$

Since there are four constraints in the primal therefore the number of dual variables are y_1, y_2, y_3 and y_4.

Now the dual problem is to maximize $(Z_y) = (14, 8, 6, 4)\ [y_1, y_2, y_3, y_4]$

i.e $14y_1 + 8y_2 + 6y_3 + 4\ y_4.$

$(Z_y) = 14\ y_1 + 8y_2 + 6y_3 + 4\ y_4.$

Subject to the constraints $A^T y \leq c^T$, $y \geq 0$

i.e.

$$\begin{bmatrix} 6 & 12 & 14 & 8 \\ 10 & 2 & 4 & 14 \\ 8 & 6 & 10 & -4 \end{bmatrix} \begin{bmatrix} y_1 \\ y_2 \\ y_3 \\ y_4 \end{bmatrix} \geq \begin{bmatrix} 2 \\ 4 \\ 3 \end{bmatrix}$$

i.e.

$6\ y_1 + 12\ y_2 + 14y_3 + 8y_4 \geq 2$

$10y_1 + 2y_2 + 4y_3 + 14\ y_4 \geq 4$

$8y_1 + 6y_2 + 10y_3 - 4\ y_4 \geq 3$

$y_1, y_2, y_3, y_4 \geq 0$

Illustration No. 5

Formulate the dual of the following L.P.P

Maximize $Z = x_1 - 3x_2$

Subject to the constraints

$3x_1 - x_2 \leq 6$

$4x_1 + 2x_2 \geq 8$

$3x_1 - 5x_2 \geq -6$

$x_1, x_2 \geq 0$

Solution:

Given

Maximize $Z = x_1 - 3x_2$

Subject to the constraints

$3x_1 - x_2 \leq 6$

$4x_1 + 2x_2 \geq 8$

$3x_1 - 5x_2 \geq -6$

$x_1, x_2 \geq 0$

But two of the constraint is greater than type (≥) i.e $4x_1 + 2x_2 \geq 8$, $3x_1 - 5x_2 \geq -6$, converting them into less than type, we get $-4x_1 - 2x_2 \leq -8$, $-3x_1 + 5x_2 \leq 6$.

The problem can be rewritten as

Maximize $Z = x_1 - 3x_2$

Subject to the constraints

$3x_1 - x_2 \leq 6$

$-4x_1 - 2x_2 \leq -8$

$-3x_1 + 5x_2 \leq 6.$

$x_1, x_2 \geq 0$

Now writing the above LPP in matrix notation, we get

Maximize $(Z_x) = cx$

Subject $Ax \leq b, x \geq 0$

Here $x = [x_1, x_2]$ and $c = [1, -3]$

$$A = \begin{bmatrix} 3 & -1 \\ -4 & -2 \\ -3 & 5 \end{bmatrix} \text{ and } b = \begin{bmatrix} 6 \\ -8 \\ 6 \end{bmatrix}$$

Since there are three constraints in the primal therefore the number of dual variables are y_1, y_2 and y_3

Now the dual problem is to minimize $(Z_y) = (6, -8, 6)\ [y_1, y_2, y_3]$

i.e $\quad 6y_1 - 8y_2 + 6y_3.$

Minimize $(Z_y) = 6y_1 - 8y_2 + 6y_3.$

Subject to the constraints $A^T y \geq c^T$, $y \geq 0$.

i.e

$$\begin{bmatrix} 3 & -4 & -3 \\ -1 & -2 & 5 \end{bmatrix}\begin{bmatrix} y_1 \\ y_2 \end{bmatrix} \geq \begin{bmatrix} 1 \\ -3 \end{bmatrix}$$

i.e.

$$3y_1 - 4y_2 - 3y_3 \geq 1$$

$$-y_1 - 2y_2 + 5y_3 \geq -3$$

$$y_1, y_2 \geq 0$$

Theorem No. 1

The dual L.P.P of the dual of a primal L.P.P is the primal itself.

Proof:

Let the given L.P.P be written in the form

Maximize $(Z_x) = c_1x_1 + c_2x_2 + c_3x_3 + \ldots + c_nx_n$

Subject to

$$a_{11}x_1 + a_{12}x_2 + a_{13}x_3 + \ldots a_{1n}x_n \leq b_1$$

$$a_{21}x_1 + a_{22}x_2 + a_{23}x_3 + \ldots a_{2n}x_n \leq b_2$$

$$\ldots$$

$$a_{m1}x_1 + a_{m2}x_2 + a_{m3}x_3 + \ldots a_{mn}x_n \leq b_m$$

and $\quad x_1, x_2, \ldots x_n \geq 0 \qquad \ldots..(1)$

Then the corresponding dual LP problem is stated as

Minimize $(Z_y) = b_1y_1 + b_2y_2 + b_3y_3 + \ldots + b_my_m$

Subject to

$$a_{11}y_1 + a_{21}y_2 + a_{31}y_3 + \ldots a_{m1}y_m \geq c_1$$

$$a_{12}y_1 + a_{22}y_2 + a_{32}y_3 + \ldots a_{m2}y_m \geq c_2$$

$$\ldots$$

$$a_{1n}x_1 + a_{2n}x_2 + a_{3n}x_3 + \ldots a_{mn}x_m \geq c_n$$

and $y_1, y_2, \ldots y_m \geq 0$ .....(2)

Consider 2 as primal; writing the dual as

Maximize $(Z_z) = c_1z_1 + c_2z_2 + c_3z_3 + \ldots + c_nz_n$

Subject to

$$a_{11}z_1 + a_{12}z_2 + a_{13}z_3 + \ldots a_{1n}z_n \leq b_1$$

$$a_{21}z_1 + a_{22}z_2 + a_{23}z_3 + \ldots a_{2n}z_n \leq b_2$$

$$\ldots$$

$$a_{m1}z_1 + a_{m2}z_2 + a_{m3}z_3 + \ldots a_{mn}z_n \leq b_m$$

and $z_1, z_2, \ldots z_n \geq 0$

Replacing (Z_z) by (Z_x) and $z_1, z_2, \ldots z_n$ by $x_1, x_2, \ldots x_n$ then, we get,

Maximize $(Z_x) = c_1x_1 + c_2x_2 + c_3x_3 + \ldots + c_nx_n$

Subject to

$$a_{11}x_1 + a_{12}x_2 + a_{13}x_3 + \ldots a_{1n}x_n \leq b_1$$

$$a_{21}x_1 + a_{22}x_2 + a_{23}x_3 + \ldots a_{2n}x_n \leq b_2$$

$$\ldots$$

$$a_{m1}x_1 + a_{m2}x_2 + a_{m3}x_3 + \ldots a_{mn}x_n \leq b_m$$

and $x_1, x_2, \ldots x_n \geq 0$

Which is the original given primal and hence the dual of dual of a primal L.P.P is the primal itself.

Illustration No. 6

Given the following L.P.P, verify that the dual of the dual is the primal itself

Maximize $Z = 4x_1 + 6x_2 + 10x_3$

Subject to the constraints

$$6x_1 + 6x_2 - 14x_3 \leq 12$$

$$10x_1 + 12x_2 + 2x_3 \leq 6$$

$$-4x_1 + 2x_2 + 8x_3 \leq 8$$

$$2x_1 - 8x_2 + 6x_3 \leq 2 \qquad \text{where } x_1, x_2, x_3 \geq 0$$

Solution: Part I

The problem can be rewritten as

Maximize $(Z_x) = cx$

Subject $Ax \le b, x \ge 0$

Here $x = [x_1, x_2, x_3]$ and $c = [4, 6, 10]$

$$A = \begin{bmatrix} 6 & 6 & -14 \\ 10 & 12 & 2 \\ -4 & 2 & 8 \\ 2 & -8 & 6 \end{bmatrix} \text{ and } b = \begin{bmatrix} 12 \\ 6 \\ 8 \\ 2 \end{bmatrix}$$

Since there are four constraints in the primal therefore the number of dual variables are y_1, y_2, y_3 and y_4.

Now the dual problem is to minimize $(Z_y) = (12, 6, 8, 2)\,[y_1, y_2, y_3, y_4]$

i.e $12y_1 + 6y_2 + 8y_3 + 2y_4$.

Minimize $(Z_y) = 12y_1 + 6y_2 + 8y_3 + 2y_4$.

Subject to the constraints $A^T y \ge c^T$, $y \ge 0$

i.e.

$$\begin{bmatrix} 6 & 10 & -4 & 2 \\ 6 & 12 & 2 & -8 \\ -14 & 2 & 8 & 6 \end{bmatrix} \begin{bmatrix} y_1 \\ y_2 \\ y_3 \\ y_4 \end{bmatrix} \ge \begin{bmatrix} 4 \\ 6 \\ 10 \end{bmatrix}$$

i.e.

$$6y_1 + 10y_2 - 4y_3 + 2y_4 \ge 4$$

$$6y_1 + 12y_2 + 2y_3 - 8y_4 \ge 6$$

$$-14y_1 + 2y_2 + 8y_3 + 6y_4 \ge 12$$

$$y_1, y_2, y_3, y_4 \ge 0$$

Part II

Now if we want to write the dual of dual of the L.P.P considering as a primal

Minimize $(Z_y) = 12y_1 + 6y_2 + 8y_3 + 2y_4$.

Subject to

$$6y_1 + 10y_2 - 4y_3 + 2y_4 \geq 4$$

$$6y_1 + 12y_2 + 2y_3 - 8y_4 \geq 6$$

$$-14y_1 + 2y_2 + 8y_3 + 6y_4 \geq 12$$

$$y_1, y_2, y_3, y_4 \geq 0$$

i.e.

The problem can be rewritten as a primal problem in matrix notation

Minimize $(Z_y) = cy$

Subject $Ay \geq b, y \geq 0$

Here $y = [y_1, y_2, y_3, y_4]$ and $c = [12, 6, 8, 2]$ i.e.

$$A = \begin{bmatrix} 6 & 10 & -4 & 2 \\ 6 & 12 & 2 & -8 \\ 14 & 2 & 8 & 6 \end{bmatrix} \text{ and } b = \begin{bmatrix} 4 \\ 6 \\ 10 \end{bmatrix}$$

Since there are three constraints in the primal therefore the number of dual variables are z_1, z_2 and z_3

Now the dual problem is to maximize $(Z_z) = (4, 6, 10)\ [z_1, z_2, z_3]$

i.e $4z_1 + 6z_2 + 10z_3.$

$(Z_z) = 4z_1 + 6z_2 + 10z_3$

Subject to the constraints $A^T z \leq c^T$, $z \geq 0$

$$\begin{bmatrix} 6 & 6 & -14 \\ 10 & 12 & 2 \\ -4 & 2 & 8 \\ 2 & -8 & 6 \end{bmatrix} \begin{bmatrix} z_1 \\ z_2 \\ z_3 \end{bmatrix} \leq \begin{bmatrix} 12 \\ 6 \\ 8 \\ 2 \end{bmatrix}$$

i.e.

$$6z_1 + 6z_2 - 14z_3 \leq 12$$

$$10z_1 + 12z_2 + 2z_3 \leq 6$$

$$-4z_1 + 2z_2 + 8z_3 \leq 8$$

$$2z_1 - 8z_2 + 6z_3 \leq 2 \qquad \text{where } z_1, z_2, z_3 \geq 0$$

Replacing the (z_1, z_2, z_3) by (x_1, x_2, x_3) we get the original primal

Maximize $Z = 4x_1 + 6x_2 + 10x_3$

Subject to the constraints

$$6x_1 + 6x_2 - 14x_3 \le 12$$

$$10x_1 + 12x_2 + 2x_3 \le 6$$

$$-4x_1 + 2x_2 + 8x_3 \le 8$$

$$2x_1 - 8x_2 + 6x_3 \le 2 \qquad \text{where } x_1, x_2, x_3 \ge 0$$

Some of the Advantages of Duality

(i) Largely reduces the computational procedure by converting into a dual where the primal may contain large number of constraints and smaller number of variables.

(ii) Provide additional information about the changes in optimal solution by changing the coefficients and the formulation.

(iii) Economic interpretation of dual helps the management in making future planning.

(iv) It helps in checking the accuracy of the primal solution.

Summary

- Objective is a desired result that needs to be achieved in form of Profit Maximization or Cost minimization etc.
- Constraints represents Limitation in resource capacity in the form of machine Hours, Labour Hours, Space constraint and Investment constraint.
- Decision Variables are variables competing against one another to grab a share of the limited resources.
- Artificial variables are fictitious in nature and used only for computational purpose with very large unit of penalty assigned in the objective function.
- Infeasible solution occurs when there exist no solution that satisfies all the constraints of LP problem.

Self Practice Exercises

Solve the following using Simplex Method

1. Maximum (Z) = 3x + 4y

   Subject to

   4x + 6y ≤ 32

   8x + 4y ≤ 32

   where x, y ≥ 0 (non-negative constraints)

2. Maximum (Z) = 4x + 9y

   Subject to

   2x + 3y ≤ 60,

   2x + y ≤ 30,

   where x, y ≥ 0 (non-negative constraints)

3. Maximum (Z) = 8x + 6y

   Subject to

   4x + 6y ≤ 24

   x ≤ 4,

   y ≤ 3,

   where x, y ≥ 0 (non-negative constraints)

4. Maximum (Z) = 6x + 4y

   Subject to

   2x + y ≤ 20

   x ≤ 8,

   y ≤ 6,

   where x, y ≥ 0 (non-negative constraints)

5. Maximum (Z) = 10x + 5y

   Subject to

   x + 2y ≤ 6,

   4x + 3y ≤ 12

   where x, y ≥ 0 (non-negative constraints)

6. Maximum (Z) = 6x + 12y + 4z

Subject to

$6x + 8y + 2z \le 4,$

$2x + 6y + 4z \le 2$

where x, y and z ≥ 0 (non-negative constraints)

7. Maximize Z = 2x + 8y + 10z

Subject to

$4x + 2y + 6z \le 72,$

$2x + 4y + 2z \le 84$

where x, y and z ≥ 0 (non-negative constraints)

Solve the following Using Big – M Method

8. Minimize Z = 3x + 4y

Subject to

$2x + 4y \ge 16,$

$6x + 4y \ge 24$

where x, y ≥ 0 (non-negative constraints)

9. Minimize Z = 10x + 5y

Subject to

$7x + 4y \ge 28,$

$2x + 4y \ge 16$

where x, y ≥ 0 (non-negative constraints)

10. Minimize Z = 4x + 3y

Subject to

$x + 2y \ge 8,$

$3x + 2y \ge 12$

where x, y ≥ 0 (non-negative constraints)

11. Minimize Z = 6x + 8y

Subject to

$2x + 3y \ge 90,$

$4x + 3y \ge 120$

where x, y ≥ 0 (non-negative constraints)

12. Formulate the dual of the following L.P.P

Maximize $Z = 6x_1 - 2x_2 + 2x_3$

Subject to the constraints

$$8x_1 - 2x_2 \leq 8$$

$$-16x_1 - 2x_2 - 2x_3 \leq -12$$

$$10x_1 - 12x_3 \leq 4$$

$$x_1, x_2, x_3 \geq 0$$

13. Formulate the dual of the following L.P.P

Maximize $Z = 6x_1 + 11x_2 + 9x_3$

Subject to the constraints

$$x_1 + 6x_2 - 2x_3 \leq 6$$

$$2x_1 + 2x_2 - 7x_3 \leq 8$$

$$x_1, x_2, x_3 \geq 0$$

14. Formulate the dual of the following L.P.P

Minimize $Z = 8x_1 + 40x_2$

Subject to the constraints

$$x_1 + x_2 \geq 40$$

$$3x_1 + 10x_2 \geq 8$$

$$4x_1 + 5x_2 \geq 8$$

$$x_1, x_2 \geq 0$$

15. Formulate the dual of the following L.P.P

Maximize $Z = 16x_1 - 20x_2$

Subject to the constraints

$$4x_1 - 10x_2 \geq 6$$

$$6x_1 - 4x_2 \geq 4$$

$$x_1, x_2 \geq 0$$

Answers

1. $x = 8$, $y = 0$ and $Z = 24$.
2. $x = 0$, $y = 20$ and $Z = 180$

3. $x = 4, y = 4/3$ and $Z = 40$
4. $x = 7, y = 6$ and $Z = 66$.
5. $x = 3, y = 0$ and $Z = 30$
6. $x = 2/5, y = 1/5, z = 0$ and $Z = 24/5$
7. $x = 0, y = 18$ and $z = 6$ and $Z = 204$
8. $x = 2, y = 3$ and $Z = 18$
9. $x = 12/5, y = 14/5$ and $Z = 36$
10. $x = 2, y = 3$ and $Z = 17$
11. $x = 15, y = 10$ and $Z = 170$
12. Minimize $(Z_y) = 8y_1 - 24y_2 + 26y_3$

    Subject to

    $$8y_1 - 16y_2 + 10y_3 \geq 6$$
    $$-2y_1 - 2y_2 \geq -2$$
    $$-6y_2 - 12y_3 \geq 2$$
    $$y_1, y_2, y_3, \geq 0.$$
13. Minimize $(Z_y) = 6y_1 + 8y_2$

    Subject to

    $$y_1 + 2y_2 \geq 6$$
    $$6y_1 + 2y_2 \geq 11$$
    $$-2y_1 - 7y_2 \geq 9$$
    $$y_1, y_2, y_3, \geq 0$$
14. Maximize $(Z_y) = 40y_1 + 60y_2 + 40y_3$

    Subject to

    $$y_1 + 3y_2 + 4y_3 \leq 6$$
    $$y_1 + 10y_2 + 5y_3 \leq 6$$

    Where $y_1, y_2, y_3, \geq 0$
15. Maximize $(Z_y) = 6y_1 + 4y_2$

    Subject to

    $$4y_1 + 6y_2 \leq 16$$
    $$-10y_1 - 4y_2 \leq -20$$

    Where $y_1, y_2 \geq 0$

CHAPTER 20

Decision Theory and Decision Tree

Introduction

In daily life we confront numerous problematic situations which call for active search for viable solution alternatives in order to solve the problem in hand. On many such occasions we tend to reach conclusion tentatively in fast and haphazard manner because of ignorance in handling a critical situation which requires finesse and scientific treatment. As we start using scientific techniques in the process of decision making we move from the zone of ignorance to a known zone thereby reaching the desired objectives which becomes smooth coupled with high success rate. Since we weight each of the alternatives with available resources and then select the best fitting one in order that the consequences are not detrimental. We have been using various scientific models in business for decision making and one such tool is decision theory or decision analysis process by which a certain course of action is selected among the set of alternatives to reach a desired objective of the problem in efficient and satisfactory manner.

According to Fishburn solving the decision model consists of finding a strategy for action, the expected relative value which is at least as great as the expected relative value of any other strategy in a specified set. The descriptive criterion of a strategy will be maximization of the decision maker's expected relative value.

General Characteristics of Decision Making

Some of the important characteristics

(i) It is the final product of various deliberations and analytical reasoning.

(ii) It results in the selection of the best fitted alternative with the constraint of resources.

(iii) It is commitment of the highest order from the beginning till the successful implementation of the project or job.

(iv) Sometimes the decision taken may go wrong in certain conditions in such scenario there should be scope for taking alternative course of actions.

Some of the Steps in Decision Making Process

The decision making process has the following steps

Identification and defining the problem

In most instances identifying and defining the problem is the most crucial and time consuming task.

Gathering and diagnosing the Facts

Once the problem is identified the next step is to gather necessary and reliable information. From the relevant facts derived should be recorded and subjected to checks and cross checks and as far as possible all conflicting data must be resolved to evolve a common consensus.

Identify all available alternatives

With the identification of the problem the next step is to look for possible set of course of action or alternatives and list them in the light of the available facts and figures. While identifying alternatives past experience, research work and brain storming and other aids can be utilized.

Evaluation of Alternatives

Alternatives which have evolved are evaluated in the light of set objectives and the resulting pay offs from each course of action. Since each course of action or alternatives have a degree of risk associated, the decision makers next course is to assess the risk and efforts are made either to eliminate risk or minimize the effect of risk. Then the alternatives are quantified in terms of risk and monetary value.

Selection of Alternatives

Those alternatives which have enlisted from the evaluation stage are then ranked according to their acceptability. And the final choice is made on the basis of effectiveness which results in optimal pay offs within acceptable limits of risk.

Implementation of the Selection Decision

In this stage the alternative which is selected as the final choice is then implemented.

Graphical Representation Decision Making Process

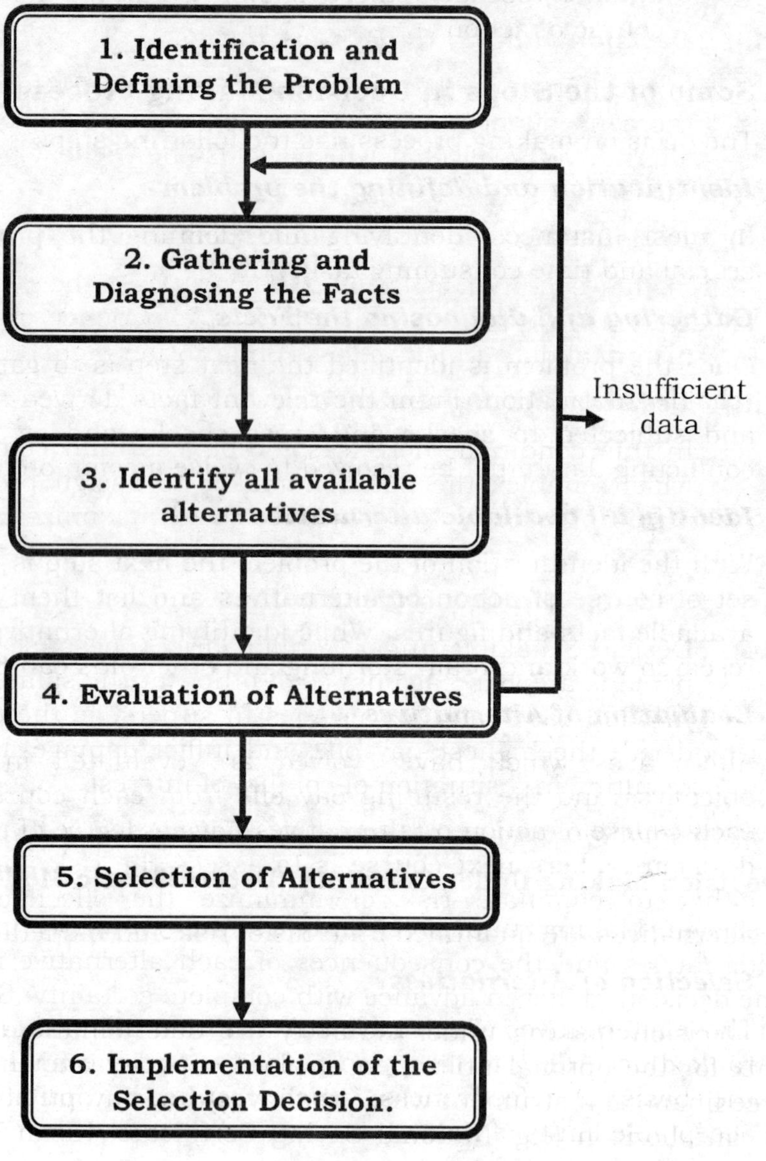

Different Categories of Decision Making Models

Decision theory helps the decision maker in selection of optimal choice from a given set of alternative courses. Depending upon the available information and different states of decision making environment, it is classified broadly into four general categories based on the predictive ability of the consequences of each alternative.

(i) ***Decision Making Under Certainty***

In this condition there is complete or perfect information regarding the consequences of each decision choice with certainty. And an alternative is selected which yields the highest possible return in the future.

(ii) ***Decision Making under Uncertainty***

In this condition it is difficult to forecast the probabilities of the different events because of lack of experience or unavailability of historical data which makes computation highly difficult.

(iii) ***Decision Making under Risk***

In this condition there is some or less than complete knowledge which enables the decision maker to assign probability to each states of nature. The objective is to maximize gain or minimize loss.

(iv) ***Decision Making Under Conflict***

Decision making under conflict arises because of two or more persons are vying against each other for the same particular thing. In such case each player tries to supersede the other in order to derive the highest pay offs and inflict damage as much possible resulting in a situation of conflict of interest.

Note: Within the purview of Game theory discussed in the next chapter.

Decision Making Under Certainty (Deterministic Model)

Decision Making under certainty is the situation in which all the alternatives and the consequences of each alternative is well known to the decision maker in advance with complete certainty. Some of the cases of decision making under certainty are determination EOQ, product – mix for the optimal utilization of plant capacity and investment in risk free financial instrument where the investors is aware of the returns after a time horizon (e.g. National Saving Certificate, Kisan Vikas Patra etc).

Some of the techniques used to deal with decision making under certainty are

(i) System of Equations
(ii) Linear Programming
(iii) Inventory Models
(iv) Queuing Models
(v) Sequencing Models
(vi) Dynamic Programming
(vii) Assignment Models

Illustration No 1

Zigma Tyres are the manufacturer tubeless radials. In order to manufacture the tyres needs processing which can be done on either one of the two machines M_1 and M_2. Following are the additional details furnished below

Cost Under Various Heads	M_1	M_2
Machining Time per Unit (Mins)	12	6
Set Up Time (Min)	40	60
Machine Cost per hour	60	80
Mis. Exp. with each machine(Rs.)	400	800

Find the selection of the machine among M_1 and M_2 if the company has received an order size of 800 and with little expectation of reorder from the vendor?

Solution:

Let

F_{M1} = The Fixed Cost per Machine M_1

F_{M2} = The Fixed Cost per Machine M_2

v_{M1} = The Variable Cost per unit in Machine M_1

v_{M2} = The Variable Cost per unit in Machine M_2

Q_{BEQ} = The Break Even Quantity.

Q = Lot size order Placed

The above illustration can be solved by Break – Even Analysis technique and the steps involved are hence furnished as

1. Collection of data

2. Calculation of Fixed Cost and Variable Cost.
3. Find The Break Even Quantity(Q_{BEQ}).
4. Compare the desired quantity of requirement (Q) with the Break Even Quantity(Q_{BEQ}),
   (i) if $Q > Q_{BEQ}$ then select the machine with lower variable cost.
   (ii) if $Q < Q_{BEQ}$ then select the machine with lower Fixed cost
   (iii) if $Q = Q_{BEQ}$ then select either of the machine.

Now by using the above information we calculate both the cost fixed and variable.

Heads	M_1	M_2
Set Up Cost (Rs.)	$\frac{40}{60} \times 60 = 40$	$\frac{60}{60} \times 80 = 80$
Misc. Cost	400	800
Fixed Cost (Set Up + Misc. Cost)	440	880
Variable Cost Per Unit (Rs.)	$\frac{12}{60} \times 60 = 12$	$\frac{6}{60} \times 80 = 8$

But the cost associated with both the machines at Break Even Quantity will be same therefore

$$F_{M1} + v_{M1}\, Q = F_{M2} + v_{M2}\, Q$$

$$440 + 12\, Q = 880 + 8\, Q$$

And hence

$$Q = \frac{880 - 440}{12 - 8} = 110.$$

As the lot size (800 > 110) therefore the second machine with lower variable cost is utilized for production.

Model I

Decision Making Under UnCertainty (Uncertain Situations)

Under this category includes those decisions where the probabilities associated with the different events cannot be estimated because of absence of knowledge secondly no past experience nor any historical data to allow computation as such the decision maker hard pressed to compute the expected pay off for his strategies. Some of the different techniques for decision making under uncertainty are

1. Maximin or Minimax Criterion (Pessimism Criterion)
2. Maximax or Minimin Criterion (Optimism Criterion)
3. Hurwicz–Alpha Criterion.
4. Laplace Criterion.
5. Regret Criterion.

Maximin or Minimax Criterion (Pessimism Criterion)

Also called as Waldian Criterion in which the decision maker undertakes a course of action which maximizes his minimum pay offs called as maximin "the best of the worst" and in case of minimax it is minimum of the maxima. The working steps of the criterion is summarized as

1. Identify the minimum pay off value in case of profit (or maximum in case of loss) in correspondence to each alternative available.
2. Select a course of action that has the best maximum pay off value among the minimum pay offs and vice–versa.

This criterion is also called as pessimistic decision criterion based on the fact that the decision maker is uncertain and conservative about the future and anticipates the worst possible outcome i.e. minimum profit and maximum loss.

Illustration No. 1

Akash Electricals manufacturer of ceiling fans brand name – Breeze and Delight has faced a typical business problem of selecting one of the brands for production which is solely determined by the market acceptance in terms of profitability which in current state is uncertain. However the state of nature is classified under four broad heads as furnished in the table below

State	Profitability of Each Model as per Market Acceptance (Rs.)	
	Breeze	Delight
Excellent	50000	60000
Good	30000	20000
Average	20000	10000
Poor	8000	4000

Find the firms decision of selecting a particular brand according to maximin criterion?

Solution:

Pay off matrix for the above problem can written as

Model Type	State of Nature				Minimum of Row	Maximin
	Excellent	Good	Average	Low		
Breeze	50000	30000	20000	8000	8000	***8000***
Delight	60000	20000	10000	4000	4000	

Since the maximum of these minimum pay off is Rs. 8000 and hence the Akash electrical should select model Breeze according to maximin criterion.

Maximax or Minimin Criterion (Optimism Criterion)

Under this category the decision maker undertakes a course of action to gain the maximum possible profit (selection of the best of the best) or the lowest possible loss. The working steps of the criterion is summarized as

1. Identify the maximum pay off value (or minimum) in correspondence to each alternative available.
2. Select a course of action that has the best anticipated pay off value, maximum for profit and minimum for loss.

This is also called as optimistic decision criterion as the decision maker selects the alternative with largest (lowest) possible pay offs.

Continuing illustration No 1:

Find what will be the course of action for selecting a brand according to maximax criterion?

Solution:

The pay off matrix is given by

Model Type	State of Nature				Maximum of Row	Maximax
	Excellent	Good	Average	Low		
Breeze	50000	30000	20000	8000	50000	***60000***
Delight	60000	20000	10000	4000	60000	

Since the maximum is Rs.60000 associated with the model Delight and hence the Akash electrical should select model Delight based on maximax criterion.

Hurwicz–Alpha Criterion

Under this criterion the decision maker is neither completely pessimistic nor optimistic but follows a midway approach which is the mixture of both. The idea is introducing coefficient of optimism denoted by α to measure the degree of optimism. The value of the coefficient lies between zero and unity where the value of zero indicates total pessimism and unity total optimism. If α represents coefficient of optimism then $(1 - \alpha)$ will represent coefficient of pessimism.

Hurwicz Criterion suggest that the decision makers course of action must maximizes criterion of realism.

$$\text{Hurwicz Criterion of Realism} = \alpha\,(\text{Maximum in Column}) + (1 - \alpha)\,(\text{minimum in column})$$

The working steps of the criterion is summarized as:

1. Identify the largest pay off and the smallest pay off for each alternative.
2. Estimate the value of the coefficient of optimism.
3. Against each alternative of largest and lowest pay off multiply with α and $1 - \alpha$ then calculate the weighted average using the above mentioned formula.
4. Select an alternative with the highest weighted average pay off.

Continuing from Illustration No 1:

Find what will Akash's preference of brand selection according to Hurwicz – Alpha criterion if the coefficient of optimism $\alpha = 0.60$?

Solution:

Model	Maximum Pay Off (i)	Minimum Pay Off (ii)	Weighted Averages (Rs.) $= \alpha \times i + (1-\alpha) \times ii$
Breeze	50000	8000	$0.60 \times 50000 + 0.4 \times 8000$ = **33200**
Delight	60000	4000	$0.60 \times 60000 + 0.4 \times 4000$ = **37600**

According to Hurwicz–Alpha Criterion the optimum solution is choosing the Delight Model as the pay off is Rs. 37600.

Laplace Criterion (Equally Likely)

Under this criterion the decision maker logically assumes that all events have equal chance of occurrence i.e. the probability assigned to each state are equal. As the occurrence of each event is equally likely and hence the probability of each of these event is

$$\frac{1}{\text{Number of States of the Nature}}$$

The working steps of the criterion is summarized as:

1. Assigning of equal probabilities by using above mentioned formula.
2. Calculate the expected pay off for each alternative by applying the following formula

   Probability of the state of nature × Pay off value for the combination of alternative.

   Or by adding all the pay –offs and dividing by the number of possible states of nature.
3. Select the best expected pay off.

Continuing illustration No 1:

Find what will be the preferred brand for Akash Electricals according to Laplace Criterion?

Solution:

According to Laplace criterion, the expected value for the two models are,

$$\text{Breeze Model} = \frac{1}{4}\times 50000 + \frac{1}{4}\times 30000 + \frac{1}{4}\times 20000 + \frac{1}{4}\times 8000$$

$$= 27000$$

$$\text{Delight Model} = \frac{1}{4}\times 60000 + \frac{1}{4}\times 20000 + \frac{1}{4}\times 10000 + \frac{1}{4}\times 4000$$

$$= 23500$$

Note: as the number states of nature are four and hence the probability of each event is $\frac{1}{4}$.

Hence Akash Electricals will select Breeze Model according to Laplace criterion.

Regret Criterion

Regret means disappointment. Under this criterion the decision maker feels regretted for taking a wrong course of action resulting in an opportunity loss in the process intends to minimize the regret.

The working steps of the criterion is summarized as:

1. Identify the largest pay off to each state of nature.
2. Subtract the all other entries from that value as shown in the solved illustration.
3. For each alternative identify the worst or maximum regret value. Record the value in a new row.
4. Select the alternative with the smallest anticipated opportunity loss value.

Illustration No. 2

The estimated profitability condition for a store–owner with various sales states for three different products are furnished below

Sales	Estimated Conditional Profits		
	X	Y	Z
High	3000	4000	3500
Average	2200	1800	2000
Low	2500	2000	2800

Find the decision course selected by the store–owner with respect to regret criterion?

Solution:

Products	Pay Off			Regret Pay Off			Maximum Regret
	High	Avg.	Low				
X	3000	2200	2500	1000	0	300	1000
Y	4000	1800	2000	0	400	800	800
Z	3500	2000	2800	500	200	0	**500**
Maximum Pay off	4000	2200	2800				

Since the alternative Z corresponds to the minimal of the maximum possible regrets, the decision maker should select the alternative Z.

Illustration No. 3

A given decision problem is expressed in the following pay off table form

Strategy	Pay Offs		
	I	II	III
X	30	60	90
Y	60	60	30
Z	60	45	90

Find the recommended decision according to

(i) Maximin Criterion?

(ii) Maximax Criterion?

(iii) Hurwicz–Alpha Criterion if $\alpha = 0.70$

(iv) Laplace Criterion?

(v) Regret Criterion

Solution:

The above pay off matrix can be written as

Maximin Criterion

Strategy	Pay Offs			Minimum Row	Maximin
	I	II	III		
X	30	60	90	30	
Y	60	60	30	30	
Z	60	45	120	45	**45**

As for the given pay off matrix the minimum pay off of each alternative strategy X: 30, Y: 30 and Z = 45 and hence the maximum of these minimum pay offs is 45 therefore the alternative Z is selected according to maximin criterion.

Maximax Criterion

Strategy	Pay Offs			Maximum Row	Maximax
	I	II	III		
X	30	60	90	90	
Y	60	60	30	60	
Z	60	45	120	120	**120**

As for the given pay off matrix the maximum pay off of each alternative strategy X: 90, Y: 60 and Z = 120 and hence the maximum of these maximum pay offs is 120 therefore the alternative Z is selected according to maximax criterion.

Hurwicz–Alpha Criterion

Strategy	Maximum Pay Off (i)	Minimum Pay Off (ii)	Weighted Averages $= \alpha \times i + (1 - \alpha) \times ii$
X	90	30	0.70 × 90 + 0.30 × 30 = 72
Y	60	30	0.70 × 60 + 0.30 × 30 = 51
Z	120	45	0.70 × 120 + 0.30 × 45 = 97.5

As for the given pay off matrix the maximum weighted averages is for the strategy Z = 97.5 and hence the strategy Z is selected according to Hurwicz–Alpha Criterion.

Laplace Criterion

The expected value for the three strategies are,

$$\text{Strategy X} = \frac{1}{3}\times 30 + \frac{1}{3}\times 60 + \frac{1}{3}\times 90 = 60$$

$$\text{Strategy Y} = \frac{1}{3}\times 60 + \frac{1}{3}\times 60 + \frac{1}{3}\times 30 = 50$$

$$\text{Strategy Z} = \frac{1}{3}\times 60 + \frac{1}{3}\times 45 + \frac{1}{3}\times 120 = 75$$

As for the given pay off matrix the maximum expected value is for the strategy Z = 75 and hence the strategy Z is selected according to Laplace Criterion.

Regret Criterion

Strategy	Pay –Off			Regret Pay Off			Maximum Regret
X	30	60	90	30	0	30	30
Y	60	60	30	0	0	90	90
Z	60	45	120	0	15	0	**15**
Maximum Pay off	60	60	120				

Since the alternative Z corresponds to the minimal of the maximum possible regret therefore the alternative Z should be selected.

Decision Making Under Risk (Stochastic Models)

Under this category of decision making the decision maker has adequate information regarding the likely occurrence of each of the events and assigning probabilities to them. Knowing the probability distribution of the state of the nature, the best decision is to select the course of action which will result in the largest expected pay off.

Probability Aspects of Decision Making

The alternative sources of probabilities are:

Past Historical Data

One of the most commonly used techniques to forecast the future course based on the available past historical data. In this method the available past data is converted into a frequency distribution form and then each individual frequency is next converted into relative frequency form by dividing individual frequency in each category by the total frequency. The following example will illustrate the concept.

Frequency		Relative Frequency	Probability
	6	$\frac{6}{100} = 0.06$	0.06
	12	$\frac{12}{100} = 0.12$	0.12
	22	$\frac{22}{100} = 0.22$	0.22
	30	$\frac{30}{100} = 0.30$	0.30
	20	$\frac{20}{100} = 0.20$	0.20
	10	$\frac{10}{100} = 0.10$	0.10
Total	**100**	1.00	1.00

Subjective Assessment

Here the probability figures are arrived by personal judgment of the decision maker with respect to the state of an event.

Theoretical Probability Distributions

Many of the events follow theoretical probability distribution namely Binomial, Poisson or Normal distribution. For example finding out the number of defective items from a given sample or estimating the number of break downs in a machine. Each following a probability distribution while in the former case Binomial distribution can be used whereas in the latter is a fitting of case Poisson distribution.

Techniques for Decision Making Under Risk

The most widely used technique for decision making under risk is "Expected Monetary Value" (EMV)

Expected Monetary Value (EMV)

Expected Monetary Value (EMV) represents the long term expected gain from and alternative course, if the decision is repeated large number of times. And it is equal to the sum of the product of the possible pay offs with their corresponding probabilities.

For example if it is known that` the strategy (say S_1) has three different states of nature with associated probabilities are p_1, p_2 and p_3. Assuming the associate pay offs of the referred states are x_{11}, x_{12} and x_{13}, then the Expected Monetary Value is

$$\text{EMV (Course of Action, } S_1) = p_1 x_{11} + p_2 x_{12} + p_3 x_{13}$$

$$= \sum_{i=1}^{3} p_i x_{ij}$$

The same can be extended to n number of states of nature

$$\text{EMV (Course of Action, } S_j) = \sum_{i=1}^{n} p_i x_{ij}$$

where n = number of states of nature.

p_i = Probability of occurrence of state of nature say N_i.

x_{ij} = pay off associated with state of nature N_i and course of action, S_j.

The working steps of the criterion is summarized as:

1. Construct the pay off matrix with possible course of action and states of nature
2. List the conditional pay off values associated with each possible combination of course of action and the state of nature along with the probabilities of the occurrence of each state of nature.
3. Calculate the EMV for each of the course of action by multiplying the conditional pay offs by the associated probabilities and add the weighted values for each course of action.
4. At last select the course of action that yields the optimum EMV.

Illustration No. 3

Vivek a modern day entrepreneur is planning a new venture into a new market with the option of either starting with snacks joint or an ice-cream - cola parlor. The viability of the project is linked to the environmental condition of the city. The city has a past record of experiencing hot and humid summers occasionally reporting heavy downpour resulting in absurd change in climatic condition then the summers are cooler. If the city has hot summer he will make an operational profit of Rs. 10000 on ice-cream - cola business otherwise a mere profit Rs 3000 if the summer is cooler. Whereas if the summer is cooler than the profit generation by snacks joint business is Rs12000 if not Rs 3500 if the summer is hot. If there is chance of 60 % that the summers will be hot then find whether Vivek choice opting for ice-cream –cola drink parlor or snacks joint?

Solution:

The conditional profit generation by undertaking alternative course of action is furnished in the table below.

Event	**Probability**	**Conditional Pay off (Rs.)**	
(E_i)	**P(E_i)**	**Ice – Cream Parlor**	**Snacks Joint**
	(i)	**(ii)**	**(iii)**
Hot Summer	0.60	10000	12000
Cool Summer	0.40	3000	3500

The expected conditional pay off are calculated as

Event	Probability	Expected Pay off (Rs.)	
(E_i)	P(E_i)	Ice – Cream Parlor	Snacks Joint
	(i)	(i) × (ii)	(i) × (iii)
Hot Summer	0.60	0.60 ×10000 = 6000	0.60 × 3500 = 2100
Cool Summer	0.40	0.40 × 3000 = 1200	0.40 × 12000 = 4800
EMV(Rs.)		**7200**	**6900**

And since the expected monetary value of running the ice cream – cola parlor is higher than the snack joint Vivek should opt for running an ice cream parlor.

Expected Opportunity Loss (EOL)

In this approach the main aim is to minimize the expected opportunity loss (EOL). The expected opportunity loss also called as expected value of regret is the difference between the highest pay off (profit) for a state of nature and the actual pay off (profit) for the particular course of action. Otherwise EOL is the pay off lost by not selecting the course of action leading to highest pay off. Thus it can be mathematically expressed as

$$\text{EOL (State of nature, } N_i) = \sum_{i=1}^{n} l_{ij} p_i$$

where

l_{ij} = opportunity loss due to the state of nature, N_i and the course of action, S_j

p_i = Probability of occurrence of nature, N_i.

The working steps of the criterion is summarized as:

1. Calculate the conditional opportunity loss (COL) for each state by subtracting the each pay off from the maximum pay off for that particular outcome.
2. Find the EOL for each course of action by multiplying the probability of each state with COL and then add the values.
3. A course of action is selected yielding minimum EOL value.

Illustration No. 4

A management of a company is fixed whether to manufacture a particular brand or not which seems to have a commercial viability. If it is known that the selling price for the product is fixed at Rs. 3,000 and all cost including cost and manufacture and selling for each unit is Rs.1,000 apart from the developmental and quality testing cost of Rs. 30,000. With a customer lot size of 100 each of whom is suppose to buy a maximum of one unit. Additionally it is known from previous experience that a probability distribution could be derived relating to the proportion of customers who will purchase the product is furnished below?

Proportion of Customers	Probability
0.04	0.15
0.10	0.20
0.12	0.20
0.20	0.25
0.22	0.20

Find the EOL value and state whether the company should go ahead with manufacturing plan or drop the idea?

Solution:

If 'x' is the proportion of customers who will purchase the new brand then the conditional profit is,

$$(3,000 - 1,000)\ 100\ x - 30,000 = 2,00,000\ x - 30,000.$$

Let A_i (i = 1, 2, 3, 4 and 5) be the possible states of nature i.e. the proportion of customers willing to buy the brand and S_1 and S_2 be the course of action.

The conditional Pay offs is furnished below

State of Nature (In terms of Proportion of Customers, x)	Conditional Profit = (2,00,000x – 30,000)	
	Course of Action	
	S_1 (Manufacture)	S_2 (Not to Manufacture)
0.04	–22,000	0
0.10	–10,000	0
0.12	–6,000	0
0.20	10,000	0
0.22	14,000	0

The opportunity loss values

State of Nature	Probability	Conditional Profit		Opportunity Loss	
		S_1	S_2	S_1	S_2
0.04	0.15	–22,000	0	22,000	0
0.10	0.20	–10,000	0	10,000	0
0.12	0.20	–6,000	0	6,000	0
0.20	0.25	10,000	0	0	10,000
0.22	0.20	14,000	0	0	14,000

From the given values of probability associated with each state of nature, the expected opportunity loss for each course of action is furnished as

$$\text{EOL}(S_1) = 0.15\,(22{,}000) + 0.20\,(10{,}000) + 0.20\,(6000) + 0.25\,(0) + 0.20\,(0) = 6{,}500$$

$$\text{EOL}(S_2) = 0.15\,(0) + 0.20\,(0) + 0.20\,(0) + 0.25\,(10{,}000) + 0.20\,(14{,}000) = 5{,}300$$

As the company goal is to minimize the expected opportunity loss, and hence the company will drop the idea of manufacturing the new brand.

Expected Value of Perfect Information (EVPI)

In this method the decision maker achieves perfect and accurate information about the occurrence of various states and then selects a course of action which yields desired pay off. It is the maximum amount that a decision maker needs to pay to acquire the information about the occurrence of various state before actual decision is made.

Mathematically expressed as

$$\text{EVPI} = \text{(Expected profit with perfect information)} - \text{(Expected profit without perfect information)}$$

$$= \sum_{i=1}^{n} p_i \max(p_{ij}) - \text{EMV}^{\#}$$

Where

p_{ij} = Best pay off when action, S_j is taken in the presence N_i state of nature

p_i = Propability of state of nature, N_i

$\text{EMV}^{\#}$ = Maximum Expected Monetary Value

Illustration No.5

Becon Electricals producer of branded electrical switches manufactures three different variety of switch A, B and C. The sales revenue generated by the three varieties is furnished in the table below for three different states of sales.

State of Nature	Sales Revenue (' 000)		
	A	B	C
High (H)	40	35	32
Medium(M)	30	45	30
Low(L)	50	40	45

Compute the EVPI if the corresponding probability of the states of the nature are P(H) = 0.4, P(M) = 0.3 and P(L) = 0.3?

Solution:

The EMV as calculated from the given table are

$$\text{EMV for A} = 40 \times 0.40 + 30 \times 0.30 + 50 \times 0.30 = 40$$

$$\text{EMV for B} = 35 \times 0.40 + 45 \times 0.30 + 40 \times 0.30 = 39.50$$

$$\text{EMV for C} = 32 \times 0.40 + 30 \times 0.30 + 45 \times 0.30 = 35.30$$

The highest EMV# is for the product A and it is Rs. 40.

Now in order to calculate the EVPI let us work out the expected value for maximum pay off under all states of nature.

	Max. Profit of Each State	Probability	Expected Value = (Probability × Profit)
High (H)	40	0.4	16
Medium(M)	45	0.3	13.5
Low(L)	50	0.3	15

Therefore the expected pay off with perfect information (16 + 13.5 + 15) = 44.5

And hence the Expected Value with Perfect Information (EVPI) = (Expected Profit with perfect information) – EMV#.

= 44.5 – 40 = 4.5.

Illustration No.6

Good Bakery a leading bakery store has faced a typical problem to bake how many cakes to meet the demand for a particular day. In order to beat the competition Good bakery prefers only to sell fresh baked cakes to its customers and any left ones are unsold. So, the store faces the problem of over production will result in loss (excess capacity) and under production resulting in disappointed customer who will buy cakes from the competitors resulting loss of sales. The owners of the baker's store has collected information from the past sales and selected 100 day period to understand the movement of sales and the details are furnished below

Sales per day	No. of Days
50	10
54	20
60	40
70	30

Construct the pay off table and the opportunity lost table. What should be the optimal number of cakes needs to be baked each day using EMV and EOL criterion. Also check the result using EPVI if the cakes sell at Rs. 100 and the cost to manufacture per unit is Rs. 70?

Solution:

Let number of units manufactured are the different courses of action represented by A_1, A_2, A_3 and A_4 and D_1, D_2, D_3 and D_4 be the number of cakes demanded.

Therefore, the conditional table is given by

Conditional pay off Table

		Demand			
		D_1	**D_2**	**D_3**	**D_4**
		50	54	60	70
A_1	50	1500	1500	1500	1500
A_2	54	1220	1620	1620	1620
A_3	60	800	1200	1800	1800
A_4	70	100	500	1100	2100
Probability		0.10	0.20	0.40	0.30

Note : the table values are computed as follows

Pay off A_1S_1 = 100 × 50 – 70 × 50 = 1500.

Explanation: 50 units demanded and 50 units are manufactured. Therefore demand matching supply.

Similarly for A_2S_1 = *100* × 50 – 70 × 54 = 1220.

Explanation: 50 units demanded and 54 units are manufactured. Therefore demand less than supply. Four units are excess manufactured thus resulting in the loss of 4 units in terms of production cost.

And hence the other table values are computed taking the clue given.

The Expected Conditional Pay – Off

	D_1	D_2	D_3	D_4	EMV
A_1	150	300	600	450	1500
A_2	122	324	648	486	**1580**
A_3	80	240	720	540	**1580**
A_4	10	100	440	630	1180

Both A_2 and A_3 have EMV value of 1580. Thus Good Bakery has the option choosing either of A_2 or A_3 courses of action according to EMV criterion.

Conditional Loss Table

	D_1	D_2	D_3	D_4
A_1	0	120	300	600
A_2	280	0	180	480
A_3	700	420	0	300
A_4	1400	1120	700	0
Probability	0.10	0.20	0.40	0.30

Expected Conditional Loss Table

	D_1	D_2	D_3	D_4	EOL
A_1	0	24	120	180	324
A_2	28	0	72	144	244
A_3	70	84	0	90	244
A_4	140	224	280	0	644

The minimum EOL corresponds to both A_2 or A_3. Thus Good Bakery has the option choosing either of A_2 or A_3 courses of action according to EOL criterion.

Note : That both EMV and EOL criterion are consistent as they give identically similar results.

Bayesian Analysis and Posterior Probabilities

In the early part of this chapter we were dealing with prior probability distribution for various states of nature in order to evaluate the expected monetary value for different states and for the selection of the optimal choice. And Bayesian theory is an extension of the above criterion where the theory is utilized to calculate the posterior probabilities knowing the prior probabilities. In our daily course of decision making we encounter numerous decision alternatives and prior probability distribution not always the best forecaster. And in cases where the identification of alternative is considered necessary it calls for revision and testing of alternatives additionally if it is to know the effects of the alternatives then the result are revisited and restated which bring out vital informations. In this section we will look into the Baye's theorem and the computation of posterior probabilities using an illustrative example.

Bayes' Theorem

Statement

Let B_i's were i = 1, 2, 3, ... n be n mutually exclusive and disjoint events with the probability of P(B_i's) ≠ 0 then for any arbitrary event A which is a subset of $B_1 \cup B_2 \cup B_3 \ldots \cup B_n$ s.t P(A)>0 then we have

$$P\left(\frac{B_i}{A}\right) = \frac{P(B_i) \times P\left(\frac{A}{B_i}\right)}{\sum_{i=1}^{n} P(B_i) \times P\left(\frac{A}{B_i}\right)}$$

Proof of the theorem:

Let B_1, B_2, B_3 ... B_n are mutually exclusive and exhaustive events s.t A occurs only when one of $B_1, B_2, B_3 \ldots\ldots\ldots\ldots B_n$ occurs i.e A occurs in combination with any one of the mutually exclusive event

$(A \cap B_1)$, $(A \cap B_2)$, ... $(A \cap B_n)$ where all $(A \cap B_1)$, $(A \cap B_2)$, ... $(A \cap B_n)$ are disjoint.

So, we can write,

$$A = (A \cap B_1) \cup (A \cap B_2) \cup \ldots \cup (A \cap B_n)$$

By applying Additional Rule of Probability, we get

$$P(A) = P(A \cap B_1) + P(A \cap B_2) + \ldots + P(A \cap B_n) \quad \ldots(i)$$

$$\text{Since } P(A \cap B_i) = P(B_i \cap A)$$

Therefore (i) can be rewritten as

$$P(A) = P(B_1 \cap A) + P(B_2 \cap A) + \ldots + P(B_n \cap A) \quad \ldots(ii)$$

Again by Conditional Probability Rule

$$P(B_i \cap A) = P(B_i) \times P\left(\frac{A}{B_i}\right)$$

Therefore

$$P(A) = P(B_1)P \times \left(\frac{A}{B_1}\right) + P(B_2) \times P\left(\frac{A}{B_2}\right) + \ldots P(B_3) \times P\left(\frac{A}{B_n}\right) + \ldots$$

$$P(A) = \sum_{i=1}^{n} P(B_i) \times P\left(\frac{A}{B_i}\right) \quad \ldots(iii)$$

But

$$P(A \cap B_i) = P(A) \times P\left(\frac{B_i}{A}\right)$$

$$P(B_i \cap A) = \times P(A)P\left(\frac{B_i}{A}\right)$$

$$(\text{Since } P(A \cap B_i) = P(B_i \cap A))$$

$$P(B_i)P\left(\frac{A}{B_i}\right) \times P(A)P\left(\frac{B_i}{A}\right)$$

$$\Rightarrow P\left(\frac{B_i}{A}\right) = \frac{P(B_i) \times P\left(\frac{A}{B_i}\right)}{P(A)}$$

Writing the value of P(A) from (iii)

$$\Rightarrow P\left(\frac{B_i}{A}\right) = \frac{P(B_i) \times P\left(\frac{A}{B_i}\right)}{\sum_{i=1}^{n} P(B_i) \times P\left(\frac{A}{B_i}\right)}$$

Hence proved

Illustration No.7

Vinayaka Agrotech a company into agro business decided to grow a particular variety of crop among alternatives X, Y and Z. The decision to grow a particular variety will solely dependent upon the amount of rainfall to be experienced in the current season. The company has made a formal estimation of profit under three categories of rainfall for each crop which are listed in the table below

Estimated Profit

Rainfall	Crop X	Crop Y	Crop Z
High	10,000	6,000	5,000
Medium	5,000	7,000	4,000
Low	2,000	4,000	3,000

Past Experience has revealed a pattern rainfall based on which the projection for the current season is made with their associated probabilities of rainfall in each category i.e. High : 0.3, Medium : 0.5 and Low :0.2

Furthermore the company has the additional option to avail the services of a forecaster in order to provide the detailed analysis of current rainfall prospects which is shown in the table below.

Rainfall	Crop X	Crop Y	Crop Z
High	0.60	0.30	0.10
Medium	0.30	0.50	0.20
Low	0.10	0.30	0.60

(a) From the available data, determine the optimal decision which crop to grow?

(b) Whether it is economical to hire the services of the forecaster?

Solution:

Calculation of EMVs

States of Nature	Prior Probability	Conditional Profit Course of Action			Expected Profit Course of Action		
		S_1	S_2	S_3	S_1	S_2	S_3
N_1	0.10	**8,000**	3,000	4,000	2,000	750	1,000
N_2	0.30	4,000	4,000	**5,000**	1,000	1,000	1,250
N_3	0.60	2,000	**5,000**	3,000	1000	2,500	1,500
				EMV	4,000	4,250	3,750

Therefore the maximum EMV value is 4,250 which is for the crop Y and the firm should execute the plan to plant the crop Y.

Now let B_i (i = 1, 2, and 3) denote the forecast outcome for High, Medium and Low rainfall and the likelihood values are given below

States of Nature(N_i)	**Forecast Likelihood**		
	$P\left(\frac{B_1}{N_i}\right)$	$P\left(\frac{B_2}{N_i}\right)$	$P\left(\frac{B_3}{N_i}\right)$
N_1	0.60	0.30	0.10
N_2	0.30	0.50	0.20
N_3	0.10	0.30	0.60

Therefore the expected profit with perfect information is given by

$$EPPI = 0.1\ (8,000) + 0.3(5,000) + 0.6\ (5,000) = \text{Rs. } 5,300$$

Thus $\quad EVPI = 5,300 - 4,250 = \text{Rs. } 1050.$

Now for each of the three forecast will result in prior and posterior probabilities which are given below

States of Nature	**Prior Probability**	**OutComes (B_i)**	**Conditional Probability** $P\left(\frac{B_i}{N_i}\right)$	**Joint Probability** $P(B_i \cap N_i) = P(N_i)P\left(\frac{B_i}{N_i}\right)$		
N_1		B_1	0.60	0.06	-	-
	0.10	B_2	0.30	-	0.03	-
		B_3	0.10	-	-	0.01
N_2	0.30	B_1	0.30	0.03		
		B_2	0.50		0.15	
		B_3	0.20			0.06
N_3	0.60	B_1	0.10	0.06		
		B_2	0.30		0.18	
		B_3	0.60			0.36
		Marginal Probability		0.15	0.36	0.43

OutComes (B_i)	Probability $P(B_i)$	States of Nature (N_i)	Posterior Probability $P(B_i \cap N_i) = P(N_i)P\left(\frac{B_i}{N_i}\right)$
B_1	0.15	N_1	$\frac{0.06}{0.15} = 0.40$
		N_2	$\frac{0.03}{0.15} = 0.20$
		N_3	$\frac{0.06}{0.15} = 0.40$
B_2	0.36	N_1	$\frac{0.03}{0.36} = 0.08$
		N_2	$\frac{0.15}{0.36} = 0.41$
		N_3	$\frac{0.18}{0.36} = 0.50$
B_3	0.43	N_1	$\frac{0.01}{0.43} = 0.02$
		N_2	$\frac{0.06}{0.43} = 0.14$
		N_3	$\frac{0.36}{0.43} = 0.84$

Now for each outcome let us recalculate the EMV with revised probability with additional details as furnished.

States of Nature (N_i)	Forecast Outcome								
	B_1			B_2			B_3		
	Prob.	COL	EOL	Prob.	COL	EOL	Prob.	COL	EOL
N_1	0.40	0	0	0.08	5000	400	0.02	4000	80
N_2	0.20	1000	200	0.41	1000	410	0.14	0	0
N_3	0.40	3000	1200	0.50	0	0	0.84	2000	1680
Posterior	**EOL**		**1400**			**810**			**1760**

Therefore the expected value of the sample information is obtained by multiplying EOL's with the revised probabilities as shown below

OutComes (B_i)	Probability $P(B_i)$	Expected Opportunity Loss (EOL)	Expected Value of Sample Information (EVSI)
B1	0.15	1400	210
B2	0.36	810	291.60
B3	0.43	1760	756.80
		Total	**1258.40**

The EVSI Rs. 1258.40 indicates charges that needs to be spend by the firm if the services of the forecaster is availed.

Decision Tree Analysis

A decision tree is a graphical representation of various alternatives and the sequence of events of a particular decision problem primarily intended for review and methodical analysis. What decision tree does is, it reduces the problem to its simplest form in similar lines of a tree with branches. It consists of nodes, branches, probability estimates and the pay offs. Nodes are two types the decision node and the chance node and while constructing the tree diagram, it is convention to represent decision nodes by the symbol square (□) and which are the places where the decision maker must select a course of action among the available options. And the chance nodes or event nodes by circle (○) are the points where the decision maker will comprehend the response to his decision.

Branches comes out of the nodes are either decisions or states of nature. Each branch coming out of decision nodes are termed as decision branches representing a course of alternatives that can be selected at the decision point available to the decision maker whereas those emanating from the chance node are termed as chance branches representing the state of nature of all situations or events.

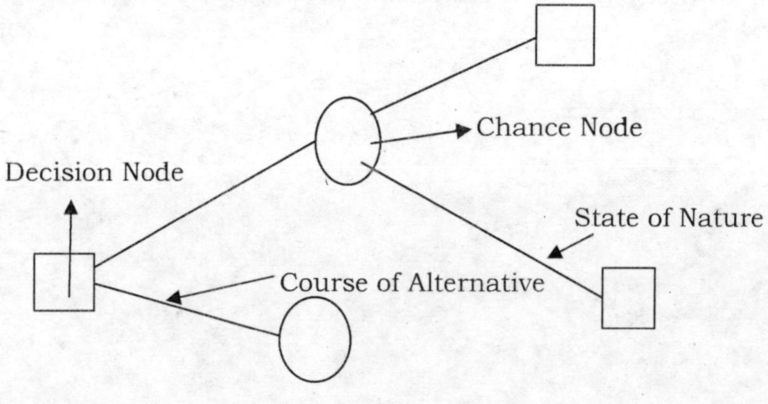

Since the decision making is made under risk therefore the probability associated with each state of nature is written along the side of the chance branch and multiplying the probabilities along the branch will yield joint probabilities of the corresponding act – event sequence. Thus the decision maker list all the possible alternatives and resulting pay off values which can be either positive or negative for each act – event sequence which help him to calculate the expected pay off and EMV for each act.

A decision tree is highly useful device in multistage sequencing situations involving series of decision where each is dependent on the other. For optimal sequence of decision in a tree is obtained by working backwards from right hand side and rolling backwards in this way eliminating all unprofitable branches with sole focus to maximize the return from the decision situation.

The following illustration will clearly display the usage of decision tree.

Illustration No.8

Animesh an investor has the option of investing in two independent investments X and Y made available to him by his service provider but because of resource constraint he can invest in any one of the two investments. First he can choose X and then stop if X is not successful and shift to Y or vice versa. The probability of success of X is 0.7 while that of Y is 0.3. Both the investment requires a minimum investment of Rs. 20,000. The risk associated with the investment is that there is no return if the venture fails whereas successful implementation of X will see a return of Rs. 40,000 over cost and that of Y is Rs. 50,000 over cost. Determine the strategy that should be adopted by Animesh and draw the decision tree?

Solution:

Decision Tree

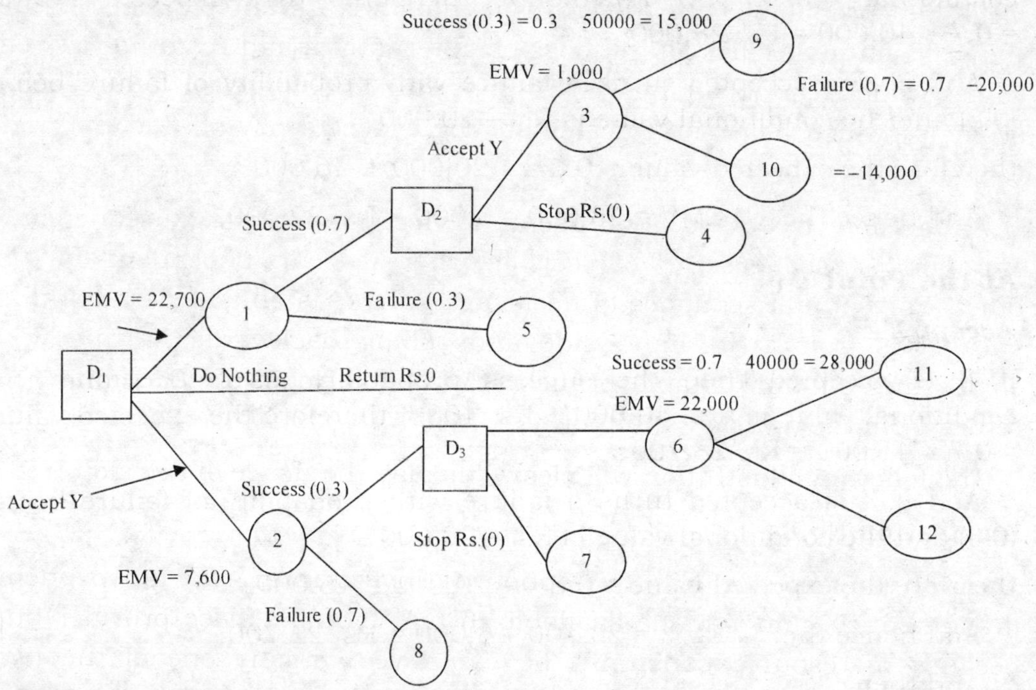

At the Point D2

Accept Y

If Y is accepted then the success probability is of (0.3) and the conditional value is Rs. 50,000

therefore the expected value = 0.3 × 50,000 = Rs. 15,000

And if Y is accepted turns a failure with probability of failure being (0.7) and the conditional value is Rs. – 20,000 therefore the expected value = 0.7 × – 20,000 = – 14,000.

And hence the EMV = 15,000 – 14,000 = Rs. 1000.

At the Point D_3

Accept X

If X is accepted then the success probability is of (0.7) and the conditional value is Rs.40,000 therefore the expected value = 0.7 × 40,000 = Rs. 28,000

And if X is accepted turns a failure with probability of failure being (0.3) and the conditional value is Rs. – 20,000

therefore the expected value = 0.3 × – 20,000 = – 6,000.

And hence the EMV = 28,000 – 6,000 = Rs. 22,000.

At the Point D_1

Accept X

If X is accepted then the success probability is of (0.7) and the conditional value is (Rs. 40,000 + Rs. 1000) therefore the expected value = 0.7 × 41,000 = Rs. 28,700

And if X is accepted turns a failure with probability of failure being (0.3) and the conditional value is Rs. – 20,000

therefore the expected value = 0.3 × – 20,000 = - 6,000.

And hence the EMV = 28,700 – 6,000 = Rs. 22,700.

And then

Accept Y

If Y is accepted then the success probability is of (0.3) and the conditional value is (Rs.50,000 + Rs 22,000)

therefore the expected value = 0.3 × 72,000 = Rs. 21,600

And if Y is accepted turns a failure with probability of failure being (0.7) and the conditional value is Rs. – 20,000

therefore the expected value = 0.7 × – 20,000 = – 14,000.

And hence the EMV = 21,600 – 14,000 = Rs. 7,600.

Do Nothing

No returns.

Since the EMV is Rs. 22,700 at node 1 is the highest and hence the best strategy is to accept the first investment proposal.

Illustration No.9

A drug manufacturing company developed a wonder drug which has received a phenomenal success in the market and the company finds that it is unable to meet the market potential resulting in backlog sales. In order to overcome the current situation the company's management has decided upon three alternatives course of action (i) Out Sourcing (ii) Sub–Contracting and (iii) Construct New Facilities. The correct choice of course of action is majorly depended upon the future demand which may be high, medium or low and the company's management has ranked their respective probabilities as 0.4, 0.3 and 0.3 respectively. The cost analysis made against each alternative reveals the profitability as shown below

Demand	Probability	Out - Sourcing	Sub – Contracting	Construct New Facilities
High	0.4	200	150	300
Medium	0.3	160	130	120
Low	0.3	80	100	-50

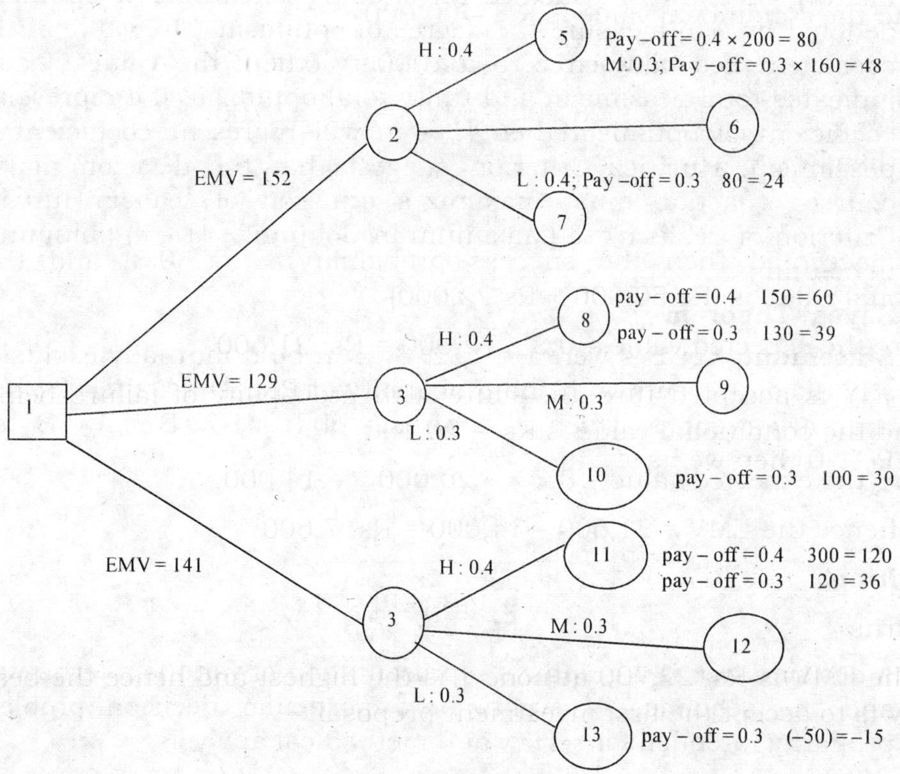

Find the most preferred alternative and the corresponding expected value and draw the decision situation using a decision tree?

Solution:

As per the calculation of EMV at nodes 2, 3 and 4, it was found that the node 2 has the highest EMV value, therefore the decision at node 1 will be to choose course of action 1 which is outsourcing.

Note: EMV at node 2 = 80 + 48 + 24 = 152

Summary

- According to Fishburn solving the decision model consists of finding a strategy for action, the expected relative value which is at least as great as the expected relative value of any other strategy in a specified set. The descriptive criterion of a strategy will be maximization of the decision maker's expected relative value.
- **Hurwicz–Alpha Criterion**

  Under this criterion the decision maker is neither completely pessimistic nor optimistic but follows a midway approach which is mixture of both. The idea is introducing coefficient of optimism denoted by α to measure the degree of optimism. The value of the coefficient lies between zero and unity where the value of zero indicates total pessimism and unity total optimism. If α represents coefficient of optimism then $(1 - \alpha)$ will represent coefficient of pessimism. Hurwicz Criterion suggest that the decision maker course of action must maximizes criterion of realism.Hurwicz Criterion of Realism = α (Maximum in Column) + $(1 - \alpha)$ (minimum in column).
- **Bayes' Theorem**

  Statement: Let B_i's were i = 1, 2, 3, ... n be n mutually exclusive and disjoint events with the probability of P(B_i's) ≠ 0 then for any arbitrary event A which is a subset of $B_1 \cup B_2 \cup B_3 \ldots \cup B_n$ s.t P(A)>0 then we have

$$P\left(\frac{B_i}{A}\right) = \frac{P(B_i) \times P\left(\frac{A}{B_i}\right)}{\sum_{i=1}^{n} P(B_i) \times P\left(\frac{A}{B_i}\right)}$$

- A decision tree is a graphical representation of various alternatives and the sequence of events of a particular decision problem primarily intended for review and methodical analysis.

Review Questions

1. Elaborate on decision making and explain the various steps involved in decision making?
2. What are the techniques used in solving problems in decision making under risk and uncertainty? Elaborate on each of these techniques?
3. State and proof Baye's theorem and write on its application on business forecasting?
4. What is decision tree? And how does it helps business firm in decision making?

Self Practice Exercises

1. Consider the following pay off (Profit) matrix

Strategy	Pay Off (Profit)		
	P_1	P_2	P_3
S_1	30	22	35
S_2	35	25	40
S_3	40	30	32

What will be the course of action according to

(i) Maximin Criterion.

(ii) Maximax Critierion.

(iii) Hurwicz–Alpha Criterion (α = 0.6).

(iv) Laplace Criterion.

(v) Regret Criterion.

2. A farmer has certain quantity of land on which he wants to make decision on which of the three crops to cultivate. The profitability of each crop is dependent on the rainfall condition during that season. He has categorized the amount of rainfall as High, Average and Scanty and his estimation of profit of each crop say X, Y and Z under different condition are furnished below

Strategy	(Expected Profit)		
	X	Y	Z
High	10000	5000	7000
Average	6000	6000	7000
Scanty	3000	7000	6000

What will be the course of action if the farmer wants to cultivate only crop according to

(i) Maximin Criterion.

(ii) Maximax Critierion.

(iii) Hurwicz–Alpha Criterion ($\alpha = 0.7$).

(iv) Laplace Criterion.

(v) Regret Criterion.

3. The probability of demand for Tempo Trucks for hiring on any particular day is given as follows

No. of Tempo Trucks Demanded	**0**	**1**	**2**	**3**	**4**
Probability Associated	0.1	0.20	0.30	0.25	0.15

The trucks have a fixed cost of Rs. 50 each for maintenance and a daily running charges of Rs. 100. If Animesh Transport owns 4 Tempo trucks, find

(i) What is the daily expectation ?

(ii) If the transport Company is about to start the business what should be the ideal number of trucks that needs to be kept?

4. Detail analysis of demand for a particular product for the last year for 52 weeks revealed the demand distribution as furnished below

Demand Distribution		
Qty. Demanded	**No. of weeks the qty. sold.**	**Probability**
20	5	0.10
21	10	0.20
22	16	0.30
23	13	0.25
24	5	0.10
25	3	0.05
	52	1.00

Selling price of the product is Rs. 5 and the acquisition cost price is Rs. 4. If the product remains unsold for a week then it is sold at loss of Rs.3.

(i) Construct the conditional profit table.

(ii) Determine the optimum number that needs to be sold in order to maximize profit.

(iii) Construct the conditional loss table(COL).

(iv) Calculate Expected Opportunity Loss (EOL).

5. A fruit vendor purchases Apple crates every morning at Rs. 100 and sells them for Rs. 150. Unsold crates at the end of the day are sold in the next day at a throw away price of Rs. 50 thereafter it has no value. Past sales of 100 days has revealed that the number of crates sold per day ranged between 20 to 23 crates.

Crates	**20**	**21**	**22**	**23**
No. of days	10	20	40	30

Then find how many crates the fruit vendor should purchase in order to maximize his profit per day using EMV?

6. Highland Steels a leading steel manufacturer has faced a typical problem to decide upon viable option based on the market conditions. One to build a large plant to accommodate any size of future demand or build a small plant initially to serve low demand and consider the possibility of further expansion based on rising demand conditions.

If the management of the firm has availed the services of a leading market research firm to analyze the market conditions, the report submitted by the research team has categorized the market into High – demand, Medium – Demand and Low – Demand with respective probabilities as 0.3 , 0.4 and 0.3 respectively.

Now if the firm builds a large plant at a cost of Rs 20 lakhs then all future demands can be accommodated and will require no further expansion. The returns of these plant for low, average and high demand is estimated at Rs.16 lakhs, Rs. 30 lakhs and Rs. 40 lakhs respectively whereas if the firm initially builds a smaller unit at a initial investment of 8 lakhs will only fulfill meeting low demand condition but, if the demand increases to higher levels further investments of Rs 8 lakhs for medium demand and Rs.16 lakhs high demand needs to be invested. Under normal condition sudden expansion of the plant is not feasible and there is loss of revenue. And the operational returns on small pant expansion is Rs. 32 lakhs and on the large plant is 38 lacs respectively. Plot all the decisions involved and their economic feasibility into a decision tree and decide on the optimal plan of action.

7. A computer manufacturing firm Zidan Infotech Ltd. is planning to develop a new touch screen 3G phone with a estimated developmental cost of Rs 10,00,000. The research team has found that the product has 70% chance to be developed successfully without any hiccups. The market potential of the new set has been assessed which is furnished below.

Market Condition	Probability	Value of Profit
High Acceptability	0.2	1,00,000
Average Acceptability	0.5	80,000
Low Acceptability	0.3	40,000

The value of profit as mentioned above does not include the developmental cost. While above plan was in consideration a second plan was also floated to upgrade the current model at the same cost but the value of profit in all the three stages are as follows

Market Condition	Probability	Value of Profit
High Acceptability	0.2	60,000
Average Acceptability	0.4	30,000
Low Acceptability	0.4	15,000

But the second plan against the former has certainty in return on investment. From the above mentioned data

1. Draw a decision tree relating to all events and choices available to the firm?
2. What should be the Co's decision with regards to investment in the tune of Rs. 10,00,000?

Answers

1.

(i) S_3

(ii) Either S_2 or S_3

(iii) S_3

(iv) S_3

(v) S_2

2.

(i) Z

(ii) X

(iii) X

(iv) Z

(v) Z

3.

(i) Rs. 15

(ii) The transport company should buy 2 Tempo Trucks as the EMV is highest for the second course of action i.e. Rs. 60.

4.

(ii) EMV is highest for when the qty. demanded is 22 units.

(iv) EOL is lowest for when the qty. demanded is 22 units

5. The fruit vendor should purchase 22 crates of apple per day to maximize profit.

6. The firm should go for small expansion for better returns.

7. The optimal decision is develop the new 3 G set for the market.

CHAPTER 21

Game Theory

Brief History

The history of game theory goes back to the early part of 20th century to the studies made by Augustine Cournot (Duopolistic Model) in the year 1838. But the momentum has gained with the works of Emile Borel (1921) and by John Von Neumann and Oskar Morgenstern with the publication of the book "Theory of the Games and Economic Behavior" which has set the basic framework that is still practiced today. Some of the other contributors are Nobel laureates John Nash, John Harsanyi and Reinhard Selten.

Introduction

Game theory is basically a study of conflicts and competition involving individuals, firms or any other combination in which the concepts are utilized to formulate, analyze and study the strategic behavior. The central assumption in the game theory is that the players are rational and their course of action is mostly driven by the one which gives the most preferred outcome (maximize gain) at the cost of its opponents.

Key Terminology

Players

The participants acting as decision makers in a game. If the game involves two players, one opposing the other then it is known as "two person game" and a game having more than 2 players is called as "n person game".

Strategy

A finite number of alternative courses of action available to a player. Mainly classified as *"Pure Strategies"* and *"Mixed Strategies"*. In pure strategy the players select the same strategy and each player is exactly aware of what the player is going to do. Whereas in mixed strategy each player uses a combination of strategies in order to keep the opponent guessing about the course of action to be selected.

Pay off

Pay off is the outcome of the game when alternative courses of action are adopted by each competing players. The pay-offs can be win, lose or draw.

Value of the Game

Value of the game is the expected pay off of a game when all the players of the game follow their optimum strategies.

Pay off Matrix

The gains or losses from a game if represented in a matrix form then it is known as pay off matrix.

For example a simple pay off matrix is represented as

$$\text{Player X}\ \overset{\text{Player Y}}{\begin{pmatrix} 2 & 1 & 4 & 2 \\ 4 & 6 & 7 & 4 \\ 5 & 2 & 3 & 3 \\ 3 & 4 & 2 & 2 \end{pmatrix}}$$

Maximin

The maximum element among the minimum elements selected from each row.

For example

$$\begin{pmatrix} 2 & 1 & 4 \\ 4 & 6 & 7 \\ 5 & 2 & 3 \end{pmatrix} \quad \begin{matrix} \text{Row Minimum} \\ 1 \\ 4 \longrightarrow \text{maximum element (Row Maximin)} \\ 2 \end{matrix}$$

Minimax

The minimum element among the maximum elements selected from each column.

For example

$$\begin{pmatrix} 2 & 1 & 4 \\ 4 & 6 & 7 \end{pmatrix}$$

Column Maximum 4 6 7 (4 : Column Minimax)

Saddle Point

A game in which "the Maximin of one player say A" is equal to "Minimax for another player say B" is called the saddle point.

Types of Games

(i) Two Person Game

A game with two competing parties one opposing the other is called as Two Person Game.

(ii) Multi Person Game

A game in which there are more than two competing parties is called as Multi Person Game.

(iii) Two Person Zero Sum Game

A game in which the gain of one player is the loss of the other is called as Two person Zero Sum Game. Otherwise the algebraic sum of the gains of both the players after the game is zero.

(iv) Two Person Non Zero Sum Game

A game in which the gain of one player is not equal to the loss of the other is called as Two person Non Zero Sum Game.

Solution to Solve Two Person Zero Sum Game (With Saddle Points)

Procedural Steps

(i) Construct the Pay off matrix.

(ii) Identify the row minimums and select the largest of the row minima.

(iii) Identify the Column maximums and select the smallest of the Column maxima.

(iv) Check for the presence of Saddle points i.e. the maximin value equals the minimax value.

(v) If there is saddle point, it is the optimum solution of the game.

Illustration No. 1

The pay off matrix of a zero sum two person is given below. Find the best strategy for each player and the value of the game?

Player II

$$\text{Player I}\begin{pmatrix} 4 & 6 \\ 7 & 8 \end{pmatrix}$$

Solution

Player II

$$\text{Player I}\begin{pmatrix} 4 & 6 \\ 7 & 8 \end{pmatrix}$$

Player I is aware that if he plays the first strategy (i.e. Row 1) his gain is either 4 or 6 depending on the strategy followed by Player II.

He is also knows that in all circumstances his gain will be at least 4 units irrespective of the strategy followed by Player II.

Similarly if Player I adopts row 2 as line of strategy, his minimum gain is 7 units whatever be the course of action (i.e. selection of column) by the second player.

	Player II	Row Minimum
Player I	$\begin{pmatrix} 4 & 6 \\ 7 & 8 \end{pmatrix}$	4 7 ← Maximin
Column Maximum	8 7 ↑ Minimax	

The pay off of 7 units is the maximin value or lower bound value of the game and the decision to play row 2 is the maximin pure strategy.

If Aim of the Player I is to maximize gain it will be natural tendency of Player II to minimize his maximum loss i.e. selection between 7 and 8 units. Therefore Player II adopts minimax pure strategy to play column I so that his loss is the least of maximum loses. Here the maximum payoff or loss of 7 units represents minimax or upper bound value of the game.

Since minimax value = maximin value = 7 represents the saddle point and also the value of the game i.e. gain of units to Player I.

Therefore the optimal strategies of both players are

Player I: (0, 1)

Player II: (1, 0)

and the optimal value = 7

Illustration No. 2

Solve the following game and determine the optimal strategies for X and Y and also determine the value of the game?

Player Y

$$\text{Player X}\begin{pmatrix} 2 & 1 & 4 & 2 \\ 4 & 6 & 7 & 4 \\ 5 & 2 & 3 & 3 \\ 3 & 4 & 2 & 2 \end{pmatrix}$$

Solution:

Player Y

$$\text{Player X}\begin{pmatrix} 2 & 1 & 4 & 2 \\ 4 & 6 & 7 & 4 \\ 5 & 2 & 3 & 3 \\ 3 & 4 & 2 & 2 \end{pmatrix}$$

	Player Y				Row Minimum	Maximin
Player X	2	1	4	2	1	
	4	6	7	4	4	4
	5	2	3	3	2	
	3	4	2	2	2	
Column Maximum	5	6	7	4		

(Saddle point)

Minimax 4

In the above game, since a saddle point exists as such it is a pure strategy and each player X and Y will stick to single strategy i.e Player X to maximize gain (Row 2 : 4 units) and Player Y to minimize loss (column 4 : 4 units)

Therefore the optimal strategies of both players are

Player X : (0, 1, 0, 0)

Player Y : (0, 0, 0, 1)

and the optimal value = 4.

Mixed Strategies (Solution of a 2 × 2 game without saddle point)

Let us consider a game without saddle points then the problem arises in the determination of probabilities with which each course of action can be chosen. In order to determine the formulation for such cases, consider a game whose pay–off matrix is of the form

$$\begin{array}{cc} & \text{Player B} \\ & \begin{array}{cc} 1 & 2 \end{array} \\ \text{Player A} \begin{array}{c} 1 \\ 2 \end{array} & \begin{pmatrix} a_{11} & a_{12} \\ a_{21} & a_{22} \end{pmatrix} \end{array}$$

If we assume that Player A chooses his i^{th} course of action with probability p_i and Player B chooses his j^{th} course of action with probability q_j. Thus $P = (p_1, p_2)$ and $Q = (q_1, q_2)$ are the mixed strategies of Player A and B respectively.

Such that

$$\sum p_i = p_1 + p_2 = 1$$

$$\sum q_j = q_1 + q_2 = 1$$

where $p_i, q_j \geq 0$ for all i, j

Now the expected gain to Player A is

$a_{11} p_1 + a_{21} p_2$, when chooses strategy 1

and $a_{12} p_1 + a_{22} p_2$, when chooses strategy 2

Similarly the expected loss to Player B is

$a_{11} q_1 + a_{12} q_2$,

and $a_{21} q_1 + a_{22} q_2$

respectively.

As we know that each game has a solution and let the value of the game be 'v' as Player A is at least suppose to get 'v'

We get,

$$\left.\begin{aligned} a_{11}p_1 + a_{21}p_2 &\geq v, \\ a_{12}p_1 + a_{22}p_2 &\geq v. \end{aligned}\right\} \quad \text{.....(i)}$$

and

Similarly the expected loss to Player B is atmost v

Therefore

$$\left.\begin{aligned} a_{11}q_1 + a_{12}q_2 &\leq v, \\ a_{21}q_1 + a_{22}q_2 &\leq v \end{aligned}\right\} \quad \text{.....(ii)}$$

and

As the values of p_1, p_2, q_1, q_2 and v needs to be determined from the above inequalities let us convert the inequalities into equation and hence we get from (i) and (ii),

$$a_{11}p_1 + a_{21}p_2 = v = a_{12}p_1 + a_{22}p_2$$

and

$$a_{11}q_1 + a_{12}q_2 = v = a_{21}q_1 + a_{22}q_2$$

Since $p_1 + p_2 = 1$ or $p_2 = 1 - p_1$, $q_1 + q_2 = 1$ or $q_2 = 1 - q_1$ in the above equation we get,

$$a_{11}p_1 + a_{21}(1 - p_1) = a_{12}p_1 + a_{22}(1 - p_1) \quad \text{.....(iii)}$$

and

$$a_{11}q_1 + a_{12}(1 - q_1) = a_{21}q_1 + a_{22}(1 - q_1) \quad \text{.....(iv)}$$

therefore rearranging (iii) and (iv) we get,

$$p_1 = \frac{a_{22} - a_{21}}{(a_{11} - a_{12}) + (a_{22} - a_{21})}$$

$$p_2 = \frac{a_{11} - a_{12}}{(a_{11} - a_{12}) + (a_{22} - a_{21})}$$

Similarly

$$q_1 = \frac{a_{22} - a_{12}}{(a_{11} - a_{12}) + (a_{22} - a_{21})}$$

$$q_2 = \frac{a_{11} - a_{21}}{(a_{11} - a_{12}) + (a_{22} - a_{21})}$$

In order to get the value of the game 'v' we substitute the value of p_1 and p_2

We get

$$v = \frac{a_{11}\, a_{22} - a_{12}\, a_{21}}{(a_{11} - a_{12}) + (a_{22} - a_{21})}$$

Since the values obtained of p_1, p_2, q_1, q_2 and v are necessarily non negative therefore they form the solution of the game.

Illustration No. 3

The pay off matrix of a game between two players I and II is given by

	Player II	Row Minima
Player I	$\begin{pmatrix} 5 & 2 \\ 3 & 4 \end{pmatrix}$	2 3

(i) Solve the game?

(ii) Find the value of the game.

Solution:

	Player II	Row Minima
Player I	$\begin{pmatrix} 5 & 2 \\ 3 & 4 \end{pmatrix}$	2 3
Column Maxima	5 4	

Viewing the above pay- off matrix there exist no saddle points therefore the optimal strategies are mixed strategies.

Let Player I strategy are p_1 and p_2 and that of the player II is q_1 and q_2

So

$$P_1 = \frac{(4-3)}{(5-2)+(4-3)} = \frac{1}{4}$$

$$P_2 = \frac{(5-2)}{(5-2)+(4-3)} = \frac{3}{4}$$

$$q_1 = \frac{(4-2)}{(5-2)+(4-3)} = \frac{2}{4} = \frac{1}{2}$$

$$q_2 = \frac{(5-3)}{(5-2)+(4-3)} = \frac{2}{4} = \frac{1}{2}$$

As we check $p_1 + p_2 = 1$, $q_1 + q_2 = 1$

And the value

$$v = \frac{(5 \times 4 - 3 \times 2)}{(5-2)+(4-3)} = \frac{14}{4} = 3\frac{1}{2}$$

Dominance Rule

It is generally observed that in some situation where some of strategies stands superior or dominates the other i.e. the outcome is always preferred overall other in all conditions. According to the principle of dominance, in such cases the size of the pay off matrix can be reduced by deleting those strategies which are dominated by the others. For instance, if each element in a particular row say i^{th} row (*Dominating Row*) is greater than or equal to the corresponding elements of another row say k^{th} row (*Dominated Row*), a player in any condition will never select k^{th} strategy, in other words the probability of selection is zero. As dominated strategies are never chosen, they are eliminated from the pay off matrix to simplify the game.

Similar situation do exist for the opponent where each element in a particular column stand superior or greater (*Dominating Column*) than the corresponding elements of another column (*Dominated Column*). The dominance technique is a viable technique in cases of mixed strategy games where the matrix does not contain a saddle point and the size of the matrix is reduced to 2 × 2 form by mere elimination of a few strategies whose removal has no effect on the result of the final solution

Dominance Rule can be applied as under.

(i) Compare the elements of a row with corresponding elements in the other row. Continue the process by proceeding from one row to another.

(ii) Identify and eliminate the dominated row(s) and construct the reduced pay off matrix.

(iii) Similarly identify and eliminate the dominating column(s).

(iv) Solve the reduced pay off matrix by either of algebraic method, sub game method or graphical method.

Illustration No. 4

The pay off matrix in the following two person zero sum game is given as under

$$\text{Player A}\begin{array}{c}\text{Player B}\\ \begin{pmatrix} 7 & 6 & 3 \\ 2 & 2 & 1 \\ 5 & 10 & 7 \end{pmatrix}\end{array}$$

(i) Use dominance rule to obtain the reduced pay off matrix?

(ii) Find the optimal strategies and the value of the game?

Solution:

Since the game has no saddle point and hence not stable. Let us consider whether dominance rule can be applied to reduce the pay off matrix. First we check for row dominance for which we need to compare each element of rows in pair wise.

$$\text{Player A}\begin{array}{c}\text{Player B}\\ \begin{pmatrix} 7 & 6 & 3 \\ 0 & 2 & -1 \\ 5 & 10 & 7 \end{pmatrix}\end{array}\quad\begin{array}{c}\text{Row Minimum}\\ 3 \\ -1 \\ 5\end{array}$$

$$\text{Column Maximum}\quad 7 \quad 10 \quad 7$$

(Note: There exists no saddle point)

(i) Comparing Row 1 elements (7, 6, 3) with the Row 2 elements (0, 2, –1) we find that each element of Row 1 is greater than the corresponding elements in *Row 2*. Thus for any strategy adopted by the Player B will see Player A selecting first strategy (Row 1) over the second (Row 2).

(ii) Similarly comparing Row 1 elements (7, 6, 3) with the Row 3 elements (5, 10, 7) we find that some element of Row 1 is greater than the corresponding elements in Row 2 (ie. 7 over 5). Whereas some element of Row 2 is greater than the corresponding elements in Row 1 (ie. 10 over 6, 7 over 3). Thus there exist no clear dominance.

(iii) By comparing the Row 2 elements (0, 2, –1) with the Row 3 elements (5, 10, 7) we find that each element of Row 2 is lesser than the corresponding elements in Row 3 and hence Row 2 is dominated by Row 3.

Hence the elimination of the dominated row (Row 2) we get the following pay off matrix.

$$\begin{array}{cc} & \text{Player B} \\ \text{Player A} & \begin{pmatrix} 7 & 6 & 3 \\ 5 & 10 & 7 \end{pmatrix} \end{array}$$

Similarly in order to analyze the dominance among the columns, we compare each element in a column with the corresponding element in the other column.

(i) Comparing Column 1 elements (7, 5) with Column 2 (6, 10) there exist no dominance i.e. (7 is superior to 6 but 5 is inferior to 10).

(ii) Comparing Column 1 elements (7, 5) with Column 3 (3, 7) there exist no dominance i.e. (7 is superior to 3 but 5 is inferior to 7).

(iii) Similarly comparing the Column 2 elements (6, 10) with Column 3 (3, 7) there exist dominance i.e. (6 is superior to 3 and 10 is inferior to 7). Given an option to Player B to select between Column 2 and 3 he will prefer column 3 as it will minimize his loses irrespective of the strategy adopted by Player A (Strategy 2 dominates Strategy 3).

Thus with the elimination of column 2 the reduced pay off matrix is as follows

$$\begin{array}{cc} & \text{Player B} \\ \text{Player A} & \begin{pmatrix} 7 & 3 \\ 5 & 7 \end{pmatrix} \end{array}$$

On verifying the reduced pay off matrix we find there exist no saddle points. A case of mixed strategy game we solve the above game

$$\begin{array}{ccc} & \text{Player B} & \text{Row Minima} \\ \text{Player A} & \begin{pmatrix} 7 & 3 \\ 5 & 7 \end{pmatrix} & \begin{matrix} 3 \\ 5 \end{matrix} \\ \text{Column Maxima} & \begin{matrix} 7 & 7 \end{matrix} & \end{array}$$

Let us presume that the Player A selects his strategies with probabilities p_1 and p_2 and Player B with q_1 and q_2.

Therefore

$$p_1 = \frac{(7-5)}{(7-3)+(7-5)} = \frac{2}{6} = \frac{1}{3}$$

$$p_2 = \frac{(7-3)}{(7-3)+(7-5)} = \frac{4}{6} = \frac{2}{3}$$

$$q_1 = \frac{(7-3)}{(7-3)+(7-5)} = \frac{4}{6} = \frac{2}{3}$$

$$q_2 = \frac{(7-5)}{(7-3)+(7-5)} = \frac{2}{6} = \frac{1}{3}$$

The value of the game

$$v = \frac{(7\times 7-5\times 3)}{(7-3)+(7-5)} = \frac{34}{6} = 17/3$$

Illustration No. 5

Solve the game whose pay off matrix is given by

Player Y

$$\text{Player X} \begin{pmatrix} 2 & 8 & 4 \\ 7 & 3 & 8 \\ 6 & 2 & 7 \end{pmatrix}$$

Find also the value of the game?

Solution:

Since the game has no saddle point let us apply dominance rule to reduce the pay off matrix. For dominance for which we need to compare each element of rows and columns pair wise.

	Player Y			Row Minima
Player X	2	8	4	2
	7	3	8	3
	6	2	7	2
Column Maxima	7	8	8	

Since it is clearly visible from the pay off matrix that Column III dominates the Column I as all elements of the Column III are greater than the corresponding elements of the Column I. And hence the pay-off matrix can be reduced as

Player Y

$$\text{Player X}\begin{pmatrix} 2 & 8 \\ 7 & 3 \\ 6 & 2 \end{pmatrix}$$

Similarly Row III is dominated by Row II and is eliminated from the pay–off matrix. And hence the game further reduces to

Player Y   Row Minima

$$\text{Player X}\begin{pmatrix} 2 & 8 \\ 7 & 3 \end{pmatrix} \quad \begin{matrix} 2 \\ 3 \end{matrix}$$

Column Maxima 7 8

Even now there exist no saddle points and the matrix cannot be further reduced. Let us presume that the Player X selects his strategies with probabilities p_1 and p_2 and Player Y withq_1 and q_2.

$$p_1 = \frac{(3-7)}{(2-8)+(3-7)} = \frac{-4}{-10} = \frac{2}{5}$$

$$p_2 = \frac{(2-8)}{(2-8)+(3-7)} = \frac{-6}{-10} = \frac{3}{5}$$

$$q_1 = \frac{(3-8)}{(2-8)+(3-7)} = \frac{-5}{-10} = \frac{1}{2}$$

$$q_2 = \frac{(2-7)}{(2-8)+(3-7)} = \frac{-5}{-10} = \frac{1}{2}$$

As we check $p_1 + p_2 = 1$, $q_1 + q_2 = 1$

And the value

$$v = \frac{(2\times 3-7\times 8)}{(2-8)+(3-7)} = \frac{-50}{-10} = 5$$

Arithmetic Method

Arithmetic Method is used in obtaining optimal strategies for each player in (2 × 2) games without saddle points. The steps involved are:

(i) Find the difference of two numbers in Row I and put it under the Row II, neglecting the negative sign if occurs.

(ii) Find the difference of two numbers in Row II and put it under the Row I, neglecting the negative sign if occurs.

(iii) Repeat the same for the two Columns.

The value thus obtained by swapping the difference represents the optimal relative frequencies of both the players. These are the frequencies can be converted to probabilities by dividing each of them by their sum. The value of the game can be obtained by any of method discussed.

Illustration No. 6

The pay off matrix in the following two person zero–sum game is given as under

$$\text{Player X} \overset{\text{Player Y}}{\begin{pmatrix} 8 & 2 & -4 \\ 11 & 10 & 11 \\ 14 & 12 & 8 \end{pmatrix}}$$

(i) Suggest the optimal strategies and the net outcome of the game?

Solution

	Player Y			Row Minima
Player X	8	2	−4	−4
	11	10	11	10
	14	12	8	8
Column Maxima	14	12	11	

The pay off matrix is without saddle points.

Hence using Dominance Rule we observe that 1st Column dominates the 2nd Column as each element of 1st Column is greater than the corresponding elements in the 2nd Column as such eliminated from the pay off matrix.

Therefore the reduced pay off matrix obtained by Dominance Property is

$$\text{Player X} \overset{\text{Player Y}}{\begin{pmatrix} 2 & -4 \\ 10 & 11 \\ 12 & 8 \end{pmatrix}}$$

Similarly 2nd Row dominates 1st Row as such 1st row can be eliminated from the pay–off matrix.

Hence we get,

Player Y

$$\text{Player X}\begin{pmatrix} 10 & 11 \\ 12 & 8 \end{pmatrix} \quad \begin{matrix} 12-8=4 & p_1=4/5 \\ 11-10=1 & p_2=1/5 \end{matrix}$$

$$\begin{matrix} 11-8=3 & 14-12=2 \\ q_1=3/5 & q_2=2/5 \end{matrix}$$

Expected Gain to Player X

(i) $10 \times 4/5 + 12 \times 1/5 = 52/5$ If Player Y adopts q_1

(ii) $11 \times 4/5 + 8 \times 1/5 = 52/5$ If Player Y adopts q_2

Expected Loss to Player Y

(i) $10 \times 3/5 + 11 \times 2/5 = 52/5$ If Player Y adopts p_1

(ii) $12 \times 3/5 + 8 \times 2/5 = 52/5$ If Player Y adopts p_2

Illustration No. 7

The pay off matrix in the following two person zero – sum game is given as under

Player B

$$\text{Player A}\begin{pmatrix} 1 & 4 & 2 \\ 4 & 7 & 6 \\ 5 & 5 & 3 \end{pmatrix}$$

(i) Suggest the optimal strategies of A and B and the net outcome of the game?

Solution:

Player B   Row Minima

$$\text{Player A}\begin{pmatrix} 1 & 4 & 2 \\ 4 & 7 & 6 \\ 5 & 5 & 3 \end{pmatrix} \quad \begin{matrix} 1 \\ 4 \\ 3 \end{matrix}$$

Column Maxima 5 7 6

The pay off matrix is without saddle points.

Hence using Dominance Rule we observe that 2nd Column dominates the 3rd Column as each element of 2nd Column is greater than the corresponding elements in the 3rd Column as such eliminated from the pay off matrix.

Therefore the reduced pay off matrix obtained by Dominance Property is

Player B

$$\text{Player A} \begin{pmatrix} 1 & 2 \\ 4 & 6 \\ 5 & 3 \end{pmatrix}$$

Similarly 2nd Row dominates 1st Row as such 1st row can be eliminated from the pay off matrix.

Hence we get,

Player B

$$\text{Player A} \begin{pmatrix} 4 & 6 \\ 5 & 3 \end{pmatrix} \quad \begin{matrix} 5-3=2 & p_1 = 1/2 \\ 6-4=2 & p_2 = 1/2 \end{matrix}$$

$$6-3=3 \quad 5-4=1$$

$$q_1 = 3/4 \quad q_2 = 1/4$$

Expected Gain to Player A

(i) $4 \times 1/2 + 5 \times 1/2 = 9/2$ If Player B adopts q_1

(ii) $6 \times 1/2 + 3 \times 1/2 = 9/2$ If Player B adopts q_2

Expected Loss to Player B

(i) $4 \times 3/4 + 6 \times 1/4 = 9/2$ If Player A adopts p_1

(ii) $5 \times 3/4 + 3 \times 1/4 = 9/2$ If Player A adopts p_2

Algebraic Method

Consider a game with pay off matrix of the form $[a_{ij}]_{m \times n}$. Let v be the value of the game and $(p_1, p_2, p_3, \ldots p_m)$ and $(q_1, q_2, q_3, \ldots q_n)$ are the respective probability of selection of optimal strategies by the players A and B.

The expected gain to be received by Player A when Player B uses any of his strategies from 1st, 2nd, ... nth strategies are

$\sum_{i=1}^{m} a_{i1}p_i$, $\sum_{i=1}^{m} a_{i2}p_i$, $\sum_{i=1}^{m} a_{i3}p_i$, ... $\sum_{i=1}^{m} a_{in}p_i$ respectively. Since the Player A is the gainer and expects at least 'v'.

Therefore

$$\sum_{i=1}^{m} a_{ij}p_i \geq v, \qquad j=1,2,3 \ldots n$$

where $$\sum_{i=1}^{m} p_i = 1,$$

Similarly by considering the losses to B when Player A uses his respective strategies we get the following the system of inequalities.

$$\sum_{j=1}^{n} a_{1j}\, q_j, \sum_{j=1}^{n} a_{2j}\, q_j, \sum_{j=1}^{n} a_{3j}\, q_j, \ldots.. \sum_{i=1}^{n} a_{mj}\, q_j$$

Since Player B is the loser

Therefore

$$\sum_{j=1}^{n} a_{ij}\, q_j \leq v, \qquad i=1, 2, 3 \ldots m$$

where $$\sum_{j=1}^{n} q_j = 1,$$

Such that p_is, q_is ≥ 0 for all i, j.

Now to get the value of p_is, q_js first of all we assume all the inequalities are considered as equations and try to solve the system of equations. But in case the system of equations so obtained is inconsistent then at least one of the inequalities must strictly hold as inequality. The solution now can be obtained using trial and error method.

Illustration No. 8

Solve the game whose pay off matrix is given by

Player N

$$\text{Player M}\begin{pmatrix} 0 & 3 & 2 \\ 2 & -1 & 3 \\ 4 & 5 & -2 \end{pmatrix}$$

Find also the value of the game?

Solution

Let the Player M advocate strategies (p_1, p_2 and p_3) and Player N choose strategies (q_1, q_2 and q_3) such that

$p_1 + p_2 + p_3 = 1$ and $q_1 + q_2 + q_3 = 1$ and v be the value of the game. Here we need to find the value of p_1, p_2, p_3, q_1, q_2, q_3 and v which satisfies the following relation

i.e.

$$0 .p_1 + 2. p_2 + 4 p_3 \geq v \quad \text{..... (1)}$$

$$3 .p_1 - 1. p_2 + 5 p_3 \geq v \quad \text{..... (2)}$$

$$2 .p_1 + 3. p_2 - 2 p_3 \geq v \quad \text{..... (3)}$$

$$0. q_1 + 3. q_2 + 2 q_3 \leq v \quad \text{..... (4)}$$

$$2. q_1 - 1. q_2 + 3 q_3 \leq v \quad \text{..... (5)}$$

$$4. q_1 + 5. q_2 - 2 q_3 \leq v \quad \text{..... (6)}$$

Along with the condition that

$$p_1 + p_2 + p_3 = 1 \quad \text{..... (7)}$$

and $$q_1 + q_2 + q_3 = 1 \quad \text{..... (8)}$$

where

$$p_1, p_2, p_3, q_1, q_2, q_3 \geq 0$$

Since it is difficult to handle inequalities we will convert the six inequalities into equations and hence we get,

$$2. p_2 + 4. p_3 = v \quad \text{..... (9)}$$

$$3. p_1 - p_2 + 5p_3 = v \quad \text{..... (10)}$$

$$2. p_1 + 3. p_2 - 2p_3 = v \quad \text{..... (11)}$$

$$3. q_2 + 2q_3 = v \quad \text{..... (12)}$$

$$2. q_1 - q_2 + 3 q_3 = v \quad \text{..... (13)}$$

$$4. q_1 + 5 q_2 - 2 q_3 = v \quad \text{..... (14)}$$

Now we have 7 unknowns and 8 equations. Solving the above set of equation, we have Player M = (17/46, 10/23, 9/46) and Player N = (7/23, 6/23, 10/23) and the value of the game v = 38/23.

Illustration No. 9

Solve the game whose pay off matrix is given by

$$\text{Firm A} \overset{\text{Firm B}}{\begin{pmatrix} 1 & 3 & 2 \\ 2 & 2 & 3 \\ 3 & 2 & 2 \end{pmatrix}}$$

Find also the value of the game?

Solution

Let the Firm A advocate strategies (p_1, p_2 and p_3) and Firm B choose strategies (q_1, q_2 and q_3) such that

$p_1 + p_2 + p_3 = 1$ and $q_1 + q_2 + q_3 = 1$ and v be the value of the game. Here we need to find the value of p_1, p_2, p_3, q_1, q_2, q_3 and v which satisfies the following relation

i.e.

$$1. p_1 + 2. p_2 + 3. p_3 \geq v \qquad \text{..... (1)}$$

$$3. p_1 + 2. p_2 + 2. p_3 \geq v \qquad \text{..... (2)}$$

$$2. p_1 + 3. p_2 + 2. p_3 \geq v \qquad \text{..... (3)}$$

$$1. q_1 + 3. q_2 + 2. q_3 \leq v \qquad \text{..... (4)}$$

$$2. q_1 + 2. q_2 + 3. q_3 \leq v \qquad \text{..... (5)}$$

$$3. q_1 + 2. q_2 + 2. q_3 \leq v \qquad \text{..... (6)}$$

Along with the condition that

$$p_1 + p_2 + p_3 = 1 \qquad \text{..... (7)}$$

and $q_1 + q_2 + q_3 = 1$ ..... (8)

where

$$p_1, p_2, p_3, q_1, q_2, q_3 \geq 0$$

Since its difficult to handle inequalities we will convert the six inequalities into equations and hence we get,

$$1. p_1 + 2. p_2 + 3. p_3 = v \qquad \text{..... (9)}$$

$$3. p_1 + 2. p_2 + 2. p_3 = v \qquad \text{..... (10)}$$

$$2. p_1 + 3. p_2 + 2. p_3 = v \qquad \text{..... (11)}$$

$$1. q_1 + 3. q_2 + 2. q_3 = v \qquad \text{..... (12)}$$

$2. q_1 + 2. q_2 + 3. q_3 = v$ ..... (13)

$3. q_1 + 2. q_2 + 2. q_3 = v$ ..... (14)

Now we have 7 unknowns and 8 equations.

Solving the above equations we get Firm A (1/4, 1/4, 1/2) and Firm B (1/4, 1/2, 1/4) and Value of the game is v = 9/4.

Matrix Method

In case of pay off matrix of a game takes a form of square matrix then the optimal solution can be obtained by Matrix method. The solution of a two person zero sum game with mixed strategies with a square pay off matrix can be calculated as

$$\text{Player A's Optimal Strategy} = \frac{[1 \quad 1]\, P_{adj}}{[1 \quad 1]\, P_{adj} \begin{bmatrix} 1 \\ 1 \end{bmatrix}}$$

$$\text{Player B's Optimal Strategy} = \frac{[1 \quad 1]\, P_{cof}}{[1 \quad 1]\, P_{adj} \begin{bmatrix} 1 \\ 1 \end{bmatrix}}$$

Therefore the value of the game is given by

v = (Player A's Optimal Strategies) × (Player B's Optimal Strategies) × (Pay off matrix P_{ij}).

where P_{adj} = adjoint matrix, P_{cof} = Cofactor matrix and Player A's optimal strategies are in the form of row vector and B's column vector.

Note: In rare cases the solution violates the non-negative conditions of probabilities i.e. p_is, q_is ≥ 0.

Illustration No. 10

The pay off matrix in the following two person zero – sum game is given as under

$$\begin{array}{cc} & \text{Player Q} \\ \text{Player P} & \begin{pmatrix} 0 & 6 & 1 \\ 5 & 1 & 6 \\ 4 & 0 & 5 \end{pmatrix} \end{array}$$

(i) Find the optimal strategies and the value of the game?

Solution:

$$\begin{array}{cc} & \text{Player Q} & \text{Row Minima} \\ \text{Player P} & \begin{pmatrix} 0 & 6 & 1 \\ 5 & 1 & 6 \\ 4 & 0 & 5 \end{pmatrix} & \begin{array}{c} 0 \\ 1 \\ 0 \end{array} \\ \text{Column Maxima} & 5 \quad 6 \quad 6 & \end{array}$$

The pay off matrix is without saddle points.

Hence using Dominance Rule we observe that 2nd Row dominates the 3rd Row as each element of 2nd Row is greater than the corresponding elements in the 3rd Row as such 3rd Row can be eliminated from the pay off matrix. Similarly 3rd Column dominates the 1st column and hence can be eliminated.

Therefore the reduced pay–off matrix obtained by Dominance Property is

$$\begin{array}{cc} & \text{Player Q} \\ \text{Player P} & \begin{pmatrix} 0 & 6 \\ 5 & 1 \end{pmatrix} \end{array}$$

Calculating the Adjoint Matrix and Cofactor Matrix of the above reduced pay matrix we get,

$$P_{adj} = \begin{pmatrix} 1 & -6 \\ -5 & 0 \end{pmatrix}$$

$$P_{cof} = \begin{pmatrix} 1 & -5 \\ -6 & 0 \end{pmatrix}$$

$$\text{Player P's Optimal Strategy} = \frac{[1 \quad 1]\begin{pmatrix} 1 & -6 \\ -5 & 0 \end{pmatrix}}{[1 \quad 1]\begin{pmatrix} 1 & -6 \\ -5 & 0 \end{pmatrix}\begin{bmatrix} 1 \\ 1 \end{bmatrix}} = \frac{[-4 \quad -6]}{-10}$$

$$= \frac{[4 \quad 6]}{10}$$

Therefore $p_1 = 4/10 = 2/5$ and $p_2 = 6/10 = 3/5$

$$\text{Player Q's Optimal Strategy} = \frac{[1\ \ 1]\begin{pmatrix} 1 & -5 \\ -6 & 0 \end{pmatrix}}{[1\ \ 1]\begin{pmatrix} 1 & -6 \\ -5 & 0 \end{pmatrix}\begin{bmatrix} 1 \\ 1 \end{bmatrix}} = \frac{[-5\ \ -5]}{-10}$$

$$= \frac{[5\ \ 5]}{10}$$

Therefore $q_1 = 5/10 = 1/2$ and $q_2 = 5/10 = 1/2$

And the value of the game is given by

v = (Player P's Optimal Strategies) × (Player Q's Optimal Strategies) × (Pay off matrix P_{ij}).

$$v = [2/5\ \ 3/5]\begin{pmatrix} 0 & 6 \\ 5 & 1 \end{pmatrix}\begin{bmatrix} 1/2 \\ 1/2 \end{bmatrix} = 3/2$$

Illustration No. 11

The pay off matrix in the following of a zero – sum game is given as under

Player II

$$\text{Player I}\begin{bmatrix} 2 & 8 & 3 \\ 1 & 3 & 8 \\ 6 & 2 & 7 \end{bmatrix}$$

(i) Find the optimal strategies of Player I and II and the value of the game?

Solution

	Player II			Row Minima
Player I	2	8	3	2
	1	3	8	1
	6	2	7	2
Column Maxima	6	8	8	

The pay off matrix is without saddle points.

Hence using Dominance Rule we observe that 3rd Column dominates the 1st Column as each element of 3rd Column is greater than the corresponding elements in the 1st Column as such eliminated from the pay off matrix.

Therefore the reduced pay off matrix obtained by Dominance Property is

$$\text{Player I} \overset{\text{Player II}}{\begin{bmatrix} 2 & 8 \\ 1 & 3 \\ 6 & 2 \end{bmatrix}}$$

Similarly 1st Row dominates 2nd Row as such 2nd row can be eliminated from the pay–off matrix.

Hence we get

$$\text{Player I} \overset{\text{Player II}}{\begin{bmatrix} 2 & 8 \\ 6 & 2 \end{bmatrix}}$$

Calculating the Adjoint Matrix and Cofactor Matrix of the above reduced pay matrix we get,

$$P_{adj} = \begin{bmatrix} 2 & -8 \\ -6 & 2 \end{bmatrix}$$

$$P_{cof} = \begin{bmatrix} 2 & -6 \\ -8 & 2 \end{bmatrix}$$

$$\text{Player's I Optimal Strategy} = \frac{[1 \;\; 1]\begin{pmatrix} 2 & -8 \\ -6 & 2 \end{pmatrix}}{[1 \;\; 1]\begin{pmatrix} 2 & -8 \\ -6 & 2 \end{pmatrix}\begin{bmatrix} 1 \\ 1 \end{bmatrix}} = \frac{[-4 \;\; -6]}{-10}$$

$$= \frac{[4 \;\; 6]}{10}$$

Therefore $p_1 = 4/10 = 2/5$ and $p_2 = 6/10 = 3/5$

$$\text{Player B's Optimal Strategy} = \frac{[1 \;\; 1]\begin{pmatrix} 2 & -6 \\ -8 & 2 \end{pmatrix}}{[1 \;\; 1]\begin{pmatrix} 2 & -8 \\ -6 & 2 \end{pmatrix}\begin{bmatrix} 1 \\ 1 \end{bmatrix}} = \frac{[-6 \;\; -4]}{-10}$$

$$= \frac{[6 \;\; 4]}{10}$$

Therefore $q_1 = 6/10 = 3/5$ and $q_2 = 4/10 = 2/5$

And the value of the game is given by

v = (Player A's Optimal Strategies) × (Player B's Optimal Strategies) × (Pay off matrix P_{ij}).

$$v = \begin{bmatrix} 2/5 & 3/5 \end{bmatrix} \begin{bmatrix} 2 & 8 \\ 6 & 2 \end{bmatrix} \begin{bmatrix} 3/5 \\ 2/5 \end{bmatrix} = 108/25$$

Graphical Method

The graphical method can used to solve pay off matrix of the form 2 × n or m × 2. It consists of two graphs

1. The pay off gains to Player I against his strategies available.
2. The pay off loses to Player II against his strategies available.

Consider the following (2 × 3) matrix assumed without a saddle point

The pay off matrix can be written as

$$\begin{array}{ccc} & & \text{Player II} & & \\ & & B_1 \quad B_2 \quad B_3 & & \text{Probability} \\ \text{Player I} & \begin{matrix} A_1 \\ A_2 \end{matrix} & \begin{bmatrix} a_{11} & a_{12} & a_{13} \\ a_{21} & a_{22} & a_{23} \end{bmatrix} & & \begin{matrix} p_1 \\ P_2 \end{matrix} \\ \text{Probability} & & q_1 \quad q_2 \quad q_3 & & \end{array}$$

Now Player I has two strategies (A_1, A_2) and Player II has three strategies (B_1, B_2, B_3). Therefore Player I can choose strategy A_1 with the probability of p_1 and strategy A_2 with the probability of p_2 such that $p_1 + p_2 = 1$ where $p_1, p_2 \geq 0$. The objective is to determine the optimal values of p_1 and p_2. Hence with the available strategies to Player II (B_1, B_2, B_3) the expected pay off to Player I is as follows

B's Strategies	A's Expected Pay-off
B_1	$a_{11} p_1 + a_{21} p_2$
B_2	$a_{12} p_1 + a_{22} p_2$
B_3	$a_{13} p_1 + a_{23} p_2$

As per the maximin criterion for mixed strategy games Player I will select (p_1, p_2) in order to maximize the minimum expected pay offs.

Thus plotting the straight lines

$$G_1 = a_{11}\, p_1 + a_{21}\, p_2$$

$$G_2 = a_{12}\, p_1 + a_{22}\, p_2$$

$$G_3 = a_{13}\, p_1 + a_{23}\, p_2$$

The lower boundary of the above lines will give the minimum expected pay off and the highest point in the lower boundary is the maximum expected pay off and hence the value of p_1 and p_2. Then we find only two strategies of Player II which passes through the maximin point in the process reducing the size of the matrix in (2×2) form. This is a case of (2×3) form; the same technique can be extended to $(2 \times n)$ form.

Similarly $(m \times 2)$ form games can treated in the similar manner except that the minimax point is the lowest point on the upper boundary of the straight lines.

Illustration No. 12

Solve the game graphically and find the value of the game

$$\begin{array}{cc} & \text{Player II} \\ \text{Player I} & \begin{pmatrix} 3 & 5 & 11 \\ 10 & 7 & 4 \end{pmatrix} \end{array}$$

Solution:

The game has no saddle points as shown below

	Player II			Row Minima	
Player I	3	5	11	3	
	10	7	4	4	→ Maximi
Column Maxima	10	7	11		
		↑ Mimimax			

Now check whether any column can be eliminated using *Dominance Rule;* on viewing the above pay – off matrix we find none. So, let us write down the constraints.

Let the Player I has two strategies with probability p_1 and p_2 respectively such that $p_1 + p_2 = 1$ (where $p_1, p_2 \geq 0$). The objective is to find the optimal values of p_1 and p_2.

Therefore for each strategies available to Player II, the expected gain for Player I is as follows

If Player II chooses the 1st strategy then the expected gain to Player I = 3 p_1 + 10 p_2.

If Player II chooses the 2nd strategy then the expected gain to Player I = 5 p_1 + 7p_2

If Player II chooses the 3rd strategy then the expected gain to Player I = 11 p_1 + 4p_2

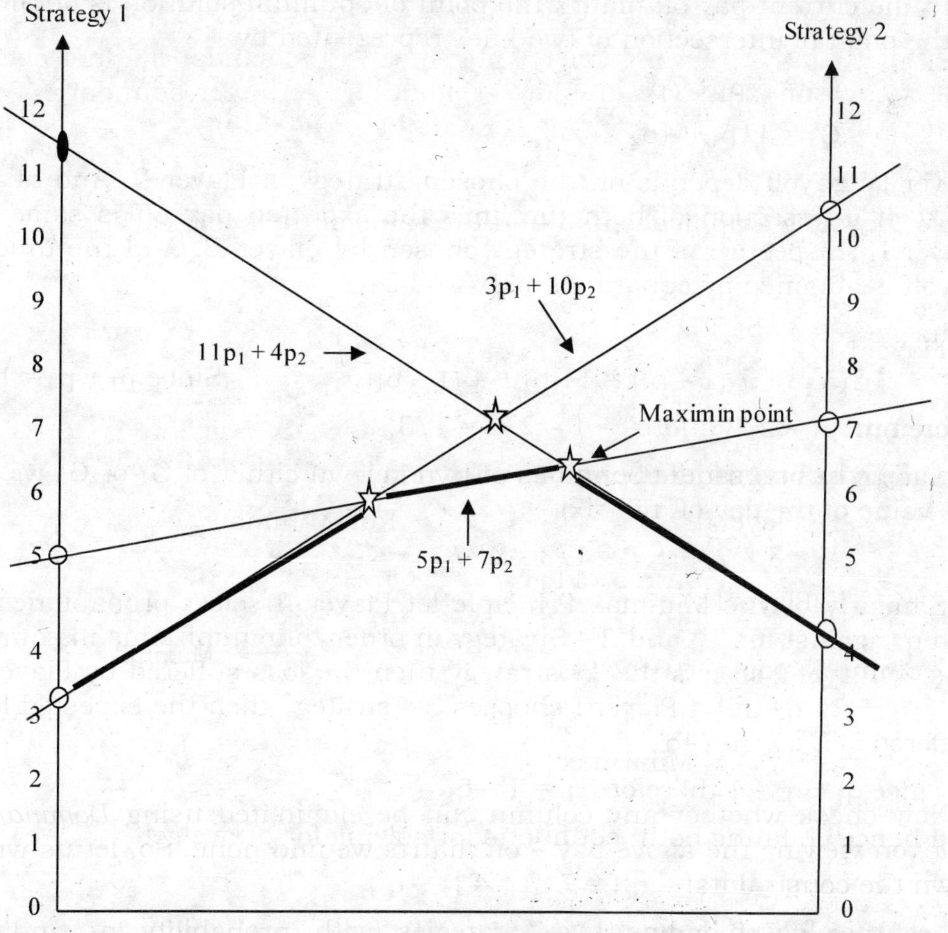

The expected pay off of the Player I is determined by the alternative strategies available. If Player II selects the 2nd strategy Players I gain is 5 by choosing 1st Strategy and 7 by choosing 2nd strategy. Thus the values

are plotted against two vertical lines and a line is drawn by joining the points representing the maximum possible pay offs. Similarly all the rest lines are drawn.

On the assumption that the player II always choose his best strategy in order to yield the worst results for Player I and thus the gains to Player I is represented by the lower boundary for any probabilistic value of Player I and the highest point on the lower boundary gives the maximum expected payoff as Player I will choose a combination of strategies to maximize his minimum gains.

In the current pay off matrix the point of optimum solution is obtained at the point of intersection of two lines represented by

$$G_2 = 5\,p_1 + 7p_2$$

$$G_3 = 11p_1 + 4p_2$$

Player I's payoff depends on the chosen strategy of Player II. And at the point of intersection of both the lines the expected pay off is same for Player I irrespective of the strategy chosen by player B. And the unique payoff is obtained by equating line L_2 and L_3

Thus we get

$$5\,p_1 + 7\,(1 - p_1) = 11\,p_1 + 4\,(1 - p_1) \qquad (\text{Since } p_1 + p_2 = 1)$$

Therefore $p_1 = 2/3$ and $p_2 = 1 - 2/3 = 1/3$.

And by substituting the values of p_1 and p_2 in either of G_1 or G_2 we get the value of the pay off matrix

$$V = 5 \times 2/3 + 7 \times 1/3 = 17/3$$

Similarly by the Minimax Principle let Player II select probabilities q_1 and q_2 against his 1st and 2nd Strategy in order to minimize his maximum loss. If Player I selects the 1st strategy then the loss suffered by Player II L_2: $5\,q_1 + 11\,q_2$ and if Player I chooses 2nd strategy then the expected loss suffered is L_3: $7\,q_1 + 4q_2$

Since $q_1 + q_2 = 1$ therefore $q_2 = 1 - q_1$

and hence equating both equations for solving for q_1, we get

$$5\,q_1 + 11\,(1 - q_1) = 7\,q_1 + 4\,(1 - q_1)$$

Therefore $q_1 = 7/9$ and $q_2 = 1 - 7/9 = 2/9$.

And by substituting the values of q_1 and q_2 in either of L_2 or L_3 we get

$$V = 5 \times 7/9 + 11 \times 2/9 = 57/9$$

Solution of Two Person Zero Sum Game by Linear Programming Method

A game with mixed strategies can be solved by transforming the problem into Linear Programming Problem. Let the pay off matrix is $[a_{ij}]_{m\times n}$ and Player A chooses strategies whose probability is defined by $(p_1, p_2, p_3, \dots p_m)$ where $\sum_{i=1}^{m} p_i = 1$ i.e $p_1 + p_2 + p_3 \dots + p_m = 1$ and Player B's probability are $(q_1, q_2, q_3, \dots q_m)$ where $\sum_{i=1}^{n} q_i = 1$ i.e $q_1 + q_2 + q_3 \dots + q_n = 1$. Then the expected gain to A when B chooses j^{th} course of action is $\sum_{i=1}^{m} a_{ij}p_i$.

Applying Maximin Criterion for Player A, we get

$\sum_{i=1}^{m} a_{ij}p_i \geq v$ i.e.

$$a_{11}p_1 + a_{21}p_2 + a_{31}p_3 + \dots a_{m1}p_m \geq v$$

$$a_{12}p_1 + a_{22}p_2 + a_{32}p_3 + \dots a_{m2}p_m \geq v$$

$$\vdots \qquad \vdots \qquad \vdots$$

$$a_{1n}p_1 + a_{2n}p_2 + a_{3n}p_3 + \dots a_{mn}p_m \geq v$$

were $p_1 + p_2 + p_3 \dots + p_m = 1$ and $p_1, p_2, p_3 \dots, p_m \geq 0$

Now divide both side of the inequalities by v we get,

$$a_{11}p_1/v + a_{21}p_2/v + a_{31}p_3/v + \dots a_{m1}p_m/v \geq 1$$

$$a_{12}p_1/v + a_{22}p_2\ /v + a_{32}p_3/v + \dots a_{m2}p_m/v \geq 1$$

$$\vdots$$

$$a_{1n}p_1/v + a_{2n}p_2/v + a_{3n}p_3/v + \dots a_{mn}p_m/v \geq 1$$

$$p_1/v + p_2/v + p_3/v \dots + p_m/v = 1/v$$

let us denote $p_1/v, p_2/v, p_3/v \dots, p_m/v$ by $x_1, x_2, x_3 \dots, x_m$

The objective of the Player A is to maximize gain which is equivalent to minimizing 1/v. Thus the resultant LPP can be written as follows

Minimize $= 1/v = Z = x_1 + x_2 + x_3 \dots + x_m$ (Max $v \leftrightarrow$ Min $1/v$)

Subject to

$$\left.\begin{array}{l} a_{11}x_1 + a_{21}x_2 + a_{31}x_3 + \ldots a_{m1}x_m \geq 1 \\ a_{12}x_1 + a_{22}x_2 + a_{32}x_3 + \ldots a_{m2}x_m \geq 1 \\ \vdots \quad \vdots \quad \vdots \\ a_{1n}x_1 + a_{2n}x_2 + a_{3n}x_3 + \ldots a_{mn}x_m \geq 1 \\ x_1, x_2, x_3 \ldots, x_m \geq 0 \end{array}\right\} \ldots\ldots \text{(i)}$$

Similarly

Considering the Minimax criterion for Player B we get,

$$a_{11}q_1 + a_{12}q_2 + a_{13}q_3 + \ldots a_{1n}q_n \leq v$$
$$a_{21}q_1 + a_{22}q_2 + a_{32}q_3 + \ldots a_{2n}\,q_n \leq v$$
$$\vdots \quad \vdots \quad \vdots \quad \vdots$$
$$a_{m1}q_1 + a_{m2}q_2 + a_{m3}q_3 + \ldots a_{mn}q_n \leq v$$

where $q_1 + q_2 + q_3 \ldots + q_m = 1$ and $q_1, q_2, q_3 \ldots, q_n \geq 0$

Now divide both side of the inequalities by v we get,

$$a_{11}q_1/v + a_{12}q_2/v + a_{13}q_3/v + \ldots a_{1n}q_n/v \leq 1$$
$$a_{21}q_1/v + a_{22}q_2/v + a_{23}q_3/v + \ldots a_{2n}q_n/v \leq 1$$
$$\vdots \quad \vdots \quad \vdots \quad \vdots$$
$$a_{m1}q_1/v + a_{m2}q_2/v + a_{m3}q_3/v + \ldots a_{mn}q_n/v \leq 1$$
$$q_1/v + q_2/v + q_3/v \ldots + q_n/v = 1/v$$

let us denote $q_1/v, q_2/v, q_3/v \ldots, q_m/v$ by $y_1, y_2, y_3 \ldots, y_n$

The objective of the Player B is to minimize loss which is equivalent to maximizing 1/v. Thus the resultant LPP can be written as follows

Minimize $= 1/v = Z = y_1 + y_2 + y_3 \ldots + y_n$ (Mini v $\leftrightarrow$ Max 1/v)

Subject to

$$a_{11}y_1 + a_{12}y_2 + a_{13}y_3 + \ldots a_{1n}y_n \leq 1$$
$$a_{21}y_1 + a_{22}y_2 + a_{23}y_3 + \ldots a_{2n}y_n \leq 1$$
$$\vdots \quad \vdots \quad \vdots$$
$$a_{m1}y_1 + a_{m2}y_2 + a_{m3}y_3 + \ldots a_{mn}y_n \leq 1$$
$$y_1, y_2, y_3 \ldots, y_m \geq 0$$

Illustration No. 13

Two Players A and B are competing against each other and their respective strategies are furnished below in the following pay off matrix

Player B

$$\text{Player A} \begin{pmatrix} 8 & 0 & 3 \\ 0 & 6 & 3 \\ 4 & 1 & 7 \end{pmatrix}$$

Use Linear Programming Method to determine the best strategies for both the players?

Solution:

Let the strategies of Player A is given by p_1, p_2 and p_3 where as q_1, q_2 and q_3 for Player B such that

$$p_1 + p_2 + p_3 = 1 \text{ and } q_1 + q_2 + q_3 = 1$$

	Player II			Row Minima
Player I	8	0	3	0
	0	6	3	0
	4	1	7	1
Column Maxima	8	6	7	

The pay off matrix is without saddle points.

Considering the pay off for the Player A the Linear Programming Problem is

$$\text{Minimize} = Z = 1/v = x_1 + x_2 + x_3$$

Subject to the constraints

$$\left.\begin{aligned} 8x_1 + 4x_3 &\geq 1 \\ 6x_2 + x_3 &\geq 1 \\ 3x_1 + 3x_2 + 7x_3 &\geq 1 \end{aligned}\right\} \quad \text{Case 1}$$

where x_1, x_2 and $x_3 \geq 0$

Similarly the pay off matrix for Player B is

$$\text{Maximize } Z' = 1/v = y_1 + y_2 + y_3$$

Subject to the constraints

$$\left.\begin{array}{l} 8\,y_1 + 3\,y_3 \le 1 \\ 6x_2 + 3y_3 \le 1 \\ 4y_1 + y_2 + 7x_3 \le 1 \end{array}\right\} \text{Case 2}$$

where y_1, y_2 and $y_3 \ge 0$

This is the dual of Case 1

Let us solve the Case 2

Maximize $Z' = y_1 + y_2 + y_3 + 0.\ s_1 + 0.s_2 + 0.s_3$

Subject to the constraints

$$\left.\begin{array}{l} 8\,y_1 + 3\,y_3 + s_1 = 1 \\ 6x_2 + 3y_3 + s_2 = 1 \\ 4y_1 + y_2 + 7x_3 + s_3 = 1 \end{array}\right\} \text{Case 2}$$

y_1, y_2, y_3, s_1, s_2 and $s_3 \ge 0$

The Simplex representation

1st Iteration

Contribution (C_j)			**1**	**1**	**1**	**0**	**0**	**0**	
C_B	Basic Variables	Solution Variables	y_1	y_2	y_3	s_1	s_2	s_3	**Min Ratio**
0	s_1	1	**8**	0	3	1	0	0	$\frac{1}{8}$ ←
0	s_2	1	0	6	3	0	1	0	$\frac{1}{0}$
0	s_3	1	4	1	7	0	0	1	$\frac{1}{4}$
(Z_j)			0	0	0	0	0	0	
($C_j - Z_j$)			1 ↑	1	1	0	0	0	

Contribution (C_j)			1	1	1	0	0	0	
C_B	Basic Variables	Solution Variables	y_1	y_2	y_3	s_1	s_2	s_3	Min Ratio
1	y_1	$\frac{1}{8}$	**1**	0	3/8	1/8	0	0	$\frac{1/8}{0}$
0	s_2	1	0	6	3	0	1	0	$\frac{1}{6}$ ←
0	s_3	1/2	0	1	11/2	-1/2	0	1	$\frac{1}{2}$
	(Z_j)		1	0	3/8	1/8	0	0	
	(C_j - Z_j)		0	1 ↑	5/8	-1/8	0	0	

Contribution (C_j)			1	1	1	0	0	0	
C_B	Basic Variables	Solution Variables	y_1	y_2	y_3	s_1	s_2	s_3	Min Ratio
1	y_1	$\frac{1}{8}$	**1**	0	3/8	1/8	0	0	$\frac{1/8}{0}$
1	y_2	1/6	0	1	1/2	0	1/6	0	$\frac{1}{6}$ ←
0	s_3	1/3	0	0	5	-1/2	-1/6	1	$\frac{1}{2}$
	(Z_j)		1	0	3/8	1/8	0	0	
	(C_j - Z_j)		0	1 ↑	5/8	-1/8	0	0	

Contribution (C_j)			1	1	1	0	0	0	
C_B	Basic Variables	Solution Variables	y_1	y_2	y_3	s_1	s_2	s_3	Min Ratio
1	y_1	$\frac{1}{8}$	**1**	0	3/8	1/8	0	0	3
1	y_2	1/6	0	1	½	0	1/6	0	$\frac{1}{3}$
0	s_3	1/3	0	0	5	-1/2	-1/6	1	$\frac{1}{15}$ ←
	(Z_j)		1	1	7/8	1/8	1/6	0	
	(C_j - Z_j)		0	0	1/8	-1/8	-1/6 ↑	0	

Contribution (C_j)			1	1	1	0	0	0	
C_B	Basic Variables	Solution Variables	y_1	y_2	y_3	s_1	s_2	s_3	Min Ratio
1	y_1	$\frac{1}{10}$	1	0	0	13/80	1/80	-3/40	
1	y_2	2/15	0	1	0	1/20	11/60	-1/10	
1	y_3	1/15	0	0	1	-1/10	-1/30	1/5	
(Z_j)			1	1	1	9/80	13/20	1/40	
(C_j - Z_j)			0	0	0	-9/80	-13/80	-1/40	

The optimality is reached at $Z' = \left(\frac{1}{10}+\frac{2}{15}+\frac{1}{15}\right)=\frac{3}{10}$

Therefore v = 10/3

And

$y_1 = 1/10 \leftrightarrow q_1/v = 1/10, \rightarrow q_1 = 1/3$

$y_2 = 2/15 \leftrightarrow q_2/v = 2/15, \rightarrow q_2 = 4/9$

$y_3 = 1/15 \leftrightarrow q_3/v = 1/15 \rightarrow q_2 = 2/9$

By Duality principle we get

$x_1 = 9/80 \leftrightarrow p_1/v = 9/80, \rightarrow p_1 = 3/8$

$y_2 = 13/80 \leftrightarrow p_2/v = 13/80, \rightarrow p_2 = 13/24$

$x_3 = 1/40 \leftrightarrow p_3/v = 1/40 \rightarrow p_3 = 1/12$

Thus we get

Strategies of the Player A = (3/8, 13/24, 1/12)

Strategies of the Player B = (1/3, 4/9, 2/9)

Summary

Game theory is basically a study of conflicts and cooperation involving individuals, firms or any other combination in which the concepts are utilized to formulate, analyze and study the strategic behavior.

Pay off is the outcome of the game when alternative courses of action are adopted by each competing players.

Maximin

The maximum element among the minimum elements selected from each row.

Minimax

The minimum element among the maximum elements selected from each column.

Saddle Point

A game in which "the Maximin of one player say A" is equal to "Minimax for another player say B" is called the saddle point.

Two Person Zero Sum Game

A game in which the gain of one player is the loss of the other is called as Two person Zero Sum Game. Otherwise the algebraic sum of the gains of both the players after the game is zero.

Matrix Method

In case of pay off matrix of a game takes a form of square matrix then the optimal solution can be obtained by Matrix method. The solution of a two person zero sum game with mixed strategies with a square pay off matrix can be calculated as

$$\text{Player A's Optimal Strategy} = \frac{[1 \;\; 1]\, p_{adj}}{[1 \;\; 1]\, p_{adj}\begin{bmatrix}1\\1\end{bmatrix}}$$

$$\text{Player B's Optimal Strategy} = \frac{[1 \;\; 1]\, p_{cof}}{[1 \;\; 1]\, p_{adj}\begin{bmatrix}1\\1\end{bmatrix}}$$

Self Practice Exercises

1. Solve the games with the following pay off matrix

(a) $\begin{pmatrix} 3 & 10 & 6 & 7 \\ 8 & 9 & 7 & 8 \\ 10 & 5 & 3 & 6 \end{pmatrix}$ (b) $\begin{pmatrix} 3 & 1 & 7 & 9 & 6 \\ 7 & 4 & 5 & 8 & 12 \\ 9 & 3 & 6 & 1 & 5 \\ 11 & 2 & 1 & 9 & 5 \end{pmatrix}$

(c) $\begin{pmatrix} 5 & 7 \\ 4 & 6 \\ 2 & 3 \\ 1 & 7 \end{pmatrix}$ (d) $\begin{pmatrix} -3 & 5 & 2 & 22 \\ 7 & 7 & 6 & 8 \\ -2 & 0 & 0 & -3 \end{pmatrix}$

(e) $\begin{pmatrix} 4 & 8 & 6 & 16 & 8 \\ 10 & 12 & 6 & 14 & 16 \\ 12 & 14 & 18 & 16 & 14 \\ 8 & 4 & 12 & 8 & 6 \end{pmatrix}$ (f) $\begin{pmatrix} 6 & 18 & 3 & 6 \\ 9 & 15 & 12 & 9 \\ 3 & 0 & 6 & 12 \end{pmatrix}$

2. Solve the following pay off matrix and calculate the value of the game

(a) $\begin{pmatrix} 2 & 14 & 4 \\ 12 & 4 & 14 \\ 10 & 2 & 12 \end{pmatrix}$ (b) $\begin{pmatrix} 7 & 5 & 7 & 3 \\ 9 & 9 & 4 & 8 \\ 6 & 2 & 3 & 5 \end{pmatrix}$

(c) $\begin{pmatrix} 9 & 6 & 11 & 0 \\ 9 & 12 & 6 & 12 \\ 13 & 14 & 12 & 4 \\ 9 & 12 & 8 & 24 \end{pmatrix}$

3. Using dominance Rule solve the following pay off matrix

(a) $\begin{pmatrix} 6 & 12 & 7 & 14 \\ 8 & 9 & 6 & 10 \\ 11 & 10 & 12 & 11 \end{pmatrix}$ (b) $\begin{pmatrix} 5 & 4 & 9 \\ 6 & 7 & 8 \\ 4 & 5 & 7 \end{pmatrix}$

4. Solve the pay off matrix by graphical method

(a) $\begin{pmatrix} 6 & 3 & 4 \\ 7 & 5 & 2 \end{pmatrix}$ (b) $\begin{pmatrix} 5 & 9 & 8 \\ 12 & 6 & 10 \end{pmatrix}$

(c) $\begin{pmatrix} 4 & 10 & 12 \\ 14 & 6 & 8 \end{pmatrix}$ (d) $\begin{pmatrix} 2 & 7 \\ 6 & 4 \\ 1 & 2 \\ -2 & -3 \end{pmatrix}$

(e) $\begin{pmatrix} 4 & 6 \\ 5 & 2 \\ 3 & 3 \end{pmatrix}$

5. Solve the following Pay off matrix by Arithmetic Method

(a) $\begin{pmatrix} 0 & 8 & 2 \\ 7 & 1 & 9 \\ 5 & 0 & 6 \end{pmatrix}$ (b) $\begin{pmatrix} 8 & 11 & 12 \\ 10 & 6 & 7 \\ 4 & 7 & 8 \end{pmatrix}$

6. Solve the following pay off matrix by algebraic method

(a) $\begin{pmatrix} -2 & 4 & 2 \\ 2 & -4 & 4 \\ 6 & 8 & -6 \end{pmatrix}$ (b) $\begin{pmatrix} 2 & 1 & 2 \\ 1 & 2 & 0 \\ 2 & 0 & 4 \end{pmatrix}$

7. Solve the following pay off matrix by linear programming method

(a) $\begin{pmatrix} 10 & 6 & 14 \\ 14 & 18 & 2 \\ 20 & 12 & 4 \end{pmatrix}$ (b) $\begin{pmatrix} 8 & 1 & 4 \\ 7 & 2 & 5 \\ 9 & 4 & 8 \end{pmatrix}$

Answer

1.

(a) Saddle Point (A_2, B_3) and the value of the game = 7
(b) Saddle Point (A_2, B_2) and the value of the game = 4
(c) Saddle Point (A_1, B_1) and the value of the game = 5
(d) Saddle Point (A_2, B_3) and the value of the game = 6
(e) Saddle Point (A_3, B_1) and the value of the game = 12
(f) Saddle Point (A_2, B_1) and the value of the game = 9

2.

(a) Probability of 1st Player = (2/5, 3/5)
Probability of 2nd Player = (1/2, 1/2)
The value of the game = 8.

(b) Probability of 1st Player = (2/3, 1/3)
Probability of 2nd Player = (5/6, 1/6)
The value of the game =32/3.

(c) Probability of 1st Player = (1/2, 1/2)

Probability of 2nd Player = (5/8, 3/8)

The value of the game =11/2.

3.

(a) Probability of 1st Player = (1/7, 6/7)

Probability of 2nd Player = (2/7, 5/7)

The value of the game = 31/10.

(b) Probability of 1st Player = (3/4, 1/4)

Probability of 2nd Player = (3/4, 1/4)

The value of the game =19/4.

4.

(a) p_1 = 3/4 and p_2 = 1/4.

q_1 = 1/2 and q_2 = 1/2.

Value of the game = 7/2

(b) p_1 = 3/5 and p_2 = 7/5

q_1 = 3/10 and q_2 =7/10

Value of the game = 29/5

(c) p_1 = 4/7 and p_2 = 3/7.

q_1 = 2/7 and q_2 = 5/7.

Value of the game = 58/7.

(d) p_1 = 2/7 and p_2 = 5/7.

q_1 = 3/7 and q_2 = 4/7.

Value of the game = 34/7

(e) p_1 = 3/5 and p_2 = 4/5.

q_1 = 4/5 and q_2 = 1/5.

Value of the game = 22/5.

5.

(a) p_1 = 3/7 and p_2 = 4/7.

q_1 = 1/2 and q_2 = 1/2.

Value of the game = 28/7.

(b) $p_1 = 4/7$ and $p_2 = 3/7$.

$q_1 = 5/7$ and $q_2 = 2/7$.

Value of the game = 62/7.

6.

(a) $S_A = (17/60, 10/23, 9/46)$

$S_B = (7/23, 6/23, 10/23)$

$V = 30/23$.

(b) $S_A = (2/3, 1/3, 0)$

$S_B = (2/3, 1/3, 0)$

$V = 8/3$

7.

(a) $S_A = (2/3, 1/3, 0)$

$S_B = (0, 1/2, 1/2)$

$V = 16$.

(b) $S_A = (0, 4/7, 3/7)$

$S_B = (0, 3/5, 2/5)$

$V = 20/7$.

Appendix

The Prisoner's Dilemma

Two accomplice were arrested for committing a robbery. There could not be booked for the crime under the law because of lack of adequate evidences against them, unless one or both of them confess for the criminal act. Each of the suspects was interrogated in isolation from the other companion so that there is no chance of communication possible in between them. The attorney (public prosecutor) gave an offer to both the accomplice (remember in isolation) promised no punishment for the suspect who confess and severe punishment for the other party (e.g. 10 years jail term). Now if both the accomplice do not confess both will escape punishment. If both confess they will be sentenced according to the law for the crime of robbery (say 5 years of imprisonment). Thus each suspect has two strategies open to him either to confess or not to confess. Now there exists a dilemma if one confesses and goes free of punishment if the other doesn't confess or get punished as according to law of land for the crime or not to confess and escape punishment if the other accomplice does not confess to crime or get heavily punished if being betrayed by the other partner. The pay off matrix of both is given below.

<table>
<tr><td rowspan="6">Accomplice A's Strategies</td><td></td><td colspan="4">Accomplice B's Strategies</td></tr>
<tr><td></td><td colspan="2">No Confession</td><td colspan="2">Confession</td></tr>
<tr><td rowspan="2">No Confession</td><td>A</td><td>B</td><td>A</td><td>B</td></tr>
<tr><td>0</td><td>0</td><td>10</td><td>0</td></tr>
<tr><td rowspan="2">Confession</td><td>A</td><td>B</td><td>A</td><td>B</td></tr>
<tr><td>10</td><td>0</td><td>5</td><td>5</td></tr>
</table>

Because of lack of communication of among the accomplice has resulted in uncertainty in loyalty against one another. Thus both of them adopts the second strategy of confessing to the crime in the process get sentenced for a 5 year term of imprisonment. Clearly it demonstrates the worst scenario chosen against the adoption of no confession strategy as they both would have pleaded not guilty for the crime and escape the penalty of punishment and thus maximizing their gain in the process.

CHAPTER 22

Mathematics of Finance

Introduction

Given an option to an individual to decide on to receive a certain sum of money now to that of an equivalent amount in the future, what will be his course of action? On all rationale count most will favor current usage than waiting a given time period to receive the same amount say a year or more, the motive behind that future has a degree of uncertainty. Thus the motivation of holding that sum for future is mostly driven by the advantage of receiving greater benefits in the future.

What is Time Value?

We say that money has a time value because that money can be invested with the expectation of earning a positive rate of return. In other words, amount of money received today is worth greater than an amount of money to be received tomorrow. That is because today's amount can be invested and gain higher returns tomorrow by systematic investment. The essence of time value concept has a great bearing in the financial world as it stands vital in sound decision making with regards to timing of investment, to a certain extent understand risk, estimate absolute cash flows and to compare the feasibility and returns of alternate scores of investment. In this chapter emphasis is placed to elaborate on the various concepts of time value and its application.

Some Key Terminology

Present Value: An amount of money today, or the current value of a future cash flow.

Future Value: An amount of money at some future time period.

Period: A length of time (often a year, but can be a month, week, day, hour, etc.)

Interest Rate: The reward paid to a lender (or saver) for the use of funds expressed as a percentage for a period (normally expressed as an annual rate).

Timelines

- A timeline is a graphical device used to illustrate the timing of the cash flows for an investment.
- Each mark a vertical straight line represents one time period

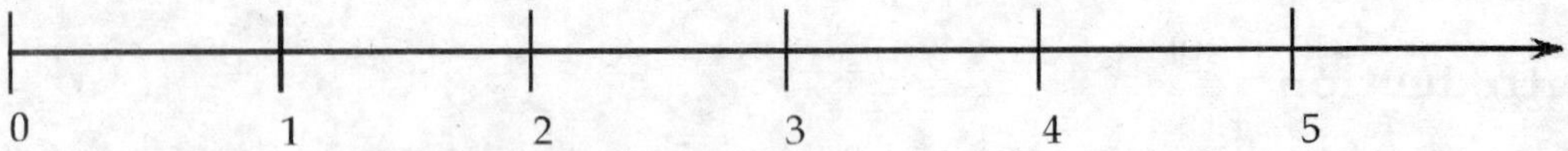

Concept of Simple and Compound Interest

Simple Interest

In case of simple interest, the interest component is calculated on the principal amount for the entire period. For example if a principal P is invested at r% per annum for a period n years then the simple interest is computed by

$$\text{Interest} = \text{Principal} \times \text{Rate} \times \text{Time period} = \text{Prn}.$$

Thus the amount that will accumulate after the end of the period is

$$\text{Amount (A)} = \text{Principal} + \text{Interest} = P + Prn = P(1 + rn).$$

Illustration No.1

If Amrita deposits Rs. 10,000 in deposit scheme drawing an annual interest rate of 10%. She withdraws Rs. 1,000 and any interest accrued on the said amount after each year. Calculate the total amount of interest received by Amrita?

Solution:

During the 1st year period Amrita earns the interest on the principal amount of Rs. 10,000.

$$= 10{,}000 \times 1 \times 0.10 = 1000.$$

But not only she withdraws the interest but also a part of the principal i.e. Rs. 1,000. And hence the net withdrawals are Rs. 2,000 out which Rs.1,000 is the interest component.

In the 2nd year the new principal is 10,000 – 1,000 = 9,000.

The interest is earned on the principal = 9,000 × 1 × 0.10 = 900

Again the interest Rs. 900 and Rs. 1,000 is withdrawn. Now the net principal for the 3rd year stands at Rs. 8,000. The interest is computed on Rs. 8,00o which is Rs. 800. Again the interest and Rs. 1000 of principal amount is withdrawn. Thus the net principal available in the beginning of the fourth year is Rs. 7000.

This procedure is continued till 10 years period and on tenth year the interest earned is Rs. 100 (1000 × 1 × 0 .10)

Thus the total interest received by Amrita during these 10 years is

= 1000 + 900 + 800 + 700 + 600 + 500 + 400 + 300 + 200 + 100.

= Rs. 5,500.

Compound Interest

If the interest earned on a principal is added after each definite time period and the whole sum is the principal for the next period on which interest is earned, then the interest calculation in this manner is called as compound interest. And the period after which interest becomes due is called interest period or conversion period.

Let us assume that an initial amount P is invested at an interest rate of r percent per annum for a period of n years.

The interest earned at the end of the year is Pr. Therefore the total amount at the end of year is given by

$$\text{Amount } (A_1) = \text{Principal} + \text{Interest} = P + Pr = P\,(1 + r).$$

Similarly the amount accumulated at the end of 2nd year is given by

$$A_2 = A_1 + A_1 r = A_1\,(1 + r) = P\,(1 + r)\,(1 + r) = P\,(1 + r)^2.$$

Extending the same principle for a time period of n years at an interest rate of r% is

$$A_n = P\,(1 + r)^n$$

The Concept of Multi-Period Compounding

In certain cases instead of interest being paid or received annually they are paid (received) any of the formats like semi-annually, quarterly, monthly or even daily in such cases the formula is adjusted in order to meet the requirement.

(i) In case of half-yearly mode of Interest Payments

$$A_n = P\left(1+\frac{r}{2}\right)^{2n}$$

(ii) In case of quarterly mode of Interest Payments

$$A_n = P\left(1+\frac{r}{4}\right)^{4n}$$

(iii) In case of monthly mode of Interest Payments

$$A_n = P\left(1+\frac{r}{12}\right)^{12n}$$

(iv) In case of daily mode of Interest Payments

$$A_n = P\left(1+\frac{r}{365}\right)^{365n}$$

The general formula is

$$A_n = P\left(1+\frac{r}{m}\right)^{mn}$$

where m stands for number of compounding in an year.

Illustration No. 2

If Rs. 1000 is invested for 5 years at an interest rate of 8% compounding every quarter then find what will be the amount accumulated at the end of the period?

Solution:

Given P =1000, and the time period n = 5 yrs. And the interest rate adjusted to quarterly mode = $\frac{8}{4}$ = 2% = 0.02.

By the application of the formula

$$A_n = P\left(1+\frac{r}{4}\right)^{4n}$$

$$= 1000\left(1+\frac{0.08}{4}\right)^{4\times 5}$$

$$= 1000\ (1+0.2)^{20}$$

$$= 1000\ (1.02)^{20}.$$

$$= 1000 \times 1.485 = \text{Rs. } 1,485$$

Nominal vs Effective Rate

Nominal rate of interest is the announced rate of interest whereas the rate at which the money grows during each year is called as the Effective Rate of interest. The mathematical formulae to calculate the effective rate of interest is given by

$$r_{eff} = \left(1+\frac{r}{k}\right)^k - 1$$

where

r = the nominal rate of interest .

k = the number of conversion period per year.

r_{eff} = the effective rate of interest

Note: In case of nominal rate compounded continuously; we have $r_{eff} = e^r - 1$.

Illustration No. 3

Calculate the effective annual rate of interest compounding equivalent to a nominal rate of 10% p.a. Compounded semi-annually?

Solution:

Here r = 0.10 and k = 2

So,

$$r_{eff} = \left(1+\frac{r}{k}\right)^k - 1$$

$$r_{eff} = \left(1+\frac{0.10}{2}\right)^2 - 1$$

$$= [1.1025] - 1 = 0.1025 \text{ or } 10.25\ \%.$$

Continuous Compounding

In certain cases the interest is calculated by continuous accrual basis. For instance banks paying interest on deposits schemes where the customers are entitled to receive interest on daily compounding basis i.e. adjustment done at the end of each day.

In such cases the formulae for calculating the future value is given by

$$FV_n = PV \times e^{(r \times n)}$$

where 'e' is the exponential function with a value of 2.7183 and 'i' the interest rate.

Illustration No. 4

Find the value at the end of 5 years for a sum of Rs. 100 bearing a rate of interest of 10% if the compounding is done continuously?

Solution:

$$FV_5 = 100 \times e^{(0.10 \times 5)}$$

As the present value (PV) = Rs. 100 and interest rate (i) = 10% = 0.10.

$$= 100 \times e^{(0.50)}$$

$$= 100 \times 1.6487$$

$$= 164.87$$

(Note; the value of $e^{0.50}$ = 1.6487 derived from table)

Thus the future value with continuous compounding has accumulated to Rs. 164.87.

Illustration No.5

A certain sum is invested at compounded annual interest of 5%. If the interest rate for the second year is Rs. 105. Find the principal at the beginning of the third year?

Solution:

(a) Let initial principal be equal to 'P'

Since the principal is invested at an interest rate of 5%, therefore the amount at the end of 1st year $= P(1+r) = P\left(1+\frac{5}{100}\right) = P\left(\frac{21}{20}\right)$.

In the second year the principal at the beginning of the second year is $P\left(\frac{21}{20}\right)$ and hence the interest earned on this amount is

$$\text{Interest } (I_2) = P\left(\frac{21}{20}\right) \times \frac{5}{100}$$

But the interest of the second year is Rs. 105

Equating both the figures i.e. $P\left(\frac{21}{20}\right) \times \frac{5}{100} = 105$

$$P = \text{Rs. } 2000$$

And hence the principal for the 3rd year

= Principal of the 2nd Year + Interest of the 2nd year

$$= 2000 \times \left(\frac{21}{20}\right) + 2000 \times \left(\frac{21}{20}\right) \times \frac{5}{100}$$

$$= 2100 + 105 = 2205$$

Illustration No. 6

If a certain sum is invested at compounded rate of 18 % per annum for 2 years will draw an amount which is Rs 190 less than the sum when invested in half yearly interest payable mode. Find the initial investment?

Solution:

Let us assume that the initial principal as P. And hence the amount after two years if the sum is invested at 18% per annum will be

Amount, if interest payable in yearly mode

$$= P\ (1 + r)\ (1 + r)$$

$$= P\left(1 + \frac{18}{100}\right)\left(1 + \frac{18}{100}\right)$$

$$= P\left(\frac{118}{100}\right)^2$$

Similarly the amount after 2 years if the interest payable in half yearly mode $= P\left(1 + \frac{9}{100}\right)^4 = P\left(\frac{109}{100}\right)^4$

But from the given conditions

$$P\left(\frac{109}{100}\right)^4 - P\left(\frac{118}{100}\right)^2 = 190$$

$$P\left\{\left(\frac{109}{100}\right)^4 - \left(\frac{118}{100}\right)^2\right\} = 190$$

$$P\ \{1.411 - 1.392) = 190$$

$$P = \text{Rs. } 10{,}000$$

Future Value of a Single Sum

Suppose an investor has invested Rs.100 in saving banks account drawing 8% interest annually what will be the sum at the end of 1st, 2nd and 3rd year if the interest income was not drawn and reinvested?

The investor's saving account will have a sum Rs 108 at the end 1st year

i.e. the initial principal	= Rs. 100.	
Plus 8% interest	= Rs. 8	(100×0.08)
Sum	= Rs. 108	

Now as the investor has not done any withdrawals, his new principal will be Rs 108

The investors saving account will show a balance of Rs. 116.64 at the end of 2nd year

i.e. the new principal	= Rs. 108	
Plus 8% interest	= Rs. 8.64	(108×0.08)
Sum	= Rs. 116.64	

Further with no withdrawals the net balance in the saving account in the end of 3rd year is Rs 125.97.

i.e. the new principal	= Rs. 116.64	
Plus 8% interest	= Rs. 9.33	(116.64×0.08)
Sum	= Rs. 125.97	

Having seen the compounding technique we can calculate the future value for any number of years.

Let us generalize the future value computation, from the above illustration we have seen the 2nd and 3rd year sum as Rs 116.64 and Rs. 125.97 which can be rewritten as

$$\begin{aligned}\text{Value at the end of the 2nd year} &= 116.64 = 108 + 8.64\\ &= 108 + 108 \times 0.08\\ &= 108 \times (1 + 0.08)\\ &= 100 \times (1 + 0.08) \times (1 + 0.08)\\ &= 100 \times (1 + 0.08)^2\end{aligned}$$

In similar counts the 3rd year value = 125.97 = $100 \times (1 + 0.08)^3$

Hence we have seen that a future sum can be expressed as a product of initial principal and a compounding factor. Therefore if $(FV)_n$ is the future value at the end of period n and P the initial principal and r represents the rate of interest, then the future value is given by

$$(FV)_n = P\,(1 + r)^n$$

where $(1 + r)^n$ is the compounding factor.

Illustration No. 7

Anmol has purchased a fixed deposit of Rs. 10,000 with a commercial bank drawing an interest rate of 10% per annum. Find how much the money will grow at the end of five years?

Solution:

The above illustration can be solved by two methods

Method - 1

The amount will grow in the following manner

Amount after the 1st year period $(FV)_1$ = $10000 + 10\% \times 10000$

= Rs 11, 000

Amount after the 2nd year period $(FV)_2$ = $11000 + 10\% \times 11000$

= Rs 12,100

Amount after the 3rd year period $(FV)_3$ = $12100 + 10\% \times 12100$

= Rs 13,100

Amount after the 4th year period $(FV)_4$ = $13100 + 10\% \times 13100$

= Rs 14, 641

Amount after the 5th year period $(FV)_5$ = $14641 + 10\% \times 14641$

= Rs 16105.1

Alternatively

By the application of formula method, we know,

$$(FV)_n = P\,(1 + r)^n$$

Therefore,

$$(FV)_5 = 10000\,(1 + 0.10)^5$$
$$= 10000(1.1)^5$$
$$= 16105.10$$

Present Value of a Single Sum

Given an option to an investor to comprehend what will be a future sum in equivalent value now so that to compare his cash flows. In order to determine the present valuation of a future sum we use discounting technique which is a method to find the present values of a single or series of future cash flows. Basically it is the current rupee value of a future sum.

Suppose we go back to the same illustration we have seen in the compounding technique what will an investor indifferent between a future sum of Rs. 108 received a year later to its equivalent value now. If the current value has grown over to Rs 108 by deploying the current sum at 8% rate of interest. The answer is simple Rs 100. How?

As we know the future value

$$108 = P(1 + 0.08)$$

Therefore $$P = \frac{108}{1.08} = 100$$

Thus an amount of Rs 100 with an opportunity cost of 8% is Rs 108 at the end of one year. Similarly any future sum can be converted to its equivalent value in present terms.

Let us generalize the present value computation. Mathematically as we know that the future value $(FV)_n$ to be received after n time period for a given present sum ' P ' is calculated using the formulae

$$(FV)_n = P(1 + r)^n$$

And hence

$$P = \frac{(FV)_n}{(1+r)_n}$$

where $$\frac{1}{(1+r)^n}$$ is called the discounting factor.

Illustration No. 8

Sanchit, an investor has invested in certain investment scheme which is suppose to grow to Rs 1,00,000 in 10 years period at an interest rate of 10%. Find what will be the present value of Sanchit's investment?

Solution:

We know that present value $P = \frac{(FV)_n}{(1+r)_n}$

Thus,

$$P = \frac{1,00,000}{(1+0.10)^{10}}$$

$$= \frac{1,00,000}{(1.10)^{10}}$$

{But the present value factor $\frac{1}{(1.10)^{10}} = 0.386$}

$= 1,00,000 \times 0.386$

$= \text{Rs.}38,600$

Annuities

Annuities are series of payments or receipts of equivalent amount made over a time horizon. For e.g . when you have availed a housing loan from a bank say for a period of 20 years where the monthly installments *(Principal + interest)* needs to be paid on regular basis. Thus you are making a commitment to the bank to honor to pay the installments till the loan terminates. Second e.g. if you have purchased a life insurance to safeguard yourself from any misfortune, you are required to pay premiums on either of the formats (*i.e. monthly, quarterly, half yearly or Annually*) to insurance co. In the first case the housing loan and in the second case the insurance premiums are some of the forms of annuities.

Annuity classification

(i) ***Ordinary or deferred annuity***: In this form the payments (receipts) are made at the end of each period – or the payments are deferred until the expiry of the period of time interval.

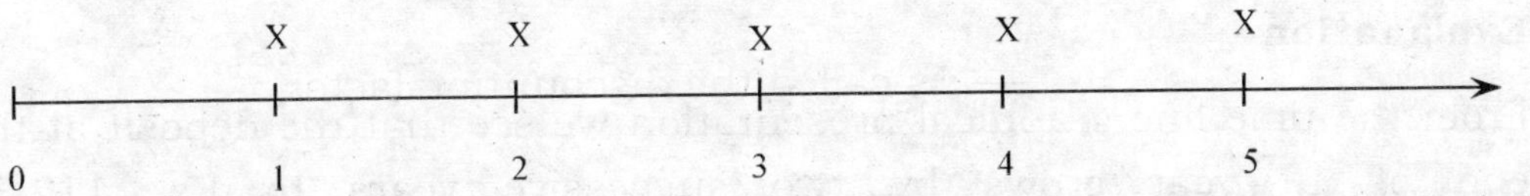

(ii) ***Annuity Due***: In this case the payments (receipts) are made at the beginning of each period.

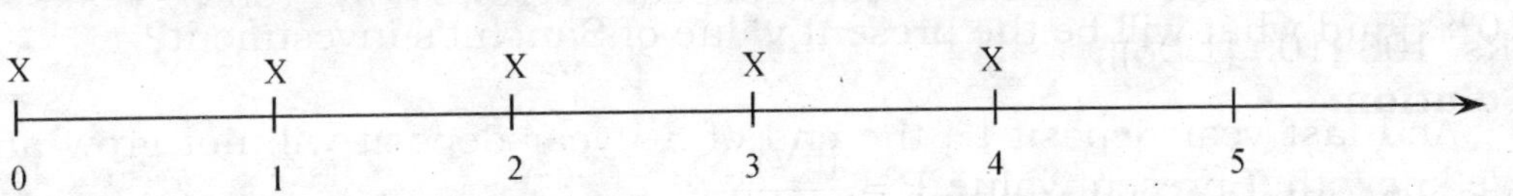

Calculation of Future Value of an Ordinary Annuity (Using Time line Analysis)

Illustration

Let us assume that a person deposits Rs. 100 at the end of each year for 3 years. Find sum accumulated at the end of third year, if the rate of interest is 6% per annum?

Representation in Time Line Axis

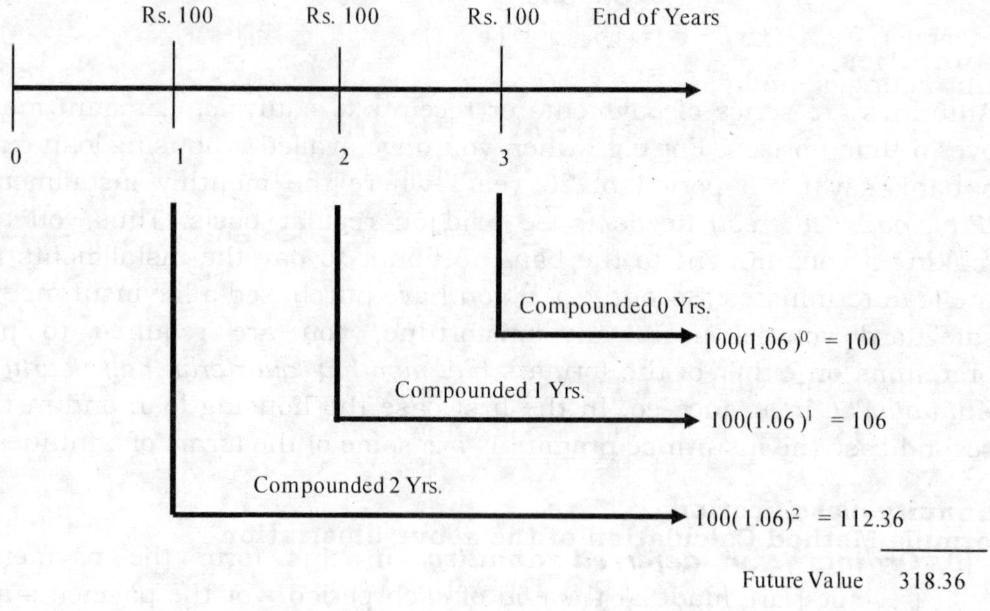

Explanation

From the time line graphical presentation we see that the deposit at the end of 1st year grows by two successive years to Rs. 112.36 {i.e. 100 $(1.06)^2$}.

The end of second year deposit grows by one year to Rs. 106 {100 (1.06)}.

And last year deposit i.e the end of 3rd year deposit will not grow and yield no interest and remains Rs. 100.

And hence Rs. 100 deposited at the end of each year for 3 consecutive years will accumulate to Rs. 318.36 {112.36 + 106 + 100}.

Generalization of Future Value of an Ordinary (Deferred) Annuity

In order to compute the future value of an Ordinary Annuity, let us assume equal receipts/payments (P) at the end of each year for a time period of n years. The computation is represented as

$FV_n = P + P(1+r) + P(1+r)^2 + P(1+r)^3 + ... + P(1+r)^{n-1}$ .....(i)

Multiply the eq. (i) by (1 + r)

$FV_n(1+r) = P(1+r) + P(1+r)^2 + P(1+r)^3 + ... + P(1+r)^n$ .....(ii)

Subtracting (ii) and (i)

$$FV_n(1+r) - FV_n = P(1+r)^n - P$$

$$= P\{(1+r)^n - 1\}$$

$$r\,FV_n = P\{(1+r)^n - 1\}$$

$$FV_n = \left(\frac{P\{1+r\}^n - 1}{r}\right)$$

Hence the future value is given by

$$FV_n = \left(\frac{P\{(1+r)\}^n - 1}{r}\right)$$

Formula Method Calculation of the above illustration

$$FV_n = \left(\frac{P\{(1+r)\}^n - 1}{r}\right)$$

$$= \frac{100\{(1+0.06)^3\} - 1}{0.06}$$

$$= \frac{100\{(1.06)^3 - 1\}}{0.06}$$

$$= 318.36$$

Illustration No. 9

Kishore has opened a Recurring Deposit account with a nationalized bank for a period of 5 years. He has decided to deposit Rs. 1500 at the end of the year for the period of deposit in the process drawing an interest rate of 8%. Find the future value to be received by Kishore on the termination of the deposit scheme?

Solution:

The future value is given by

$$FV_n = \left(\frac{P\{1+r\}^n - 1}{r}\right)$$

Here P = 1500, n = 5 years and r = 8% or 0.08

$$= \left(\frac{1500(1.08^5 - 1)}{0.08}\right)$$

= Rs. 8799.90.

Alternatively

Year	Cash Deposits	Future Value Factor	Future Value
1	1500	1.36	2040.73
2	1500	1.25	1889.56
3	1500	1.16	1749.60
4	1500	1.08	1620.00
5	1500	-	1500.00
		Future Value of Annuity	= 8799.90

Calculation of Present Value of an Ordinary Annuity (Using Time line Analysis)

Illustration

Representation in Time Line Axis

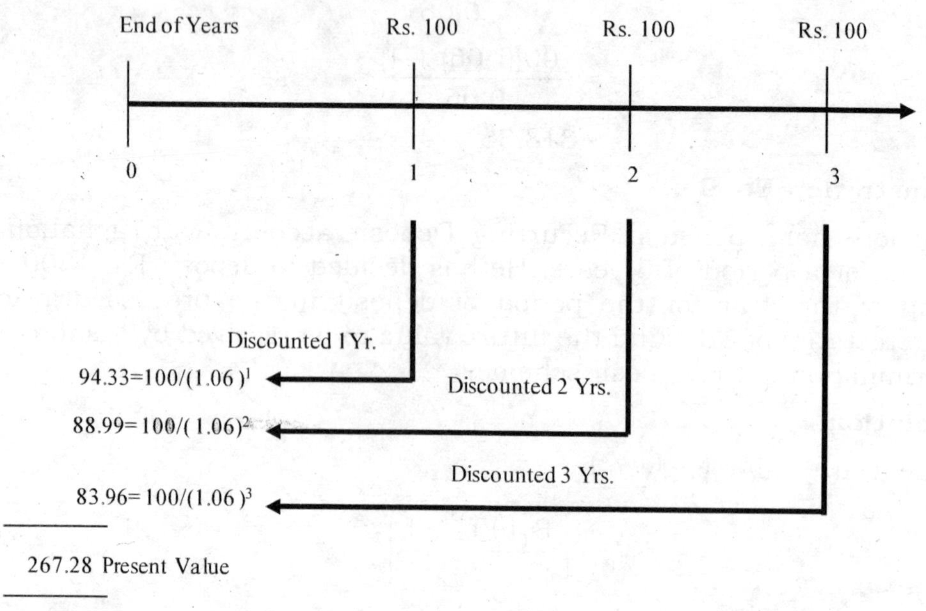

Explanation

We have seen from the graphical representation that Rs 100 one year from now when discounted by 6% equals to Rs. 94.33 { i.e $\frac{100}{1.06} = 94.33$}. So an individual is indifferent between receiving Rs 94.33 now and Rs.100 after a year period.

The second year amount if discounted by two years equals to Rs. 88.99 {100/(1.06)²}.

And last year deposit i.e the 3rd year deposit when discounted by three years equals to Rs. 83.96 {100/(1.06)³}.

And hence the total present value of the annuity is equal to Rs. 267.36 {112.36 + 106 + 100}.

Generalization of Present Value of an Ordinary Annuity

Understanding the present value of a stream of cash flows is immense importance in financial world and its computation is in similar lines as discussed above. Let P be the cash inflow discounted at a given rate say r for a period of n years, then the stream can represented in present term as

$$PV_n = \left\{\frac{P}{(1+r)} + \frac{P}{(1+r)^2} + \frac{P}{(1+r)^3} + \ldots + \frac{P}{(1+r)^n}\right\}$$

$$PV_n = P\left\{\frac{1}{(1+r)} + \frac{1}{(1+r)^2} + \frac{1}{(1+r)^3} + \ldots + \frac{1}{(1+r)^n}\right\} \qquad \ldots..(i)$$

Multiply (i) by (1 + r)

$$PV_n(1+r) = P\left\{1 + \frac{1}{(1+r)} + \frac{1}{(1+r)^2} + \ldots + \frac{1}{(1+r)^{n-1}}\right\} \qquad \ldots..(ii)$$

Subtracting (ii) and (i)

$$PV_n(1+r) - PV = P\left\{1 - \frac{1}{(1+r)^n}\right\}$$

$$r\ PV_n = P\left\{1 - \frac{1}{(1+r)^n}\right\}$$

$$PV_n = \frac{P}{r}\left\{1 - \frac{1}{(1+r)^n}\right\}$$

or $$PV_n = P\left\{\frac{1}{r} - \frac{1}{r(1+r)^n}\right\}$$

The Present Value of Ordinary annuity is given by

$$PV = P\left\{\frac{1}{r} - \frac{1}{r(1+r)^n}\right\}$$

Formula method calculation of the above illustration

Since the present value of ordinary annuity is given by

$$PV = P\left\{\frac{1}{r} - \frac{1}{r(1+r)^n}\right\}$$

$$= 100\left\{\frac{1}{0.06} - \frac{1}{0.06(1.06)^3}\right\}$$

$$= 267.28$$

Illustration No. 10

A heavy machinery is purchased on hire purchase basis with Rs. 10,000 as down payment on signing the deal and the rest is paid in next five yearly installments of Rs. 4000 at the end of each year. If the interest charged is 8% per annum then calculate the cash down price of the machinery?

Solution:

We need to calculate the present value of the five installments as

$$PV = P\left\{\frac{1}{r} - \frac{1}{r(1+r)^n}\right\}$$

as stated P = 4000, r = 8% or 0.08 and n = 5

$$= 4000\left\{\frac{1}{0.08} - \frac{1}{0.08(1+0.08)^5}\right\}$$

$$= 4000\left\{\frac{1}{0.08} - \frac{1}{0.08(1.08)^5}\right\}$$

$$= 4000 \times (12.5 - 8.54)$$

$$= \text{Rs. } 15{,}840.$$

And hence the cash down price of the machinery would be

Rs. (10,000 + 15840) = Rs. 25,840

Sinking Fund, Amortization and Capital Recovery

Sinking Fund

One of the vital applications of the concept of annuities is to understand the repayment of interest bearing debt instruments like bonds. For example, a firm accumulating sufficient funds to honor the liability of settling the maturity amount on expiry period. These types of funds created to settle dues are sinking funds and are used to pay the principal portion of the debt. In order to arrive the calculation of sinking fund as we know that the future sum is calculated by

$$FV_n = \left(\frac{P(1+r)^n - 1}{r}\right) \text{ or}$$

$$= P\left(\frac{(1+r)^n - 1}{r}\right)$$

Therefore,

$$P = \frac{FV_n \times r}{[(1+r)^n - 1]}$$

where $\frac{r}{[(1+r)^n - 1]}$ is the sinking fund factor used to calculate the annuity of a future sum.

Illustration No. 11

A machine costing Rs. 1,00,000 has shelf life of 10 years. If the scrap realized is Rs. 10,000, what should be the amount set aside from the profits at the end of each year so as to replace the machine if the compounding rate being 5% per annum?

Solution:

The machine replacement can be considered as a case of Sinking Fund that needs to be created for machine replacement.

The cost of the machine after 10 years = 100000 – 10000 = Rs. 90,000

Let P be the amount set aside each year

Since

$$FV_n = \left(\frac{P(1+r)^n - 1}{r}\right)$$

Here FV_{10} = 90,000, r = 5% or 0.05 and n = 10

$$90{,}000 = \left(\frac{P(1+0.05)^{10} - 1\}}{0.05}\right)$$

$$P = \frac{90{,}000 \times 0.05}{(1.05)^{10} - 1}$$

$$= \frac{4500}{(1.6288) - 1} = \text{Rs.}7156.48$$

(*Note:* the table value of $(1.05)^{10}$ = 1.6288)

Annually the firm has to set aside Rs. 7156.48. for acquiring the machinery at the end of 10th year period.

Amortization

A debt is said to be amortized when both the principal and interest are repaid sequentially over equal time periods with equal installments or payments. Against sinking fund which is primarily meant to discharge the principal of the debt whereas amortization is used to cover the both the outstanding interest and part of the principal. Loan availed to purchase house or car for a given period of time are some of the common examples. Let us understand the concept using an illustration.

Illustration No. 12

Jyothi has availed a loan of Rs 30,000 from a commercial bank to purchase a two wheeler at 12% rate of interest for a period of 5 years. Find what shall be the annual installments so that the loan is completely repaid within the time period?

Solution:

Since

$$PV = P\left\{\frac{1}{r} - \frac{1}{r(1+r)^n}\right\}$$

But the outstanding amount taken in the form of loan is Rs. 30,000 availed at 12% rate for 5 years period.

Therefore PV =30,000, r= 12% and n = 5 years.

$$30{,}000 = P\left\{\frac{1}{0.12} - \frac{1}{0.12(1.12)^5}\right\}$$

And hence 30, 000 = P × 3.613.

P = Rs. 8303.34

Thus annual installment of Rs 8303.34 each year paid for five years will completely square of the loan.

The details of the loan amortization schedule is furnished below

End of Year	Installments	Principal Component Adjusted	Interest Paid	Outstanding Balance
0				30,000
1	8303.34	4703.34	3600	25296.66
2	8303.34	5267.74	3035.55	20028.91
3	8303.34	5899.86	2403.47	14129.04
4	8303.34	6607.86	1695.48	7521.18
5	8303.34	7400.79	902.54	-

Note: the outstanding balance may show some amount because of decimals consideration.

Capital Recovery

Capital recovery helps in understanding the income generation by an investment. It is basically the annuity of an investment which is invested at specified rate for a specified time period. For example if an individual is saving certain sum keeping in mind for meeting any future requirements in the form squaring his liabilities or any other responsibilities, he will be obvious to know that what amount of investment today will create a fund that meets his desired objectives. Capital recovery can be computed from the formula of present value calculation of an ordinary annuity since,

$$PV = P\left\{\frac{1}{r} - \frac{1}{r(1+r)^n}\right\}$$

Therefore,

$$P = \frac{PV}{\left\{\frac{1}{r} - \frac{1}{r(1+r)^n}\right\}} = PV \times CRF$$

where CRF is capital recovery factor which is equal to $\left\{\frac{1}{r} - \frac{1}{r(1+r)^n}\right\}$

Illustration No. 13

Shyam Prasad has invested Rs. 5000 for a period of 5 years at an interest rate of 11%. How much income should he receive annually to recover his investment?

Solution:

Let P be the amount recovered each year

Since

$$P = \frac{PV}{\left\{\frac{1}{r} - \frac{1}{r(1+r)^n}\right\}}$$

Here PV = 5,000, r = 0.11

And hence

$$P = \frac{PV}{\left\{\frac{1}{r} - \frac{1}{r(1+r)^n}\right\}}$$

$$= \frac{5000}{\left\{\frac{1}{0.11} - \frac{1}{0.11(1.11)^5}\right\}}$$

$$= 5000 \times \left\{\frac{1}{3.685}\right\} = \text{Rs. } 1356.85$$

Future Value of an Annuity Due

Time Line Representation

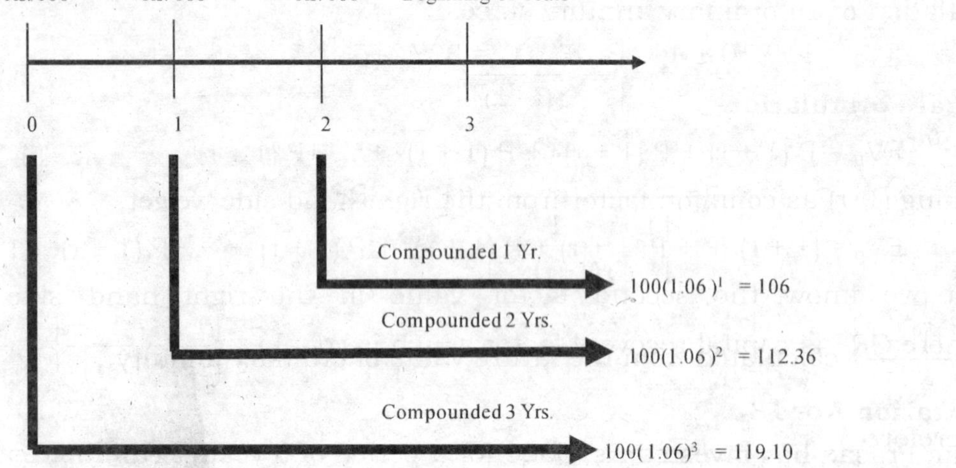

Explanation

As stated earlier the difference between ordinary annuity and annuity due is the period of inflow or outflow of cash. In case of annuity due instead of cash flow happening at the end of the period it takes place at the beginning of the period. The following table will illustrate the difference in the timing of cash inflow or out flow in both the cases for the benefit of the students.

Comparison of Cash Flow of Ordinary Annuity and Annuity Due		
Year	**Ordinary Annuity**	**Annuity Due**
0	0	100
1	100	100
2	100	100
3	100	0

From the illustration above we see that the 1st inflow/outflow of Rs.100 is compounded for 3 yrs to 119.10 at 6% rate. The 2nd year sum of Rs. 100 is compounded by 2 years to 112.36 and the 3rd year sum is compounded by one year to Rs. 106. We see that an annuity due earn an additional year of interest against an annuity ordinary. And the future value(FV)$_3$

$$FV_3 = 119.10 + 112.36 + 106 = 337.46.$$

General Formulation

$$FV_n = P\,(1 + r) + P\,(1 + r)^2 + P\,(1 + r)^3 + \ldots + P\,(1 + r)^n \qquad \ldots\ldots(i)$$

Taking (1+r) as common factor from the right hand side we get,

$$FV_n = (1 + r)\,\{P + P\,(1 + r) + P\,(1 + r)^2 + P\,(1 + r)^3 + \ldots + P\,(1 + r)^{n-1}\}$$

But we know the second factor value in the right hand side $P\left[\frac{(1+r)^n - 1}{r}\right]$ calculated from the future value of ordinary annuity.

Therefore

$$FV_n = (1 + r) \times P\left[\frac{(1+r)^n - 1}{r}\right]$$

or

$$FV_n = P\left[\frac{(1+r)^n - 1}{r}\right] \times (1+r)$$

Formula Method

$$FV_n = \left[\frac{P\{(1+r)^n - 1\}}{r}\right] \times (1+r)$$

$$= \frac{100\{(1+0.06)^3 - 1\}}{(0.06)} \times (1+0.06)$$

$$= \frac{100\{(1.06)^3 - 1\}}{(0.06)} \times (1.06)$$

$$= 337.46.$$

Illustration No. 14

Gautam receives lease rental of Rs. 10,000 annually. As per the agreement of the lease the lessee has to pay the rental every year on the first day of the beginning of the year. If the agreement is for 5 years and if Gaurav has an opportunity to invest at 10% per annum. Find the future sum of the investment?

Solution

We know that

$$FV_n = P\left[\frac{(1+r)^n - 1}{r}\right] \times (1+r)$$

$$= 10000 \times \left[\frac{(1+0.1)^5 - 1}{0.1}\right] \times (1.1)$$

$$= \text{Rs } 67{,}156.1$$

Alternatively

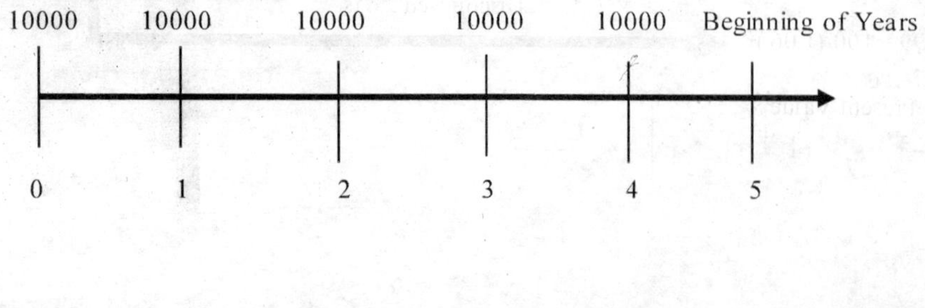

The 1st year sum received at the beginning of the 1st year i.e at t = 0 is compounded by 5 years i.e. 10000 $(1.1)^5$ = Rs.16,105.1

The 2nd year amount is compounded by 4 years $10000(1.1)^4$

= Rs 14,641.

The 3rd year amount is compounded by 3 years $10000(1.1)^3$

= Rs 13,310.

The 4th year amount is compounded by 2 years $10000(1.1)^2$

= Rs 12,100.

The 5th year amount is compounded by a year $10000(1.1)^1$

= Rs 11,000.

FV = 16,105.1 + 14,641 + 13,310 + 12,100 + 11,000

= Rs. 67156.1

Present Value of an Annuity Due

Time Line Representation

Beginning of the Year

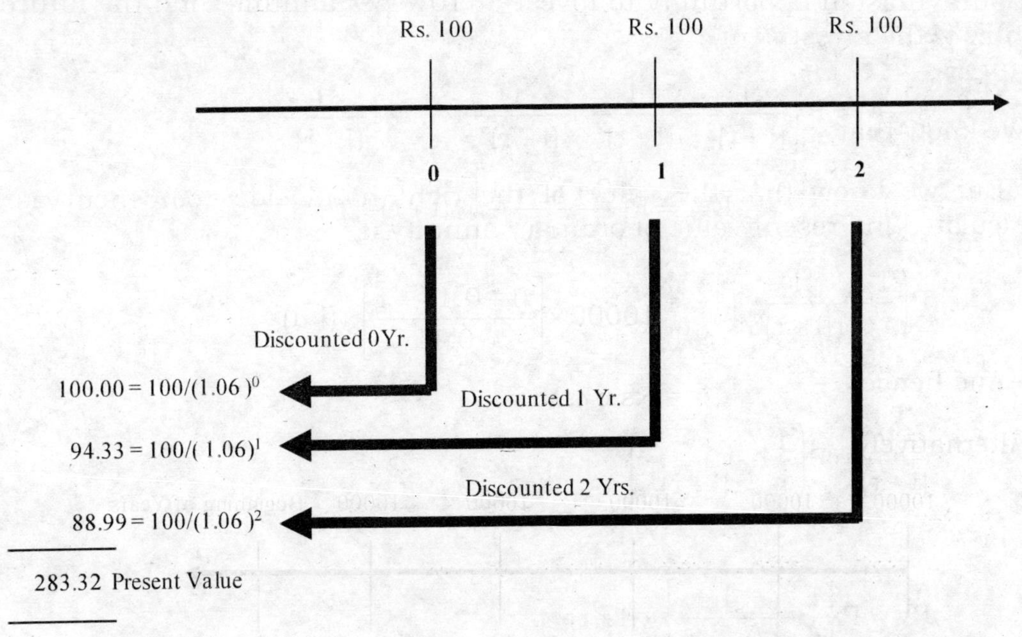

Explanation

The first sum of Rs. 100 present value remain the same whereas the 1st year amount is discounted by a year and the 2nd year sum is discounted by two years. Thus the present value of the series with payment starting with the beginning of the period is given by

$$PV = \frac{100}{1.06^0} + \frac{100}{1.06^1} + \frac{100}{1.06^2} = 100 + 94.33 + 88.99 = \text{Rs.}283.32.$$

General Formulation

Present value

$$PV = \left\{P + \frac{P}{(1+r)} + \frac{P}{(1+r)^2} + \frac{P}{(1+r)^3} + \ldots\ldots + \frac{P}{(1+r)^{n-1}}\right\}$$

$$PV = P\{1 + \frac{1}{(1+r)} + \frac{1}{(1+r)^2} + \frac{1}{(1+r)^3} + \ldots + \frac{1}{(1+r)^{n-1}}$$

$$PV = \left\{P + \frac{P}{(1+r)} + \frac{P}{(1+r)^2} + \frac{P}{(1+r)^3} + \ldots\ldots + \frac{P}{(1+r)^{n-1}}\right\} \qquad \ldots\ldots\text{(i)}$$

Divide both side by (1+r)

We get,

$$\frac{PV}{(1+r)} = P\left\{\frac{1}{(1+r)^1} + \frac{1}{(1+r)^2} + \frac{1}{(1+r)^3} + \ldots\ldots + \frac{1}{(1+r)^n}\right\}$$

But we know that the value of the right hand side component as calculated in Present value of ordinary annuity is

$$P\left(\frac{1}{r} - \frac{1}{r(1+r)^n}\right)$$

And hence

$$\frac{PV}{(1+r)} = P\left(\frac{1}{r} - \frac{1}{r(1+r)^n}\right)$$

or

$$PV = P\left\{\frac{1}{r} - \frac{1}{r(1+r)^n}\right\} \times (1+r)$$

Formula Method Calculation

$$PV = P\left\{\frac{1}{r} - \frac{1}{r(1+r)^n}\right\} \times (1+r)$$

$$= 100\left\{\frac{1}{0.06} - \frac{1}{0.06(1.06)^2}\right\} \times (1.06) = 193.98$$

Illustration No. 15

Rakesh receives lease rentals every year at the beginning of each year and the first installment is received from the day the lease agreement gets implemented. He receives regularly payments of Rs. 10,000 for the lease period of 5 years. Find the present value of Rakesh's rental income if the interest rate is 8% ?

Solution:

We know that the present value of annuity due is given by

$$PV = P\left\{\frac{1}{r} - \frac{1}{r(1+r)^n}\right\} \times (1+r)$$

But the value of P = 10,000, r = 8% per annum and time period n = 5 years

$$= 10{,}000\left\{\frac{1}{0.08} - \frac{1}{0.08(1.08)^5}\right\} \times (1.08)$$

= Rs. 43,200.

Bonds Basics

A bond is a long term debt instrument where the issuer agrees to pay periodical interest rate at specified point of time and principal at maturity.

Some Common Types of Bonds

(i) ***Treasury or Government Bonds:*** Basically issued by government and governmental bodies like central bank in behalf of the government.

(ii) ***Municipal Bonds:*** Mostly issued by municipalities, corporation and semi government organization to public, this bonds may bear sum default risk.

(iii) ***Mortgage Bonds:*** These are issued against which assets are pledge as security for obligation to make the payments. It is a form of contract in which the rights and obligation of both the parties are indentured.

(iv) ***Deep Discount Bonds:*** Deep Discounted bonds are issued at deep discounted price to the public and does not carry any explicit rate of interest.

(v) ***Convertible Bonds:*** Issued by companies that can be converted into equities after specific period of time on the option executed by the bond holder.

Characteristics of Bonds

(i) It's a long term debt instrument.

(ii) Carries an interest or coupon rates.

(iii) Issued for a specific period of time and repaid on maturity.

(iv) Has a market value.

Bond Valuation

The value of a bond is the present value of the contractual payments made by the issuer from the beginning till maturity. It is calculated as

$$B = \sum_{t=1}^{n} \frac{I_t}{(1+r)^t} + \frac{M.V}{(1+r)^n}$$

where

I = Coupon or Interest Rate.

r = Bonds required rate of return.

n = Number of years to maturity.

M.V = Maturity Value.

Illustration No. 16

Sanchit an investor is considering to purchase a five year Rs 1000 par value bond, bearing a normal rate of interest of 8% per annum. If the Sanchit required rate of return is 10%. What will be ready to pay now to purchase the bond if it matures at par?

Solution:

Sanchit will receive Rs. 80 as interest for five years and Rs. 1000 on maturity at the end of the fifth year. We need to determine the present value of the bond

Therefore,

$$B = \sum_{t=1}^{n} \frac{I_t}{(1+r)^t} + \frac{M.V}{(1+r)^n}$$

i.e. $$\frac{80}{(1.10)^1} + \frac{80}{(1.10)^2} + \frac{80}{(1.10)^3} + \frac{80}{(1.10)^4} + \frac{80}{(1.10)^5} + \frac{1000}{(1.10)^5}$$

$= 72.72 + 66.11 + 60.10 + 54.64 + 49.67 + 620.92.$

$=$ Rs. 924.16.

Thus Rs. 1000 bond is worth Rs. 924.16 today if the required rate of return is 10% per annum.

Summary

- Motivation for an investor holding a sum for future is mostly driven by the advantage of receiving greater benefits in the future.
- Present Value - An amount of money today, or the current value of a future cash flow.
- **Interest Rate** - The reward paid to a lender (or saver) for the use of funds expressed as a percentage for a period (normally expressed as an annual rate).
- **Timelines**
  - A timeline is a graphical device used to illustrate the timing of the cash flows for an investment.
  - Each mark a vertical straight line represents one time period
- If $(FV)_n$ is the future value of a given sum at the end of period n and P being the initial principal and r representing the rate of interest , then the future value is given by

  $$(FV)_n = P(1 + r)^n$$

  where $(1 + r)^n$ is the compounding factor.
- Annuities are series of payments or receipts of equivalent amount made over a time horizon. Classified as Ordinary Annuity and Annuity due.

- The future value of an annuity for a period n at a rate of interest r is mathematically represented as

$$FV_n = P\left[\frac{(1+r)^n - 1}{r}\right]$$

where $\frac{(1+r)^n - 1}{r}$ is the compounding factor of an annuity.

- The Present Value of Ordinary annuity is given by

$$PV = P\left\{\frac{1}{r} - \frac{1}{r(1+r)^n}\right\}$$

- A debt is said to be amortized when both the principal and interest are repaid sequentially over equal time periods with equal installments or payments.
- The Present Value of annuity due is given by

$$PV = P\left\{\frac{1}{r} - \frac{1}{r(1+r)^n}\right\} \times (1+r)$$

- If the interest earned on a principal is added after each definite time period and the whole sum is the principal for the next period on which interest is earned, then the interest calculation in this manner is called as compound interest.

Review Questions

1. What do you understand by time value of money? Elaborate?
2. What is the significance of time value concepts in financial decision making?
3. Write about annuities? What are the different types annuities, explain with detail graphical presentation?
4. Deduce the mathematical formulation for calculation the of present value of growing annuities ordinary?
5. Write short notes
   (i) Amortization?
   (ii) Sinking Fund?
   (iii) Continuous Compounding?

6. Write about the concept of simple and compound interest with examples?
7. What is capital recovery and explain its importance by an example?
8. What is multi period compounding? How does it affect the annual rate of interest by a solved example?
9. What are bonds? Write about the types and characteristics of bonds?

Self Practice Exercises

1. Ramesh deposits Rs. 2000 at the each year for 5 years in the saving bank account earning 5% interest compounded annually. Find how much sum he will have at the end of 5th year?
2. An investor deposits Rs. 1000 in his recurring deposit for 5 years at 8% interest. Find out the amount he will have in his account, if his interest compounded
   (i) Annually?
   (ii) Semi-annually?
   (iii) Quarterly?
   (iv) Continuously?
3. A bond of face value of Rs. 1,00,000 is payable 10 years later. There is an option to equal yearly installments at the end of each year at 10% annual interest. What would be the sum received each year?
4. A sinking fund is created by the redemption of Rs. 1,00,000 at the end of 25 years. How much amount should be set aside out of profits each year so that the sinking fund is created, if the investment can draw a normal interest of 4%?
5. Nidhi an investor has a regular practice of depositing Rs 5000 at the end of each year in a deposit scheme with a commercial bank in the process drawing an interest rate of 11% compounded annually. Find the amount in her account at the end of 6 year?
6. The sums of Rs 4000, Rs 6,000 and Rs 8,000 are due at the end of 2nd, 4th and 6th year respectively. If there is a proposal to replace the entire sum by a single amount of Rs 20,000 payable after a

time period n. If all are drawing 10% rate of interest compounded yearly then find n?

7. Shyam has availed a personal loan of Rs 20,000 which needs to be repaid in 5 equal annual installments. If the interest rate charged is 10% annually then calculate the yearly installments?

8. Praveen planning to sent his daughter abroad for higher education as such he is saving a certain sum each year from his income so that the amount builds to Rs. 10,00,000 after a period of 20 years. If the investment can earn 10 % rate per annum find fund set aside yearly?

9. Brijesh has borrowed Rs. 30,000 at 6% rate of interest and agreed to pay off the entire loan in 20 equal installments each beginning at the end of each year. Find what will be his annual installment?

10. A firm has two business plans. Plan A where his initial investment is Rs. 80,000 and whereas Plan B is Rs. 45,000. If Plan A will yield an income of Rs. 20,000 per year for the next 6 years while Plan B will yield an income of Rs. 16,000 for the next 4 years. If the discounting rate of the future income is 10% per annum then find which plan is better in terms of profitability?

Answers

1. Rs. 11054.
2. i) Rs. 1469 (Approx), ii) Rs.1480 (Approx), iii) Rs. 1486 (Approx), iv) 1492 (Approx)
3. Rs. 6273.50.
4. Rs. 2408 (Approx).
5. Rs. 35, 071.
6. n = 5.45 yrs.
7. Rs. 5275.65
8. Rs. 17,461
9. Rs. 2616
10. Plan a
11. a

Logarithms Table - A

	0	1	2	3	4	5	6	7	8	9	1	2	3	4	5	6	7	8	9
10	0000	0043	0086	0128	0170	0212	0253	0294	0334	0374	4	8	12	17	21	25	29	33	37
11	0414	0453	0492	0531	0569	0607	0645	0682	0719	0775	4	8	11	15	15	23	27	30	34
12	0792	0828	0864	0899	0934	0969	1004	1038	1072	1106	3	7	10	14	14	21	24	28	31
13	1139	1173	1206	1239	1271	1303	1335	1367	1399	1430	3	6	10	13	13	19	23	26	29
14	1461	1492	1523	1553	1584	1614	1614	1673	1703	1732	3	6	9	12	12	18	21	24	27
15	1761	1790	1818	1847	1875	1903	1931	1959	1987	2014	3	6	8	11	14	17	20	22	25
16	2041	2068	2095	2122	2148	2175	2201	2227	2253	2279	3	5	8	11	13	16	19	21	24
17	2304	2330	2355	2380	2405	2430	2455	2480	2504	2529	3	5	8	10	13	15	18	20	23
18	2553	2577	2601	2625	2648	2672	2695	2718	2742	2765	2	5	7	9	12	14	16	19	21
19	2788	2810	2833	2856	2878	2900	2923	2945	2967	2989	2	4	7	9	11	13	16	18	20
20	3010	3032	3054	3075	3090	3118	3139	3160	3181	3201	2	4	6	8	11	13	15	17	19
21	3222	3243	3263	3284	3304	3324	3345	3365	3385	3404	2	4	6	8	10	12	14	16	18
22	3424	3444	3464	3483	3502	3522	3541	3560	3579	3598	2	4	6	8	10	12	14	15	17
23	3617	3636	3655	3674	3692	3711	3729	3747	3766	3784	2	4	6	7	9	11	13	15	17
24	3802	3820	3838	3856	3874	3892	3909	3927	3945	3962	2	4	5	7	9	11	12	14	16
25	3979	3997	4014	4031	4048	4065	4082	4099	4116	4133	2	3	5	7	9	10	12	14	15
26	4150	4165	4183	4200	4216	4232	4249	4265	4281	4298	2	3	5	7	8	10	11	13	15
27	4314	4330	4346	4362	4378	4393	4409	4425	4440	4456	2	3	5	6	8	9	11	13	14
28	4472	4487	4502	4518	4533	4548	4564	4579	4594	4609	2	3	5	6	8	9	11	12	14
29	4624	4639	4654	4669	4683	4698	4713	4728	4742	4757	1	3	4	6	7	9	10	12	13
30	4771	4786	4800	4814	4829	4843	4857	4871	4886	4900	1	3	4	6	7	9	10	11	13
31	4914	4928	4942	4855	4869	4983	4997	5011	5024	5038	1	3	4	6	7	8	10	11	12
32	5051	5065	5079	5092	5105	5119	5132	5145	5159	5172	1	3	4	5	7	8	9	11	12
33	5185	5198	5211	5224	5235	5250	5263	5276	5289	5302	1	3	4	5	6	8	9	10	12
34	5315	5328	5340	5353	5366	5378	5391	5403	5416	5428	1	3	4	5	6	8	9	10	11
35	5441	5453	5465	5478	5490	5502	5514	5527	5539	5551	1	2	4	5	6	7	9	10	11
36	5563	5575	5587	5599	5611	5623	5635	5647	5658	5670	1	2	4	5	6	7	8	10	11
37	5682	5694	5705	5717	5729	5740	5752	5763	5775	5786	1	2	3	5	6	7	8	9	10
38	5798	5809	5821	5832	5843	5855	5866	5877	5888	5899	1	2	3	5	6	7	8	9	10
39	5911	5922	5933	5944	5955	5966	5977	5988	5999	6010	1	2	3	4	5	7	8	9	10
40	6021	6031	6042	6053	6064	6075	6085	6096	6107	6117	1	2	3	4	5	6	8	9	10
41	6128	6138	6149	6160	6170	6180	6191	6201	6212	6222	1	2	3	4	5	6	7	8	9
42	6232	6243	6253	6263	6274	6284	6294	6304	6314	6325	1	2	3	4	5	6	7	8	9
43	6335	6345	6355	6365	6375	6385	6395	6405	6415	6425	1	2	3	4	5	6	7	8	9
44	6435	6444	6454	6464	6474	6484	6493	6503	6513	6522	1	2	3	4	5	6	7	8	9
45	6532	6542	6551	6561	6571	6580	6590	6599	6609	6618	1	2	3	4	5	6	8	8	9
46	6628	6637	6646	6656	6665	6675	6684	6684	6702	6712	1	2	3	4	5	6	7	7	8
47	6721	6730	6739	6749	6758	6767	6776	6776	6794	6803	1	2	3	4	5	5	7	7	8
48	6812	6821	6830	6839	6848	6857	6866	6866	6884	6893	1	2	3	4	4	5	7	7	8
49	6902	6911	6920	6928	6937	6946	6955	6955	6972	6981	1	2	3	4	4	6	7	7	8

Contd.....

	0	1	2	3	4	5	6	7	8	9	1	2	3	4	5	6	7	8	9
50	6990	6998	7007	7016	7024	7033	7042	7050	7059	7067	1	2	3	3	4	5	6	7	8
51	7076	7084	7093	7101	7110	7118	7126	7315	7143	7152	1	2	3	3	4	5	6	7	8
52	7160	7168	7177	7185	7193	7202	7210	7218	7226	7235	1	2	2	3	4	5	6	7	7
53	7243	7251	7259	7267	7275	7284	7292	7300	7308	7316	1	2	2	3	4	5	6	6	7
54	7324	7332	7340	7348	7356	7364	7372	7380	7388	7396	1	2	2	3	4	5	6	6	7
55	7404	7412	7419	7427	7435	7443	7451	7459	7466	7474	1	2	2	3	4	5	5	6	7
56	7482	7490	7497	7505	7513	7520	7528	7536	7543	7551	1	2	2	3	4	5	5	6	7
57	7559	7566	7574	7582	7589	7597	7604	7612	7619	7627	1	2	2	3	4	5	5	6	7
58	7634	7642	7649	7657	7664	7672	7679	7686	7694	7701	1	1	2	3	4	4	5	6	7
59	7709	7716	7733	7731	7738	7745	7752	7760	7767	7774	1	1	2	3	4	4	5	6	7
60	7782	7789	7796	7803	7810	7818	7825	7832	7839	7846	1	1	2	3	4	4	5	6	6
61	7853	7860	7868	7875	7882	7889	7896	7903	7910	7917	1	1	2	3	4	4	5	6	6
62	7924	7931	7938	7945	7952	7959	7966	7973	7980	7987	1	1	2	3	3	4	5	6	6
63	7993	8000	8007	8014	8021	8028	8035	8041	8048	8055	1	1	2	3	3	4	5	5	6
64	8062	8069	8075	8082	8089	8096	8102	8109	8116	8122	1	1	2	3	3	4	5	5	6
65	8129	8136	8142	8149	8156	8162	8169	8176	8182	8189	1	1	2	3	3	4	5	5	6
66	8195	8202	8209	8215	8222	8228	8235	8241	8248	8254	1	1	2	3	3	4	5	5	6
67	8261	8267	8274	8280	8287	8293	8299	8306	8312	8319	1	1	2	3	3	4	5	5	6
68	8325	8331	8338	8344	8351	8357	8363	8370	8376	8382	1	1	2	3	3	4	4	5	6
69	8388	8395	8401	8407	8414	8420	8426	8432	8439	8445	1	1	2	2	3	4	4	5	6
70	8451	8457	8463	8470	8476	8482	8488	8494	8500	8506	1	1	2	2	3	4	4	5	6
71	8513	8519	8525	8531	8537	8543	8549	8555	8561	8567	1	1	2	2	3	4	4	5	5
72	8573	8579	8585	8591	8597	8603	8609	8615	8621	8627	1	1	2	2	3	4	4	5	5
73	8633	8639	8645	8651	8657	8663	8669	8675	8681	8686	1	1	2	2	3	4	4	5	5
74	8692	8698	8704	8710	8716	8722	8727	8733	8739	8745	1	1	2	2	3	4	4	5	5
75	8751	8756	8762	8768	8774	8779	8785	8791	8797	8802	1	1	2	2	3	3	4	5	5
76	8808	8814	8820	8825	8831	8837	8842	8848	8854	8859	1	1	2	2	3	3	4	5	5
77	8865	8871	8876	8882	8887	8893	8899	8904	8910	8915	1	1	2	2	3	3	4	4	5
78	8921	8927	8932	8938	8943	8949	8954	8960	8965	8971	1	1	2	2	3	3	4	4	5
79	8976	8932	8937	8993	8998	9004	9009	9015	9020	9025	1	1	2	2	3	3	4	4	5
80	9031	9036	9042	9047	9053	9058	9063	9069	9074	9079	1	1	2	2	3	3	4	4	5
81	9085	9090	9096	9101	9106	9112	9117	9122	9128	9133	1	1	2	2	3	3	4	4	5
82	9138	9143	9149	9154	9159	9165	9170	9175	9180	9196	1	1	2	2	3	3	4	4	5
83	9191	9196	9201	9206	9212	9217	9222	9227	9232	9238	1	1	2	2	3	3	4	4	5
84	9243	9248	9253	9258	9263	9269	9274	9279	9284	9289	1	1	2	2	3	3	4	4	5
85	9294	9299	9304	9309	9315	9320	9325	9330	9335	9340	1	1	2	2	3	3	4	4	5
86	9345	9350	9355	9360	9365	9370	9375	9380	9385	9390	1	1	2	2	3	3	4	4	5
87	9395	9400	9405	9410	9415	9420	9425	9430	9435	9440	0	1	1	2	2	3	4	4	4
88	9445	9450	9455	9460	9465	9469	9474	9479	9484	9489	0	1	1	2	2	3	3	4	4
89	9494	9499	9504	9509	9513	9518	9523	9528	9533	9538	0	1	1	2	2	3	3	4	4
90	9542	9547	9552	9557	9562	9566	9571	9576	9581	9586	0	1	1	2	2	3	3	4	4
91	9590	9595	9600	9605	9609	9614	9619	9624	9628	9633	0	1	1	2	2	3	3	4	4
92	9638	9643	9647	9652	9657	9661	9666	9671	9675	9680	0	1	1	2	2	3	3	4	4
93	9685	9689	9694	9699	9703	9708	9713	9717	9722	9727	0	1	1	2	2	3	3	4	4
94	9731	9736	9741	9745	9750	9754	9759	9763	9768	9773	0	1	1	2	2	3	3	4	4
95	9777	9782	9786	9791	9795	9800	9805	9809	9814	9818	0	1	1	2	2	3	3	4	4
96	9823	9827	9832	9836	9841	9845	9850	9854	9859	9863	0	1	1	2	2	3	3	4	4
97	9868	9872	9877	9881	9886	9890	9894	9899	9903	9908	0	1	1	2	2	3	3	4	4
98	9912	9917	9921	9936	9930	9934	9939	9943	9948	9952	0	1	1	2	2	3	3	4	4
99	9956	9961	9965	9969	9974	9978	9983	9987	9991	9996	0	1	1	2	2	3	3	3	4

Antilogarithms: Table - B

	0	1	2	3	4	5	6	7	8	9	1	2	3	4	5	6	7	8	9
.00	1000	1002	1005	1007	1009	1012	1014	1016	1019	1021	0	0	1	1	1	1	2	2	2
.01	1023	1026	1028	1030	1033	1035	1038	1040	1042	1045	0	0	1	1	1	1	2	2	2
.02	1047	1050	1052	1054	1057	1059	1062	1064	1067	1069	0	0	1	1	1	1	2	2	2
.03	1072	1074	1076	1079	1081	1084	1086	1089	1091	1094	0	0	1	1	1	1	2	2	2
.04	1096	1099	1102	1104	1107	1109	1112	1114	1117	1119	0	1	1	1	1	2	2	2	2
.05	1122	1125	1127	1130	1132	1135	1138	1140	1143	1146	0	1	1	1	1	2	2	2	2
.06	1148	1151	1153	1156	1159	1161	1164	1167	1169	1172	0	1	1	1	1	2	2	2	2
.07	1175	1178	1180	1183	1186	1189	1191	1194	1197	1199	0	1	1	1	1	2	2	2	2
.08	1202	1205	1208	1211	1213	1216	1219	1222	1225	1227	0	1	1	1	1	2	2	2	3
.09	1230	1233	1236	1239	1242	1245	1247	1250	1253	1256	0	1	1	1	1	2	2	2	3
.10	1259	1262	1265	1268	1271	1274	1276	1279	1282	1285	0	1	1	1	1	2	2	2	3
.11	1288	1291	1294	1297	1300	1303	1306	1309	1312	1315	0	1	1	1	2	2	2	2	3
.12	1318	1321	1324	1327	1330	1334	1337	1340	1343	1346	0	1	1	1	2	2	2	2	3
.13	1349	1352	1355	1358	1361	1365	1368	1371	1374	1377	0	1	1	1	2	2	2	3	3
.14	1380	1384	1387	1390	1393	1396	1400	1403	1406	1409	0	1	1	1	2	2	2	3	3
.15	1413	1426	1419	1422	1426	1429	1432	1435	1439	1442	0	1	1	1	2	2	2	3	3
.16	1445	1449	1452	1455	1459	1462	1466	1469	1472	1476	0	1	1	1	2	2	2	3	3
.17	1479	1483	1486	1489	1493	1496	1500	1503	1507	1510	0	1	1	1	2	2	2	3	3
.18	1514	1517	1521	1524	1528	1531	1535	1538	1542	1545	0	1	1	1	2	2	2	3	3
.19	1549	1552	1556	1560	1563	1567	1570	1574	1578	1581	0	1	1	1	2	2	3	3	3
.20	1585	1589	1592	1596	1600	1603	1607	1611	1614	1618	0	1	1	1	2	2	3	3	3
.21	1622	1626	1629	1633	1637	1641	1644	1648	1652	1656	0	1	1	2	2	2	3	3	3
.22	1660	1663	1667	1671	1675	1679	1683	1687	1690	1694	0	1	1	2	2	2	3	3	3
.23	1698	1702	1706	1710	1714	1718	1722	1726	1730	1734	0	1	1	2	2	2	3	3	4
.24	1738	1742	1746	1750	1754	1758	1762	1766	1770	1774	0	1	1	2	2	2	3	3	4
.25	1778	1782	1786	1791	1795	1799	1803	1807	1811	1816	0	1	1	2	2	2	3	3	4
.26	1820	1824	1828	1832	1837	1841	1845	1849	1854	1858	0	1	1	2	2	3	3	3	4
.27	1862	1866	1871	1875	1879	1884	1888	1892	1887	1901	0	1	1	2	2	3	3	3	4
.28	1905	1910	1914	1919	1923	1928	1932	1936	1941	1945	0	1	1	2	2	3	3	4	4
.29	1950	1954	1959	1963	1968	1972	1977	1982	1986	1991	0	1	1	2	2	3	3	4	4
.30	1995	2000	2004	2009	2014	2018	2023	2028	2032	2037	0	1	1	2	2	3	3	4	4
.31	2042	2046	2051	2056	2061	2065	2070	2075	2080	2084	0	1	1	2	2	3	3	4	4
.32	2089	2094	2099	2104	2109	2113	2118	2123	2128	2133	0	1	1	2	2	3	3	4	4
.33	2138	2143	2148	2153	2158	2163	2168	2173	2178	2183	0	1	1	2	2	3	3	4	4
.34	2188	2193	2198	2203	2208	2213	2218	2223	2228	2234	1	1	2	2	3	3	4	4	5
.35	2239	2244	2249	2254	2259	2265	2270	2275	2280	2286	1	1	2	2	3	3	4	4	5
.36	2291	2296	2301	2307	2312	2317	2323	2328	2333	2339	1	1	2	2	3	3	4	4	5
.37	2344	2350	2355	2360	2366	2371	2377	2382	2388	2393	1	1	2	2	3	3	4	4	5
.38	2399	2404	2410	2415	2421	2427	2432	2438	2443	2449	1	1	2	2	3	3	4	4	5
.39	2455	2460	2466	2472	2477	2483	2498	2495	2500	2506	1	1	2	2	3	3	4	5	5
.40	2512	2518	2523	2529	2535	2541	2457	2553	2559	2564	1	1	2	2	3	4	4	5	5
.41	2570	2576	2582	2588	2594	2600	2606	2612	2618	2624	1	1	2	2	3	4	4	5	5
.42	2630	2636	2642	2649	2655	2661	2667	2673	2679	2685	1	1	2	2	3	4	4	5	6
.43	2692	2698	2704	2710	2716	2723	2729	2735	2742	2748	1	1	2	3	3	4	4	5	6
.44	2754	2761	2762	2773	2780	2786	2793	2799	2805	2812	1	1	2	3	3	4	4	5	6
.45	2818	2825	2831	2838	2844	2851	2858	2864	2871	2877	1	1	2	3	3	4	5	5	6
.46	2884	2891	2897	2904	2911	2951	2924	2931	2938	2944	1	1	2	3	3	4	5	5	6
.47	2951	2958	2965	2972	2979	2985	2992	2999	3006	3013	1	1	2	3	3	4	5	5	6
.48	3020	3027	3034	3041	3048	3055	3062	3069	3076	3083	1	1	2	3	4	4	5	6	6
.49	3090	3097	3105	3112	3119	3126	3133	3141	3148	3155	1	1	2	3	4	4	5	6	6

Contd.....

	0	1	2	3	4	5	6	7	8	9	1	2	3	4	5	6	7	8	9
.50	3162	3170	3177	3184	3192	3199	3206	3214	3221	3228	1	1	2	3	4	4	5	6	7
.51	3236	3243	3251	3258	3266	3273	3281	3289	3296	3304	1	2	2	3	4	5	5	6	7
.52	3311	3319	3327	3334	3342	3350	3357	3365	3373	3381	1	2	2	3	4	5	5	6	7
.53	3388	3396	3404	3412	3420	3428	3436	3443	3451	3459	1	2	2	3	4	5	6	6	7
.54	3447	3475	3483	3491	3499	3508	3516	3524	3532	3540	1	2	2	3	4	5	6	6	7
.55	3548	3556	3565	3573	3581	3589	3597	3606	3614	3622	1	2	2	3	4	5	6	7	7
.56	3631	3639	3648	3656	3664	3673	3681	3690	3698	3707	1	2	3	3	4	5	6	7	8
.57	3715	3724	3733	3741	3750	3758	3767	3776	3784	3793	1	2	3	3	4	5	6	7	8
.58	3802	3811	3819	3828	3837	3846	3855	3864	3873	3882	1	2	3	4	4	5	6	7	8
.59	3890	3899	3908	3917	3926	3936	3945	3954	3963	3972	1	2	3	4	5	5	6	7	8
.60	3981	3990	3999	4009	4018	4027	4036	4036	4055	4064	1	2	3	4	5	6	6	7	8
.61	4074	4083	4093	4102	4111	4121	4130	4140	4150	4159	1	2	3	4	5	6	7	8	9
.62	4169	4178	4188	4198	4207	4217	4227	4236	4246	4256	1	2	3	4	5	6	7	8	9
.63	4266	4276	4285	4295	4305	4315	4325	4335	4345	4355	1	2	3	4	5	6	7	8	9
.64	4365	4375	4385	4395	4406	4416	4426	4436	4446	4457	1	2	3	4	5	6	7	8	9
.65	4467	4477	4487	4498	4508	4519	4529	4539	4550	4560	1	2	3	4	5	6	7	8	9
.66	4571	4581	4592	4603	4613	4624	4624	4645	4656	4667	1	2	3	4	5	6	7	9	10
.67	4677	4688	4699	4710	4721	4732	4732	4753	4764	4775	1	2	3	4	5	7	8	9	10
.68	4786	4797	4808	4819	4831	4842	4842	4864	4875	4887	1	2	3	4	6	7	8	9	10
.69	4898	4909	4920	4932	4943	4955	4955	4977	4989	5000	1	2	3	5	6	7	8	9	10
.70	5012	5023	5035	5047	5058	5070	5082	5093	5105	5117	1	2	4	5	6	7	8	9	11
.71	5129	5140	5152	5164	5176	5188	5200	5212	5224	5236	1	2	4	5	6	7	8	10	11
.72	5248	5260	5272	5284	5297	5309	5321	5333	5346	5358	1	2	4	5	6	7	8	10	11
.73	5370	5383	5395	5408	5420	5433	5445	5458	5470	5483	1	3	4	5	6	8	9	10	11
.74	5495	5508	5521	5534	5546	5559	5572	5585	5598	5610	1	3	4	5	6	8	9	10	12
.75	5623	5636	5649	5662	5675	5689	5702	5715	5728	5741	1	3	4	5	7	8	9	10	12
.76	5754	5768	5781	5794	5808	5821	5834	5848	5861	5875	1	3	4	5	7	8	9	11	12
.77	5888	5902	5916	5929	5943	5967	5972	5984	5998	6010	1	3	4	5	7	8	10	11	12
.78	6026	6039	6053	6067	6081	6095	6109	6124	6138	6152	1	3	4	6	7	8	10	11	12
.79	6166	6180	6194	6209	6223	6237	6252	6266	6281	6295	1	3	4	6	7	9	10	11	13
.80	6310	6324	6339	6353	6368	6383	6397	6412	6427	6442	1	3	4	6	7	9	10	12	13
.81	6457	6471	6486	6501	6516	6531	6546	6561	6577	6592	2	3	5	6	8	9	11	12	14
.82	6607	6622	6637	6653	6668	6683	6699	6714	6730	6745	2	3	5	6	8	9	11	12	14
.83	6761	6776	6792	6808	6823	6839	6855	6871	6887	6902	2	3	5	6	8	9	11	13	14
.84	6918	6934	6950	6966	6982	6998	7015	7031	7047	7063	2	3	5	6	8	10	11	13	15
.85	7079	7096	7112	7129	7145	7161	7178	7194	7211	7228	2	3	5	7	8	10	12	13	15
.86	7244	7261	7278	7295	7311	7328	7345	7362	7379	7396	2	3	5	7	8	10	12	13	15
.87	7413	7430	7447	7464	7482	7499	7516	7534	7551	7568	2	3	5	7	9	10	12	14	16
.88	7586	7603	7621	7638	7656	7674	7691	7709	7727	7745	2	4	5	7	9	10	12	14	16
.89	7762	7780	7798	7816	7834	7852	7870	7889	7907	7925	2	4	5	7	9	10	13	14	16
.90	7943	7962	7980	7998	8017	8035	8054	8072	8091	8110	2	4	6	7	9	11	13	15	17
.91	8128	8147	8166	8185	8204	8222	8241	8260	8279	8299	2	4	6	8	9	11	13	15	17
.92	8318	8337	8356	8375	8395	8414	8433	8453	8472	8492	2	4	6	8	10	12	14	15	17
.93	8511	8531	8551	8570	8590	8610	8630	8650	8670	8690	2	4	6	8	10	12	14	16	18
.94	8710	8730	8750	8770	8790	8810	8831	8851	8872	8892	2	4	6	8	10	12	14	16	18
.95	8913	8933	8954	8974	8995	9016	9036	9057	9078	9099	2	4	6	8	10	12	15	17	19
.96	9120	9141	9162	9183	9204	9226	9247	9268	9290	9311	2	4	6	8	11	13	15	17	19
.97	9333	9354	9376	9397	9419	9441	9462	9484	9506	9528	2	4	7	9	11	13	15	17	20
.98	9550	9572	9594	9616	9638	9661	9683	9705	9727	9750	2	4	7	9	11	13	16	18	20
.99	9772	9795	9817	9840	9863	9886	9908	9931	9954	9977	2	5	7	9	11	14	16	18	20

Value of e^x and e^{-x} : Table - C

x	e^x	e^{-x}	x	e^x	e^{-x}
0.00	1.000	1.000	3.00	20.086	0.050
0.10	1.105	0.905	3.10	22.198	0.045
0.20	1.221	0.819	3.20	24.533	0.041
0.30	1.350	0.741	3.30	27.113	0.037
0.40	1.492	0.670	3.40	29.964	0.033
0.50	1.649	0.607	3.50	33.115	0.030
0.60	1.822	0.549	3.60	36.598	0.027
0.70	2.014	0.497	3.70	40.447	0.025
0.80	2.226	0.449	3.80	44.701	0.022
0.90	2.460	0.407	3.90	49.402	0.020
1.00	2.718	0.368	4.00	54.598	0.018
1.10	3.004	0.333	4.10	60.340	0.017
1.20	3.320	0.301	4.20	66.686	0.015
1.30	3.669	0.273	4.30	73.700	0.014
1.40	4.055	0.247	4.40	81.451	0.012
1.50	4.482	0.223	4.50	90.017	0.011
1.60	4.953	0.202	4.60	99.484	0.010
1.70	5.474	0.183	4.70	109.95	0.009
1.80	6.050	0.165	4.80	121.51	0.008
1.90	6.686	0.150	4.90	134.29	0.007
2.00	7.389	0.135	5.00	148.41	0.007
2.10	8.166	0.122	5.10	164.02	0.006
2.20	9.025	0.111	5.20	181.27	0.006
2.30	9.974	0.100	5.30	200.34	0.005
2.40	11.023	0.091	5.40	221.41	0.005
2.50	12.182	0.082	5.50	244.69	0.004
2.60	13.464	0.074	5.60	270.43	0.004
2.70	14.880	0.067	5.70	298.87	0.003
2.80	13.445	0.061	5.80	330.30	0.003
2.90	18.174	0.055	5.90	365.04	0.003
3.00	20.086	0.500	6.00	403.43	0.002

Poisson Distribution : Table - D

r \ λ = np	0.1	0.2	0.3	0.4	0.5	0.6	0.7	0.8	0.9	0.10
0	.9048	.8187	.7408	.6708	.6065	.5488	.4966	.4493	.4066	.3679
1	.0905	.1637	.2222	.2681	.3033	.3293	.3476	.3595	.3659	.3679
2	.0045	.0164	.0333	.0536	.0758	.0988	.1217	.1438	.1647	.1839
3	.0002	.0011	.0033	.0072	.0126	.0198	.0284	.0383	.0494	.0613
4	.0000	.0001	.0002	.0007	.0016	.0030	.0050	.0077	.0111	.0153
5	.0000	.0000	.0000	.0001	.0002	.0004	.0007	.0012	.0020	.0031
6	.0000	.0000	.0000	.0000	.0000	.0000	.0001	.0002	.0003	.0005
7	.0000	.0000	.0000	.0000	.0000	.0000	.0000	.0000	.0000	.0001
	1.1	1.2	1.3	1.4	1.5	1.6	1.7	1.8	1.9	2.0
0	.3329	.3012	.2725	.2466	.2231	.2019	.1827	.1653	.1496	.1353
1	.3662	.3614	.3543	.3452	.3347	.3230	.3106	.2975	.2842	.2707
2	.2014	.2169	.2303	.2417	.2510	.2584	.2640	.2678	.2700	.2707
3	.0738	.0867	.0998	.1128	.1255	.1378	.1496	.1607	.1710	.1804
4	.0233	.0260	.0324	.0395	.0471	.0551	.0636	.0723	.0812	.0902
5	.0045	.0062	.0084	.0111	.0141	0176	.0216	.0260	.0309	.0361
6	.0008	.0012	.0018	.0026	.0035	.0047	.0061	.0078	.0098	.0120
7	.0001	.0002	.0013	.0005	.0008	.0011	.0015	.0020	.0027	.0034
8	.0000	.0000	.0001	.0001	.0001	.0002	.0003	.0005	.0006	.0009
9	.0000	.0000	.0000	.0000	.0000	.0000	.0001	.0001	.0001	.0002
	2.1	2.2	2.3	2.4	2.5	2.6	2.7	2.8	2.9	3.0
0	.1225	.1108	.1003	.0907	.0821	.0743	.0672	.0608	.0550	.0498
1	.2572	.2438	.2306	.2177	.2052	.1931	.1815	.1703	.1596	.1494
2	.2700	.2681	.2652	.2613	.2565	.2510	.2450	.2384	.2314	.2240
3	.1890	.1966	.2033	.2090	.2138	.2176	.2205	.2225	.2237	.2240
4	.0992	.1082	.1169	.1254	.1336	.1414	.1488	.1557	.1622	.1680
5	.0417	.0476	.0538	.0602	.0668	.0735	.0804	.0872	.0940	.1008
6	.0146	.0174	.0206	.0241	.0278	.0319	.0362	.0407	.0455	.0504
7	.0044	.0055	.0068	.0083	.0099	.0118	.0139	.0163	.0188	.0216
8	.0011	.0015	.0019	.0025	.0031	.0038	.0047	.0057	.0068	.0081
	.0003	.0004	.0005	.0007	.0009	.0011	.0014	.0018	.0022	.0027
10	.0001	.0001	.0001	.0002	.0002	.0003	.0004	.0005	.0006	.0008
11	.0000	.0000	.0000	.0000	.0000	.0001	.0001	.0001	.0002	.0002
12	.0000	.0000	.0000	.0000	.0000	.0000	.0000	.0000	.0000	.0001

r \ λ = np	3.1	3.2	3.3	3.4	3.5	3.6	3.7	3.8	3.9	4.0
0	.0450	.0408	.0369	.0334	.0302	.0273	.0247	.0224	.0202	.0183
1	.1397	.1304	.1217	.1135	.1057	.0084	.0915	.0850	.0789	.0733
2	.2165	2087	.2008	.1929	.1850	.1771	.1692	.1615	.1539	.1465
3	.2237	.3226	.2209	.2186	.2158	.2125	.2087	.2046	.2001	.1954
4	.1734	.1781	.1823	.1858	.1888	.1912	.1931	.1944	.1951	.1954
5	.1075	.1140	.1203	.1264	.1322	.1377	.1429	.1477	.1522	.1563
6	.0555	.0608	.0662	.0716	.0771	.0826	.0881	.0936	.0989	.1042
7	.0246	.0278	.0312	.0348	.0385	.0425	.0466	.0508	.0551	.0595
8	.0095	.0111	.0129	.0148	.0169	.0191	.0215	.0241	.0269	.0298
9	.0033	.0049	.0047	.0056	.0066	.0076	.0089	.0102	.0116	.0132
10	.0010	.0013	.0016	.0019	.0023	.0028	.0033	.0039	.0045	.0053
11	.0003	.0004	.0005	.0006	.0007	.0009	.0011	.0013	.0016	.0019
12	.0001	.0001	.0001	.0002	.0002	.0003	.0003	.0004	.0005	.0006
13	.0000	.0000	.0000	.0000	.0001	.0001	.0001	.0001	.0002	.0002
14	.0000	.0000	.0000	.0000	.0000	.0000	.0000	.0000	.0000	.0001
	4.1	4.2	4.3	4.4	4.5	4.6	4.7	4.8	4.9	5.0
0	.0166	.0150	.0136	.0123	.0111	.0101	.0091	.0082	.0074	.0067
1	.0679	.0630	.0583	.0540	.0500	.0462	.0427	.0395	.0365	.0337
2	.1393	.1323	.1254	.1188	.1125	.1063	.1005	.0948	.0894	.0842
3	.1904	.1852	.1798	.1743	.1687	.1631	.1517	.1517	.1460	.1404
4	.1951	.1944	.1933	.1917	.1898	.1875	.1849	.1820	.1789	.1755
5	.1600	.1633	.1662	.1687	.1708	.1725	.1738	.1747	.1753	.1755
6	.1093	.1143	.1191	.1237	.1281	.1323	.1362	.1432	.1432	.1462
7	.0640	.0686	.0732	.0778	.0824	.0869	.0914	.0959	.1002	.1044
8	.0328	.0360	.0393	.0428	.0463	.0500	.0537	.0575	.0614	.0653
9	.0150	.0168	.0188	.0209	.0232	.0255	.0280	.0307	.0334	.0363
10	.0061	.0071	.0081	.0092	.0104	.0118	.0132	.0147	.0164	.0181
11	.0023	.0027	.0032	.0037	.0043	.0049	.0056	.0064	.0073	.0082
12	.0008	.0009	.0011	.0014	.0016	.0019	.0022	.0026	.0030	.0034
13	.0002	.0003	.0004	.0005	.0006	.0007	.0008	.0009	.0011	.0013
14	.0001	.0001	.0001	.0001	.0002	.0001	.0003	.0003	.0004	.0005
15	.0000	.0000	.0000	.0000	.0001	.0001	.0001	.0001	.0001	.0001

Standard Normal Distribution: Table - E

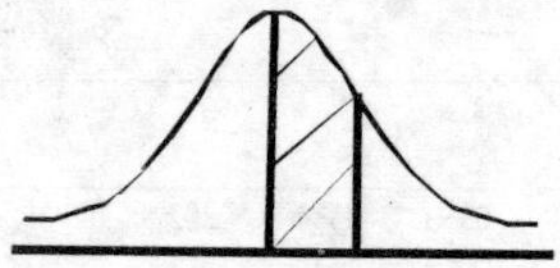

Z = 0 z

Z	0	1	2	3	4	5	6	7	8	9
0.0	.0000	.0040	.0080	.0120	.0160	.0199	.0239	.0279	.0319	.0359
0.1	.0398	.0438	.0478	.0571	.0557	.0596	.0636	.0675	.0714	.0754
0.2	.0793	.0832	.0871	.0910	.0948	.0987	.1026	.1064	.1103	.1141
0.3	.1179	.1217	.1255	.1293	.1331	.1368	.1406	.1443	.1480	.1517
0.4	.1554	.1591	.1628	.1664	.1700	.1736	.1772	.1808	.1844	.1879
0.5	.1915	.1950	.1985	.2019	.2054	.2088	.2123	.2157	.2190	.2224
0.6	.2258	.2291	.2324	.2357	.2389	.2442	.2454	.2486	.2518	.2549
0.7	.2580	.2612	.2642	.2673	.2704	.2734	.2764	.2794	.2823	.2852
0.8	.2881	.2910	.2939	.2967	.2996	.3023	.3051	.3078	.3106	.3133
0.9	.3159	.3186	.3212	.3228	.3264	.3289	.3315	.3340	.3365	.3389
1.0	.3413	.3438	.3461	.3485	.3508	.3531	.3554	.3577	.3599	.3621
1.1	.3643	.3665	.3686	.3708	.3729	.3749	.3770	.3790	.3810	.3830
1.2	.3849	.3869	.3888	.3907	.3925	.3944	.3962	.3980	.3997	.4015
1.3	.4032	.4049	.4066	.4082	.4099	.4115	.4131	.4147	.4162	.4177
1.4	.4192	.4207	.4222	.4236	.4251	.4265	.4279	.4292	.4306	.4319
1.5	.4332	.4345	.4357	.4370	.4382	.4394	.4406	.4418	.4429	.4441
1.6	.4452	.4463	.4474	.4484	.4495	.4505	.4515	.4525	.4535	.4545
1.7	.4554	.4564	.4573	.4582	.4591	.4599	.4608	.4616	.4625	.4633
1.8	.4641	.4649	.4656	.4664	.4671	.4678	.4686	.4693	.4699	.4706
1.9	.4713	.4719	.4726	.4732	.4738	.4744	.4750	.4756	.4761	.4767
2.0	.4772	.4778	.4783	.4788	.4783	.4708	.4803	.4808	.4812	.4867
2.1	.4821	.4826	.4830	.4834	.4838	.4812	.4846	.4850	.4854	.4857
2.2	.4861	.4864	.4868	.4871	.4875	.4878	.4881	.4884	.4887	.4890
2.3	.4893	.4896	.4898	.4901	.4904	.4906	.4909	.4911	.4913	.4916
2.4	.4918	.4920	.4922	.4925	.4927	.4929	.4931	.4932	.4934	.4936
2.5	.4938	.4940	.4941	.4943	.4945	.4946	.4948	.4949	.4951	.4952
2.6	.4953	.4955	.4956	.4957	.4959	.4960	.4961	.4962	.4963	.4964
2.7	.4965	.4966	.4967	.4968	.4969	.4970	.4971	.4972	.4973	.4974
2.8	.4974	.4975	.4976	.4977	.4977	.4978	.4979	.4979	.4980	.4981
2.9	.4981	.4982	.4982	.4983	.4984	.4984	.4985	.4985	.4986	.4986
3.0	.4987	.4987	.4987	.4988	.4988	.4989	.4989	.4989	.4990	.4490
3.1	.4990	.4991	.4991	.4991	.4992	.4992	.4992	.4992	.4993	.4993
3.2	.4993	.4993	.4994	.4994	.4994	.4994	.4994	.4995	.4995	.4995
3.3	.4995	.4995	.4995	.4996	.4996	.4996	.4996	.4996	.4996	.4997
3.4	.4997	.4997	.4997	.4997	.4997	.4997	.4997	.4997	.4997	.4998
3.5	.4998	.4998	.4998	.4998	.4998	.4998	.4998	.4998	.4998	.4998
3.6	.4998	.4998	.4999	.4999	.4999	.4999	.4999	.4999	.4999	.4999
3.7	.4999	.4999	.4999	.4999	.4999	.4999	.4999	.4999	.4999	.4999
3.8	.4999	.4999	.4999	.4999	.4999	.4999	.4999	.4999	.4999	.4999
3.9	.5000	.5000	.5000	.5000	.5000	.5000	.5000	.5000	.5000	.5000

Note: The table values are in proportion to area in between Z = 0 to z as shown in the figure. As normal curve is symmetrical curve the negative values are derived using the same principle.

Compound Value Factor of an Annuity (CVFA) of Re 1 : Table -F

Year	Interest Rate													
	1%	*2%*	*3%*	*4%*	*5%*	*6%*	*7%*	*8%*	*9%*	*10%*	*11%*	*12%*	*13%*	*14%*
1	1.000	1.000	1.000	1.000	1.000	1.000	1.000	1.000	1.000	1.000	1.000	1.000	1.000	1.000
2	2.010	2.020	2.030	2.040	2.050	2.060	2.070	2.080	2.090	2.100	2.110	2.120	2.130	2.140
3	3.030	3.060	3.091	3.122	3.153	3.184	3.215	3.246	3.278	3.310	3.342	3.374	3.407	3.440
4	4.060	4.122	4.184	4.246	4.310	4.375	4.440	4.506	4.573	4.641	4.710	4.779	4.850	4.921
5	5.101	5.204	5.309	5.416	5.526	5.637	5.751	5.867	5.985	6.105	6.228	6.353	6.480	6.610
6	6.152	6.308	6.468	6.633	6.802	6.975	7.153	7.336	7.523	7.716	7.913	8.115	8.323	8.536
7	7.214	7.434	7.662	7. 898	8.142	8.394	8.654	8.923	9.200	9.487	9.783	10.089	10.405	10.730
8	8.286	8.583	8.892	9.214	9.549	9.897	10.260	10.637	11.028	11.436	11.859	12.300	12.757	13.233
9	9.369	9.755	10.159	10.583	11.027	11.491	11.978	12.488	13.021	13.579	14.164	14.776	15.416	16.085
10	10.462	10.950	11.464	12.006	12.578	13.181	13.816	14.487	15.193	15.937	16.722	17.549	18.420	19.337
11	11.567	12.169	12.808	13.486	14.207	14.972	15.784	16.645	17.560	18.531	19.561	20.655	21.814	23.045
12	12.683	13.412	14.192	15.026	15.917	16.870	17.888	18.977	20.141	21.384	22.713	24.133	25.650	27.271
13	13.809	14.680	15.618	16.627	17.713	18.882	20.141	21.495	22.953	24.523	26.212	28.029	29.985	32.089
14	14.947	15.974	17.086	18.292	19.599	21.015	22.550	24.215	26.019	27.975	30.095	32.393	34.883	37.581
15	16.097	17.293	18.599	20.024	21.579	23.276	25.129	27.152	29.361	31.772	34.405	37.280	40.417	43.842
16	17.258	18.639	20.157	21.825	23.657	25.673	27.888	30.324	33.003	35.950	39.190	42.753	46.672	50.980
17	18.430	20.012	21.762	23.698	25.840	28.213	30.840	33.750	36.974	40.545	44.501	48.884	53.739	59.118
18	19.615	21.412	23.414	25.645	28.132	30.906	33.999	37.450	41.301	45.599	50.396	55.750	61.725	68.394
19	20.811	22.841	25.117	27.671	30.539	33.760	37.379	41.446	46.018	51.159	56.939	63.440	70.749	78.696
20	22.019	24.297	26.870	29.778	33.066	36.786	40.995	45.762	51.160	57.275	64.203	72.052	80.947	91.025
25	28.243	32.030	36.459	41.646	47.727	54.865	63.249	73.106	84.701	98.347	114.413	133.334	155.620	181.871
30	34.785	40.568	47.575	56.085	66.439	79.058	94.461	113.283	136.308	164.494	199.021	241.333	293.199	356.787
40	48.886	60.402	75.401	95.026	120.800	154.762	199.635	259.057	337.882	442.593	581.826	767.091	1013.704	1342.025
50	64.463	84.579	112.797	152.667	209.348	290.336	406.529	573.770	815.084	1163.909	1668.771	2400.018	3459.507	4994.521

Contd.....

	Interest Rate											
Year	15%	16%	17%	18%	19%	20%	21%	22%	23%	24%	25%	30%
1	1.000	1.000	1.000	1.000	1.000	1.000	1.000	1.000	1.000	1.000	1.000	1.000
2	2.150	2.160	2.170	2.180	2.190	2.200	2.210	2.220	2.230	2.240	2.250	2.300
3	3.473	3.506	3.539	3.572	3.606	3.640	3.674	3.708	3.743	3.778	3.813	3.990
4	4.993	5.066	5.141	5.215	5.291	5.368	5.446	5.524	5.604	5.684	5.766	6.187
5	6.742	6.877	7.014	7.154	7.297	7.442	7.589	7.740	7.893	8.048	8.207	9.043
6	8.754	8.977	9.207	9.442	9.683	9.930	10.183	10.442	10.708	10.980	11.259	12.756
7	11.067	11.414	11.772	12.142	12.523	12.916	13.321	13.740	14.171	14.615	15.073	17.583
8	13.727	14.420	14.773	15.327	15.902	16.499	17.119	17.762	18.430	19.123	19.842	23.858
9	16.786	17.519	18.285	19.086	19.923	20.799	21.714	22.670	23.669	24.712	25.802	32.015
10	20.304	21.231	22.393	23.521	24.709	25.959	27.274	28.657	30.113	31.643	33.253	42.619
11	24.349	25.733	27.200	28.755	30.404	32.150	34.001	35.962	38.039	40.238	42.566	56.405
12	29.002	30.850	32.824	34.931	37.180	39.581	42.142	44.874	47.788	50.895	54.208	74.327
13	34.352	36.786	39.404	42.219	45.244	48.497	51.991	55.746	59.779	64.110	68.760	97.625
14	40.505	43.672	47.103	50.818	54.841	59.196	63.909	69.010	74.528	80.496	86.949	127.913
15	47.580	51.660	56.110	60.965	66.261	72.035	78.330	85.192	92.669	100.815	109.687	167.286
16	55.717	60.925	66.649	72.939	79.850	87.442	95.780	104.935	114.983	126.011	138.109	218.472
17	65.075	71.673	78.979	87.068	96.022	105.931	116.894	129.020	142.430	157.253	173.636	285.014
18	75.836	84.141	93.406	103.740	115.266	128.117	142.441	158.405	176.188	195.994	218.045	371.518
19	88.212	98.603	110.285	123.414	138.166	154.740	173.354	194.254	217.712	244.033	273.556	483.973
20	102.444	115.380	130.033	146.628	165.418	186.688	210.758	237.989	268.785	303.601	342.945	630.165
25	212.793	249.214	292.105	342.603	402.042	471.981	554.242	650.955	764.605	898.092	1054.791	2348.803
30	434.745	530.312	647.439	790.948	966.712	1181.882	1445.151	1767.081	2160.491	2640.916	3227.174	8729.985
40	1779.090	2360.757	3134.522	4163.213	5529.829	7343.858	9749.525	12936.535	17154.046	22728.803	30088.655	120392.883
50	7217.716	10435.649	15089.502	21813.094	31515.336	45497.191	65617.202	94525.279	135992.154	195372.644	280255.693	1659760.743

Present Value Factor of an Annuity (PVFA) of Re 1: Table G

Year	Interest Rate													
	1%	2%	3%	4%	5%	6%	7%	8%	9%	10%	11%	12%	13%	14%
1	0.990	0.980	0.971	0.962	0.952	0.943	0.935	0.926	0.917	0.909	0.901	0.893	0.885	0.877
2	1.970	1.942	1.913	1.886	1.859	1.833	1.808	1.783	1.759	1.736	1.713	1.690	1.668	1.647
3	2.941	2.884	2.829	2.775	2.723	2.673	2.624	2.577	2.531	2.487	2.444	2.402	2.361	2.322
4	3.902	3.808	3.717	3.630	3.546	3.465	3.387	3.312	3.240	3.170	3.102	3.037	2.974	2.914
5	4.853	4.713	4.580	4.452	4.329	4.212	4.100	3.993	3.890	3.791	3.696	3.605	3.517	3.433
6	5.795	5.601	5.417	5.242	5.076	4.917	4.767	4.623	4.486	4.355	4.231	4.111	3.998	3.889
7	6.728	6.472	6.230	6.002	5.786	5.582	5.389	5.206	5.033	4.868	4.712	4.564	4.423	4.288
8	7.652	7.325	7.020	6.733	6.463	6.210	5.971	5.747	5.535	5.335	5.146	4.968	4.799	4.639
9	8.566	8.162	7.786	7.435	7.108	6.802	6.515	6.247	5.995	5.759	5.537	5.328	5.132	4.946
10	9.471	8.983	8.530	8.111	7.722	7.360	7.024	6.710	6.418	6.145	5.889	5.650	5.426	5.216
11	10.368	9.787	9.253	8.760	8.306	7.887	7.499	7.139	6.805	6.495	6.207	5.938	5.687	5.453
12	11.255	10.575	9.954	9.385	8.863	8.384	7.943	7.536	7.161	6.814	6.492	6.194	5.918	5.660
13	12.134	11.348	10.635	9.986	9.394	8.853	8.358	7.904	7.487	7.103	6.750	6.424	6.122	5.842
14	13.004	12.106	11.296	10.563	9.899	9.295	8.745	8.244	7.786	7.367	6.982	6.628	6.302	6.002
15	13.865	12.849	11.938	11.118	10.380	9.712	9.108	8.559	8.061	7.606	7.191	6.811	6.462	6.142
16	14.718	13.578	12.561	11.652	10.838	10.106	9.447	8.851	8.313	7.824	7.379	6.974	6.604	6.265
17	15.562	14.292	13.166	12.166	11.274	10.477	9.763	9.122	8.544	8.022	7.549	7.120	6.729	6.373
18	16.398	14.992	13.754	12.659	11.690	10.828	10.059	9.372	8.756	8.201	7.702	7.250	6.840	6.467
19	17.226	15.678	14.324	13.134	12.085	11.158	10.336	9.604	8.950	8.365	7.839	7.366	6.938	6.550
20	18.046	16.351	14.877	13.590	12.462	11.470	10.594	9.818	9.129	8.514	7.963	7.469	7.025	6.623
25	22.023	19.523	17.413	15.622	14.094	12.783	11.654	10.675	9.823	9.077	8.422	7.843	7.330	6.873
30	25.808	22.396	19.600	17.292	15.372	13.765	12.409	11.258	10.274	9.427	8.694	8.055	7.496	7.003
40	32.835	27.355	23.115	19.793	17.159	15.046	13.332	11.925	10.757	9.779	8.951	8.244	7.634	7.105
50	39.196	31.424	25.730	21.482	18.256	15.762	13.801	12.233	10.962	9.915	9.042	8.304	7.675	7.133

Contd....

	Interest Rate											
Year	***15%***	***16%***	***17%***	***18%***	***19%***	***20%***	***21%***	***22%***	***23%***	***24%***	***25%***	***30%***
1	0.870	0.826	0.855	0.847	0.840	0.833	0.826	0.820	0.813	0.806	0.800	0.769
2	1.626	1.605	1.585	1.566	1.547	1.528	1.509	1.492	1.474	1.457	1.440	1.361
3	2.283	2.246	2.210	2.174	2.140	2.106	2.074	2.042	2.011	1.981	1.952	1.816
4	2.855	2.798	2.743	2.690	2.639	2.589	2.540	2.494	2.448	2.404	2.362	2.166
5	3.352	3.274	3.199	3.127	3.058	2.991	3.926	2.864	2.803	2.745	2.689	2.436
6	3.784	3.685	3.589	3.498	3.410	3.326	3.245	3.167	3.092	3.020	2.951	2.643
7	4.160	4.039	4.922	3.812	3.706	3.605	3.508	3.416	3.327	3.242	3.161	2.802
8	4.487	4.344	4.207	4.078	3.954	3.837	3.726	3.619	3.518	3.421	3.329	2.925
9	4.772	4.607	4.451	4.303	4.163	4.031	4.905	3.786	3.673	3.566	3.463	3.019
10	5.019	4.833	4.659	4.494	4.339	4.192	4.054	3.923	3.799	3.682	3.571	3.092
11	5.234	5.029	4.836	4.656	4.486	4.327	4.177	4.035	3.902	3.776	3.656	3.147
12	5.421	5.197	5.988	4.793	4.611	4.439	4.278	4.127	3.985	3.851	3.725	3.190
13	5.583	5.342	5.118	4.910	4.715	4.533	4.362	4.203	4.053	3.912	3.780	3.223
14	5.724	5.468	5.229	5.008	4.802	4.611	4.432	4.265	4.108	3.962	3.824	3.249
15	5.847	5.575	5.324	5.092	4.876	4.675	4.489	4.315	4.153	4.001	3.859	3.268
16	5.954	5.668	5.405	5.162	4.938	4.730	4.536	4.357	4.189	4.033	3.887	3.283
17	6.047	5.749	5.475	5.222	4.990	4.775	4.576	4.391	4.219	4.059	3.910	3.295
18	6.128	5.818	5.534	5.273	5.033	4.812	4.608	4.419	4.243	4.080	3.928	3.304
19	6.198	5.877	5.584	5.316	5.070	4.843	4.635	4.442	4.263	4.097	3.942	3.311
20	6.259	5.929	5.628	5.353	5.101	4.870	4.657	4.460	4.279	4.110	3.954	3.316
25	6.464	6.097	5.766	5.467	5.195	4.948	4.721	4.514	4.323	4.147	3.985	3.329
30	6.566	6.177	5.829	5.517	5.235	4.979	4.746	4.534	4.339	4.160	3.995	3.332
40	6.642	6.233	5.871	5.548	5.258	4.997	4.760	4.544	4.347	4.166	3.999	3.333
50	6.661	6.246	5.880	5.554	5.262	4.999	4.762	4.545	4.348	4.167	3.000	3.333

INDEX

A

B

C

D

E

F

H

I

L

M

O

P

R

S

U